INTRODUCTION TO FLAVONOIDS

Chemistry and Biochemistry of Organic Natural Products

A series of books bringing into perspective the available information on the occurrence and distribution of natural products, their structural diversity and biogenetic interrelationships, chemical syntheses as well as their possible role in nature and their physiological activities.

Edited by B. Ravindranath, Bharavi Laboratories Ltd, Bangalore, India

---

Volume 1
The Isoquinoline Alkaloids
by Kenneth W. Bentley

Volume 2
Introduction to Flavonoids
by Bruce A. Bohm

Other Volume in Preparation

The Coumarins
A. M. Zobel

This book is part of a series. The publisher will accept continuation orders which may be cancelled at any time and which provide for automatic billing and shipping of each title in the series upon publication. Please write for details.

INTRODUCTION TO FLAVONOIDS

Bruce A. Bohm

Department of Botany, University of British Columbia
Vancouver, Canada

harwood academic publishers
Australia • Canada • China • France • Germany • India
Japan • Luxembourg • Malaysia • The Netherlands • Russia
Singapore • Switzerland

Printed in Singapore.

Amsteldijk 166
1st Floor
1079 LH Amsterdam
The Netherlands

British Library Cataloguing in Publication Data

Bohm, Bruce A.
Introduction to flavonoids. – (Chemistry and biochemistry of organic natural products; v. 2)
1. Flavonoids
I. Title
572.5'92

ISBN 90-5702-353-9
ISSN 1027-7498

The illustration on the cover: an *Aristolochia* species (Aristolochiaceae) showing the pattern of anthocyanin deposition in the floral tube. The combination of pigment pattern and distinctive aroma produced by the flower serve to entice insects into the tube where they effect pollination. In some species trapped insects can escape after the floral tube has expanded while in others it is a one-way trip.

Pen and ink drawing by Lesley R. Bohm (Vancouver)

This work is dedicated to the memory of E. C. Bate-Smith,
Hans Grisebach and Tony Swain,
giants upon whose shoulders so many of us have stood.

CONTENTS

PREFACE TO THE SERIES

Development of organic chemistry has been closely associated with that of the chemistry of natural products. It is common knowledge that many of the techniques of extraction, separation, structure determination and synthesis have been developed to understand the structural variation among the natural products. Intellectual curiosity, which has been the driving force for natural products research during the earlier years, has given way to the search for physiologically active compounds for medicinal use. In fact, potential use in medicine has long been the justification for research in this field. Realisation of this objective, however, has been less than satisfactory, the success stories of several antibiotics and the likes of taxol notwithstanding. Nevertheless, research in this area has certainly led to better understanding of the structural requirements for a variety of physiological activities, leading to the synthesis and modification of several lead compounds and analogues. Many fascinating developments may be expected in the future for, while several hundreds of thousands of compounds have been isolated and identified, many more remain. By some estimates, over 95% of the flora remain untouched and the growing realisation that several species of the plant as well as animal kingdom are endangered has added the needed impetus for more vigorous screening programs the world over. With the currently available tools for extraction, chromatographic separation and structural identification, a fresh look at the well-studied plants may also be rewarding.

Two other aspects of natural products chemistry may be noted. One is the biochemical basis of their formation and their role in nature, which is still subject to speculation. Among the theories are that the secondary metabolites are, probably, (a) the waste (or end) products of bioenergetics and metabolism, (b) required for self protection and (c) needed for supporting other organisms (allelochemicals or chemical signals). An understanding of how and why these compounds are made in nature would certainly lead to their better utilisation.

More recently, natural products have assumed additional significance as renewable raw materials in place of the fast-depleting petroleum reserves. This is particularly true of the by-products of the agricultural and food industries, which only need to continue to grow in view of the increasing world population. While some natural products can be used as such, several may need to be converted to value-added products. In some cases, newer uses may be found for abundantly available natural products. Study of the structural relationships among natural products would go a long way in these endeavours.

The present series aims to bring into perspective the available information on the occurrence and distribution of the natural products, their structural diversity and biogenetic interrelationships, their possible role in nature and their physiological activities. Detailed account of the methods of their extraction, separation, structure determination and synthesis would probably make the work more comprehensive, but it is difficult to do justice, given the space limitations; only general methods and distinguishing features are discussed in order to help identification of the compounds and reference to the most recent reviews are made wherever available. Also, discussion on syntheses essentially aimed at chemical or biological interconversions are included in order to bring forth the potential of the natural products as raw materials for value-added products.

Organisation of the vast and diverse information on the subject of natural products is a problem in itself. Classification of the natural products by origin, chemical structure, physiological activity or biosynthetic pathways all have their limitations. The traditional classification of the compounds, which reflects the structural and biogenetic relationships to a large extent, is followed in this work. Each volume is dedicated to a major class of natural products and each volume is further split into several parts to adequately cover smaller groups of closely related compounds. Each volume or part thereof is designed to be an essentially independent and self-sufficient book, authored or edited by an expert.

I would like to take this opportunity to express my deep sense of appreciation to the authors and editors and to the staff of Harwood Academic Publishers for their unfailing support in bringing out the series.

B. Ravindranath

ACKNOWLEDGEMENTS

I wish to thank several colleagues and friends (not mutually exclusive) who have shown interest in my various writing projects over the years. In particular, Pam and Doug Soltis, at Washington State University, have been very supportive, as well as showing great patience in explaining some of the more mysterious aspects of molecular genetics to me. Dan Crawford, at the Ohio State University, has been a good friend and supporter for many years. Tod Stuessy, sequentially at The Ohio State University and Los Angeles County Museum, and now at Universität Wien, has been a most enthusiastic supporter in this project, as well as co-author on several others. I have also enjoyed very much my correspondence with Eckard Wollenweber, Darmstadt, who carries on the great German tradition of flavonoid chemistry almost single-handedly, it seems! I should also like to express my appreciation to Prof. J. B. Harborne of Reading University who gave me the opportunity to contribute on a regular basis to *The Flavonoid* series.

In the late 1950s, I became acquainted with E. W. (Ted) Underhill, who was a doctoral candidate in the Faculty of Pharmacy at the University of Rhode Island at the time I was working toward my degree in the Department of Chemistry. My interest in plant biochemistry came about as a result of that meeting as I witnessed for the first time the use of a radio-labeled precursor in Ted's study of tropic acid biosynthesis. It was he who suggested that I try for a National Research Council (Canada) postdoctoral fellowship to work at McGill University with G. H. Neil Towers. It was a memorable experience to see an entirely new field open up before me. Many thanks go to Neil who, in addition to giving me a job at U.B.C. in 1966, continues to amaze with his broad knowledge of plant chemistry. Dr Elijah Tannen, now retired, has provided support over the years in ways that are difficult to describe.

My research career in Canada has been supported exclusively by programs of the Natural Sciences and Engineering Research Council (Formerly, the National Research Council) of Canada. These have taken the form of regular operating and equipment grants, as well as a strategic grant. My sincerest thanks go to these programs.

The fine line drawings that appear in Chapter Seven and on the cover of this book are the work of my wife Lesley Bohm, whose grudging tolerance of my seemingly endless writing activities is hereby acknowledged as well.

Last, but by no means least, I should like to acknowledge the help provided to me by the staff of Harwood Academic Publishers.

CHAPTER ONE

INTRODUCTION AND HISTORICAL PERSPECTIVE

A detailed history of the discovery of flavonoids, giving full recognition to all those whose contributions have provided the foundation upon which we stand today, is far beyond the scope of this book. It is necessary, however, to give some idea of where current ideas came from and to identify some of the main players. It should come as no surprise to learn that some very eminent scientists have left their mark upon the subject. Unfortunately, superficial views of any subject tend to victimize many individuals whose contributions would be worthy of comment in a larger treatment. To those so offended (or their descendants), I offer my apologies. The best attempt at a comprehensive treatment of the subject was Muriel (Wheldale) Onslow's (1925) scholarly effort *The Anthocyanin Pigments of Plants*, 2nd ed. (from which many of the older citations below were taken). As the name suggests, the work was aimed at assembling into one place as much of the information on anthocyanins as was possible. Although dedicated to the one class of flavonoids, the work nonetheless touches on many aspects of flavonoids in general. It is, of course, not surprising that the anthocyanins should attract the lion's share of attention since they were the flavonoids most readily observed in Nature. The variation in flower color among members of the same species was also widely recognized and provided the opportunity for studying the patterns of inheritance of the pigments. As the reader may recall, Mendel's experiments with peas included, among other features, pod color, seed color, and, relevant to the subject at hand, flower color (purple vs. white). We will return to the contribution to genetics played by anthocyanins below.

The earliest suggestion that substances existed in Nature that would eventually be recognized as flavonoids can be traced back to the 17th century. Nehemiah Grew (1682), in papers read to the Royal Society in London, discussed differences in solubility properties of plant pigments; some were soluble in oil, others in water. He also commented on their different behaviors toward acids and alkalis. Autumn coloration of leaves has attracted the attention of many workers over the years, one of the earliest being Macaire-Princep who, in 1828, advanced the idea that red pigments are oxidation products of chlorophyll. The term "anthocyanin" was first used by Marquart (1835) to describe water soluble red, blue and violet pigments from plants. (He referred to yellow pigments as "anthoxanthins.") Marquart suggested that anthocyanins were formed by dehydration of chlorophyll. This view was not universally held, however. von Mohl (1838), for instance, argued that the origin of anthocyanins had nothing to do with chlorophyll. Some years later, Wigand (1862) argued that anthocyanin has no relationship with chlorophyll but is formed instead from some colorless compound, which he called a "tannin." This was an important idea that set the stage for a serious and dedicated search for the chromogenic compound or compounds from which anthocyanins were formed. Maumené (1882) provided further support for the existence of a colorless precursor for anthocyanins. Grapes dried over sulfuric acid did not change color, but upon exposure to air they developed a red color. This observation led him to conclude that oxidation, or possibly hydration, of some chromogenic compound must occur. Interest in the nature of the chromogens continued to attract attention as can be seen in a paper by Wheldale (1909a) where she made the suggestion that the chromogens were glycosides of either flavones or xanthones and that their conversion to anthocyanins resulted from the action of an oxidase. Research in various laboratories over the years led to the observation that "leucoanthocyanidins" or "proanthocyanidins" exist in some plants, either (or both) of which are converted to anthocyanidins under acidic conditions. The true nature of these compounds, whether mono-, di- or oligomeric, was not fully appreciated until fairly recently when it became clear that flavanol derivatives were responsible. The ultimate biosynthetic precursor of anthocyanidins remains unresolved, although we know with reasonable certainty that flavan-3,4-diols are involved.

Other observations that proved to be prophetic involved co-pigments and cell acidity. The idea of co-pigmentation, although it was not expressly identified as such, came from the work of Filhol (1860) who felt that only one kind of pigment existed and that the variety of colors observed in flowers arose from its interaction with other substances in the cell. We know, of course, that Filhol's single pigment idea is wrong, but the interaction of anthocyanins with other cell components, co-pigments, is well documented. A hint of the direct involvement of acid in color production came from an experimental study by Molisch (1889) who boiled *Coleus* and *Perilla* leaves in water, which resulted in loss of their red color. The red pigmentation returned after acidification. He suggested that acidity of the cell sap was necessary for color. We now know, of course, that subtle changes in cell pH can have significant effects upon flower color (Chapter Two).

Definitive work on anthocyanin structure came from the pioneering studies by R. Willstätter and his co-workers (see Willstätter and Everest, 1913, and subsequent publications). Many papers from that group were published during the period 1913 to 1916. A summary of Willstätter's work can be found in a review written by Everest (1915). The general field was reviewed by Perkin and Everest (1918) in a book entitled *The Natural Organic Colouring Matters*. Willstätter's structural elucidation studies set the scene for the equally impressive synthetic efforts of Robert Robinson and his many co-workers in subsequent years (ca. 1922–1934). An early paper from the series, entitled "A synthesis of pyrylium salts of anthocyanidin type," is highly instructive of the methods utilized for making a wide variety of anthocyanin derivatives (Pratt and Robinson, 1922). One of the synthetic sequences used to make anthocyanins is discussed in Chapter Five.

As we will see in Chapter Seven, flavonoids play a variety of roles in plants. Although most of the experimental work in this area is of fairly recent vintage, there are many suggestions in the older literature concerning possible functions of anthocyanins. An association of anthocyanins with wounding response was implied by the observation of Beyerinck (1885) that color developed in regions of roots of the grass *Poa nemoralis* that had been injured by the gall insect *Cecidomyia poae*. Similar conclusions were drawn by Sorauer (1886) who discussed the possible connection between anthocyanin formation and infection of plants by parasites. More recently,

Walker (1923) noted that pigmented onions were more resistant to disease than unpigmented ones. One of the more imaginative suggestions was that of Kuntze (1877) who thought that the red color of plants might be protective against animals because it resembled blood. Ludwig (1891) suggested that the red color of certain plants was a warning to animals, particularly snails. The involvement of a variety of flavonoids, including anthocyanins, as part of the response of a plant to infection by microorganisms or in protecting plants against herbivores has been well documented, as will be discussed in Chapter Seven. So far as I am aware, haemomimicry has not been examined experimentally.

The participation of various flavonoid derivatives in attraction of pollinators has been well documented in the contemporary literature. The possibility that anthocyanins functioned in this manner was suggested by several workers toward the end of last century. For example, Ludwig (1889) concluded that the location of red pigment in flowers of *Impatiens*, *Sambucus*, and *Viburnum* was such that it would lead insects to the nectaries. This point was reiterated by Macchiati (1899) who studied the nectaries of *Prunus laurocerasus*. Anthocyanin was also put forward as protection for leaves against strong light (Johow, 1884), while Kny (1892) also thought that they might act as light shields, that in so doing could transform light into heat. Bringing this idea to the present, it is interesting to note that Sturgeon and Mitten (1980) have shown that the purple cones (presumably anthocyanic) of *Abies concolor* in southern Colorado have a higher internal temperature compared to green cones when placed in direct sunlight. They suggest that the darker cone color is an adaptation to the lower temperatures experienced by trees at higher elevations.

It is probably safe to say that anthocyanins played a major role in the development of genetics. Flower color variation was a well known and particularly appealing phenomenon and seemed a logical choice for studying patterns of inheritance. The early days of anthocyanin genetics included studies by such pioneers in the field as Bateson, Correns, de Vries, Emerson, Punnett, and Wheldale. Some of the early studies involved attempts to relate the inheritance of flower color in snapdragons with specific chemical characters (e.g., Wheldale, 1909b, 1914, 1915). This was a difficult task because the chemistry of the complex mixture of pigments in this species was not yet known. Wheldale's work was nonetheless significant because it helped focus attention on the potential of combining genetic with chemical studies, an approach that ultimately proved to be extremely powerful. These studies were done at the John Innes Horticultural Research Institute (Bayfordbury, Hertford, now Norwich) whose director at the time (1910–1926) was W. Bateson. In his biography of J. B. S. Haldane, Ronald Clark (1984) described the Institute under Bateson as having "... been transformed into a nursery for the growing science of genetics..." Following Bateson's death in 1926, Sir Daniel Hall was appointed to the directorship of the Institute. Taking advice from Julian Huxley as to whom he might turn for scientific advice, Hall managed to acquire the "loan" of Haldane, on a part-time basis, from Cambridge. One of Haldane's interests concerned the relationship between chemical cause and physiological effect in humans, which led him to think about similar relationships in plants with flower color as a possible experimental tool. The person chosen to do the chemical genetic work was Rose Scott-Moncrieff whose contributions to the new discipline of biochemical genetics were of major significance (Lawrence and Scott-Moncrieff, 1935; Scott-Moncrieff, 1936; Lawrence and Price, 1940). It seems reasonable to take Clark's lead (1984, p. 103) in quoting Scott-Moncrieff directly on the significance of the chemico-genetic approach. "By crossing plants with known chemical characteristics we can now combine and rearrange these qualities in their progeny just as surely as we can mix chemicals together in the laboratory, and when these controlled biological reactions diverge from the expected, we obtain important evidence of other unknown processes which must also be involved." For additional insights into this important phase of genetic history, the entire story as presented by Clark makes fascinating reading.

Before we finish off our this brief historical tour, a few comments on the early days of some of the other flavonoid types is in order. The first dihydrochalcone, phloridzin, for example, was isolated from apple root bark in 1835 by de Koninck. The term "flavone" first appeared in a paper by von Kostanecki and Tambor (1895) whose studies pioneered the structural work on this class of flavonoids. de Laire and Tiemann (1893) obtained the first isoflavone from the rhizomes of *Iris florentina*, but the first structurally defined isoflavone, prunetin, was not described until some years later by Finnemore (1910). The first chalcone described as a naturally occurring compound was apparently "carthamin" from safflower (*Carthamus tinctorius*) a dye plant widely used in India (Kametaka and Perkin, 1910). The isomeric relationship between flavanones and chalcones, involving ring opening of the former to yield the latter, was described by Dean and Nierenstein in 1925. The first flavanone recognized as naturally occurring was butrin, the 7-*O*-glucoside of butin, which Lal and Dutt (1935) described from the legume *Butea frondosa*. Aurones were comparative new comers to the flavonoid family being described originally by Geissman and Heaton in 1944. *C*-Glycosylflavones and biflavonoids are also recent additions. The Geissman volume (1962) contained only a few comments on the former and none on the latter.

The term "leucoanthocyanin" was first used by Rosenheim (1920) to identify compounds that he isolated from young grape leaves. Treatment with acid easily converted these compounds to anthocyanins. He postulated a pseudobase structure (a cyclic hemiketal) to account for their properties. Robinson and Robinson (1933) thought the pseudobase structure was unlikely owing to the ease with which it can be converted to the parent anthocyanin. Recent studies of anthocyanin chemistry, however, suggest that at pH values above 3, anthocyanins do indeed react with water to form hemiacetals, which will be discussed in Chapter Two.

The pigments of Centrospermae posed a major challenge to phytochemists for many years. These so-called "nitrogenous anthocyanins" represented a range of pigments that superficially resembled anthocyanins but seemed to constitute their own class of compounds. An early observation of this phenomenon was recorded by Braconnot (1807) who was concerned with the properties of *Phytolacca americana*, the "vulgaire raisin d'amerique" in his terms. Extracts of the fruits did not behave toward acids, alkalis, and metal salts as expected for extracts of other fruits. Hilger (1879) came to much the same conclusion using extracts of related taxa, *Amaranthus*, *Beta*, and *Chenopodium* as well as *Phytolacca*. Structures of these compounds, which are not anthocyanins at all, were determined with certainty only much more recently. Structures for some of these compounds appear in Chapters Two and Three along with references to recent reviews describing their chemistry and biology.

CHAPTER TWO

STRUCTURAL VARIATION AND THE FLAVONOID LITERATURE

Before embarking on our survey of flavonoid structures it is necessary to comment on the source of the information below. The literature on flavonoid chemistry and occurence is vast, well beyond anything that could be handled in a single volume. The problem of access to the flavonoid literature has been addressed by J. B. Harborne, of Reading University, who, with the help of editorial associates and a number of contributing authors, has managed to bring the mass of information available in the primary literature into an easily accessible form. The product of these efforts is the series *The Flavonoids*, which first appeared in 1975 and has appeared roughly every six years since, with the *Advances in Research* volumes appearing in 1982, 1988, and 1994. Each of these volumes consists of chapters written by specialists who deal in depth with some aspect of the subject, whether it is a class of flavonoids, analytical techniques, or some other relevant topic. I have made frequent use of these very important sources in the preparation of this book.

But it did not begin with *The Flavonoids* series. *The Chemistry of Flavonoid Compounds*, edited by T. A. Geissman, appeared in 1962 and served as the first comprehensive review of the flavonoid literature. This was followed in 1963 by F. M. Dean's masterful treatment entitled *Naturally Occurring Oxygen Ring Compounds*, which, although not specializing in flavonoids, contained important chemical information about them. In 1967, Prof. Harborne published his *Comparative Biochemistry of the Flavonoids*, in which he described not only the chemistry and distribution of flavonoids but also their genetics and functions. Other books of note include Edwin Haslam's *The Shikimate Pathway* (1974) and his later work, *Plant Polyphenols. Vegetable Tannins Revisited* (1989), and Helen Stafford's *Flavonoid Metabolism* (1990).

Much of interest is to be found in the older literature but access is a serious problem. A major review of information from that era can be found in *The Anthocyanin Pigments of Plants* by Muriel Wheldale Onslow, the second edition of which was published in 1925. Many of the historical references in Chapter One came from that source.

A GENERAL INTRODUCTION TO FLAVONOID STRUCTURE

As we saw briefly in Chapter One, flavonoids are based upon a fifteen-carbon skeleton. At the simplest level, the skeleton consists of two phenyl rings connected by a three-carbon bridge. Since most flavonoids have a carbonyl function located at one end of the bridge they can be referred to as derivatives of 1,3-diphenylpropan-1-one [2-1]. Compound [2-1], commonly known as dihydrochalcone, is one of the simplest naturally occurring flavonoids, having been isolated from the stinkhorn fungus (*Phallus impudicus*) by List and Freund (1968). The related chalcone, compound [2-2], was isolated from *Centaurea calcitropa* (Asteraceae, sunflower family) by Dawidar and coworkers (1989). Structure [2-3], whose common name is flavone and which shows the numbering system used for flavonoids, forms the skeleton for a large of class of flavonoids and is itself a naturally occurring compound, having been found on the leaf surfaces of some species of *Primula* (Primulaceae, primrose family) in the early part of the century (Muller, 1915), and more recently in members of the genus *Pimelea* (Thymelaeaceae) (Freeman *et al.*, 1981). But simplicity is decidedly not a feature of flavonoid chemistry. It is the vastly rich array of structural variation seen in the group that as attracted the attention of natural product chemists and plant biochemists over the years. Let's begin by looking at some fundamental structural features.

Chalcones and dihydrochalcones represent the two classes of flavonoids in which the three-carbon bridge is said to be "open." In the remaining classes, the three-carbon bridge is part of a heterocyclic ring involving a phenolic group on the adjacent ring. Since this is the first mention of phenolic groups it is time to introduce them because it is their existence, number, placement, and degree of substitution that provide much of the structural variation that we see in flavonoids in Nature. Structures [2-4], [2-5], and [2-6] will serve as model compounds for this purpose. The biosynthetic pathway leading to flavonoids is such that hydroxyl groups are usually found in specific arrangements on the two rings as well as at one of the end carbons of the three-carbon bridge. The normal positions of oxygenation are shown in the common flavone "apigenin" [2-4], the common flavonol "quercetin" [2-5], and the common isoflavone "genistein" [2-6]. (We will see where these names come from later in this chapter.) The numbering convention used for identifying the positions in these compounds is indicated on the flavone system [2-3]. First, it is necessary to establish that the left hand ring (which we will see is derived from the acetate/malonate pathway) is referred to as the A-ring. The right hand ring, which is derived from the ring carbons of phenylalanine, is referred to as the B-ring. A heterocyclic ring, when present between the two, is referred to as the C-ring. According to convention, the heterocyclic oxygen is designated as position-1. Numbering of the carbons starts with the one to which the B-ring is attached, C-2, and continues around the C-ring so that the carbonyl carbon is C-4. The four available positions on the A-ring are numbered 5, 6, 7, and 8. Thus, we can see so far that apigenin [2-4] has hydroxyl groups at positions C-5 and C-7. [The carbons shared by the A- and C-rings are identified as 4a and 8a. These carbons do not figure in our discussion of substitutions but need to be identified when we look at the ^{13}C NMR spectra of flavonoids (Chapter Four)]. The B-ring carbons are given primed numbers such that the hydroxyl group in apigenin is at C-4′. Since apigenin belongs to the class of flavonoids known as flavones, its name according to the system commonly used would be 5,7,4′-trihydroxyflavone. Although proper chemical numbering would place the 4′ first (because it is the lowest number), the convention used in this book lists A-ring substituents first followed by B-ring substituents. In the case

2-1

2-2

2-3

2-4

Aryl group at position 2

OHs at positions 5,7 and 4'

2-5

Aryl group at position 2

OHs at positions 3,5,7,3' and 4'

2-6

Aryl group at position 3

OHs at positions 5, 7 and 4'

2-7

Phenylmethylene group at position 2

OHs at positions 4, 6, 3', 4' and 5'

of flavonols we have two options. We can name structure [2-5], quercetin, as a flavone derivative, i.e., 3,5,7,3′,4′-pentahydroxyflavone, or if we wish to use the flavonol option, where there is already an hydroxyl group at C-3 by definition, it could be called 5,7,3′,4′-tetrahydroxyflavonol. This second option can lead to confusion at times, so, unless special circumstances warrant, flavonols will be numbered as derivatives of flavone. A simple isoflavone, genistein, is shown as structure [2-6]. Isoflavones are isomeric to flavones by virtue of their having the B-ring attached at position-3 rather than at position-2 as in flavones proper. In all other respects, the nomenclature of isoflavones parallels that of flavones. Structure [2-7] represents a class of flavonoids known as aurones wherein the heterocyclic ring (C-ring) consists of five carbons rather than six. The numbering convention takes into consideration this C-ring shrinkage so that the A-ring carbons are numbered 4, 5, 6, and 7. These are biosynthetically equivalent to the A-ring carbons of other flavonoids. There is no change in the B-ring numbering. The numbering convention for chalcones and dihydrochalcones will be introduced in the sections dealing with these flavonoid classes.

THE COMMON LANGUAGE OF FLAVONOIDS

Common or trivial names are routinely used in much of the flavonoid literature. Many of the common names used relate in some way to the specific epithet or genus from which the compound was originally obtained. The name apigenin, for instance, is derived from *Apium*, a genus in Apiaceae (carrot family, also called Umbelliferae). "Apiin," a glycosidic derivative of apigenin, was originally isolated from *Apium graveolens* var. *dulce*, the common celery. Apigenin 4′-methyl ether was originally obtained from a species of *Acacia*, hence its common name "acacetin." Similarly, the common name "kaempferol" comes from *Kaempferia* (a relative of ginger), while "quercetin" is derived from *Quercus*, the generic name for oaks, which, incidentally, provide excellent sources of several quercetin glycosides. "Patuletin" comes from the specific epithet of *Tagetes patula*, a marigold; "morin" from the mulberry genus *Morus*; "centaureidin" from the genus *Centaurea*; "jaceidin" from the specific epithet of *Centaurea jacea*, "artemetin" from *Artemisia*; and "gossypetin" from the cotton plant, *Gossypium*. Pairs of isomeric compounds often combine part of the plant name with the "iso" term as in the case of the flavonols "rhamnetin" and "isorhamnetin" whose root derives from *Rhamnus*., the buckthorn. The "iso" preface refers to the location of the methyl ether group in these compounds and tells us that rhamnetin was the first of the pair to be identified. Other plants have contributed in several different ways. For example, the flavanone "sakuranetin" derives its common name from "sakura," the Japanese word for cherry (*Prunus cerasus*, widely cultivated). Sakuranetin can be obtained from the root bark of the cherry. One would suspect that such common names as "prunin" (an isoflavone) and "cerasin" (a chalcone) might owe their names to the cherry as well. The origin of the names "hesperitin" and "hesperidin," well known compounds from species of *Citrus*, is more complex. According to mythology, the Hesperides were the nymphs who guarded the golden apples that were given by Gaea to Hera as a wedding gift. Since both Romans and Greeks tended to call all fruits by the term *malum* (which we associate with apples, i.e., the genus *Malus*) the description of an orange as a "golden apple" makes some sort of sense. [How oranges entered the scene in the first place is another story, one engagingly told by John McPhee (1967) in his book entitled *Oranges*.] The Hesperides were immortalized in botanical terminology by means of the technical term "hesperidium" used to identify the fruit of *Citrus*, which is a type of berry. It was not too great a jump from there to flavonoids, which occur in abundance in the skins of the hesperidia.

Situations similar to that seen in *Centaurea*, where several compounds have been isolated from one genus (or one species of a genus), are fairly common. For instance, the common names "chrysosplenetin," "chrysosplenin," "chrysosplenol-C," and "chrysosplenol-D" all derive from the genus *Chrysosplenium* (Saxifragaceae), although some of these compounds have been obtained from other plants in the mean time. Another way of handling this situation is seen in the case of several flavanones obtained from the legume genus *Euchresta* (Fabaceae) which the authors simply called "euchrestaflavanone-A," "-B," and "-C" (Shiritaki *et al.*, 1981). The first-time reader should not be discouraged by the fact that these names appear to convey little structural information; most don't. Other than supposing that the compounds from *Euchresta* are flavanones, which they are, the name does not say anything about their complex structures.

Perhaps the most colorful examples of common names among the flavonoids are to be found in the class of pigments known as anthocyanidins (anthocyanins are the corresponding glycosides). Although the individual names give no clue as to the structures of the compounds, the plants from which they originally came are easily recognized. We see "delphinidin" from *Delphinium*, "malvidin" from *Malva*, "pelargonidin" from *Pelargonium*, "peonidin" from *Paeonia*, and "petunidin" from the *Petunia*. The original generic source of "rosinidin" should not prove too difficult a puzzle! A somewhat unusual anthocyanidin named "apigeninidin" does not give any information about its plant source but a worker acquainted with the structure of apigenin should infer some structural resemblance between the two compounds.

The *Chemical Abstracts* system of nomenclature, which we will meet below, is precise in conveying its message, but is cumbersome and requires a knowledge of chemistry beyond that necessary to understand and appreciate the biological significance of flavonoids. Conversely, the use of common names is convenient but conveys little meaning to those individuals unfamiliar with "flavojargon." Adding to the problem is the fact that workers in each specialist laboratory are comfortable with a certain subset of common names. For example, workers familiar with isoflavones (common in legumes) would have some difficulty communicating with people working with Asteraceae where isoflavones are rare. Some accommodation is obviously necessary. This is achieved in a haphazard fashion. It is generally assumed that most workers are familiar with commonly occurring compounds: apigenin, luteolin, and tricin among the flavones, kaempferol, quercetin, myricetin, gossypetin, and quercetagetin among the flavonols, naringenin and eriodictyol among the flavanones, aromadendrin (dihydrokaempferol) and taxifolin (dihydroquercetin) among dihydroflavonols. It is useful to know that "genistein" [2-6] from the legume genus *Genista*) and "orobol" (from *Orobus*, also a legume) are isoflavones with substitution patterns equivalent to apigenin and luteolin (5,7,3′,4′-tetrahydroxyflavone), respectively. Other common names will appear in the following text.

When we deal with compounds that are not likely to be widely known by their common names, it is necessary to know the flavonoid class involved and its associated numbering system. Thus, for workers who are familiar with the flavonoids of citrus fruits, the name "nobiletin" is a meaningful term. For the rest of us, however, it will be necessary to write this compound as 5,6,7,8,3′,4′-hexamethoxyflavone. Likewise, "tlatlancuayin" would likely be a mystery to all but a very few. The general reader might appreciate the latter compound better by learning that it is 5,2′-dimethoxy-6,7-methylenedioxyisoflavone.

A FORMAL NOMENCLATURE SYSTEM

As we have seen, it is common practice in flavonoid studies to use a class name, i.e., chalcone, dihydrochalcone, flavone, or flavonol (and others) to represent the type of compound under consideration. A visit to *Chemical Abstracts*, however, will not turn up much information under these names. The basic structural unit in flavonoids having a six-membered C-ring is 4*H*-1-benzopyran-4-one [2-8]. A phenyl group normally occurs at C-2 so that flavones are considered derivatives of 2-phenyl-4*H*-1-benzopyran-4-one. Apigenin, which we saw above as 5,7,4′-trihydroxyflavone [2-4], would be 2-(4-hydroxyphenyl)-5,7-dihydroxy-4*H*-1-benzopyran-4-one. The actual entry in *Chemical Abstracts* would be based upon the root compound, benzopyran-4-one in this case, so that the entry would be: 4*H*-1-Benzopyran-4-one, 2-(4-hydroxyphenyl). Information on the equivalent isoflavone, genistein [2-6], would be found under 4*H*-1-benzopyran-4-one, 3-(4-hydroxyphenyl)-5,7-dihydroxy. Quercetin [2-5] would be called 2-(3,4-dihydroxyphenyl)-3,5,7-trihydroxy-4*H*-1-benzofuran-4-one. Flavanones, which look like flavones except that they do not have the C-2/C-3 double bond, would add the "2,3-dihydro" term to the appropriate name. The flavanone equivalent to apigenin, naringenin [2-10], would be known formally as 2-(4-hydroxyphenyl)-2,3-dihydro-5,7-dihydroxy-4*H*-1-benzofuran-4-one. Dihydroflavonols, which bear the same structural relationship to flavonols as flavanones do to flavones, are named in this system by adding the "2,3-dihydro" term to the appropriate flavonol name. Thus, dihydroquercetin [2-11] would be found under 4*H*-1-benzofuran-4-one, 2-(3,4-dihydroxyphenyl)-2,3-dihydro-3,5,7-trihydroxy. Additional terms necessary to describe stereochemical features of flavanones and dihydroflavonols, which are illustrated in structures [2-10] and [2-11], will be introduced below.

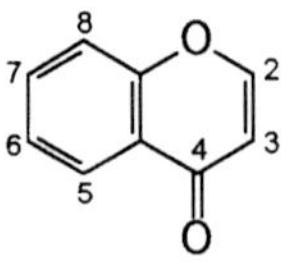

2-8

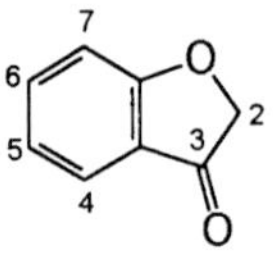

2-9

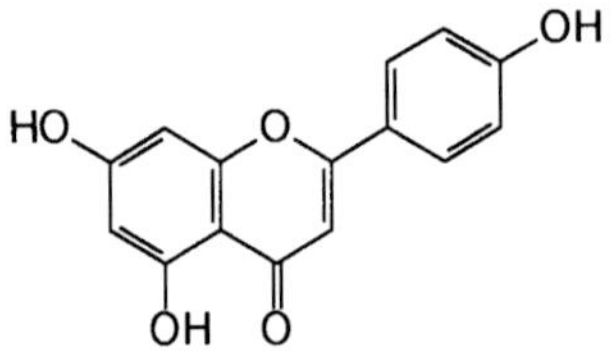

2-4

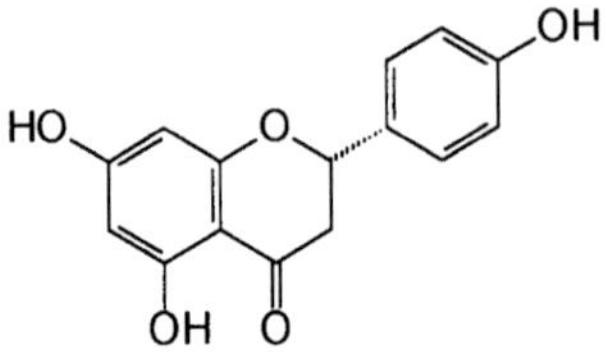

2-10

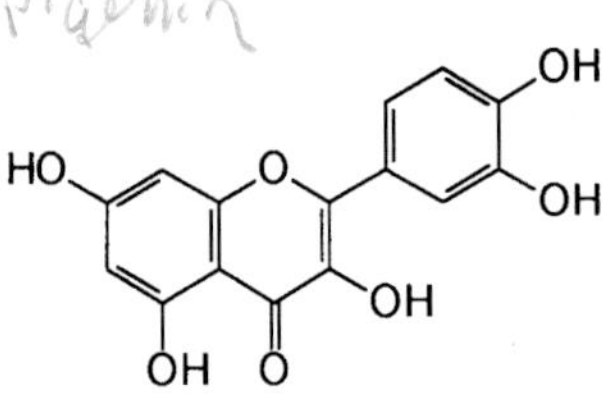

2-5

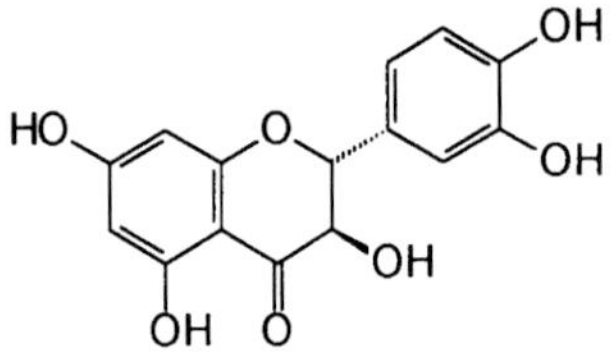

2-11

Flavonoids with a five-membered C-ring, the aurones and closely related auronols, can be found in *Chemical Abstracts* under the general entry 3(2*H*)-benzofuranone

[2-9]. Derivatives of aurone would be known formally as 2-phenylmethylene- 3(2*H*)-benzofuranone. Substituents are added in the same fashion as we saw for the benzopyranone derivatives above. Thus, the aurone bracteatin [2-9] would appear as 3(2*H*)benzofuranone, 2-(3,4,5-trihydroxyphenyl)methylene-4,6-dihydroxy.

GENERAL STRUCTURAL FEATURES OF FLAVONOIDS

This section is intended to give a very brief overview of the major structural features that characterize flavonoids; detailed information on each class of compounds will appear later in the chapter. Our initial goal is an examination of the structural features that define the classes of naturally occurring flavonoids: (1) whether the three-carbon bridge is open or whether it is involved as part of a heterocyclic ring; (2) if a heterocyclic ring exists; whether it is five- or six-membered; (3) the position of attachment of the B-ring; (4) the level of oxidation of the C-ring; and (5) whether the compound is monomeric, dimeric, or is a higher oligomer. The following material repeats some of the points made above, in somewhat different context, and can be skipped by the more experienced reader. I feel that some repetition is useful at this stage for the newcomer to flavonoid chemistry.

Two classes of flavonoids lack a C-ring, chalcones and dihydrochalcones. The parent compounds appeared above as [2-2] and [2-1], respectively. Examples of naturally occurring members of these classes are the chalcone isoliquiritigenin [2-12] and the dihydrochalcone phloretin [2-13] both of which exhibit comparatively simple substitution patterns. Many chalcones and dihydrochalcones are known that have additional oxygen-containing functions on one or both rings as well as on the remaining carbons. (Don't worry about the common names of these two compounds! The name "phloretin" tells us nothing, while the best we can get from "isoliquiritigenin" is that it is likely an isomer of a compound called "liquiritigenin.") It is important to recognize that the numbering convention for chalcones and dihydrochalcones differs from that used for other flavonoid classes. In these two classes, the primed numbers refer to the A-ring positions and the unprimed numbers refer to the B-ring. Thus, isoliquiritigenin [2-12] is 2′,4′,4-trihydroxychalcone and phloretin [2-13] is 2′,4′,6′,4-tetrahydroxydihydrochalcone. This apparent numbering switch is an artifact of chalcone chemistry that reflects their kinship with acetophenones whose ring carbons were often identified with primed numbers. The sequence of numbering used for both chalcones and dihydrochalcones adheres to our rule that the A-ring positions will always be presented first.

All other flavonoids are based upon three rings, the A- and B-rings plus a heterocyclic ring involving two or three of the bridge carbons. (We ignore heterocyclic systems produced by cyclization of sidechain prenyl functions, which do not involve bridge carbons.) Aurones, and the closely related auronols, are characterized by the presence of a five-membered heterocyclic ring, the rest by having six-membered heterocycles. A typical member of this class, bracteatin [2-7], was seen above. Another commonly encountered aurone, particularly in certain groups of Asteraceae, is "sulfuretin" [2-14], whose aurone-based name is 6,3′,4′-trihydroxyaurone. Note that there is no hydroxyl group at position-4. This is not a mistake; some

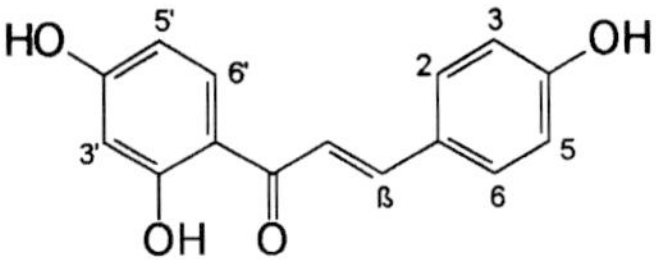

2-12

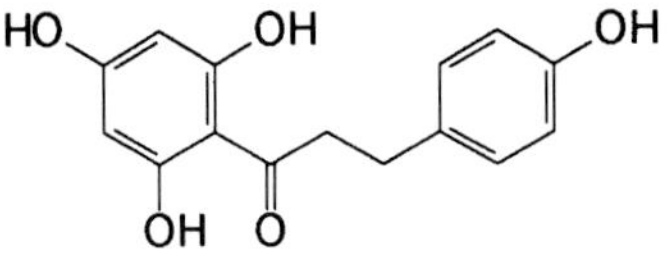

2-13

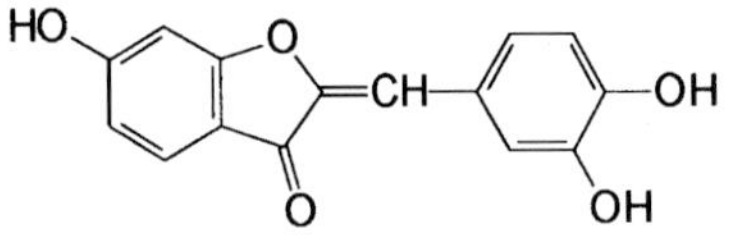

2-14

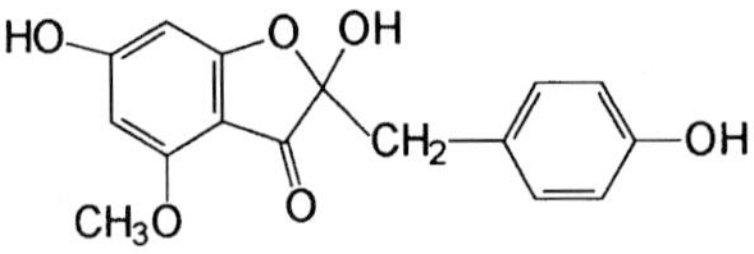

2-15

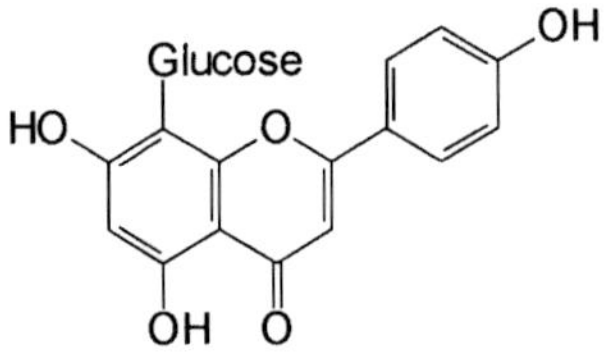

2-16

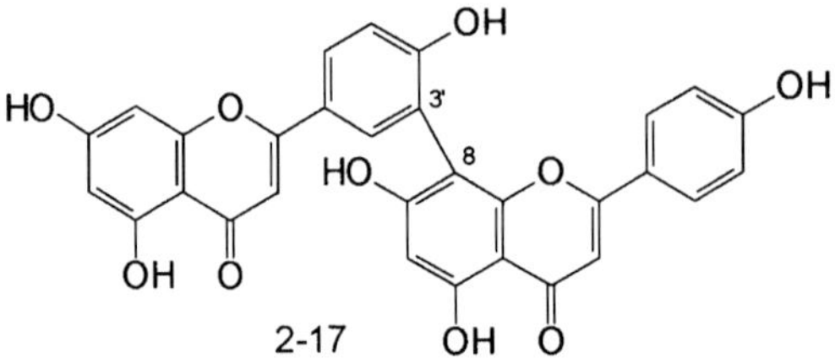

2-17

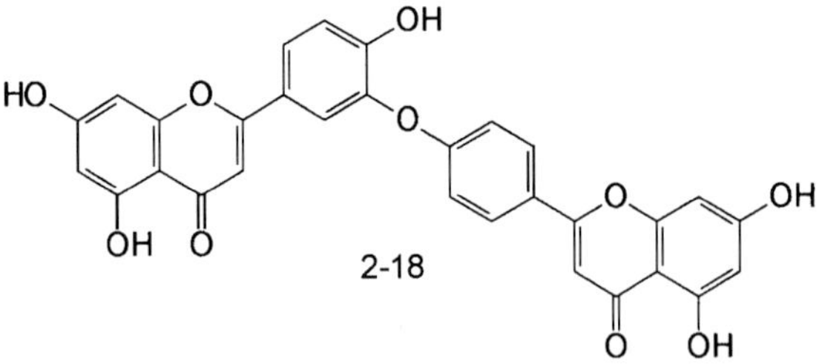

2-18

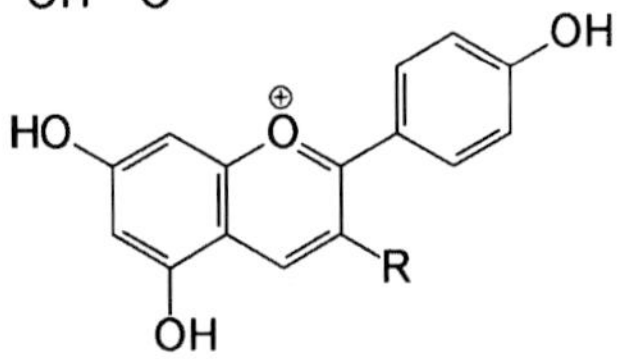

2-19 R = OH

2-20 R = H

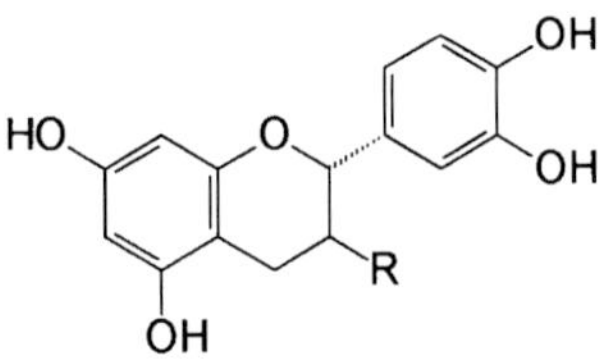

2-21 R = —OH

2-22 R = ·····OH

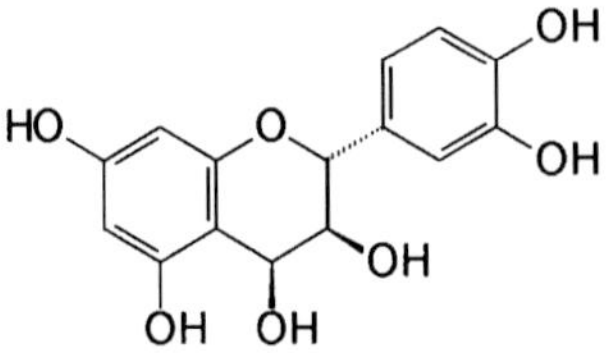

2-23

flavonoids lack one or more hydroxyl groups. 5-Deoxyflavonoids (equivalent to 4-deoxyaurones), are, in fact, quite common in legumes. Structure [2-15] is the auronol "carpasin," whose formal (*Chemical Abstracts*) name would be based upon 2-hydroxy-2-benzylcoumaranone. Carpasin would be more familiarly known as 6,4′-dihydroxy-4-methoxyauronol.

The remaining flavonoid classes all have six-membered C-rings, but vary with regard to the point of attachment of the B-ring, the oxidation level of the C-ring carbons, and how many monomeric units are involved. Naringenin [2-10], one of whose glycosides imparts the bitter taste to grapefruit, is representative of the class of flavonoids known as flavanones. These compounds are characterized by having a saturated C-ring with the B-ring attached at C-2. Flavones, typified by apigenin [2-4] above, differ from flavanones in having a double bond between carbons C-2 and C-3. The B-ring is attached at C-2 and there is no substituent (usually) at C-3. Isoflavones differ from flavones in having their B-rings attached at C-3, as seen in the structure of genistein [2-6] above.

Flavonols are characterized by having their B-ring attached at C-2, as with flavones and flavanones, but differ from the latter by possessing an hydroxyl group at C-3. Quercetin [2-5] is typical of this class of flavonoids. Dihydroflavonols constitute a related class that differs in lacking the double bond between C-2 and C-3. Dihydroquercetin, whose common name is taxifolin, is shown as structure [2-11].

Many representatives of most classes of flavonoids occur in the natural state in the form of *O*-glycosides. There is a group of glycosidic derivatives, however, that is accorded status as a class of flavonoids in their own right. These *C*-glycosylflavonoids are characterized by having one or two sugar units linked directly to the aromatic nucleus through carbon–carbon bonds. Vitexin [2-16] is shown as an example of this class of compounds. Others are known that have two *C*-linked sugars and further elaboration of the compound may involve *O*-methylation or *O*-glycosylation. The majority of *C*-glycosylflavonoids are flavones, but a few *C*-glycosides of other classes of flavonoids are known.

Biflavonoids consist of two flavonoids linked through either a carbon–carbon or carbon–oxygen bond. The compound shown in [2-17] is "amentoflavone," one of the earliest of this class of flavonoids to be described. It consists of two apigenin units linked via the C-3′ position of one and the C-8 position of the other. In "ochnaflavone" [2-18], luteolin, a flavone with two hydroxyl groups on the B-ring, is linked to apigenin through oxygen. In addition to the existence of such heterodimers, as opposed to homodimers (amentoflavone), biflavonoids can involve members of different classes of flavonoid. They can also involve linkage via bridge carbons. The special nomenclatural rules needed for these compounds will be discussed below. A few triflavonoids are also known.

Anthocyanins, the main source of red, mauve, purple, and blue flower colors in most plants consist of an anthocyanidin coupled with one or more sugar units. The flavonoid ring system in anthocyanins is often written as the flavylium ion, an example of which, "cyanidin," is shown as [2-19]. These compounds carry a formal positive charge on the flavonoid moiety. The structures actually responsible for flower color, which invariably are *O*-glycosides, can be very complex and will be dealt with in some detail below. 3-Deoxyanthocyanins, which obviously lack the 3-OH group normally seen in anthocyanins, and here represented by apigeninidin [2-20], are of more limited occurrence in plants.

The term flavanol refers to a group of flavonoids that lack the C-4 carbonyl but have one or two hydroxyl groups on the C-ring. The most familiar of these are the flavan-3-ols, typical examples of which are (+)-catechin [2-21] and (−)-epicatechin [2-22]. Compounds with 3′,4′,5′-hydroxylation and corresponding stereochemistry are called (+)-gallocatechin and (−)-epigallocatechin, respectively. Flavan-3,4-diols are important compounds that are involved in the formation of the so-called condensed tannins. As can be seen in [2-23], which can be referred to as 2,3-*trans*-3,4-*trans*-leucocyanidin, these compounds contain three chiral centers. Owing to the possession of a saturated C-ring, these compounds are often collectively referred to as flavans: flavan-3-ols, flavan-3,4-diols, flavans (with no substitution other than the phenyl group at C-2), and flav-3-enes.

STRUCTURAL ELABORATIONS

Oxygenation Pattern

The oxygen atoms at C-4, C-5, C-7, C-4′, and in the heterocyclic ring of flavones, flavonols, flavanones, isoflavones, etc., at C-3, C-4, C-6, C-4′, and in the heterocyclic ring of aurones, and at C-2′, C-4′, C-6′, and C-4, as well as the carbonyl oxygen, of chalcones (and dihydrochalcones) arise from the precursors of the flavonoid skeleton. The A-ring oxygens come from malonyl coenzyme-A; the B-ring oxygen and the carbonyl oxygen from *p*-coumaroyl coenzyme-A (Chapter Six). Thus, 2′,4′,6′,4-tetrahydroxychalcone, the corresponding dihydrochalcone, the flavanone naringenin, the flavone apigenin, the isoflavone genistein, and the 3-deoxyanthocyanidin apigeninidin are compounds whose oxygenation pattern represents the fundamental biosynthetic condition. In all other situations, which include the 3-hydroxyl group of dihydroflavonols and flavonols, additional oxygenation on the bridge carbons of chalcones and dihydrochalcones, extra oxygenation on the A-ring, and oxygens on the B-ring other than C-4′, are the result of reactions that occur after the flavonoid ring system has been formed.

Many naturally occurring flavonoids exhibit oxygenation patterns with at least one additional oxygenation, relative to the basic pattern. Quercetin and its derivatives, arguably among the most widely occurring of all flavonoids, belong to the "extra" oxygenation group, although most casual workers in the field probably don't consider them thus. Two of quercetin's oxygens, the ones at C-3 and C-3′, are introduced after the formation of the flavonoid rings has been completed. A third oxygenation, parallel to the establishment of the quercetin 3′-substitution, results in the formation of the 3′,4′,5′-trioxygenated B-ring pattern characteristic of myricetin. The three flavonols, kaempferol [2-24], quercetin [2-25], and myricetin [2-26], in the form of a myriad of glycosidic derivatives, are the most widely occurring members of this class. Several other flavonol B-ring oxygenation patterns are known, but they are much less frequently encountered and, if not restricted to, are at least more often found in certain families than in others. For example, 3,5,7,2′,4′-pentahydroxyflavone [2-27], commonly known as morin, occurs widely in Moraceae (mulberry family) and sporadically in other families.

Extra oxygenation on the A-ring is possible at two positions, C-6 and C-8. Flavonols are known that exhibit oxygenation at one or the other and a few are known in which both positions are substituted. One of the more frequently encountered flavonols with extra A-ring oxygenation is 3,5,6,7,3′,4′-hexahydroxyflavone [2-28], which can be considered as 6-hydroxyquercetin, but is more commonly referred to as

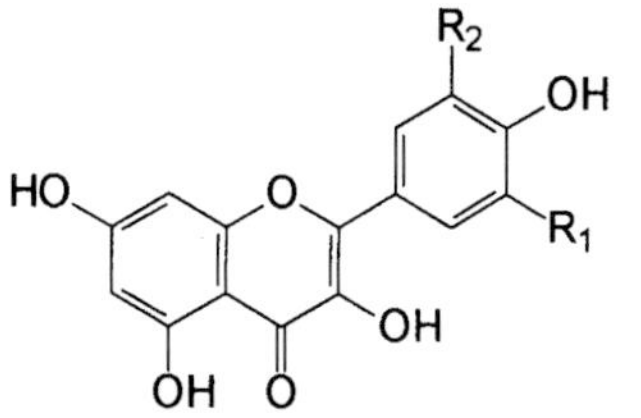

2-24 $R_1 = R_2 = H$ (K)
2-25 $R_1 = OH$, $R_2 = H$ (Q)
2-26 $R_1 = R_2 = OH$ (M)

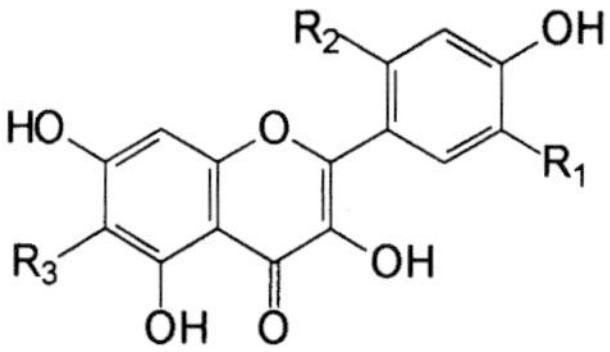

2-27 $R_1 = R_3 = H$, $R_2 = OH$
2-28 $R_1 = R_3 = OH$, $R_2 = H$ (Qt)

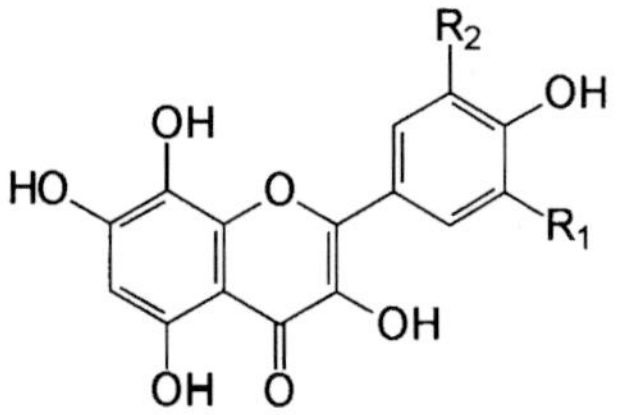

2-29 $R_1 = R_2 = H$ (Herb)
2-30 $R_1 = OH$, $R_2 = H$ (Goss)
2-31 $R_1 = R_2 = OH$ (Hib)

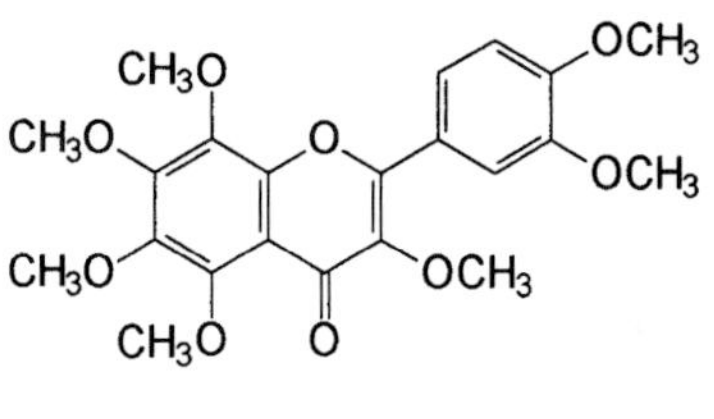

2-32

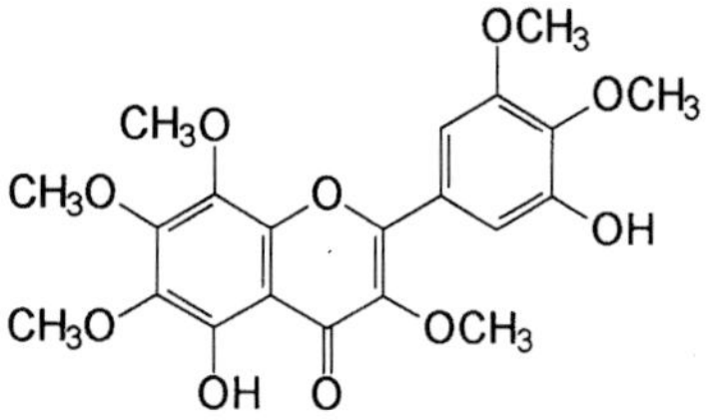

2-33

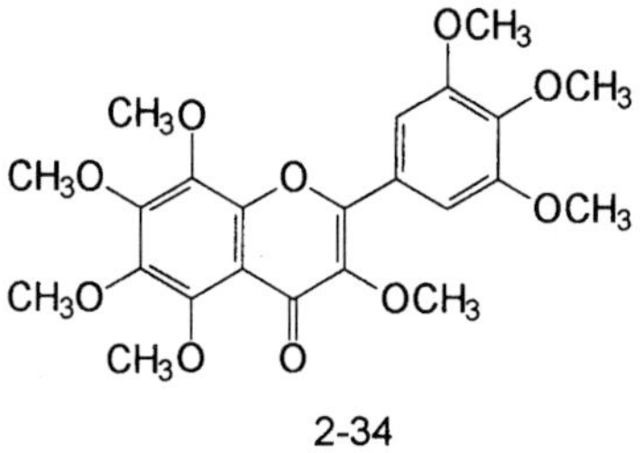

2-34

quercetagetin. Quercetagetin derivatives, both methyl ethers and glycosides as well as combinations thereof, are widely distributed in members of Asteraceae, but occur in several other families as well. The equivalent 6-hydroxy- and 6-methoxy-kaempferol derivatives, which seem not to have acquired common names, are also known, both in Asteraceae and in other plant families. The 8-oxygenated derivatives of kaempferol, quercetin, and myricetin, known respectively as herbacetin [2-29], gossypetin [2-30], and hibiscetin [2-31], are well known flavonols with somewhat more restricted distributions than we see for the parent flavonols. All three of the 8-hydroxy compounds occur in members of Malvaceae (the mallow family), herbacetin and gossypetin from species of *Gossypium* (cotton) and hibiscetin from species of *Hibiscus*, among others. (The common name for herbacetin derives from

G. herbaceum, a diploid African cotton species.) Although these compounds can be thought of as characteristic of Malvaceae, it is interesting to note that glycosidic derivatives of herbacetin are also known from members of Crassulaceae, Papaveraceae (poppy family), Primulaceae (primrose family), Rosaceae (rose family), Scrophulariaceae (snapdragon family), as well as *Equisetum* (horsetails). Naturally occurring *O*-methylated derivatives of herbacetin, as well as of the other 8-hydroxy flavonols, are also well known natural products.

6,8-Dioxygenated derivatives of kaempferol, quercetin, and myricetin are also known compounds, but are represented in the plant kingdom much more frequently as *O*-methylated derivatives. Examples of these highly substituted flavonols are 3,5,6,7,8,3′,4′-heptamethoxyflavone [2-32], known from several *Citrus* species, 5,3′-dihydroxy-3,6,7,8,4′,5′-hexamethoxyflavone (digicitrin) [2-33] from *Digitalis purpurea*, and 3,5,6,7,8,3′,4′,5′-octamethoxyflavone (digicitrin dimethyl ether) [2-34] from *Murraya exotica*. *Citrus* and *Murraya* are both members of Rutaceae (rue family); *Digitalis* belongs to Scrophulariaceae. These are some of the more extreme examples of *O*-methylation; many flavonoids with these of oxygenation patterns exist that have lower levels of *O*-methylation.

We find a similar situation in the realm of the flavones *per se*. The five oxygen atoms present in apigenin [2-4] are derived from original precursors, whereas luteolin [2-35] is homologous to quercetin in that its 3′-oxygen enters the picture after the flavonoid ring system has been established. Tricetin [2-36] has oxygen at C-5′ as well and represents a situation parallel to myricetin. As in the case of flavonols, other flavone B-ring patterns are known, a few of which are highly unusual. Extra oxygenation on the A-ring of apigenin or luteolin, followed by *O*-methylation and/or glycosylation, results in the formation of a large number of compounds. 6-Hydroxyapigenin [2-37], known commonly as scutellarein, a name it acquired from the mint genus *Scutellaria*, and its derivatives are widespread in Lamiaceae as well as in many other families, e.g. Asteraceae, Betulaceae (birch family), Verbenaceae. Scutellarein 6-methyl ether (hispidulin) [2-38] its 6,7-dimethyl ether (cirsimaritin) [2-39], and its 6,4′-dimethyl ether (pectolinarigenin) [2-40] are present in many members of Asteraceae and Lamiaceae, and occur in a number of other plant groups as well including several ferns. "Isoscutellarein," the 8-hydroxy isomer of scutellarein, is also known but does not appear to be as widely distributed in Nature as scutellarein itself. Apigenin with extra oxygenation at both C-6 and C-8 represents a fairly rare substitution . Most of the naturally occurring derivatives of this flavone appear to be *O*-methylated and are known from a variety of sources including Asteraceae, Lamiaceae, and ferns.

6-Hydroxyluteolin [2-41], the flavone analog of quercetagetin, is known from species of *Lippia* (Verbenaceae) (Skaltsa and Shammas, 1988), *Thymus* (Lamiaceae) (Adzet *et al*., 1986b, cited by Wollenweber, 1994), and *Viguiera* (Asteraceae) (Valant-Vetschera and Wollenweber, 1988). Its 6-methyl ether, "nepetin" or "eupafolin" [2-42], however, is widely distributed in members of several families. The 8-hydroxy isomer, common name "hypolaetin," occurs as the 7-*O*-glucoside [2-43] in *Juniperus macropoda* (Cupressaceae) (Siddiqui and Sen, 1971), and in several angiosperm species as *O*-methyl derivatives. 5,6,7,8,3′,4′-Hexahydroxyflavone exists as the base molecule for a variety of *O*-methylated derivatives, again from several plant families.

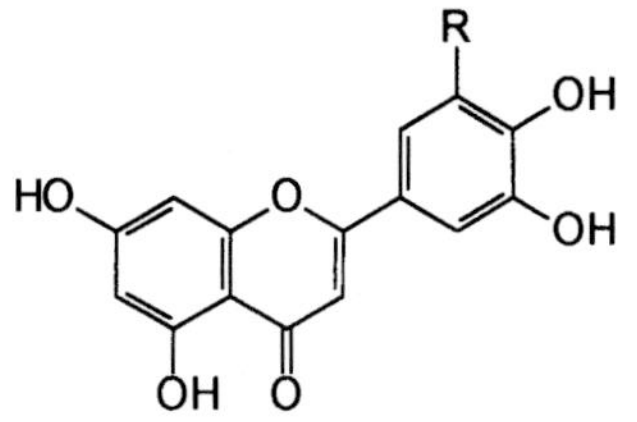

2-35 R = H
2-36 R = OH

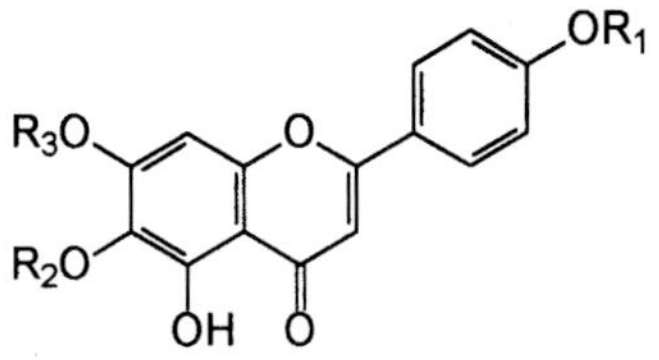

2-37 $R_1 = R_2 = R_3 = H$
2-38 $R_1 = R_3 = H, R_2 = CH_3$
2-39 $R_1 = H, R_2 = R_3 = CH_3$
2-40 $R_1 = R_2 = CH_3, R_3 = H$

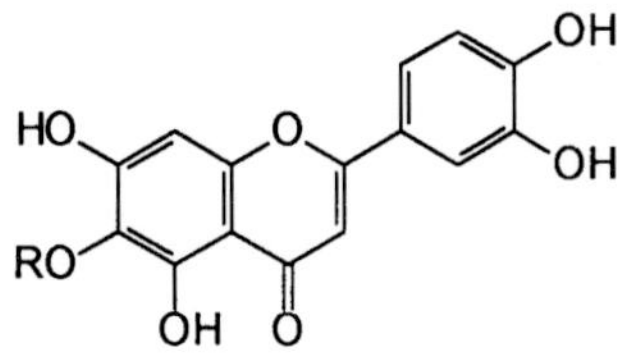

2-41 R = H
2-42 $R = CH_3$

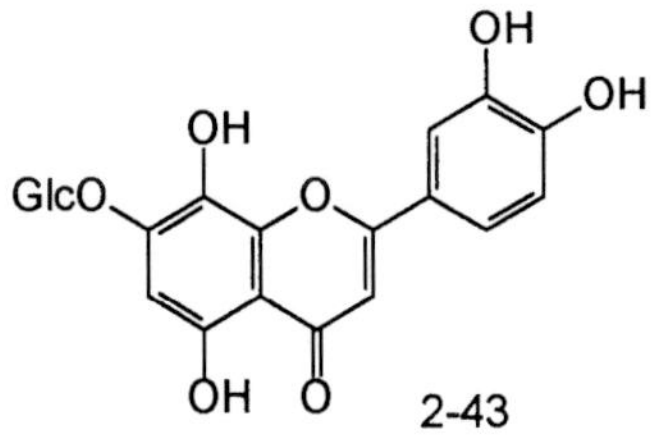

2-43

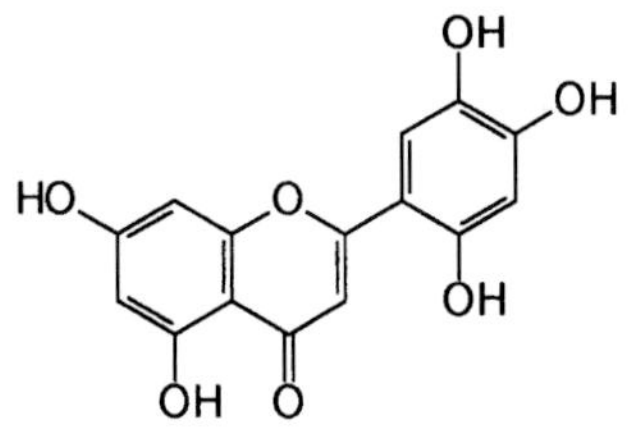

2-44

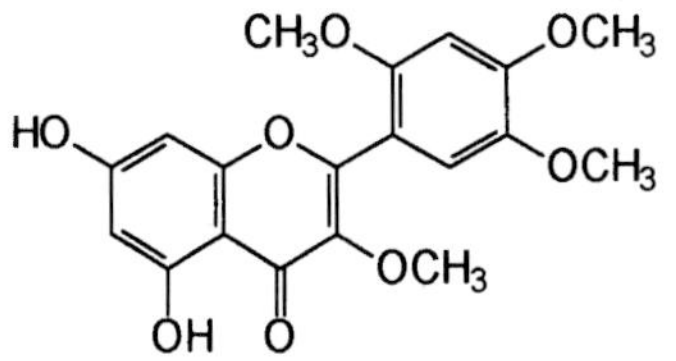

2-45

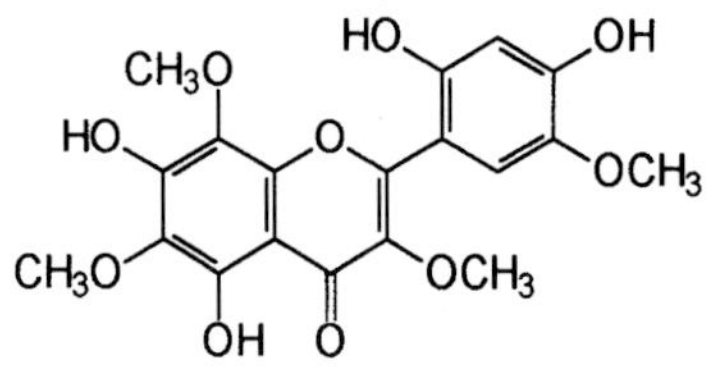

2-46

Tricetin-based flavones with 6-, 8-, or 6,8-oxygenation exist but, as seen repeatedly in this highly oxygenated series, appear to occur mainly as methyl ethers.

The foregoing has dealt with a repeating pattern of increasing oxygenation in both rings, but has not strayed too far into the, admittedly arbitrary, arena of unusual oxygenation patterns. Let's look at some of these now. Several groups of plants make flavonoids with a 2′,4′,5′-substituted B-ring. "Isoëtin," 5,7,2′,4′,5′-pentahydroxyflavone [2-44], derives its name from *Isoëtes*, (Isoëtaceae, the quillworts) (Voirin *et al*., 1975). Glycosides of isoëtin have been reported from several genera of Asteraceae (Marco *et al*., 1988; Gluchoff-Fiasson *et al*., 1991; Harborne, 1991). Isoëtin methyl ethers have been reported from members of four genera of Asteraceae, *Ageratum* (Quijano *et al*., 1987), *Brickellia* (Iinuma *et al*., 1985 in a correction of Roberts *et al*., 1984), *Gutierrezia* (Fang *et al*., 1985a,b; Li *et al*., 1987), and *Gymnosperma* (Yu *et al*., 1988). An example from the last named citation is 5,7,4′,5′-tetrahydroxy-3,2′-dimethoxyflavone [2-45] from *Gymnosperma glutinosum*. Additional A-ring oxygenation coupled with the 2′,4′,5′-system in the latter two genera lead to 3,5,6,7,8,2′,4′,5′-octaoxygenated flavonoids, e.g., [2-46].

Several complex flavonoids characterized by the presence of 2′,4′,6′-trioxygenation have been reported during the past few years. Iinuma and colleagues (1992, 1994 and citations therein) isolated *C*-alkylated flavanone derivatives from *Echinosophora koreensis* and from *Sophora exigua* (both Fabaceae). Typical of these are (2*S*)-5,7,2′,6′-tetrahydroxy-4′-methoxy-8-*C*-prenylflavanone [2-47] from the former and (2*S*)-5,7,2′,4′,6′-pentahydroxy-8-*C*-lavandulylflavanone [2-48] from the latter. Compounds with this kind of B-ring are not unique to legumes, but are also known from *Artocarpus heterophyllus* (Moraceae) (Lu and Lin, 1993), examples of which are 5-hydroxy-7,2′,4′,6′-tetramethoxyflavanone [2-49] and the corresponding 8-*C*-prenyl derivative [2-50]. The phloroglucinol (1,3,5-trihydroxybenzene) oxygenation pattern is generally taken as indicating biosynthesis *via* the acetyl coenzyme-A/malonyl coenzyme-A route which, as we will see in Chapter Six, is the route for formation of flavonoid A-rings. The B-ring pattern in *Sophora* is an A-ring mimic, but it is highly unlikely that these compounds are formed by a pathway other than the normal flavonoid pathway, i.e., the B-ring is derived from *p*-coumaroyl CoA.

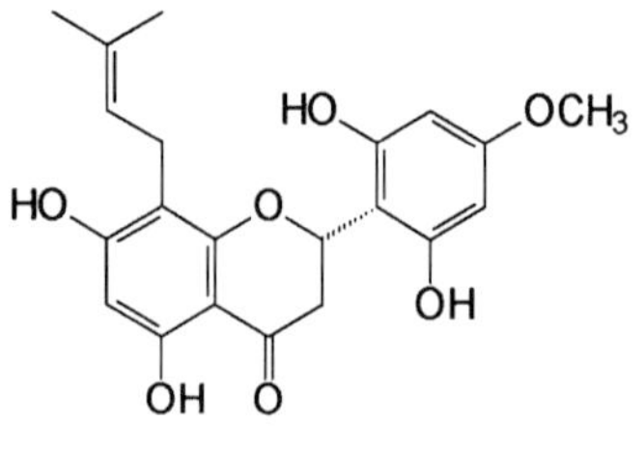

2-47

(Group at C-8 is prenyl)

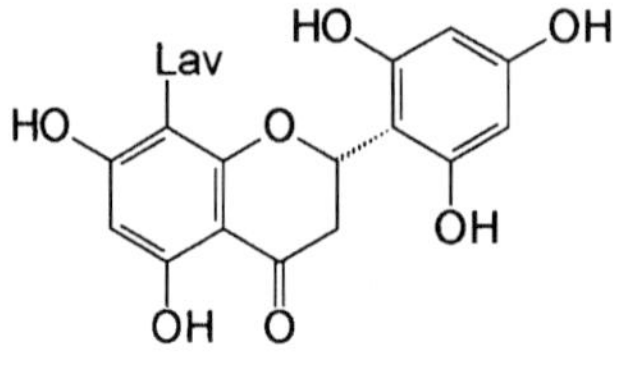

2-48

Lav = Lavandulyl

(5-methyl-2-isopropenylhex-4-enyl)

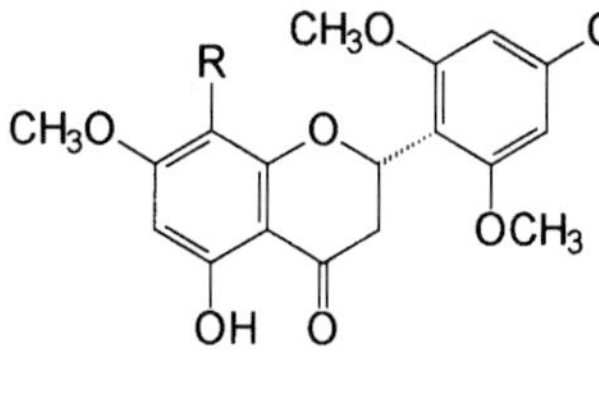

2-49 R = H

2-50 R = Prenyl

Three possible configurations are possible for arranging four oxygens on the B-ring. Several compounds are known that exhibit a 2′,3′,4′,5′-tetraoxygenated B-ring. Flavonoids with this B-ring configuration have been reported from two members of Asteraceae, *Brickellia glutinosa* (Goodwin *et al*., 1984), and *Ageratum houstonianum* (Quijano *et al*., 1987). The former afforded 5,3′-dihydroxy-6,7,2′,4′,5′-pentamethoxyflavone [2-51] while the latter gave hexa-, hepta-, and octa-*O*-methylated derivatives of 5,6,7,8,2′,3′,4′,5′-octahydroxyflavone, e.g., [2-52]. One of the other configurations, a 2′,3′,4′,6′-oxygenated B-ring, is seen in compound [2-53], which Malan (1993) obtained from the yellow heartwood of *Distemonanthus benthamianus* (Fabaceae). Flavonoids with 2′,3′,5′,6′-oxygenation do not appear to have been found to occur in Nature.

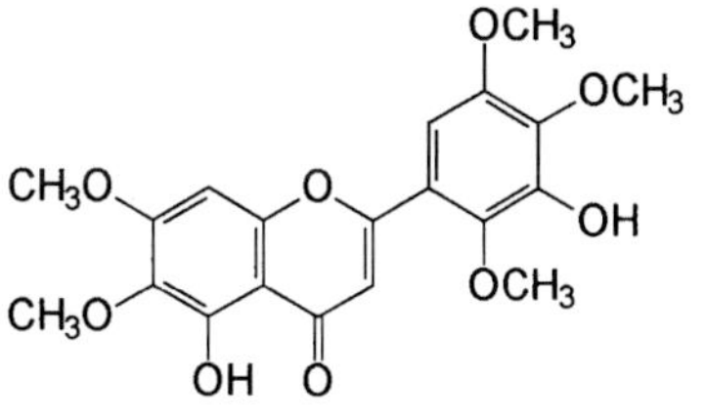

2-51

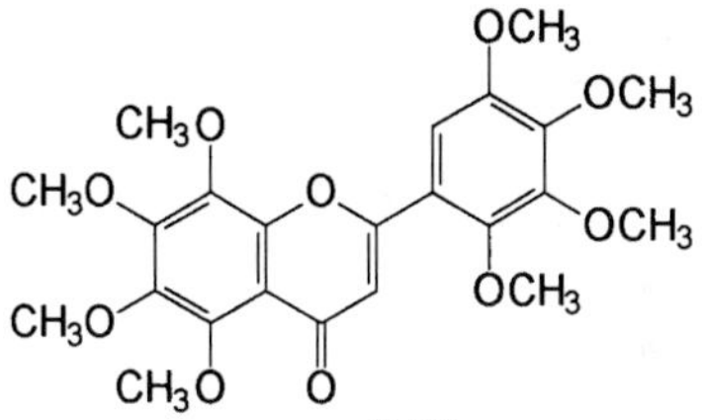

2-52

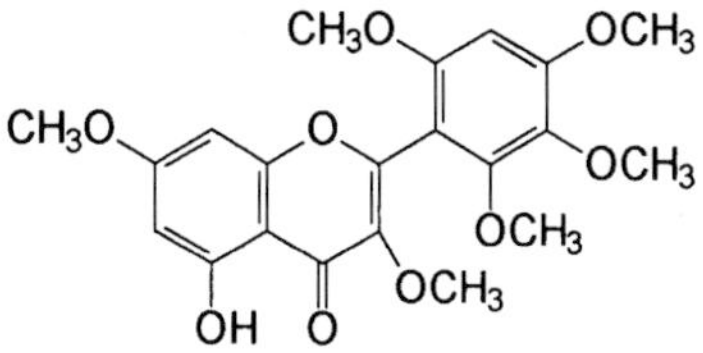

2-53

Many of the remaining examples of flavonoids with unusual oxygenation patterns appear to occur in no particular taxonomically clear-cut grouping. Let's look at one that does. Foremost in this category are the flavone derivatives produced by species of *Primula* and other genera of Primulaceae. In addition to flavonoids with normal oxygenation patterns, members of this genus produce an array of flavones with few or no hydroxyl groups. Flavone itself [2-3] was first isolated from the granular exudate (farina) on leaves of *P. pulverulenta* and *P. japonica* (Muller, 1915) and subsequently found to occur in (or rather, on) many other species (Blasdale, 1945). 5-Hydroxyflavone (primuletin) [2-54] has also been reported, as have 2′-methoxyflavone [2-55], 6-methoxyflavone [2-56], 2,2′-dihydroxyflavone [2-57], and 6,2′-dimethoxyflavone [2-58] (Wollenweber and Mann, 1986). These compounds are not restricted to the leaf surface as shown by Ahmad and coworkers (1991) who described 7,2′-dihydroxyflavone 7-*O*-glucoside [2-59] from the polar fraction of *P. macrophylla*. 2′-Hydroxyflavone (not shown) has been reported from the unrelated plant *Daphnopsis selloniana* (Thymelaeaceae) (Blasko *et al*., 1988). Other unusual flavonoid oxygenation patterns have been identified from a wide array of species. Table 2.1 lists some of these along with their sources. Additional examples have been compiled by Wollenweber (1994).

Many flavonoids are known that lack oxygen atoms from one or more of the positions normally occupied by them. Commonly seen in this category are flavonoids lacking oxygen at the C-5 position (equivalent to the C-4 position in aurones and the C-2′ position in chalcones and dihydrochalcones). Evidence suggests that oxygen atoms are lost from these positions during the early stages of the biosynthetic pathway before the first fifteen-carbon product is freed from the synthetic enzyme(s). Examples of naturally occurring A-ring deoxyflavonoids are

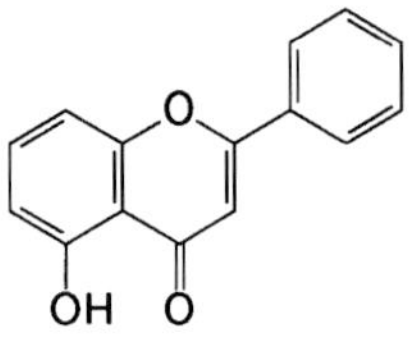

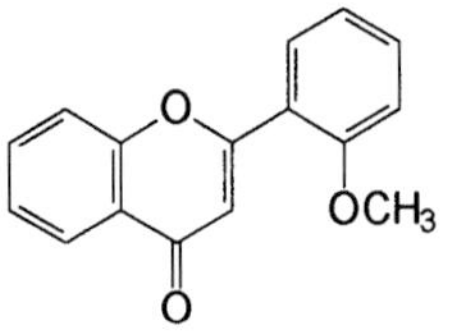

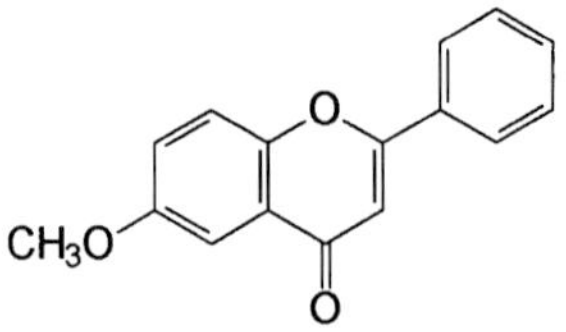

Table 2.1 A sampling of unusual flavonoid oxygenation patterns

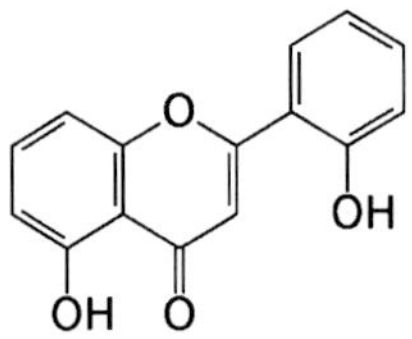

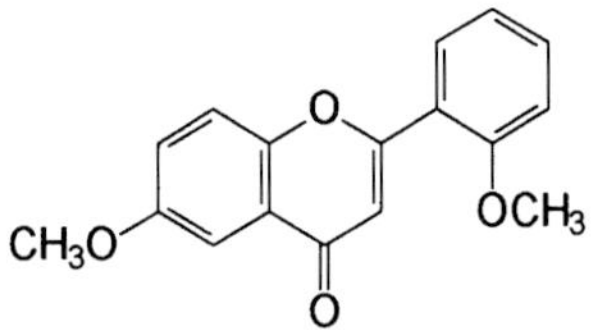

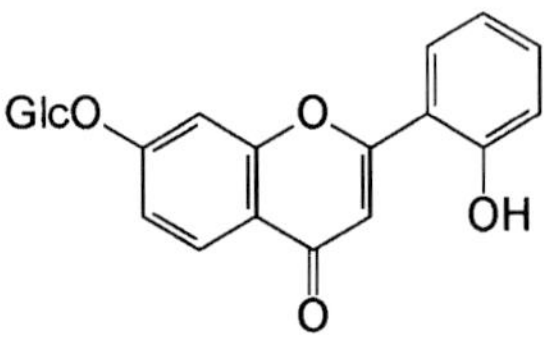

Oxygenation pattern	Species	Family	Reference
3,3′ (chalcone)	*Primula macrophylla*	Primulaceae	Ahmad *et al.*, 1992
5,6	*Primula pulverulenta*	Primulaceae	Wollenweber and Mann, 1986
5,7,3′	*Flourensia retinophylla*	Asteraceae	Stuppner and Müller, 1994
5,6,4′	*Artemisia campestris*	Asteraceae	Rauter *et al.*, 1989
5,8,4′	*Artemisia campestris*	Asteraceae	Rauter *et al.*, 1989
3,5,7,2′	*Datisca cannabina*	Datiscaceae	Grisebach and Grambow, 1968
5,7,2′	*Iris pseudoacorus*	Iridaceae	Hanawa *et al.*, 1991
5,6,7,2′	*Iris pseudoacorus*	Iridaceae	Hanawa *et al.*, 1991
5,8,3′,5′	*Artemisia subdigitata*	Asteraceae	Shi *et al.*, 1992
6,7,3′,5′	*Inula grantioides*	Asteraceae	Ahmad and Ismail, 1991
3,6,7,3′,4′	*Millingtonia hortensis*	Bignoniaceae	Bunyapraphatsara *et al.*, 1989
5,7,8,2′,3′	*Andrographis paniculata*	Acanthaceae	Kuroyanagi *et al.*, 1987
3,5,7,2′,5′	*Scutellaria planipes*	Lamiaceae	Zhang *et al.*, 1994
3,5,7,3′,5′	*Baileya multiradiata*	Asteraceae	Dominguez *et al.*, 1976
3,5,6,7,8,3′	*Gymnosperma glutinosa*	Asteraceae	Dominguez and Torre, 1974
3,5,7,8,2′,3′	*Notholaena* species	Pteridophyte	Wollenweber *et al.*, 1988c
5,6,7,8,2′,3′,6′	*Scutellaria planipes*	Lamiaceae	Zhang *et al.*, 1994
5,7,8,2′,3′,4′	*Andrographis paniculata*	Acanthaceae	Kuroyanagi *et al.*, 1987

2′,4′,4-trihydroxychalcone (isoliquiritigenin) [2-12], the isomeric 7,4′-dihydroxyflavanone (liquiritigenin), 3′-hydroxy-7,4′-dimethoxyflavone (tithionine) [2-60], 6,3′,4′-trihydroxyaurone (sulfuretin) [2-14], 7,4′-dihydroxyisoflavone (daidzein) [2-61], 7-hydroxy-4′-methoxyisoflavone (formononetin) [2-62], and the related pair [2-63] and [2-64]. Isoliquiritigenin and liquiritigenin are widespread in several plant families, e.g. Asteraceae and Fabaceae, sulfuretin is a common constituent of many genera of Asteraceae, and the two isoflavones daidzein and formononetin are common components of many legumes. Of much more limited occurrence are "tithionine," which Correa and Cervera (1971) isolated from *Tithonia tubeformis* (Asteraceae), and the pair of compounds [2-63] and [2-64], isolated from *Graziella mollissima* (Asteraceae) by Nakashima and coworkers (1994).

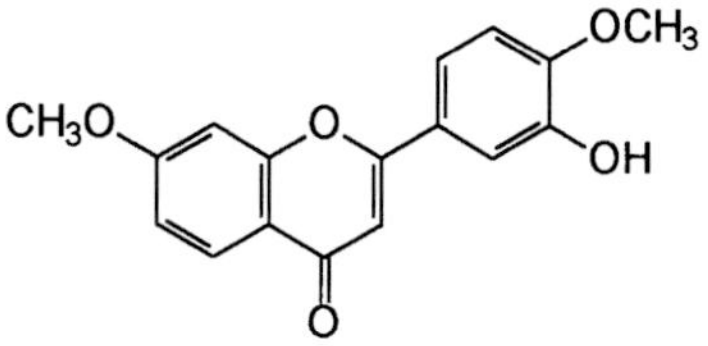

CH_3O, O, OCH_3, OH, O — 2-60

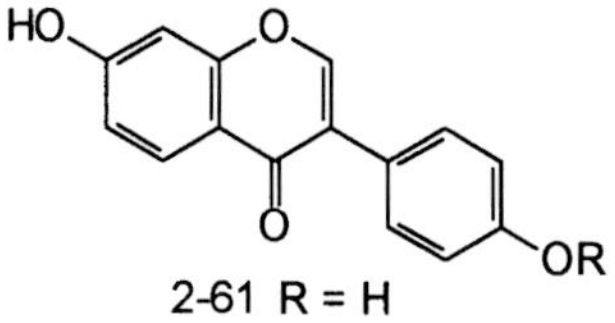

HO, O, O, OR — 2-61 R = H; 2-62 R = CH_3

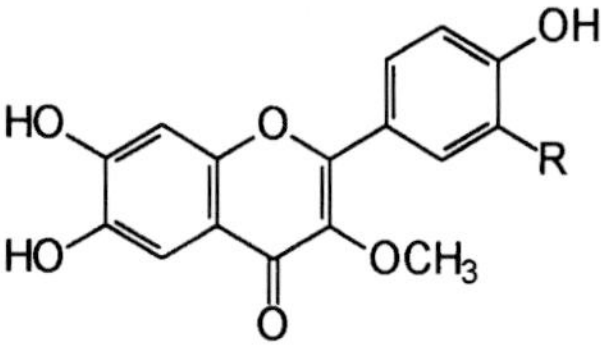

HO, HO, O, OH, R, OCH_3, O — 2-63 R = H; 2-64 R = OH

B-Ring deoxyflavonoids are also widely distributed in the plant kingdom. A survey of the literature revealed their occurrence in some 50 families of flowering plants (Bohm and Chan, 1992). Commonly occurring B-ring deoxyflavonoids are 5,7-dihydroxyflavanone (pinocembrin) [2-65], 3,5,7-trihydroxyflavanone (pinobanksin) [2-66], 5,7-dihydroxyflavone (chrysin) [2-67] and 3,5,7-trihydroxyflavone (galangin) [2-68]. Methyl ethers of some of these are also well known, e.g., 5-hydroxy-7-methoxyflavone (tectochrysin) [2-69]. Chalcones without B-ring substitution, e.g., 2′,4′,6′-trihydroxychalcone [2-70] and its 4′- [2-71] and 6′-methyl [2-72] ethers, have been reported from several sources. Extra oxygenation on the A-ring of flavones yields such compounds as 5,6,7-trihydroxyflavone (baicalein) [2-73], whose derivatives are widespread in the mints (Lamiaceae), its 6-methyl ether, oroxylin A [2-74]. The 8-oxygenated isomer of baicalein, norwogonin [2-75], and its derivatives are also well known naturally occurring flavonoids.

O-Alkylation

The most common *O*-alkyl group is methyl and it is found in all classes of flavonoids. Singly *O*-methylated flavonoids are extremely common, examples of which are the

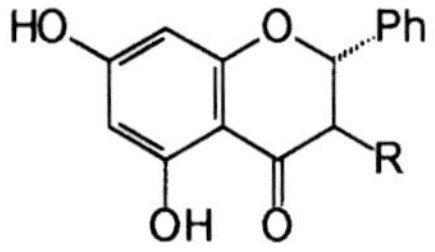

2-65 R = H
2-66 R = OH

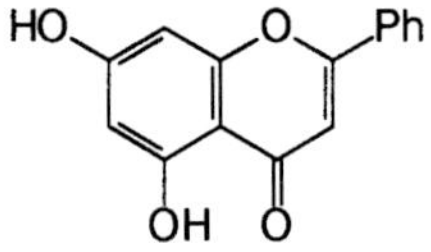

2-67 R = H
2-68 R = OH

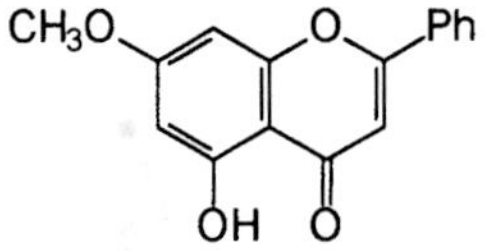

2-69

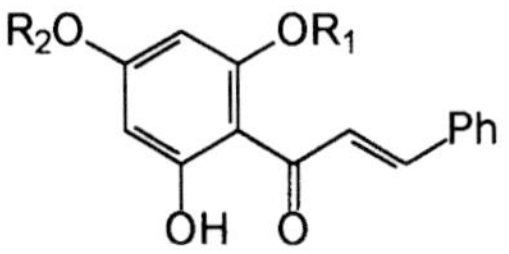

2-70 $R_1 = R_2 = H$
2-71 $R_1 = H, R_2 = CH_3$
2-72 $R_1 = CH_3, R_2 = H$

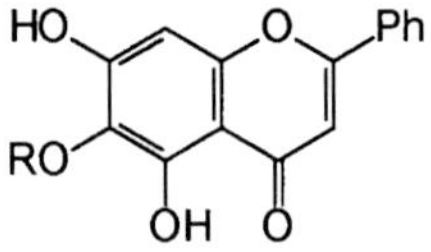

2-73 R = H
2-74 $R = CH_3$

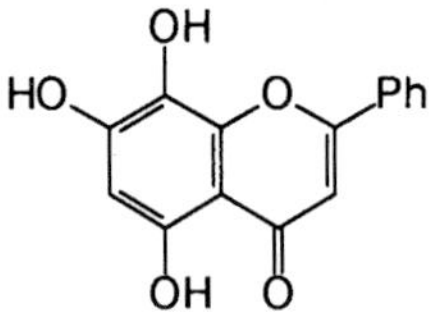

2-75

flavone methyl ethers apigenin 4′-methyl ether (acacetin) [2-76], apigenin 7-methyl ether (genkwanin) [2-77], luteolin 3′-methyl ether (chrysoeriol) [2-78], luteolin 4′-methyl ether (diosmetin) [2-79], and 5,7,4′-trihydroxy-3′,5′-dimethoxyflavone (tricin) [2-80]. Commonly encountered flavonol methyl ethers include quercetin 3′-methyl ether (isorhamnetin) [2-81], quercetin 7-methyl ether (rhamnetin) [2-82], and 3,5,7,3′,4′-pentahydroxy-6-methoxyflavone (patuletin or quercetagetin 6-methyl ether) [2-83]. Sakuranetin is the 7-methyl ether of naringenin [2-84]; isosakuranetin is the corresponding 4′-methyl ether [2-85]. Hesperitin [2-86] is the 4′-methyl ether of the flavanone eriodictyol. Peonidin is the 3′-methyl ether of cyanidin [2-87] while malvidin is the 3′,5′-dimethyl ether of delphinidin [2-88].

Several compounds have all, or almost all, of their hydroxyl groups methylated. A widely occurring compound in Asteraceae is 5-hydroxy-3,6,7,3′,4′-pentamethoxy-flavone (artemetin) [2-89]. Other highly *O*-methylated flavonoids were encountered above, 3,5,6,7,8,3′,4′-heptamethoxyflavone [2-32], 5,3′-dihydroxy-3,6,7,8,4′,5′-hexa-methoxyflavone [2-33], and the octamethoxyflavone [2-34]. Many flavonoids also exhibit the methylenedioxy function, along with a variety of other substituents. One of the more highly substituted flavonoids obtained from a natural source is the octa-substituted chalcone [2-90], which Miyaichi and coworkers (1989) isolated from *Scutellaria indica* (Lamiaceae). Simpler, but no less interesting, is the light-sensitive aurone [2-91] isolated by Pare and coworkers (1991) from *Cephalocereus senilis* (Cactaceae). 2′,4-Dihydroxy-6′-methoxy-3′,4′-methylenedioxychalcone [2-92] has been identified as a component of *Helichrysum glomeratum* (Bohlmann and Suwita, 1979), while 2′,6′-dimethoxy-3′,4′-methylenedioxychalcone [2-93] and the corresponding dihydrochalcone have been isolated from *H. sutherlandii* (Asteraceae) (Bohlmann *et al.*, 1978). A final example comes from the work of Jaques and colleagues (1987) on *Cicer arietinum* (Fabaceae) from which they obtained 7-hydroxy-3′,4′-methylenedioxy-isoflavone (pseudobaptigenin) [2-94].

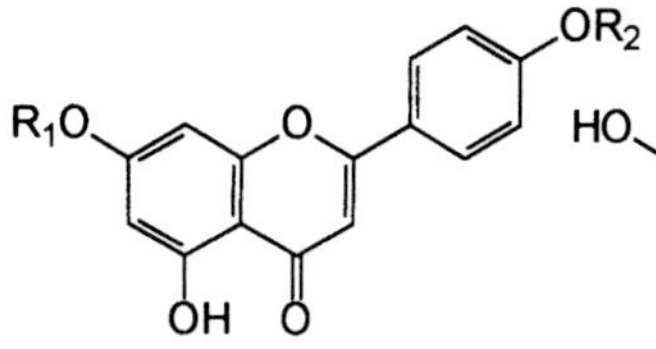

2-76 $R_1 = H, R_2 = CH_3$
2-77 $R_1 = CH_3, R_2 = H$

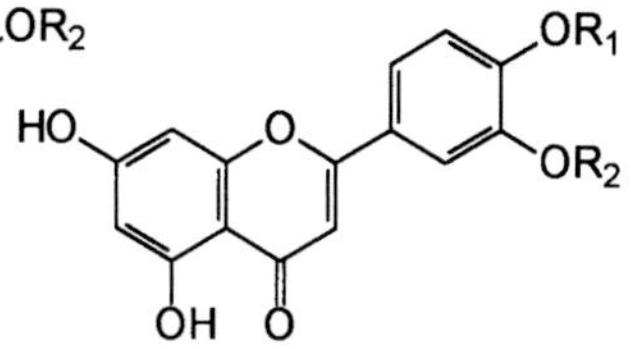

2-78 $R_1 = H, R_2 = CH_3$
2-79 $R_1 = CH_3, R_2 = H$

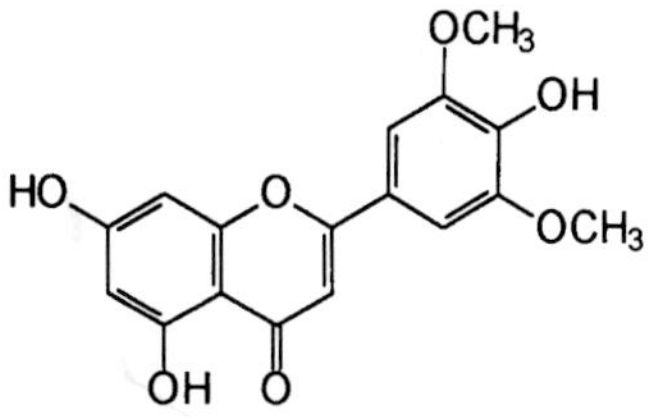

2-80

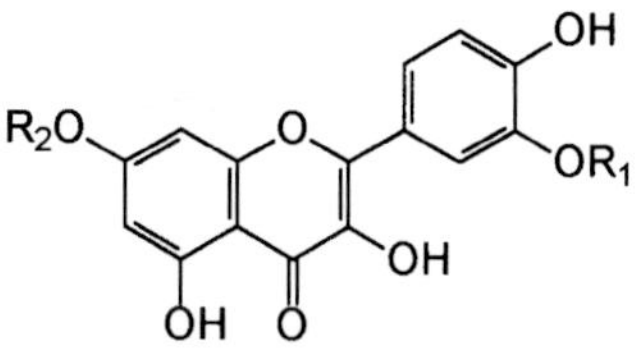

2-81 $R_1 = CH_3, R_2 = H$
2-82 $R_1 = H, R_2 = CH_3$

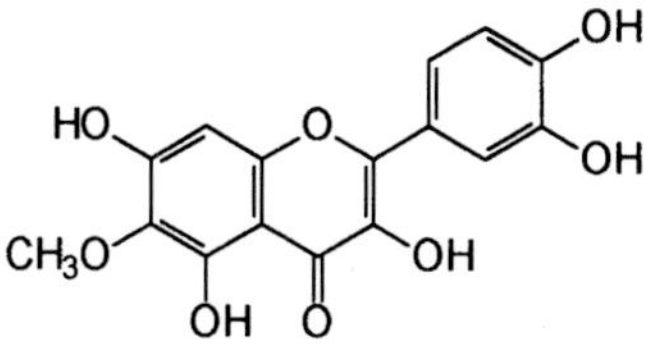

2-83

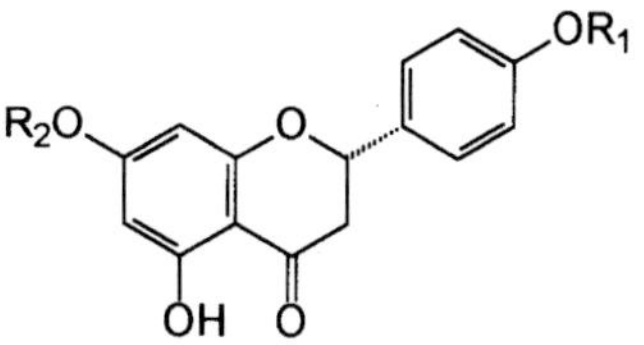

2-84 $R_1 = H, R_2 = CH_3$
2-85 $R_1 = CH_3, R_2 = H$

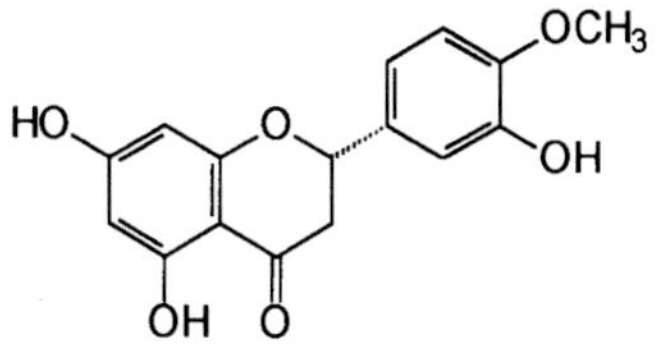

2-86

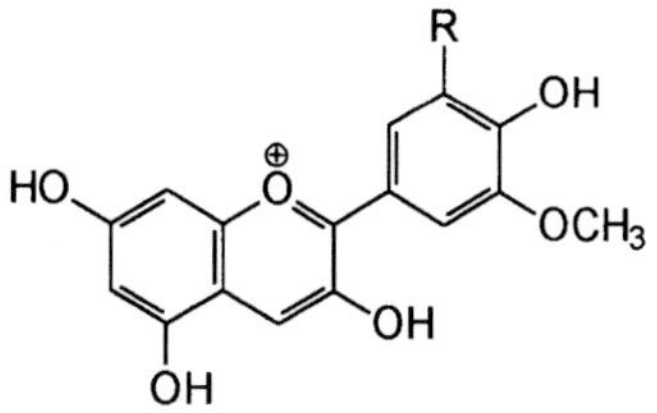

2-87 R = H
2-88 $R = OCH_3$

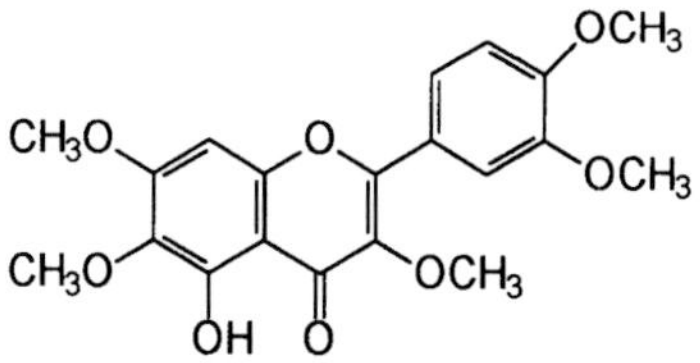

2-89

The 3-methylbuten-2-yl group (or γ,γ-dimethylallyl, commonly referred to as prenyl) is frequently seen as a *C*-alkylating function, as we will discover a little later. However, prenyl ethers of flavonoids have been identified as components of certain plants as well. For example, the 4′-prenyl ethers of 2′,4′,6′-trihydroxychalcone [2-95] and 2′,4′,6′,4-tetrahydroxychalcone [2-96] were identified as constituents of *Helichrysum athrixiifolium* [Bohlmann and Ates (Gören), 1984]. The 7-prenyl ethers of

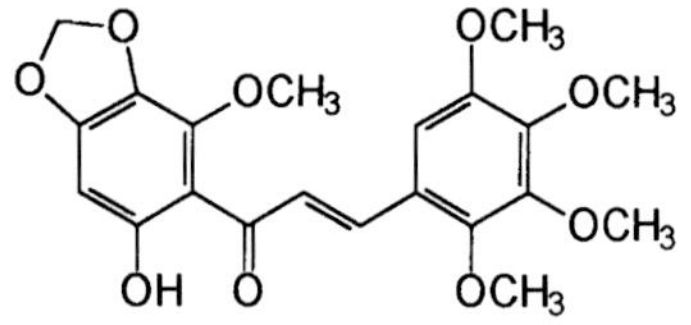

2-90

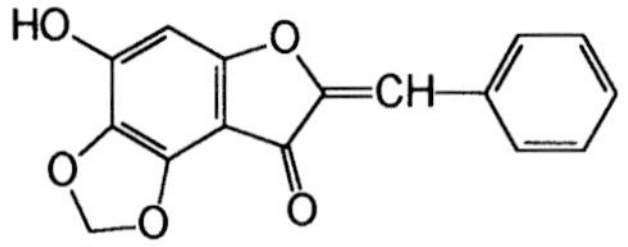

2-91

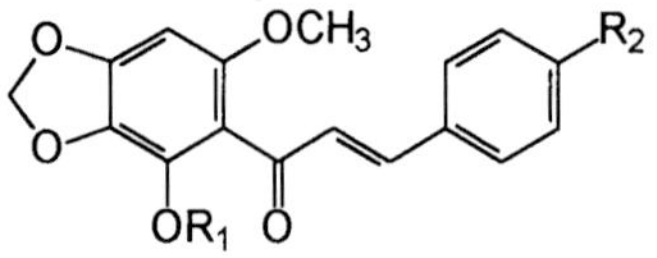

2-92 R_1 = H, R_2 = OH

2-93 R_1 = CH_3, R_2 = H

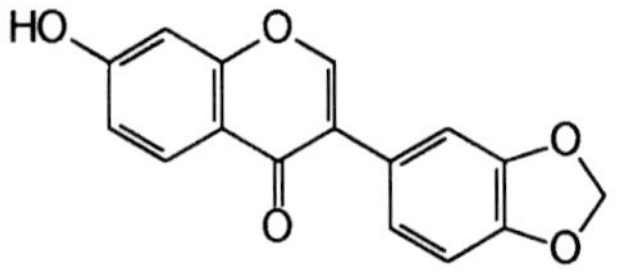

2-94

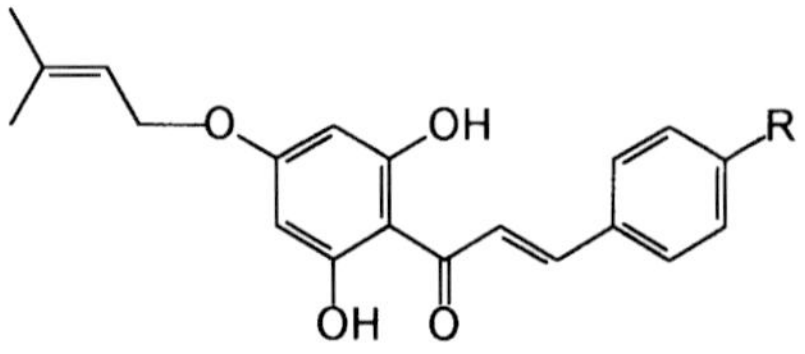

2-95 R = H

2-96 R = OH

5,7-dihydroxyflavanone and 5,7,4′-trihydroxyflavanone were also described in that report, but it is possible that they were artifacts of the isolation procedure, i.e., the products of cyclization of the corresponding chalcones.

Biflavonoids linked through oxygen, such as ochnaflavone [2-18], could formally be included in this section because they are diaryl ethers, but discussion of this group of compounds will be deferred until later in this chapter.

The flavonolignans, an unusual group of *O*-alkylated flavonoids, were first isolated from fruits of the European milk thistle, *Silybum marianum* (Asteraceae), a species reputed from the time of Dioscorides to protect the liver from hepatitis. Early studies by Pelter and Hänsel (1968) on the fruits revealed compounds that appeared to have been formed by the condensation of a flavonoid with 3-methoxy-4-hydroxycinnamyl alcohol (coniferyl alcohol). [This was ultimately shown to be the case by Schrall and Becker (1977) who observed a free radical condensation of dihydroquercetin (taxifolin) and coniferyl alcohol using cultured *S. marianum*.] Problems met in the early stages of structural determinations of these compounds were finally resolved using ^{13}C NMR, which led to structure [2-97] for "silychristin," a typical member of the group (Pelter *et al*., 1977; Wagner *et al*., 1978). Other members of the group exhibit structures built on the *p*-dioxane system as illustrated in the partial structure [2-98].

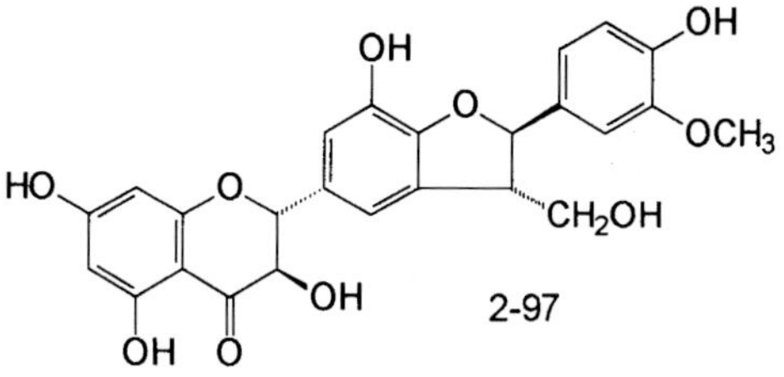

2-97

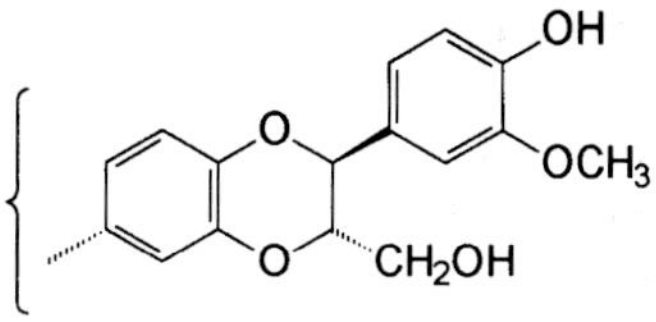

2-98

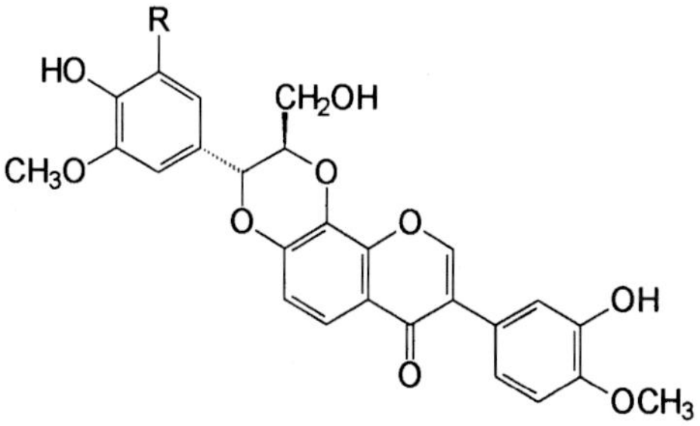

2-99 R = CH_3

2-100 R = H

Additional information on compounds from *Silybum* can be found in a report by Fiebig and Wagner (1984). X-Ray analysis has also proven to be a very useful tool in establishing structures of some of the more refractory members of this group (Abraham *et al.*, 1970).

It also became evident, as discussed in several of the papers, particularly that of Fiebig and Wagner (1984), that specimens of *Silybum marianum* from different locations, or that exhibited different flower color (or both), differed in their flavonolignans, both qualitatively and quantitatively. Such infraspecific variation is common in plants [see review by Bohm (1987) and the case study involving *Lasthenia* in Chapter Three]. In fact, the argument has been made that infraspecific chemical differences be accorded formal recognition (Tétényi, 1970). Although Tétényi was primarily concerned with drug plants, the principle could apply equally well to any situation. His suggestion has never attracted serious attention, however.

The compounds from *Silybum* illustrated above are based upon a flavanone (eriodictyol) or a dihydroflavonol (dihydroquercetin), but other classes of flavonoids can be involved. A luteolin-based flavonolignan, similar in other respects to the *p*-dioxane-based compounds, has been reported from *Hydnocarpus wightiana* (Flacourtiaceae) (Ranganathan and Seshadri, 1973; Parthasarathy *et al.*, 1979), from the legume genus *Cassia* by Kostova and Rangaswami (1977), and more recently from *Onopordum corymbosum* (Asteraceae) by Cardona and coworkers (1990). *Xanthocercis zambesiaca* (Fabaceae) afforded a pair of isoflavone derivatives, xanthocerin-A [2-99] and xanthocerin-B [2-100] (Bezuidenhout *et al.*, 1988).

C-Alkylation

C-Alkylation of two general sorts is seen in flavonoids, one involving the linkage of a sugar directly to the flavonoid nucleus, and the other involving a wide variety of different alkyl groups. Flavonoids of the first group comprise their own class, the *C*-glycosylflavonoids, and are treated separately below. The others, which can involve several other classes of flavonoids, will be lumped together in the present section.

The simplest *C*-alkylation seen in flavonoids involves the methyl group. The earliest reports of *C*-methylated flavonoids included 6-*C*-methylchrysin (strobochrysin) [2-101] from *Pinus* species (Pinaceae) and from the fern *Lonchitis tisserantii* (Voirin and Lebreton, 1967), the 6,8-di-*C*-methylflavanone matteucinol [2-102] from the ostrich fern genus *Matteuccia* (Fujise, 1929, cited by Harborne, 1967; see also Jiang *et al*., 1994b), and eucalyptin [2-103] and sideroxylin [2-104] from *Eucalyptus* species (Wollenweber and Kohorst, 1981). Continuing investigation of *Matteuccia orientalis* by Jiang and colleagues (1994a) resulted in finding, in addition to matteucinol, its 3′-methoxyisomer "isomatteucinol."

Wollenweber and Jay (1988) listed a dozen families from which *C*-methylflavonoids had been obtained. The main contributing family continued to be Myrtaceae, to which *Eucalyptus* belongs. Ericaceae has also been a contributor of these flavonoids. Other families include the fern family Pteridaceae (*Pityrogramma*), Pinaceae (the pines) representing the gymnosperms, the monocot families Agavaceae (agave family) and Liliaceae (lilies), as well as Annonaceae (sweetsop or custard apple family), Clusiaceae (Guttiferae,), Didieriaceae and Nyctaginaceae (both betacyanin families), Fabaceae, and Meliaceae (neem tree family). In his 1994 chapter, Wollenweber added Platanaceae (plane trees) and Asteraceae, although he expressed doubts about the identification of 5-hydroxy-7,4′-dimethoxy-6-*C*-methylflavone, which had been reported from *Artemisia austriaca* (Adekenov *et al*., 1987). In most of these cases, Myrtaceae and Ericaceae excluded, reports of *C*-methylated flavonoids usually involve only a single genus and only a few species at most.

Examination of the list of *C*-methylated flavonoids compiled by Wollenweber and Jay (1988) and Wollenweber (1994) clearly shows a total exclusion of *C*-methylation of the B-ring. *C*-Methylation at *C*-6, *C*-8, and at both positions abound, with the single report of methylation at *C*-3 coming from a study of *Eugenia kurzii* (Myrtaceae) from which Painuly and Tandon (1983) obtained glycosides of 3-*C*-methylapigenin [2-105] and 3-*C*-methylluteolin [2-106]. There are several examples of compounds in the flavonoid literature demonstrating that "over-methylation" of the A-ring occurs. One of the early examples of this phenomenon is the chalcone "ceroptene" [2-107] from the leaf exudate of the gold-back fern *Pityrogramma* (*Ceropteris*) *triangularis*. [Nomenclatural problems are a perpetual nuisance! *Pityrogramma triangularis* has now been given a new name, *Pentagramma triangularis* (Yatskievych *et al*., 1990)]. *Myrica gale* (Myricaceae), common in bogs of the Northern Hemisphere, was shown by Anthonsen and colleagues (1971) to accumulate the cyclohexatrione derivative [2-108], which contains three *C*-methyl groups. It is accompanied by the normal dihydrochalcone [2-109], and the highly unusual shrunken A-ring compound [2-110]. Leaves of *Myrica* were used at one time to improve the flavor and foaming properties of beer (Mabberley, 1987). The interested reader might look up the

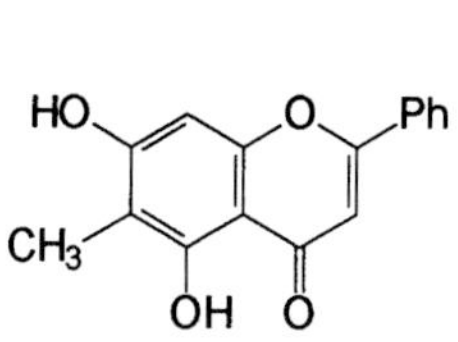

2-101

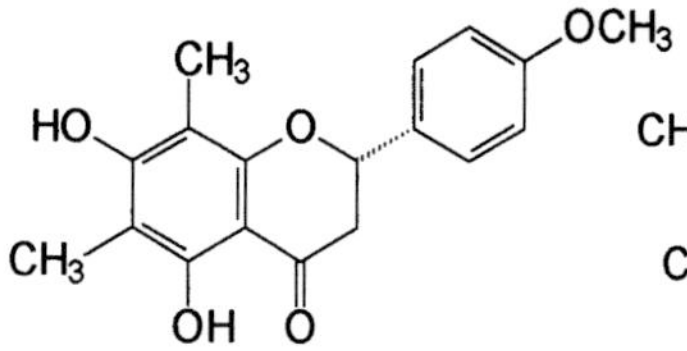

2-102

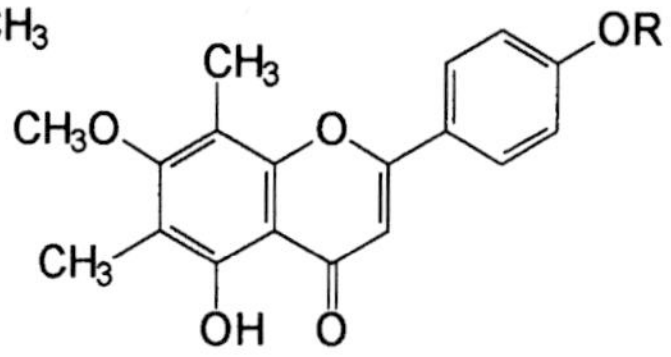

2-103 R = CH_3

2-104 R = H

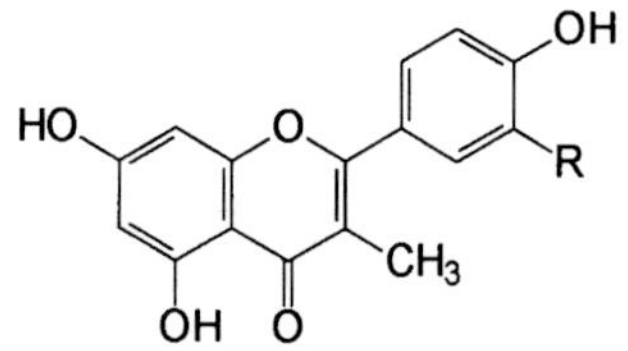

2-105 R = H

2-106 R = OH

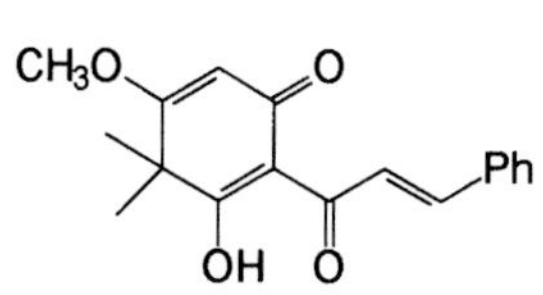

2-107

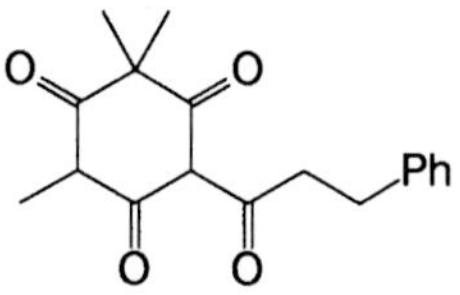

2-108

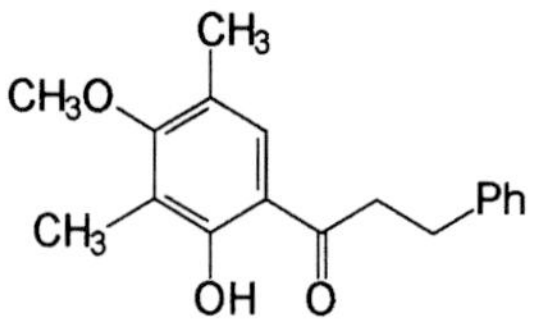

2-109

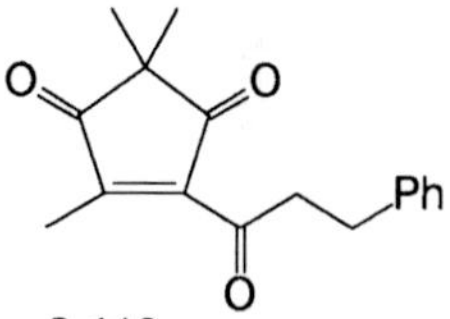

2-110

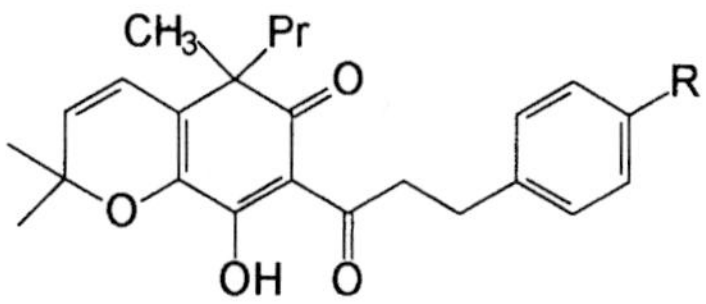

2-112 R = H

2-113 R = OH

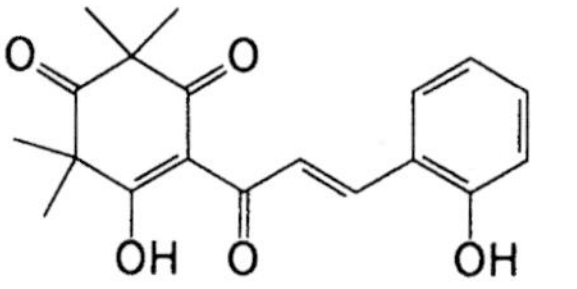

2-111

structure of humulone, the flavoring principal of hops, and compare its structure with the *Myrica* compounds. A recent re-examination of *M. gale* by Uyar and coworkers (1978) confirmed the presence of [2-108] and [2-109], but failed to find any of the shrunken A-ring compound. Two possibilities exist: (1) the shrunken-ring compound is seasonal; or (2) it was an artifact of the isolation procedure. Ring contraction of a (non-flavonoid) cyclohexatrione to a cyclopentendione has been reported (Siaens *et al.*, 1977). Syntheses of the *Myrica* compounds have been described (Uyar *et al.*, 1978; Misirlioglu *et al.*, 1978).

Double *C*-methylation at two positions on the A-ring resulted in the production of the highly unusual hexenedione-A-ring flavonoid seen in structure [2-111] which Hufford and coworkers (1981) reported from *Uvaria afzelii* (Annonaceae). Double alkylations need not always involve the methyl group as evidenced by the existence of the *gem*-dialkyl compounds "grenoblone" [2-112] and its 4′-hydroxyl derivative

[2-113] which Kaoudji and his coworkers (Kaoudji, 1986; Kaoudji *et al.*, 1986) isolated from bud excretions of *Platanus acerifolia* (Platanaceae), the plane tree common in urban horticulture.

A-Ring *C*-alkylation is easy to explain when one considers the nature of the precursor units of which that ring is built. Condensation of the cinnamoyl coenzyme-A starter unit with three malonyl coenzyme-A units produces an enzyme-bound polyketide whose methylene carbons, activated by the adjacent carbonyl groups, provide ideal targets for alkylation. The same distinction can be made in considering the mechanism of *C*-glycosylation, appears to be restricted to the A-ring (3-C-glycosylflavonoids are known). Since *C*-prenylation occurs with apparent equal facility on either ring, there would appear to be some fundamentally different mechanism involved.

C-Formylflavonoids have been reported from a few species representing several unrelated families in the plant kingdom. Early reports by S. R. Gupta and coworkers (1975, 1977) included a description of 2′,4-dihydroxy-4′-methoxy-5′-*C*-formylchalcone, "neobavachalcone" [2-114] from *Psoralea corylifolia* (Fabaceae). The isomeric "isoneobavachalcone," with the methoxy group at the 2′-position, has been reported from the same species by B. K. Gupta and coworkers (1980). 2′,4′,6′-Trihydroxy-3′-*C*-formylchalcone [2-115] has been reported from the guava relative *Psidium acutangulum* (Myrtaceae) (Miles *et al.*, 1991). Extracts of *P. acutangulum* were active against the pathogenic fungi *Rhizoctonia solani*, *Helminthosporium teres*, and *Pythium ultimum*. Compound [2-115] was active against *R. solani* and *H. teres.* The retrochalcone derivative [2-116] was obtained from *Anredera scandens* (Basellaceae) (Calzada *et al.*, 1990). Wollenweber and Jay (1988) listed three *C*-formylflavones based on chrysin, e.g., 5,7-dihydroxy-6-*C*-formyl-8-*C*-methylflavone [2-117], its 7-methyl ether, and 5,7-dihydroxy-6-*C*-methyl-8-*C*-formylflavone. So far as I am aware at the moment, the biosynthetic origin of the *C*-formyl group in flavonoids has not been addressed. Is direct formylation possible or does this group arise from oxidation of a *C*-methyl group?

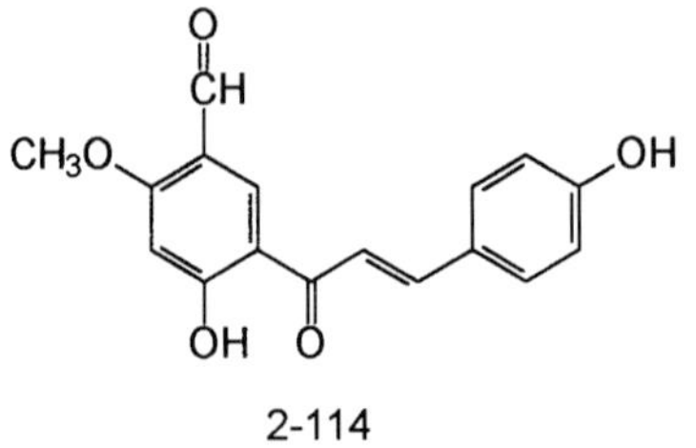

2-114

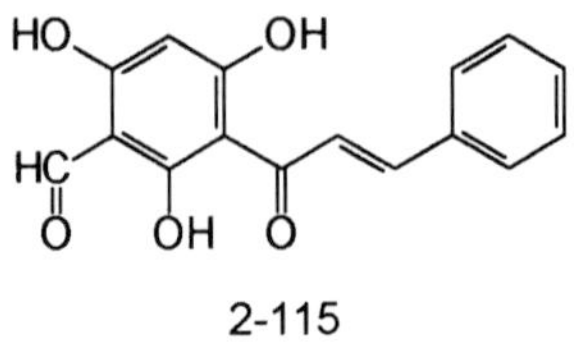

2-115

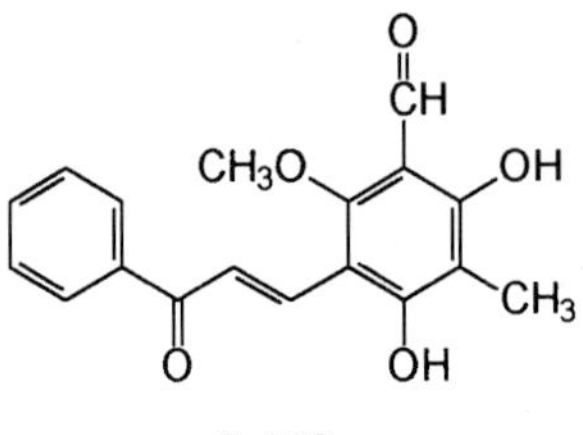

2-116

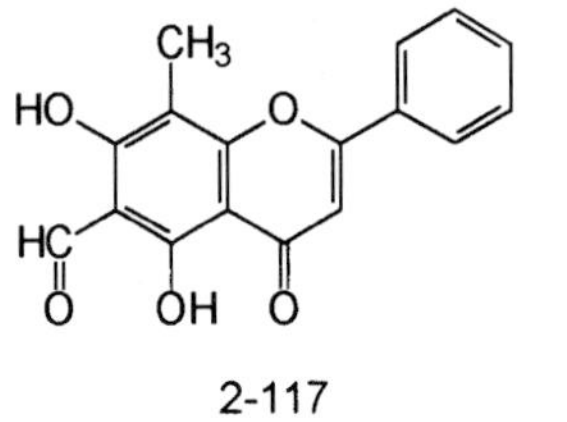

2-117

The most frequently seen *C*-alkyl group in flavonoids is 3-methylbuten-2-yl. This function is often referred to as the γ,γ-dimethylallyl, or, more informally as we saw above, as the prenyl function. We will use the term prenyl throughout this book. *C*-Prenylated flavonoids belonging to several structural classes are known, although isoflavones probably lead in this category. Owing to the presence of a double bond in the prenyl group, several structural modifications can be effected after initial prenylation has occurred. These modifications include reduction, isomerization, hydroxylation, epoxidation, and cyclization to five-, six,- or seven- membered rings. As mentioned above, in contrast to the situation seen with *C*-methyl groups, flavonoids with *C*-prenyl groups on their B-rings are common. One often finds prenylated A- and B-rings in the same compound. Many other combinations, including C_{10}-sidechains, are known.

Most commonly encountered *C*-prenylated flavonoids contain a single such group, a typical example of which is [2-118]. This isoflavone, known as "prebarbigerone," is 7,2′,4′,5′-tetramethoxy-8-*C*-prenylisoflavone and was isolated from seeds of *Millettia ferruginea* ssp. *ferruginea* (Fabaceae) (Dagne and Bekele, 1990). The closely related "barbigerone," [2-119], described in the same paper, was obtained from seeds of *M. ferruginea* ssp. *darassana*. Barbigerone contains a new heterocyclic system that arises through reaction of the prenyl double bond with the adjacent phenolic group. This compound would be formally named 2′,4′,5′-trimethoxy-6″,6″-dimethylpyrano(2″,3″ : 7,8)isoflavone. For those unfamiliar with this nomenclatural convention, the numbers within the parentheses indicate the bond shared by the chromene ring and the isoflavone A-ring. (This function is sometimes referred to as the dimethylpyrano or DNP group.) Compound [2-120] is a typical example of a di-*C*-prenylated flavonoid while [2-121] will represent flavonoids that have undergone three *C*-prenylations. Both of these compounds have been obtained from species of the legume genus *Lespedeza* (Wang *et al.*, 1987; Li *et al.*, 1990).

Several flavonoids are known that feature a C_{10}-sidechain, two isomeric forms of which exist, the geranyl (3,7-dimethylocta-2,6-dienyl) and lavandulyl (2-isopropylene-5-methylhex-4-enyl) isomers. Examples of the occurrence of geranyl derivatives include 6-*C*-geranylnaringenin [2-122] from *Bonannia graeca* (Apiaceae) (Bruno *et al.*, 1985) and *Diplacus* (*Mimulus*) *aurantiaca* (Scrophulariaceae), the 8-*C*-isomer from *Sophora tomentosa* (Shirataki *et al.*, 1983), and the 3′-isomer from *Morus lhou* (Fukai *et al.*, 1985). Not intending to slight any other flavonoid class in this respect, it is appropriate to note the existence of 5,7,4′-trihydroxy-3′-methoxy-6-*C*-geranylflavone (6-*C*-geranylchrysoeriol) as a component of *Cannabis sativa* (Cannabidaceae) (Barrett *et al.*, 1986). An example of a flavonoid with a *C*-lavandulyl sidechain is "kurarinone" [2-123], which Hatayama and Komatsu (1971) isolated from a species of *Sophora*. A variant of the *C*-geranyl group of compounds can be seen in [2-124] where cyclization has occurred, as in the case with a prenyl group, to give a chromeno derivative. The compound shown comes from a species of *Boesenbergia* (Zingiberaceae, the gingers) and was identified by Tuntiwachwuttikul and coworkers (1988 and citations therein). [Note that I drew the structure for this compound, boesenbergin-B, incorrectly in Bohm, 1994, page 395, where it had grown an extra methylene group!]. Following the notation described above for cyclized prenyl groups, boesenbergin would be named 2′-hydroxy-6′-methoxy-6″-methyl-6″-(4-methylpenten-3-yl)(2″,3″: 4′,3′)pyranochalcone.

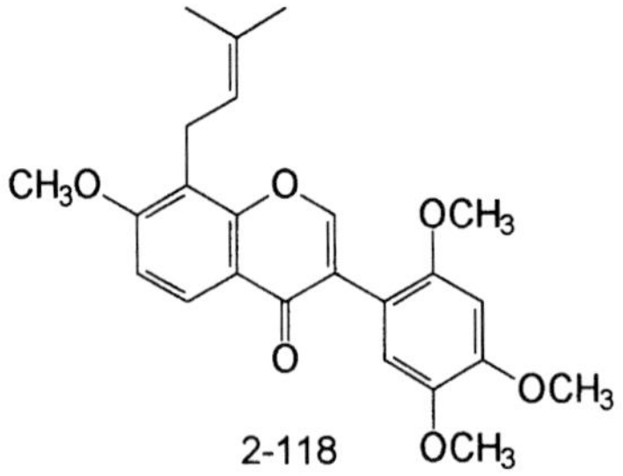

2-118 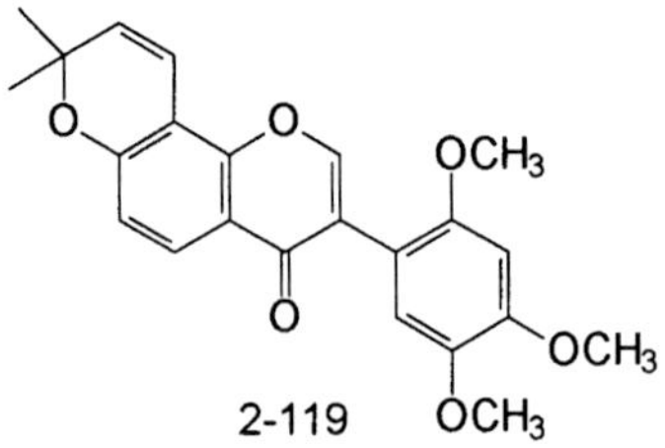 2-119

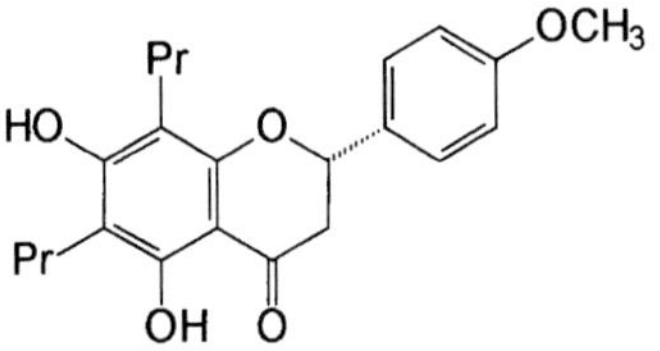

2-120 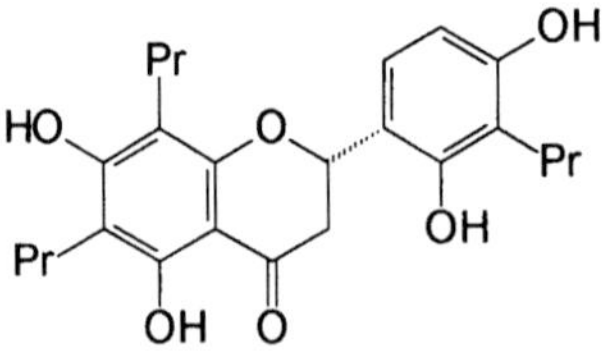 2-121

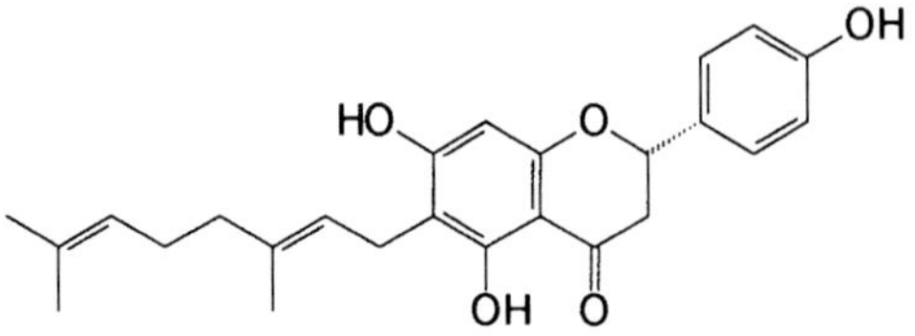

2-122 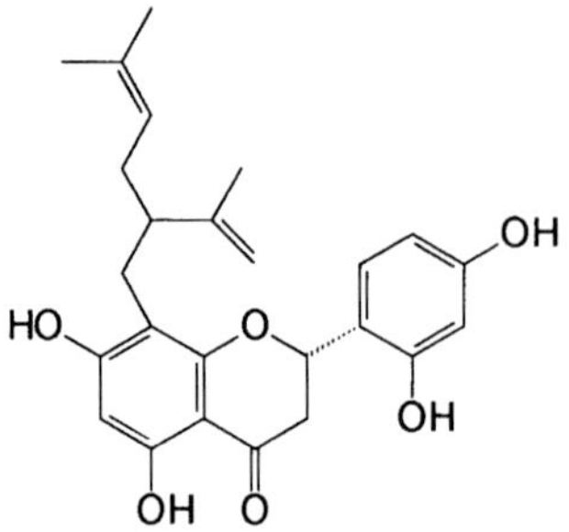 2-123

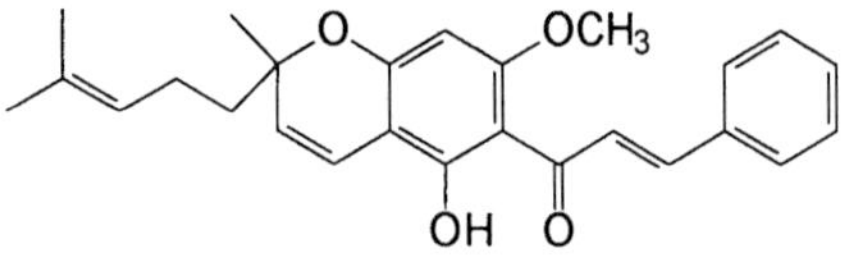

2-124

A further variation on the C_{10} theme involves monoterpene units as alkylating groups. A case in point is the existence of the dihydrochalcones linderatin [2-125] and methyllinderatin [2-126], components of *Lindera umbellata* var. *lancea* (Lauraceae, cinnamon family) (Ichino, 1989; Ichino *et al*., 1988a, 1989). Equivalent chalcones and flavanones were also described. A more complicated situation existed with regard to the chalcone "rubranine" [2-127]. This compound was originally isolated from another lauraceous taxon, *Aniba rosaeodora*, by Combes and coworkers (1970). Alkylation of the flavonoid nucleus was followed by two additional cyclizations with the adjacent phenolic hydroxyls to provide the unique pentacyclic structure. The existence of rubranine as a natural compound was challenged by

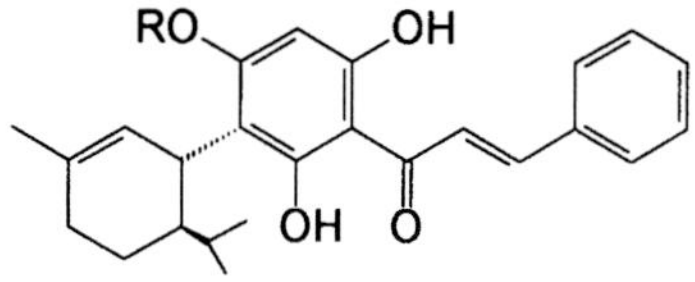

2-125 R = H
2-126 R = CH_3

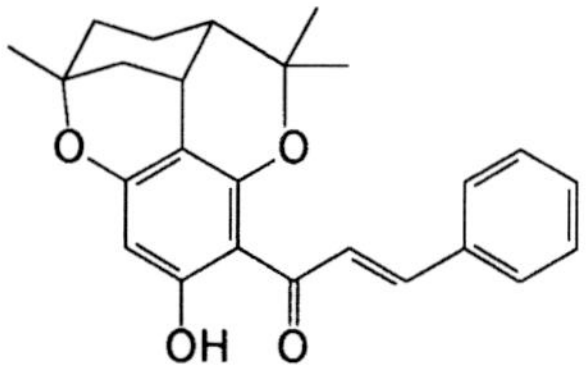

2-127

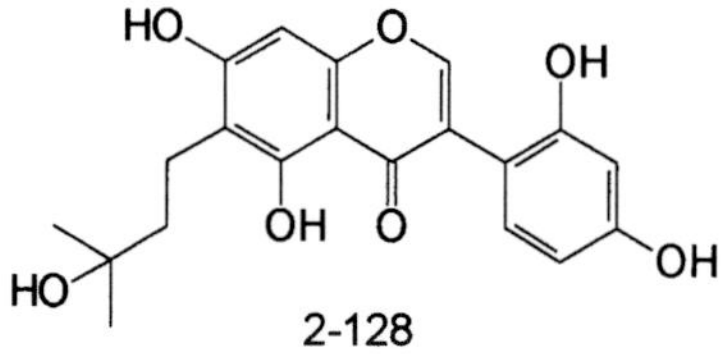

2-128

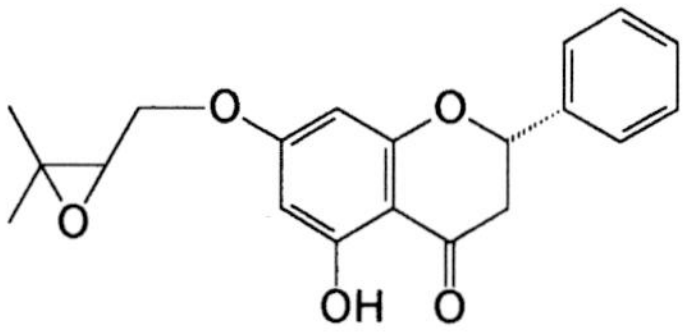

2-129

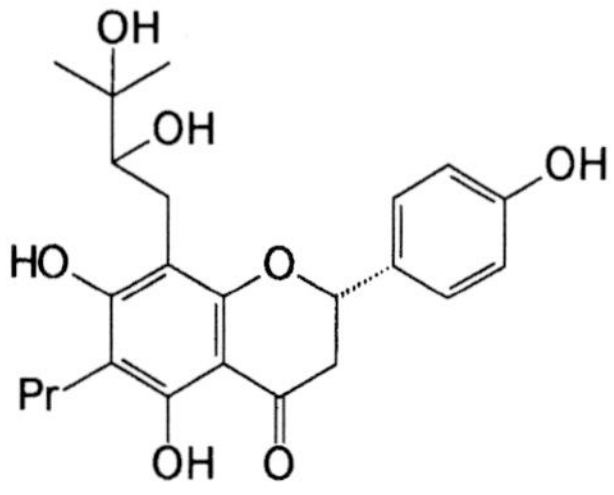

2-130

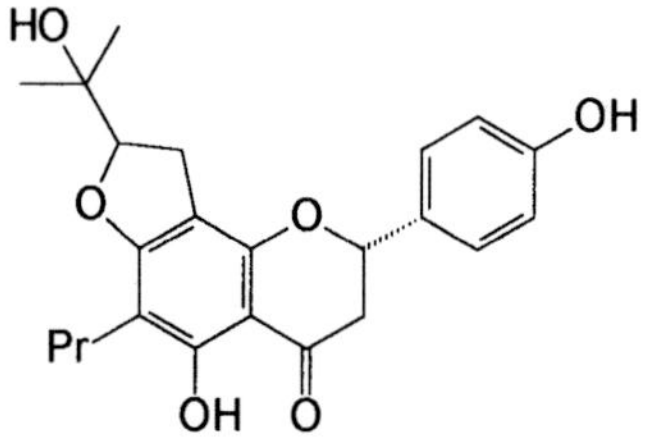

2-131

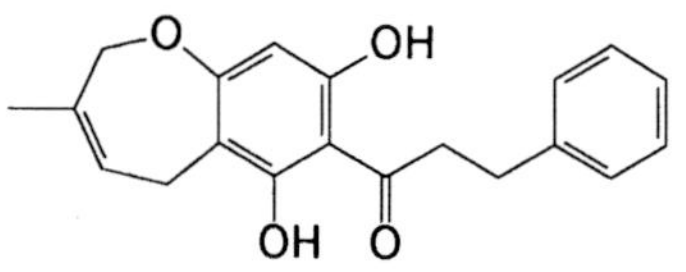

2-132

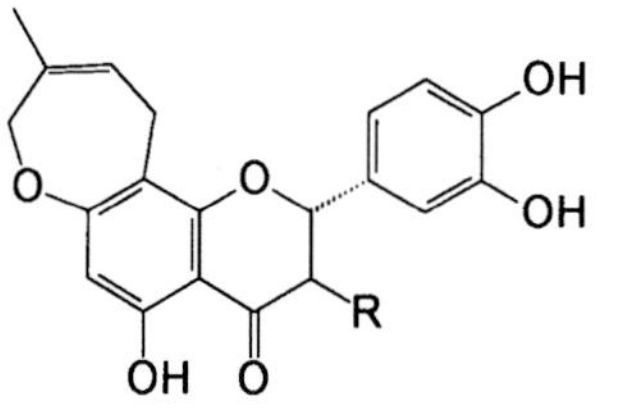

2-133 R = H
2-134 R = —OH

Montero and Winternitz (1973) who suggested that it may have been an artifact of isolation. They supported their view by effecting the condensation of pinocembrin (the flavanone corresponding to the foundation chalcone) with citral in the presence of anabine, a known alkaloidal constituent of *Aniba*. Proof that rubranine is indeed a naturally occurring constituent was based on two lines of evidence (de Alleluia *et al.*, 1978). These workers isolated (−)-rubranine from wood of *A. rosaeodora* by brief extraction with hexane, conditions under which base-catalyzed condensation is unlikely, and they showed that citral does not occur in the plant.

Hydration of the prenyl double bond can occur to yield an alcohol such as luteone hydrate [2-128] from *Lupinus luteus* root (Hashidoko *et al.*, 1986). Hydration is not limited to prenyl groups *per se*, but can occur with the cyclized forms as well. Epoxide derivatives such as [2-129] are also known. This one was obtained from *Achyrocline flaccida* (Asteraceae) by Norbedo and coworkers (1984). *Lonchocarpus minimiflorus* (Fabaceae) has provided a number of interesting flavonoids two of which serve to illustrate further structural variations on the prenyl side chain. In addition to lonchocarpol (6,8-di-*C*-prenylnaringenin), the dihydroxy compound [2-130] and furan derivatives of the sort illustrated as [2-131] occur in the plant (Roussis *et al.*, 1987). These compounds were active as feeding deterrents against the leaf-cutting ant *Atta cephalotes*.

Cyclization of a prenyl group to yield a seven-membered ring also occurs. Asakawa and coworkers (1978) identified the dihydrochalcone [2-132], and other similarly substituted compounds, from the liverwort *Radula variabilis*. Some of the related compounds in that plant were bibenzyls with the same seven-carbon heterocyclic attachment. In a study of *Wyethia angustifolia* (Asteraceae) McCormick and coworkers (1986) reported similar dihydroöxepine derivatives of eriodictyol [2-133] and dihydroquercetin [2-134].

A novel group of *C*-alkylated flavonoids has been identified from several members of the genus *Uvaria* of the tropical family Annonaceae. The alkyl group in question, 2-hydroxybenzyl, is seen in 2′,4′-dihydroxy-6′-methoxy-3′-*C*-(2-hydroxybenzyl)dihydrochalcone [2-135], which was obtained from *U. acuminata* (Cole *et al.*, 1976). The 5′-mono- and 3′,5′-disubstituted have also been identified as constituents of species of *Uvaria*. The 3′,5′-di-*C*-(2-hydroxybenzyl)dihydrochalcone derivative [2-136] has also been described from *U. chamae* (Okorie, 1977). Note that in the latter compound a secondary reaction between the hydroxyl group on the benzyl ring and the hydroxyl group at C-4′ has occurred to yield a compound that contains a diphenyl ether linkage. Similarly substituted flavanones have also been identified from species of *Uvaria* by Hufford and coworkers (1976, 1978a,b, 1980). Many of these compounds showed activity in lymphocytic leukemia test systems.

We saw in the section above on *O*-alkylation that a some species have the capacity to bring about the condensation of 3-methoxy-4-hydroxyconiferyl alcohol with flavonoids having hydroxyl groups at C-3′ and C-4′ to produce flavonolignans, e.g., [2-97] to [2-100]. An interesting compound has been reported by Dean *et al.* (1993) that has some similarities to the flavonolignans. An isoflavanoid derivative, shown to have structure [2-137], was obtained from *Aglaia ferruginaea* (Meliaceae). As diagrammed, the flavonoid C-ring lies above the plane of the paper with C-2 and C-3, of the presumed phenylpropanoid precursor, along with the attached aromatic ring, lying somewhat below. It is possible to visualize a condensation of some *p*-methoxyphenylpropanoid derivative, possibly *p*-methoxycinnamic acid, with the parent isoflavonoid followed by decarboxylation.

O-Glycosylation

Flavonoids can occur in two forms in plants, as such, and as glycosides that can be either *O*- or *C*-linked. Glandular exudates, which have been the subject of intensive

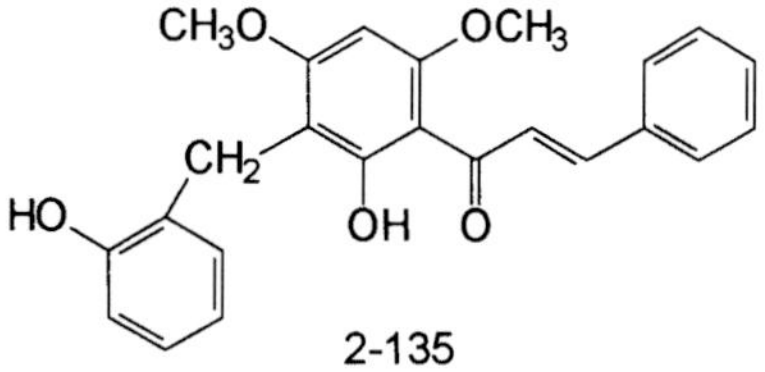

2-135

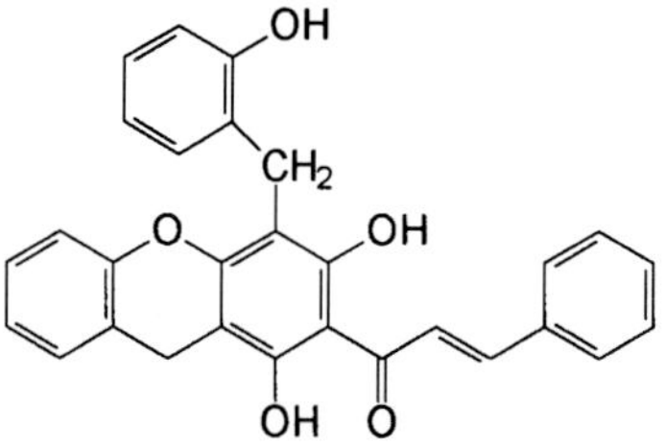

2-136

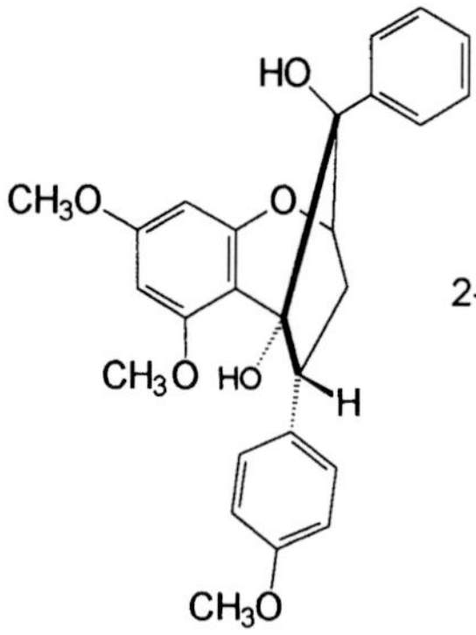

2-137

study by Eckhard Wollenweber and his associates in Darmstadt, involve non-polar compounds that are associated with and exuded by various glandular structures on the leaves and other organs of many plants. The polar flavonoid fraction, however, consists of mainly of glycosides (sulfates will be discussed below). Flavonoid glycosides of virtually every flavonoid class are known. The variety of flavonoid glycosides is based on several factors: the number of positions on the flavonoid for glycosylation, the number of different sugars that can be involved, the level of glycosylation, whether mono-, di, or higher, and combinations of these. The main monosaccharides involved are glucose, galactose, rhamnose, galacturonic acid, glucuronic acid, arabinose, and xylose. Other monosaccharides seen occasionally are allose, apiose, and mannose. Apiose is a branched pentose that attracted attention during studies of the biosynthesis of apiin, apigenin 7-*O*-apiosylglucoside, the main flavonoid glycoside in parsley (see Chapter Six).

Two additional factors contributing to complexity are the possible existence of sugars in either the furanose or pyranose ring forms, and stereochemical possibilities at the anomeric carbon. In most cases not all possible isomeric are not known, but likely do exist in Nature. This prediction is based upon the knowledge that all four isomers of kaempferol 3-*O*-arabinoside are known, i.e., α- and β-pyranosides and α- and β-furanosides. Classical methods of carbohydrate analysis provide information on the existence of these forms; carbohydrate identifications based upon simple acid hydrolysis and chromatography generally do not. The use of ^{1}H and ^{13}C NMR techniques provide nondestructive means of determining all of these features. Several Case Studies in Chapter Four demonstrate the power of these analytical methods.

Glycosylation can occur at any hydroxyl group but certain ones seem favored. Although most flavonols occur as 3-*O*-glycosides, 7-*O*-glycosides are also widespread. Flavones mostly occur as 7-*O*-glycosides, but substitution can occur at any other position(s), including positions-6 or -8 when those positions are hydroxylated. For example, Soltis (1980a) identified the 6-*O*-galactoside and 6-*O*-galactosylglucoside of 5,6,3′,4′-tetrahydroxy-7-methoxyflavone (pedalitin) [2-138] from species of *Sullivantia* (Saxifragaceae). More recently, Allais and coworkers (1995) reported the first flavanone 8-*O*-glucoside to be described from a plant source. The compound, isolated from *Calluna vulgaris* (Ericaceae), was identified as 8-hydroxynaringenin 8-*O*-glucoside [2-139]. It was accompanied by the corresponding dihydroflavonol derivative, 8-hydroxydihydrokaempferol 8-*O*-glucoside [2-140].

Glycosylation at two positions is common. The family of naturally occurring flavonol di-*O*-glycosides includes such compounds as: kaempferol and myricetin 3,4′-di-*O*-glucosides, [2-141] and [2-142], respectively, from the Norway spruce (*Picea abies*) (Slimestad *et al*., 1993); kaempferol 3-*O*-glucoside-7-*O*-rhamnoside from *Arabis caucasica* (Brassicaceae, Cruciferae, cabbage family) (Matlawska *et al*., 1991); myricetin 3-*O*-galactoside-3′-*O*-rhamnoside [2-143] from *Buchanania lanzan* (Anacardiaceae, sumac family) (Arya *et al*., 1992); and myricetin 4′-methyl ether 3,7-di-*O*-rhamnoside [2-144] from the fern *Asplenium antiquum* (Mizuno *et al*., 1991). Flavonoids with more than two glycosidic linkages have also been reported, as in the case of rhamnetin 3,3′,4′-tri-*O*-glucoside [2-145] and rhamnetin 3-*O*-arabinoside-3′,4′-di-*O*-glucoside [2-146] from species of *Pyrola* (Ericaceae) (Averett and Bohm, 1986). Three examples illustrate flavonoids that have a monosaccharide at one position and a disaccharide at another: isorhamnetin 3-*O*-rutinoside-4′-*O*-glucoside [2-147] from *Conyza linifolia* (Asteraceae) (El-Sayed *et al*., 1991); apigenin 7-*O*-rhamnoside-4′-*O*-(4′-*O*-glucosyl-α-rhamnoside) from *Asplenium normale* (Iwashina *et al*., 1993); and quercetin 4′-methyl ether (tamarixetin) 3-*O*-rutinoside-7-*O*-rhamnoside from the legume *Cassia italica* (El-Sayed *et al*., 1992).

Several naturally occurring di- and trisaccharides that have been identified as components of flavonoid glycosides are listed in Table 2.2. The entries represent only a sampling; many other examples are known (Harborne and Williams, 1982, 1988). Examples of the several isomeric permutations of diglycoside structures can be seen in the case of flavonoid dirhamnosides and diglucosides. The three positional isomers involving two rhamnose units are known: isorhamnetin 3-*O*-rhamnosyl(1 → 2)rhamnoside from *Calendula officinalis* (Asteraceae) (Kommissarenko *et al*., 1988); 8-prenylkaempferol 4′-methyl ether 3-*O*-rhamnosyl(1 → 3)rhamnoside 7-*O*-glucoside from *Vancouveria hexandra* (Berberidaceae) (Mizuno *et al*., 1991a); and kaempferol 3-*O*-rhamnosyl(1 → 4)rhamnoside from *Cassinopsis madagascariensis* (Icacinaceae) (Rasoanaivo *et al*., 1990). Three groups of isomeric flavonoid diglucosides have been identified from plants. Glucosyl(1 → 6)glucosides (gentiobiosides) and glucosyl(1 → 2) (sophorosides) are quite common, whereas glucosyl(1 → 3)glucosides, known trivially as laminaribiosides (name from the 1,3-glucan storage polysaccharide from the macroalga *Laminaria*) are known from only a handful of plant species, including the recent report of quercetin and isorhamnetin 3-*O*-laminaribiosides in *Pteridium aquilinum*, the common bracken fern (Imperato, 1995). Using the flavonoid glycoside

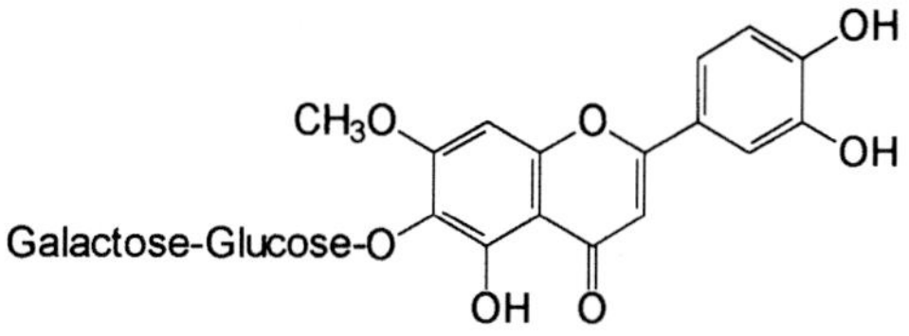

2-138

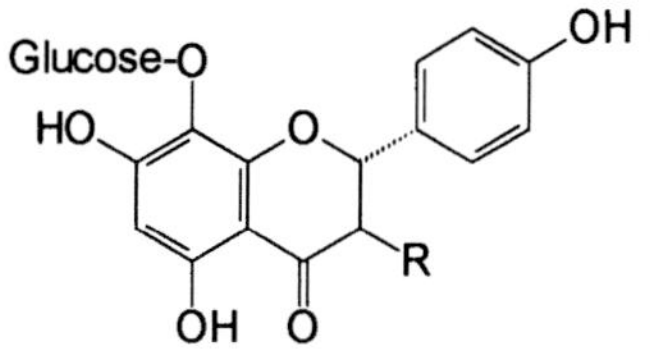

2-139 R = H
2-140 R = OH

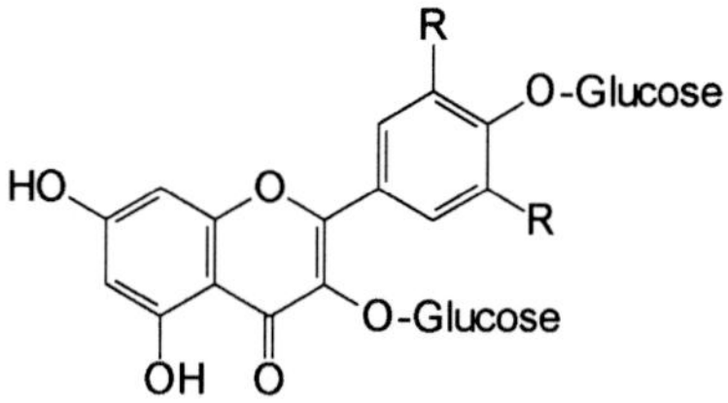

2-141 R = H
2-142 R = OCH_3

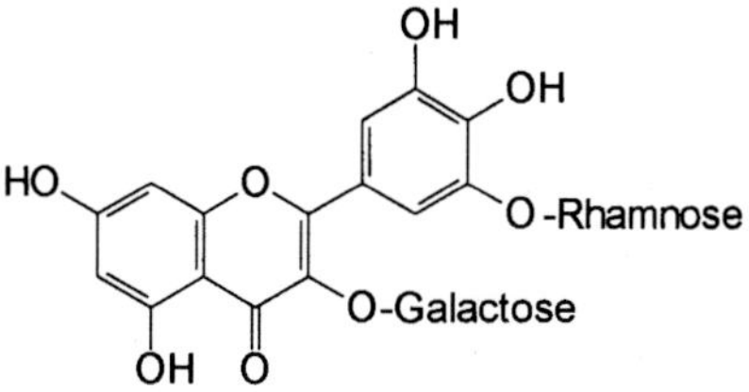

2-143

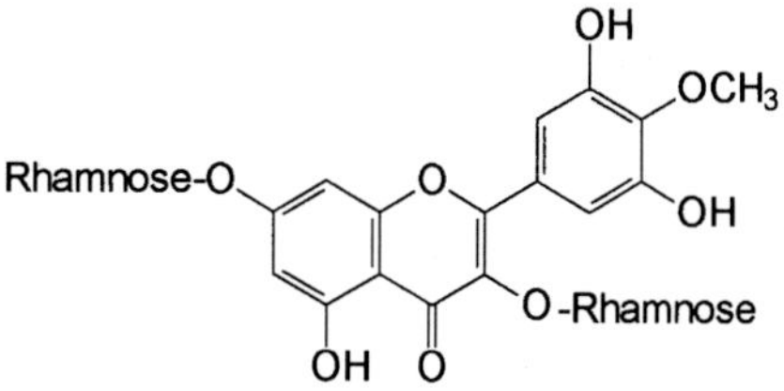

2-144

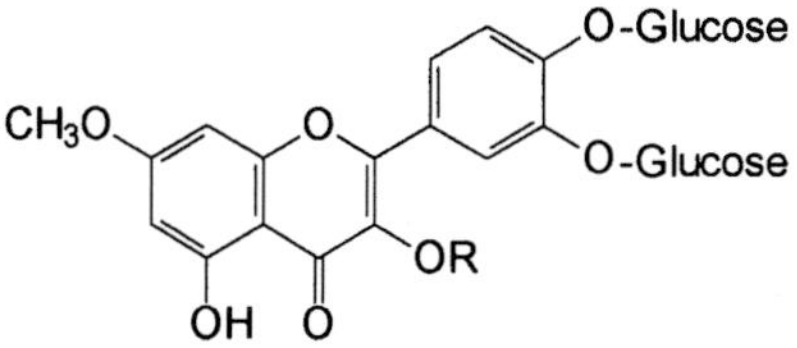

2-145 R = Glucose
2-146 R = Arabinose

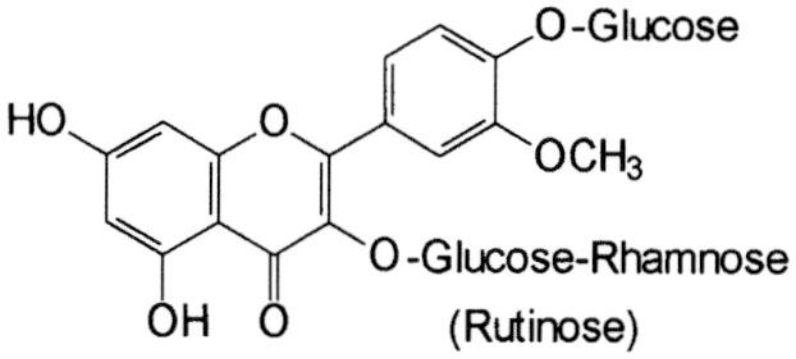

O-Glucose-Rhamnose (Rutinose)

2-147

surveys of Harborne and Williams (1975, 1982, 1988, 1994) it appears that the glucosyl(1 → 4)glucoside isomer (cellobioside) is not known in the flavonoid literature.

At our current level of knowledge, it is probably safe to say that the most commonly encountered flavonoid diglycosides are those that involve glucose and rhamnose, in various combinations, as shown above. Many other sugars and combinations of sugars are known, of course, and others are sure to be identified

Table 2.2 A sampling of flavonoid di- and trisaccharides

Common name	Structure
DIGLYCOSIDES	
Gentiobiose	6-*O*-β-D-Glucosyl-D-glucose
Lactose	4-*O*-β-D-Galactosyl-D-glucose
Laminaribiose	3-*O*-β-D-Glucosyl-D-glucose
Lathyrose	2-*O*-β-D-Xylosyl-D-galactose
Neohesperidose	2-*O*-α-L-Rhamnosyl-D-glucose
Robinobiose	6-*O*-α-L-Rhamnosyl-D-galactose
Rungiose	3-*O*-α-L-Rhamnosyl-D-glucose
Rutinose	6-*O*-α-L-Rhamnosyl-D-glucose
Sambubiose	2-*O*-β-D-Xylosyl-D-glucose
Sophorose	2-*O*-β-D-Glucosyl-D-glucose
Vicianose	6-*O*-α-L-Arabinosyl-D-glucose
TRIGLYCOSIDES	
Gentiotriose	*O*-β-Glucosyl-(1 → 6)-*O*-β-glucosyl-(1 → 6)-glucose
2′-Glucosylgentiobiose	*O*-β-Glucosyl-(1 → 2)-*O*-β-glucosyl-(1 → 6)-glucose
2′-Rhamnosylrutinose	*O*-α-Rhamnosyl-(1 → 2)-*O*-α-rhamnosyl-(1 → 6)-glucose
Sophorotriose	*O*-β-Glucosyl-(1 → 2)-*O*-β-glucosyl-(1 → 2)-glucose
2G-Apiosylrutinose	*O*-β-Apiosyl-(1 → 2)-*O*-[α-rhamnosyl-(1 → 6)-glucose]
2G-Glucosylgentiobiose	*O*-β-Glucosyl-(1 → 2)-*O*-[β-glucosyl-(1 → 6)-glucose]

with the availability of magnetic resonance methods. Table 2.3 presents a sampling of some of less often reported flavonoid di- and triglycosides. A tetraglycoside has been included to represent the level of complexity possible when one is confronted by the combination of different sugars and different points of attachment of the sugars on the flavonoid nucleus. In the tetraglycoside from *Alangium*, the 3-*O*-galactose unit has rhamnose moieties attached at positions 2″ and 6″ (Kijima *et al.*, 1995). This compound is represented semidiagramatically as [2-148]. A similar compound has been identified from *Rhazya stricta* (Apocynaceae), and shown to consist of a branched triglycoside, consisting of two rhamnose residues and a galactose, at position-3, as well as another rhamnose at position-7 (Andersen *et al.*, 1987). The "old man" cactus, *Cephalocereus senilis*, yielded a similar compound that had a branched triglycoside at position-3 consisting of galactose, glucose, and rhamnose, and a second rhamnose at position-7 (Liu *et al.*, 1994). It is useful to point out, using the old man cactus as an example, that sets of related glycosides are often found together. In this case, the flavonoid glycoside fraction consisted, along with the tetraglycoside, kaempferol 7-*O*-rhamnoside and kaempferol 3-*O*-(6-*O*-α-L-rhamnosylgalactoside). Clues to the structures of the more complex glycosides can often be gleaned from the simpler ones. A recent report of a flavone tetraglycoside deserves comment because, unlike the tetraglycosides just mentioned, it has all of its sugars attached through one flavonoid position. Sharaf (1996) described acacetin 7-*O*-(2⁗-*O*-α-rhamnosyl- 2‴-*O*-β-glucosyl-2″-*O*-β-glucosyl)-glucoside [2-149] from *Peganum harmala*.

Several *C*-glycosylflavonoids are known that also bear additional sugar units, sometimes bound to the *C*-sugar itself, and sometimes linked through one of the phenolic oxygens. In the former category are compounds of the sort recently described from *Galipea trifoliata* (Rutaceae) (Bakhtiar *et al.*, 1994). Two of the compounds identified were the known 2″-*O*-glucosyl [2-150] and 2″-*O*-xylosyl [2-151]

Table 2.3 A sampling of unusual naturally occurring flavonoid glycosides

Glycoside	Source plant	Family	Reference
Diglycosides . . .			
Anhydroicaritin 3-*O*-rhamnosyl(1 → 2)rhamnoside	*Epimedium* spp.	Berberidaceae	Li *et al*., 1996
Apigenin, luteolin and scutellarein 7-*O*-glucuronyl(1 → 2) glucuronides	*Perilla ocimoides*	Lamiaceae	Yoshida *et al*., 1993
Apigenin 7-*O*-rhamnoside-4′-*O*-glucosyl(1 → 4)rhamnoside	*Asplenium normale*	Pteridophyte	Iwashina *et al*., 1993
Chrysoeriol 7-*O*-mannosyl(1 → 2)alloside	*Cassia alata*	Fabaceae	Gupta and Singh, 1991
Isoscutellarein 7-*O*-allosyl(1 → 2)glucoside (acetate)	*Lycopus europaeus*	Lamiaceae	Bucar *et al*., 1995
Kaempferol 3-*O*-glucosyl(1 → 2)rhamnoside	*Ginkgo biloba*	Ginkgoaceae	Markham *et al*., 1992
Luteolin 7-*O*-glucuronyl(1 → 2)glucuronide	*Aloysia triphylla*	Verbenaceae	Darnat *et al*., 1995
6-Hydroxyluteolin 7-*O*-sambubioside	*Hebe stricta*	Scrophulariaceae	Kellam *et al*., 1993
Prunetin 4′-O-apiosyl(1 → 6)glucoside	*Dalbergia coromandeliana*	Fabaceae	Ramesh and Yuvarajan, 1995
Rhamnetin 3-*O*-mannosyl(1 → 2)alloside	*Cassia alata*	Fabaceae	Gupta and Singh, 1991
Strobopinin 7-*O*-xylosyl(1 → 3)xyloside	*Mosla chinensis*	Lamiaceae	Zheng *et al*., 1996
Tamarixetin 3-*O*-glucosyl(1 → 2)galactoside	*Cynanchum thesioides*	Asclepiadaceae	Yuan and Zuo, 1992
Triglycosides . . .			
Isorhamnetin 3-*O*- glucosyl(1 → 2)rhamnosyl(1 → 6)galactoside	*Anthyllis sericea*	Fabaceae	Marco *et al*., 1989
Isorhamnetin 3-*O*-glucosyl(1 → 2)galactoside 7-*O*-glucoside	*Anthyllis sericea*	Fabaceae	Marco *et al*., 1989
Kaempferol 3-*O*- rhamnosyl(1 → 4)rhamnosyl(1 → 6)galactoside	*Actinidia* spp.	Actinidiaceae	Webby and Markham, 1990
Rhamnocitrin 3-*O*-apiosyl(1 → 5)apioside 4′-*O*-glucoside	*Mosla chinensis*	Lamiaceae	Zheng *et al*., 1996
Tetraglycoside. . .			
Kaempferol 3-*O*-dirhamnosyl(1 → 2, 1 → 6)galactoside-7-*O*-glucoside	*Alangium premnifolium*	Alangiaceae	Kijima *et al*., 1995

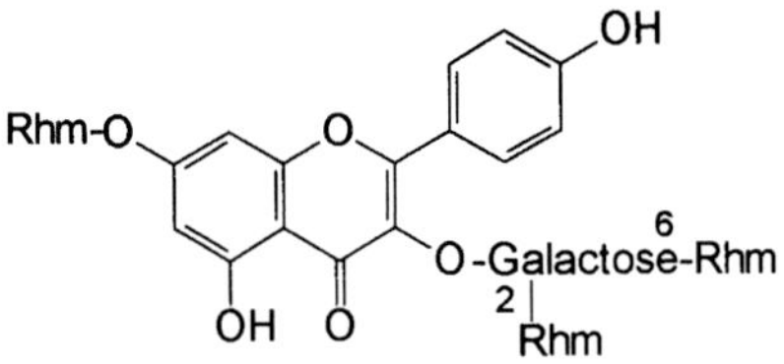

2-148

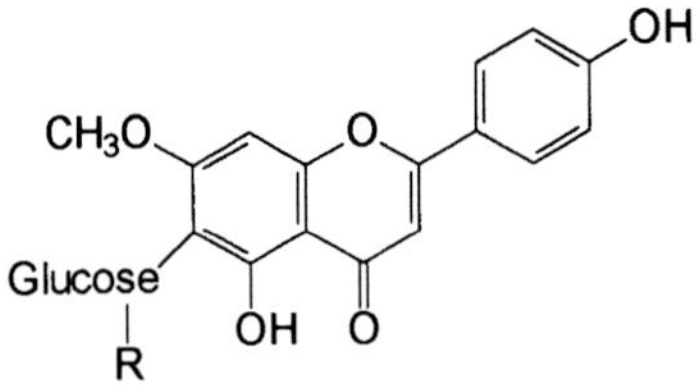

2-150 R = Glucose
2-151 R = Xylose

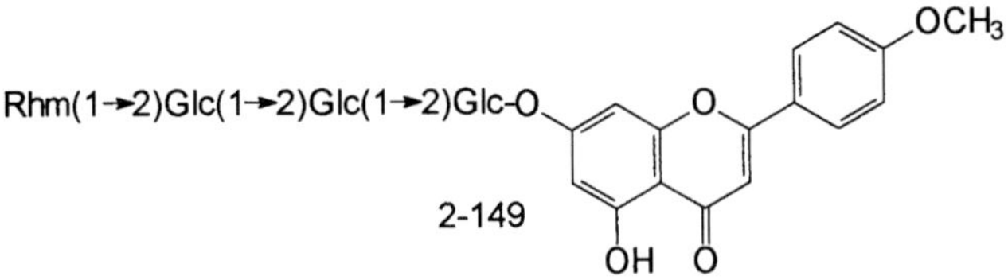

2-149

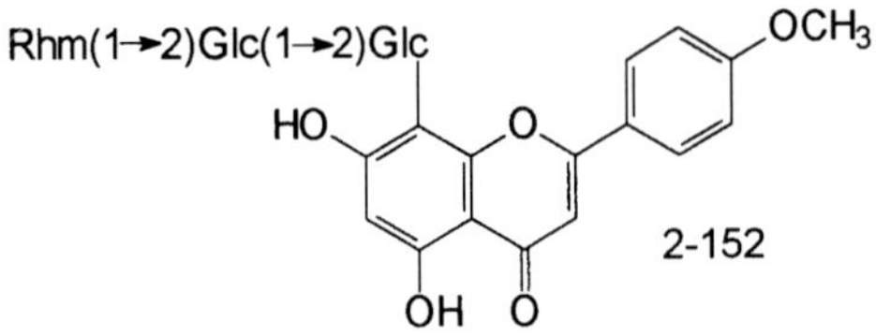

2-152

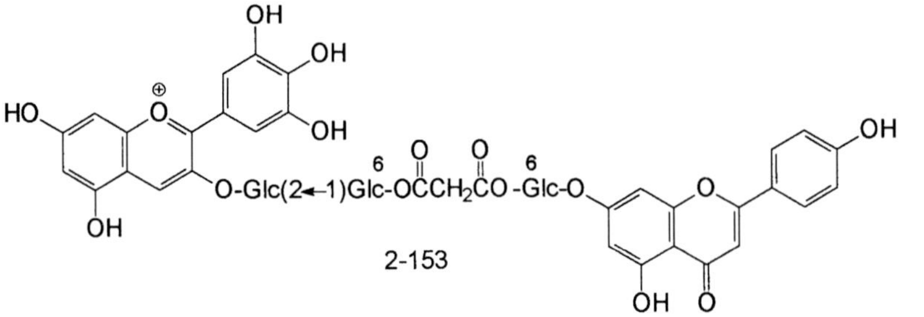

2-153

derivatives of swertisin (6-*C*-glucosylapigenin 7-methyl ether). The third compound, also known, was the 2″-*O*-glucosyl derivative of isoswertisin (8-*C*-glucosylapigenin 7-methyl ether). The fourth compound, new as a natural product at the time of discovery, was identified as 2″-*O*-β-xylosyl-8-*C*-β-galactosylapigenin. Sharaf and coworkers (1997) have recently described a suite of flavone glycosides from *Peganum harmala* (Zygophyllaceae) among which was a *C*-glycosylflavone identified as 2‴-*O*-α-rhamnosyl-2″-*O*-β-glucosylcytisoside [2-152]. Cytisoside is the 8-*C*-glucosyl derivative of apigenin 4′-methyl ether (acacetin). The genetic control of glycosylation has been studied in detail in *Silene* in which plants *C*-glucosylflavones can be secondarily glycosylated either at position 2″ of the *C*-glucosyl group or at position-7 of the flavone. The compounds generated by this system can arise from transfer of arabinose, glucose, rhamnose, or xylose to the 2″-hydroxyl group, or galactose, glucose, or xylose to the 7-position of isovitexin (Steyns and van Brederode, 1986; van Brederode *et al.*, 1987). This study of glycosylation in *Silene* by the Dutch geneticists stands as one of the best efforts to understand the genetic basis, and evolutionary significance, of flavonoid biosynthesis.

Several anthocyanins are known that are characterized by complex glycosylation (and acylation) patterns. These compounds are discussed in some detail later in this chapter. Several examples from the literature can be found in Table 2.4.

Acylation

Several different organic acids have been found esterified to one or more hydroxyl groups of flavonoid glycosides or directly to a flavonoid hydroxyl. Aliphatic acids involved include acetic, malonic, lactic, succinic, butyric, 2-methylbutyric, 3-methylbutyric, 2-butenoic, tiglic (*cis*-2-methylbut-2-enoic acid), 3-methyl-3-hydroxy-glutaric, and quinic acids. Aromatic acids include benzoic, *p*-hydroxybenzoic, gallic

Table 2.4 A sampling of acylated anthocyanins

Plant species	Family	Antho[a]	Sugars	Acids[b]	Reference
Aconitum chinense	Ranunculaceae	Dp	3-Glucose 7-Glucose	HB	Takeda *et al.*, 1994
Ajuga pyramidalis	Lamiaceae	Cy	3-Glucose 5-Glucose	M, F	Madhavi *et al.*, 1996
Allium victorialis	Liliaceae	Cy	3-Glucose	M	Andersen and Fossen, 1995
Clitoria ternatea	Leguminosae	Dp	3-Glucose 3′-Glucose 5′-Glucose	M, C	Terahara *et al.*, 1996
Consolida armeniaca	Ranunculaceae	Dp	3-Glucose 7-Glucose	M, C	Saito *et al.*, 1996c
Evolvulus pilosus	Convolvulaceae	Dp	3-Glucose 5-Glucose	M, CF	Toki *et al.*, 1994a
Gentiana cultivar	Gentianaceae	Cy	3-Glucose 5-Glucose	CF	Hosokawa *et al.*, 1995
Ipomoea batatas	Convolvulaceae	Cy, PE	3-Sophorose 5-Glucose	CF	Goda *et al.*, 1997
Ipomoea purpurea	Convolvulaceae	Cy	3-Sophorose 5-Glucose	CF	Saito *et al.*, 1995
x*Laeliocattleya* cv	Orchidaceae	Cy	3-Glucose 7-Glucose 3′-Glucose	M, C	Tatsuzawa *et al.*, 1994
Matthiola incana	Brassicaceae	Pg	3-Sambubiose 5-Glucose	M, C, F, S	Saito *et al.*, 1996a
Meconopsis spp.	Papaveraceae	Cy	3-Sambubiose 5-Glucose	M	Takeda *et al.*, 1996
Pharbitis nil	Convolvulaceae	Cy, Pn	3-Glucose 5-Glucose	CF	Saito *et al.*, 1996b
Ranunculus asiaticus	Ranunculaceae		3-Glu-Xyl	M	Toki *et al.*, 1996
Senecio cruentus	Asteraceae	Pg	3-Glucose 7-Glucose	M, CF	Toki *et al.*, 1995a

[a]Anthocyanidin abbreviations as in text
[b]Acylating acids: C = *p*-coumaric; CF = caffeic; F = ferulic; HB = *p*-hydroxybenzoic; M = malonic; S = sinapic
The acylating acid follows the sugar to which it is attached

(3,4,5-trihydroxybenzoic), cinnamic, *p*-coumaric (4-hydroxycinnamic), caffeic (3,4-dihydroxycinnamic), ferulic (3-methoxy-4-hydroxycinnamic), isoferulic (3-hydroxy-4-methoxycinnamic), and sinapic acids (3,5-dimethoxy-4-hydroxycinnamic). Two or more acyl groups may be present on any given sugar, or on different sugars in a di- or triglycoside, and it is not uncommon to find different acids involved. Examples of acylated glycosides showing a range of diversity and acylating acids are shown in Table 2.5.

Complex acylated flavonoids have been identified recently that consist of two monomeric flavonoid glycosides whose saccharide units are acylated jointly by a bifunctional acid, such as malonic acid. An example of this type of compound was isolated from flowers of the water hyacinth *Eichhornia crassipes* (Pontederiaceae) by Toki and colleagues (1994b) and shown to consist of delphinidin 3-*O*-gentiobioside and apigenin 7-*O*-glucoside linked by malonic acid via the 6‴-hydroxyl group of the gentiobiose unit and the 6″-hydroxyl group of the apigenin glucoside. This compound is represented diagramatically as [2-153]. Other examples involving flavone glucosides linked by bifunctional acids include compounds from *Agastache rugosa* (Lamiaceae) (Itokawa *et al.*,1981) and *Citrus sudachi* (Rutaceae) (Horie *et al.*, 1986). The latter example involves 3-hydroxy-3-methylglutaric as the linking acid.

Species belonging to the mint genus *Stachys* have provided a rich array of flavonoids over the years, including flavones with interesting B-ring oxygenation patterns and several acylated glycosides (e.g., see El-Ansari *et al.*, 1991). A recent study of *S. aegyptica* from the same laboratory has added an entirely new member to the roster of acylated glycosides. The new compound was shown to consist of two apigenin 7-*O*-glucoside units acylated at the 6″-oxygens with *p,p*′-dihydroxytruxinic acid [2-154] (El-Ansari *et al.*, 1995). The new compound, which the authors named "stachysetin," was accompanied in the plant by the known apigenin 7-*O*-(6″-*p*-coumaroyl)glucoside [2-155]. Dimerization of two molecules of the 6″-*p*-coumarate through interaction of the cinnamyl double bonds yields the cyclobutyl system characteristic of truxillic acid and its isomers.

Two highly unusual compounds, named "poriolide" [2-156] and "isoporiolide" [2-157], were obtained from *Leucothoë keiskei* (Ericaceae) where they occurred with the known *C*-methylflavanones "poriol" [2-158] and its glucoside "poriolin" [2-159] (Ogiso and Koshida, 1972). Poriolide and isoporiolide differ with respect to the position of the hydroxyl group on the salicylic acid residue, whether it is "inside" or "outside" the macrolide structure. Formation of these compounds can be rationalized by assuming that the flavanone 7-*O*-glucoside poriolin is first acylated with a salicylic acid (2-hydroxybenzoic acid) [2-160] residue. This is reasonable in view of the fairly common occurrence of salicylic acid derivatives in Ericaceae (Harborne and Williams, 1973). Phenolic coupling would then account for the bond between the salicylic acid group and the B-ring of the flavanone. Both inside and outside orientations are possible because the salicylic acid unit is free to rotate.

A unique acyl derivative was reported recently from the bracken fern (*Pteridium aquilinum*) by Imperato (1996). Hydrolysis with acid afforded kaempferol, apiose, *p*-coumaric acid, the amino acid glycine, and several intermediate products each of which yielded only glycine upon further treatment with acid. Alkaline hydrolysis yielded kaempferol 3-*O*-apioside, *p*-coumaric acid, glycine, and intermediates.

Table 2.5 A Sampling of acylated flavonoid glycosides

Compound	Source	Family	Reference
Apigenin 5-*O*-(6″-malonyl) glucoside	*Equisetum arvense*	Equisetaceae	Veit *et al.*, 1990
Apigenin 7-*O*-[2″-(2-hydroxypropanoyl)] glucoside	*Marrubium vulgare*	Lamiaceae	Nawwar *et al.*, 1989
Luteolin 7-*O*-(6″-*E*-cinnamoyl) glucoside	*Salix gilgiana*	Salicaceae	Mizuno *et al.*, 1987a
Pelargonidin 3-*O*-glucoside-5-*O*-(6″-acetylglucoside)	*Verbena hybrida*	Verbenaceae	Toki *et al.*, 1995
Cyanidin 3-*O*-(6″-oxalyl) glucoside	Orchids	Orchidaceae	Strack *et al.*, 1989
Kaempferol 3-*O*-(3″,6″-di-*p*-coumaroyl) glucoside	*Pinus sylvestris*	Pinaceae	Jungblut *et al.*, 1995b
Quercetin 3-*O*-(3″,6″-di-*p*-coumaroyl) glucoside	*Pinus sylvestris*	Pinaceae	Jungblut *et al.*, 1995b
Okanin 4-*O*-(2″-caffeoyl-6″-acetyl) glucoside	*Bidens frondosa*	Asteraceae	Karikome *et al.*, 1992
Okanin 4′-*O*-(2″,4″-diacetyl-6″-*p*-coumaroyl) glucoside	*Bidens pilosa*	Asteraceae	Hoffmann and Hölzl, 1988
4-*O*-Methylokanin 4′-*O*-(6″-*p*-coumaroyl) glucoside	*Bidens frondosa*	Asteraceae	Karikome *et al.*, 1992
Maritimetin 6-*O*-(2″,4″,6″-triacetyl) glucoside	*Bidens pilosa*	Asteraceae	Hoffmann and Hölzl, 1989
Kaempferol 3-*O*-(2″-*p*-coumaroyl) rhamnoside 7-*O*-rhamnoside	*Cheilanthes fragrans*	Adiantaceae	Imperato, 1992b
Kaempferol 3-*O*-(2″-galloyl) arabinoside	*Eucalyptus rostrata*	Myrtaceae	Okamura *et al.*, 1993
Kaempferol 3-*O*-(5″-feruloyl)apioside	*Pteridium aquilinum*	Dennstaedtiaceae	Imperato, 1996
Quercetin 4′-*O*-(6″-galloyl) glucoside	*Eucalyptus rostrata*	Myrtaceae	Okamura *et al.*, 1993
Quercetin 3-*O*-(2″-*p*-hydroxybenzoyl-4″-*p*-coumaroyl) glucoside	*Libocedrus bidwillii*	Cupressaceae	Franke and Markham, 1989
Patuletin 3-*O*-rhamnoside 7-*O*-(3″,4″-diacetyl) rhamnoside	*Kalanchoë gracilis*	Crassulaceae	Liu *et al.*, 1989
Patuletin 3-*O*-(4″-acetylrhamnoside)-7-*O*-(2″,4″-diacetylrhamnoside)	*Kalanchoë gracilis*	Crassulaceae	Liu *et al.*, 1989
Quercetin 3-*O*-[6‴-sinapoylglucosyl](1 → 2)-galactoside	*Thevetia peruviana*	Apocynaceae	Abe *et al.*, 1995
Myricetin 3-*O*-(6″-galloyl)-*β*-D-galactoside	*Myrica gale*	Myricaceae	Nagai *et al.*, 1995
Myricetin 5′-methyl ether 3-*O*-(6″-acetyl)-glucoside	*Picea abies*	Pinaceae	Slimestad and Hostettmann, 1996

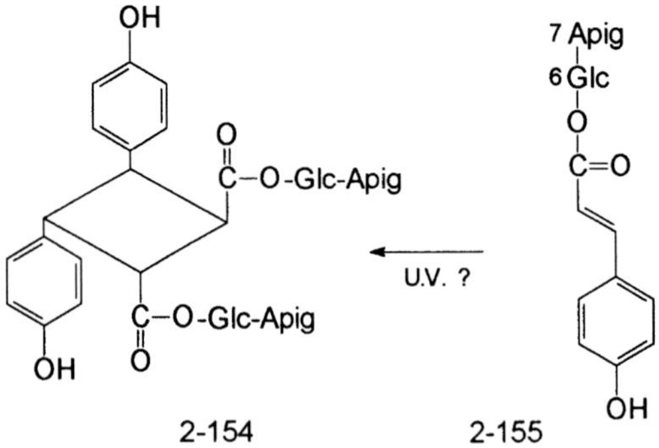

2-154 2-155

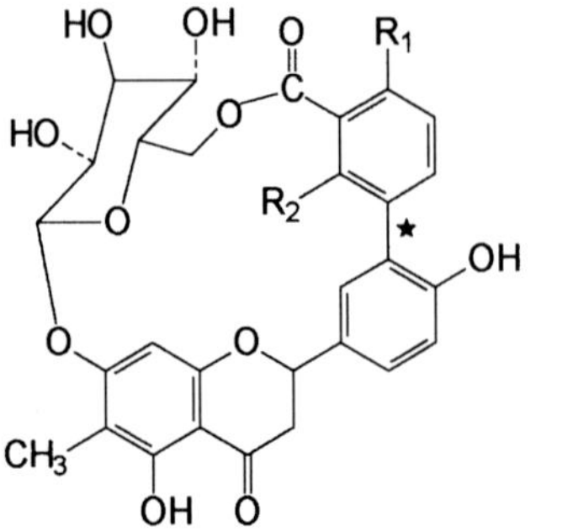

2-156 R_1 = OH, R_2 = H

2-157 R_1 = H, R_2 = OH

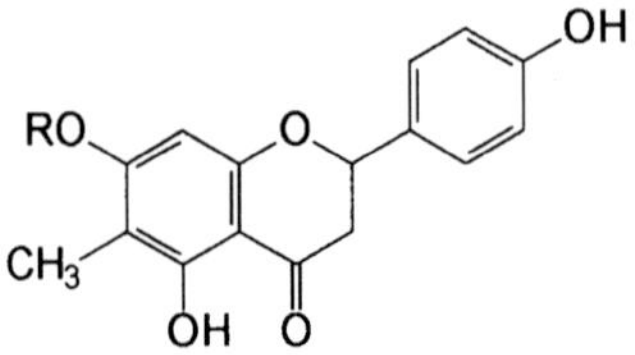

2-158 R = H

2-159 R = Glucose

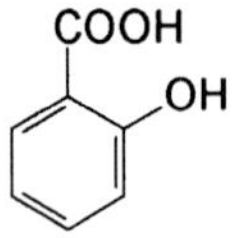

2-160

Detailed study of the compound using fast-atom bombardment mass spectrometry (FAB-MS, see Chapter Four for comments on this technique) showed that the compound was kaempferol 3-*O*-apioside acylated at one site with *p*-coumaric acid and at the other with a polyglycine chain consisting of seven amino acid residues [2-161]. The positions of acylation of apiose were not determined.

Flavonoids acylated directly through a phenolic group, or through the *C*-3 hydroxyl group in flavonols or dihydroflavonols, have been reported in recent years. These compounds appear to be components of the non-polar exudates of certain plants. An unusual addition to the list of acylated flavonoids involves a compound that exhibits both direct phenolic acylation as well as acylation of sugar hydroxyls. Bilia and coworkers (1991) identified an eriodictyol 7-*O*-glucoside derivative that was acetylated at all four of the glucoside hydroxyl groups and at the 3′,4′-hydroxyl groups as well [2-162]. Most of the acids listed above have been identified in directly acylated flavonoids. It is highly likely that many of these compounds have been

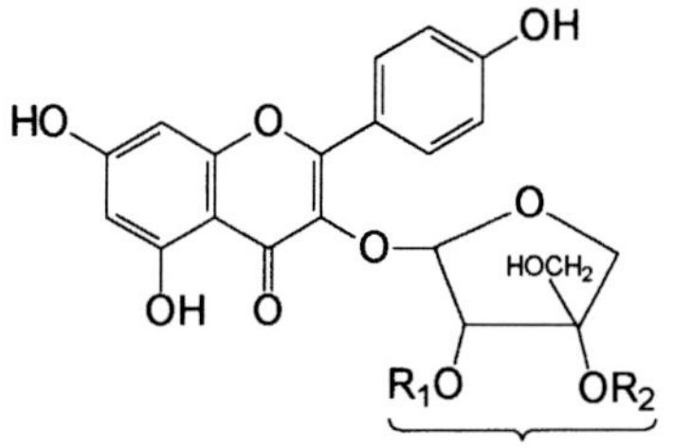

R_1/R_2 = *p*-coumaroyl 2-161

R_2/R_1 = $-(\overset{O}{\overset{\|}{C}}CH_2\overset{H}{\overset{|}{N}})_6\overset{O}{\overset{\|}{C}}CH_2NH_2$

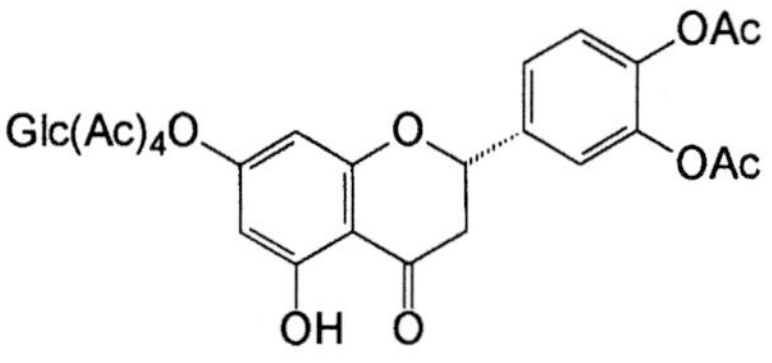

2-162

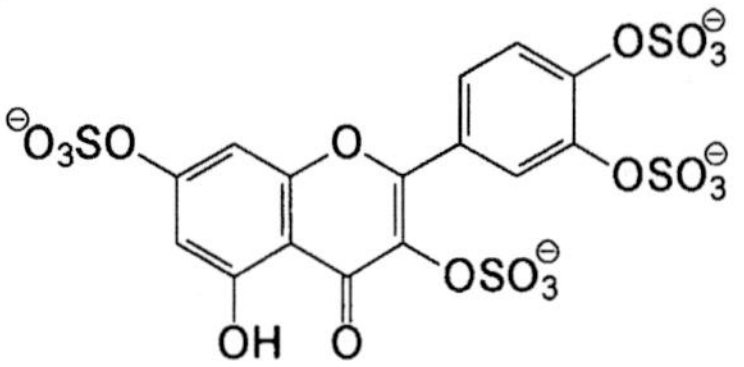

2-163

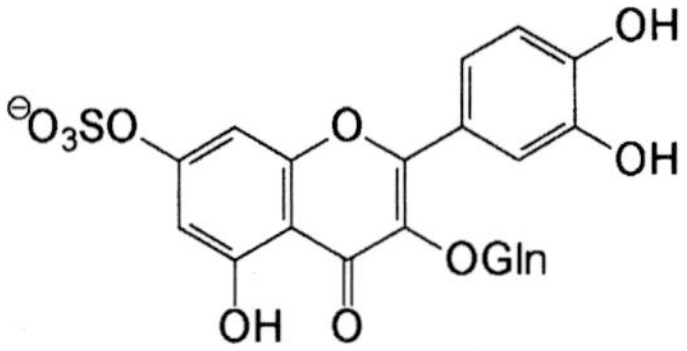

2-164

overlooked in the past owing to their sensitivity to hydrolysis. Table 2.6 lists examples of these compounds along with the plants from which they were obtained.

Sulfation

As in the case of acylation, sulfation can involve either sugar or ring hydroxyl groups or both in the same molecule. Sulfation of flavonoid *C*-5 hydroxyl groups does not appear to occur but sulfation of any other hydroxyl group, or several of them, can. 7-Sulfates of flavones and 3-Sulfates of flavonols are the most commonly encountered derivatives of these classes. Detailed studies of *Flaveria* (Asteraceae) by Barron and coworkers (1988, and citations therein) have resulted in finding several di- and trisulfates of quercetin as well as the 3,7,3′,4′-tetrasulfate [2-163] derivative. Many mixed compounds, e.g., quercetin 3-*O*-glucuronide-7-sulfate [2-164], have been reported. A list of flavonoid sulfates was compiled by Williams and Harborne (1994, Table 8.5).

FLAVONOID CLASSES

In this section we will examine the structures of each class of flavonoids in a bit more detail. There may be significant differences in the size of treatment of the individual classes, however. This apparent favoritism is occasioned by intense recent activity in

Table 2.6 Examples of directly acylated flavonoids

Compound	Species	Family	Reference
5,7-Dihydroxyflavone 7-benzoate	*Baccharis bigelovii*	Asteraceae	Arriaga-Giner *et al.*, 1986
2′,5′-Dihydroxyflavone 5′-acetate	*Primula pulverulenta*	Primulaceae	Wollenweber *et al.*, 1988b
3,5,7,3′-Tetrahydroxyflavone 3-isobutyrate	*Flourensia retinophylla*	Asteraceae	Stuppner and Müller, 1994
5,7,3′,4′-Tetrahydroxy-6-C-glucosyl flavone 7-caffeate	*Vellozia* species	Velloziaceae	Harborne *et al.*, 1994
3,5,7,3′,5′-Pentahydroxyflavone 3-, 3′-, 4′-isobutyrates	*Traversia bacchariodes*	Asteraceae	Kulanthaivel and Benn, 1986
3,5,7-Trihydroxy-8,3′, 4′-trimethoxyflavone 3-angelate	*Polygonum flaccidum*	Polygonaceae	Ahmed *et al.*, 1991
5,2′,5′-Trihydroxy-3,7, 8-trimethoxyflavone 2′-acetate	*Notholaena californica*	Pteridaceae	Wollenweber *et al.*, 1988c
5,7,8-Trihydroxy-8-methoxyflavone 8-(*Z*-2-methyl-2-butenoate)	*Gnaphalium robustum*	Asteraceae	Urzua and Cuadra, 1990
5,7,8-Trihydroxy-8-methoxyflavone 8-(2-methylbutyrate)	*Gnaphalium robustum*	Asteraceae	Urzua and Cuadra, 1990

certain areas. A case in point involves the anthocyanins. In recent years the level of understanding of how flower color is produced in plants has come under detailed scrutiny by both organic and physical chemists. It is essential that the reader be given an opportunity to see the results of as much of that work as is practicable. Proanthocyanidins, or tannins in the broader sense, have also been the subject of intensified work of late owing to the availability of ever more powerful analytical instrumentation. Isoflavonoids continue to attract a good deal of interest owing to their biological activities as well as their prominence as major components of legumes, not only a large and conspicuous family, but also an economically important one. Other flavonoid classes are not necessarily any less biologically important or chemically less interesting but they, as groups, may not have attained the level of "fame" enjoyed by isoflavones and anthocyanins. To be sure, representatives of all classes will be encountered in Chapter Seven where the role of individual compounds in Nature will be discussed.

In addition to common members of each class of compounds, rare or unusual compounds will also be described. These compounds are structurally interesting, but they often give useful insights into the biosynthetic capabilities of the organisms under study as well. It is obvious that a full reckoning of each class of compounds is beyond the scope of this book, but enough examples will be presented to provide the reader with an appreciation of the structural diversity within each class. Overlap is inevitable since some of the compounds encountered in one class may be closely related to members of another class, e.g., the β-ketodihydrochalcones (or dibenzoylmethanes) provide an opportunity to develop a link between dihydrochalcones and flavones. Similarly, "retrochalcones" may appear to be aberrant chalcones, but their structural relationship with corresponding "anthocyanin water 2-adducts" is significant (Brouillard and Dangles, 1994; see below). Additional information concerning interesting distributional or functional aspects of particular flavonoids will be provided whenever appropriate.

Anthocyanins

The anthocyanins constitute the most conspicuous group of flavonoids. They are the compounds responsible for the red, mauve, violet, and blue flower colors seen in most plants. The qualification 'most' is necessary because the red-flowered members of Caryophyllales (Centrospermae, e.g., beets, *Bougainvillea*, *Cactus*) are caused not by anthocyanins at all but by a group of pigments called betacyanins. These compounds, part of a larger group of called betalains, are interesting from both structural and evolutionary perspectives and will be discussed in a bit more detail later (Chapter Three).

Anthocyanins are not restricted to flowers; they can be found in leaves, stems, fruits, seeds, and even root tissue. Although most commercially available plants are attractive because of their flower color, several owe their attractiveness to patterned deposition of anthocyanins in leaves. The "prayer plant," *Maranta* sp. (Marantaceae) is an excellent example of this phenomenon. The many-hued varieties of *Coleus* owe their leaf colors to anthocyanins as well. Less spectacular, perhaps, but interesting nonetheless, is the existence of leaf chevrons in the clover (*Trifolium tridentatum*). These anthocyanin deposition patterns are under single gene control with the allele for spots being dominant (Griffiths and Ganders, 1983).

Anthocyanins consist of an anthocyanidin bound to one or more sugar moieties. There is always a sugar at position-3 and frequently additional sugars at positions-5 and -7. Further modification of the glycosides through acylation and through complexation with noncyanic flavonoids and metal ions also can occur. Subtle differences in cell pH, along with the possibilities of copigmentation, and ionic complexation, lead to a virtually unlimited degree of color variation. All of this variation is based upon a limited number of fundamental flavonoid structures when compared to most of the other classes of flavonoids. The commonly occurring anthocyanidins, with their standard abbreviations, are pelargonidin (Pg) [2-165], cyanidin (Cy) [2-166], peonidin (Pn) [2-167], delphinidin (Dp) [2-168], petunidin (Pt) [2-169], and malvidin (Mv) [2-170].

Peonidin, petunidin, and malvidin, all *O*-methylated anthocyanidins, have been recognized as common flower pigments for many years. More recently, several additional *O*-methylated pigments have been reported: rosinidin is the 7,3′-dimethyl ether of cyanidin [2-171], while several delphinidin derivatives join the list: pulchellin is the 5-methyl ether, europinidin is the 5,3′-dimethyl ether, capensidin is the 5,3′,5′-trimethyl ether (malvidin 5-methyl ether), and hirsutidin is the 7,3′,5′-trimethyl ether (malvidin 7-methyl ether) [2-172]. Some of the sources of these unusual pigments are listed in Table 2.7 (data from Harborne, 1967 and *The Flavonoids* series).

5-Deoxyanthocyanidins, such as fisetinidin (5-deoxycyanidin) [2-173] and robinetinidin (5-deoxydelphinidin) [2-174] can be formed from proanthocyanidins (Harborne, 1967). Extra A-ring oxygenation is also seen in a few naturally occurring anthocyanins. Aurantinidin, 6-hydroxypelargonidin [2-175], occurs in the petals of

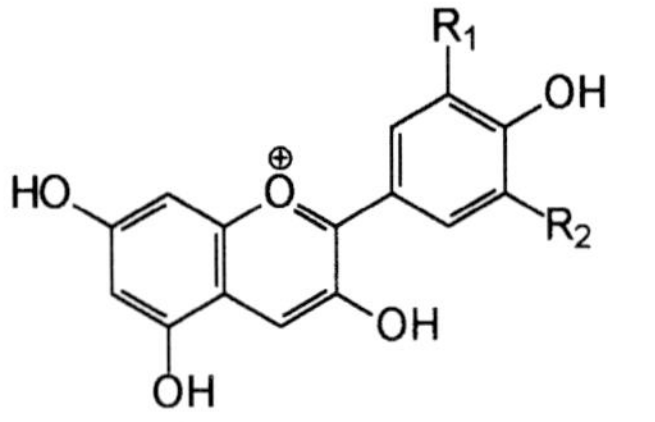

2-165 $R_1 = R_2 = H$
2-166 $R_1 = OH, R_2 = H$
2-167 $R_1 = OCH_3, R_2 = H$
2-168 $R_1 = R_2 = OH$
2-169 $R_1 = OH, R_2 = OCH_3$
2-170 $R_1 = R_2 = OCH_3$

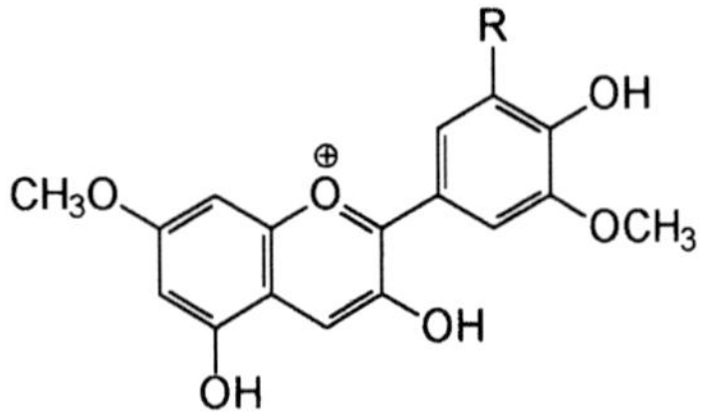

2-171 R = H
2-172 $R = OCH_3$

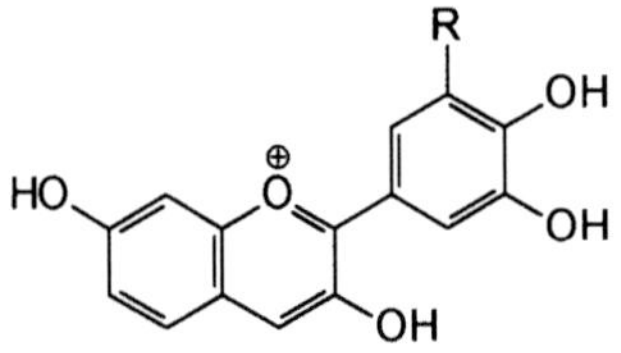

2-173 R = H
2-174 R = OH

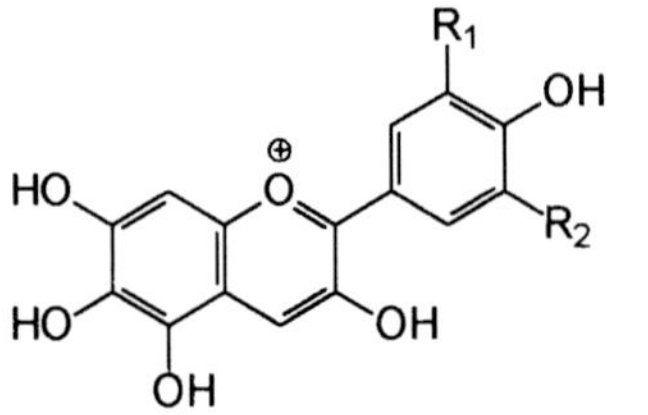

2-175 $R_1 = R_2 = H$
2-176 $R_1 = OH, R_2 = H$
2-177 $R_1 = R_2 = OH$

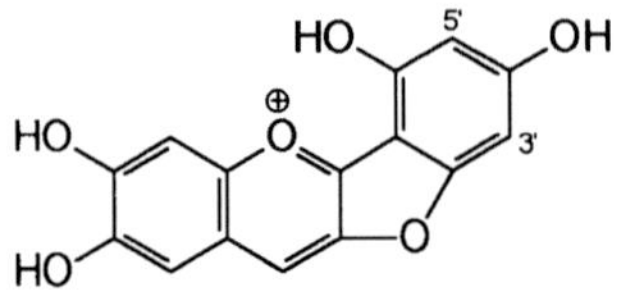

2-178

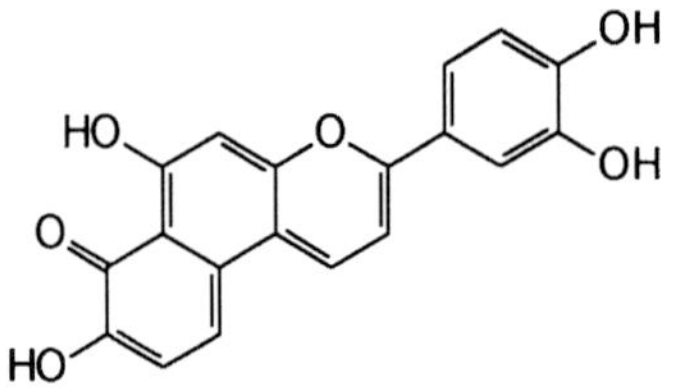

2-179

Impatiens aurantiaca (Balsaminaceae) (Clevenger, 1964), while 6-hydroxycyanidin [2-176] and 6-hydroxydelphinidin [2-177] (as 3-*O*-rutinosides), along with cyanidin and delphinidin glycosides, have been reported from Chilean species of *Alstroemeria* (Alstroemeriaceae) (Norbaek *et al.*, 1996).

Anthocyanidins occur in a variety of glycosylated forms, but all of them have at least a glucose moiety at the C-3 position. The formation of anthocyanidin 3-*O*-glucosides, in fact, is considered one of the obligate steps in the biosynthesis of anthocyanins (see Chapter Six for details). Further glycosylations and acylations depend upon genetic factors in individual groups of plants. Glycosylation at position-5 is common and occasionally one finds sugars at other positions as well. Extensive lists of anthocyanin occurrences appear in each volume of *The Flavonoids*.

Table 2.7 Structures and sources of some unusual anthocyanidins

Anthocyanidin	No. in text	Structure	Source	Family	Reference
Cyanidin 5-methyl ether	–		*Egeria densa*	Elodeaceae	Momose *et al*., 1977
Rosinidin	2-162	Cyanidin 7,3′-dimethyl ether	*Primula rosea*	Primulaceae	Harborne, 1968
Pulchellin	2-163	Delphinidin 5-methyl ether	*Plumbago pulchella*	Plumbaginaceae	Harborne, 1967b
Europinidin	2-164	Delphinidin 5,3′-dimethyl ether	*Plumbago europaea*		Harborne, 1966
Capensidin	2-165	Delphinidin 5,3′,5′-trimethyl ether	*Plumbago capensis*		Harborne, 1962
Hirsutinidin	2-166	Delphinidin 7,3′,5′-trimethyl ether	*Catharanthus roseus*	Apocynaceae	Carew and Krueger, 1976
Aurantinidin	2-169	6-Hydroxypelargonidin	*Impatiens aurantiaca*	Balsaminaceae	Clevenger, 1964
6-Hydroxycyanidin	–		*Alstroemeria* species	Alstroemeriaceae	Norbaek *et al*., 1996
6-Hydroxydelphinidin	–		*Alstroemeria* species		Norbaek *et al*., 1996
Fisetinidin	2-167	5-Deoxycyanidin	From proanthocyanidins		Harborne, 1967a, p. 4
Robinetinidin	2-168	5-Deoxydelphinidin	From proanthocyanidins		Harborne, 1967a, p. 4

A number of bryophytes (mosses, liverworts, and hornworts) are known that have reddish or purplish colors associated with their cell walls. The reddish coloration of the liverwort *Ricciocarpos natans* was the subject of a recent study by Kunz and collaborators (1994) who reported two novel anthocyanidins, "riccionidin A" [2-178] and its apparent dimer "riccionidin B." Several features of these compounds are noteworthy among which is the ether linkage between the B- and C-rings. The 6,7-dioxy-5-deoxy pattern is unique among anthocyanidins and the 2′,4′,6′-trioxygenation pattern of the B-ring is a feature not seen in other pigments of this general class. The second compound, "riccionidin B," was shown to have a molecular ion (FABMS) $m/z = 569.0783$ $[M+H]^+$, as compared to $m/z = 285.0399$ $[M+H]^+$ for riccionidin A, indicating that it is a dimer. The two monomeric units are linked through C-3′ or C-5′. Both riccionidins were detected chromatographically in extracts of the liverworts *Marchantia polymorpha*, *Riccia duplex*, and *Scapania undulata*.

Chemical studies of *Sphagnum* some years ago resulted in the identification of the red pigment [2-179] (Vowinkel, 1975). Although there is a resemblance between "sphagnorubin" and flavonoids in its ring system, the absence of a charge and the presence of the α-hydroxy carbonyl system clearly distinguishes between them. So far as I am aware, there have been no biosynthetic studies of "sphagnorubin" nor any attempts to establish its breadth of occurrence.

Acylated anthocyanins involving caffeic or acetic acid had been reported from a few plant species by a number of workers but no comprehensive surveys had been undertaken until fairly recently. Reports of anthocyanins acylated with malonic acid in *Papaver nudicaule* (Papaveraceae) (Cornuz *et al.*, 1981) and *Cichorium intybus* (Asteraceae, corn flower) (Bridle *et al.*, 1984), however, prompted Harborne and Boardley (1985) to undertake a survey of anthocyanins across a wide spectrum of flowering plant families. Since acylated glycosides are sensitive to the HCl-methanolic extraction procedure usually used for anthocyanins, mineral acid was omitted from their procedure. Electrophoretic analysis of 81 species representing 27 plant families indicated that zwitterionic compounds existed in half the sample with nearly all of the members from Asteraceae and Lamiaceae tested being positive. It soon became evident that malonylated anthocyanins are very common, and that in some instances both malonic acid and an aromatic acid were present in the same molecule. Often several such compounds were observed to occur in a single species. An example of this is seen in the anthocyanin chemistry of *Salvia splendens* (Lamiaceae). Kondo and coworkers (1989b) isolated a total of 11 anthocyanins from different flower color variants and showed them to be derivatives of pelargonidin, delphinidin, or malvidin. All were 3,5-diglycosides and had either *p*-coumaric or caffeic acid, and most (9 of 11) had one or two malonic acid residues as well. Table 2.4 lists some examples of moderately complex acylated anthocyanins described in recent years. The anthocyanidin is identified along with the positions and levels of glycosylation, and the acylating acid or acids.

In addition to the existence of sets of analogous pigments in a given species, sets of structurally related anthocyanins can often co-occur as well. A comparatively simple example of this phenomenon involves pigments identified from the violet-blue flowers of the Japanese morning glory (*Pharbitis nil* syn. *Ipomoea nil*) (Convolvulaceae) by Lu and coworkers (1991). The most complex pigment of the set was identifed as peonidin

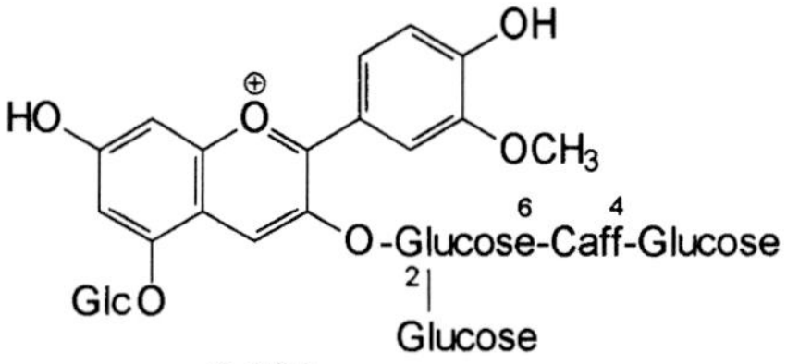

2-180

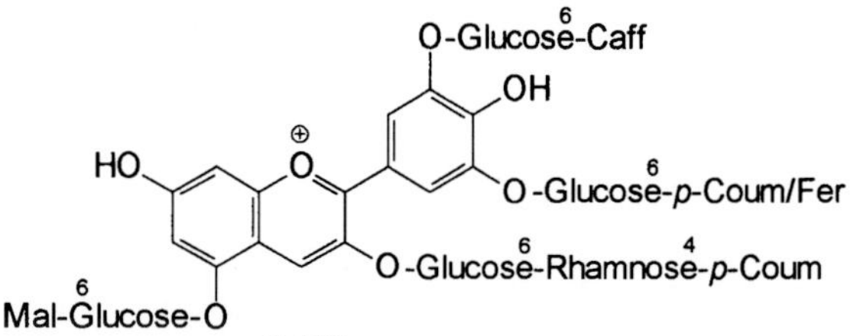

2-181

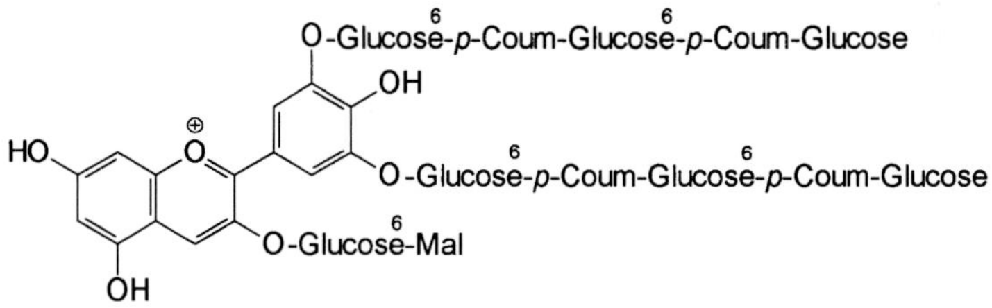

2-182

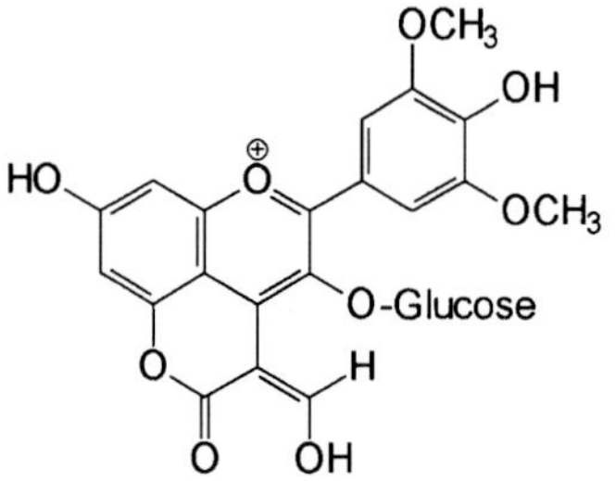

2-183

3-*O*-(2-*O*-(β-D-glucopyranosyl)-6-*O*-(*trans*-4-*O*-(β-D-glucopyranosyl)-caffeoyl)-β-D-glucopyranoside)-5-*O*-(β-D-glucopyranoside), shown diagramatically as [2-180]. Also identified in the pigment fraction from this species was the non-acylated peonidin 3-*O*-gentiobioside-5-*O*-glucoside. In situations such as this it would not be uncommon to see the corresponding 3-*O*-glucoside-5-*O*-glucoside or the 3-*O*-gentiobioside present as well. The exact nature of the pigment mix would obviously rely on the specificities of the glycosyltransferases involved.

Two examples will illustrate the level of complexity attained by some species. Two major pigments from have been isolated from flowers of *Lobelia erinus* (Campanulaceae, Lobeliaceae) and identified (Kondo *et al.*, 1989). Abbreviated structures for "lobelinin A" and "lobelinin B" are shown in [2-181, with either *p*-coumaroyl or feruoyl as the variable acylating unit]. These compounds, cumulatively, consist of two different sugars (glucose and rhamnose), malonic acid, three different cinnamic acids (*p*-coumaric, caffeic, and ferulic), and have glycosidic linkages involving four positions on the anthocyanidin (3-, 5-, 3′-, and 4′). More complicated still, and the largest of these compounds at the time of its discovery (Terahara *et al.*, 1990), was "ternatin A1" [2-182] from *Clitoria ternatea* (Fabaceae). It consists of delphinidin, seven molecules of glucose, four molecules of *p*-coumaric acid, and one molecule of malonic acid. Strack and Wray (1994, page 11) pointed out that the FAB-MS spectrum of ternatin A1 had a molecular-ion peak at *m*/*z* 2107, which corresponds to $C_{96}H_{107}O_{53}^{+}$!

A convenient nomenclatural shortcut, in which the components of these complicated acylated glycosides are represented in a linear fashion, was demonstrated by Terahara and colleagues (1996). This shortcut is particularly useful in cases where sugar and acyl units alternate. A simple set of abbreviations, wherein "G" stands glucose, "M" stands for malonyl, and "C" stands for *p*-coumaroyl, allows complicated structures to be represented easily, e.g., delphinidin 3-GM-3′-GCG-5′-GCG or delphinidin 3-GM-3′-GCGCG-5′-GC should be readily interpreted. Other acids could be handled in the same way: "F" for ferulic acid, "CF" for caffeic acid, and "S" for sinapic acid. In the case of *p*-coumaroyl, feruloyl, and sinapoyl residues, there is only one site for glycosylation so further information is not required. In the case of caffeoyl residues, however, where two hydroxyl groups are available for attachment of a sugar, it would be necessary to indicate which hydroxyl group is involved.

An entirely different type of anthocyanin derivative was described recently (Bakker *et al.*, 1997) that appears to be present in red wines (normal and fortified) but not in the grapes from which they are made. The compound had a higher then expected molecular weight for a normal malvidin monoglucoside and exhibited chemical features suggestive of an anthocyanin that was substituted at C-4. The flavylium ion, stabilized at low pH, was shown to have structure [2-183] by extensive NMR analyses. By adjustment of the pH of the medium these workers were able to detect a total of five isomeric forms of the new compound the significance of which will become clear in the following paragraphs. The origin of the three-carbon unit was not established although a pyruvate derivative might be involved. The use of 2D NMR methods in arriving at the structure of the new compound will be discussed in Chapter Four.

As we have just seen, some anthocyanins are characterized by very complex levels of glycosylation and acylation. Early studies of naturally occurring blue pigments suggested that non-anthocyanic flavonoids as well as metal ions, along with organic acids, might be involved in producing color. In fact, an association between anthocyanins and metal ions has been recognized for many years (Shibata *et al.*, 1919). An extremely important review entitled "Structure, stability and color variation of natural anthocyanins," by Goto (1987) provides a thorough view of the complexity of this family of pigments including detailed analyses of spectroscopic information.

With an appreciation of the levels of complexity that exist in floral pigments, it is time to ask how these complex molecules contribute to the production of color.

The flavylium ion [2-184], the ionic form most often given as the typical structure of naturally occurring anthocyanins, is stable, and therefore brightly colored, only in very acid conditions, less than pH=3 [see Brouillard and Dangles (1994) for a detailed discussion]. At higher pH values, the flavylium cation becomes deprotonated so that in the pH range 3–6 the molecule can exist in a quinonoidal form, represented by [2-185]. Two other neutral quinonoidal forms can exist that involve the A-ring oxygens. These are highly colored species and are the predominant forms of anthocyanins in many pigment cells. An alternative decoloration process may occur in the pH range 2–7 whereby the flavylium ion reacts with a molecule of water to form a 2-hydroxy-flav-3-ene [2-186]. Compound [2-186] is a hemiketal, which can be written in equilibrium with the *cis*-retrochalcone [2-187]. At pH values above 7 three ionized quinonoidal forms can exist, e.g., [2-188]. These species exhibit a blue color, which, on standing in alkaline conditions, changes to green, and ultimately fades owing to degradation of the flavonoid skeleton.

Santos and colleagues (1993), using one- and two-dimensional NMR techniques, studied the forms of malvin (4′-hydroxy-3′,5′-dimethoxy B-ring) present in aqueous solution in the pH range 0.3–4.5. There was a dramatic drop-off of flavylium ion

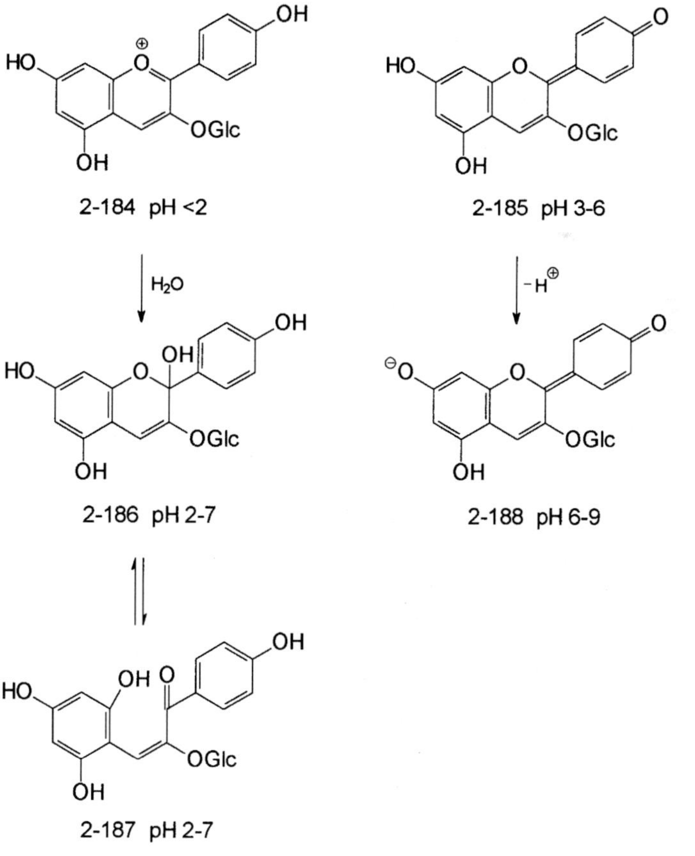

HO, OH, OGlc, OH

2-184 pH <2

HO, O, OGlc, OH

2-185 pH 3-6

H_2O

$-H^{\oplus}$

HO, OH, OH, OGlc, OH

2-186 pH 2-7

$^{\ominus}O$, O, OGlc, OH

2-188 pH 6-9

HO, OH, O, OH, OGlc, OH

2-187 pH 2-7

content as the medium reached and exceeded pH = 1 with a concomitant appearance and increase in the content of the hemiacetal forms . The presence of both the *cis*- and *trans*-chalcone forms could be clearly discerned. At pH < 1 only the flavylium ion was present in solution. At about pH = 2.2 both flavylium ion and the summed hemiacetals were present in equal amounts (ca. 0.42 mole fraction), but at pH = 4 the flavylium ion accounted for less than 10% of the species in solution while the hemiacetals accounted for over 60%. The chalcone forms accounted for the rest.

Before going on to an examination of the significance of these various forms, it is useful to note that their occurrence within plant tissues does not have to be an all or none situation. This was established by Merlin and coworkers (1985) who showed, using resonance Raman microspectrometry, that the pigments in *Vitis vinifera* cv. 'Pinot noir' (Vitaceae) berries were differentially located. Malvidin 3-*O*-glucoside occurred inside the skin, essentially in the quinonoid base form, while in the outer face of the skin of the fruit it occurred primarily in the flavylium ion form. This microprobe method allows one to investigate the nature of a pigment *in vivo* with only minimal harm being done to the cell.

Returning to the matter of the different forms in which anthocyanins can exist, we must consider an apparent paradox. If the flavylium ion is only stable at very low pHs, how do we explain the existence of highly colored pigments in cells where the pH is near neutrality? Several means exist by which the flavylium ion can be protected from nucleophilic attack by water molecules. What is required is some entity that has strong affinity with the flavylium ion and can form a protecting complex with it. The flavylium ion is essentially planar and is characterized by extensive electronic delocalization which makes it an ideal candidate for complexation. The simplest solution would be self complexation, in which several anthocyanin units become associated with each other. Self association involving neutral quinonoid bases is much stronger than that involving flavylium ions (lower pH) or ionized quinonoid bases (higher pH) presumably because of lack of charge repulsion. Self association is favored when the concentration of an anthocyanin is high and in some cases can lead to precipitation of pigment complex, which has been observed to occur with cyanin (cyanidin 3-*O*-glucoside). It is interesting that concentrations of anthocyanins within cells can reach 2.4×10^{-2} M, as reported by Asen and coworkers (1971) for cyanin in petal epidermal cells of the rose cultivar 'Better Times.'

Another factor involved in self association is the nature of glycosylation of the anthocyanin. Anthocyanins with sugars at both the 3- and 5-positions are more stable than those with only a 3-*O*-glycosidic linkage (Hoshino *et al.*, 1981). More recently, Hoshino (1991) described ^{1}H NMR studies of malvidin 3,5-di-*O*-glucoside designed to shed light on the nature of the pigment's self association. He observed that the most strongly shielded protons (in the NMR sense) were those at C-2′ and C-6′ suggesting that these protons are most nearly located over adjacent nuclei. The resulting model shows the quinonoidal form of the pigment self-associated in a helically stacked fashion. The model has the anthocyanidin rings in the center of the helix, thus establishing a hydrophobic domain, surrounded by the hydrophilic glycoside units.

A second possible source of complexation involves interaction between the flavylium unit and the aromatic part of an acyl group. Lending support to this possibility is the observation that pigments with two or more aromatic acyl groups are exceptionally resistant to loss of color through hydration in slightly acidic or neutral media. Several of the blue pigments, e.g. "heavenly blue anthocyanin" (HBA), are among the most stable. Heavenly blue anthocyanin consists of a peonidin unit, six glucoses, and three caffeic acid residues. An interesting experiment that demonstrates the stability of these complexes involved adding rutin (quercetin 3-*O*-rutinoside) to a neutral solution of HBA. Rutin is one of the best copigments known and might be expected to disrupt the anthocyanin complex. However, rutin, had no effect on the visible spectrum of HBA, suggesting that competition between intra- and intermolecular copigmentation strongly favors the former (Asen *et al*., 1977; Ishikura and Yamamoto, 1980). Further evidence for interaction between acyl functions and the anthocyanin unit comes from kinetic and equilibrium studies of anthocyanin hydration using pigments from *Pharbitis nil* (Dangles *et al*., 1993). The experimental system consisted of pelargonidin 3-*O*-sophoroside-5-*O*-glucoside and four of its caffeoyl esters, one with three, one with two, and two with one each. The investigators observed that both the rate and extent of hydration of pelargonidin are reduced in the presence of the caffeoyl groups, suggesting an interaction between the anthocyanin chromophore and the caffeoyl groups. The saccharide group serves as a spacer. Structure [2-189] represents a simple modeling of this kind of interaction

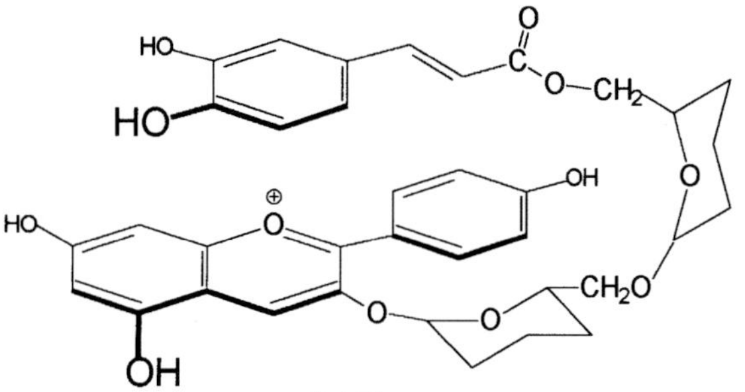

2-189

where the aromatic ring of the caffeic acid unit lies parallel with and over the anthocyanin C-ring. The caffeic acid ring is positioned in such a way that (1) the flavylium ion charge is effectively countered by the π-electron cloud of the aromatic ring, and (2) the approach of a water molecule, required in the conversion of 2-184 to 2-186, is prevented. That this effect is indeed a general one is indicated by recent work on anthocyanins from *Eichhornia crassipes* (Pontederiaceae) and *Evolvulus pilosus* (Convolvulaceae) by Figueiredo and coworkers (1996). The *Eichhornia* pigment, which consists of a delphinidin glycoside and apigenin 7-*O*-glucoside joined by a malonic acid unit [2-153], and a set of delphinidin glycosides from *Evolvulus*, were subjected to kinetic analysis of hydration. As in the examples above, acylated anthocyanins, including the "double" compound from *Eichhornia*, showed a significant reduction in the extent of hydration compared to simple delphinidin glycosides.

Brouillard and Dangles (1994) comment that, allowing for pH, intermolecular copigmentation is probably the single most important factor influencing flower color. Metal ions are known to complex with catecholic functions in flavonoids, but blue pigments tend not to have the catechol system. Thus, some other explanation is required. Some experimental observations with the blue pigment from *Hydrangea macrophylla* will indicate that involvement of metal ions alone is not sufficient in itself to ensure stable pigmentation. Addition of aluminum ions to a solution of the *Hydrangea* anthocyanin, delphinidin 3-*O*-glucoside [2-190], was not sufficient to produce a stable blue pigment. Addition of a copigment, however, yielded a significantly different outcome. The copigment in this case could be either 3-*p*-coumaroyl- or 3-caffeoylquinic acid, [2-191] and [2-192], respectively. Addition of either of these compounds resulted in appearance of a stable blue pigment. Addition of 5-caffeoylquinic acid (caffeic acid esterified at position-5 in 2-192) however, failed to produce blue pigment (solution remained purple-red) (Takeda *et al*., 1985a,b). The stability of color produced from delphinidin 3-*O*-glucoside, aluminum ions, and either of the 3-cinnamoylquinic acids rivals that of the polyacylated anthocyanin glycosides discussed above (Takeda *et al*., 1990). The issue of color stability was also addressed by Lu and coworkers (1992a,b) in their work with the complex pigments of *Pharbitis nil*. These investigators showed that caffeic acid residues were responsible for the stability of the *Pharbitis* pigments, with the degree of stability increasing as a function of the number of acid residues present.

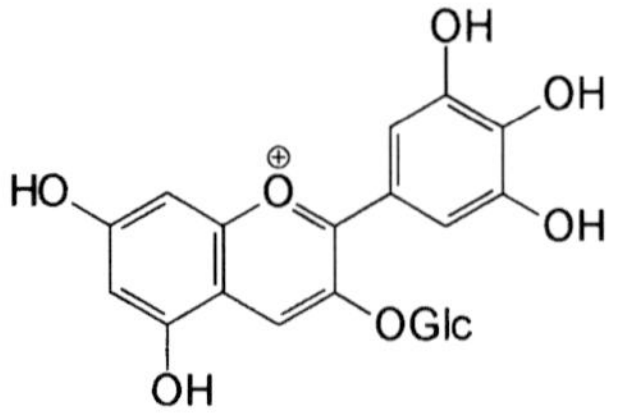

2-190

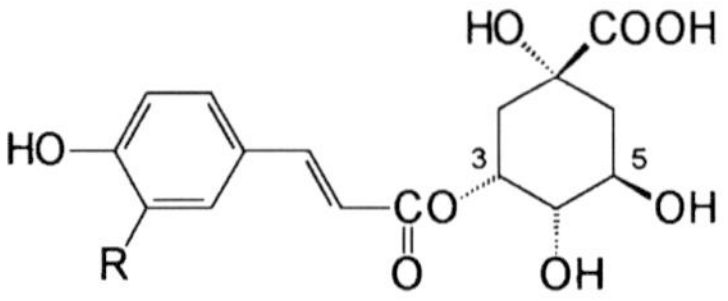

2-191 R = H
2-192 R = OH

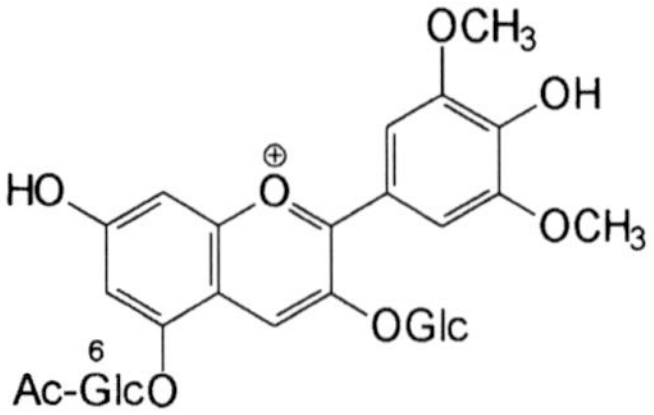

2-193

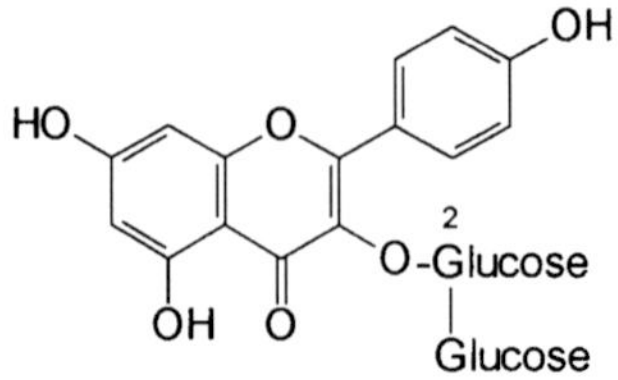

2-194

The combined effect of copigmentation and pH is very well documented in a study of flower color in three geraniums, *Geranium* 'Johnson's Blue' (*G. himalayense* × *G. pratense*), *G. pratense*, *and G. sanguineum*. *Geranium* 'Johnson's Blue' was described as having bluish purple flowers, while *G. pratense* has purplish blue, and *G. sanguineum* bluish magenta flowers. The major pigments and their interactions were described by Markham and coworkers (1997) who attempted to regenerate the respective pigments under laboratory conditions. Despite the different colors exhibited by the three species, these workers found that the major flavonoids in each were the same, malvidin

3-*O*-glucoside-5-*O*-(6-acetylglucoside) [2-193] and kaempferol 3-*O*-sophoroside [2-194]. Lesser amounts of other flavonol 3-*O*-glycosides and anthocyanins were also present. A major difference, however, lay in the ratios of the major components. *Geranium* 'Johnson's Blue' had a flavonol/anthocyanin ratio of approximately 2 : 1, while the value for *G. pratense* was 4 : 1 and that for *G. sanguineum* was 1: 5. Differences in the pH of expressed flower juice were also noted: *Geranium* 'Johnson's Blue,' 5.40; *G. pratense*, 5.43; and *G. sanguineum*, 4.63. Attempts to reconstitute the natural color of each of these species using the observed ratio of flavonol and anthocyanin at the pH of the expressed juice failed. Increasing the amounts of flavonols, however, produced a blue shift, but the natural color was elusive until the pH was altered by the careful addition of dilute base. At a pH in the range of 6.6–6.8 the reconstituted colors approached the natural color of both *Geranium* 'Johnson's Blue' and *G. pratense*, while a pH of 5.6 was necessary for regeneration of the *G. sanguineum* color. As those workers pointed out, the pH of the epidermal cell vacuoles, where the pigments reside, must be 1–1.4 units higher than that of the expressed juice. Minor differences between observed natural pigment absorbances and those of the reconsitituted pigments may lie in the fact that only the major flavonol and anthocyanin were used in the reconstitution experiments.

One of the most interesting systems involving both a copigment and a metal ion involves "commelinin," the pigment responsible for the blue flower color of *Commelina communis* (Commelinaceae) (Kondo *et al*., 1992). Commelinin consists of the anthocyanin "malonylawobanin" (M), the flavone "flavocommelin" (F), and magnesium ions (Tamura *et al*., 1986). The anthocyanin is delphinidin 3-[6″-(*p*-coumaroyl)glucoside]-5-(6″-malonylglucoside) [2-195]; the copigment is the *C*-glucosylflavone isovitexin 7-methyl ether 4′-*O*-glucoside [2-196]. Structure [2-197], adapted from Goto (1987), shows one way in which magnesium can function to join these two compounds. Electrophoresis of an aged solution of the natural pigment showed an array of evenly spaced blue bands the fastest moving one of which was identical to synthesized commelinin, indicated as M_6F_6, while the slowest one was the totally demalonylated A_6F_6($A=M$ less its malonyl group). The intermediate bands were attributed to pigments that had undergone successive loss of malonate on standing, e.g., M_5AF_6 would have lost one malonate group, $M_4A_2F_6$ would have lost two, and so forth. X-Ray crystallographic studies and analytical ultracentrifugation provided information from which it was possible to deduce the native structure of commelinin as consisting of six equivalents of malonylawobanin, six equivalents of flavocommelin, and two atoms of magnesium. The question then turned to the physical arrangement of these pigment associations. An important clue came from X-ray diffraction studies, which showed that the structure has a three-fold axis of symmetry. This can be accommodated by placing a magnesium ion in the center around which three flavonoid units are arrayed radially. The second magnesium ion would be located below the first with its three flavonoid units arrayed so that hydrophobic anthocyanin rings in one layer could interact with anthocyanin rings in the next.

Two additional points are worth noting. The first deals with the exterior positioning of the malonyl carboxyl groups in this model. These groups exist in the ionized carboxylate form near neutrality and provide a polar face for the surrounding aqueous medium. The second point deals with the complexation of magnesium with

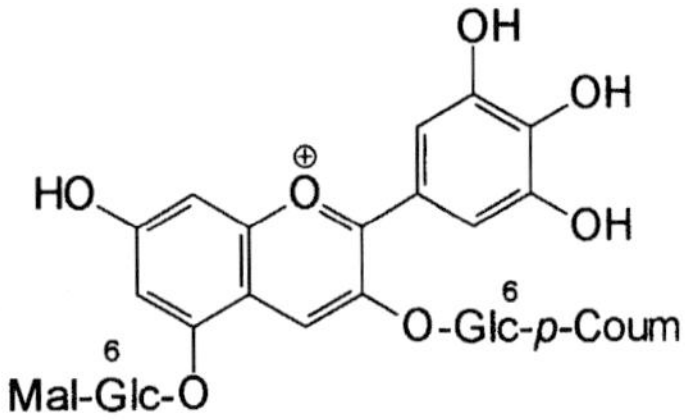

2-195
"Malonylawobanin"

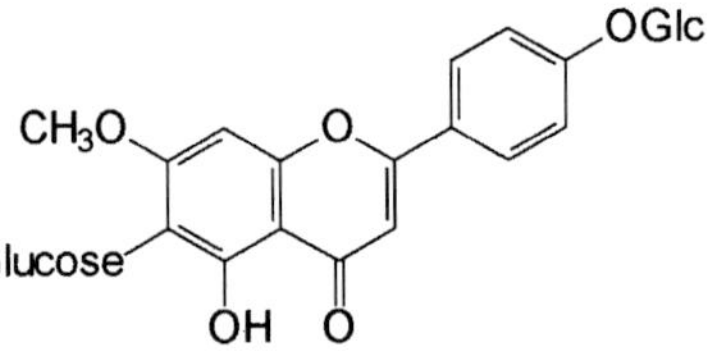

2-196
"Flavocommelin"

2-197
"M_2F_2Mg"

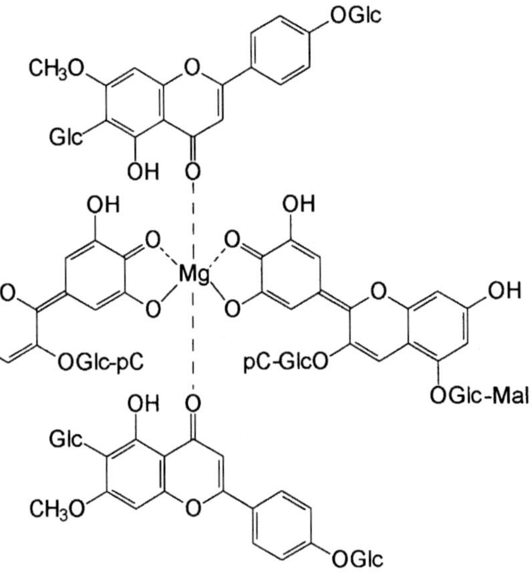

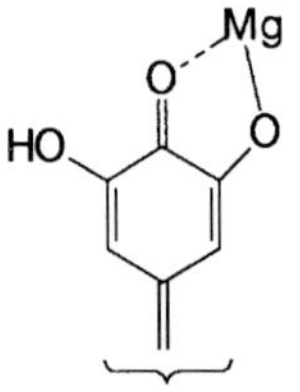

2-198

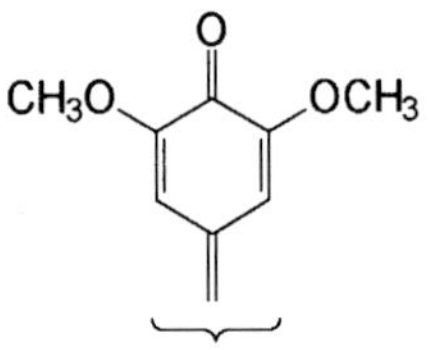

2-199

the quinonoidal form of the anthocyanin required by the model. The quinone-carbonyl with adjacent phenolic group on the B-ring of delphinidin provides an ideal site for complexation with the metal ion as shown in partial structure [2-198]. It was determined that complexation with metal ions failed with anthocyanidins that lacked these functional groups, e.g., malvidin, where the quinone is possible but no hydroxyl group is available [2-199]. A similar system apparently accounts for the blue color of corn flowers, *Centaurea cyanus*. This pigment consists of six molecules each of cyanidin 3-*O*-(6″-succinylglucoside)-5-*O*-glucoside and apigenin 4′-*O*-(6″-malonylglucoside) and one atom each of iron and magnesium (Goto *et al.*, 1986; Goto and Kondo, 1991).

The availability of a reliable source of tissue for biochemical studies of flavonoids, anthocyanins included, has been answered in many instances in recent years through the use of axenically cultured cells. An example of such a system for study of complex anthocyanin derivatives involves cultured cells of *Daucus carota* (Apiaceae), which make cyanidin 3-*O*-(2″-*O*-β-xylopyranosyl-6″-*O*-β-glucopyranosyl-β-galactopyranoside) acylated at the 6-OH of the glucose residue with either *p*-hydroxybenzoic, *p*-coumaric, ferulic, or sinapic acid (Glässgen *et al.*, 1992). Many features of pigment biosynthesis and complexation could be addressed using axenic cultures of this sort.

The 3-deoxyanthocyanins are, as their name indicates, anthocyanins that lack oxygen at C-3. They occur as glycosides bound through the 5-OH group. These pigments are much less widely distributed in the plant kingdom than are the anthocyanins proper. 3-Deoxyanthocyanins have been reported from members of subfamily Gesnerioideae of Gesneriaceae (see Chapter Three) as well as from some ferns, mosses, and a scattering of other species (Harborne, 1966, 1967). The three main members of this subclass, apigeninidin (Ap) [2-200], luteolinidin (Lt) [2-201], and tricetinidin (Tr) [2-202], take their common names from the flavones with corresponding oxygenation patterns, apigenin, luteolin, and tricetin, respectively. Apigeninidin 7-methyl ether was recently reported from *Sorghum caudatum* by Pale and coworkers (1997). The discovery of luteolinidin 5-*O*-glucoside as a component of species of the moss genus *Bryum* (Bendz *et al.*, 1962) provided the first incontestable report of a flavonoid in a moss. The significance of the presence of these compounds in Gesneriaceae will be discussed in Chapter Three.

Aurones (Including Auronols)

Aurones are comparatively late arrivals on the flavonoid scene. During a study of the flower pigments of *Coreopsis grandiflora* (Asteraceae), Geissman and Heaton (1943, 1944) obtained two compounds, related as aglycone and glycoside. Although they gave the color reaction associated with chalcones (development of deep red color with base) they were clearly not polyhydroxychalcones. The possibility was considered that they were benzalcoumaranones, which in fact they were shown to be by synthesis (Geissman and Heaton, 1944). The structure of "leptosidin" and its glucoside "leptosin" are shown below as [2-203] and [2-204], respectively. A few years later Seikel and Geissman (1953) identified another member of this new group of pigments, this time from flowers of the snapdragon, *Antirrhinum majus* (Scrophulariaceae). The pigment, which they named called "aureusin," yielded glucose and an aglycone, "aureusidin," that was identified as 4,6,3′,4′-tetrahydroxybenzalcoumaranone [2-205]. The point of attachment of the glucose was undetermined. At about this same time, workers in Italy described a different glucoside of aureusidin from flowers of *Oxalis cernua* (Oxalidaceae), but they did not determine the site of glycosylation (Ballio *et al.*, 1953). It fell to Harborne and Geissman (1955) to determine that aureusin was glycosylated at the C-6 hydroxyl while the compound

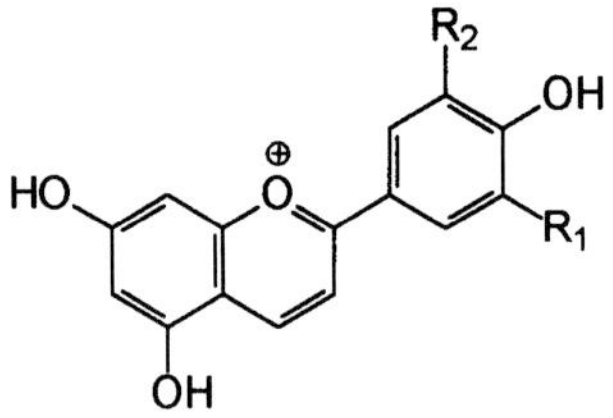

2-200 $R_1 = R_2 = H$
2-201 $R_1 = OH, R_2 = H$
2-202 $R_1 = R_2 = OH$

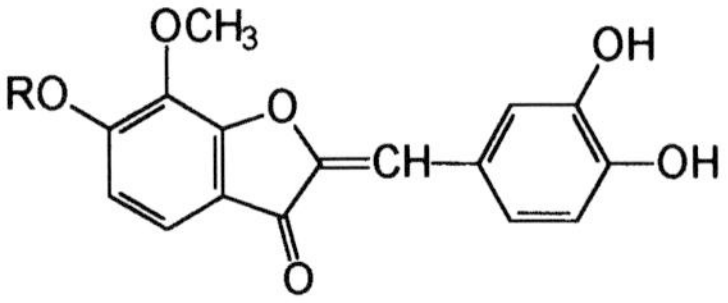

2-203 R = H
2-204 R = Glc

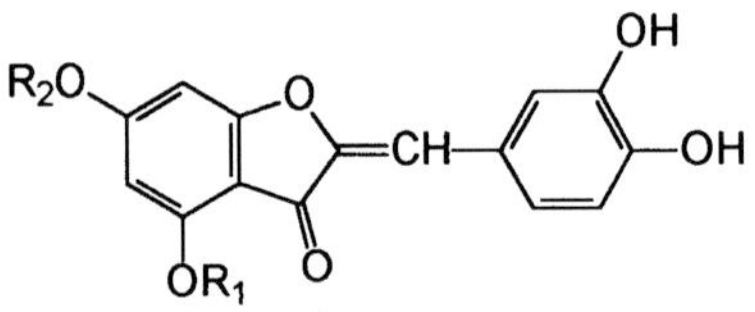

2-205 $R_1 = R_2 = H$
2-206 $R_1 = H, R_2 = Glc$
2-207 $R_1 = Glc, R_2 = H$

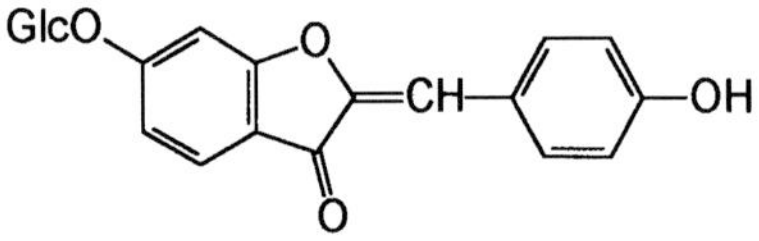

2-208

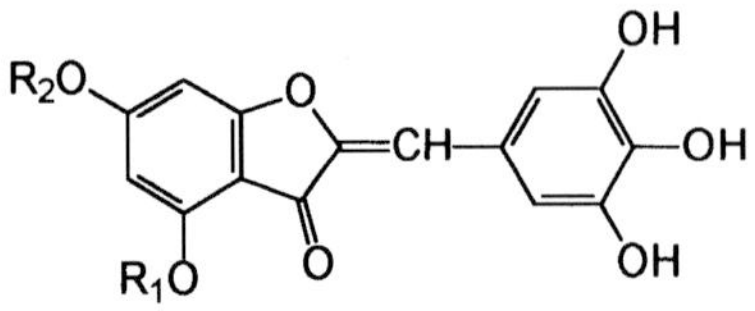

2-209 $R_1 = R_2 = H$
2-210 $R_1 = Glc, R_2 = H$
2-211 $R_1 = H, R_2 = Gln$

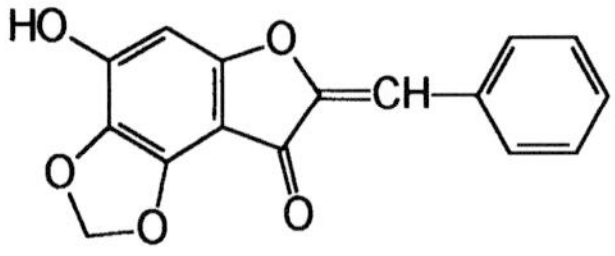

2-212

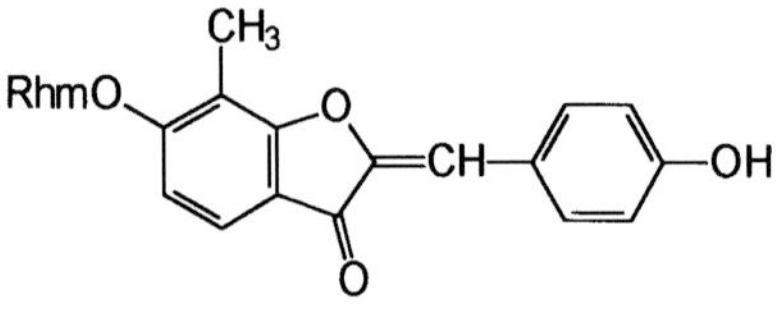

2-213

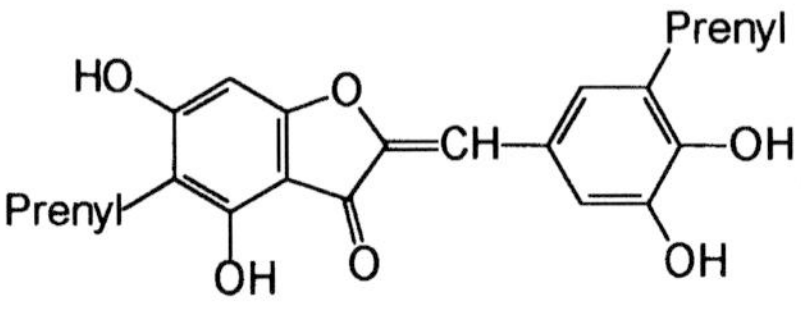

2-214

from *Oxalis*, "cernuoside," was glycosylated at the C-4 hydroxyl. These compounds are illustrated as [2-206] and [2-207], respectively. In order to accommodate this new group of compounds, and make communication a little easier, Bate-Smith and Geissman (1951) coined the term "aurone."

Along with identification of the structure of the aurones came the realization that their optical properties were very similar to those of the much better known compounds, the chalcones. As time passed, it also became widely appreciated that these two classes of flavonoids often occur together. It was also noted that chalcones could be converted to aurones in the presence of weak base and atmospheric oxygen, and that chalcone spots or bands on chromatograms left to dry in the air slowly accumulated aurones. Conversion of chalcones to aurones by enzyme extracts from plant tissues have also been demonstrated (Shimokoriyama and Hattori, 1953). (It is interesting to note that, despite early recognition that these two classes of pigments are closely related, the biosynthesis of aurones remains one of the least studied aspects of flavonoid formation. The biosynthesis of chalcones, on the other hand, has been examined in great depth as we shall see in Chapter Six.) The apparent ease of conversion of chalcones to aurones has prompted workers dealing with flavonoid surveys to report the presence of chalcone/aurone pairs in recognition of the possibility that the aurones observed might have been artifacts of chromatographic analysis (e.g., Crawford and Stuessy, 1981; Schilling and Panero, 1988). It seems reasonable to this writer, at least, that most reported sightings of aurones can be attributed to Nature rather than to atmospheric oxidation. The reader should also note that chalcones and aurones are frequently referred to collectively as "anthochlors." Table 2.8 lists some of the more common aurones, their glucosides, and corresponding chalcones and their glucosides.

In Shimokoriyama's chapter on anthochlors in 1962, aurones were reported from members of three plant families, Asteraceae, Oxalidaceae, and Scrophulariaceae, with Asteraceae being most strongly represented. In the 1975 volume of *The Flavonoids* aurones were reported from members of several new families: Cyperaceae, Fabaceae, Gesneriaceae, and Plumbaginaceae. Members of Asteraceae were still the most frequently cited, however. Among the more interesting compounds was the simplest aurone yet seen in Nature. Wong (1966) described "hispidol" from a species of *Glycine* (Fabaceae) and identified it as 6,4′-dihydroxyaurone 6-*O*-glucoside [2-208]. The

Table 2.8 Common chalcones, aurones and their glucosides

Structure	Common name
2′,4′,4-Trihydroxychalcone	Isoliquiritigenin
2′,4′,4-Trihydroxychalcone 4-*O*-glucoside	
2′,4′,6′,4-Tetrahydroxychalcone	Chalcononaringenin
2′,4′,6′,4-Tetrahydroxychalcone 2′-*O*-glucoside	Isosalipurposide
2′,4′,3,4-Tetrahydroxychalcone	Butein
2′,4′,3,4-Tetrahydroxychalcone 4′-*O*-glucoside	Coreopsin
2′,3′,4′,3,4-Pentahydroxychalcone	Okanin
2′,3′,4′,3,4-Pentahydroxychalcone 4′-*O*-glucoside	Marein
2′,4′,3,4-Tetrahydroxy-3′-methoxychalcone	Lanceoletin
2′,4′,3,4-Tetrahydroxy-3′-methoxychalcone 4′-*O*-glucoside	Lanceolin
6,3′,4′-Trihydroxyaurone	Sulfuretin
6,3′,4′-Trihydroxyaurone 6-*O*-glucoside	Sulfurein
6,7,3′,4′-Tetrahydroxyaurone	Maritimetin
6,7,3′,4′-Tetrahydroxyaurone 6-*O*-glucoside	Maritimein
6,3′,4′-Trihydroxy-7-methoxyaurone	Leptosidin
6,3′,4′-Trihydroxy-7-methoxyaurone 6-*O*-glucoside	Leptosin
4,6,3′,4′-Tetrahydroxyaurone	Aureusidin
4,6,3′,4′-Tetrahydroxyaurone 4-*O*-glucoside	Cernuoside
4,6,3′,4′-Tetrahydroxyaurone 6-*O*-glucoside	Aureusin
4,6,3′,4′,5′-Pentahydroxyaurone	Bracteatin
4,6,3′,4′,5′-Pentahydroxyaurone 4-*O*-glucoside	Bractein

following volume (Bohm, 1982) included two new flowering plant families as sources of aurones, Anacardiaceae and Zingiberaceae. The most interesting finds, however, involved several liverworts and a moss. Bracteatin [2-209], originally isolated from *Helichrysum bracteatum* (the strawflower) as the 4-*O*-glucoside [2-210], was reported from sporophytes of the moss *Funaria hygrometrica* where it occurred as the free phenol (Weitz and Ikan, 1977). This apparent reproductively-linked occurrence is not unique. Markham and Porter (1978) identified aureusidin 6-*O*-glucuronide [2-211] from antheridiophores of three liverworts, *Marchantia berteroana, M. polymorpha*, and *Conocephalum supradecompositum*. These workers failed to detect it in thallus (the leafy portion of the plant) or in female plants. That this compound appears to be performing some specific function within the reproductive cycle of these organisms is indicated by its presence in the organism only for the four to eight weeks during which the reproductive structures are functional. What the compound is doing, however, remains a mystery.

Examples of somewhat unusual aurones include the B-ring deoxyaurone [2-212], which was reported as a constituent of the cactus *Cephalocereus senilis* by Pare and coworkers (1991). This compound was observed in mature plant tissues as well as cultured cells that had been challenged with chitin. The first *C*-methylated aurone, 6,4′-dihydroxy-7-*C*-methyl 6-*O*-rhamnoside [2-213], was isolated from *Pterocarpus marsupium* (Fabaceae) by Mohan and Joshi (1989). Another *C*-methylated aurone was added to the list by Seabra and coworkers (1995) who identified 6,3′,4′-trihydroxy-4-methoxy-5-*C*-methylaurone from *Cyperus capitatus* (Cyperaceae). Although aurones have been reported from other species of *Cyperus* (Harborne *et al*., 1985), this is the first report of one from underground parts of the plant. Two di-*C*-prenylaurones, obtained from *Antiaris toxicaria* (Moraceae), represent the first *C*-prenylated members of this class to be reported (Hano *et al*., 1990). The structure of "antiarone B" is shown as [2-214]. "Antiarone A" is the 2′,5′-di-C-prenyl isomer. The description of bracteatin [2-209] as a component of *Asplenium kaulfussii* by Imperato (1989) marked the first report of an aurone from a fern. El Tohamy and coworkers (1993) reported isolation of "5,7-dihydroxyaurone" from *Lonchocarpus speciosus* along with other, known flavonoids. It seems highly unlikely, however, that the A-ring hydroxylation pattern is correct as written. Although chemically possible, the 5,7-dihydroxylation pattern is biosynthetically highly improbable as it would require loss of oxygens at both C-4 and C-6 with subsequent gain at both C-5 and C-7. I suspect that the proper designation would be 4,6-dihydroxyaurone which is the numbering system for phloroglucinol-type A-rings in aurones.

A recent paper report by Seabra and coworkers (1997) describes isolation and identification of aureusidin, 6,3′-dihydroxy-4,4′-dimethoxy-5-*C*-methylaurone, and the novel 4,6,3′,4′-tetramethoxyaurone from *Cyperus capitatus*. It was noted in that paper that both the *cis*- [2-215] and *trans*- [2-216] isomers of the tetramethoxyaurone were present and that they tend to isomerize in light. The reader will have noticed that the stereochemical configuration of other aurones has not been included; rarely is this feature included in descriptions of these compounds.

Auronols are 2-hydroxy-2-benzylcoumaranones. The first one to fall to structural analysis was a compound called "alphitonin" that had been isolated from wood of *Alphitonia excelsa* (Rhamnaceae) by Smith and Read (1922). It was not until 1960,

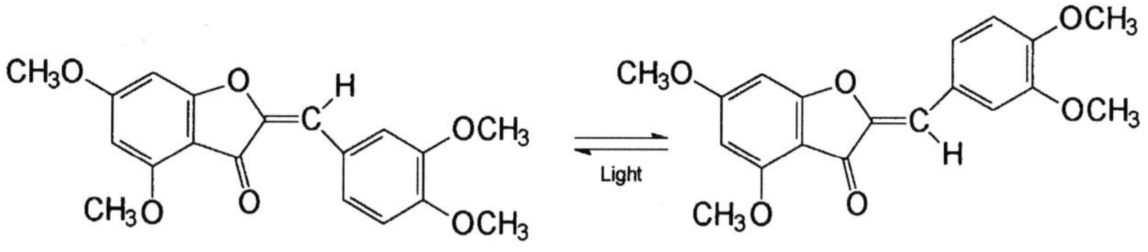

2-215 ⇌ (Light) 2-216

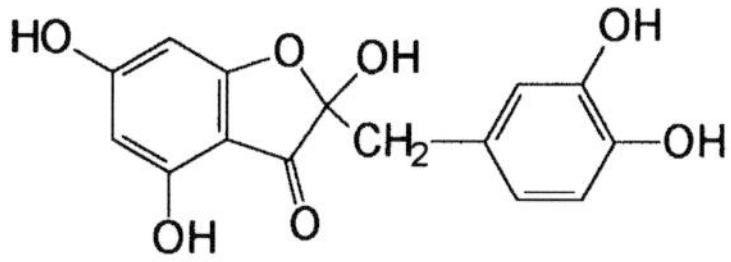

2-217

2-218

2-219

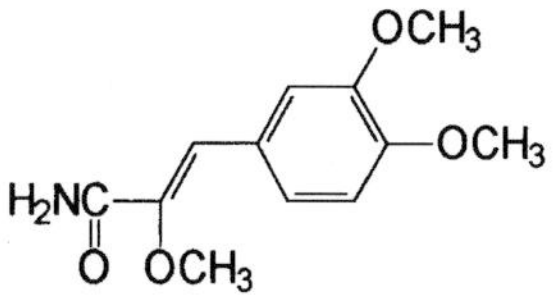

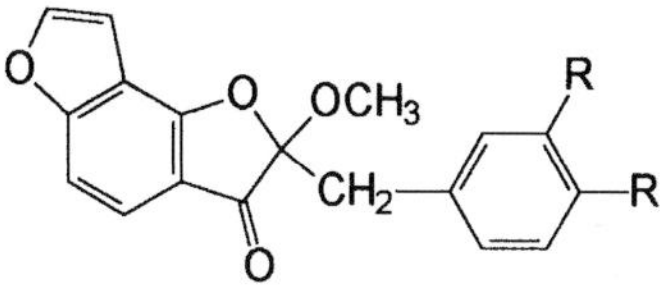

2-220 R's = H

2-221 R's = OCH_2O

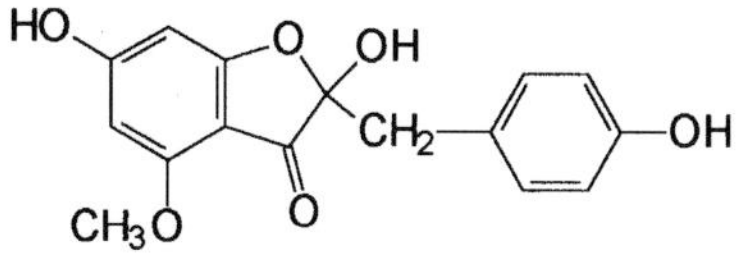

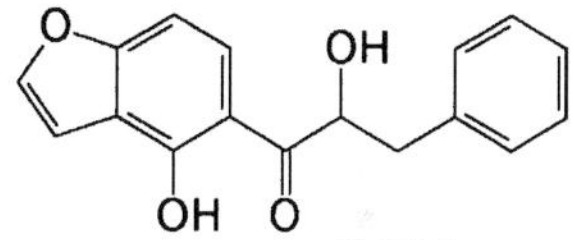

2-222 R's = H

2-223 R's = OCH_2O

2-224

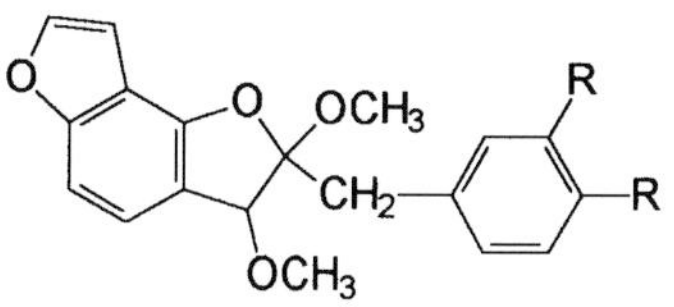

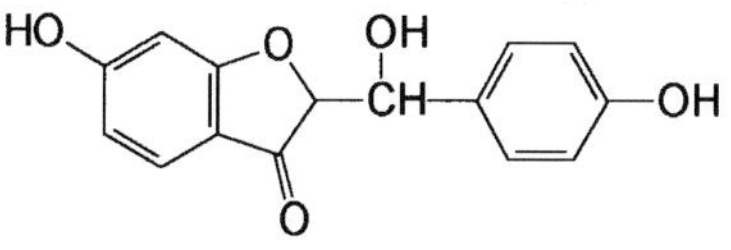

2-225

however, that its structure was established by Birch and coworkers (1960) as 4,6,3′,4′-tetrahydroxyauronol [2-217]. Two key reactions in the determination of its structure depended on low temperature reductions. The first involved reduction of alphitonin pentamethyl ether with sodium in liquid ammonia to yield 4,6,3′,4′-tetramethoxydihydroaurone, which was readily identified by comparing it with authentic material, and established the location of four of the hydroxyl groups. The position of the remaining hydroxyl group was determined by reduction of the pentamethyl ether with potassium amide/ammonia (KNH_2/NH_3) which yielded α,3,4-trimethoxycinnamide [2-218], thus establishing the hydroxyl group at C-2 of the auronol. Other auronols known from Nature include 6,3′,4′-trihydroxyauronol and its 4′-methyl ether from *Schinopus* (Anacardiaceae), 4,6,4′-trihydroxyauronol from *Maesopsis* and *Phyllogeiton* (both Rhamnaceae), and 6,7,3′,4′-tetrahydroxyauronol form an *Acacia* species (for details see Bohm, 1994, page 474). Two recent reports of auronols are 6,4′-dihydroxy-4-methoxyauronol ("marsupin") from the legume *Pterocarpus marsupium* (Manickam *et al*., 1997) and 4,6,4′-trihydroxyauronol and its 6-*O*-glucoside from *Ceanothus americanum* (Rhamnaceae) (Li *et al*., 1997). The *Pterocarpus* compound was identified as one of the agents responsible for the hypoglycemic activity of extracts of heartwood extracts, while the second compound showed some activity toward oral pathogens.

An interesting mix of modified auronols plus an unusual α-hydroxydihydrochalcone was obtained from heartwood of the Mexican *Lonchocarpus castilloi* by Gómez-Garibay and coworkers (1990). Two of the modified auronols, [2-220] and [2-221], were methylated on the 2-hydroxyl group. The other two had undergone reduction of the carbonyl group and *O*-methylation to yield [2-222] and [2-223]. The dihydrochalcone was shown to have structure [2-224]. The known resistance of *Lonchocarpus castilloi* heartwood to fungal attack led these workers to test the five compounds against the pathogen *Lenzites trabea*. All five were active with the dihydrochalcone exhibiting the highest antifungal activity.

Before we leave aurones and auronols, it is instructive to mention briefly some experiments performed by Wong and Wilson (1972) on the effect of cell-free preparations from garbanzo beans (*Cicer arietinum*) or soybeans (*Glycine max*) on 2′,4′,4-trihydroxychalcone. In addition to a flavonol, a dihydroflavonol, the aurone, and an odd spirodiketone, these investigators identified the "hydrated aurone" 2-(α,4-dihydroxybenzyl)-6-hydroxyaurone [2-225] from the reaction mixture. The biosynthetic significance of the hydroxy dihydrochalcone or hydrated aurone, if any, has not been investigated.

Biflavonoids

The autumn leaves of *Ginkgo biloba* (Ginkgoaceae, maidenhair tree) yielded a compound that gave an elemental analysis of $C_{32}H_{22}O_{10}$, which, along with color tests and other chemical characteristics, suggested to Nakazawa (1941) that the compound was flavonoid in nature. Chemical degradations and intercoversions (see Dean, 1962, for historical details) led to the recognition that the compound was based on the linkage of two flavone nuclei. The nature of the linkage was eventually found to involve a bond between the A-ring of one unit and the B-ring of the other (Baker *et al*., 1959). The parent compound, called "amentoflavone," consists of two apigenin units linked as shown in [2-226]. The *Ginkgo* compound actually turned out to be a mixture of dimethyl ethers, ginkgetin [2-227] , with methoxy groups at C-7 and C-4′, and isoginkgetin [2-228] with methoxy groups at C-4′ and C-4‴. These compounds could also be called, respectively, amentoflavone 7,4′-dimethyl ether and amentoflavone 4′,4‴-dimethyl ether. Notice in these compounds that the numbering of positions reflects the fact that the second apigenin unit (lower right hand unit) can be considered to be a substituent of the parent unit. Amentoflavone might also be called 3′-(8-apigeninyl)apigenin, if one were so inclined. The reader

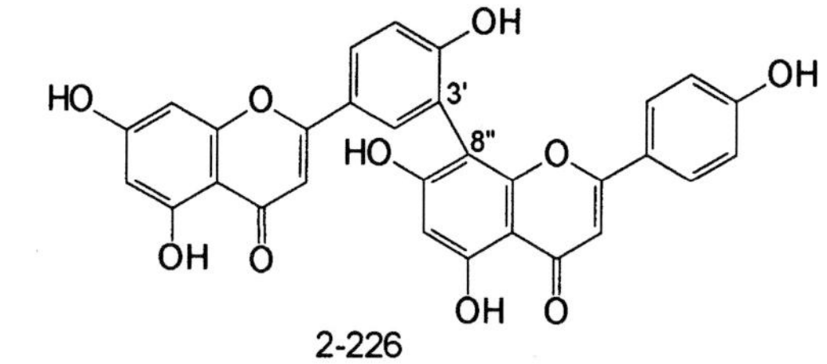

2-226

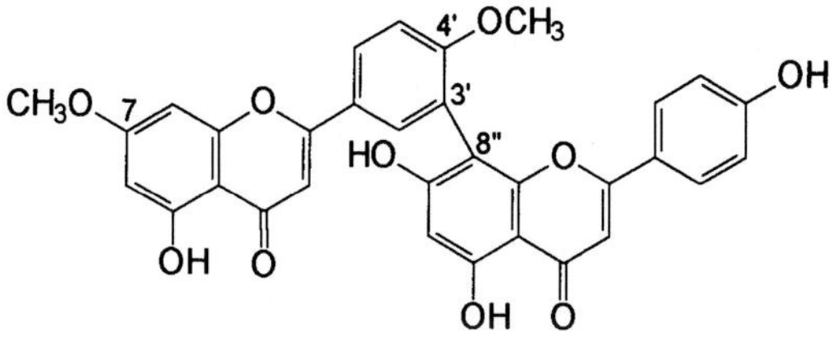

2-227

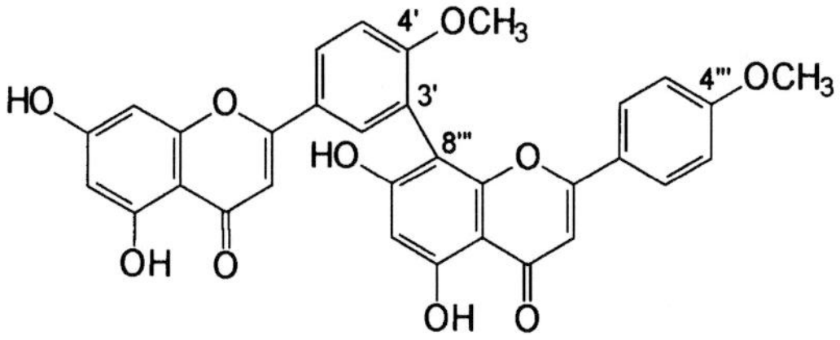

2-228

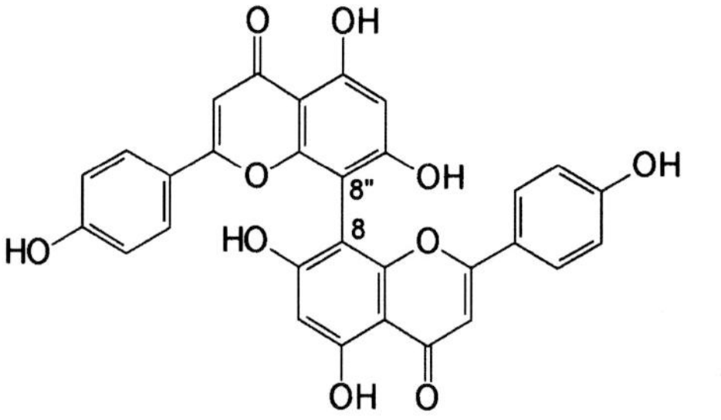

2-229

would be correct in concluding that nomenclature in this group of flavonoids can be a complicated business! It was not long after the description of ginkgetin and isoginkgetin appeared in the literature that other biflavonoids began to be reported on a regular basis. The new compounds were based on a variety of inter-flavone linkages, including compounds such as "cupressoflavone," linked through the C-8 positions of the respective units, as illustrated in [2-229], "agathisflavone" linked through C-6 of one unit and C-8 of the other, and "robustiflavone," joined through C-3′ and C-6 of the respective units. In "garcinia" type biflavonoids the linkage involves C-8 of one unit and C-3 (a C-ring carbon) of the other. Inter-flavonoid bonds may involve oxygen as illustrated in [2-230] for "hinokiflavone." Other oxygen-linked biflavonoids are also known such as "ochnaflavone," members of which group are linked by oxygen between C-4′ of one unit and C-3′ of the second. Any nomenclature system based upon common names tends to become unwieldy and increasingly difficult to use by non-specialists. Adding to the problem is the fact that a newly isolated compound tends to acquire a common name before its structure is fully established and a more or less systematic name can be assigned to it. Use of a polysaccharide-style bond notation can facilitate communication in many instances. Amentoflavone would simply be identified as apigenin-(3′ → 8)-apigenin, ginkgetin as 7,4′-di-*O*-methylapigenin-(3′ → 8)-apigenin, and isoginkgetin as 4′-*O*-methylapigenin-(3′ → 8)-4′-*O*-methylapigenin. Since 4′-*O*-methylapigenin is also known as acacetin, this latter compound could also be called acacetin(3′ → 8)acacetin. Although this system eliminates the use of double and triple primed numbers, it still requires knowing the common names of the flavones themselves.

A nomenclatural convention for biflavonoids was discussed by Locksley (1973) in his review of the chemistry of these compounds. In consultation with Professor Ollis and other natural product chemists, a system was developed that avoided reliance on common names. Each monomeric unit is assigned a Roman numeral running in sequence from the beginning of the molecule to the end (essentially left to right as these compounds are usually written). The points of connection between the monomeric units are indicated by a Roman numeral to identify the monomer and an Arabic numeral to indicate the position of linkage. The linkage is then be written as a pair of Roman numeral–Arabic numeral combinations separated by a comma and enclosed within brackets. For example, the interunit linkage in amentoflavone would be written [I-3′, II-8]. The full name of amentoflavone, locating the positions of substituents, would be: I-4′, II-4′, I-5, II-5, I-7, II-7-hexahydroxy [I-3′, II-8] biflavone. Ginkgetin is I-4′, I-5, II-5, II-7-tetrahydroxy-I-4′, I-7-dimethoxy [I-3′, II-8] biflavone in this system; isoginkgetin is I-5, II-5, I-7, II-7-tetrahydroxy-I-4′,II-4′-dimethoxy [I-3′, II-8] biflavone. Biflavones joined by oxygen linkages include the "*O*" term in the bracketed linkage term. Hinokiflavone would be represented as II-4′, I-5, II-5, I-7, II-7-pentahydroxy [I-4′-*O*-II-6] biflavone.

Although most biflavonoids discovered in the early days of research in this area involved flavones, it soon became clear that other classes of flavonoid could also be involved. In fact, most possible combinations have been found to occur naturally. Thorough reviews of the literature by Hans Geiger and Christopher Quinn (Geiger and Quinn, 1975, 1982, 1988; Geiger, 1994) have been a regular part of *The Flavonoids* series. The sampling of compounds presented next come from the recent literature and should give an idea of the variation possible in this class. From the New Zealand moss species *Campylopus clavatus* and *C. holomitrium* Geiger and Markham (1992) isolated an unusual biflavonoid that consisted of 4,6,3′,4′-tetrahydroxyaurone (aureusidin) and the flavanone eriodictyol. The compound, called "campylopusaurone" by those authors, and illustrated here as [2-231] was shown to be aureusidin-(5′ → 6″)-eriodictyol. Other anthochlors (aurones and chalcones) have been found as part of biflavonoids as typified for our purposes by the compound "lophirone C" [2-232] obtained from *Ochna calodendron* (Ochnaceae), a tree prized for its medicinal properties by

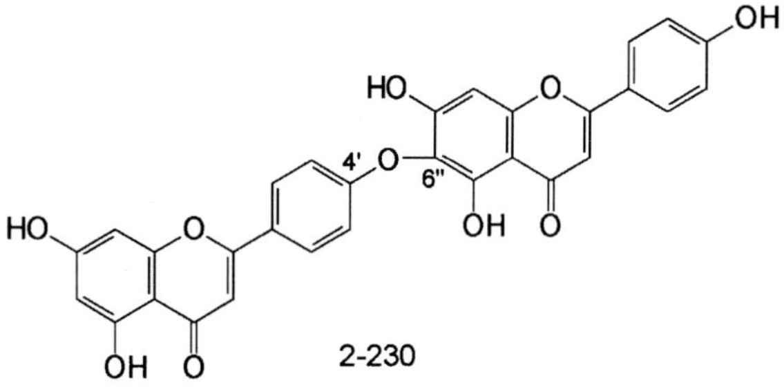

2-230

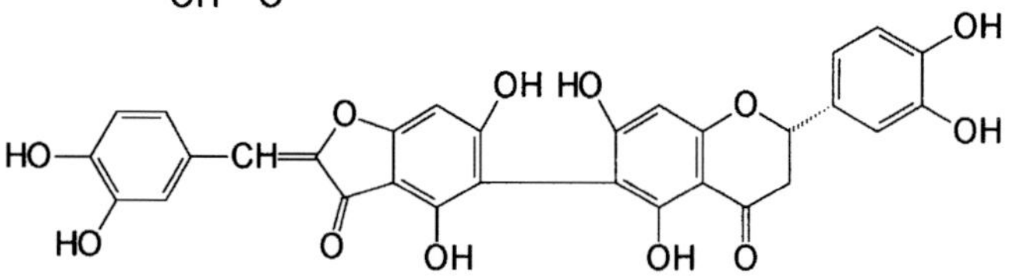

2-231

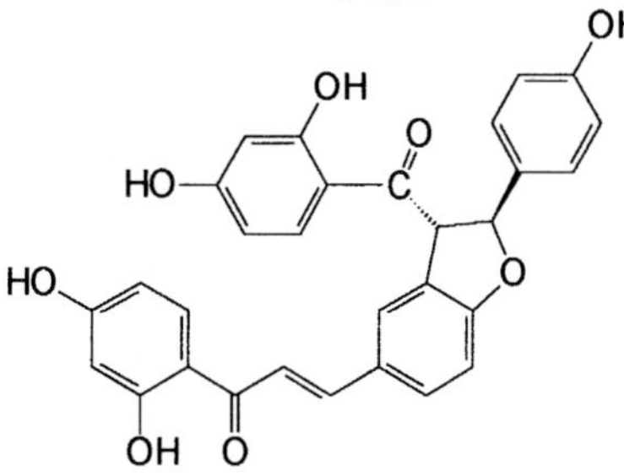

2-232

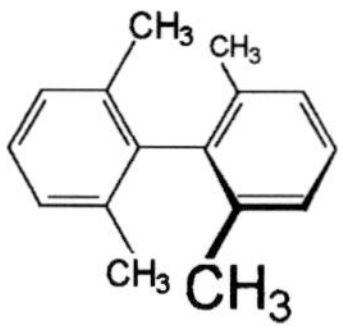

2-233

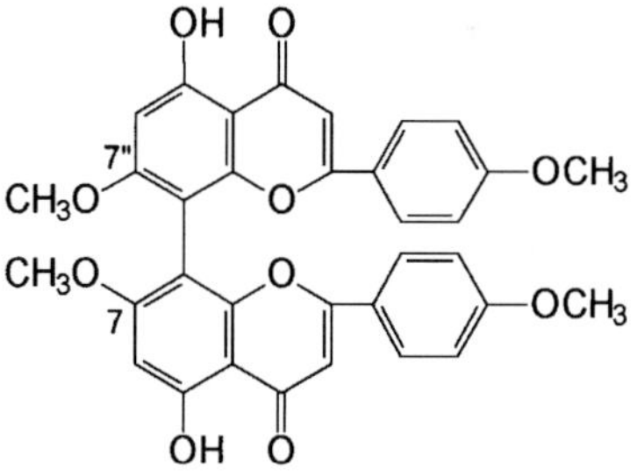

2-234

the peoples of southern Cameroon (Messanga *et al.*, 1994). Note that one of the chalcone units is linked twice with the other, once through oxygen and once by a carbon–carbon bond. The resulting pyran ring system is a frequently seen feature of di-, tri-, and higher flavonoids in which a chalcone plays a part. This arises from the reactivity of the α,β-unsaturated carbonyl function. The first biaurone was recently described from the moss *Aulacomnium palustre* (Aulacomniaceae) and shown to be 5,5′-biaureusidin [substitute aureusidin for eriodictyol in 2-231] (Hahn *et al.*, 1995). The UV spectrum and characteristic color developed with diphenylboric acid

β-aminoethyl ester (Naturstoffreagenz A) suggested that the compound was an aurone, while the FAB-MS spectrum gave $m/z = 569$ for $[M\text{-}H]^-$. ^{13}C NMR confirmed the existence of only aureusidin (no other monomeric unit) but also that the linkage involved C-5 of both units. In addition to the new biflavonoid, those workers reported seven known biflavonoids and triluteolin, a member of a small group of triflavonoids about which more will be said below. Isoflavonoids can also be involved in biflavonoids as evidenced by the presence of biisoflavanones in roots of the Brazilian *Ouratea hexasperma* (Ochnaceae) (Moreira *et al.*, 1994).

Certain biflavonoids have the capacity to exist in atropisomeric forms. It is well known that biphenyl derivatives with substituents at positions 2,2′,6, and 6′, e.g., [2-233], can exist as atropisomers owing to restricted rotation about the single bond joining the two rings. Biflavonoids linked by a C-C bond can be considered to belong to a special class of biphenyl derivatives. The existence of atropisomerism in biflavonoids can be appreciated by looking at two examples, one representing the family of biflavones *per se*, and one representing biflavonoids that consist of two different flavonoid types. An example of the first type was featured in the study of (−)-4′,4‴,7,7″-tetra-*O*-methylcupressuflavone [2-234] by Harada and coworkers (1992). These workers determined the compound's absolute stereochemistry by theoretical calculation of its circular dichroism (CD) spectra. The second example involves a novel biflavonoid isolated from *Garcinia kola* roots by Terashima and coworkers (1995). The compound was identified as comprising naringenin linked to kaempferol via a C-3/C-8″ bond as shown in [2-235]. At 25° this compound exists as two atropisomers in the ratio of 1:1.2. If the reader has access to space-filling molecular models, it is instructive to build models of these compounds to see for oneself the crowding that prevents free rotation. A further example of atropisomerism involving isoflavones with bulky groups at C-2′ and C-6′ will be discussed below.

It was pointed out above in reference to lophirone-C [2-232], above, that some biflavonoids are joined by more than one inter-unit bond. Further examples will demonstrate this additional level of complexity. The first involves a study "anhydrobartramiaflavone" from the moss genus *Bartramia*, which Seeger and coworkers (1991, 1992) found to consist of two luteolin units joined by bonds involving, reciprocally, positions C-8 and C-2′ in each unit. This doubly linked dimer is shown as structure [2-236]. A more complicated situation exists in compounds linked in non-symmetric fashions, which may additionally involve different classes of flavonoids. Examples of this kind of biflavonoid come from a recent study of anti-inflammatory constituents of *Sarcophyte piriei* (Balanophoraceae), a parasitic plant that grows on roots of *Acacia* species in east Africa (Ogundaini *et al.*, 1996). Two compounds were identified that were based on a flavan–flavanone pairing with linkages involving the C-ring of the former and the A-ring of the latter. Both exist in the plant as 7-*O*-glucosides on the flavanyl unit. The compounds, illustrated as [2-237] and [2-238], are named 5,7,3′,4′-tetrahydroxyflavanyl-7-*O*-β-D-glucosyl-(4β-8; 2β-*O*-7)-naringenin and -eriodictyol, respectively.

It was necessary for Professor Geiger (1994) to rename his chapter to accommodate the appearance in the literature of the first triflavonoids, reports of which came from his group (Seeger *et al.*, 1992, 1995)! The first member of this group, called "bartramiatriluteolin," which name recognizes the compound's source, the moss genus *Bartramia*,

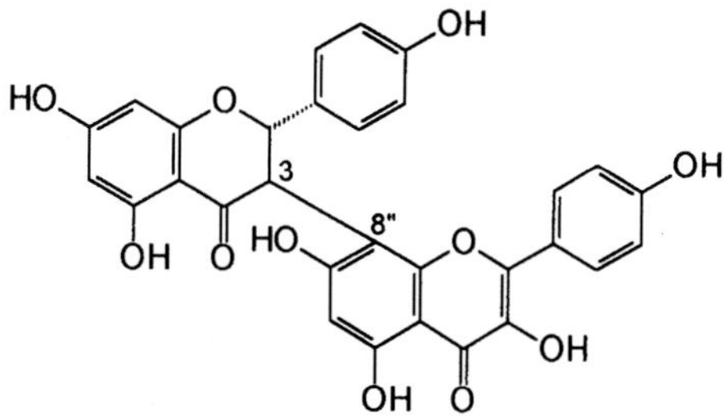

2-235

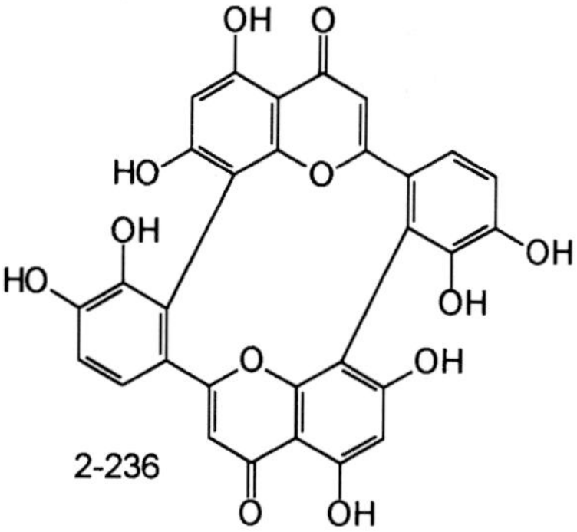

2-236

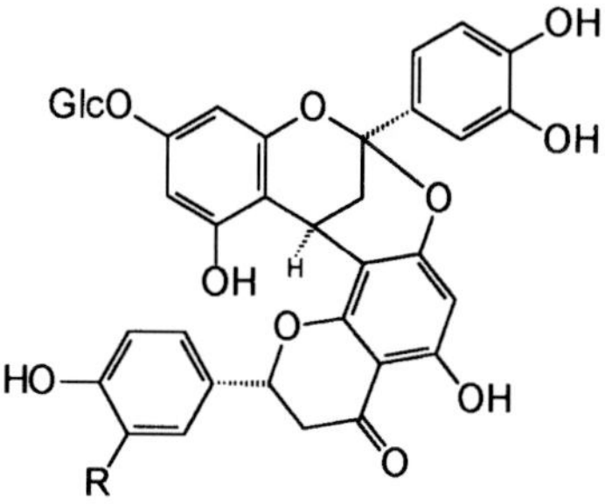

2-237 R = H

2-238 R = OH

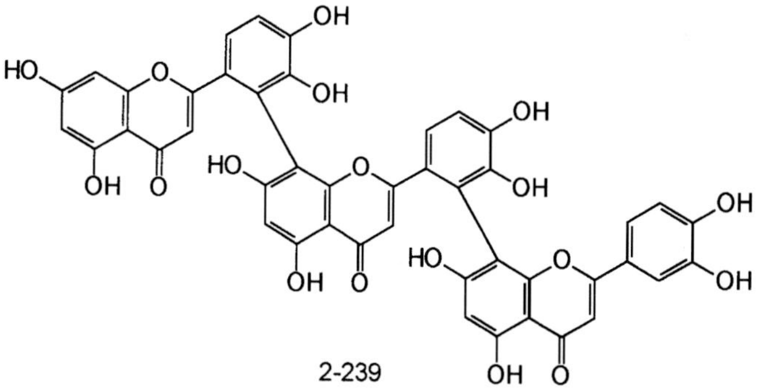

2-239

as well as its unique tri-flavone structure, is shown as [2-239]. Using one of the nomenclatural devices from above, we can call this novel molecule luteolin-(2′ → 8)-luteolin-(2′ → 8)-luteolin. An isomeric triluteolin called "aulacomniumtriluteolin" (Hahn, 1993) has the first two luteolins linked 2′ → 8, but the third is connected *via* the C-6 position of the terminal luteolin unit. It is obvious that the number of isomeric possibilities within this group is quite large, but the moss had other surprises in store. A further level of complexity was documented by Geiger and coworkers (1995) in their description of "cyclobartramiatriluteolin," also obtained from *Bartramia stricta*. In this compound, three luteolin units are linked head-to-tail to form an 18-membered macrocycle. The molecular weight of linear triluteolin [2-239] is 854, whereas the newly isolated compound gave a value of 852 (3×284). Since the molecular weight of luteolin is 286, the new compound has to have three luteolins, each less two hydrogen atoms. As well, the ^{13}C NMR exhibited only 15 signals, not the expected number for a trimer. The inter-luteolin linkages are all $8 \rightarrow 2'$ and the compound possesses a three-fold axis of symmetry. Structure [2-240] represents a distorted view of the compound. Again, molecular models would be necessary to gain a more accurate view of what such a compound might look like. The extent of occurrence of oligomeric flavone in mosses will be of continuing interest.

As we shall see in Chapter Six, the route to the formation of monomeric flavonoids is well established and appears to function much the same across the plant kingdom. The formation of biflavonoids, however, takes us into a somewhat different realm. Flavonoids are phenolic compounds, of course, and are thus susceptible to one-electron oxidation. Abstraction of hydrogen from the phenolic function leaves an oxygen free radical that can be stabilized by delocalization, as shown in the model reactions in Figure 2.1. Removal of the phenolic proton from [a] affords the oxygen radical [b], which in turn can yield radicals [c] and [d]. Each of the canonical forms represents a point at which radical coupling may occur. This is demonstrated in the lower part of the figure where several of the possible reactions occur by which new inter-unit carbon–carbon or carbon–oxygen bonds are formed. In the first three examples the initial coupling product undergoes keto–enol tautomerization to yield the coupled phenolic products.

Theoretically, any phenolic group can undergo one-electron oxidation, which means that flavonoid dimerization (and beyond) could occur at any stage. Figure 2.2 illustrates the formation of reactive species that would lead to couplings involving A-rings (top sequence) and B-rings (bottom sequence). The three flavone forms would lead to the formation of O-linked, C-8-linked, and C-6-linked biflavonoids, respectively. The lower sequence shows the possibilities when the cinnamoyl system of a chalcone is involved in delocalization of the radical.

Phenolic coupling reactions may occur between different classes of flavonoids, of course, but there is now proof that coupling between major classes of phenolic compounds is possible. In a search for HIV-reverse transcriptase inhibitors, Wang and coworkers (1994) encountered a compound from *Swertia franchetiana* (Gentianaceae) that had properties suggestive of a flavone as well as a xanthone. Detailed spectroscopic analysis showed that the compound was a flavone–xanthone dimer, the first of its kind reported from Nature. The flavone portion consisted of isoörientin joined at its 8-position to the 3-position of 1,4,5-trihydroxy-7-methoxyxanthone [2-241].

Chalcones

The first isolable compound formed in flavonoid biosynthesis is a chalcone. Although chalcones are thus considered as obligate intermediates in the pathway, they do not necessarily accumulate to any appreciable degree in most plants. This phenomenon might be better appreciated by looking at a plant that normally accumulates more highly colored flavonoids in its flowers but is not known to

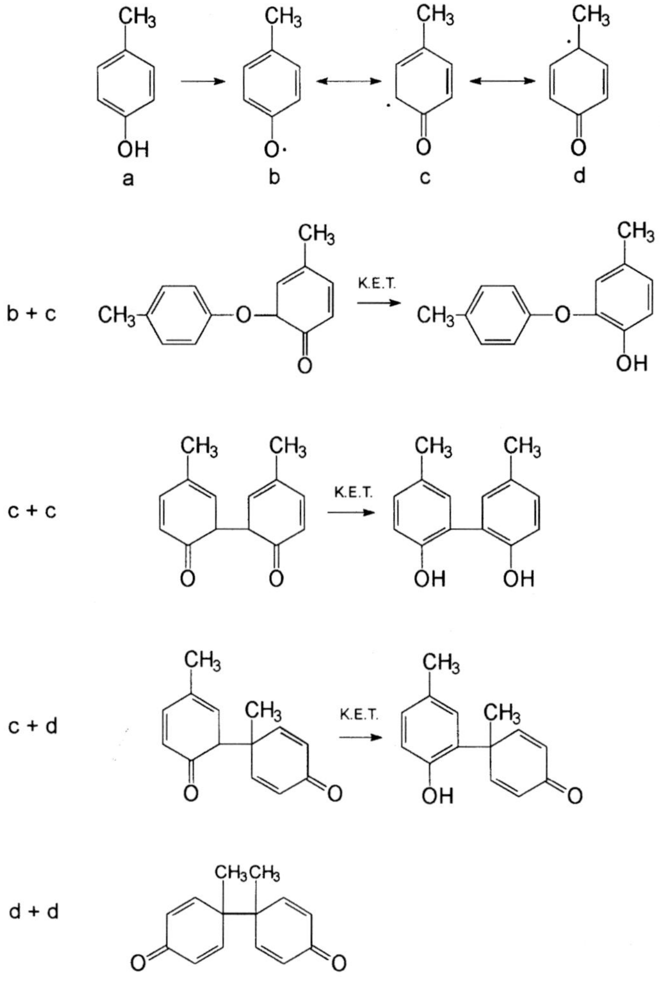

Figure 2.1 Phenolic coupling reactions. K.E.T. = Keto–enol tautomerization.

accumulate chalcones. Kuhn and his colleagues (1978) found that colorless flowers of *Callistephus chinensis* (Asteraceae) accumulated the chalcone isosalipurposide (2′,4′,6′,4-tetrahydroxychalcone-2′-*O*-glucoside) [2-242] but had no other flavonoids. The accumulation of this compound correlated with the *chch* genotype, in which the next enzyme in line was absent. Plants with the *Ch*__ genotype have chalcone isomerase (CHI), which acts by catalyzing the cyclization of chalcone to the flavanone stage setting the stage for subsequent steps in the flavonoid pathway.

In many plants, however, chalcone derivatives occur as major components of the flavonoid profiles. We will examine the diversity of naturally occurring chalcones in this chapter. Although chalcones have been found to accumulate in most plant

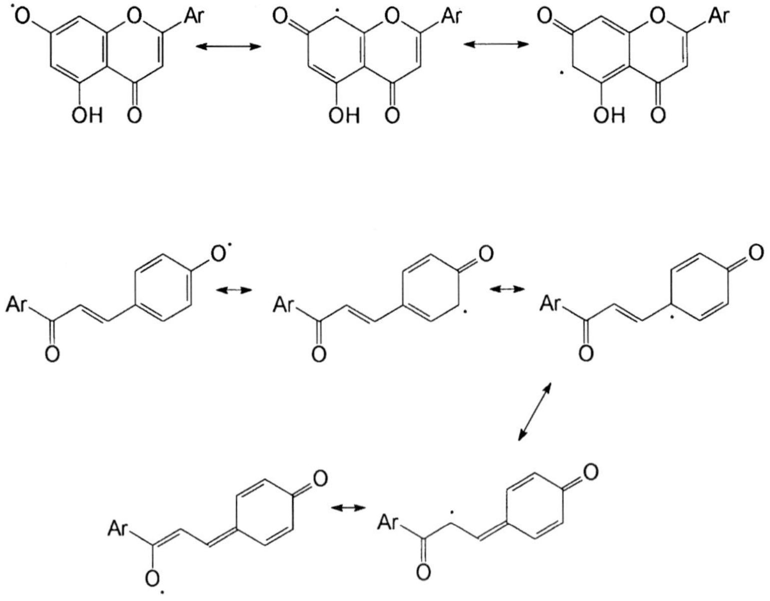

Figure 2.2 Canonical forms involved in free-radical coupling reactions (phenolic couplings) in biflavonoid formation. A-ring forms shown in top set, B-ring forms showing chalcone involvement in bottom set.

tissues, they have probably attracted most attention through the part they play in floral pigmentation. They are often localized in specific patterns in petals (or ray florets in Asteraceae) where, because of their capacity to absorb ultraviolet radiation, they contribute to the attraction of pollinators. Details of this function will be presented in Chapter Seven.

Chalcone itself has been identified as a component of *Centaurea calcitropa* (Asteraceae) by Dawidar and coworkers (1989). NMR showed the presence of bridge protons belonging to both the *cis*- [2-244] and *trans*- [2-245] isomers along with two unsubstituted phenyl rings. High resolution MS gave $C_{15}H_{12}O$ as the molecular formula. That two isomers were present in the purified material was evident by the presence of two sharp peaks in the IR spectrum, one at 1690 and one at 1630 cm^{-1} representing the two α,β-unsaturated carbonyl groups. The isomers proved inseparable. After one week standing in methanol solution Z → E (*cis* to *trans*) isomerization was complete.

Table 2.8 listed several of the more common naturally occurring chalcones and the aurones with which they are most closely associated. Detailed accounts of flavonoid structure and occurrences can be found in *The Flavonoids* series (Bohm, 1975, 1982, 1988, 1994). Our purpose will be served here by looking at a selection of chalcones representing the range of structures known in Nature. The finding that chalcone itself occurs naturally sets the obvious lower limit on flavonoid substitution. A chalcone with one hydroxyl group was isolated from *Shorea robusta* (Dipterocarpaceae) and shown to be the glucoside of 4′-hydroxychalcone [2-245] (Jain *et al.*,

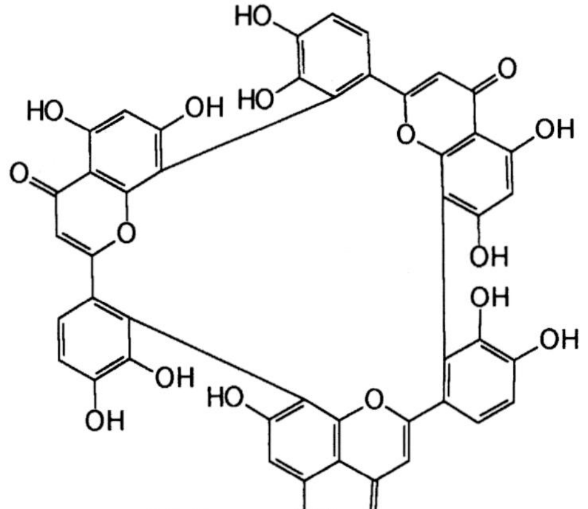

2-240

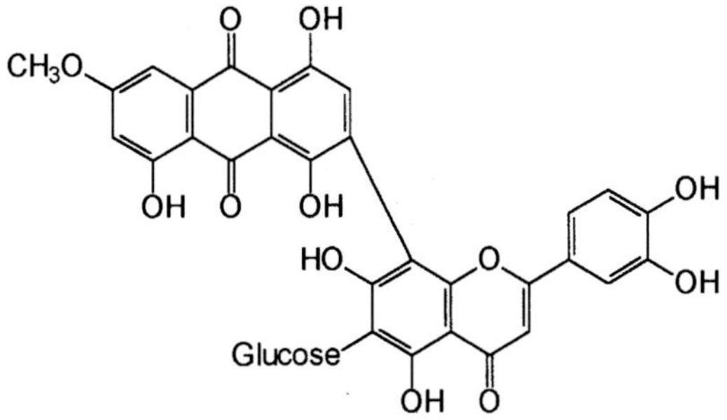

2-241

1982). Several chalcones are known whose A-rings are based on resorcinol and whose B-ring lacks substitution. Three of these were identified as components of *Acacia neovernicosa* by Wollenweber and Seigler (1982) and identified as 2′,4′-dihydroxychalcone [2-246], 2′-methoxy-4′-hydroxychalcone [2-247], and the extra-oxygenated 2′,4′-dihydroxy-3′-methoxychalcone [2-248]. Elaborations on this base molecule have been observed, for example, in studies of *Helichrysum rugulosum* from which source Bohlmann *et al.* (1984) obtained 2′-methoxy-4′-prenyloxychalcone [2-249], 2′,4′-dihydroxy-3′-*C*-prenylchalcone [2-250], and the corresponding 4′-methyl ether. A further variation on *C*-prenylation, this time involving a compound with a 4-methoxy group, is seen in [2-251] which Singhal and coworkers (1983) obtained from *Millettia pachycarpa*. It was noted in the section above on substitution patterns that *Primula* species have yielded several flavones have been unusual substitution patterns. An unusual chalcone was isolated from *Primula macrophylla* by Ahmad and coworkers (1992) and shown to be 3′,3-dihydroxychalcone [2-252].

The most common chalcone, although it doesn't necessarily accumulate in any given plant, is 2′,4′,6′,4-tetrahydroxychalcone [2-253], which is often referred to as "chalcononaringenin." This is the first compound formed (in most instances)

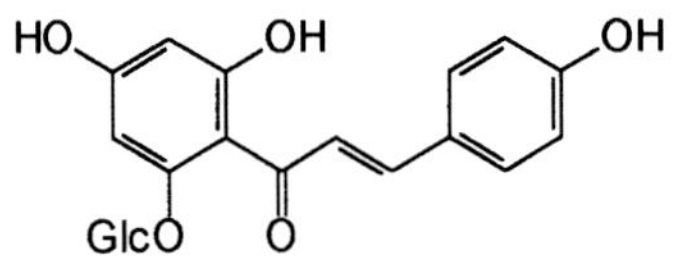

2-242 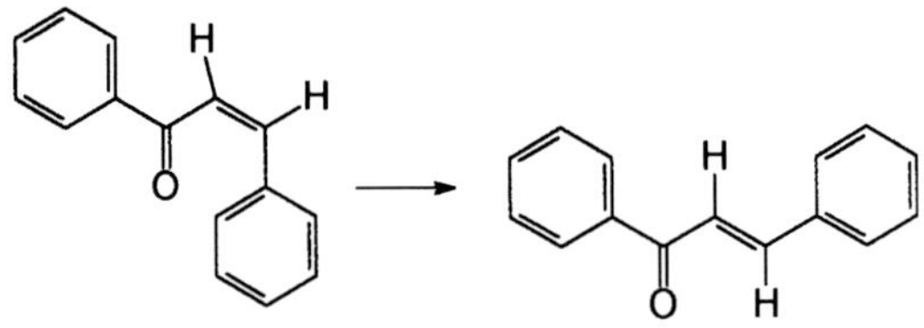 2-243 2-244

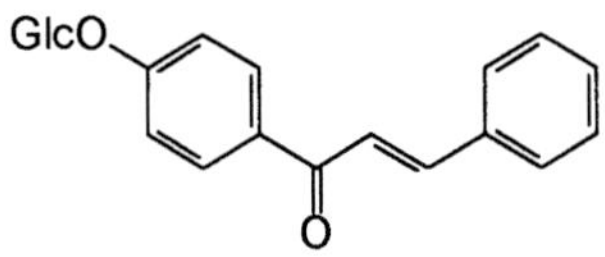

2-245

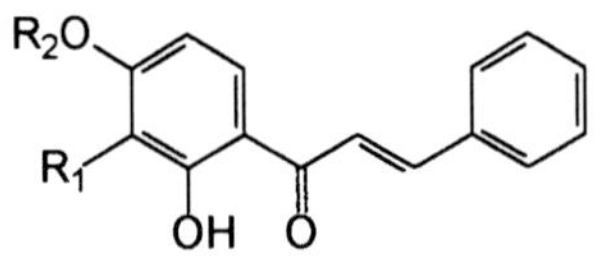

2-246 $R_1 = R_2 = H$

2-247 $R_1 = H$, $R_2 = CH_3$

2-248 $R_1 = OCH_3$, $R_2 = H$

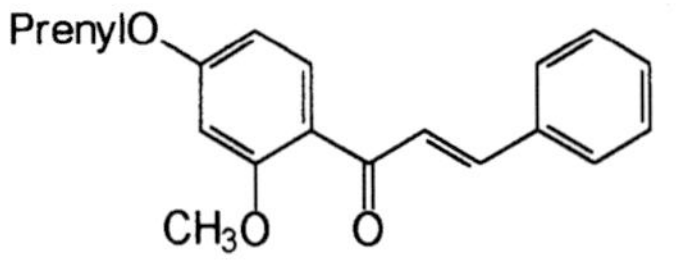

2-249 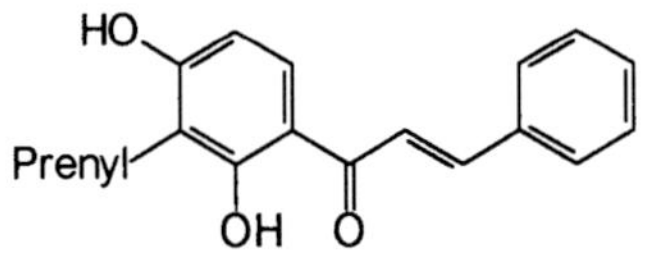 2-250

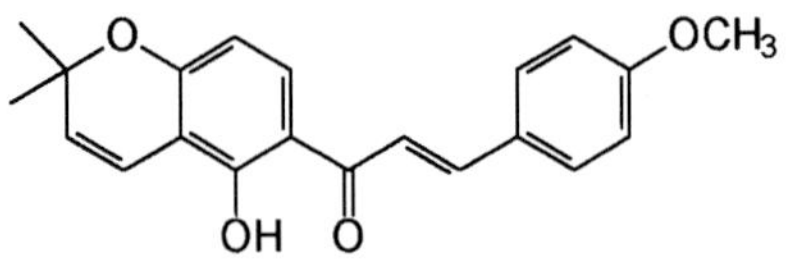

2-251 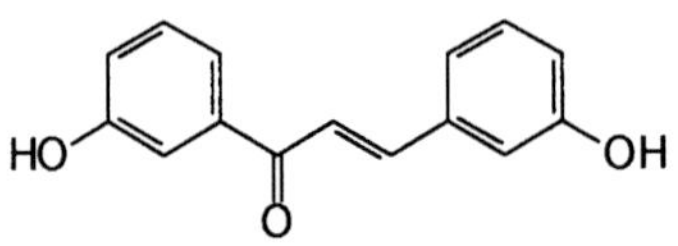 2-252

through the action of chalcone synthase (CHS). Reports of this compound occurring as the aglycone are rare owing to the ease with which it is cyclized to naringenin. Two such reports, however, involve pollen from unrelated species, *Petunia hybrida* (de Vlaming and Kho, 1976) and *Nothofagus antartica* (Fagaceae) (Wollenweber and Wiermann, 1979). Many glycosidic and *O*-alkylated derivatives of chalcononaringenin have been tabulated (Bohm, 1982).

As in the case of most other flavonoid classes, many chalcones are known that are characterized by high levels of substitution involving hydroxyl, methoxyl, alkyl, and combinations of these. Examples are 2′,4′-dihydroxy-3′,5′-di-*C*-prenylchalcone [2-254] from roots of *Tephrosia spinosa* (Sharma and Rao, 1992),

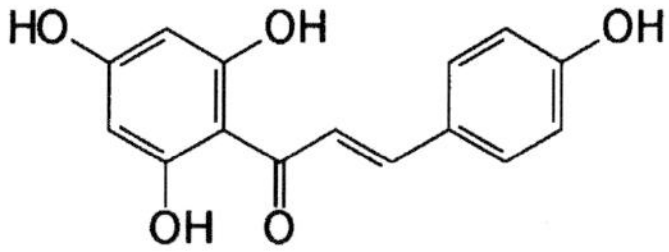

2-253 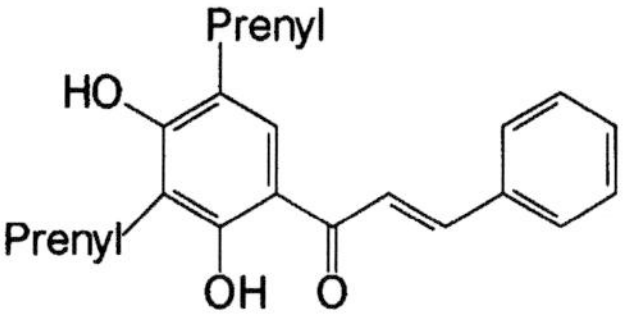 2-254

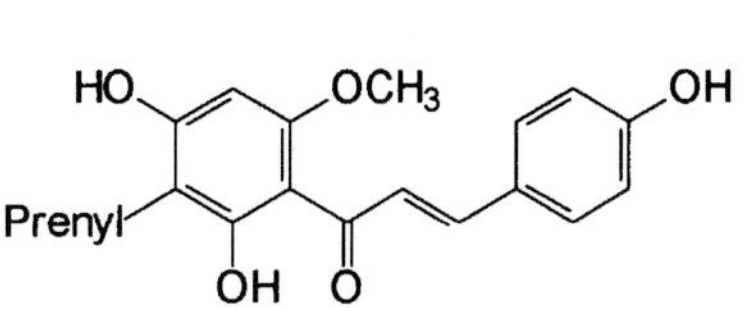

2-255 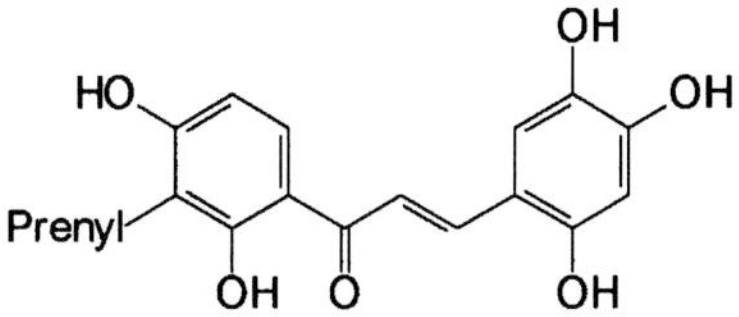 2-256

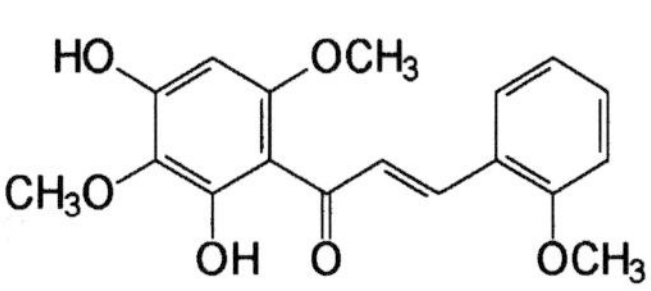

2-257

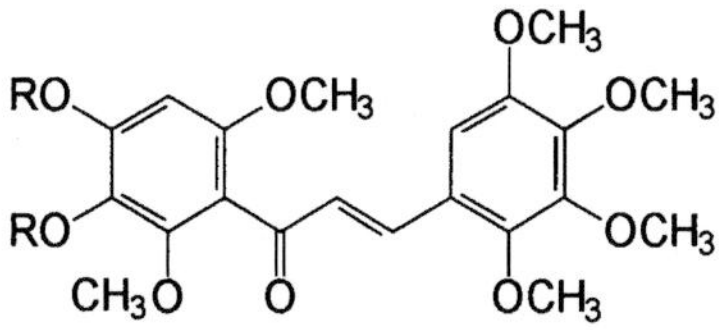

2-258 Rs = H

2-259 Rs = OCH_2O

2′,4′,4-trihydroxy-6′-methoxy-3′-*C*-prenylchalcone [2-255] from *Humulus lupulus* (Cannabidaceae, hops) (Mizobuchi and Sato, 1985), and 2′,4′,2,4,5-pentahydroxy-3′-*C*-prenylchalcone [2-256] from *Crotalaria ramosissima* (Fabaceae) (Khalilullah *et al*., 1993). Members of the genus *Scutellaria* have yielded a number of highly substituted chalcones, some of them among the most highly substituted flavonoids known. *Scutellaria luzonica*, for example, yielded 2′,4′-dihydroxy-3′,6′,2-trimethoxychalcone [2-257] (Lin *et al*., 1991), but the winner in this category may well be *S. indica* from which Miyaichi and coworkers (1989) obtained, among others, the two compounds whose structures are shown as [2-258] and [2-259].

Among the complex flavonoids obtained from members of Moraceae [see Nomura (1988) for review] are compounds that appear to have been formed by a condensation process that is reminiscent of the Diels–Alder reaction. As indicated in the model reaction below, the Diels–Alder condensations a 1,3-diene [2-260] condenses with a dienophile [2-261] to form a new six-membered ring compound [2-262]. The dienophile is most effective if it possesses an electron withdrawing group (R in 2-261) in conjugation with the double bond. In the case of flavonoids from Moraceae, it is thought that a doubly unsaturated prenyl group on one flavonoid unit [2-263] serves as the diene while the bridge double bond of a chalcone [2-264] serves as the dienophile. In this case the carbonyl group of the chalcone functions as the electron withdrawing group. This entire process is illustrated below in the formation of "kuwanon-J" [2-265] from callus cultures of *Morus alba* (Ueda *et al.*, 1982). Additional information on Diels–Alder type compounds and the reactions leading to them can be found in Nomura (1988).

A small group of chalcones is known whose rings appear to be on backward. That is to say, the A-ring and B-ring oxygenation patterns appear reversed. The first

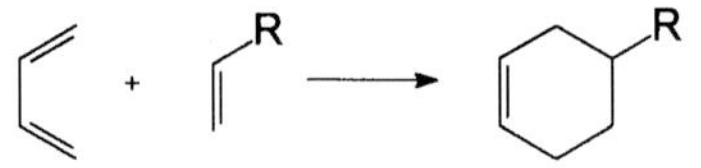

2-260 2-261 2-262

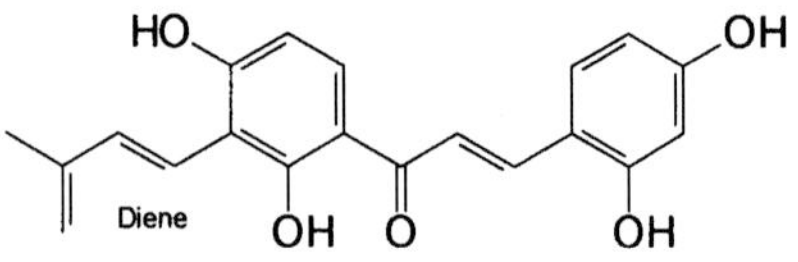

Diene

2-263

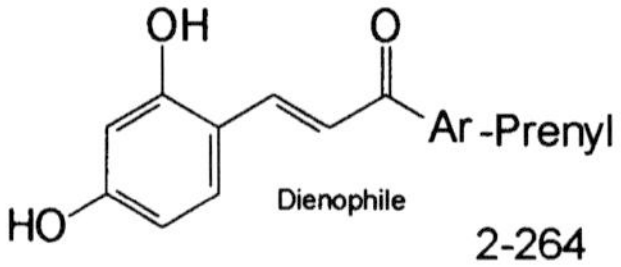

Dienophile

Ar-Prenyl

2-264

2-265

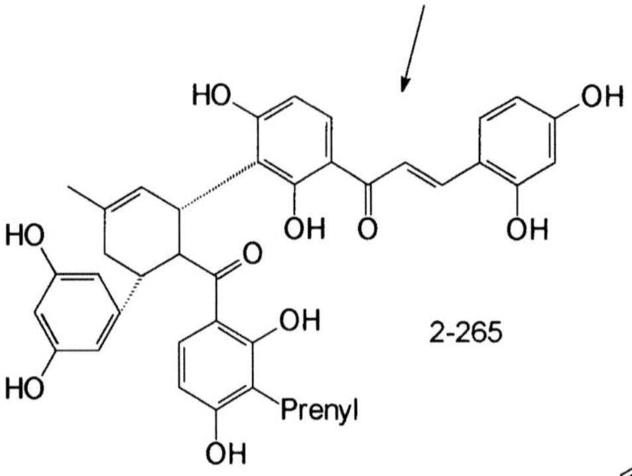

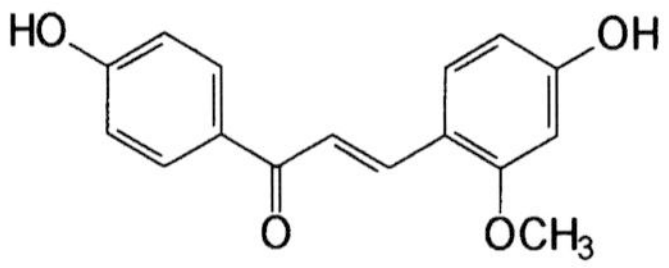

2-266

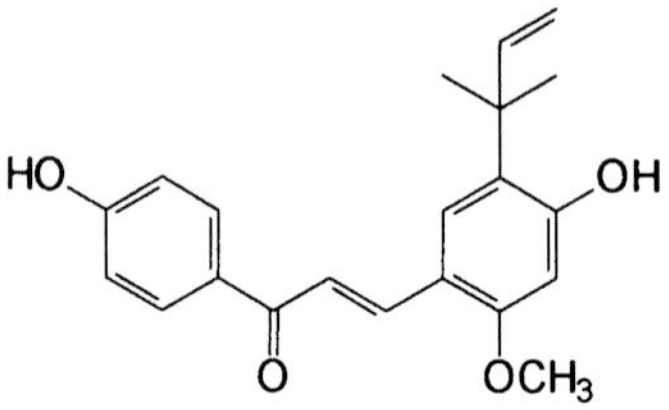

2-267

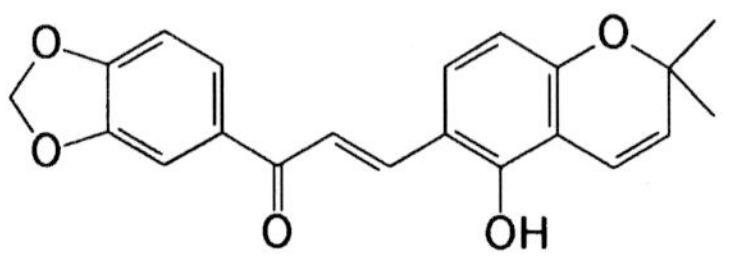

2-268

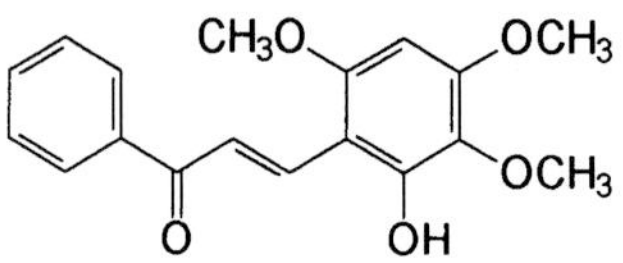

2-269

member of this group to be identified was 4′,4-dihydroxy-2-methoxychalcone (echinatin) [2-266] from cultures of *Glycyrrhiza echinata* (Fabaceae) (Saitoh *et al.*, 1975). This was followed in the same year by a report of two compounds from roots of *G. glabra* (Saitoh and Shibata, 1975), "licochalcone-A" [2-267], which also features the α,α-dimethylallyl group, and 4′,3,4-trihydroxy-2-methoxychalcone, "licochalcone-B." The term "retroflavonoid" was coined to identify the group. Biosynthetic studies (Saitoh *et al.*, 1975) demonstrated that the A-ring, as written in these compounds, was formed from phenylpropanoid precursors as would be expected for a phenolic compound with a single oxygen at the *para*-position, and that the B-ring, bearing oxygens on alternate carbons suggesting a malonyl-CoA derived system, was indeed formed from acetate. Ayabe and coworkers (1980a,b) suggested that a dibenzoylmethane (β-ketodihydrochalcone) intermediate may be involved. As we shall see in the discussions of dihydrochalcones below, a β-ketodihydrochalcone offers a mechanistic opportunity for equilibration of the terminal carbons of the three-carbon bridge to occur.

Additional, recently described members of the retrochalcone group include compound [2-268] from *Pongamia glabra* (Fabaceae) (Saini *et al.*, 1983), the unusual *C*-formyl chalcone [2-116] from *Anredera scandens* (Basellaceae) (Calzada *et al.*, 1990), and 2-hydroxy-3,4,6-trimethoxychalcone [2-269] from the antimalarial *Uvaria dependens* (Nkunya *et al.*, 1993).

Retrochalcones have received comparatively little attention as a group owing undoubtedly to their scarcity in Nature. In the section on anthocyanins above, however, we met a *cis*-retrochalcone as one of the possible forms that may exist as a result of the reaction of an anthocyanin [2-270] with water (Brouillard and Dangles, 1994). The reaction involves formation of a hemiacetal [2-271] which can be considered to be in equilibrium with the *cis*-retrochalcone [2-272]. The reactions are repeated below. Although the significance of anthocyanin hydration has been thoroughly discussed in the literature, the part played by the retrochalcone has not been pursued in any depth.

A small group of compounds exists that can be considered either as chalcones with a β-hydroxyl group or as β-ketodihydrochalcones as indicated in the case of a compound isolated from *Baccharis salicifolia* (Wollenweber *et al.*, 1989). These structures represent the keto and enol forms which can presumably exist as an equilibrium mixture. In the case of the *Baccharis* material, the keto [2-273] and enol [2-274] forms were found to exist in the ratio of 7 : 3. In contrast, the keto–enol pair [2-275] and [2-276] isolated from *Glycyrrhiza inflata* (Kajiyama *et al.*, 1992) appeared to exist in ratio of 3 : 7. Two β-hydroxychalcones, "pongagallone-A" [2-277] and "pongagallone-B" [2-278], have been reported from the galls on leaves of *Pongamia glabra* (Gandhidasan *et al.*, 1987). The lack of a triplet for a β-diketone carbon near δ 55.0 in the ^{13}C NMR spectrum indicates that the compounds occur in the β-hydroxy forms illustrated. A further complicating factor with β-hydroxychalcones is that they can also exist in the cyclic hemiketal form. This is shown in the case of a compound from a species of *Malus* for which Williams (1967) proposed structure [2-279]. Chadenson and coworkers (1972), who had obtained a similar compound from *Populus niger*, preferred cyclic hemiketal form [2-280].

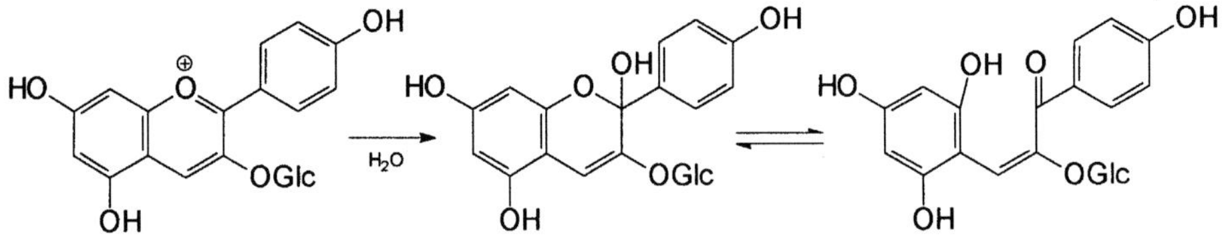

HO OH OH OGlc 2-270 H_2O HO OH OH OGlc OH 2-271 HO OH O OH OGlc OH 2-272

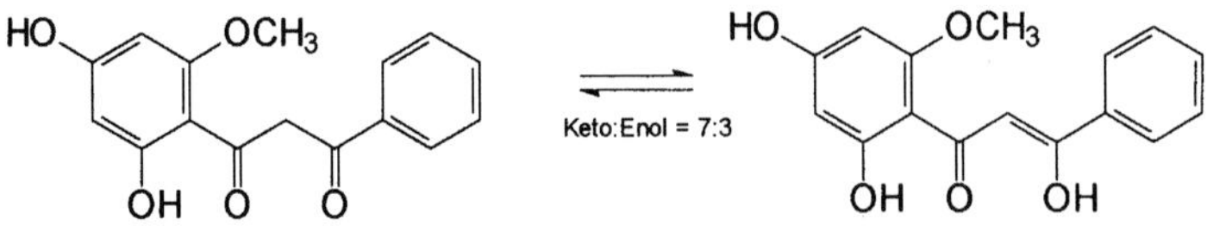

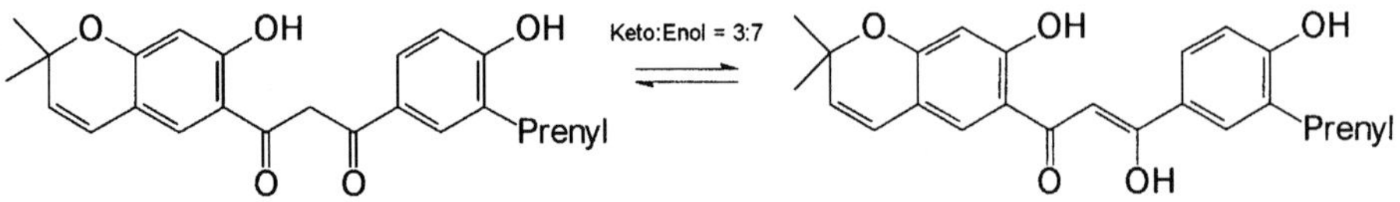

HO OCH_3 OH O O 2-273 Keto:Enol = 7:3 HO OCH_3 OH O OH 2-274

O OH OH Prenyl O O 2-275 Keto:Enol = 3:7 O OH OH Prenyl O OH 2-276

Prenyl CH_3O R R CH_3 O O OH

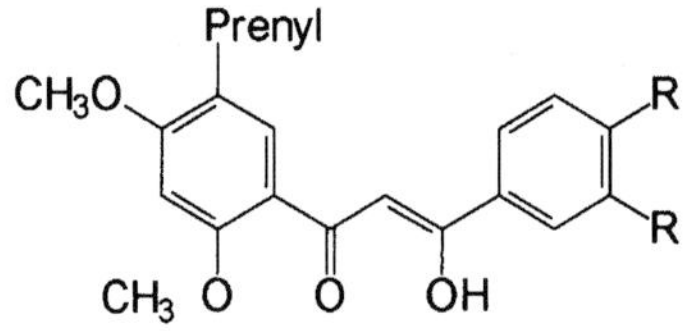

2-277 Rs = H

2-278 Rs = OCH_2O

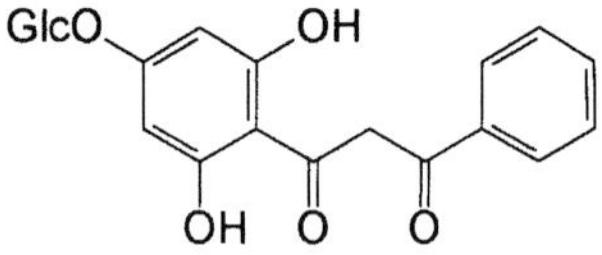

GlcO OH OH O O 2-279

GlcO O OH OH O 2-280

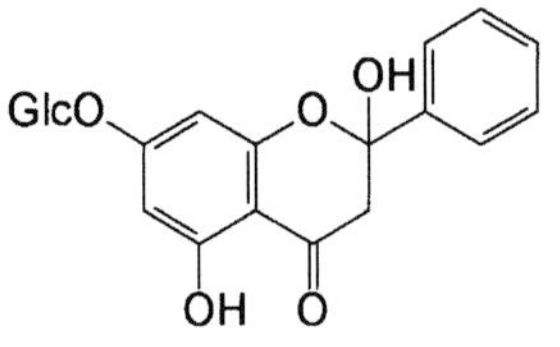

Dihydrochalcones

Dihydrochalcones comprise a small group of flavonoids in which the three-carbon bridge double bond has been reduced. This reduction destroys the chalcone chromophore with the result that dihydrochalcones are not as readily apparent as are their parent molecules, particularly on paper or thin layer chromatograms. Their ultraviolet spectral characteristics are similarly dulled. Nonetheless, several dihydrochalcones are known some of which possess interesting properties. The earliest dihydrochalcone to be identified as a natural product was 2′,4′,6′,4-tetrahydroxydihydrochalcone 2′-*O*-glucoside [2-281], which de Koninck (1835) reported from bark of apple, cherry, plum, and pear. The report of "phloridzin" in apple bark has been authenticated, but it does not occur in the other three (Williams, 1966). Phloridzin is thought to be responsible for the apple replant problem or "old orchard disease." Apple seedlings planted in an older orchard grow poorly and often have discolored roots. The unhealthy roots severely restrict productivity. Börner (1959) investigated several aspects of this phenomenon including the involvement of phloridzin. Five phenolic compounds were identified, either from growth experiments with cultured apple seedlings (with and added apple root bark) or from soil samples: phloridzin, phloretin (the aglycone), *p*-hydroxydihydrocinnamic acid, phloroglucinol, and *p*-hydroxybenzoic acid, all of which inhibited growth. Of this set of compounds, only phloridzin was identified as a natural constituent of apple root bark. He concluded that phloridzin is broken down by soil microorganisms and, until the soil is cleared of these compounds, successful replanting is unlikely. Subsequent experiments by Holowczak *et al.* (1960) showed that numerous isolates of the soil microorganism *Venturia inaequalis* were capable of decomposing phloridzin to phloretin and glucose and phloretin into *p*-hydroxydihydrocinnamic acid and phloroglucinol. Other examples of regrowth problems, though not necessarily involving flavonoids, can be found in Rice's (1987) monograph entitled *Allelopathy*.

At the time of preparation of the first volume of *The Flavonoids* series (Bohm, 1975), there were reports in the literature of only a handful of dihydrochalcones, although the hint of normal flavonoid structural variation was there. B-Ring deoxydihydrochalcones were represented by 2′,6′-dihydroxy-4′-methoxydihydrochalcone [2-282] from poplar (*Populus*, Salicaceae) (Goris and Canal, 1935) and from the fern genus *Pityrogramma* (Star and Mabry, 1971). The first *C*-glucosyl member of the class, 2′,4′,6′,3,4-pentahydroxy-3′-*C*-glucosyldihydrochalcone [2-283], had been identified from *Aspalathus linearis* (Fabaceae) (Koeppen and Roux, 1965). All but one of the compounds listed in the first survey were based upon the 2′,4′,6′-trihydroxy- or phloroglucinol pattern. The first dihydrochalcone based on the 2′,4′-dihydroxy-, or resorcinol, pattern was 2′,4′,4-trihydroxydihydrochalcone 2′-*O*-glucoside from *Viburnum davidii* (Caprifoliaceae) (Bohm and Glennie, 1969). It is interesting to note that a survey of 52 additional species from Caprifoliaceae, including a total of 15 species of *Viburnum*, failed to disclose any additional source of the compound in the family. By the time of appearance of the second volume (Bohm, 1982), the list of known compounds and the number of plant families involved had grown significantly. Among the newly reported compounds were those that were *C*-methylated, such as 2′,4′-dimethoxy-5′-*C*-methyl-6′-hydroxydihydrochalcone [2-284] from *Myrica gale* (Malterud *et al.*, 1977), *C*-prenylated, as in the case of [2-285] and [2-286] from *Helichrysum* species (Bohlmann *et al.*, 1979), and *C*-benzylated, represented by [2-287], 2′,4′-dihydroxy-3′-(2-hydroxybenzyl)-6′-methoxydihydrochalcone from *Uvaria acuminata* (Cole *et al.*, 1976). One additional compound has to be commented here, not because of its structure, which is entirely ordinary, but because of its name: "confusoside," 2′,4′,4-trihydroxydihydrochalcone

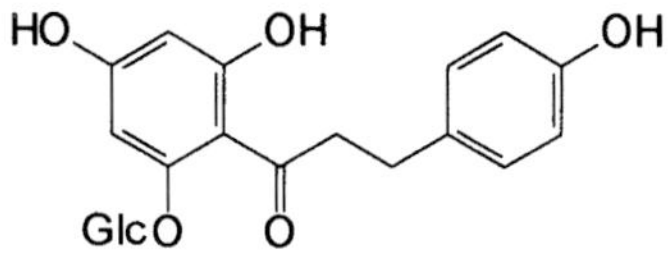

2-281

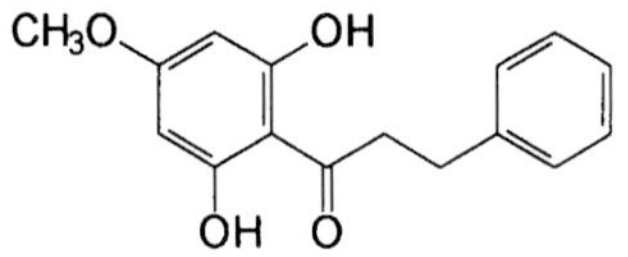

2-282

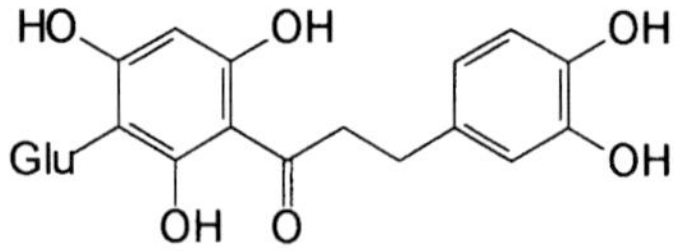

2-283

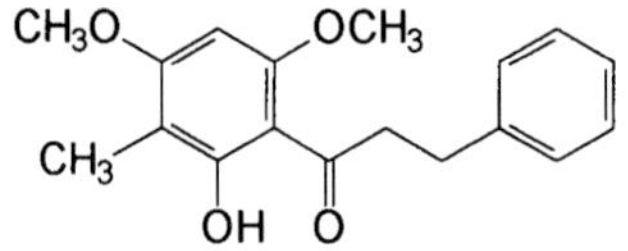

2-284

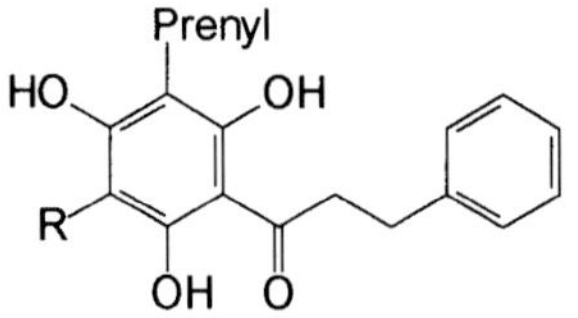

2-285 R = OCH_3

2-286 R = Prenyl

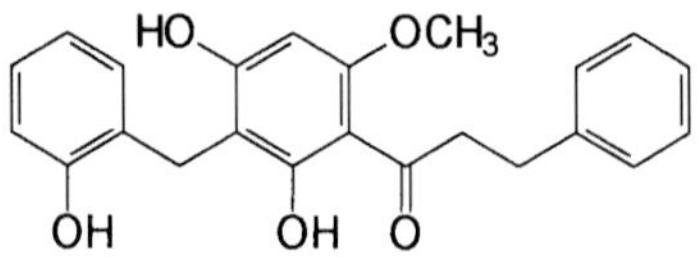

2-287

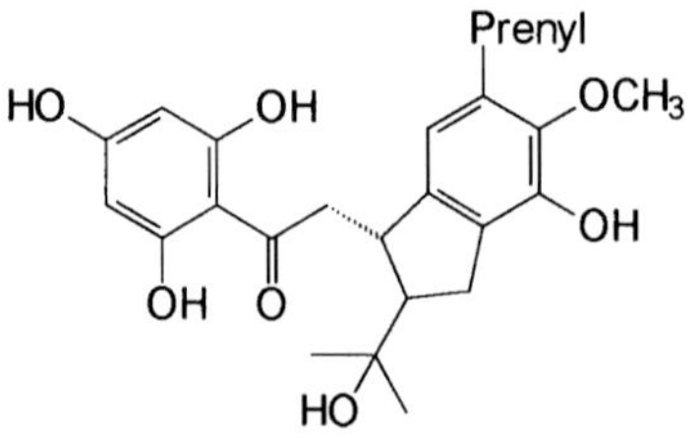

2-288

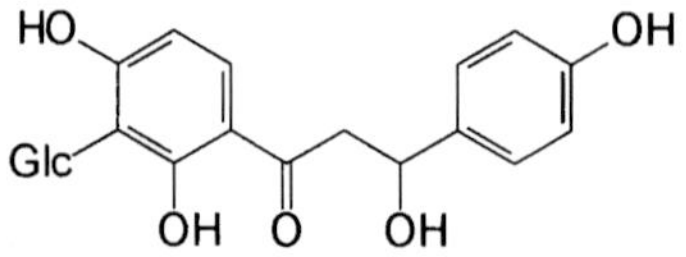

2-289

4′-*O*-glucoside, was isolated from *Symplocus confusa* (Symplocaceae) by Tanaka and coworkers (1982). The trivial name is indeed an unfortunate choice considering all of the examples we have seen that so richly deserve such descriptive attention!

One of the structural characteristics of many of the flavonoids obtained from members of Moraceae is the presence of *C*-prenyl groups, which may occur on any of the rings. Examples of *C*-prenylation in the flavone series will be seen below, but *C*-alkylated dihydrochalcones are also known from this family. Two such compounds, called "antiarone-J" [2-288] and "antiarone-K," were identified as constituents of the root bark of *Antiaris toxicaria*, an Indonesian plant used for preparation of arrow poison (its latex contains cardiac glycosides). Hano and coworkers (1991) showed that both antiarones contained a prenyl function at C-2 (B-ring) which has undergone secondary reaction at the bridge β-position to yield the five-membered rings seen in the illustration. Antiarone-J carries an additional prenyl function at C-5 and both are hydroxylated on their isopropyl groups. Antiarone K lacks the C-prenyl group and has a second B-ring O-methyl. The flavonoid chemistry of Moraceae has been discussed in detail by Nomura (1988).

Several dihydrochalcones are known that are distinguished by having an oxygen function at one or the other of the bridge carbons. "Coatline-A" [2-289] is an example of the small group of compounds that have an hydroxyl function adjacent to the carbonyl. This particular example comes from the legume *Eysenhardtia polystachya* where it occurs with the 3′,4′-dihydroxy analogue (Beltrami *et al.*, 1982). Compounds with β-oxygenation are also known. A dihydrochalcone unusual for its bridge hydroxylation as well as for its rare B-ring *O*-methylation pattern was identified as a constituent of *Gliricidia sepium* (Fabaceae) by Manners and Jurd (1979). 2′,4′,3,5,β-pentahydroxy-4-methoxydihydrochalcone is illustrated in [2-290]. A pair of β-methoxydihydrochalcones, shown as [2-291] and [2-292], were identified as components of *Millettia hemsleyana* by Mahmoud and Waterman (1985). These compounds are isomers that differ in the relative positions of their carbonyl and methoxy groups. They can be thought of as representing the two keto–enol tautomers of milletenone [2-293] that have been reduced and trapped as methyl ethers. A few dihydrochalcones are also known that have oxygen on the carbon adjacent to the carbonyl carbon, an example of which is α,4′,3,4-tetrahydroxy-2′-methoxydihydrochalcone [2-294] from *Xanthocercis zambesiaca* (Bezuidenhout *et al.*, 1988).

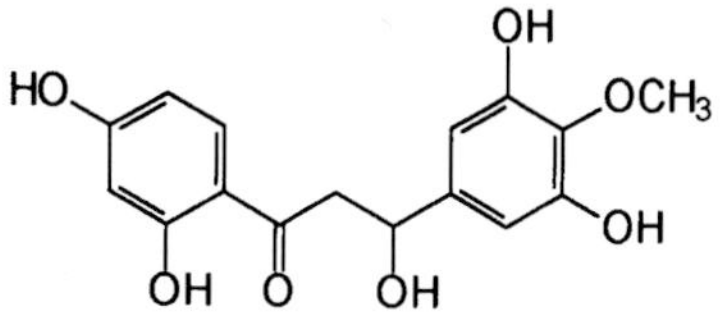

OH HO OCH3 OH OH O OH

2-290

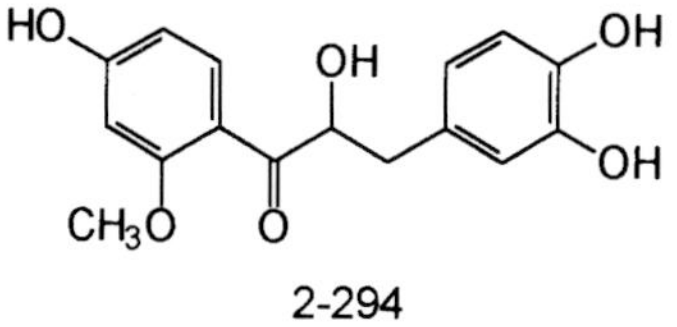

HO OH OH OH CH3O O

2-294

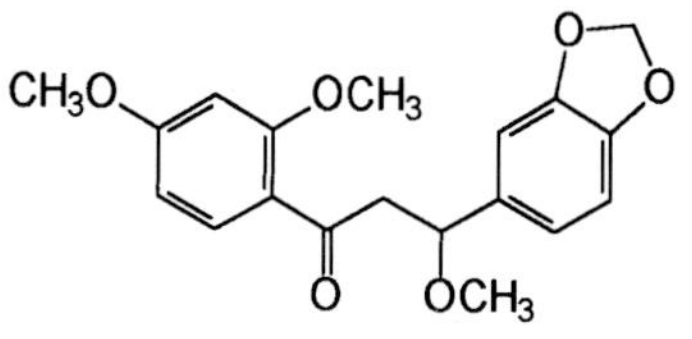

CH3O OCH3 O O O OCH3

2-291

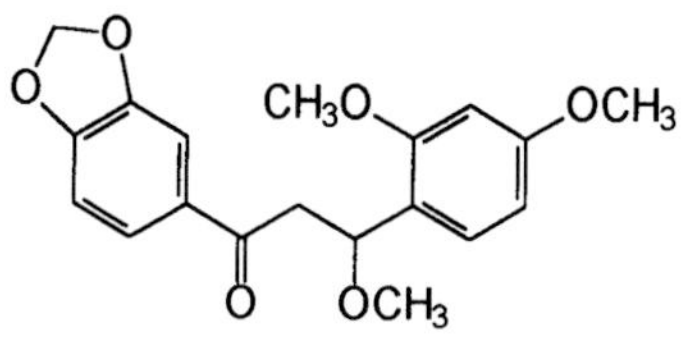

O O CH3O OCH3 O OCH3

2-292

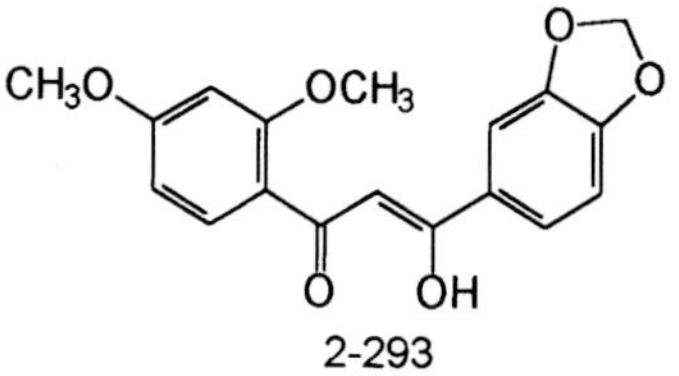

CH3O OCH3 O O O OH

2-293

Compounds related to phloretin, neohesperidin dihydrochalcone, and naringenin dihydrochalcone, possess antifeeding activity against the aphids *Schizaphis graminum* and *Myzus persicae* (Dreyer and Jones, 1981). A key feature from that work is that the compounds showed the feeding deterrent activity at concentrations that approximate natural concentrations in the plant.

Sweet-tasting dihydrochalcones will be discussed in Chapter Eight.

Dihydroflavonols

Dihydroflavonols, or 3-hydroxyflavanones as they may also be called, are obligate intermediates on the pathway to flavonols by one route, and to anthocyanidins *via* flavan-3,4-diols by another. In these roles, there is no apparent need for the intermediates to accumulate. Nonetheless, dihydroflavonols exhibiting a wide variety of structural modifications are well known and major components of the flavonoid profiles of many plants. The most commonly encountered members of this class correspond to the common flavonols: dihydrokaempferol (aromadendrin) [2-295], dihydroquercetin (taxifolin, or distylin in the older literature) [2-296], and dihydromyricetin (ampelopsin) [2-297]. A corresponding series of 5-deoxydihydroflavonols also exists: 5-deoxydihydrokaempferol (garbanzol) [2-298], 5-deoxydihydroquercetin (fustin or dihydrofisetin) [2-299], and 5-deoxydihydromyricetin (dihydrorobinetin) [2-300]. Among the simplest dihydroflavonols are pinobanksin (3,5,7-trihydroxyflavanone) [2-301] and its 5,7-dimethyl ether [2-302], and 3,7-dihydroxyflavanone [2-303]. These are all widely distributed in the plant kingdom (see Bohm, 1975b, 1982, 1988, 1994 for extensive lists of sources).

As in the case of other flavonoid classes, modified dihydroflavonols occur in a variety of plant species. Most of the compounds listed above also occur as glycosides in some plants, and several have been identified that have methyl ether groups; some have both. Dihydroisorhamnetin (taxifolin 3′-methyl ether) [2-304] has been reported from *Artemisia dracunculus* (tarragon) (Balza *et al.*, 1985a) while the dihydroflavonol corresponding to rhamnetin (quercetin 7-methyl ether) [2-305] was reported from *A. glutinosa* (Gonzalez *et al.*, 1983). The latter compound, whose trivial name is "padmatin," occurs as the 3-acetate [2-306] in *Inula viscosa* (Grande *et al.*, 1985).

Quite complex mixtures of dihydroflavonol derivatives are known to accumulate in some species. A case in point is the legume *Sophora flavescens* from which Wu and coworkers (1985) obtained a series of *C*-alkylated dihydroflavonols, called "kushenols" that are based on the 5,7,2′,4′-oxygenation pattern. Kushenol-L [2-307], for example, is the 6,8-di-*C*-prenyl derivative, while kushenols-I and -N are the (2*R*,3*R*) [2-308] and (2*R*,3*S*) [2-309] isomers, respectively, of 7,2′,4′-trihydroxy-5-methoxy-8-*C*-lavandulyldihydroflavonol. *C*-Methylation is seen in 5,7,3′,4′,5′-pentahydroxy-6,8-di-*C*-methyldihydroflavonol [2-310] from *Alluaudia humbertii* (Didieriaceae) (Voirin *et al.*, 1986). A dihydroflavonol from *Myrica rubra* was identified by Nonaka and coworkers (1983) as 5,7,2′,4′,5′-pentahydroxydihydroflavonol 3-*O*-gallate-2′-sulfate (potassium salt) [2-311].

Dihydroflavonols have chiral centers at C-3 and C-4, which translates into four possible orientations of the two groups. The common stereochemistry, (2*R*,3*R*), was seen in all but one of the above structures. There are a few cases where all four

2-295 $R_1 = R_2 = H$
2-296 $R_1 = OH, R_2 = H$
2-297 $R_1 = R_2 = OH$

2-298 $R_1 = R_2 = H$
2-299 $R_1 = OH, R_2 = H$
2-300 $R_1 = R_2 = OH$

2-301 $R_1 = OH, R_2 = H$
2-302 $R_1 = OCH_3, R_2 = CH_3$
2-303 $R_1 = R_2 = H$

2-304 $R_1 = R_3 = H, R_2 = CH_3$
2-305 $R_1 = R_2 = H, R_3 = CH_3$
2-306 $R_1 = C(=O)CH_3, R_2 = H, R_3 = CH_3$

2-307

2-308 R = —OH
2-309 R = ┅OH

isomers of a given compound are known, although one of them usually predominates in Nature. The Chinese use a preparation of *Engelhardtia chrysolepis* (Juglandaceae), called "Hung-qi," in their traditional medicine. This drug is sweet which prompted Kasai and coworkers (1988) to search for the compound or compounds responsible. Four taxifolin 3-*O*-rhamnosides were isolated and tested for sweetness. The common isomer, "astilbin," with (2*R*,3*R*) stereochemistry [2-312] was not sweet. Its isomer with (2*S*,3*S*) stereochemistry, "neoastilbin" [2-313], does have a sweet taste and is thought to be the compound responsible for the sweetness of the drug. The remaining two isomers, "isoastilbin" with (2*R*,3*S*) stereochemistry [2-314], and "neoisoastilbin" with (2*S*,3*R*) stereochemistry [2-315], were not sweet. [N. B. The stereochemistry given for these four compounds reflects recent work of Gaffield

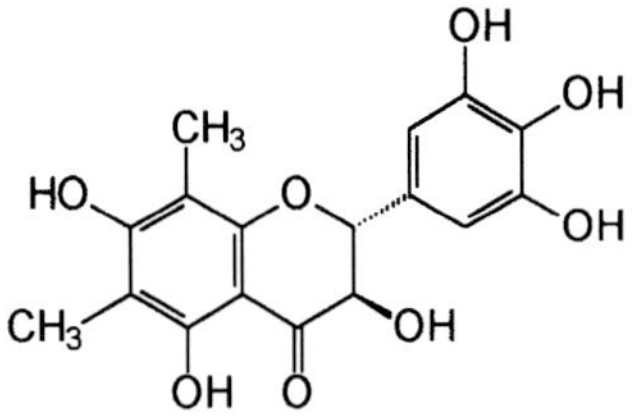

2-310

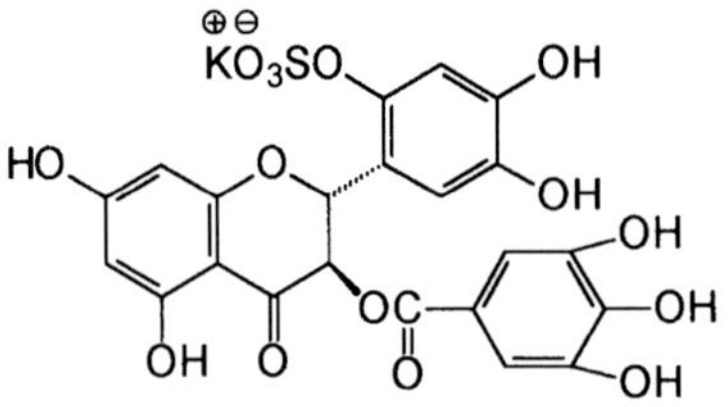

2-311

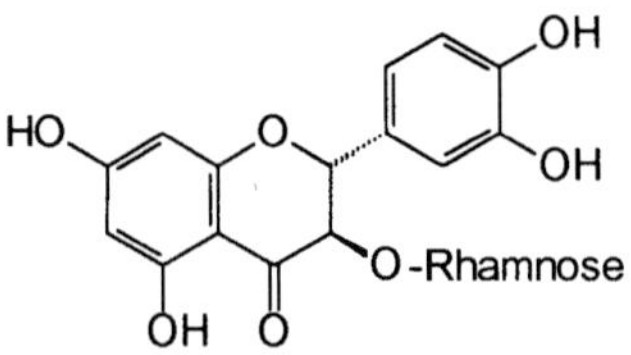

2-312 (2*R*,3*R*)

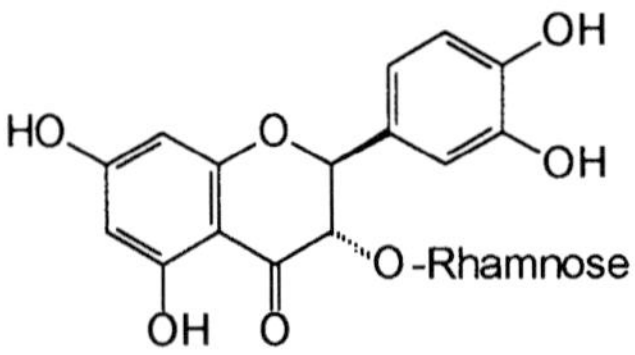

2-313 (2*S*,3*S*)

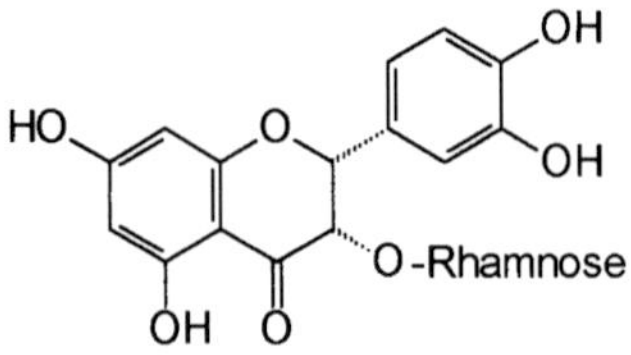

2-314 (2*R*,3*S*)

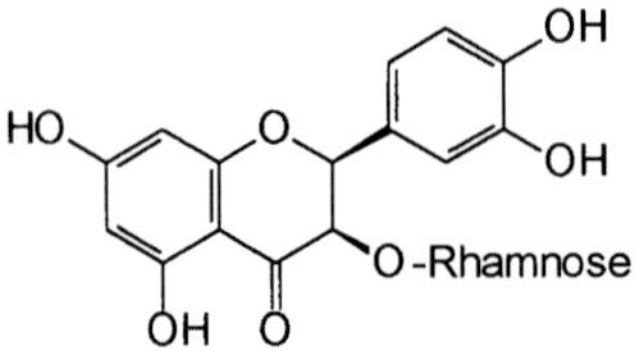

2-315 (2*S*,3*R*)

(1996) who corrected erroneous configurations for the latter two isomers that had been published earlier.] An extension of this story involves the fact that heating the drug increases its sweetness. This is interesting since it had been shown some years earlier by Tominaga (1960) that astilbin could be isomerized to a mixture of the other isomers. Other studies of sweet compounds in Nature have resulted in observations similar to those from *Engelhardtia*. Thus, Kinghorn (1987) described (2*R*,3*R*)-taxifolin 3-acetate [2-296 acetate] as the agent responsible for the sweet taste of young leaves of *Tessaria dodoneifolia* (Asteraceae). It was estimated that the compound was about 50 times sweeter than a 2% solution of sucrose. Chemical *O*-methylation of the C-4′ hydroxyl group increased the sweetness of the dihydroflavonol even more.

Flavanones

Naringenin, a common member of this class, and the first cyclized flavonoid to be formed in the normal flavonoid biosynthetic pathway, is shown as structure [2-316]. Two structural features characterize flavanones, the absence of the C2-C3 double bond, and the presence of a chiral center at C-2. The majority of naturally occurring flavanones have the C-2 phenyl group oriented down, or beneath the plane of the paper (the α-configuration). This requires that we prefix the name of the compound in question with the designation [2*S*], as opposed to [2*R*], in which case the phenyl group would be oriented above the plane of the paper (β-configuration).

In addition to providing a useful physical characteristic that uniquely identifies a given compound (along with such features as melting point, etc.), the determination of optical activity of a newly isolated flavanone establishes it as a true natural product and not an artifact of isolation or purification. As stated above, chalcones and flavanones with the same substitution pattern are isomeric. It is also characteristic of these compounds that they can be interconverted. Chemical conversion of a chalcone to the corresponding flavanone involves reaction of the *o*-hydroxyl group (the oxygen atom actually) on the A-ring of the chalcone with the bridge double bond. Since this reaction can occur from either side (top or bottom) of the double bond with equal likelihood, the product consists of equal amounts of the two epimers. The formation of a flavanone from a chalcone by natural processes, on the other hand, involves binding the chalcone to the surface of the isomerizing enzyme (chalcone isomerase, CHI) in such a way that attack of the hydroxyl group can occur from only one side of the chalcone double bond. This results in a single epimer being formed. A mixture of the two epimers would be optically neutral whereas the enzyme-produced single epimer would exhibit optical activity.

The example shown as structure [2-316], naringenin, represents a very common substitution pattern. Eriodictyol, compound [2-316] with a second B-ring hydroxyl, and its derivatives are also very common. Many other substitution patterns are known within the family of flavanones, ranging from the very simple 7-hydroxyflavanone [2-317] known from many legume species, to elaborate molecules such as 3′-hydroxy-5,7,4′-trimethoxy-8-*C*-methylflavanone [2-318] from *Adina cordifolia* (Rubiaceae) (Srivastava and Gupta, 1983). Extensive tabulations of occurrences of naringenin, eriodictyol, their glycosides, and various *O*-methyl derivatives can be found in *The Flavonoids* series (Bohm, 1975b, 1982, 1988, 1994).

The least common position for a hydroxyl group to occur in flavonoids is C-2. Such a location is possible, of course, only in flavanones and dihydroflavonols (and their isoflavonoid equivalents) where C-2 is not involved in a C2-C3 double bond. 2-Hydroxyflavanones have been put forward as possible intermediates in flavonoid biosynthesis, and 2-hydroxyisoflavanones have been directly implicated in isoflavone biosynthesis (See Chapter Six for further information). Only a few reports of these compounds as natural products have appeared in the literature, however. This may be due to their scarcity, or it may be due to the relative ease with which they undergo loss of water to form the corresponding flavones. The earliest appears to be 2,5,7-trihydroxyflavanone 7-*O*-glucoside [2-319] from *Malus* species (Williams, 1967). 2,5-Dihydroxy-7-methoxyflavanone [2-320] has been reported from *Populus nigra* (Chadenson *et al*., 1972) and from *Uvaria rufus* (Chantrapromma *et al*., 1989). Chopin and coworkers (1978) identified the isomeric pair 8-*C*-methyl-6-*C*-formyl- [2-321] and 6-*C*-methyl-8-*C*-formyl-2,5,7-trihydroxyflavanones [2-322] from *Unona lawii* (Annonaceae). 2-Hydroxynaringenin (2,5,7,4′-tetrahydroxyflavanone) [2-323] and its 7-*O*-glucoside [2-324] have been identified as constituents of *Berchemia*

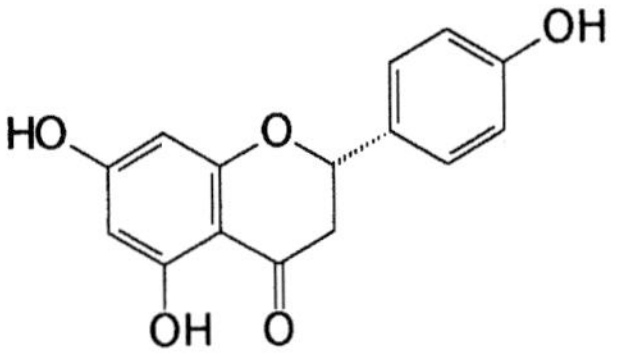

2-316

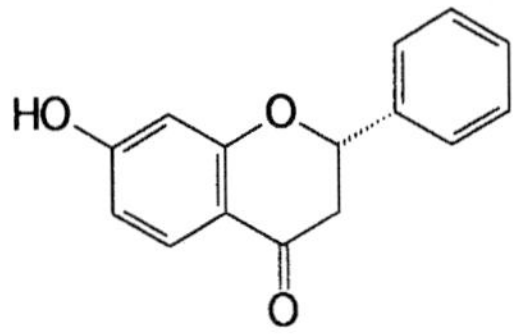

2-317

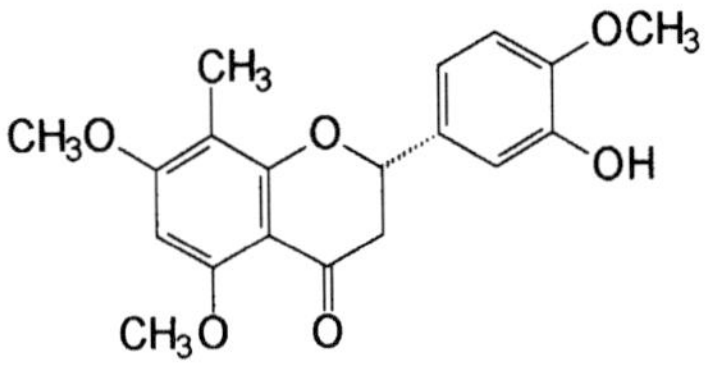

2-318

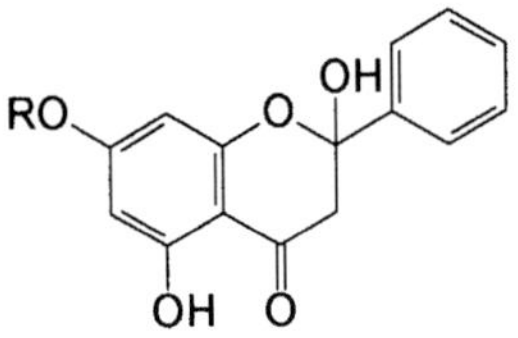

2-319 R = Glucose

2-320 R = CH_3

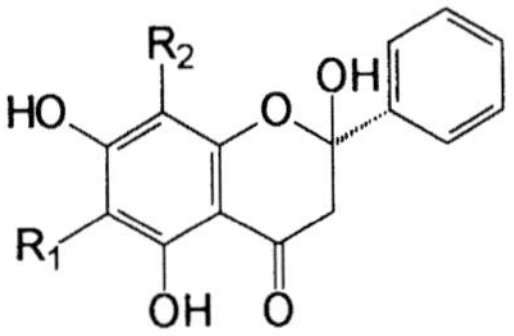

2-321 $R_1 = CH_3$, $R_2 = CHO$

2-322 $R_1 = CHO$, $R_2 = CH_3$

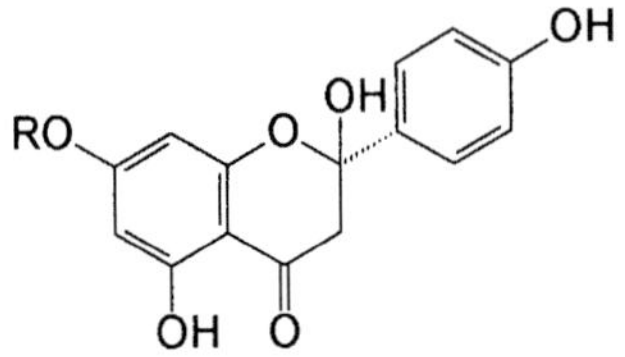

2-323 R = H

2-324 R = Glucose

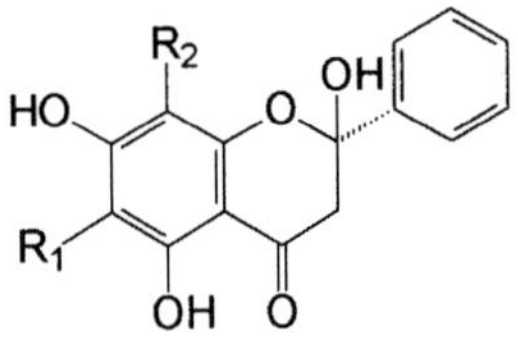

2-325 $R_1 = CH_3$, $R_2 = H$

2-326 $R_1 = H$, $R_2 = CH_3$

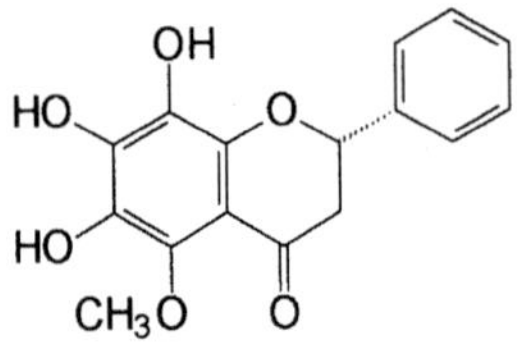

2-327

formosana (Rhamnaceae) (Lee *et al.*, 1995). The aglycone was optically active, thus arguing for its recognition as a natural product, although it could have arisen from hydrolysis of the glucoside. The most recent report of 2-hydroxyflavanones comes from the work of Fleischer and coworkers (1997) who identified the 6-*C*- [2-325] and 8-*C*-methyl [2-326] isomers of 2,5-dihydroxy-7-methoxyflavanone, along with the known 2,5-dihydroxy-7-methoxyflavanone and a series of known, structurally related flavones, from *Friesodielsia enghiana* (Annonaceae) (details appear in Case Study No. 11 in Chapter Four).

2-Hydroxyflavanones would be expected to be quite labile since they are hemiketals capable of losing a molecule of water to form the corresponding flavone. Fleischer and coworkers (1997) point out that the co-occurrence of 2,5-dihydroxyflavanones and the corresponding 5-hydroxyflavones in *Friesodielsia* is, thus, noteworthy. Hemiketals of this sort could also undergo ring opening to yield the β-ketodihydrochalcone.

Although it doesn't represent an unusual oxygenation pattern, 6,7,8-trihydroxy-5-methoxyflavanone [2-327], isolated from *Isodon oresbius* (Lamiaceae) by Hao *et al.* (1996) does represents an unusual methylation pattern.

Recent X-ray crystallographic studies have shown that the conformation of (R,S)-hesperitin (the racemic mixture) in the anhydrous crystal form is quite different from that of the hydrated compound (Fujii *et al.*, 1994). Hesperetin is 5,7,3′-trihydroxy-4′-methoxyflavanone. It had been shown that in the hydrated hesperetin molecule the 2-phenyl group was twisted out of the plane of the pyrone ring only slightly (dihedral angle, $\Phi = 0.6°$) (Shin *et al.*, 1987), whereas the present work showed that in the anhydrous crystal the phenyl group is twisted by a much larger amount ($\Phi = 53.1(3)°$). Additional examples of X-ray crystallographic studies of flavonoids can be found in Chapter Four (analytical methods), and in Chapter Eight, where studies of flavonoids as potential medicinal agents will be discussed.

Flavones and Flavonols

Although they are biosynthetically distinct, flavones and flavonols can be lumped conveniently under one heading owing to their close chemical relationship. Flavones have substitutions on the A- and B-rings but lack oxygenation at C-3. Chemically, flavonols are 3-hydroxyflavones. The widespread flavone apigenin [2-328] and flavonol kaempferol [2-329] exemplify these classes. These two classes were treated together in the two most recent volumes of the *Flavonoids* series (Wollenweber and Jay, 1988; Wollenweber, 1994). An extremely rich and diverse array of these compounds occurs in Nature, and a large number of others have been prepared synthetically. Iinuma and Mizuno (1989) calculated that there are 38,627 different possible combinations of hydroxyl and/or methoxyl substitutions! If one takes into consideration other substitution possibilities, e.g., *O*-alkylations other than methyl, *C*-alkylations, *C*-glycosylations, and combinations of these, the number of possibilities becomes astronomical. Even adding a single sugar substitution to the Iinuma-Mizuno number leads to an unimaginable list of possibilities. Not all combinations are known in Nature, of course, while some are extremely common. To give an idea of how many are known, we can turn to Wollenweber's (1994) Tables 7.1 and 7.2 which show, respectively, some 300 flavones and 380 flavonols having only hydroxyl/methoxyl substitution. Approximately 10% of those numbers represented additions to the lists since the 1988 compilation. He also listed 225 aglycones, flavones and flavonols combined, that bore some other substitution, e.g. *C*-methyl, *C*-prenyl, methylenedioxy. Of that number, 57 were newly reported in the literature since the 1988 list had been compiled. An added feature of the flavone/flavonol chapter in the 1988 volume was a set of tables (Table 7.8 to 7.14) in which frequencies of occurrence of substitutions and combinations of substitutions were presented. For example,

Table 7.9 informs us that in the case of flavones *per se*, 57% of reported compounds have hydrogen at C-8, 8% have hydroxy, 34% have methoxy, and 1% have *C*-methyl. In the flavonol series these percentages are 56, 11, 28, and 5, respectively. Table 7.14 tells us that in Asteraceae (sunflower family) 80% of reported flavones have an hydroxy group at C-5, 19% have a methoxy group at that position, and only 1% have no oxygenation at that position (5-deoxyflavonoid), whereas in Fabaceae (the legumes) 55% have 5-hydroxy, 25% have 5-methoxy, and 20% lack substitution at C-5. In both Solanaceae (potato family) and in ferns the C-5 position invariably has an hydroxyl group. These tables contain an enormous amount of information and deserve careful study.

It is clear that we can not begin to do justice to all of the workers who have reported interesting and novel compounds. What we can do is become familiar with the more or less commonly encountered members of these classes and then sample the less common ones to provide an idea of the structural variety that exists. We will start with flavones *per se*. By far the most common members of this class of flavonoids are 5,7,4′-trihydroxyflavone (apigenin) [2-328] and 5,7,3′,4′-tetrahydroxyflavone (luteolin) [2-330]. They are most often encountered bound to some sugar at position-7 but can be derivatized in a wide variety of forms including *O*-methylation, *C*-alkylation, and combinations thereof. The widespread occurrence parallels the occurrence of the flavonols having the same A-ring and B-ring oxygenations, i.e., kaempferol [2-329] and quercetin [2-331], respectively. The flavone with three B-ring hydroxyl groups, called tricetin [2-332], which is analogous to the flavonol myricetin [2-333], is known but is of somewhat limited occurrence. More common is its 3′,5′-dimethyl ether tricin [2-334]. The flavonol analogous to tricin is syringetin [2-335]. B-Ring deoxyflavones, e.g. chrysin [2-336] and tectochrysin [2-337], are known but are not common. The corresponding B-ring deoxyflavonols based on [2-338] are known but not commonly encountered. Of somewhat wider distribution are and flavonols with extra A-ring oxygenation, particularly at C-6. 5,6,7,4′-Tetrahydroxyflavone, scutellarein [2-339], occurs widely in Lamiaceae (mint family), while its *O*-methyl derivatives are common components in many plant groups, e.g. Asteraceae. Among the latter compounds are the 6-methyl ether (hispidulin) [2-340], the 6,7-dimethyl ether (cirsimaritin) [2-341], and the 6,4′-dimethyl ether (pectolinarigenin) [2-342]. The corresponding flavonols, based on 3,5,6,7,4′-pentahydroxy- (6-hydroxykaempferol, no common name) [2-343] and 3,5,6,7,3′,4′-hexahydroxyflavone (quercetagetin) [2-344] are commonly encountered compounds that occur most frequently as *O*-methylated derivatives. Again, Asteraceae is a good source of many of these.

More complex flavone derivatives are also listed in the two chapters referred to above. Among these are compounds that bear one or more methylenedioxy groups such as is seen in 7-methoxy-3′,4′-methylenedioxyflavone from stem bark of *Millettia hemsleyana* (Fabaceae) (Mahmoud and Waterman, 1985) and 7-methoxy-5,6,3′,4′-dimethylenedioxyflavone [2-345] from the wood of *Bauhinia splendens* (Fabaceae) (Laux *et al*., 1985). *C*-Methylated flavones have been reported from several plants such as 5,7-dihydroxy-8-methoxy-6-*C*-methylflavone [2-346] from *Platanus acerifolia* buds (Kaouadji and Ravanal, 1990), 5-hydroxy-7,4′-dimethoxy-6,8-di-*C*-methylflavone

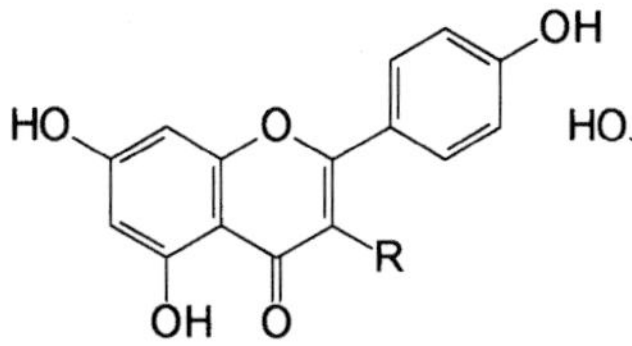

2-328 R = H
2-329 R = OH

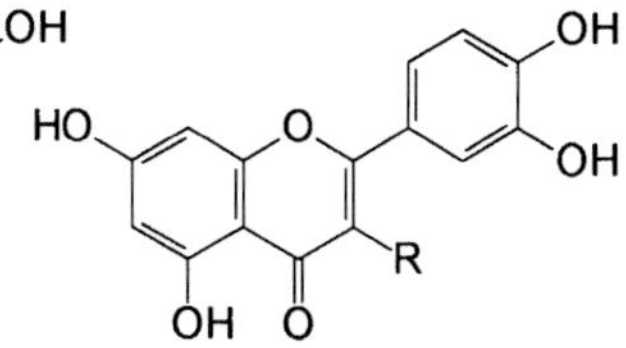

2-330 R = H
2-331 R = OH

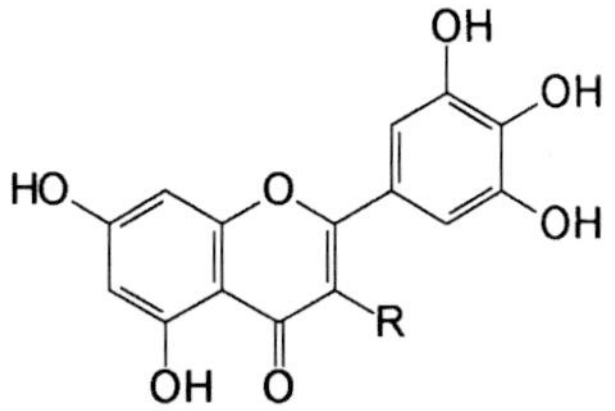

2-332 R = H
2-333 R = OH

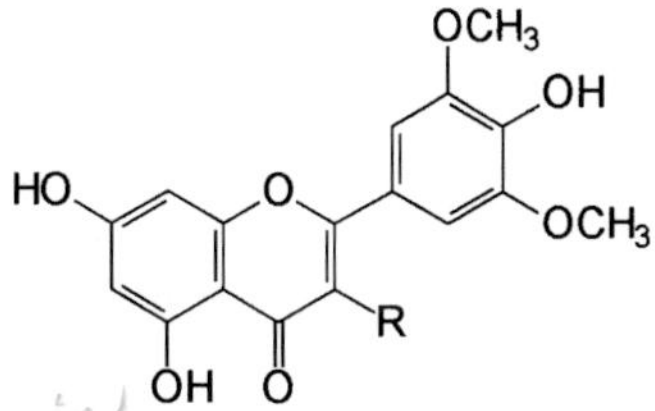

2-334 R = H
2-335 R = OH

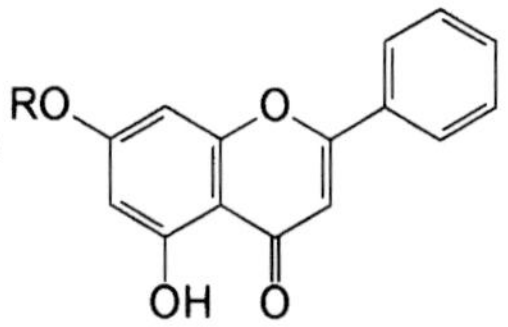

2-336 R = H
2-337 R = CH_3

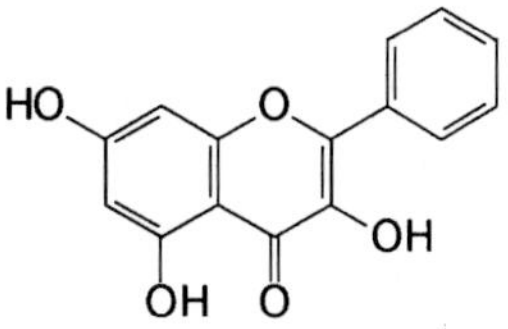

2-338

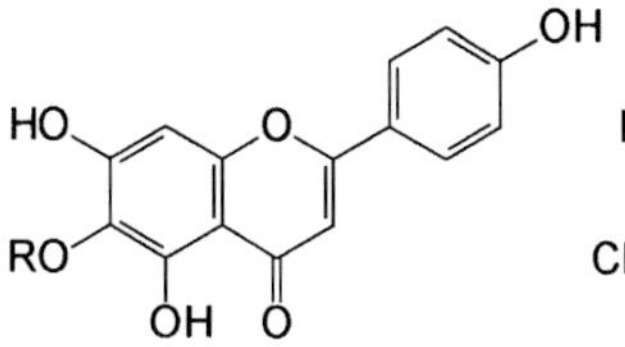

2-339 R = H
2-340 R = CH_3

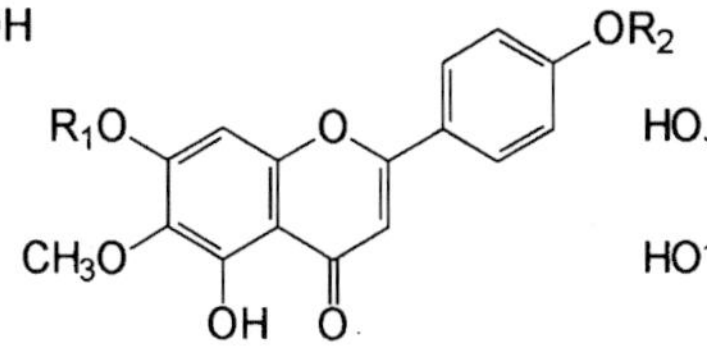

2-341 R_1 = CH_3, R_2 = H
2-342 R_1 = H, R_2 = CH_3

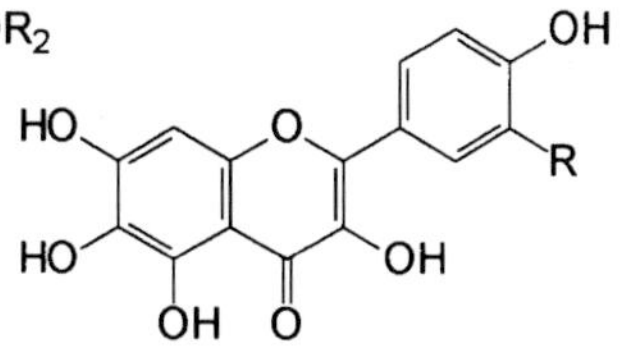

2-343 R = H
2-344 R = OH

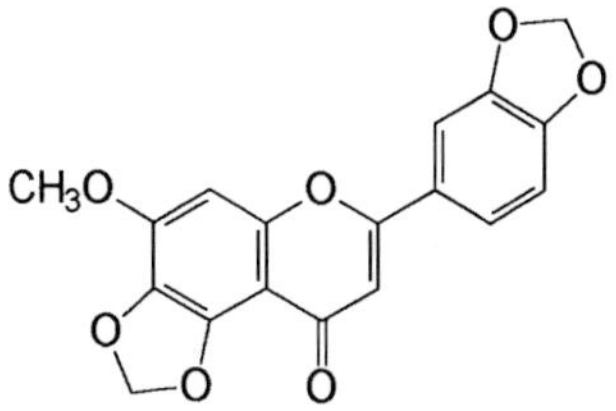

2-345

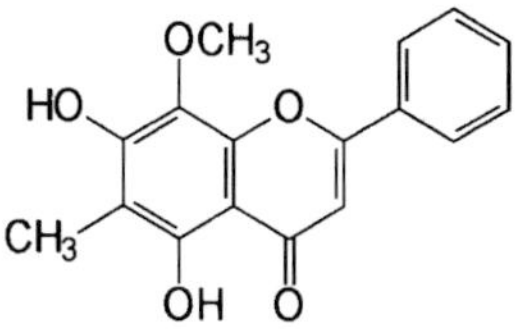

2-346

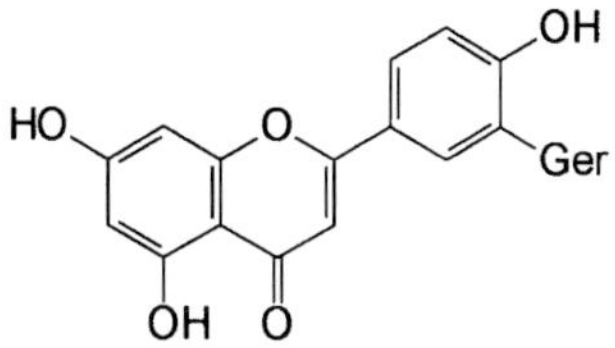

2-347

(eucalyptin) from the surface of *Eucalyptus* leaves (Wollenweber and Kohorst, 1981), and 5-hydroxy-7-methoxy-6,8-di-*C*-methylflavone from *Desmos cochinchinensis* (Annonaceae) (Wu *et al.*, 1994). Several flavones with larger *C*-alkyl groups, such as 5,7,4′-trihydroxy-3′-*C*-geranylflavone (kuwanone-S; 3′-geranylapigenin) [2-347] have also been identified.

C-Glycosylflavonoids

C-Glycosylflavonoids did not gain significant attention until the mid 1960s when their unique structure was established by NMR methods. Nor did they merit a chapter in the book edited by Geissman (1962), although a few references to *C*-glycosylflavonoids, particularly the isoflavonoid derivative puerarin, were made in that book. In his treatment of flavonoids in 1967, Harborne could report the existence of *C*-glycosyl derivatives of flavones, an isoflavone, a flavanone, a dihydroflavonol, and two flavonols. In the first volume of *The Flavonoids* series, Chopin and Bouillant (1975) presented an impressive list of naturally occurring *C*-glycosylflavonoids. The chemistry, occurrence, and biosynthesis of *C*-glycosyl compounds, including *C*-glycosylflavonoids, were reviewed by Franz and Grün (1983). With over 100 references, this treatment offers excellent entry to the original literature. One should also consult a review by Valent-Vetschera (1985).

That a new class of flavonoids existed became evident in the early to mid 1960s when compounds, which appeared to be flavones as judged by chromatographic and solubility properties, gave un-flavonelike behavior with acid. Normally, flavone glycosides are cleaved by heating with mineral acid for varying periods of time (depending upon which flavonoid oxygen is involved in the linkage). All flavonoid glycosides known at that time could be completely hydrolyzed into their component aglycone and sugar(s) by heating with 2N ethanolic-HCl (1:1) for four hours (Harborne 1967, p. 50). The anomalous flavone glycosides did not respond in the usual manner, but did often yield additional compounds that had chromatographic properties (R_f and color reactions) similar to the starting material. Analytical tools normally used for sugars, cleavage by periodic acid, for example, gave results that were difficult to analyze. The solution to the problem came from the application of nuclear magnetic resonance methods (NMR). Thus, Horowitz and Gentili (1964) showed that vitexin, obtained from the wood of *Vitex lucens* (Verbenaceae), and isovitexin were, respectively, the 8-*C*-β-D-glucopyranosyl [2-348] and 6-*C*-β-D-glucopyranosyl [2-349] derivatives of apigenin. Koeppen (1964) determined that orientin [2-350], obtained from *Polygonum orientale* (Polygonaceae) by Hörhammer and coworkers (1958), and isoörientin [2-351] were the corresponding derivatives of luteolin. In the following year, Seikel and Mabry (1965) demonstrated, again using NMR, that an unidentified compound, also from *Vitex lucens*, and called lucenin-1, had sugar substituents at both the C-6 and C-8 positions of luteolin, although it was not possible to establish what the sugars were from their data. Lucenin-1 was shown to be the 6-*C*-β-D-xylopyranosyl-8-*C*-β-D-glucopyranosyl [2-352] derivative of luteolin. Vicenin-3 [2-353] has the sugar substitutions reversed. The greatest difficulty in determining the structures of *C*-glycosylflavonoids lies in establishing the nature of the sugar residues. It is also necessary to establish the stereochemistry of their linkages to the flavonoid nucleus. Schaftoside itself was isolated from *Silene schafta* (Caryophyllaceae) and shown to be 6-*C*-β-D-glucopyranosyl-8-*C*-α-L-arabinopyranosylapigenin [2-354] (Chopin *et al*., 1974). Its isomerization product, isoschaftoside [2-355], has the opposite arrangement of its sugars but their stereochemistries are unchanged. Isoshaftoside was subsequently isolated from *Flourensia cernua* (Asteraceae) (Dillon *et al*., 1976). A third isomer, neoschaftoside, was eventually

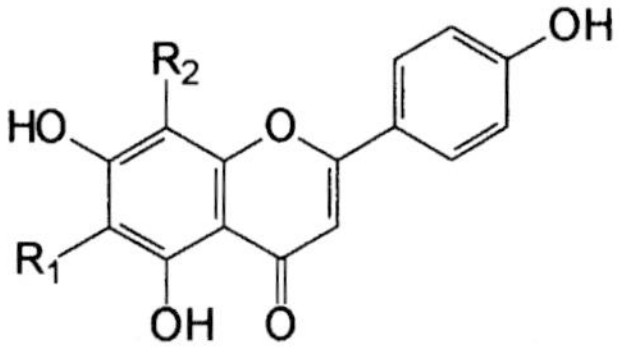

2-348 R_1 = H, R_2 = Glucose

2-349 R_1 = Glucose, R_2 = H

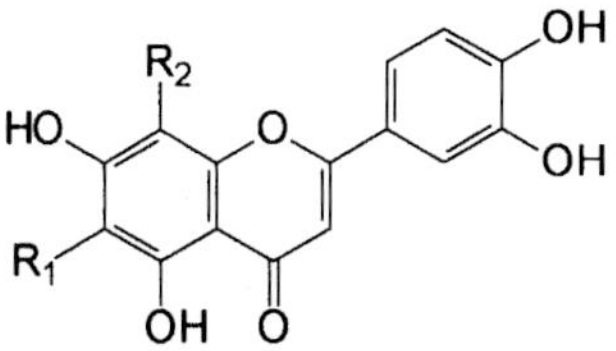

2-350 R_1 = H, R_2 = Glucose

2-351 R_1 = Glucose, R_2 = H

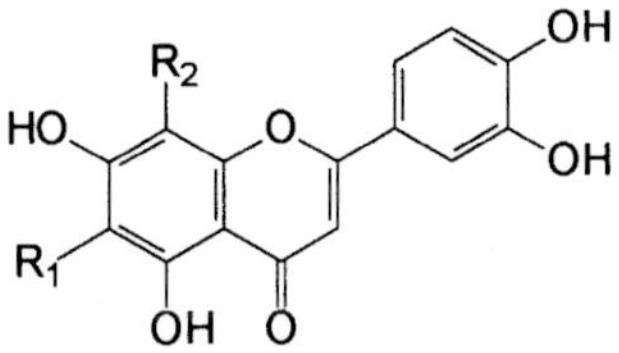

2-352 R_1 = Xylose, R_2 = Glucose

2-353 R_1 = Glucose, R_2 = Xylose

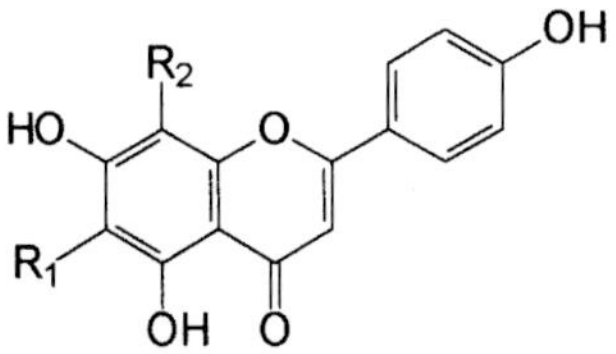

2-354 R_1 = Glucose, R_2 = Arabinose

2-355 R_1 = Arabinose, R_2 = Glucose

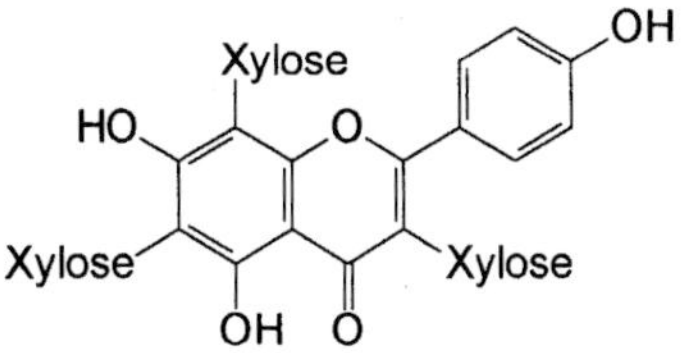

2-356

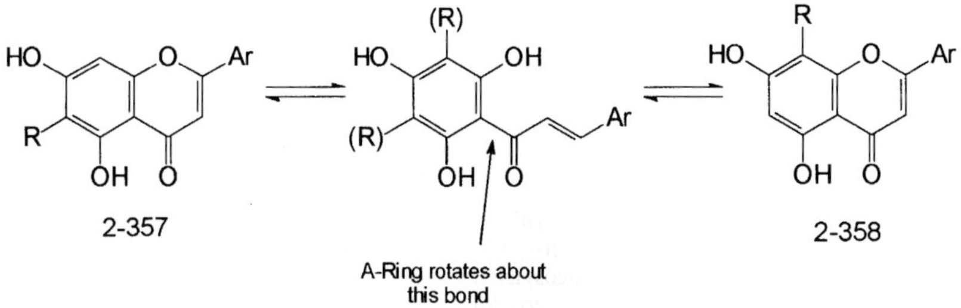

2-357

2-358

A-Ring rotates about this bond

shown to be isomeric to schaftoside (6-glucosyl-8-arabinosylapigenin) except that the arabinose is in the β-L configuration. The fourth isomer, neoisoschaftoside, isolated from the moss *Mnium undulatum*, was determined to be 6-*C*-α-L-arabinopyranosyl-8-*C*-β-D-glucopyranosylapigenin (Österdahl, 1979). The use of NMR and mass spectral techniques have proved to be invaluable for dealing with these compounds. Table 2.9 lists some of the more commonly encountered *C*-glycosylflavones, while Table 2.10 lists some *C*-glycosylflavonoids involving flavonoids other than flavones.

Table 2.9 Some common *C*-glycosylflavones

Common name	Aglycone	Substituents at positions[a]				
		6	7	8	3′	4′
Vitexin	Apigenin	Glc	OH	H	H	OH
Isovitexin	Apigenin	H	OH	Glc	H	OH
Orientin	Luteolin	Glc	OH	H	OH	OH
Isoörientin	Luteolin	H	OH	Glc	OH	OH
Swertisin	Genkwanin	Glc	OCH_3	H	H	OH
Isoswertisin	Genkwanin	H	OCH_3	Glc	H	OH
Cytisoside	Acacetin	H	OH	Glc	H	OCH_3
Isocytisoside	Acacetin	Glc	OH	H	H	OCH_3
Swertiajaponin	7-Methylluteolin	Glc	OCH_3	H	OH	OH
Isoswertiajaponin	7-Methylluteolin	H	OCH_3	Glc	OH	OH
Scoparin	Chrysoeriol	H	OH	Glc	OCH_3	OH
Isoscoparin	Chrysoeriol	Glc	OH	OH	OCH_3	OH
Vicenin-1	Apigenin	Xyl	OH	Glc	H	OH
Vicenin-2	Apigenin	Glc	OH	Glc	H	OH
Vicenin-3	Apigenin	Glc	OH	Xyl	H	OH
Violanthin	Apigenin	Glc	OH	Rhm	H	OH
Isoviolanthin	Apigenin	Rhm	OH	Glc	H	OH
Schaftoside (α-L)[b]	Apigenin	Glc	OH	Ara	H	OH
Isoschaftoside (α-L)	Apigenin	Ara	OH	Glc	H	OH
Neoschaftoside (β-L)	Apigenin	Glc	OH	Ara	H	OH
Neoisoschaftoside (β-L)	Apigenin	Ara	OH	Glc	H	OH
Lucenin-1	Luteolin	Xyl	OH	Glc	OH	OH
Lucenin-2	Luteolin	Glc	OH	Glc	OH	OH
Lucenin-3	Luteolin	Glc	OH	Xyl	OH	OH

[a]Ara = arabinose, Glc = glucose, Rhm = rhamnose, Xyl = xylose
[b]α-L and β-L refer to the configuration of the arabinose moiety (all glucose moieties have the β-D configuration)

Table 2.10 Examples of *C*-glucosylflavonoids not involving flavones

Trivial name	Chemical name[a]	Common name
Keyakinin	3,5,4′-Trihydroxy-6-*C*-glucosyl-7-methoxyflavone	6-Glucosylrhamnocitrin
	3,5,7,3′,4′-Pentahydroxy-6-*C*-glucosylflavone	6-Glucosylquercetin
Keyakinin B	3,5,3′,4′-Tetrahydroxy-6-*C*-glucosyl-7-methoxyflavone	6-Glucosylrhamnetin
Keyakinol	3,5,7,3′,4′-Pentahydroxy-6-*C*-glucosylnaringenin	6-Glucosyldihydroquercetin
Hemiphloin	5,7,4′-Trihydroxy-6-*C*-glucosylflavanone	6-Glucosylnaringenin
Puerarin	7,4′-Dihydroxy-8-*C*-glucosylisoflavone	8-Glucosyldaidzein
	2′,4′,4-Trihydroxy-3′-*C*-glucosylchalcone	3′-Glucosylisoliquiritigenin
Nothofagin	2′,4′,6′,4-Tetrahydroxy-*C*-glucosyldihydrochalcone	Glucosylphloridzin[b]
Aspalathin	2′,4′,6′,3,4-Pentahydroxy-3′-*C*-glucosyldihydrochalcone	
Coatline A	2′,4′,4,a-Tetrahydroxy-3′-*C*-glucosyldihydrochalcone	

[a]The full expression β-D-glucopyranosyl was omitted for brevity
[b]Position of *C*-linkage undetermined

One of the more unusual members of this class of compounds came from the work of Imperato (1993) who reported the novel 3,6,8-tri-*C*-xylosylapigenin [2-356] from the fern *Asplenium viviparum*. He had reported 6,8-di-*C*-rhamnosylluteolin from the same species in an earlier paper (Imperato, 1992).

Other than possessing one or more *C*-linked sugars, *C*-glycosylflavonoids are known that have one or more of the many other structural features common to flavonoids in general. These include additional sugar residues attached through any, or several, of the phenolic hydroxyl groups as well as sugars attached to the *C*-linked sugar. Acylation involving aliphatic or aromatic acids is known and sulfate derivatives have been reported from a few species. *O*-Methylation is a common feature as well. Jay (1994) compiled a very thorough list of these kinds of derivatives. He also illustrates the sugar residues found in naturally occurring *C*-glycosylflavonoids. In addition to the commonly occurring sugars glucose, galactose, xylose, arabinose, and rhamnose, he shows structures for the following residues: 6-deoxy-*xylo*-hexos-4-ulosyl, *β*-L-fucopyranosyl, *α*-D-mannopyranosyl, *β*-D-oliopyranosyl, *β*-L-boivinopyranosyl, *β*-D-chinovopyranosyl, and D-apiofuranosyl.

As mentioned in the introduction to this section, acid treatment of *C*-glycosylflavonoids often produces a small amount of product with properties similar to starting compound. This is the result of the Wessely-Moser rearrangement wherein a 5-hydroxy-*C*-glycosylflavone [2-357] undergoes acid catalyzed opening of the pyran ring, rotation of the A-ring, and recyclization to yield [2-358]. Since cyclization can occur with either of the hydroxyl groups *ortho* to the carbonyl, a mixture of the two isomers results. This behavior of an unknown flavone glycoside with acid can be taken as evidence for the existence of a *C*-linked sugar. This reaction is shown diagramatically below.

Isoflavonoids

The title of this section is intentionally general in to accommodate examples of most of the groups of flavonoids that have the B-ring attached at C-3 of a phenylchromane skeleton. Ollis (in Geissman, 1962) pointed out that isoflavonoids, in the broad sense, constitute not just one of the larger classes of flavonoids but of natural products in general. This situation appears not to have changed as the contributors to all four volumes of *The Flavonoids* series continue to point out. It's possible to get an idea of how significant a part these compounds play in natural product chemistry by looking at the number of times the subject has been reviewed prior to the appearance of the Geissman book: Seshadri (1951), Warburton (1954), Venkataraman (1959), and Dean (1963). *The Flavonoids* series includes reviews by Wong (1975) and Dewick (1982, 1988, 1994). More recently, Donnelly and Boland (1995) reviewed the literature published between 1991 and 1993.

In each treatment in *The Flavonoids* series, the respective authors have stressed that Fabaceae (Leguminosae in most of the earlier treatments) are the principal sources of isoflavonoids, primarily the Papilionoideae, although in each volume the number of other plant families from which isoflavonoids have been isolated has grown. Isoflavones also occur abundantly in Iridaceae, but members of some 20 other plant families have yielded at least one isoflavone derivative. The list is broad and includes monocots, dicots, gymnosperms, and even a moss (Dewick, 1994, Table 5.1). Occasionally, an isoflavonoid derivative will appear in a wholly unexpected place as for example the finding of "rotenone," well known as a component of many legumes (and marketed as "Derris dust"), in a member of Asteraceae, *Balduina angustifolia* (Lee *et al.*, 1972).

Dewick has compiled lists of HPLC conditions that have been used by workers for separation and purification of a variety of isoflavonoids (Dewick, 1988, Table 5.1; 1994, Table 5.2). Owing to the location of the B-ring at C-3 in isoflavonoids, the proton(s) at C-2 are easily distinguished in ^{1}H NMR spectra and the ^{13}C NMR signal for the carbon at position-2 is also readily discernible. The UV absorption spectra of isoflavones are also characteristic. Examples of these features will be discussed in more detail in Chapter Four. General techniques for studying isoflavonoids have been reviewed by Williams and Harborne (1989).

Isoflavones

Isoflavones exhibit the same level of oxidation in the heterocyclic ring as do flavones and exhibit a high level of complexity with regard to substitution on the other two rings. Oxygenation patterns range from as few as two to situations where nearly all positions are substituted, often with combinations of hydroxyl, methoxy, and prenyl groups. For an appreciation of the full richness of isoflavonoid structural variation it is convenient to consult the chapters by Dewick (1988, 1994) or the review by Donnelly and Boland (1995). In this section we will look at the structures of some of the most common isoflavonoids and proceed to examples from the recent literature that are more highly substituted. Isoflavone itself does not appear to have been found as a naturally occurring compound nor have isoflavones with only a single oxygenation. 7,4′-Dihydroxyisoflavone (daidzein) [2-359] and its 7-methyl ether (formononetin) [2-360] are very commonly encountered, however. Similarly, 5,7,4′-trihydroxyisoflavone (genistein) [2-361], its 4′-methyl ether (biochanin A) [2-362], and its 7-methyl ether (prunetin) [2-363] are also very common compounds. Prunetin derives its common name from the genus *Prunus* (Rosaceae) from which it was obtained many years ago. Prunetin, and most of the other isoflavones listed, are almost routinely reported from members of Fabaceae. The isoflavone equivalent to luteolin, 5,7,3′,4′-tetrahydroxyisoflavone [2-364], known as "orobol," occurs widely. Some of its methyl ether and *C*-prenyl derivatives have been reported from members of Asteraceae recently (McCormick *et al.*, 1986). Isoflavones with extra-oxygenation in the A-ring are also known; 5,7,4′-trihydroxy-6-methoxyisoflavone (tectorigenin) [2-365] has been obtained from several legumes. A few examples of rather unusual structures show the diversity of isoflavone biosyntheses. The compound, known as "irisone-B" [2-366], has been described from *Iris missouriensis*, where it occurs with its 2′-methyl ether, "irisone-A" (Wong *et al.*, 1987), and from *Beta vulgaris* (Chenopodiaceae) (Takahashi *et al.*, 1987). Chenopodiaceae is also represented by the discovery that *Salsola somalensis* accumulates the peculiarly substituted 5,3′-dihydroxy-6,7,8,2′-tetramethoxyisoflavone [2-367] (Woldu and Abegaz, 1990). An isoflavone from *Sopubia delphiniifolia* (Scrophulariaceae) exhibits *C*-prenylation and exists as a glycoside [2-368] (Saxena and Bhadoria, 1990). Isoflavones lacking B-ring substitution are also known. 5-Hydroxy-7-methoxyisoflavone [2-369], which is equivalent to the flavone tectochrysin, has been isolated from seeds of *Derris robusta*,

(Chibber and Sharma, 1979). The doubly *O*-methylated derivative was found in immature fruits of *Aspergillus*-infected *Arachis hypogaea* (Fabaceae), the peanut (Turner *et al.*, 1975).

Turning to examples from the recent literature, we begin to see some of the more highly substituted members of the group. Two new isoflavones came from an

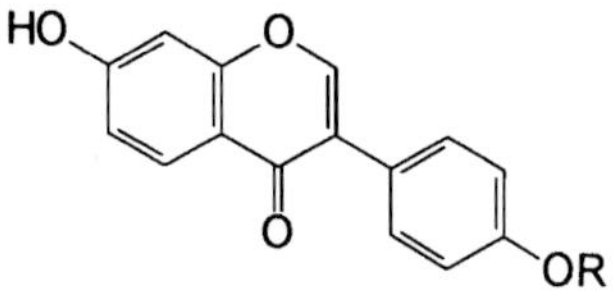

2-359 R = H
2-360 R = CH_3

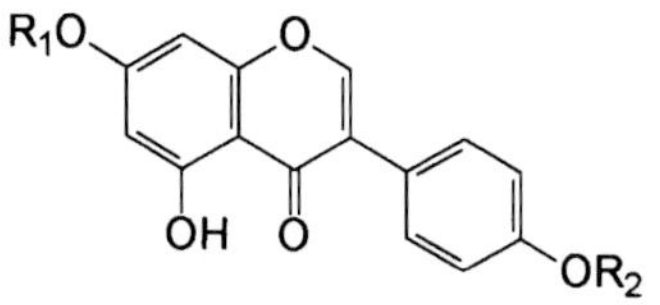

2-361 $R_1 = R_2 = H$
2-362 $R_1 = H$, $R_2 = CH_3$
2-363 $R_1 = CH_3$, $R_2 = H$

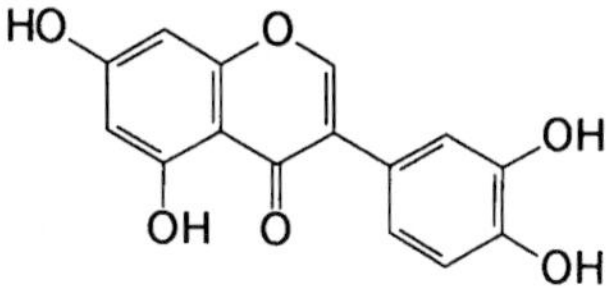

2-364

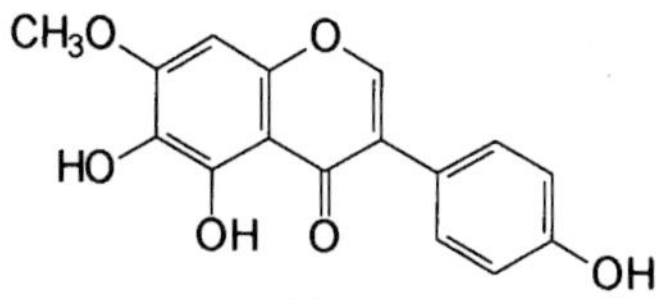

2-365

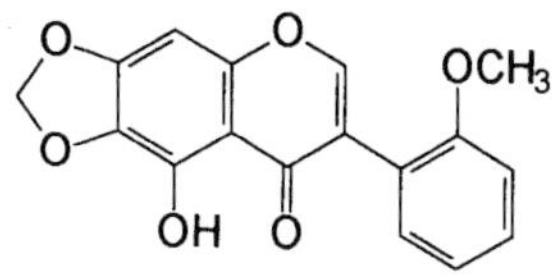

2-366

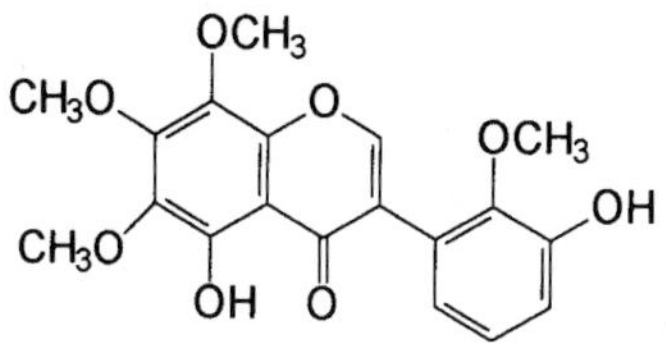

2-367

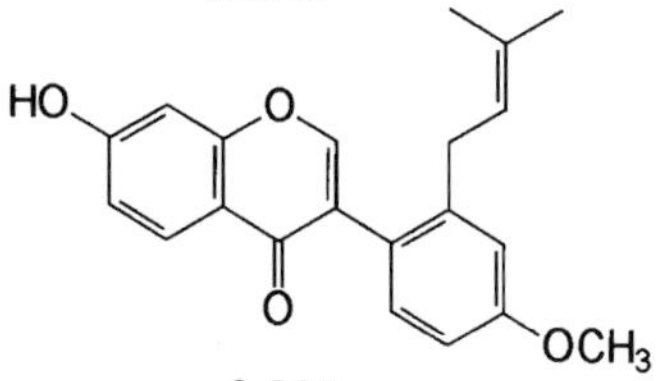

2-368

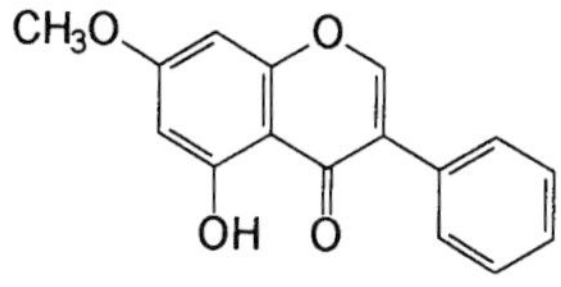

2-369

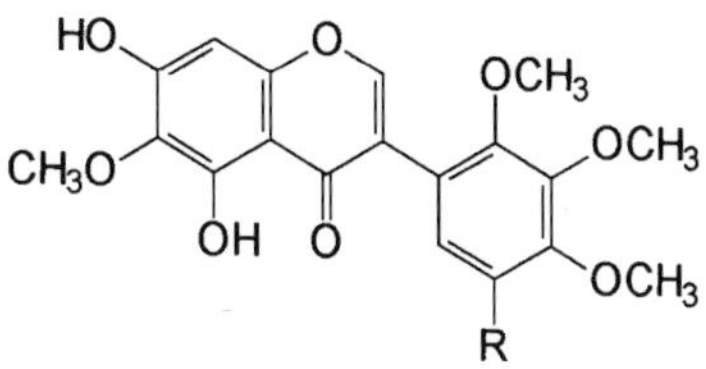

2-370 R = H
2-371 R = OCH_3

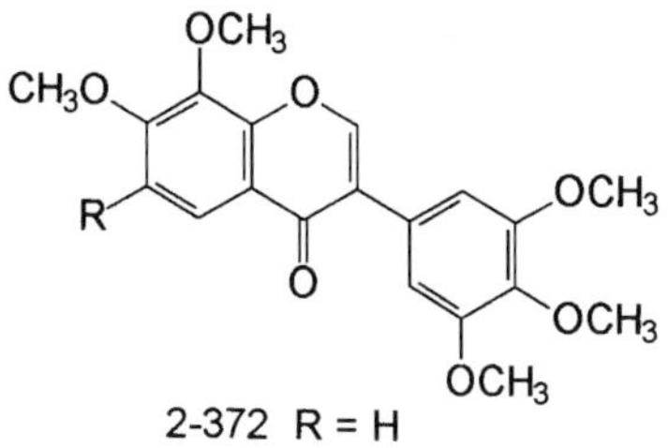

2-372 R = H
2-373 R = OCH_3

examination of *Iris japonica* by Minami *et al.* (1996), 5,7-dihydroxy-6,2′,3′,4′-tetramethoxy- [2-370] and 5,7-dihydroxy-6,2′,3′,4′,5′-pentamethoxyisoflavones [2-371]. Two highly *O*-methylated compounds were obtained from the legume *Petalostemon purpurescens* by Chaudhuri and coworkers (1996): 7,8,3′,4′,5′-pentamethoxyisoflavone [2-372] and 6,7,8,3′,4′,5′-hexamethoxyisoflavone [2-373].

Roots of *Millettia auriculata* (Fabaceae) afforded a number of prenylated compounds among which were the novel 7,8-methylenedioxy-4′-*O*-prenylisoflavone [2-374] and the rotenoid "sumatrol," whose structure can be found below [2-415] (Venkata Rao *et al.*, 1992). Further examples of structural elaboration involving prenyl groups are seen in studies of two legume genera, *Derris* and *Erythrina*. New compounds reported from *Derris scandens* (Narayana Rao *et al.*, 1994) include the doubly *C*-prenylated compound [2-375], which shows two of the modifications often seen in, but not restricted to, isoflavonoids. A common modification involves cyclization of a prenyl function with a neighboring phenolic group to give the chromene ring system. In [2-375] the 6-*C*-prenyl function has interacted with the 7-hydroxyl group to form a linear dimethylchromeno isoflavonoid. The second prenyl modification in [2-375] involves the methoxylated side-chain, which is a feature much less often encountered. A variation on oxygenated *C*-prenyl groups can be seen in the closely related epoxyisoflavone, a partial structure of which is shown as [2-376]. Wandji and coworkers (1994) obtained this compound from the Cameroonian medicinal plant *Erythrina senegalensis*.

It is well known that the production of flavonoids by plants can be significantly affected by microbial infection or, in some cases, treatment with an inorganic salt (more in Chapter Seven). This phenomenon was observed by Hakamatsuka and colleagues (1991) in a study of *Pueraria lobata* (Fabaceae) treated with cupric chloride ($CuCl_2$) solution. Treatment resulted in a 5–10 fold increase in the concentration of three compounds known to be biosynthesized by this species, 7,4′-dihydroxyisoflavone [2-359], 5,7,4′-trihydroxyisoflavone [2-361] and coumestrol [2-377], and the *de novo* formation of three additional compounds. The compounds induced by the cupric chloride treatment were identified as the known pterocarpan derivative tuberosin [2-378] and 5,7,4′-trihydroxy-8-*C*-prenylisoflavone (8-prenylgenistein) [2-379], and the novel 7,4′-dihydroxy-8-*C*-prenylisoflavone (8-*C*-prenyldaidzein) [2-380].

An interesting stereochemical situation exists when isoflavones have bulky groups at both C-2′ and C-6′. Tahara and colleagues (1993) discussed the possibility that atropisomers (rotational isomers) might exist for compounds such as 2′,6′-di-*C*-prenyl flavonoids owing to restricted rotation about the C_3-$C_{1'}$ axis. Erythbigenin, from *Piscidia erythrina*, has such an arrangement, as shown in structures [2-381] and [2-382]. These isomers probably exist but they are impossible to detect using ^1H NMR because they are enantiomers. Additional studies of *P. erythrina* bark, however, disclosed the presence of four additional compounds that offered the opportunity to resolve the atropisomer question (Tahara *et al.*, 1993). For the following discussion it is more convenient to think of these compounds in terms of two pairs and to focus our attention on one of them; whatever was observed for one pair applies equally to the other. The pair of immediate interest, then, are erythbigenols A and B in which one of the prenyl functions has undergone cyclization with the 3′-hydroxyl group to give six-membered pyrano-derivatives. These are illustrated

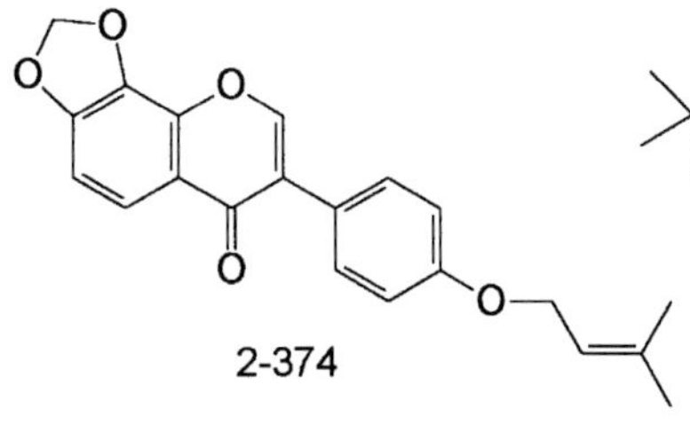

2-374 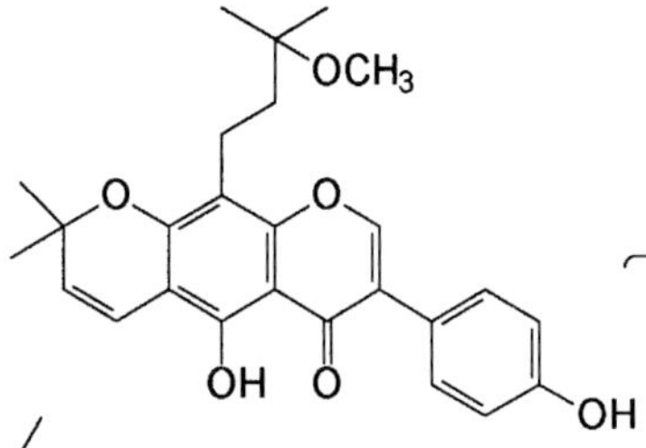 2-375 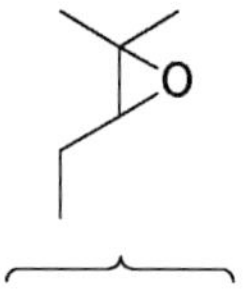 2-376

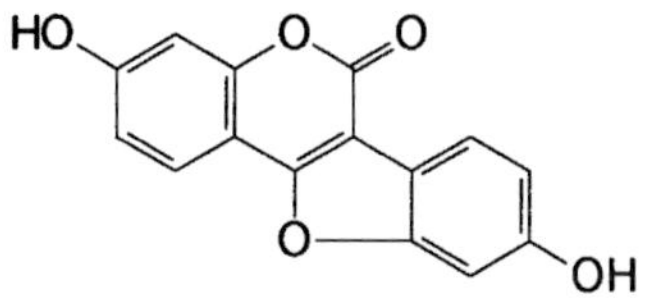

2-377 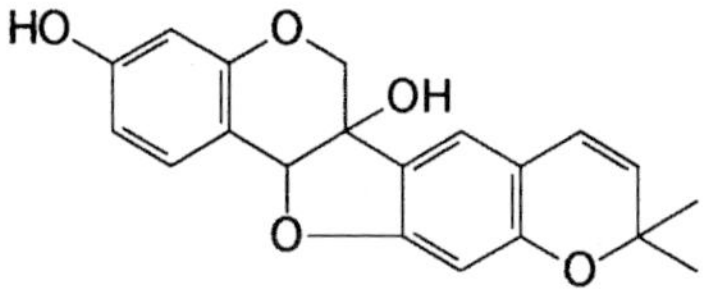 2-378

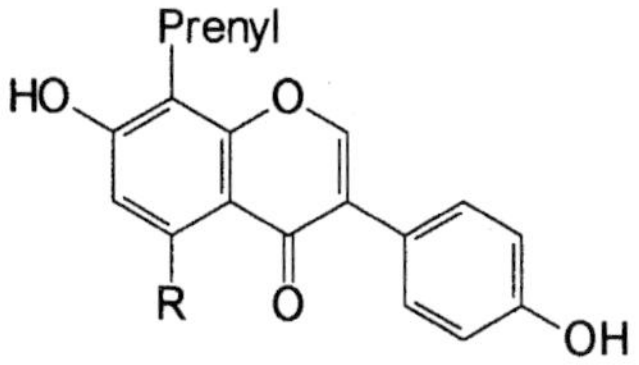

2-379 R = OH

2-380 R = H

here as [2-383, front] and [2-384, back] the terms "front" and "back" being my way of locating the furano ring system as lying either on the front side of the molecule (toward the reader), or on the back side of the molecule (away from the reader). These atropisomers are identified as erythbigenol-A and erythbigenol-B, respectively. Because each of these atropisomers has a chiral center each should be separable into its component diastereoisomers. This was accomplished using the chiral reagent 2-methoxy-2-trifluoromethyl-2-phenylacetic acid (MPTA). After conversion of the erythbigenols to their trimethyl ethers, they were esterified with the acid chloride of either [*S*]-(−)-MPTA or [*R*]-(+)-MPTA. The esters arising from erythbigenol-A were resolved chromatographically (TLC on silica gel). Tahara and colleagues referred to these esters simply as "slow" and "fast." Thus, all four possible erythbigenol isomers were obtained. Risking criticism for oversimplification, they can be listed as: [2-383, front-fast], [2-383, front-slow], [2-384, back-fast], and [2-384, back-slow]. The erythbigenone system, which consists of the pyrano-ring analogues, would also be expected to comprise four isomers. The detailed structural analysis of the erythbigenones and erythbigenols using one and two-dimensional ^{1}H and ^{13}C NMR presented in this paper is worthy of detailed study but is beyond the scope of the present discussion.

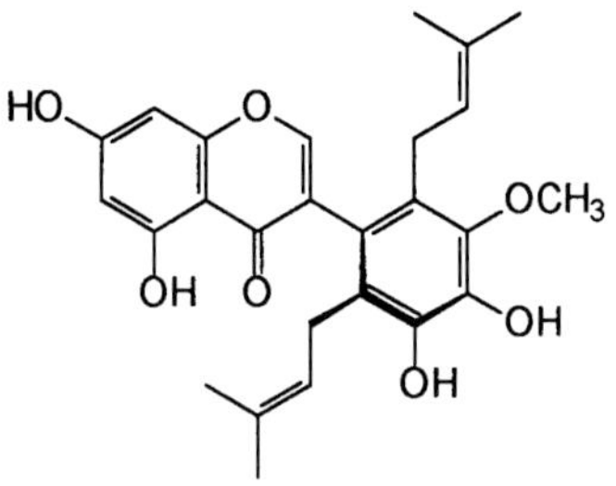

2-381

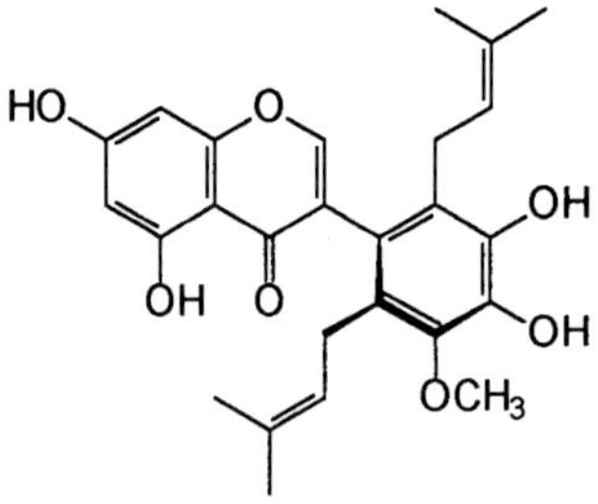

2-382

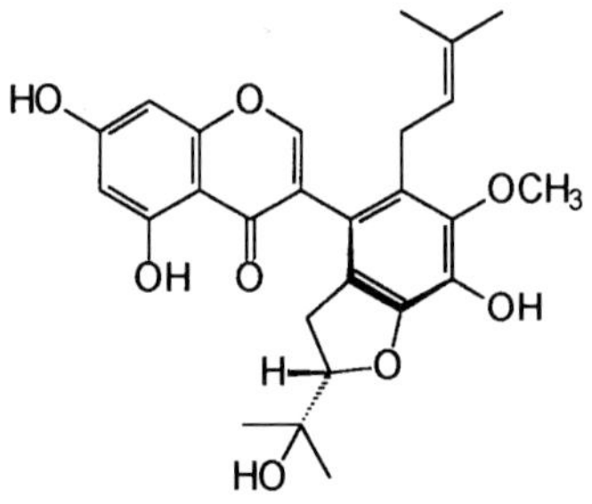

2-383

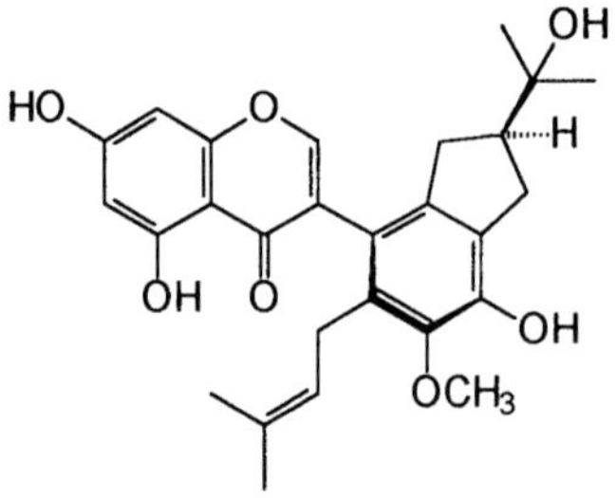

2-384

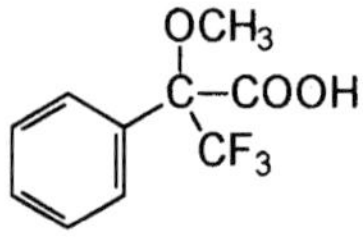

2-385

Isoflavanones

Isoflavanones bear the same relationship to isoflavones as flavanones do to flavones. And, as in the case of flavanones, isoflavanones have a chiral center (C-3 in isoflavanones). The number of known isoflavanones is smaller than isoflavones but their structural complexity is in no way reduced as we shall see with some examples from the recent literature. Once again, legumes seem to be the major source of isoflavanones. Within Fabaceae, several genera have attracted attention owing to their widespread use in indigenous medicine. For example, Nkengfack and collaborators have described numerous compounds from species of *Erythrina* from West Africa. Preparations of roots and root bark of *Erythrina sigmoidea* yielded two new isoflavanones, compound [2-386], identified as 7′-hydroxy-2′-methoxy-2″,2″-dimethylpyrano[5″,6″ : 5′,4′]isoflavanone (Nkengfack *et al.*, 1994a), and isoflavanone [2-387], whose structure was determined to be 7,4′-dihydroxy-2′,5′-dimethoxy-6-*C*-prenylisoflavanone (Nkengfack *et al.*, 1994b). *Erythrina eriotricha*, a species endemic to Cameroon, also yielded a new isoflavanone, identified as 7,2′,4′-trihydroxy-6,8-di-*C*-prenylisoflavanone [2-388], as well as five known pterocarpans (see below)

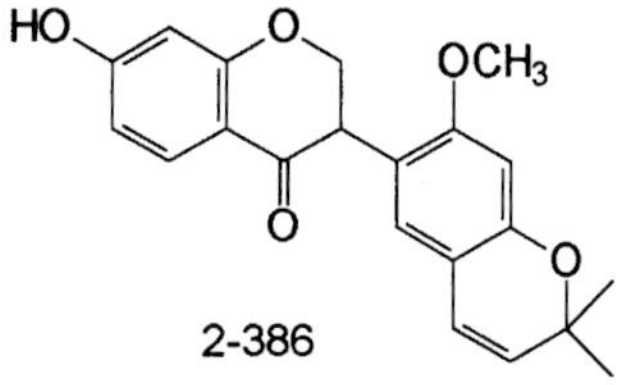

2-386

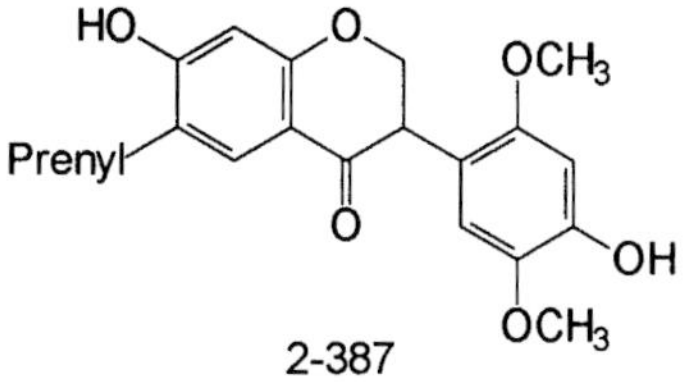

2-387

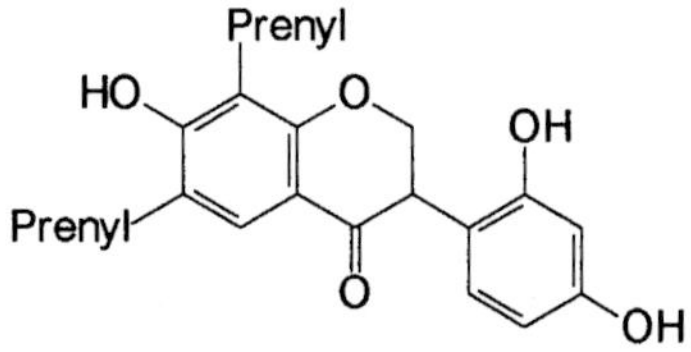

2-388

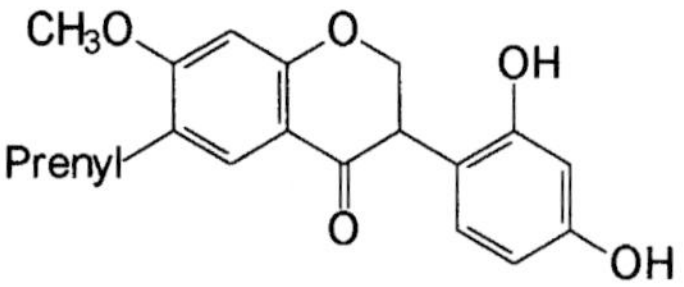

2-389

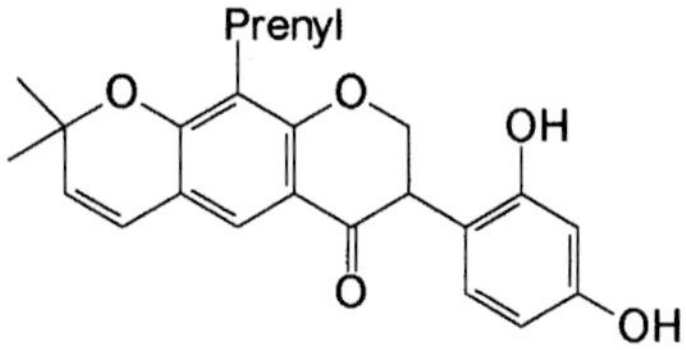

2-390

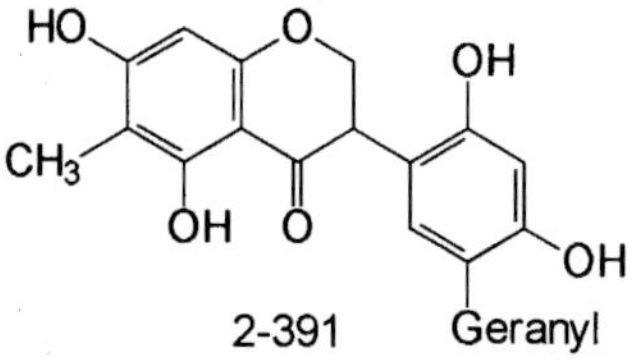

2-391

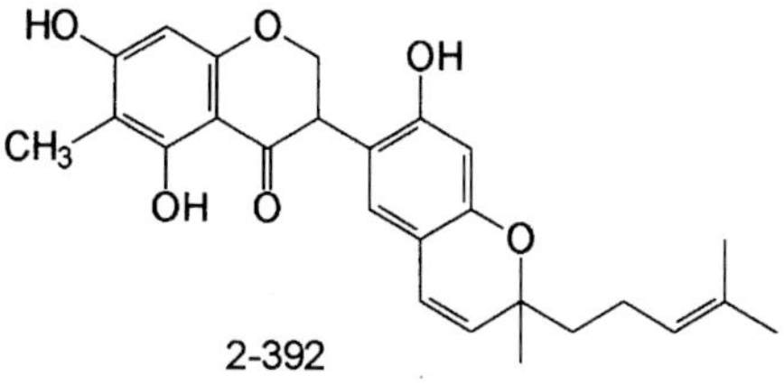

2-392

(Nkengfack *et al.*, 1995). Stem bark of *E. vellutina*, a South American species that also finds use in traditional medicine, yielded the novel 2′,4′-dihydroxy-7-methoxy-6-*C*-prenylisoflavanone [2-389] (Da-Cunha *et al.*, 1996). A structurally related compound, structure [2-390], was obtained from the Asian legume *Ormosia monosperma*, where it occurred with a number of isoflavones. It was shown to have moderate activity against selected oral bacteria (Iinuma *et al.*, 1994a). A study of the South American legume *Desmodium canum* afforded several compounds that possess a ten-carbon side chain as opposed to the five-carbon side chains (the prenyl groups) seen so far. In addition, a *C*-methyl group on the A-ring represents a somewhat unusual, but by no means unique, situation as well. The functional group of interest in these compounds is the *C*-geranyl (3,7-dimethylocta-2,6-dienyl) group. It exists as such in [2-391], 5,7,2′,4′-tetrahydroxy-6-*C*-methyl-5′-*C*-geranylisoflavanone, but has undergone cyclization to yield [2-392]. The cyclization is analogous to cyclization of the prenyl group except that the substituents on the final product are a methyl group and a 4-methylpent-3-enyl group rather than two methyl groups. All of the isoflavanones isolated in this study were optically active and were shown to have the 3*R*-configuration. Compound [2-392] had significant inhibitory activity toward a number of pathogenic bacteria in the range of 1–30 $\mu g\,ml^{-1}$. It was also active against *Candida albicans* at a concentration of 50–100 $\mu g\,ml^{-1}$. It was inactive against both *E. coli* and *Neurospora crassa*.

Isoflavans and Isoflavenes

One of the simplest members of these subclasses, 7,4′-dihydroxyisoflavan (equol) [2-393], whose configuration was established to be (3*S*) by Kurosawa and coworkers (1968), is probably the only flavonoid that was first obtained as an animal metabolite (Wong, 1975). Dewick (1988) expands on this with information to the effect that equol, which has been isolated from the urine of several animals, including humans, is likely a product of metabolism of dietary isoflavones such as daidzein and formononetin. It is also of significance that all isoflavans of plant origin have oxygen at C-2′! This is thought to reflect biosynthetic relationships with pterocarpans. An apparent consequence of possessing a 2′-hydroxy group is the potential for oxidation to quinones several examples of which are known based upon the isoflavan system. An example of this sort of compound is “mucroquinone” [2-394] obtained from *Machaerium mucronulatum* (Fabaceae) by Kurosawa and colleagues (1968). Another interesting observation concerns the very uncommon occurrence of isoflavans with oxygenation at C-5; only two appeared in the Dewick’s 1988 list, and no new ones in his list of 1994. Four new ones, all from *Glycyrrhiza aspera*, were listed by Donnelly and Boland (1995). “Glyasperin-G” [2-395] (Zeng *et al*., 1992) is typical. Two isoflavans from the recent literature will indicate that structural variation, beyond the features already mentioned, are much the same as one might expect from other isoflavonoids from legumes. Two isoflavans with rearranged prenyl functions, [2-396] and [2-397], were reported from *Millettia racemosa* and shown to have moderate activity against *Staphylococcus aureus* at 0.1 $\mu g\,ml^{-1}$ (Prakash Rao *et al*., 1996). An additional example, also featuring a rearranged prenyl group [2-398], was recently identified as one of the components of *Maackia tenuifolia* (Zeng *et al*., 1996). A brief summary of earlier reports of naturally occurring isoflavans and related isoflavanquinones was given by Alves (1968).

Isoflav-3-enes are among the rarest of flavonoids and have only recently been added to the roster of naturally occurring compounds, although they have been known as synthetic intermediates for some time. There were no naturally occurring isoflavenes listed in either of the first two volumes of *The Flavonoids*, only four listed in the third (Dewick, 1988), and one further example appeared in the most recent volume (Dewick, 1994). Their scarcity may be a reflection of their chemical reactivity. An example of a naturally occurring member of this elite group is 6,7,3′-trihydroxy-2′,4′-dimethoxyisoflav-3-ene [2-399] from the heartwood of *Baphia nitida* (Fabaceae) (Arone *et al*., 1981), whose wood, incidentally, is used for the construction of, among

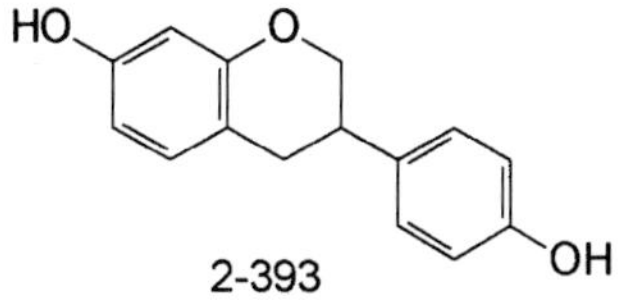

2-393

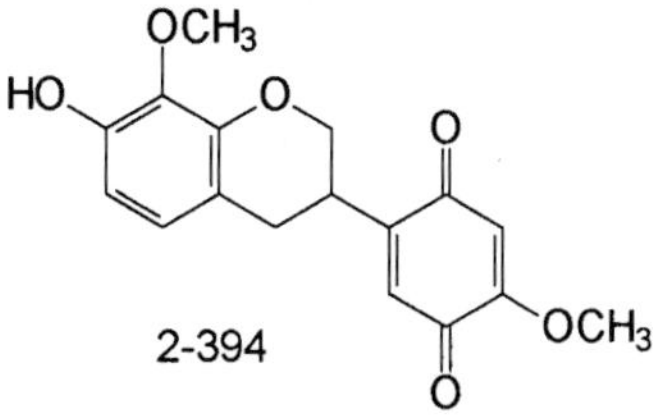

2-394

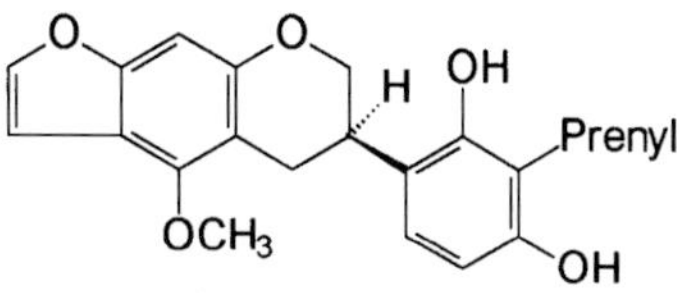

2-395

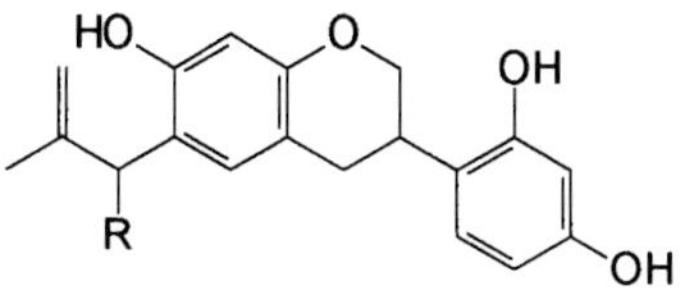

2-396 R = CH_3

2-397 R = CH_2OH

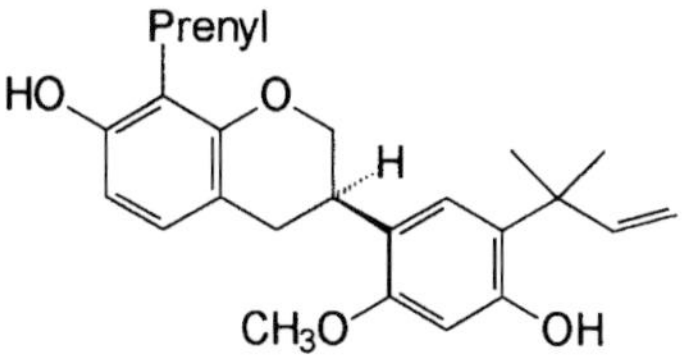

2-398

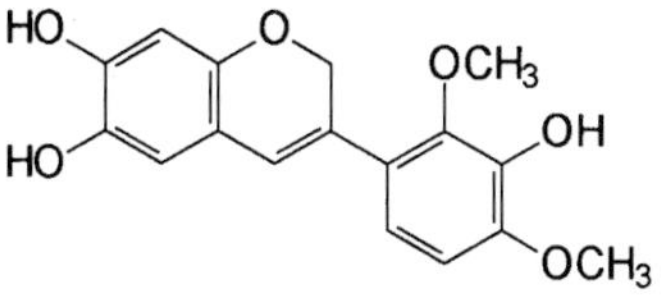

2-399

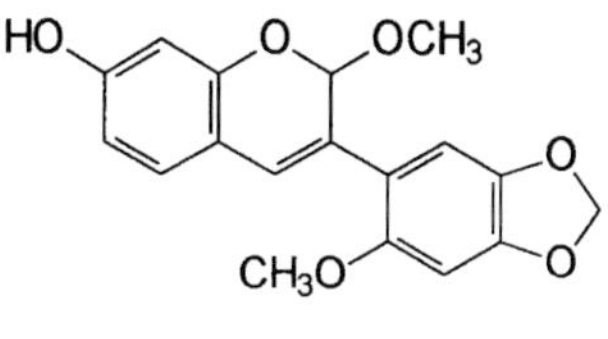

2-400

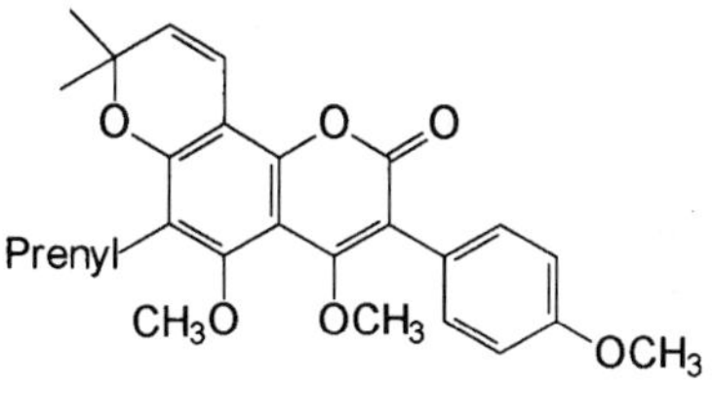

2-401

other things, violin bows. The structure of the compound has been confirmed by synthesis (Shoukry *et al.*, 1982). Very recently, Veitch and Stevenson (1997) isolated a new isoflav-3-ene from roots of *Cicer bijugum* and determined its structure to be 7-hydroxy-2,2′-dimethoxy-4′,5′-methylenedioxyisoflav-3-ene [2-400]. There was some indication that a glycoside of the isoflavene was also present in the root extracts.

3-Arylcoumarins

The arylcoumarins are characterized by the presence of a carbonyl function at C-2 and they may or may not have oxygenation at C-4. Fewer than a dozen of these compounds have been described in the literature. A recent report described the identification of a member of this small family of compounds from *Derris scandens* (Fabaceae) (Narayana Rao *et al.*, 1994). The compound is called 4,4′-di-*O*-methylscandenin [2-401]. Dewick (1994) listed only six 3-arylcoumarins and three 3-aryl-4-hydroxycoumarins, of which scandenin was one.

Tetracyclic Isoflavonoids

Members of this general group are characterized by the presence of an additional heterocyclic ring (other than those formed by cyclization of prenyl groups). Some authorities make further distinctions based upon details of hydroxylation and the presence or absence of certain unsaturations, but for the purposes of the present treatment, we will consider only four types, pointing out the distinguishing features as required: pterocarpans [2-402], coumestans [2-403] , rotenoids [2-404], and coumaronochromones [2-405]. Note that the numbering systems for coumestans, pterocarpans, and rotenoids are different from the isoflavone numbering system. The isoflavone numbering system is used in the coumaronochromones [2-405].

Pterocarpans have a saturated six membered heterocyclic ring, which is their characteristic feature, and a five-membered heterocyclic ring identical to that seen in coumestans. Pterocarpans appear to be fairly common having been isolated from a wide variety of legume species (Dewick, 1994). Two of the most frequently reported are "medicarpin" [2-406] and "maackiain" [2-407]. Pterocarpans bearing *C*-prenyl functions are also common and are represented here by "phaseolin" [2-408]. Phaseolin is a phytoalexin obtained originally from *Phaseolus* species following infection by fungal spores.

Pterocarpans that have an hydroxyl group at position 6a are often referred to their own group known, logically enough, as the 6a-hydroxypterocarpans. The first member of this group to be identified as a natural product came from pods of the common garden pea (*Pisum sativum*) that had been infected with fungal spores. Cruickshank and Perrin (1964) identified the compound as [2-409] and named it "pisatin." This finding provided the first proof that the phytoalexin theory, advocated some 20 years earlier by Müller and Börger (1941), had substance. One of the main points in the theory requires that the active compound be formed or activated only when the host plant comes in contact with the infecting agent (Harborne, 1988, pp. 315–316). The discovery that these isoflavone derivatives had such vital functions to perform led to intensive research on the subject the result of which was the discovery of many new isoflavone-based phytoalexins (Dewick, 1988, 1994), as well as members of other classes of natural products that served their respective plants in the same way. The chemical synthesis of pisatin will be described in Chapter Five.

Loss of water from a 6a-hydroxypterocarpan results in the establishment of a double bond between carbons 6a and 11a; the product is called a pterocarpene. There is convincing evidence, however, that a few of these compounds do occur naturally (Dewick, 1988, 1994). An example of this group is 3-hydroxy-8,9-methylenedioxypterocarpene [2-410], which Malan and Swinny (1990) obtained from the heartwood

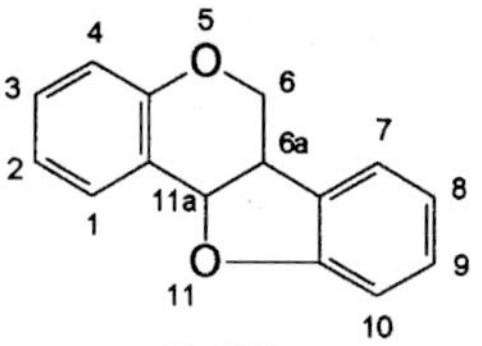
2-402

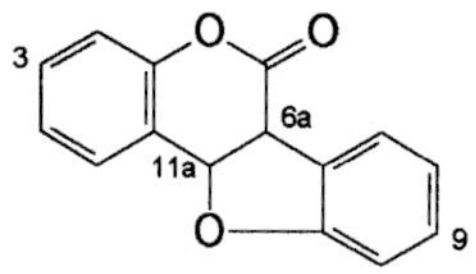
2-403

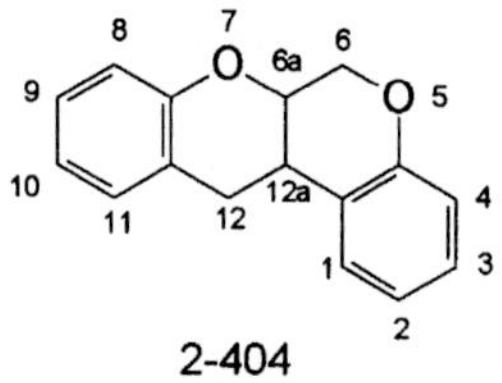
2-404

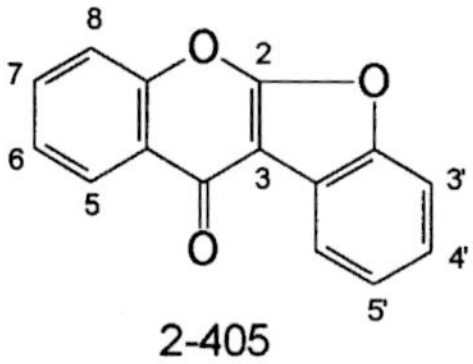
2-405

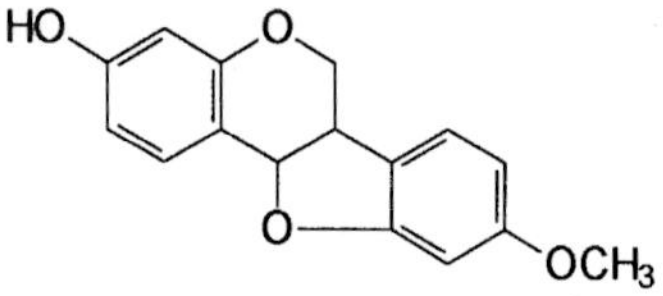
2-406

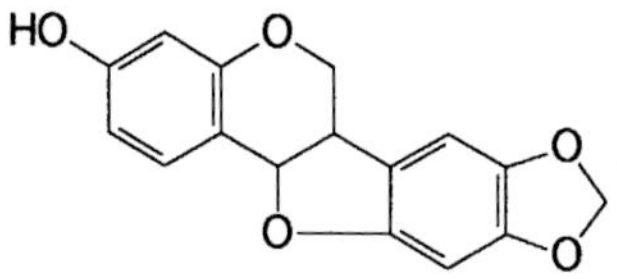
2-407

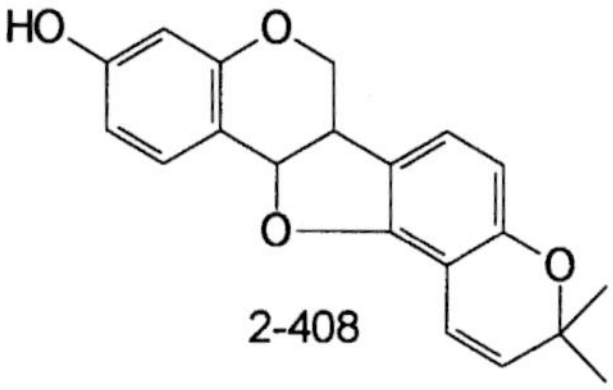
2-408

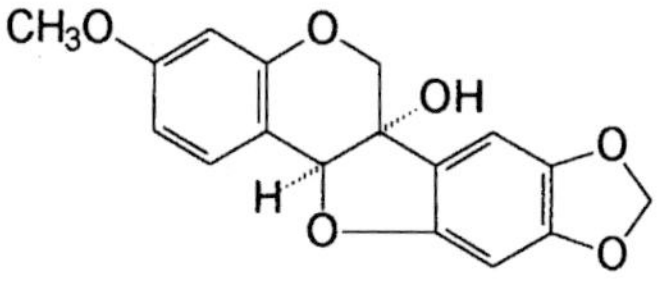
2-409

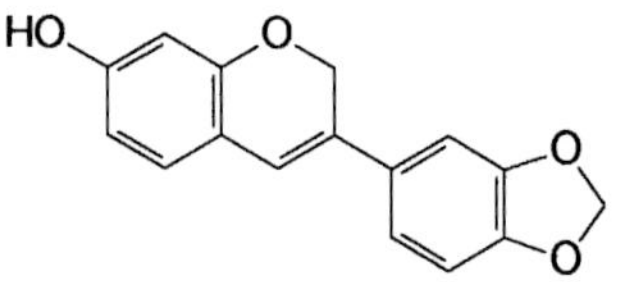
2-410

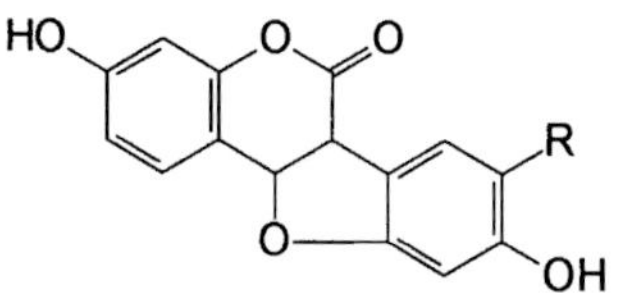
2-411 R = H

2-412 R = OH

of the South African timber tree *Virgilia oroboides* (Fabaceae) where it occurs with the common pterocarpans maackiain [2-407] and its methyl ether (pterocarpin).

Coumestans [2-403] resemble 3-arylcoumarins in having a carbonyl function adjacent to the C-ring oxygen, which is equivalent to C-2 in the isoflavone system, position-6 in the coumestan system. This group of flavonoids is distinguished by the five-membered heterocyclic ring involving the oxygen at C-4 and the 2′-carbon of the B-ring (isoflavone numbering). In the coumestan numbering system the positions would be C-11a and C-10a, respectively. Typical examples are structure [2-411], which is one of the cupric chloride-induced compounds from *Pueraria lobata* (Hakamatsuka *et al*., 1991), and compound [2-412], which was obtained from *Erythrina sigmoidea* by Nkengfack and coworkers (1994a). These compounds would be called 3,9-di- and 3,4,9-trihydroxycoumestan, respectively.

The next group of tetracyclic isoflavonoids are the rotenoids [2-404]. A comparatively simple representative of this group is "munduserone" [2-413] from the legume genus *Mundulea*, most of whose dozen or so species occur (or occurred!) in Madagascar. Using the accepted convention, this compound would be named 2,3,9-trimethoxychromanochromone, which is not much easier to remember than is the common name! Perhaps the most well known member of this group of compounds is "rotenone" [2-414], the active principle in the well known insecticide "Derris dust." Several of the plant species from which this compound has been isolated are used by indigenous peoples as a fish poison. It is safe for use by humans which makes it an ideal natural toxin. The 11-hydroxy derivative of rotenone [2-415], known as sumatrol, was mentioned above as one of the recently discovered constituents of *Millettia auriculata* (Venkata Rao *et al*., 1992).

The vast majority of tetracyclic isoflavonoid derivatives occur in members of Fabaceae. This certainly indicates that legumes have special biosynthetic capacities, but these capacities are not restricted to legumes. There have been reports of rotenoids from several other plant families none of them particularly closely related to Fabaceae. One of the earliest reports of a rotenoid occurring in a non-legume came from the study of Lee and coworkers (1972) who were trying to identify an antitumor agent from *Balduina angustifolia*, a member of Asteraceae. The compound identified was rotenone, which, incidentally, was not responsible for the antitumor activity of the plant. Another sighting of a rotenoid beyond the realm of legumes has been recorded. Sultana and Ilyas (1987) identified sumatrol [2-415] in leaf extracts of *Macaranga indica*, which is a member of Euphorbiaceae (the spurge family).

Rotenoid derivatives are also known from members of Nyctaginaceae (the 4-O'clock family), from a species of *Iris*, and from the southeastern Asian and tropical Australian monocot family Stemonaceae. There is a fundamental structural difference, however, between the rotenoids present in these species and those described above. Members of this group possess an hydroxyl function at C-12a as seen in compounds [2-416] and [2-417], which were identified as components of *Boerhavia coccinea* (Nyctaginaceae) (Messana *et al*., 1986). Compound [2-418] was obtained from rhizomes of *Iris spuria* (Shawl *et al*., 1988). With reference to its isoflavonoid foundation, this compound lacks B-ring oxygenation but exhibits extra oxygenation at C-6 of the A-ring.

Loss of a molecule of water from 6a-hydroxyrotenoids yields dehydrorotenoids a few of which have been reported as natural products. Interestingly, four of the six members of this subgroup listed by Dewick (1994) were found in species of *Boerhavia* (Kadota *et al*., 1988, 1989; Ahmed *et al*., 1990). An example of this small group is "repenone" [2-419] from *B. repens* (Ahmed *et al*., 1990).

The last group to be considered in this section, the coumaronochromones, are comparative newcomers. As Dewick (1994) pointed out, only one such compound was

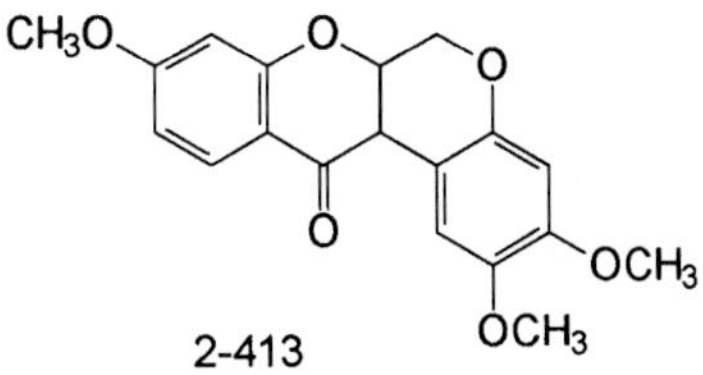

2-413

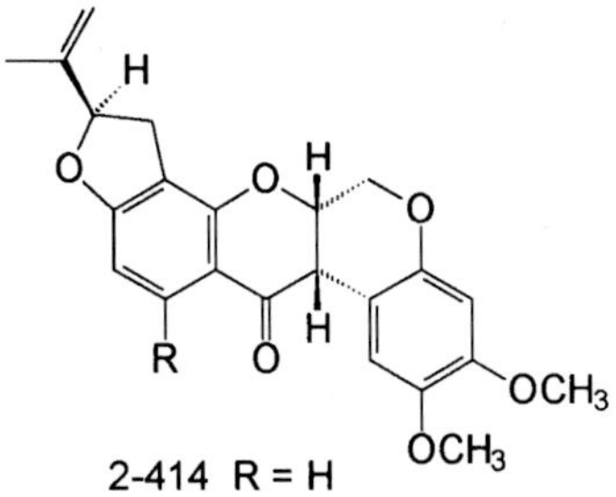

2-414 R = H

2-415 R = OH

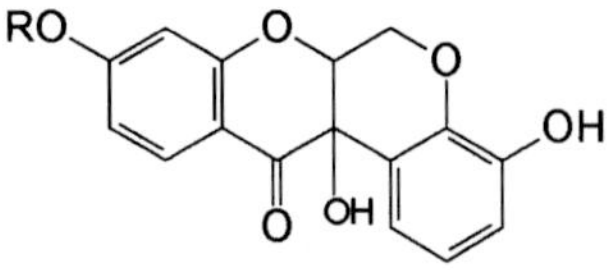

2-416 R = H

2-417 R = CH_3

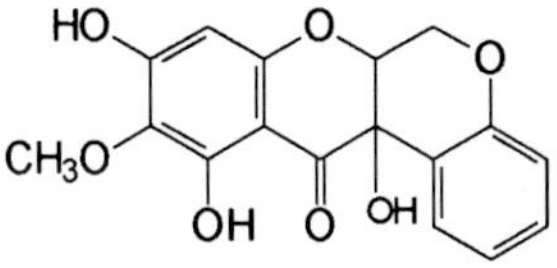

2-418

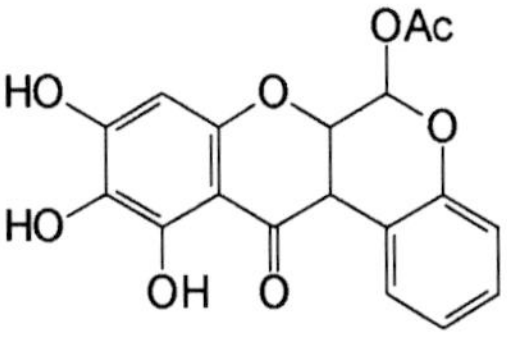

2-419

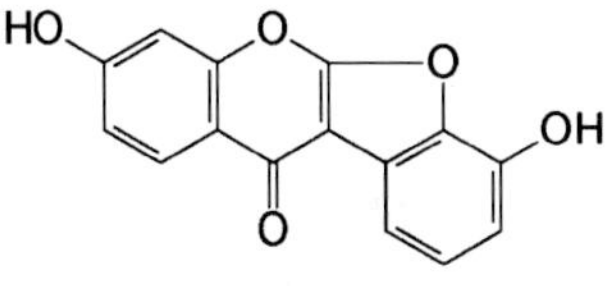

2-420

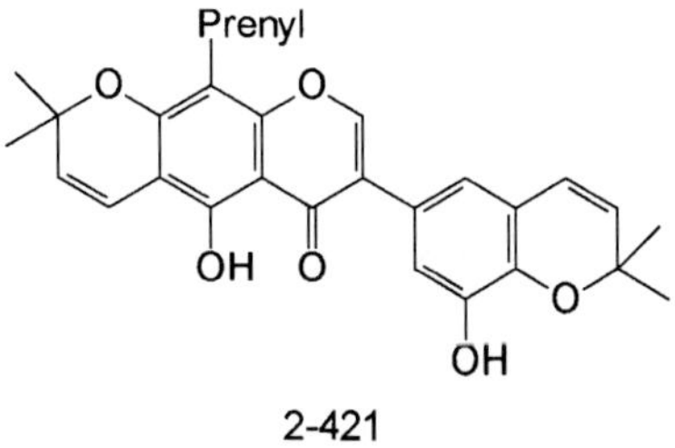

2-421

known until fairly recently. The number is increasing rapidly undoubtedly aided by the power of current spectroscopic instrumentation. It should not be surprising to learn that all but one of the reports of coumaronochromones involves a legume. It is also worth noting that the only non-legume to yield one of these compounds, at least to the end of 1993 (Donnelly and Boland, 1995), is none other than an acquaintance, *Boerhavia coccinea*! The *Boerhavia* compound, whose structure was shown as [2-420] (Ferrari *et al*., 1991), is comparatively simple, whereas the legume representative [2-421], which comes from *Euchresta formosana*, is a more complex molecule, as one might have come to expect from this family (Mizuno *et al.*, 1991).

The foregoing discussion of isoflavonoid structure and distribution offers a chance to comment on a potentially interesting experimental opportunity. When two groups of organisms with no close phylogenetic (evolutionary) relationship share a set of characters, whether morphological or biochemical, it is normal to ask if these features come about by similar processes. Are they homologous, in other words, or do they represent the end products of quite different processes and thus represent the results of convergent evolution? Since Fabaceae and Iridaceae are not sister groups, that is, they do not share an immediate ancestor, it would appear that the capacity to make the isoflavonoid molecule has developed independently in the two. Since the crucial step in the biosynthesis of isoflavones involves aryl migration, it seems logical to ask about the enzyme(s) that catalyze this step in the two families. Probing further still, how do the genes controlling these enzymes differ? Another logical direction would be to expand the question to include the other families that make these compounds, Euphorbiaceae, Nyctaginaceae, and Stemonaceae. It ought to be borne in mind that Nyctaginaceae is already known to have an unusual flavonoid chemistry in that it is one of the families within Caryophyllales (Centrospermae) that lacks the capacity to produce anthocyanidins (it makes betacyanins instead).

Proanthocyanidins and Flavans

In this section we will encounter perhaps the most complex members of the flavonoid family. Although this class of compounds contains some common and comparatively simple compounds, catechin and epicatechin in particular, the overall structural complexity of the group is impressive. Besides its complex chemistry, this group is interesting because of the major role it has played in the development of flavonoid chemistry in the broadest sense. Interest in compounds we now call the proanthocyanidins arose because members of the group are among those natural substances known as "tannins." The art (only more recently a science?) of tanning leather has had a long and impressive history that parallels mankind's association with animals as providers of companionship, clothing, as well as something to eat.

The development of our current ideas of proanthocyanidin structure began, it is claimed by some, as early as the middle of the 19th century (Haslam, 1975) when it was noted that certain colorless plant materials yielded bright red colors upon treatment with mineral acid. Since anthocyanidins were known at the time, the term "leucoanthocyanidin" for these substances seemed a logical one. The tangled history of the subject, fascinating, but beyond the scope of the present introduction, has been very well told by Haslam's (1975) contribution to *The Flavonoids* and in his *Plant Polyphenols – Vegetable Tannins Revisited* (Haslam, 1989). One of the critical observations in the field, which Haslam points out in those treatments, was the realization that leucoanthocyanidins and many of the substances responsible for the tanning action of plants belonged, essentially, to the same family of "polyphenols." But, as we will see below, not all tannic substances are flavonoids, although they certainly do qualify as polyphenolic. There was, in a word, a communication problem. Once it was recognized that two fundamentally different types of "tannins" existed, one based on flavonoid units and one based on gallic acid glucose esters, the problem resolved itself into what the flavonoid-based tannins should be called. Porter (1988) suggested that leucoanthocyanidin be used to define monomeric compounds that require breakage of a carbon–oxygen bond for the formation of an anthocyanidin. Porter also included flavan-4-ols. A proanthocyanidin, on the other hand, was defined as an oligomeric compound that required breakage of a carbon–carbon bond for the formation of color. Haslam (1989) sees the distinction between monomers (flavan-3-ols or flavan-3,4-diols) and oligomers (two to several monomeric flavan-3-ol units joined by C-C bonds) as less clear-cut. He has suggested that the term leucoanthocyanidin not be used. What we deal with below, then, are proanthocyanidins of two sorts, little ones and bigger ones. What are some of these compounds?

In order to answer that question we must first acquaint ourselves with the nomenclatural system currently in use. The system was first suggested by Hemmingway and coworkers (1982) and subsequently modified by Porter (1988). The rules (slightly expanded and modified from Porter, 1994) are as follows: (1) The system assumes the normal flavonoid skeleton numbering. (2) The system uses an agreed upon list of names for the basic flavan units, e.g., afzelechin [2-422], catechin [2-423], gallocatechin [2-424], and fisetinidol [2-425]. The proanthocyanidin class to

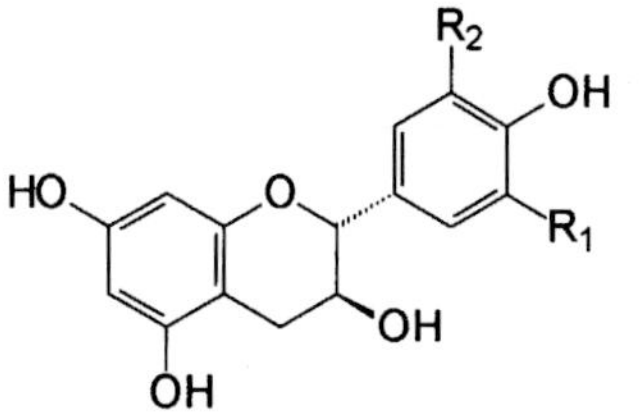

2-422 $R_1 = R_2 = H$

2-423 $R_1 = OH$, $R_2 = H$

2-424 $R_1 = R_2 = OH$

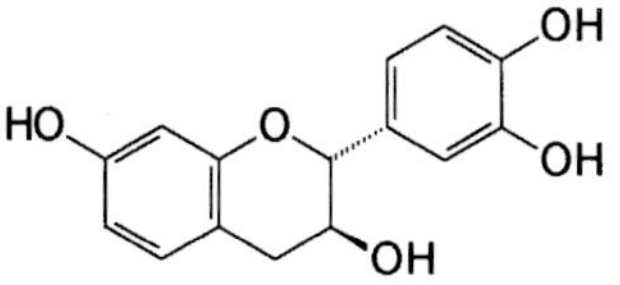

2-425

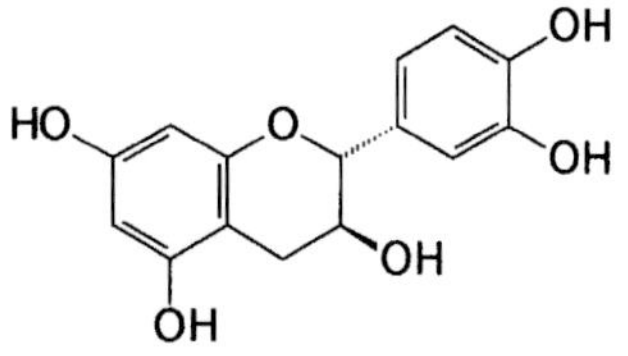

2-426 (+)-Catechin (2*R*:3*S*)

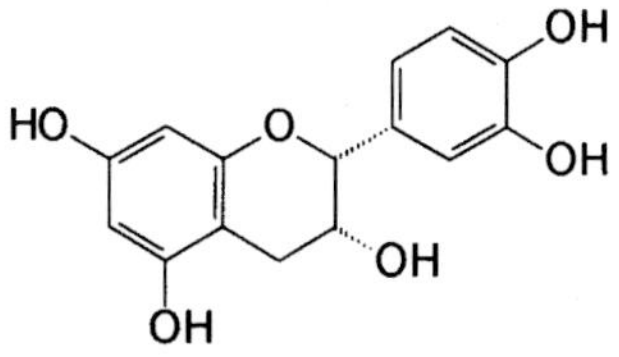

2-427 (–)-Epicatechin (2*R*:3*R*)

which these compounds belong are, respectively, propelargonidin, procyanidin, prodelphinidin, and profisetinidin. 3-Deoxy analogues are treated in the same fashsion as exemplified by apigeniflavan/proapigeninidin and luteoliflavan/proluteolinidin. The anthocyanidins, or 3-deoxyanthocyanidins in the case of the latter two, produced from these proanthocyanidins should be obvious from their names (just remove the "pro" prefix). (3) The flavan-3-ols listed in Rule (2) all have the (2*R*, 3*S*) configuration. Those with the (2*R*, 3*R*) configuration carry the "epi" prefix. Thus, (+)-catechin has the (2*R*, 3*S*) configuration while (−)-epicatechin has the (2*R*, 3*R*) configuration. Flavan-3-ols with the (2*S*) configuration would be identified with the prefix "ent" to indicate that it is an enantiomer. (4) The interflavanoid linkage is indicated in the same way as is done with oligosaccharides: the positions involved in the bond and the bond direction are indicated in parentheses and stereochemistry is expressed using the alpha–beta system. These rules can best be appreciated by examining a real-life situation. It is convenient to follow Haslam's (1989) example and describe the four dimeric procyanidins, B1, B2, B3, and B4. In short, these compounds represent all possible combinations of catechin and epicatechin. Dimer B1 is epicatechin-($4\beta \rightarrow 8$)-catechin [2-428]; B2 is epicatechin-($4\beta \rightarrow 8$)-epicatechin [2-429]; B3 is catechin-($4\alpha \rightarrow 8$)-catechin [2-430]; and B4 is catechin-($4\alpha \rightarrow 8$)-epicatechin [2-431]. Trimeric proanthocyanidins are also known and can involve the same monomers, or different ones, and can exist in an almost infinite combination of configurations and inter-unit linkages. A comparatively straightforward example of a trimeric proanthocyanidin is compound [2-432], epiafzelechin-($2\beta \rightarrow 7$; $4\beta \rightarrow 8$)-epiafzelechin-($4\beta \rightarrow 8$)-epiafzelechin, which Kashiwada and coworkers (1990) obtained from leaves of the fern *Dicranopteris pedata*. Biflavonoids joined by a single carbon–carbon bond, as in the case of the procyanidins, are said to be "B-linked," whereas oligomers such as the afzelechin trimer, which include a carbon–carbon as well as a carbon–oxygen bond, are referred to as being "A-linked." The chemistry of oligomeric proanthocyanidins has been reviewed recently by Ferreira and Bekker (1996).

The determination of the structures of monomeric proanthocyanidins requires answers to two fundamental questions: (1) what is the substitution pattern of the A- and B-rings; and (2) what is the stereochemistry of the C-ring? Determination of oxygenation patterns is comparatively straightforward owing to the ease of conversion of proanthocyanidins to anthocyanidins, or other flavonoids, whose structures were well known. Classical methods applied to C-ring structure, including the use of reagents supposedly specific for *cis*-diols, were reviewed critically by Haslam (1975). A recent study of constituents of *Cinnamomum camphora* (Lauraceae, camphor tree) provides a useful example of structural analysis of flavan-3-ols, two of which were studied using ^{1}H and ^{13}C NMR (Mukherjee *et al.*, 1994). They were shown to be (2*R*, 3*R*)-4′-hydroxy-5,7,3′-trimethoxyflavan-3-ol [2-433] and the corresponding 3′,4′-methylenedioxy derivative. The determination that these compounds have epicatechin-type (2,3-*cis*) stereochemistry was made on the basis of the C-2 proton signal which exhibited a J value of less than 1 Hz. Absolute stereochemistry, (2*R*, 3*R*), was assigned on the basis of negative rotations observed for (−)-(2*R*, 3*R*)-epicatechin and its tetramethyl ether (Birch *et al.*, 1957). A recent addition to the list of flavan-3-ols with this stereochemistry comes from the work of Chung *et al.* (1997) who reported (2*R*, 3*R*)-5,7,2′,5′-tetrahydroxyflavan-3-ol [2-434] from *Hypericum geminiflorum* (Guttiferae).

Before leaving the monomeric proanthocyanidins, it is important to point out that several of these compounds occur widely esterified to gallic acid through the C-3 hydroxyl group. Common examples of these esters are (+)-catechin gallate [2-435] and (−)-epigallocatechin gallate [2-436].

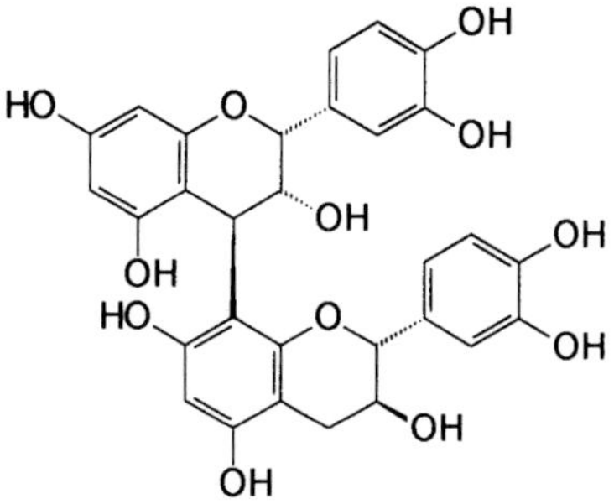

2-428 (B1)

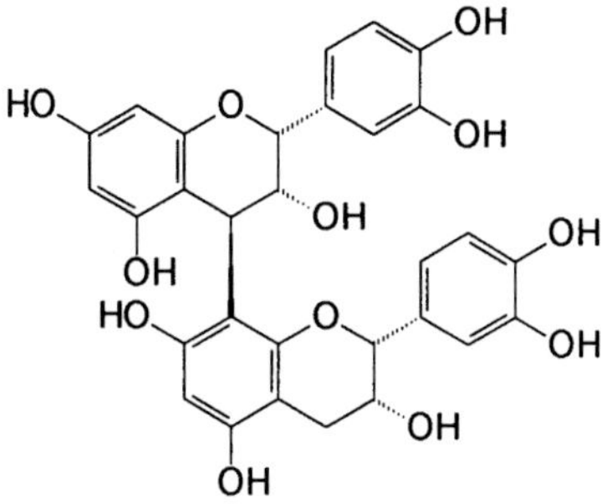

2-429 (B2)

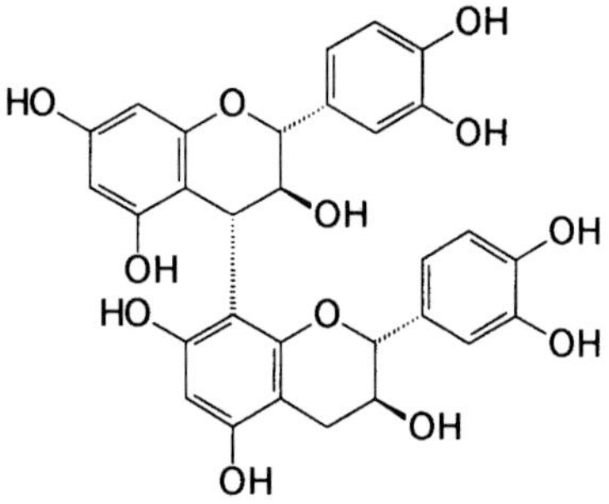

2-430 (B3)

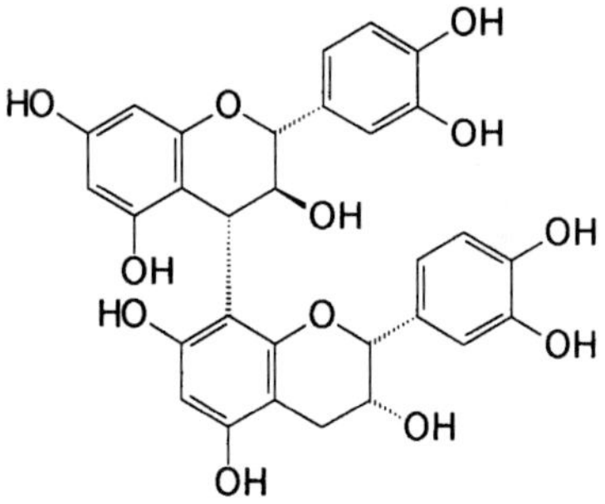

2-431 (B4)

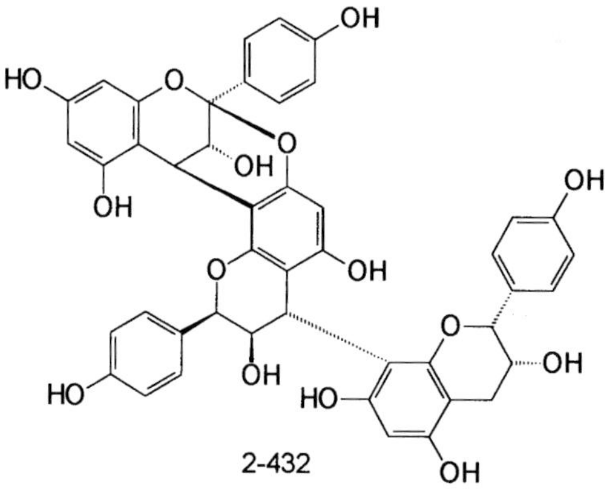

2-432

In determining the structures of polymeric proanthocyanidins, several questions have to be answered. First, one needs to know what the constituent monomers are, and equally critically, how many there are. The size of the polymer (or degree of polymerization, DP) must be determined [see Koupai-Abyazani *et al*. (1993a,b) for a comparison of methods]. Several techniques have been developed to obtain information on the polymer sequence based upon the reactivity of the C-4 (benzylic) position. Treatment of the unknown with acid in the presence of a thiol, such as

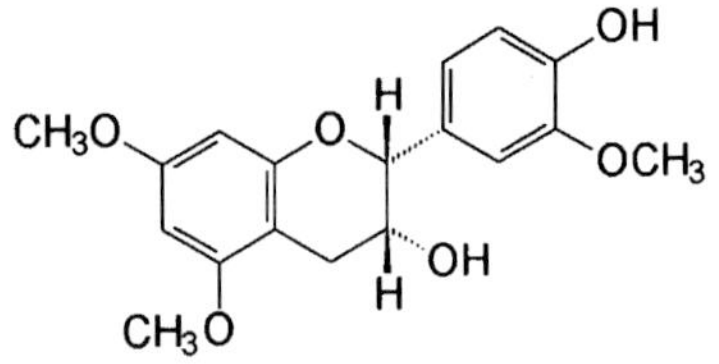

2-433 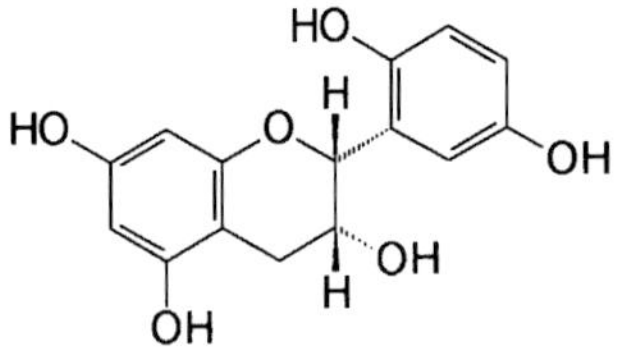 2-434

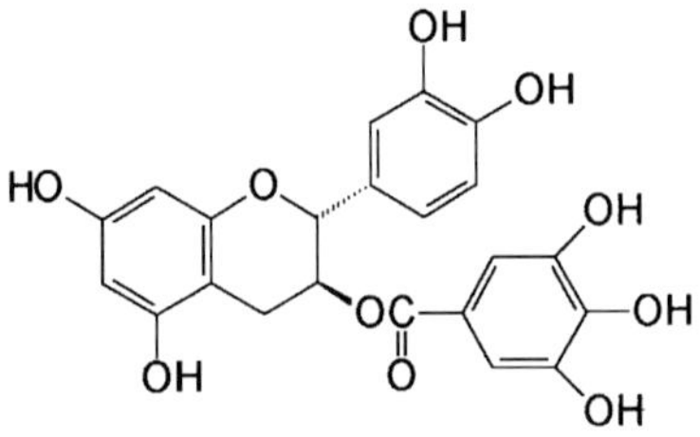

2-435 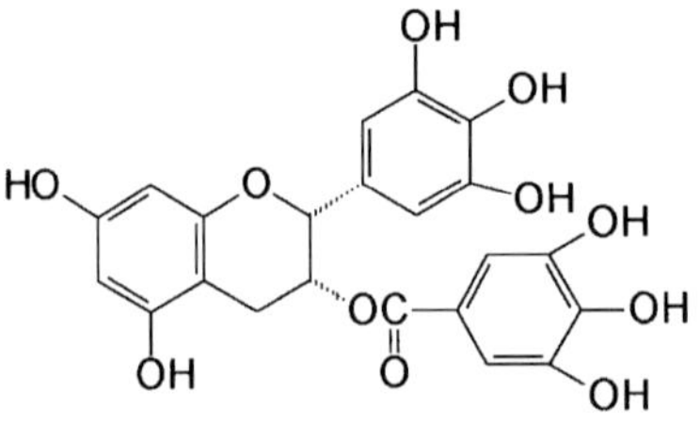 2-436

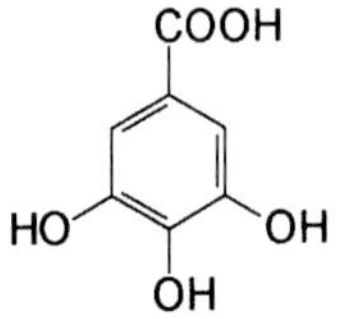

2-437 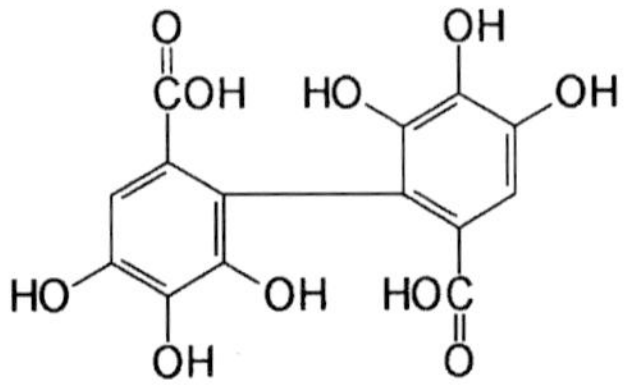 2-438 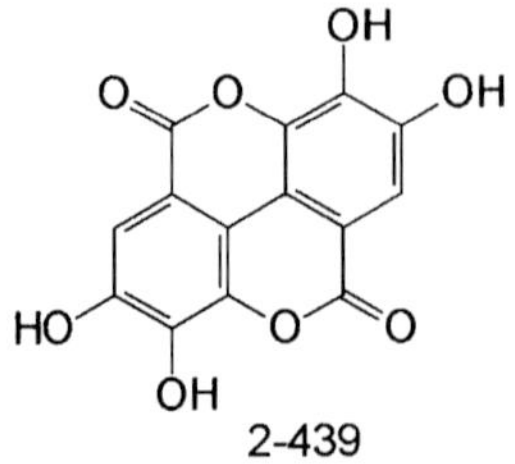 2-439

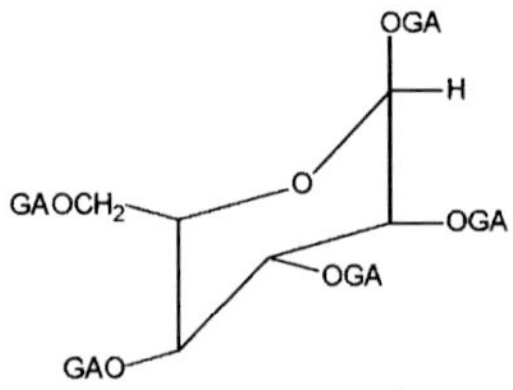

2-440 GA = Galloyl 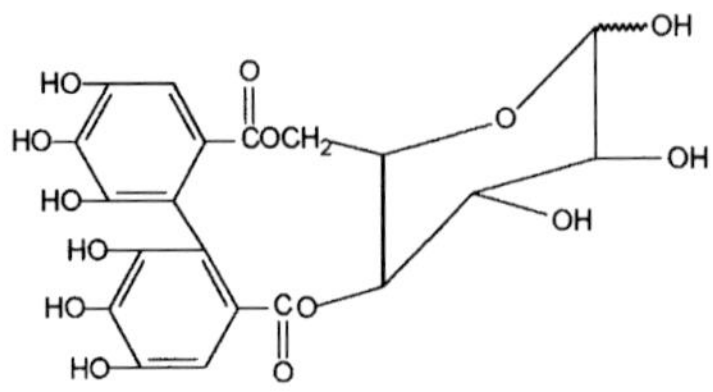 2-441

thioglycolic acid or benzene thiol, yield monomeric products whose stereochemistry at C-4 can be established. Equally effective (and less odoriferous!) is a procedure using phloroglucinol or a phloroglucinol derivative to capture the monomeric unit. Koupai-Abyazani and coworkers (1993a,b) coupled this method with high performance liquid chromatography in a study of proanthocyanidin structure in some forage legumes. These hydrolytic procedures allow one to assess the nature of the terminal unit in the polymer since it is obtained as a flavan-3-ol, the extension units all appearing as the phloroglucinol derivative(s).

Much of the ambiguity of assignment of position and stereochemistry of interflavanoid linkages can be avoided by the application of sophisticated methods of 2D NMR. Balas and coworkers (1994, 1995), whose papers serve well as leading references, applied 2D NMR methods to arrive at complete and unambiguous assignments of ^{1}H and ^{13}C NMR spectra of peracetylated catechin-(4α-8)-catechin-(4α-8)-catechin trimer. That work marked the first time that the sites of condensation between the units of a trimer had been unequivocally established.

Rossouw and coworkers (1994) addressed difficulties that may arise from interpretation of circular dichroism studies of oligomeric proanthocyanidins. They applied the Mosher Method (Dale and Mosher, 1973; Sullivan *et al.*, 1973) which involves forming the *R*-(+)- and *S*-(−)-2-methoxy-2-trifluromethyl-2-phenylacetic acid (MTPA) [2-385] esters of the flavan-3-ols and 4-arylflavan-3-ols under study. ^{1}H NMR of the resulting esters provides reliable information on the C-3 proton environment and thus the stereochemistry at that carbon. The interested reader should consult the Rossouw *et al.* (1994) paper for detailed illustrations of the stereochemistry of the model compounds used.

Many oligomeric proanthocyanidins have been isolated from heartwood of various tree species, which has led some individuals, unaware of the subtle nature of these compounds, their formation, and their physiology, to suggest that they are essentially end products of phenolic metabolism. Recent evidence from a study of proanthocyanidins in the forage herb *Onobrychis viciifolia* (Fabaceae, sainfoin) demonstrates, to the contrary, that they can be very much involved in dynamic processes. Preliminary studies had shown that the sainfoin leaf polymers consisted of (+)-catechin, (+)-gallocatechin, (−)-epicatechin, and (−)-epigallocatechin and that the last three listed compounds occurred as both extension and terminal units while (+)-catechin was present only as a terminal unit (Koupai-Abyazani *et al.*, 1993a,b). A study of proanthocyanidin formation in leaves of sainfoin required that the time of most active biosynthesis, and therefore enzyme activity, be established. Toward this goal, leaves were sampled at several developmental stages for proanthocyanidin polymers, which were isolated, purified and subjected to a phloroglucinol-based degradation procedure followed by analysis of the fragments by high performance liquid chromatography. The degree of polymerization (DP) of the polymer (see 1993 paper for details of DP) showed an overall change from 5.4 to 6.9 with a peak of 8.3 in the newly unfolding leaflets. This peak coincided with a sharp reduction in the rate of polymer synthesis. Most striking, however, was the decrease in the percentage of *cis*-isomers (epicatechin and epigallocatechin) from 83% to 48% coupled with the increase in the total percentage of prodelphinidin monomers (gallocatechin and epigallocatechin) during leaf maturation. Whatever metabolic and/or catabolic processes are occurring it is clear that the situation with regard to polymeric proanthocyanidin in sainfoin is anything but static.

Tannin and related terms were used above without much in the way of definition other than to suggest that they are very general ones. The process of tanning involves interaction of the protein in animal hides with polyphenolic compounds of plant origin (we ignore tanning by chromium salts in this work). Although several excellent sources of tannins were known to the early practitioners of the art, the chemical nature of these substances was not a major factor in their successful utilization, although it was known that tannic substances tended to give blue colors with iron salts and that they were astringent (which is actually a manifestation of their capacity to bind proteins). In time, natural product chemists began to unravel the structures of these compounds. One of the major observations was the fact that two sorts of compounds were involved. One of these types yielded flavonoids upon degradation, which led to them being called "condensed tannins," while the other type yielded 3,4,5-trihydroxybenzoic acid (gallic acid) [2-437] and glucose and became known as "hydrolyzable tannins." If gallic acid was the only aromatic acid obtained upon hydrolysis the particular hydrolyzable tannin was often referred to as a "gallocatechin." Gallic acid can undergo dimerization *via* phenolic coupling to give a product in which the two monomeric units are joined by a carbon–carbon bond to form ellagic acid [2-438] with secondary reciprocal esterification between hydroxyls and carboxyl groups on the two rings to yield [2-439]. Typical hydrolyzable tannins might have a glucose molecule esterified to several gallic acid residues or a glucose molecule esterified to one or two ellagic acids. In the latter case, the particular hydrolyzable tannins would be called "ellagitannins." Mixed tannins are also possible. Pentagalloyl glucose (the β isomer) and the 4,6-hexahydroxydiphenoyl ester of glucose are illustrated diagramatically as [2-440] and [2-441], respectively.

Gallic acid is occasionally seen as an acylating acid in flavonoid glycosides as in the case of quercetin-3-(6″-galloylglucoside) from *Tellima grandiflora* (Saxifragaceae) (Collins *et al*., 1975) or quercetin 3-(2″-galloylglucoside) from *Lasiobema japonica* (Fabaceae) (Iwagawa *et al*., 1990). It is more often encountered as an acylating unit involving catechin and related flavan-3-ols. Catechin gallate [2-435] and gallocatechin gallate [C-3 epimer of 2-436] are examples of this phenomenon. Gallic acid esters of proanthocyanidin oligomers also occur as reported by Nonaka and coworkers (1983) who obtained prodelphinidin gallate from leaves of green tea. The oligomeric unit in that compound is [2-436] linked *via* positions-8 and -4.

The last compounds to be examined in this section are flavan derivatives other than flavan-3-ols. This small group comprises three subgroups, flavans *per se*, flavan-4-ols, and flavan-3,4-diols. Only a few flavans have been reported to occur naturally, a situation that is certain to change as awareness of the existence of such comparatively simple compounds grows. One of the major reasons why they are not more often reported is that they are not readily visible on chromatograms and they have simple spectroscopic profiles compared to most other flavonoid classes. A few examples of naturally occurring flavans should give an idea of the structural variation to be expected. Among the earliest reports were 3′,4′-dihydroxy-5,7-dimethoxyflavan [2-442] and 2′-hydroxy-7-methoxy-3′,4′-methylenedioxyflavan [2-443] from *Iryanthera* species (Myristicaceae) (Franca *et al*., 1974). Another example, which has insect growth inhibitory properties, is a compound from *Viscum tuberculatum* (Viscaceae), one of the tropical mistletoes. Kubo and coworkers (1987) identified luteoliflavan-5-*O*-β-xyloside [2-444], along with its 2″-*p*-hydroxybenzoyl and 2″-caffeoyl derivatives. "Kazinol E" [2-445], which Ikuta and colleagues (1986) obtained from *Broussonetia kazinoki* (Moraceae), is interesting in that it bears three *C*-alkyl groups, two of which are normal prenyls while the third is the isomeric 1,1-dimethylpropen-2-yl (or α,α-dimethylallyl) function. In "kazinol H" the 3′-*C*-prenyl group has cyclized with the 4′-hydroxyl group. Five flavans, along with three

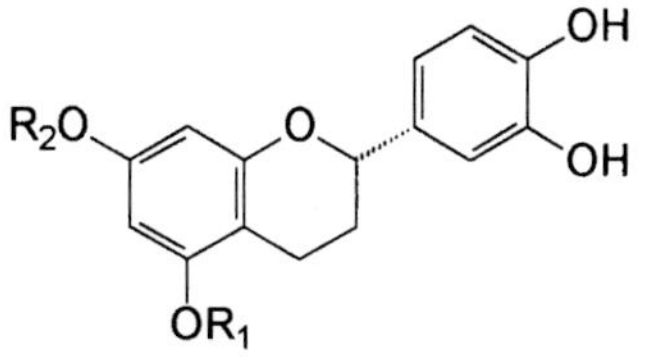

2-442 $R_1 = R_2 = CH_3$

2-444 R_1 = Xylose(Acyl), R_2 = H

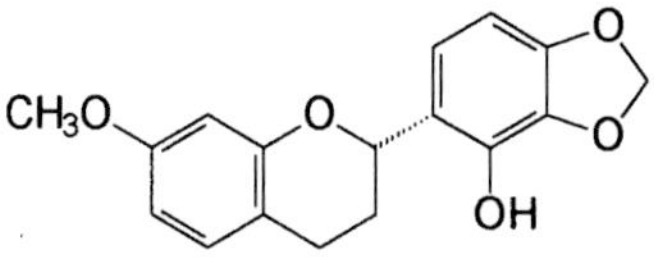

2-443

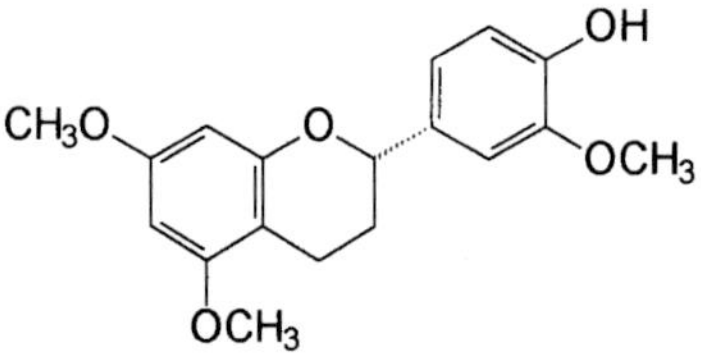

2-446

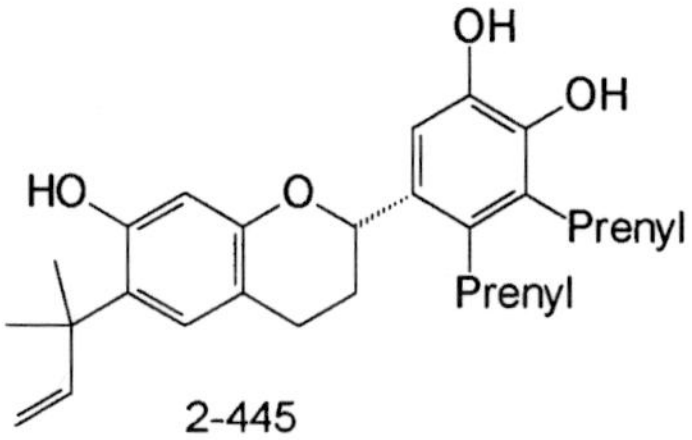

2-445

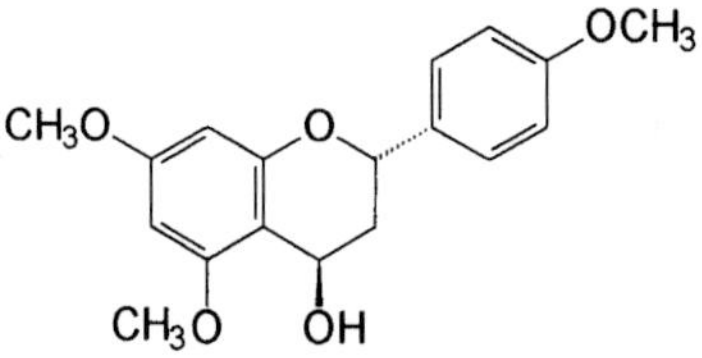

2-447

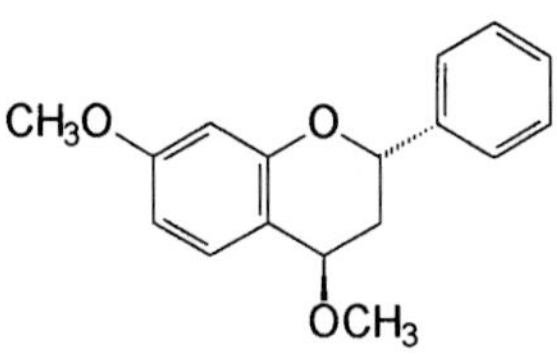

2-448

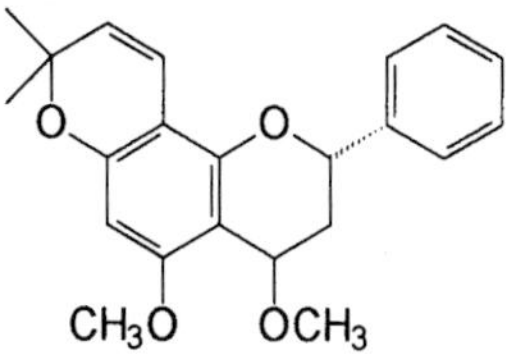

2-450

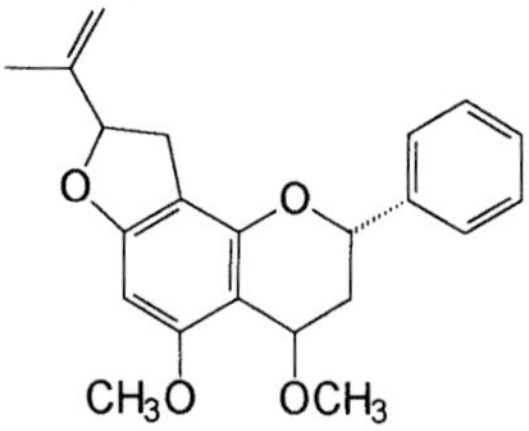

2-449

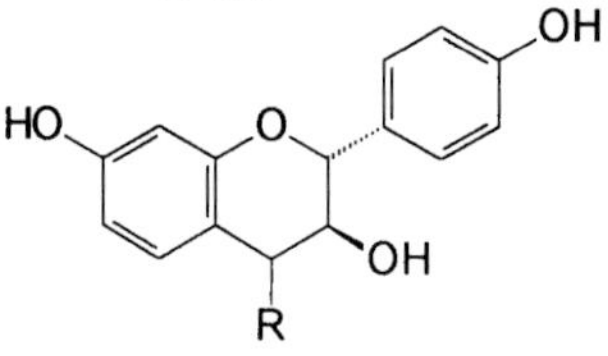

2-451 R = —OH

2-452 R = ·····OH

flavanones, were isolated from extracts of *Mariscus psilostachys* (syn. *Cyperus*, Cyperaceae) by Garo and co-workers (1996). Two of the flavans, (2*S*)-4′-hydroxy-5,7,3′-trimethoxyflavan [2-446] and (±)-5,4′-dihydroxy-7,3′-dimethoxyflavan, were newly described. The others were the known 3′,4′-dihydroxy-5,7-dimethoxyflavan, 7,3′-dihydroxy-4′-methoxyflavan, and 7,4′-dihydroxy-3′-methoxyflavan. The flavanones, all known compounds, were identified as eriodictyol, its 3′-methyl ether, and its 7,3′-dimethyl ether.

Although not a large group, the flavan-4-ols nonetheless exhibit a range of structural features that one comes to expect in any class or subclass of flavonoids. A few examples will make this clear. The first flavan-4-ol was described by Lam and Wrang (1975) in a study of phenolic compounds in *Dahlia tenuicaulis* (Asteraceae, the garden dahlia). One of the compounds identified was 5,7,4′-trimethoxy-2,4-*trans*-flavan-4-ol [2-447]. Fabaceae has provided its examples to this subclass of flavonoids (as it has to most others). Three of these are "tephrowatsin A," [2-448], isolated from *Tephrosia watsoniana* (Gomez *et al*., 1985), and "hilgardtol A" [2-449] and "hilgardtol B"[2-450], an isomeric pair isolated from *T. hildebrandtii* by Delle Monache and coworkers (1986). Note that these compounds differ by the nature of their 8-C-prenyl cyclization product, the five-ring in [2-449] and the six-ring in [2-450].

The flavan-3,4-diols have attracted more attention than the flavans and flavan-4-ols owing to their established function as intermediates in the biosynthetic path leading to anthocyanidins (Chapter Six). Haslam (1982) listed about a dozen naturally occurring 3,4-diols, while Porter (1988) listed about 30. There was little new to add in this arena for the latest volume (Porter, 1994). Flavan-3,4-diols are chemically interesting owing to their ready conversion to anthocyanidins by treatment with hot mineral acid, and their possession of three chiral centers with the attendant possibility of six isomeric configurations. Among the naturally occurring members of this subgroup, several cases exist where compounds having the same A- and B-ring oxygenation patterns have different C-ring stereochemistries. For example, *Acacia cultriformis* accumulates two isomers of 7,4′-dihydroxyflavan-3,4-diol, one with 2*R*:3*S*:4*S* stereochemistry, and one with 2*R*:3*S*:4*R* stereochemistry [2-451] and [2-452], respectively (du Preez and Roux, 1970). When dealing with molecules with this level of chirality, and especially considering that C-4 is benzylic and thus highly sensitive to solvolysis (see Haslam 1975 for review), extreme care must be exercised in extraction and purification procedures. It had been shown some years ago (Drewes and Roux, 1965) that two hours autoclaving at 15 psi of (+)-mollisacacidin (7,3′,4′-trihydroxyflavan-3,4-diol), whose stereochemistry is (2*R*:3*S*:4*R*), afforded a mixture of isomers containing 11% of the (2*S*:3*S*:4*R*) isomer, 12% of the (2*R*:3*S*:4*S*) isomer, and 2.8% of the (2*S*:3*S*:4*S*) isomer. Classic means of determining the absolute stereochemistry of flavan-3,4-diols by systematic degradation and comparison of products with compounds of known structure were reviewed by Haslam (1975). Early studies of the mass spectral disintegration of 3,4-diols were important because they demonstrated unique fragmentation patterns for these compounds (Clark-Lewis, 1968; Drewes, 1968). Application of ^{1}H and ^{13}C NMR methods have proved extremely important in this field as well.

Flavonoids Containing Nitrogen

Flavonoid alkaloids first appeared in the most recent volume of *The Flavonoids* series (Wollenweber, 1994). There are only about a half dozen of these "cross-over" molecules but their existence raises an interesting question, i.e., at what point do flavonoid and alkaloid biosyntheses converge? Since a diverse range of plant groups, and flavonoid classes, is involved, it is clear that the existence of these compounds does not represent a local aberration in flavonoid biosynthesis. In addition to the five compounds listed by Wollenweber (1994), one additional "true" alkaloid (presence of heterocyclic nitrogen) plus two flavonoid monoamines have been reported. The earliest report of flavonoid alkaloids appears to have been that of Johns *et al.* (1965) who identified "ficine" [2-453] and its 6-isomer "isoficine" from *Ficus pantoniana* (Moraceae). Investigation of the seagrass *Phyllospadix iwatensis* (Zosteraceae) by Takagi and coworkers (1980) revealed the existence of "phyllospadine" which is in fact 6-methoxyficine [2-454]. A flavan derivative [2-455] has been reported from *Vochysia guaianensis* (Vochysiaceae) (Baudouin *et al.*, 1983). Floral tissue of *Lilium caudatum* (Liliaceae) afforded a flavonoid alkaloid that was shown to have structure [2-456] (Masterova *et al.*, 1987). More recently, three nitrogen-containing isoflavone derivatives were obtained from the legume *Piscidia erythrina* (Moriyama *et al.*, 1993). Two of the compounds were simple amines, 5,7,3′-trihydroxy-5′-methoxy-2′,6′-di-*C*-prenyl-4′-aminoisoflavone and the 8,2′-di-*C*-prenyl isomer [2-457], while the third member of this group, "piscerythroxazole," was shown to possess the oxazole ring system. The structure of an unusual nitrogen-containing flavonoid derivative from *Carthamnus tinctorius* (Asteraceae, safflower) is shown as [2-458] (Meselhy *et al.*, 1992). This compound, called "tinctoramine," which acts as a calcium antagonist, is based upon a quinochalcone *C*-glucoside structure similar to others that been reported from this species (Onodera *et al.*, 1979, 1981; Takahashi *et al.*, 1984).

A novel nitrogen-containing flavonoid derivative was obtained from leaves of *Pteridium aquilinum* (bracken fern) by Imperato (1993) and shown to be a flavonoid glycoside acylated with a hexapeptide consisting of seven glycine units. This compound appeared above as structure [2-161].

As a point of historical nomenclatural caution, it should be noted that in the earlier plant chemical literature there are references to the red pigments of beets and related plants (Caryophyllales, Centrospermae) as "nitrogenous anthocyanins." They gained this descriptive title because their colors approximate those of known anthocyanin-bearing species and, indeed, they afforded nitrogen on elemental analysis. In time it was realized that these pigments belong to their own class of natural products, the "betacyanins." "Betanin," the principal pigment of beet root (*Beta vulgaris*), is shown as [2-459]. The interested reader should consult Piattelli (1981) for additional information on these compounds.

Neoflavonoids

Neoflavonoids constitute a group of flavonoid derivatives that have their aryl group attached to C-4 as opposed to C-2 in flavonoids and C-3 in isoflavonoids. The carbon skeleton is shown in [2-460] with the familiar numbering convention used in other common flavonoids. "Calophyllolide" [2-461], isolated from *Calophyllum inophyllum* (Guttiferae, Clusiaceae), was the first of these compounds to be identified (Polonsky, 1955). The key degradation product from that study was the 4-phenylcoumarin derivative [2-462], which established the fundamental ring structure of this class of flavonoids. Subsequently, many additional neoflavonoids were obtained from other members of Guttiferae almost all of which exhibited *C*-prenyl substituents on the A-ring. Neoflavonoids also occur in members of Fabaceae, but they differ from

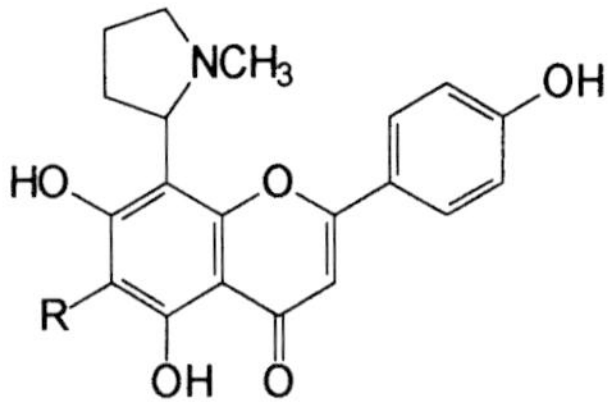

2-453 R = H

2-454 R = OCH_3

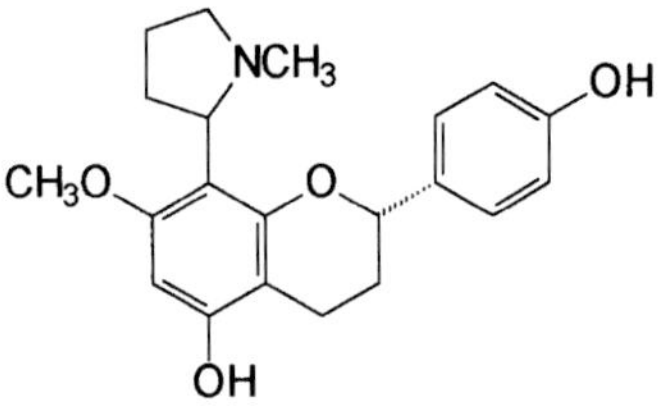

2-455

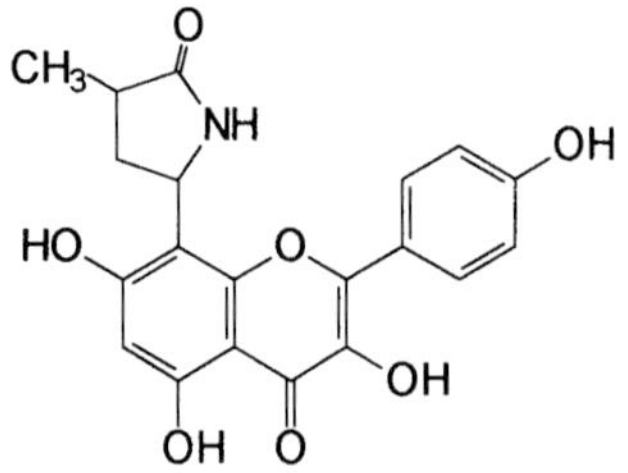

2-456

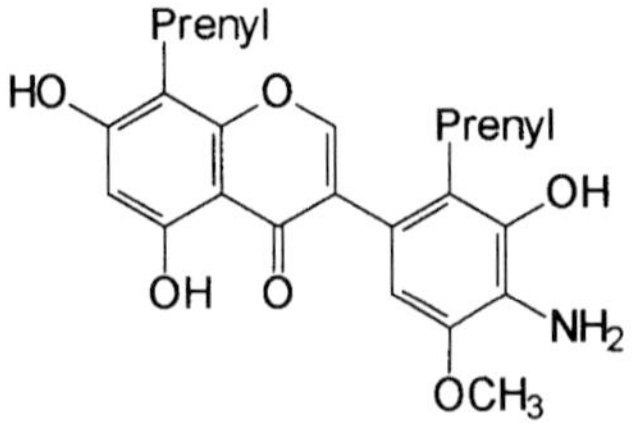

2-457

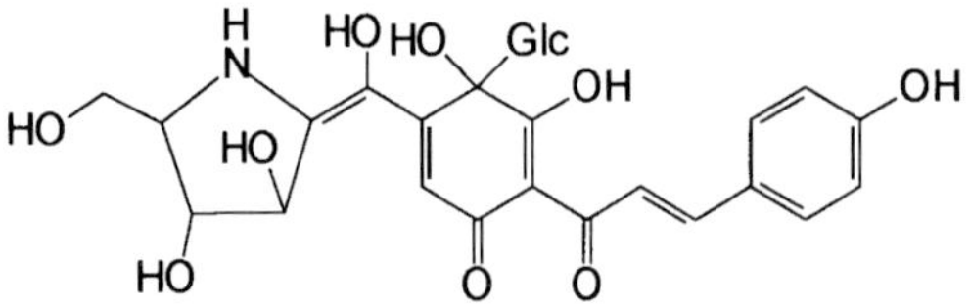

2-458

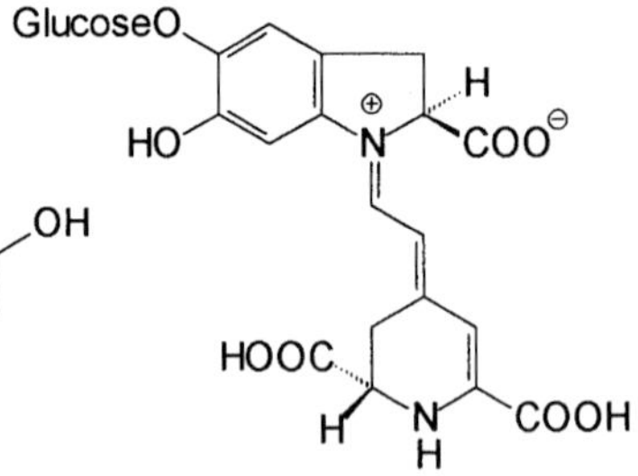

2-459

the Guttiferae compounds in the absence of the variety of *C*-alkyl groups seen in the latter, but in place are characterized by the presence of 6,7-dioxygenation. One of the early neoflavonoid derivatives isolated from a legume were the "dalbergione" derivatives [2-463] and [2-464] which are known from several species of *Dalbergia* (see Donnelly, 1975 for early references). Neoflavonoid derivatives have more recently been obtained from other plants including the fern *Pityrogramma* (see entry in Chapter Three). Neoflavonoids from members of Rubiaceae are also known and are characterized by 5,7-oxygenation with a variety of B-ring patterns much akin to what one expects to see in flavonoids in general. Structure [2-465] represents the small group of neoflavonoids called neoflavenes. These compounds have been found in a few legumes and may be biosynthesized from dalbergiones. Recognition of the 4-arylchroman ring system goes back to the early days of the 20th century in the studies of the composition of pigments from "Brazil wood," *Caesalpinia braziliensis*,

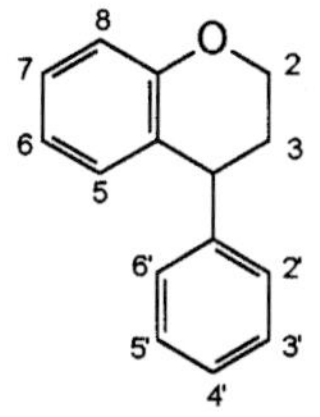

2-460 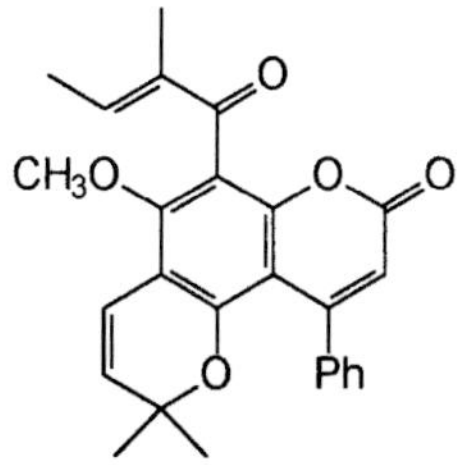 2-461 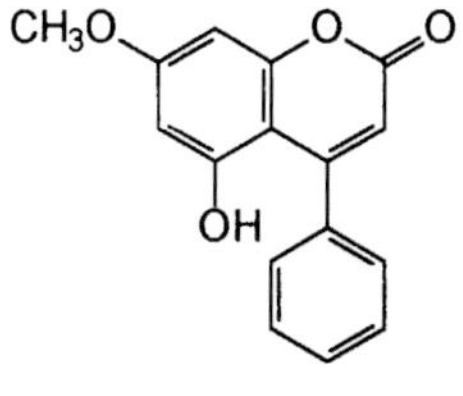 2-462

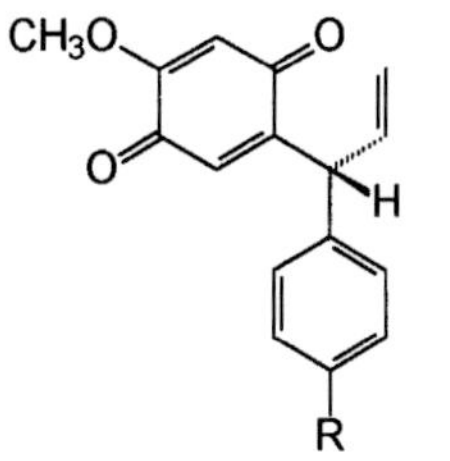

2-463 R = H
2-464 R = OCH_3

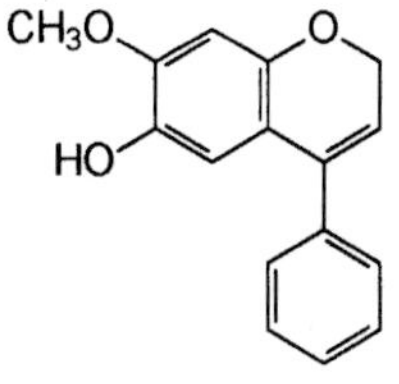

2-465

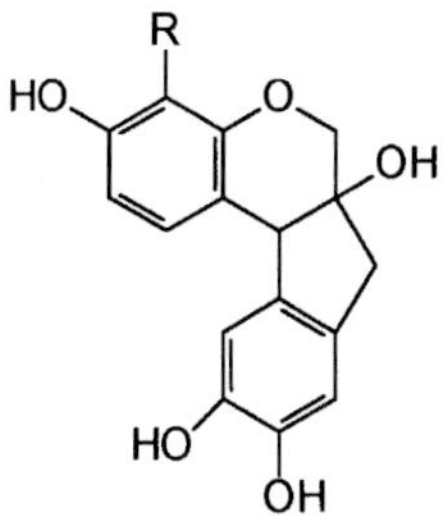

2-466 R = H
2-467 R = OH

and *Haematoxylon camapechianum.* The major components of these trees are shown below; [2-466] is "brazilin" and [2-467] is its hydroxy derivative "haematoxylin." The chemistry of these interesting compounds was reviewed by Robinson (1962). Detailed reviews of neoflavonoids, including structural variation, spectroscopic characteristics, and syntheses can be found in Donnelly (1985), Donnelly and Sheridan (1988), and Donnelly and Boland (1994, 1995).

It is important to add that the biosynthesis of neoflavonoids involves the expected precursors. Experiments using young plants of *Calophyllum inophyllum* showed that the label from 3-^{14}C-phenylalanine was found almost exclusively (92%) at C-4 of calophyllolide [2-461], while labelled acetate was incorporated into the A-ring as expected (Kanesch and Polonsky, 1967; Gautier *et al.*, 1972). Owing to the predominant occurrence of neoflavonoids as heartwood constiutents, more detailed experiments on the formation of these compounds have not been done.

CHAPTER THREE

OCCURRENCE AND DISTRIBUTION OF FLAVONOIDS

INTRODUCTION

Flavonoids, along with a variety of other secondary metabolites including acetylenic compounds, alkaloids, cyanogenic glycosides, glucosinolates, and terpenoids, have played a major role in the development of the general subject of plant chemotaxonomy that flourished from the late 1950s well into the 1970s. The study of secondary metabolites and their application to systematic problems continues today, despite increased attention given to macromolecular systematics at the present time. In the early days of the subject, much effort was devoted to flavonoid surveys, some of which were quite extensive (e.g., Bate-Smith, 1962). This kind of broad scale search for chemical information, which would not have been feasible using the methods of classical natural product chemistry, was made possible through the newly available technique of paper chromatography. The researcher required little more than some glass tanks, chromatography paper, and appropriate solvents. The technique was sufficiently straightforward that it could be run in almost any laboratory providing a darkroom was available and, preferably, some sort of ventilation system! It was immediately obvious that, with minimal chemical sophistication, a person could accumulate a good deal of information from the observed chromatographic spot patterns. A mass of presence/absence data could then be subjected to multivariate analysis which provided an assortment of similarity coefficients from which relationships among the test organisms could be made. Chemical taxonomy was well on its way to becoming a major botanical growth industry!

The stage was set for the next step in the process. It was not long before critics pointed out the hazard of comparing patterns of spots the chemical nature of which were largely unknown. The increasing availability of compounds of known structure that could be used as chromatographic standards and the development of various more or less specific chromogenic sprays allowed workers to say with some certainty that the compounds with which they were dealing were flavonoids, or at least phenols. As we will see in Chapter Four, standardization of both chromatographic solvent systems and ultraviolet spectroscopic (UV) methods had a profound impact upon the subject. The combination of improved separation systems with a standardized analytical protocol brought about another significant change often overlooked. It became possible to obtain reliable data from very small amounts of plant material. No longer was it necessary to threaten a taxon with extinction in order to get enough material for structural determination of many of its components. It also became practicable to study individual pieces of a plant, or individuals in a population, or sample the same organism repeatedly to document diurnal or other environmental changes.

With the availability of improved spectroscopic tools, ^{1}H NMR and mass spectroscopy (MS), it was possible to determine the total structure of nearly all flavonoids. The availability of complete structures led the way to speculation on how the compounds were related in terms of numbers of biosynthetic steps. This was becoming an increasingly powerful means of analyzing a given set of data because major advances in establishing the flavonoid biosynthetic pathway were also being made during this period. With knowledge of the biosynthetic steps and associated enzymes it was a logical step to infer genetic similarities between flavonoid arrays.

CHEMOTAXONOMIC LITERATURE

The plant chemotaxonomic literature is vast. Since chemotaxonomy, or biochemical systematics, is a multidimensional, synthetic subject, practitioners must be acquainted with many separate disciplines, including systematics, genetics, evolution, plant biochemistry, and natural product chemistry. The botanical literature regularly contains papers in which chemicals, micromolecular or macromolecular, figure significantly. The natural product chemical literature, including such journals as *Journal of Organic Chemistry, Tetrahedron*, or *Tetrahedron Letters*, can be an important source of information for micromolecular studies. Pharmaceutical journals, such as the *Journal of Natural Products* (formerly *Lloydia*), *Planta Medica*, or *Plantes Medicinales et Phytotherapie*, contain much of value to the chemotaxonomist. Agricultural journals such as *Agricultural and Biological Chemistry* and the *Journal of Food Chemistry and Agriculture* often have information of use. *Phytochemistry* is an immensely important source of plant chemical information which, along with its sister publication *Biochemical Systematics and Ecology*, serve the chemotaxonomic community directly. Although not directly aimed at a biosystematic audience, *The Journal of Chemical Ecology* often has comparative information of interest to the more systematically oriented reader. Its ecological impact, of course, is significant.

The most comprehensive treatment of plant chemistry from the point of view of taxonomic application is the monumental multivolume *Chemotaxonomie der Pflanzen* by Robert Hegnauer. Begun in the early 1960s, this series reports essentially all chemical data on a family by family basis for all plant groups. Current taxonomic thinking on most of the groups is another feature of the treatment that is invaluable. For North American workers, Hegnauer's volumes are especially useful because of the difficulty of access of much of the European literature covered in his work. (If program directors in universities need a reason for suggesting German as a working tool for botany students, the Hegnauer series is it!) The reader should also be aware of the four volume set entitled *Chemotaxonomy of Flowering Plants* by R. D. Gibbs (1974) formerly of McGill University in Montreal. This publication records results of the author's personal research of many years and is valuable in that regard as well as for the detailed taxonomic information available. Unfortunately, information from the literature is not well referenced. Several books on various aspects of the subject have appeared and it is only possible to mention a few of them here.

The first book to address the subject of chemotaxonomy in a broad and introductory fashion, *Biochemical Systematics*, was written by Ralph Alston and

B. L. Turner (1963a) of the University of Texas, which has been and remains one of the major centers of biosystematic research in North America. The volume, although a comparatively slim one, masterfully combined the insights of a geneticist (Alston) with those of an experienced field botanist/taxonomist (Turner). Many of the ideas first aired in that book were re-examined by Alston (1967) in a long essay that included, among many other topics, a thoughtful discussion of variation in occurrence of natural products. Two additional comprehensive treatments of chemotaxonomy are *The Chemotaxonomy of Plants* by P. M. Smith (1976) and *Plant Chemosystematics* by J. B. Harborne and B. L. Turner (1984). Either of the latter two titles would serve admirably as textbooks on the subject. An important, and very personal, treatment entitled *Micromolecular Evolution, Systematics and Ecology* by O. R. Gottlieb (1982) should be read by everyone interested in the potential usefulness of small molecules as indicators of evolutionary relationships.

The last mentioned work by Gottlieb (1982b) presents a powerfully argued philosophical position on the overall subject of comparative small molecule biology. Other essays, before and since, have addressed the subject and provide thought provoking insights. These include a discussion of the role of chemistry in plant systematics by V. Heywood (1973) in which he reminds phytochemists of the tentativeness of most phylogenetic proposals, the need for understanding infraspecific variation, and the need for a better understanding of taxonomic classification systems. Hegnauer (1986) reviewed modern methods of plant classification and discussed the problems associated with name changes in plant taxonomy. He also discussed the importance of secondary chemicals in resolving infraspecific problems. The much broader aspects of angiosperm origins and evolution and the importance of secondary chemicals as sources of information were discussed by Kubitzki and Gottlieb (1984). They argued that the successful origin of angiosperms and their subsequent radiation was facilitated, along with other factors, by a rich chemical versatility of the new organisms.

There have been several conferences and symposia at which various aspects of chemotaxonomy have been discussed with the proceedings appearing eventually in book form. A few of these worth noting are *Chemotaxonomy and Serotaxonomy* edited by Hawkes (1968), *Phytochemical Phylogeny*, edited by Harborne (1970), *Chemistry in Botanical Classification*, which arose from a Nobel Symposium on the subject, edited by Bendz and Santesson (1973), *Chemistry in Evolution and Systematics*, from an International Union of Pure and Applied Chemistry symposium, edited by T. Swain (1973), *Phytochemistry and Angiosperm Phylogeny* (Young and Seigler, 1981), and the succinctly titled *Polyphenolic Phenomena*, edited by A. Scalbert (1993). This last title is the product of the Groupe Polyphénols, perhaps the most active group in the world at the moment dedicated to studying all aspects of, as the title says, polyphenolic phenomena.

Review articles too numerous to list have appeared over the years addressing general or very specific issues relating to all aspects of biochemical systematics. For example, Giannasi (1979) and Crawford (1979) published back-to-back reviews in which they debated several current issues including the question as to the limitations of flavonoid data at upper levels of the taxonomic hierarchy. Giannasi and Crawford (1986) combined their efforts in a reprise of most aspects of biochemical systematics discussed by Ralph Alston in 1967. Among the specialized problems discussed in the literature are such topics as intraspecific flavonoid variation (Bohm, 1987) and the question of how one determines whether a given flavonoid, or flavonoid profile, is evolutionarily primitive or advanced (Gornell and Bohm, 1978; Gornall *et al.*, 1979). The interested reader will find much of value in the four *Flavonoids* volumes on the significance of occurrence of certain compounds or groups of compounds. The third volume has chapters on flavonoids as taxonomic indicators in major taxa written by specialists: algae, bryophytes, ferns, and fern allies (Markham, 1988), gymnosperms (Niemann, 1988), dicotyledons (Giannasi, 1988), and monocotyledons (Williams and Harborne, 1988).

Over the past two decades or so several books have appeared that deal with individual flowering plant families or orders in great detail. Most of these have chapters devoted to biochemical information, including flavonoids. Families covered include Apiaceae (Heywood, 1971), Asteraceae (Heywood *et al.*, 1977), Fabaceae (Harborne *et al.*, 1971), Labiatae (Harley and Reynolds, 1992), and Solanaceae (Hawkes *et al.*, 1979). Two orders were treated in detail: Euphorbiales (Jury *et al.*, 1987) and Rutales (Waterman and Grundon, 1983). The bryophytes have been treated to their own volume as well (Zinsmeister and Mues, 1990).

CRITERIA FOR COMPOUNDS TO BE USEFUL AS TAXONOMIC MARKERS

In his 1967 treatment of the flavonoids, Harborne listed several criteria that should be met for a group of naturally occurring compounds to be useful in discussions of taxonomic relationships: (1) structural diversity; (2) widespread occurrence; (3) stability; and (4) ease of identification. A few comments on each of these with specific reference to flavonoids are in order. The matter of structural diversity was covered in some detail in Chapter Two where the different classes of flavonoids and accompanying structural modifiers were discussed. Flavonoids are nearly ideal molecules in this regard simply because not all plant groups have all classes of flavonoids and, in the cases where the plants under consideration do share the same flavonoid classes, not all species within these groups necessarily have the same individual compounds. Flavonoids meet the criterion of widespread occurrence very well. It is generally understood that flavonoids occur ubiquitously in ferns and fern allies (Pteridophyta), conifers (Gymnospermae), dicots, and monocots. In a few cases where flavonoids have not been found to occur in a given species, or group of species, it is generally considered that the capacity to make these compounds has been lost in these plants. Flavonoids have been reported from some green algae but the extent of the distribution in this group of plants is not known with certainty. One of the advantages of the occurrence of flavonoids over such a large range of plant groups is that it allows one to speculate on the evolution of these compounds in a very broad context. We will look at the overall subject of flavonoid evolution at the end this chapter.

The third criterion for usefulness deals with stability, which involves two distinct aspects. First, flavonoids are very stable chemicals that require no special equipment or procedures for handling. They are solids, have reasonable solubility properties, and can be stored for considerable lengths of time in solution in a refrigerator or in the dry state on the shelf. Solutions have to be protected from excess exposure to atmospheric oxygen and direct sunlight, but normal laboratory operations are sufficient for the vast majority of compounds. Extremes in pH should be avoided during isolation and purification of flavonoids. Since they are phenolic compounds they are sensitive to oxidation under alkaline conditions. Many flavonoids occur as glycosides and are thus prone to acid hydrolysis. Under neutral conditions, however, flavonoids are very forgiving compounds. One of the advantages of using flavonoids as taxonomic indicators is the possibility of getting highly dependable results from analysis of herbarium specimens. The second aspect of stability concerns physiological, developmental, and ecological stability of flavonoids. Ideally, these factors should be considered in each study, but in most circumstances it is not possible to do so, and often the work required is not considered worth the effort. The simplest way to avoid most of the problems (and criticisms!) is to work with plant material of more or less the same age. Under ideal conditions, of course, all work should be done with plants maintained under controlled conditions in a common garden.

The final criterion has to do with the comparative ease of detection, isolation, and structural analysis of flavonoids. Paper chromatographic methods along with readily available chromogenic reagents made analysis of flavonoid profiles easy to do and the development of standardized methods of ultraviolet (UV) spectroscopy (Mabry *et al.*, 1970; Markham, 1982) allowed the determination of structures. Simple surveys of plants using two-dimensional chromatography provided a good deal of information even if the structures of the compounds were not established. If structures were identified it became possible to judge relative genetic differences between species pairs or among groups of taxa in terms of steps in the flavonoid pathway, information on which was accumulating in parallel with the survey work. Although very sophisticated instrumentation (e.g., NMR, MS, X-Ray analysis) is currently used in flavonoid research, one should never disparage the early studies where the most basic methods were employed. Research should be judged, ideally, by the significance of its outcome and not on the sophistication of the machinery used.

THE ANALYSIS OF FLAVONOID DATA

Someone once told me that collecting data is the easy part of science; making sense out of it all is the challenging part. This is especially true in the study of evolutionary relationships and no less so in the application of chemical information to such problems. While it is quite feasible to arrive at the "correct" answer to the question, "What is the structure of compound Q that we have obtained from species X, Y, and Z?" it is another matter entirely to interpret the result in terms of the relationships of the three with the supposedly related species M and N, which do *not* have the compound. The structure of compound Q will have been determined through the correct interpretation of the usual spectroscopic data (UV, MS, NMR) (journal referees will check on this to be sure!) and corroborated through conversion to known compounds and possibly through total synthesis as well. However, when we attempt to infer relationships among the five species in question based upon the presence/absence data, we will certainly encounter some "opposition." The opposition could come from an expert in the field who doesn't think that the five species are really all that closely related in the first place, or from someone who thinks that M and N are so closely related that they shouldn't be recognized as distinct species at all and that their similarity to X, Y, and Z is entirely superficial. Or, criticism could come from a molecular biologist who rejects attempts to apply micromolecular data to evolutionary problems because these compounds can easily be influenced by environmental factors. A cladist (syn. phylogenetic systematist) might criticize the work because no attempts were made to "polarize" the data, that is, to establish whether the compound in question represents an advanced character (apomorphy in cladistic terminology) or a primitive one (plesiomorphy). A common criticism from phylogenetic systematists is that describing a compound, even one that is new and has a restricted distribution, represents little more than an exercise in natural history, and for the observation to make evolutionary sense, a more sophisticated, objective analysis is required. This requires the polarization step mentioned plus utilization of "outgroup comparison," wherein the worker uses a more or less related group of species (or whatever) of known evolutionary relationships as a comparison standard. A further problem, which might be pointed out by someone acquainted with the biosynthesis and enzymology of the compound in question, involves the possibility that it might have arisen by different pathways in the different species, and that this uncertainty could only be removed by examining the biosynthesis of the compound in the different species. One could, of course, take this to the limit and suggest that until the DNA sequence of the gene(s) coding for the enzymes involved have been determined there would always be doubt as to the "true" nature of the compound's origin.

All of the above criticisms are valid, depending upon one's point of view, and each of the approaches represents a legitimate means of addressing the question of relationships. The means by which problems are approached depend as much upon philosophy, the "narrative" as opposed to the cladistic approach, as they do upon the availability of research funding, considering that DNA sequence analysis tends to be more costly than studies of small molecules. Ideally, of course, as many different techniques as possible should be brought to bear on a problem. Since many laboratories are not equipped to look at all aspects of a problem, it is common to find collaborations involving several different specialists.

The history of flavonoid-based chemotaxonomy has seen all of the criticisms mentioned above (and then some)! The subject has survived, however, and matured to the point where today data from a flavonoid survey are nearly always discussed in relation to other information whether the source is cytogenetics, geography, micro- or macromorphology, breeding behavior, or DNA sequences. Owing to readily available DNA sequence data bases, it is now possible to superimpose flavonoid (or any other secondary plant product) occurrence data on a gene sequence-based phylogeny and arrive at an idea of how changes in the structures of the secondary compounds have occurred during evolution of the particular taxa. This approach can be seen below in the case involving the flavonoids of *Itea* and *Pterostemon* and the relationship of these genera to other members of Saxifragales (Bohm, unpubl. data).

In the early days of chemosystematics, many workers simply compared spot patterns obtained from two-dimensional paper chromatography as a way of judging relationships. This "phenetic" approach is conceptually straightforward being based on the idea that the more features that taxa have in common the more likely it is that the taxa are related. Consider the situation where chromatograms of species A, B, and C exhibit a total of five spots with A and B having four in common and with C having three spots, two of which are held in common with A and B as shown in Table 3.1. Using the phenetic approach, simple inspection suggests that A and B are more closely related to each other than either is to C. Although OTUs (OTUs are "operational taxonomic units" in the jargon of numerical taxonomy) A and B have identical scores (four), this does not automatically mean that they are identical, however, although the natural tendency would be to suggest that they are. It is entirely possible that unrelated compounds could have similar chromatographic behavior and give the impression that the observed patterns are identical. One of the major criticisms leveled at "spot counting" is that one knows little about the compounds responsible for the spots other than that they have responded to some visualization technique.

In cases where there are a greater number of variables (spots) and several taxa (OTUs), as shown in the data set in Table 3.2, some means other than simple inspection is obviously needed to analyze the data. Simply counting spots immediately gives us a value for each OTU; OTU A has the fewest spots with four while OTU B has the most with seven. Although this data set is comparatively simple, it is not possible to conclude from simple inspection, at least with any degree of certainty, that OTU A is more closely related to OTU B than it is to OTU C. Several factors make it difficult to assess relationships among members of even such a small group as this. There is the very real possibility that the compounds (spots) missing in OTU A are actually intermediates in the flavonoid pathway that accumulate in the other OTUs. Thus, all taxa might have the capacity to make most of the

Table 3.1 Flavonoid data set 1

	Character				
OTU	1	2	3	4	5
A	1	1	1	1	0
B	1	1	1	1	0
C	0	0	1	1	1

Table 3.2 Flavonoid data set 2

	Character										
OTU	1	2	3	4	5	6	7	8	9	10	(FS)[1]
A	1	1	1	0	0	1	1	0	0	1	6
B	1	1	0	0	1	1	1	0	1	1	7
C	0	0	1	1	1	0	0	1	0	0	4

[1]FS = Flavonoid score

compounds and it is only the differences in accumulation tendencies of the OTUs that we observe. Another factor that must be borne in mind involves the possibility that in any given OTU a compound may be present but at a concentration below the level of detection. The only way to deal with this problem is to use equivalent amounts of plant material from each OTU.

The calculation of "similarity indices" to assess degree of relatedness among individuals in a set had been successfully applied in other fields of science where complex arrays of data existed. It seemed logical to use this approach in analyzing chromatographic spot data so as to avoid the subjectivity of simple spot counting. The calculation of similarity indices and their use in taxonomic applications was treated in detail by Stuessy (1990, pp. 63–65), in his discussion of phenetic analysis. We will demonstrate the calculation of the *simple matching coefficient* using data in Table 3.2, which involves presence and absence values for ten flavonoids from three OTUs. The first step in this approach is to determine the number of character states that pairs of OTUs have in common and then divide that number by the total number of characters. (There are two character states in this analysis, present and absent, 1 or 0.) Thus, comparing OTUs A and B one sees that they have the same values for seven of the characters (both 1s or both 0s) resulting in a value of 0.7. Comparing A and C gives a value of 0.2, comparing B and C gives a value of 0.1. After all pairwise comparisons have been made, a data matrix is set up as shown in Table 3.3. The next step is to construct a phenogram (the general term is dendrogram) which presents a graphic view of relationships among the OTUs (Figure 3.1). OTUs A and B have the highest coefficient of similarity (0.7). The similarity between the A/B pair and OTU C is calculated by determining the unweighted arithmetic (= simple) average of values for the OTU pairs A/B, A/C, and B/C: $0.7 + 0.2 + 0.1 = 1.0/3 = 0.33$.

Manual calculation of coefficients of similarity for large data sets is tedious, error-prone, and, fortunately, unnecessary owing to the availability of computer

Table 3.3 Similarity matrix

	A	B	C
A	1.0	–	–
B	0.7	1.0	–
C	0.2	0.1	1.0

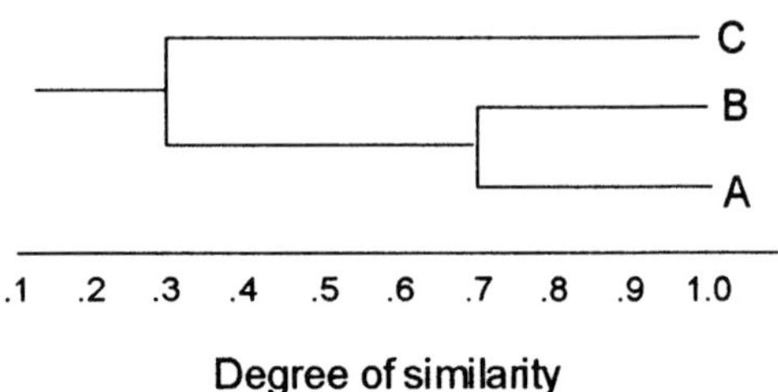

Figure 3.1 Phenogram derived from data set number 1.

programs, such as NTSYS-pc (Rohlf, 1992). Stuessy (loc. cit.) lists numerous sources of information concerning calculation of a variety of similarity measures and the assumptions underlying their use in phenetic analysis. For a particularly readable account of similarity indices, *An Introduction to Numerical Classification* by Clifford and Stephenson (1975) is highly recommended. An in-depth treatment of the subject can be found in *Principles of Numerical Taxonomy* by Sokal and Sneath (1963). It probably goes without saying that there has been extensive debate in the literature on the virtues of various similarity indices as they apply to different types of data. Some familiarity with the problems likely to be encountered in applying numerical methods will help the newcomer avoid some of the obvious pitfalls of the business. A comparison of different numerical taxonomic methods applied to a set of flavonoid data from Limnanthaceae was described by Parker (1976) who discussed the advantages, disadvantages, and assumptions associated with some of the more commonly used techniques including factor analyses. The examples used in the current chapter, however, represent artificially simple ones intended solely to demonstrate general methods.

The need to identify as many of the components of a flavonoid array as possible has been stressed as essential for arriving at a reasonable conclusion. A lively debate on the general subject, statistical analysis of spot patterns vs. the need for chemical identification, enlivened the pages of several journals in the late 1960s and early 1970s (Runemark, 1968; Adams, 1972, 1974; Weimarck, 1972; Crawford, 1979). Steady improvement in techniques for isolation and identification of flavonoids (including standardization of methods!), coupled with increasing knowledge of flavonoid biosynthesis and genetics resulted in the near elimination of spot counting. It was still necessary, however, to assign some measure of primitiveness or advancement to individual compounds or suites of compounds in order to assess the relative positions of the taxa in an evolutionary scenario. For example, Bate-Smith and Richens (1973) used a point count method to assess evolutionary advancement in *Ulmus*. To a large measure, their approach came from Bate-Smith's (1962) broad survey of dicotyledons for phenolic compounds which was based on the observations that the frequency of occurrence of myricetin and "leucoanthocyanidins" was highest in more primitive taxa. The *Ulmus* point system involved scoring +1.0 for the presence of myricetin or leucodelphinidin, +1.0 for the presence of leucocyanidin, and +1.0 when quercetin was present in higher concentration than kaempferol. A score of −1.0 was given if kaempferol concentration exceeded that of quercetin. Thus, with this system, higher numbers indicate a more primitive condition. Bate-Smith and Richens pointed out that any conclusions concerning a group as large and complex as *Ulmus* would have to be advanced with caution. With that caveat, they concluded that their flavonoid data supported the view that the genus was of east Asian origin (higher scores) with more advanced members distributed in western Eurasia, northeastern United States, and Canada (lower scores) with a second center of primitive species in southeastern United States and Mexico. This is in general agreement with the origin and dispersal of *Ulmus* as surmised by others on the basis of traditional data (see the Bate-Smith and Richens paper for leading references).

Harborne (1977) took the idea of primitive and advanced flavonoids somewhat farther by identifying the character state (primitive vs. advanced) for several additional flavonoid classes and structural modifications (Table 3.4). It should be appreciated that these "polarizations" represent very broad generalizations and that many "primitive" flavonoids occur in advanced plant groups and that the converse is true as well. It is useful to bear in mind that these values are used to best advantage when dealing with more or less closely related taxa; attempts to apply them in a broader evolutionary context runs a serious risk of encountering like compounds in taxa obviously not closely related. The issue of convergence will be met in a specific example below.

In answer to criticisms that simple point scores put too much emphasis on the presence or absence of compounds and not enough (none, actually) on their biosynthesis, Levy (1977) proposed a system that uses the "Minimum Biosynthetic Step Distance" (MBSD) and "Biosynthetic Step Identity" (BSI) as measures of similarity between pairs of taxa. The MBSD value is the sum of biosynthetic steps that distinguish a pair of flavonoid profiles. (The profiles could be among individuals in population, among populations, or among any collection of taxa of interest.) Put another way, MBSD values represent the number of biosynthetic steps necessary to convert two profiles to identity. The BSI value represents the number of biosynthetic steps in common beyond the last common precursor. We can illustrate the determination of these values using information that Levy (1977) used in his paper, with a slight elaboration. Consider three populations, P-1, P-2, and P-3, that are characterized by having, respectively, an apigenin 6-*C*-glucoside and *O*-glucoside thereof, only the *C*-glucoside, and neither compound. This simple array is seen in Table 3.5 using Levy's notation. In comparing P-1 and P-2 we see that the arrays are distinguished by the occurrence of both the mono- and the diglycoside in the former and only the

Table 3.4 Flavonoid character states after Harborne (1977)

Character	Primitive condition	Advanced condition
Anthocyanidins	Cyanidin	Pelargonidin or delphinidin
Anthocyanins	Simple glycosides	Complex glycosides
Leaf proanthocyanidin	Present	Absent
Flavonols	Myricetin common	Kaempferol and quercetin common
Flavones	Less common	More common
O-Methylation	Less common	More common
C-Glycosylflavones	Common	Uncommon
Biflavonoids	Present	Absent
A-Ring extra hydroxyl	At C-8	At C-6
Anthochlors	Chalcones	Aurones

Table 3.5 Flavonoid[1] data from Levy (1977)

Population 1:	Precursor →	A6Cg →	A6CgOg
Population 2:	Precursor →	A6Cg	–
Population 3:	Precursor	–	–
Population 4:	Precursor →	[ ] →	A6CgOg

[1] A = Apigenin; Cg = *C*-glucoside; Og = *O*-glucoside

monoglycoside in the latter. The MBSD value is "1" in this case because one additional biosynthetic step would be required to achieve identical profiles. The BSI value would also be "1" because these two populations have one compound in common beyond the last common precursor in the pathway. Comparison of P-1 and P-3 results in MBSD = 2, because two steps would be required to produce identical profiles, and BSI = 0 because they have no compounds in common beyond the common precursor. Similarly, a comparison between P-2 and P-3 gives MBSD = 1 and BSI = 0. One of the important features of this analytical technique is that the presence of an end product means that all of its precursors must be scored as being present as well. Although this statement may sound a bit peculiar, it recognizes that the presence of a compound in a profile represents the presence of all of the genetic material necessary to make that compound. Levy (1977) refers to the missing intermediate compounds as the "unsequestered precursors." With this idea in mind, we can return to the data in Table 3.5 and look at P-4. For the purpose of this discussion, this population has only a single compound, the diglycoside "A6CgOg," which we already encountered in P-1. Since we are working on the assumption that formation of A6CgOg requires the intermediacy of A6Cg, we are led to conclude that this precursor, although not accumulated (sequestered) in this population, is nonetheless present. Comparisons of P-4 with P-2 and P-3 yields the same MBSD and BSI values as those determined for comparisons of P-1 with P-2 and P-3. All pairwise comparisons are made following this procedure. The procedure can be quite complicated as evidenced by Levy's system wherein the 25 compounds represented a minimum of 19 pathways. In order to compare the degree of flavonoid similarity among *Phlox carolina* populations, Levy (1977) compared their standard Jaccard coefficients (J_{std}), based on compound identity (compounds in common divided by total number of compounds between the pair of taxa), with a Jaccard coefficient based on biosynthetic step identify (J_{bs}), defined as $J_{bs} = BSI + MBSD$. The outcome of these comparisons was that the biosynthetic-step indices, as compared to the use of the Jaccard coefficient "... proved chemically more definitive and systematically more consistent with other lines of evidence." A second example of the use of BSI and MBSD values can be found in the discussion of relationships among the sections of North American *Coreopsis* which formed part of a study of phylogenetic reconstruction using flavonoid data (Stuessy and Crawford, 1983).

It is felt by many workers that the closer one gets to the gene, the "better" the information. (No better seen than in the current DNA sequence frenzy.) The use of genetic data, as seen in the BSI/MBSD application, can have its drawbacks, however. The problem has been discussed in some detail by Crawford and Levy (1978) who based their critical appraisal on work done by van Brederode and colleagues on glycosylation of the *C*-glycosylflavone isovitexin in *Silene* (cited as *Melandrium*). The Dutch workers found that the isovitexin 7-*O*-glycosylation locus is allelic consisting of: (1) g^G which transfers glucose; (2) g^X and $g^{X'}$ which transfer xylose; and (3) the null allele g. The allele g^G is dominant over g^X while g^G and $g^{X'}$ are codominant. Three combinations yield isovitexin 7-*O*-glucoside, g^Gg^G, g^Gg, and g^Gg^X, while five combinations yield the 7-*O*-xyloside, g^Xg^X, $g^Xg^{X'}$, $g^{X'}g^{X'}$, g^Xg, and $g^{X'}g$. The combination $g^Gg^{X'}$, because these two alleles are codominant, yield a mixture of isovitexin 7-*O*-glucoside and 7-*O*-xyloside. It is obvious from this exercise that having identical phenotypes does not necessarily indicate identical genotypes. The situation becomes strikingly more complex if one considers the hypothetical situation where isovitexin is xylosylated at position-7, as above, and kaempferol is xylosylated at position-3 with a similar set of xylosyltransferases. In this system there are 25 different combinations that would afford a mixture of isovitexin 7-*O*-xyloside and kaempferol 3-*O*-xyloside! Despite these theoretical drawbacks, analysis of flavonoid genotypes can provide important insights into most situations. Indeed, much work is required, but the results often repay the efforts handsomely, as was the case with *Silene*.

Before going on to the next topic, it seems appropriate to express regrets at the apparent lack of interest by the chemosystematic community in using the BSI/MBSD approach. It is one of the few numerical methods that requires workers to focus on events, i.e., a sequence of biosynthetic reactions, rather than simply on an array of products. This approach requires that attention be paid to events that are actually happening in the organism, which should lead to a greater appreciation of process and less attention being paid simply on pattern.

Since most of the original work with flavonoids involved searching for patterns of similarity in large data bases (phenetic approach), it is not surprising that the more recent appearance on the scene of cladistically oriented people, who discounted the work as merely "narrative" having no theoretical basis, would engender some opposition, if not outright hostility. This was often the case when proponents of the two schools of thought squared off at meetings or in print! This conflict was not limited to chemotaxonomy, of course, but encompassed the whole range of systematic research. The pages of *Systematic Zoology* and *Taxon*, for example, fairly bristled with pointed disputes for several years. One of the problems with cladistic thinking, as seen by pheneticists, is that decisions about relationships are made on the basis of certain uncompromising *a priori* rules rather than on the familiar and more intuitive approach with which most non-cladistic workers were comfortable. Further problems involved the need to learn a new set of terms (some of which were seen parenthetically above), the question of polarization of features (chemical structures in our context), and the requirement for outgroup comparisons. Many comparative phytochemists, unfamiliar with taxonomic nuances, found the newer ideas too restrictive and continued reporting flavonoid distribution patterns and making comparisons along traditional lines. Further friction developed when comparative phytochemical manuscripts were given to cladistically oriented reviewers for comment (the reverse situation also existed, to be sure). With the passage of time, however, most of the ruffled feathers fell back into place, most tempers cooled, and, for the most part, objectivity returned. Current workers in the field recognize that analysis of a set of data by a variety of techniques will offer greater insight into the problem of relationships, just as the use of different sorts of data do. Cladistic analysis has taken its place along with other multivariate analytical techniques in the armamentorium of modern biosystematists. It might also be pointed out that a significant boost to the use of cladistic analysis for chemical data has come from the treatment of DNA sequence information where losses and gains of base pairs are easier to see in terms of primitive (plesiomorphic) or derived (apomorphic) characters.

The subject of cladistic analysis of flavonoids has been well discussed by two groups of workers, both of which have been active in the field from its early days. Richardson and Young (1982), in a paper entitled "The phylogenetic content of flavonoid point scores," review earlier efforts to use flavonoid point scores, examples of which were mentioned above, and present the case for cladistic analysis as an alternative approach. They discuss two examples, one using a family of their invention, the Imaginaceae (Humorales?), and one from the flavonoid literature involving the liverwort family Marchantiaceae, that demonstrate the overall analytical process including character state polarization. It was interesting to learn that relationships based on flavonoid point scores in the Marchantiaceae work (Campbell *et al.*, 1979) agreed quite closely with the results of a cladistic re-analysis by Richardson and Young. The other major contribution in the field is a paper entitled "Flavonoids and phylogenetic reconstruction" by Stuessy and Crawford (1983) who use their extensive experience with the genus *Coreopsis* (Asteraceae) to demonstrate the setting up of a data matrix, the calculation of MBSD and BSI values, and the construction of a phenogram showing evolutionary relationships among the several sections within the genus.

For the purposes of demonstrating a cladistic relationship I have invented a set of four taxa whose flavonoid array is shown in Table 3.6. All four share an array of flavonol 3-*O*-glucosides with additional structural features, diglycosides, 7-*O*-rhamnosides, acylated glucosides, and methyl ethers, distributed as shown. It is often easier to understand the analytical process by looking at the results first. A "cladogram" resulting from analysis of these data is shown in Figure 3.2a. Remember that in a cladistic analysis we are interested in the appearance of a new character (apomorphy) and whether it is shared with any other OUT (if so, it becomes a synapomorphy). We note that in the example, OTUs A and B share the capacity to make flavonol diglucosides (a synapomorphy), which became possible as a result of "evolutionary event 1," as marked on the cladogram. Owing to their sharing of the derived character, they are referred to as "sister taxa." OTU A, however, has also gained the capacity to make acylated glycosides through "evolutionary event 2," a feature not seen elsewhere in the set. (Although of no help cladistically, this feature is nonetheless useful in helping to distinguishing between OTUs A and B.) "Evolutionary event 3" involved gain of the capacity to make flavonol methyl ethers and clearly distinguishes the A/B branch of the cladogram from the C/D branch. Since both OTUs C and D share this feature they are also

Table 3.6 Flavonoid data set 4

	Structural features[1]				
OTU	3-Gly	3-DiG	7-Rha	Acyl	*O*-MeO
A	+	+	+	+	–
B	+	+	–	–	–
C	+	–	–	–	+
D	+	–	+	–	+

[1]3-Gly = 3-Glycosides; 3-DiG = 3-Diglycosides; 7-Rha = 7-Rhamnosides; Acyl = Acylation; *O*-MeO = *O*-Methylation

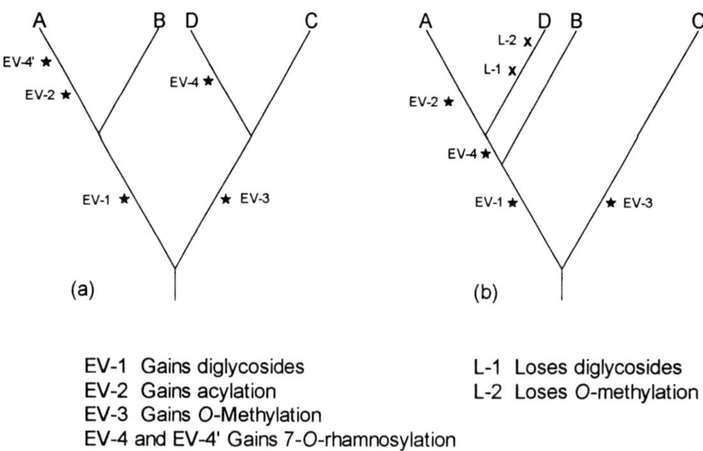

Figure 3.2. a/b Cladograms arising from data set 4.

considered as sister taxa. A fourth evolutionary event has occurred such that OTUs A and D have gained the capacity to make flavonol 7-*O*-rhamnosides. One explanation of this is that the appearance of 7-*O*-rhamnosylation represents convergent evolution (a homoplasy in cladistic terminology) in the two lines. In other words, these two OTUs have attained this new synthetic ability independently and not from a common ancestor. It would be possible, of course, to consider the alternative explanation, namely, that OTUs A and D are sister taxa because of their shared capacity to make 7-*O*-rhamnosides (a synapomorphy in this scenario). Such a relationship is represented as Figure 3.2b. There is a problem with this, however, in that if we choose to ally OTUs A and D on the basis of their shared capacity to make 7-*O*-rhamnosides, then OTU D must have lost the capacity to make both diglucosides and methyl ethers. This scenario, which involves two losses, is more drastic than the first, which involves only the convergent appearance of one feature. According to the guiding principle of "parsimony," the simpler scenario is the preferred one, which leads us to the conclusion that Figure 3.2a is the more likely relationship. In real-life situations, where quite large data bases can be involved, "reversals" of this sort are quite common. Cladistic analysis of large data sets by hand is a daunting proposition. Several computer programs have been developed that perform the necessary analyses many of which are discussed in a book entitled "*Cladistics: Perspectives on the Reconstruction of Evolutionary History*" (Duncan and Stuessy, 1984). Additional information on many aspects of cladistic analysis, including a chapter on flavonoids and phylogenetic systematics by Richardson (1983), appear in a volume edited by Platnick and Funk (1983).

Despite criticism from the cladistic school, there continue to be discussions of evolutionary relationships based upon absence/presence data. Most of the examples below involve straightforward comparisons of flavonoid profiles in taxa in question with known flavonoid profiles of putatively related taxa. Examples include the presence of the unusual 2′-hydroxyflavonols in *Datisca*, which help to distinguish it from *Octomeles* and *Tetrameles* within Datiscaceae (Bohm, 1988), the presence of *C*-glycosylflavones in *Itea* and *Pterostemon* in contrast to their absence in other members of Saxifragales (Bohm, unpubl.), the presence of 5-deoxyflavonoids in *Amphipterygium*, which argues for inclusion of the genus within Anacardiaceae (Young, 1981), the higher frequency of occurrence of anthochlors (aurones and chalcones) in subtribe Coreopsidinae than in any other subtribe of Heliantheae and, in fact, other comparably sized taxon in Asteraceae (Crawford and Stuessy, 1981), and the observation that 3-deoxyanthocyanins occur in members of subfamily Gesnerioideae of Gesneriaceae and not in Cyrtandroideae (Harborne, 1966). If one is at least willing to consider the possibility that the presence of these unusual compounds or groups of compounds may represent derived features within their respective taxa, then one is not too far from satisfying a cladistic criterion. I should imagine that judicious choice of outgroups might confirm this. In point of fact, there are precious few data to support, unequivocally, that these compounds are either primitive (plesiomorphic) or advanced (apomorphic) features (Gornall and Bohm, 1978). Most workers are content to recognize that the differences exist and that they correlate well with other features of the respective taxa.

CASE STUDIES

In this section we will examine the application of flavonoid data to a variety of systematic problems starting with the "plant kingdom" itself. The examples have been chosen to illustrate that flavonoids can be useful in addressing problems involving interfamily relationships, represented here by Asteraceae–Calyceraceae, or patterns within very large taxonomic groups such as Psilophytales and Caryophyllales, or establishing pattern of species migrations as in the case of *Silene*. Flavonoids have been instrumental in detecting evolutionary patterns within and among populations of such genera as *Lasthenia* (Asteraceae) and have proved very instructive in helping to establish the parentage of individuals within hybrid swarms, as in the classic case of *Baptisia*. Useful information has even emerged from studies of fossil flavonoids. Examples will also be presented in which the systematic problems addressed have noteworthy geographical significance with specific with reference to oceanic islands. The examples below are presented more or less from higher to lower levels in the taxonomic hierarchy starting with a few comments on the plant kingdom itself followed by interfamily relationships, on down the list, ending with intrapopulational variation.

The Plant Kingdom

In the introductory paragraph to this section, the term plant kingdom was set off by quotation marks by which device one recognizes problems in defining so a grand body of organisms. The classical view of plants included the fungi; more current thinking recognizes that this "lumping" is not acceptable for a number of reasons, which are beyond the scope of this book. Although the biochemical capabilities of fungi are formidable, their capacity to make flavonoids seems to be very limited.

Only a few reports of flavonoids from fungi have appeared and these were met with some skepticism. A compound with antibiotic properties isolated from cultures of *Aspergillus candidus* (BRL 274) (Richards *et al*., 1969) was shown by subsequent work to be 5,2′-dihydroxy-3,7,8-trimethoxy-3′-chloroflavone [3-1], and named "chlorflavonin," by (Bird and Marshall, 1969). The microorganism was cultured on a

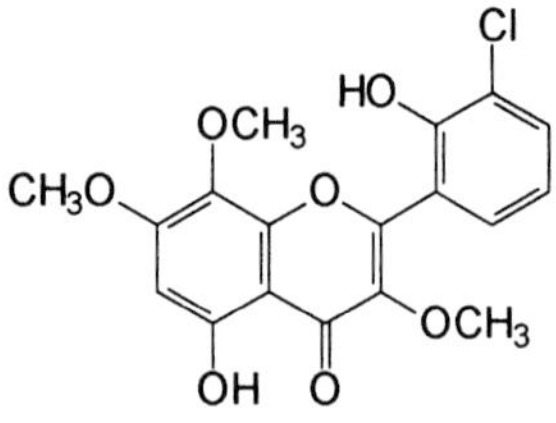

3-1

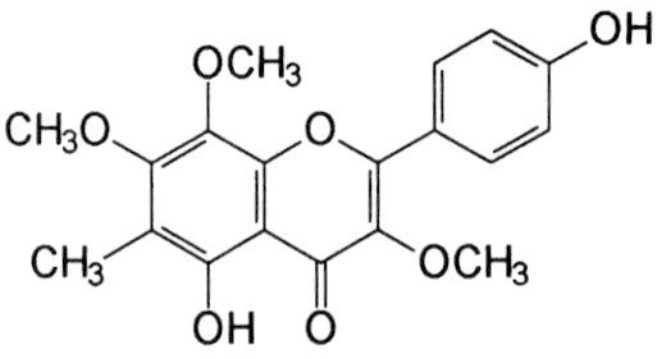

3-2

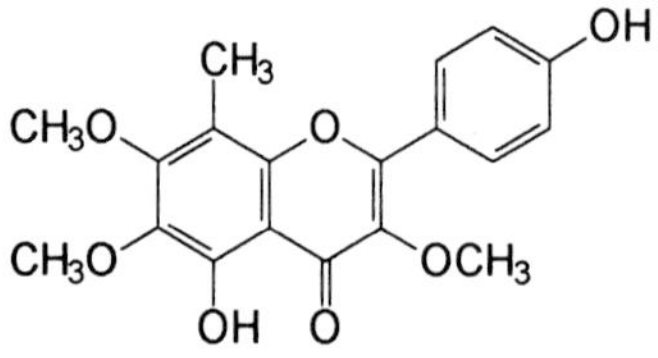

3-3

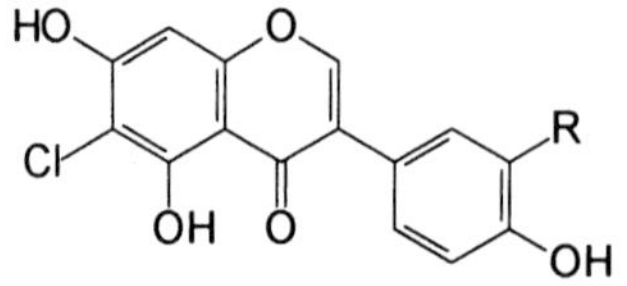

3-4 R = H

3-5 R = Cl

corn-steep liquor dextrose broth. There was no discussion as to whether the flavonoid might have originated from the nutrient mixture. This issue was addressed by Marchelli and Vining (1973), however. In addition to reporting the dechloro congener of chlorflavonin in this organism, they showed that β-^{14}C-phenylalanine efficiently labeled C-2 of chlorflavonin which led them to the conclusion that the flavonoid was biosynthesized *de novo* by the fungus. The recent report of *C*-methylflavonols from *Colletotrichum dematium* f. sp. *epilobii* by Abou-Zaid and coworkers (1997) also appears to be firmly based. 5,4′-Dihydroxy-3,7,8-trimethoxy-6-*C*-methylflavone [3-2] and 5,4′-dihydroxy-3,6,7-trimethoxy-8-*C*-methylflavone [3-3], whose structures were established by UV, ^{1}H NMR and ^{13}C NMR, were not observed in exhaustive analyses of sterile culture medium. Two chlorinated isoflavones were isolated from cultures of *Streptomyces griseus* and identified as 6-chloro- [3-4] and 6,3′-dichlorogenistein [3-5] (König *et al*., 1977). These compounds were shown to arise by chlorination of genistein present in the soybean-based nutrient medium.

The presence of flavonoids in algae has been a topic of some interest over the years in view of evolutionary links that have been suggested between certain algal groups and terrestrial plants that are known to produce flavonoids. Markham (1988) discussed earlier reports of flavonoids from *Chlamydomonas* and *Nitella*, pointed out the problems associated with those studies, and documented his failure to find any trace of flavonoids in a study of 15 genera. The question as to whether algae have the capacity to make flavonoids remains unanswered at this time.

While on the subject of algal flavonoids, it is appropriate to describe what Stafford (1990) refers to as "The flavonoid fraud of 1940 to 1955." The problem surrounds the claim by the German geneticist F. Moewus that certain flavonoids served as female and male sex hormones in *Chlamydomonas* (Birch *et al.*, 1953), isorhamnetin (quercetin 3′-methyl ether) in the former and cyanidin 3′-methyl ether diglycoside in the latter. Subsequent notes (Moewus, 1954a,b) reported that an alteration in the pathway from production of the above compounds to rutin (quercetin 3-*O*-rutinoside) resulted in sterility. Attempts by others to reproduce what would have been an extremely important finding failed (Ryan, 1955). Stafford added that she was unaware of any official retraction of the idea; I know of none either. In a recent paper by Musgrave and van den Ende (1987) on reproductive behavior of *Chlamydomonas*, no mention of this bizarre episode was made.

All other groups of plants have at least some members that have been conclusively shown to synthesize flavonoids. Again citing Markham's (1988) review, we learn that about 41% of liverworts (Class Hepaticae) and 48% of mosses (Class Musci) tested were flavonoid-positive. One of the problems that manifests itself in such surveys, however, is sampling error as evidenced by the fact that (at that time) only 6% of liverwort species had been examined. Flavonoids had not been reported in members of Class Anthocerotae at the time of that review. Flavonoids are otherwise a common feature in members of the plant kingdom, narrowly defined if one chooses.

Psilophytales

This order consists of two genera, *Psilotum* with two or three species, and *Tmesipteris* with from two to six species, depending upon taxonomic opinion. Earlier views of this general group saw Psilophytales as comprising only fossil genera (*Psilophyton*, etc.) while the two extant genera comprised Psilotales. We follow the treatment of Bold (1973) here in which the five associated orders that constitute the so-called fern allies group are given ordinal status: Equisetales, Isoëtales, Lycopodiales, Psilophytales, and Selaginellales. Details of the distribution of flavonoids in these taxa can be found in Markham (1988). The predominant flavonoids in *Psilotum* and *Tmesipteris* are biflavonoids of the amentoflavone group (apigenin–apigenin derivatives) [3-6] , which occur along with small amounts of apigenin and amentoflavone glycosides (Wallace and Markham, 1978; Markham, 1984). Amentoflavone glycosides appear to be unique to this group. Flavonols have not been found and proanthocyanidins also appear to be absent. This flavonoid profile is unique within the fern allies group. The only other biflavonyl-bearing member of the group are Selaginellales, which accumulate hinokiflavone (oxygen-linked apigenin dimers) [3-7] derivatives as well as amentoflavones. Markham (1988) refers to this apparent specialization in amentoflavone by Psilophytales as a mark of "biosynthetic purity" and that it suggests a long period of isolation.

In addition to the unusual flavonoid array presented by these genera, another compound type has also attracted some attention. Psilotin [3-8] and its hydroxy derivative [3-9] are *C*-glucosyl phenols known only from members of these two genera (McInnes *et al.*, 1965; Tse and Towers, 1967; Balza *et al.*, 1985b).

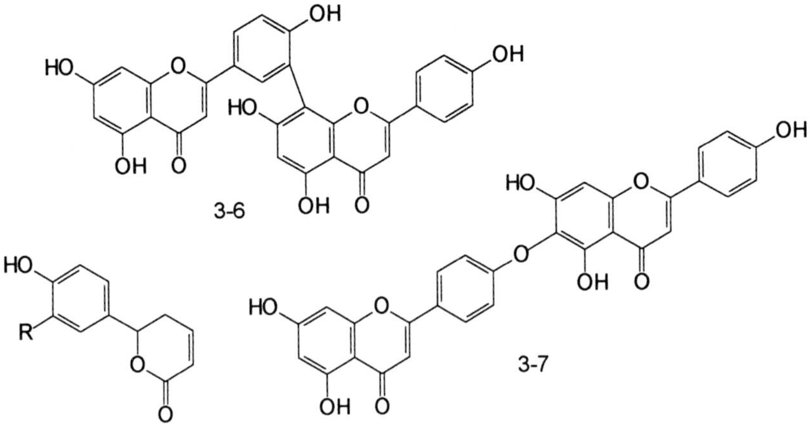

3-6

3-7

3-8 R = H
3-9 R = OH

Flavonoids and the Liliiflorae *sensu* Dahlgren

One of the most sweeping efforts within recent taxonomic history has been a re-evaluation of the monocotyledons under the leadership of the late Rolf Dahlgren (Dahlgren and Clifford, 1982; Dahlgren and Rasmussen, 1983; Dahlgren *et al.*, 1985). The result of these efforts is a completely redefined system comprising five orders: Asparagales, Burmaniales, Dioscoreales, Liliales, and Melanthiales. Many new families are recognized and several older ones have been resurrected. The impact of the new treatment is most clearly seen in the reduction of the formerly large Liliaceae [294 genera with 4550 species (Mabberley, 1987)] to a much smaller assemblage comprising only 13 genera and 385 species. Liliales, by contrast, is expanded to incorporate Orchidaceae [Orchidales in Cronquist (1981)] as well as Iridaceae. These two families, Orchidaceae and Iridaceae, are singled out for comment because of the large amount of information available concerning their flavonoids.

Earlier flavonoid surveys of monocots had shown a good correlation between frequency of occurrence of certain pigment types and family taxonomy (Williams and Harborne, 1988). It was considered timely by Williams and coworkers (1988) to reinvestigate flavonoid patterns from the point of view of the new treatment. To this end, they compiled information on 181 species representing 29 of the 52 families. Other than a few odd compounds whose significance is not certain owing to limited sampling in their respective source families, flavonoid profiles fell into two groups, those yielding only flavonols (23 families) and those yielding both flavonols and flavones (6 families). The occurrence of these two profile types cuts across the ordinal boundaries of the new system and, thus, provides no additional support for its recognition. This is not surprising in view of the number of families being considered and the limited amount of variation in the observed flavonoid profiles. The authors (Williams *et al.*, 1988) suggest that a careful study of the glycosylated forms of the flavonoids may reveal insights not forthcoming from the aglycones.

A problem in reconciling the new system is the way in which Orchidaceae and Iridaceae are included in Liliales. The flavonoid profiles of these two families are based upon *C*-glycosylflavones, which are otherwise rare in Liliiflorae. Biflavonoids are known from more primitive members of Iridaceae and isoflavones are common components of *Iris* species, along with quinones and xanthones. It has been suggested that these two families might well be placed in their own orders on the basis of their unique morphologies and supporting chemistries (Williams, 1979; Williams *et al.*, 1986).

One of problems encountered in very large surveys, and the monocot survey is certainly a case in point, is the lack of reliable data for many families or key genera. Only a little over 44% of the families recognized by the Dahlgren school were available for flavonoid analysis. This represents a major sampling error, although the problem is likely lessened in this case because of the lack of variation in the kinds of flavonoid classes involved (only flavones and flavonols). The surveys of both Iridaceae and Orchidaceae, along with information already in the literature, were based on a better sampling of those families, hence the conclusion that they are chemically quite different from the Liliales *sensu* Dahlgren stands on comparatively firm ground.

Relationship between Calyceraceae and Barnadesioideae (Asteraceae)

This case study combines investigations into two groups of plants, one involving Calyceraceae and one involving Barnadesioideae, the most recently described subfamily within Asteraceae. The stories are combined here because of the role these major taxa have played in addressing the question of the origin of Asteraceae. It will be useful to say something about each group before we consider the chemical information. Calyceraceae is a South American family comprising six genera and 50–60 species. The genera are *Acicarpha, Boöpis, Calycera, Gamocarpha, Moschopsis*, and *Nastanthus*. A review of the taxonomic literature (for leading references see Lammers, 1992) shows that Calyceraceae has generally been considered to lie near Asteraceae. Attention to the family increased markedly as it became increasingly clear that Calyceraceae may be the sister group to Asteraceae (Hansen, 1992; Gustafsson and Bremer, 1995; DeVore and Stuessy, 1995). A study of the flavonoids of 11 species representing five genera of Calyceraceae showed the presence of kaempferol and quercetin 3-*O*-glucosides, quercetin 3-*O*-galactoside, kaempferol and quercetin 3-*O*-rutinosides, 6-methoxykaempferol, and 6-methoxyquercetin (patuletin). The latter two compounds were widespread in the five genera and were seen only as aglycones. The simplest profile, consisting of kaempferol and quercetin 3-*O*-glucosides and 3-*O*-rutinosides, was seen in *Acicarpha tribuloides, Gamocarpha alpina*, and one specimen (of two) of *Calycera leucanthema*. Owing to the size of Asteraceae and the amount of information compiled for the family, it is not surprising that no clear-cut comparison can be made between Calyceraceae and Asteraceae as a whole. The significance of the Calyceraceae flavonoids emerges only when we consider those data in relation to a particular group within Asteraceae, which brings us to the second part of this case study.

The finding by Jansen and Palmer (1987) that all members of Asteraceae, except three genera in tribe Mutisieae subtribe Barnadesiinae, possessed a 22 kb inversion in the chloroplast DNA led to a series of investigations that has resulted in a reassessment of phylogenetic relationships within Asteraceae. They judged the inversion to be a derived trait (synapomorphy) in the family. The absence of the inversion in the genera belonging to Barnadesiinae, *Barnadesia, Chuquiraga*, and *Dasyphyllum*, was taken to indicate that that group represented a primitive lineage in Asteraceae. The absence of the chloroplast DNA inversion coupled with morphological information led to the elevation of subtribe Barnadesiinae to the status of subfamily, Barnadesioideae (Bremer and Jansen, 1992). [Currently, only three subfamilies are recognized, Asteroideae, Barnadesioideae, and Cichorioideae (Bremer, 1994).]

The flavonoid profiles of seven of the nine genera (31 of 90 species) of Barnadesioideae have been determined (Bohm and Stuessy, 1995) (number of species studied in parentheses), *Arnaldoa* (1), *Barnadesia* (12), *Chuquiraga* (9), *Dasyphyllum* (6), *Doniophyton* (1), *Fulcaldea* (1), and *Schlectendahlia* (1). Compounds reported were kaempferol and quercetin 3-*O*-glucosides, kaempferol and quercetin 3-*O*-glucuronides, quercetin 3-*O*-rhamnoside, kaempferol and quercetin 3-*O*-rutinosides, isorhamnetin 3-*O*-glucoside, and eriodictyol 7-*O*-glucoside (sugar tentative). Kaempferol and quercetin 3-*O*-glucosides and 3-*O*-rutinosides were nearly ubiquitous, the glucuronides were present in about half of the specimens, eriodictyol was seen in four species, and isorhamnetin was detected in only one species, *Barnadesia parviflora*. Our conclusion in that paper (Bohm and Stuessy, 1995) was that "... no other major taxon within the Asteraceae (i.e., other subfamily, tribe, or subtribe) has such a consistently simple pigment profile..."

With the flavonoid chemistry of both Calyceraceae and Barnadesioideae in hand, meaningful comparisons were possible. It was noted that the simple flavonoid profile observed for Barnadesioideae was also present in *Acicarpha tribuloides*, *Gamocarpha alpina*, and in a single specimen of *Barnadesia*. The inclusion of *Acicarpha* in this short list is significant since it is the genus that has been advanced as representative of prototypical Calyceraceae (DeVore, 1994). This view is based on several features, centripetal floral development, diploid chromosome number ($n=8$), which is the only diploid known so far in the family, and its range of occurrence in the grasslands of Uruguay, which is similar to the range of *Schlectendahlia* of Barnadesioideae (see Bohm *et al*., 1995 for details). A simple and presumed primitive array of flavonoids can be added to that list of linking characters.

Caryophyllales

Caryophyllales, or Centrospermae as the taxon was more commonly referred to in the older literature, present us with both one of the most interesting studies in flavonoid systematics as well as one of the most perplexing. Caryophyllales represents a large assemblage of species assorted into 12 families 10 of which do *not* have anthocyanins despite their strongly red to purple colored flowers or other pigmented organs. The nitrogen-containing pigments responsible for these colors were originally called "nitrogenous anthocyanins" but were ultimately shown to belong to a class of natural products in their own right, the "betalains." Betalains have in common a structural unit based on betalamic acid [3-10] whose formation through ring cleavage of 3,4-dihydroxyphenylalanine (L-dopa) and recyclization has been well documented (Mabry, 1977; Piattelli, 1981). "Betacyanins" consist of the betalamic acid unit and a cyclodopa [3-11] unit to form the isomeric pair "betanidin" [3-12] and "isobetanidin" [3-13]. These two bright red pigments were originally isolated from the common beet, *Beta vulgaris* (Chenopodiaceae). "Betaxanthins," consist of the betalamic acid moiety and some protein amino acid; "indicaxanthin," pictured in [3-14], is based upon proline.

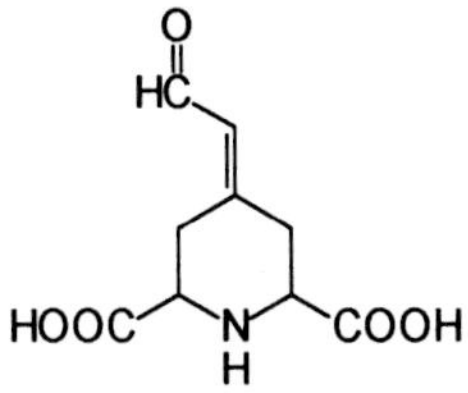

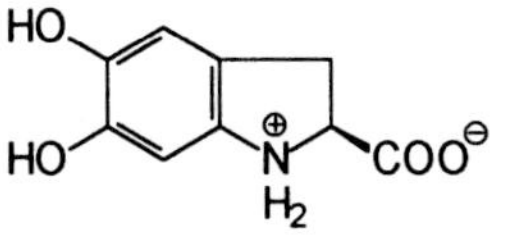

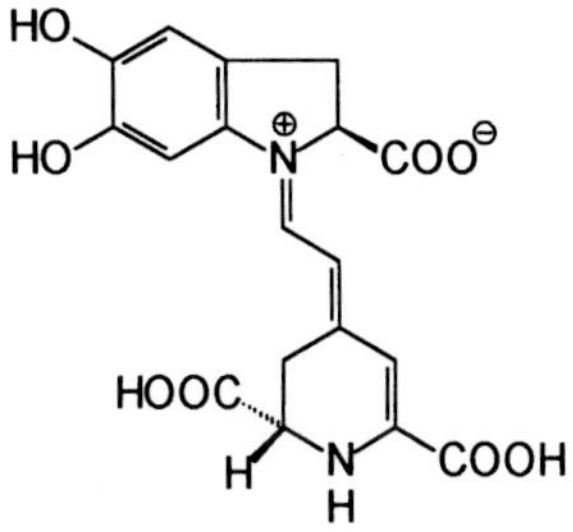

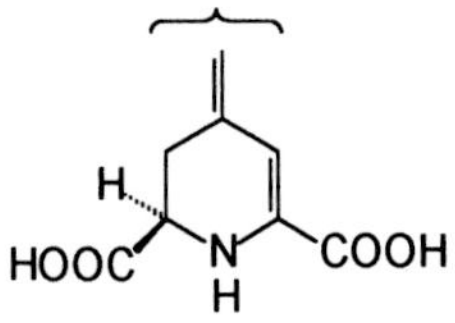

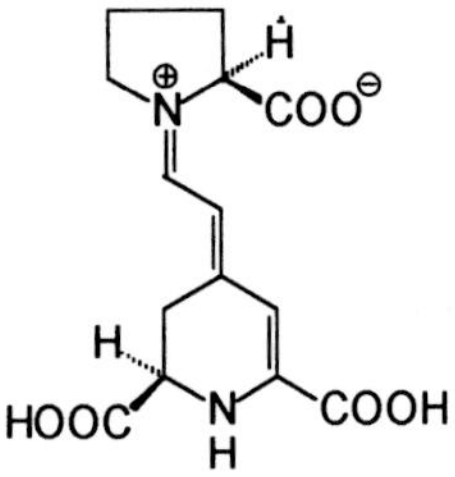

Betacyanins and/or betaxanthins have been identified in species belonging to the following families: Achatocarpaceae, Aizoaceae (ice-plants), Amaranthaceae (grain amaranth), Basellaceae, Cactaceae, Chenopodiaceae (beets, pigweed), Didiereaceae, Nyctaginaceae (4-O'clocks; *Bougainvillea*), Phytolaccaceae (pokeweed), and Portulacaceae (*Lewisia*). The two remaining families, Caryophyllaceae (pinks) and Molluginaceae, do not make betalains but do make anthocyanins. With regard to the rest of the flavonoid classes, the entire order appears to be completely normal in that they accumulate a wide array of flavones and *C*-glycosylflavones. It is the anthocyanin/betacyanin replacement phenomenon that has been the focus of attention in the order. The chemistry and systematic botany of Caryophyllales has been discussed in detail (Mabry, 1977, 1980; Piattelli, 1981). The main botanical argument has centered on how to recognize the betalain-positive taxa, or, in fact, whether they deserve separate recognition on the basis of the presence of one chemical type (accompanied by the absence of another, to be sure). Mabry (1973, 1977) raised the issue as to whether the order is monophyletic and whether the two groups of families, those with betalains and those without, should be recognized at the subordinal level. This would have the betalain-positive taxa constitute suborder Chenopodiinae while the anthocyanin-positive taxa would reside in Caryophyllinae.

Although there had been scattered reports in the literature, it fell to Richardson (1978) to provide the definitive statement concerning the non-anthocyanic flavonoids in Caryophyllales. We now enjoy the results of his survey of 112 species representing 61 genera and 11 families. Flavonols, mainly kaempferol and quercetin but occasionally myricetin and isorhamnetin, were observed in all 11 families sampled, while *C*-glycosylflavones were observed in only nine, being absent from Aizoaceae and Cactaceae. Richardson pointed out an interesting feature of the myricetin sightings, which involved Didieraceae and Chenopodiaceae. The report from a single species of *Chenopodium* (six surveyed), out of some 120 (Mabberley, 1987) highlights the need for a broader sampling of the larger genera. The positive sighting in Didieraceae, in two species of *Alluaudia* [six species tested; six known (Mabberley, 1987)], is more likely to represent a useful character. Didieraceae are endemic to Madagascar and have morphological similarities with members of *Pereskia* (Cactaceae). The two species of *Pereskia* (perhaps 20 are known) examined had only kaempferol and quercetin. The occurrence of myricetin in these two families may represent an example of convergence/parallelism, as pointed out by Richardson (1978).

Ultrastructural studies of members of this order have revealed an important feature that reinforces the idea that the group is a natural and coherent one. The sieve-element plastids in all members of the order, betalain and anthocyanin groups alike, contain ring-like inclusions composed of proteinaceous strands not known from any other group in the angiosperms (see Behnke, 1976 for review). Another interesting feature of the group is the high percentage of taxa that possess the so-called C_4 photosynthetic machinery (Kranz syndrome). This level of occurrence of the Kranz syndrome in both betalain and anthocyanin families and its sporadic occurrence in other dicot groups provides support for the idea that Caryophyllales (Centrospermae) is an old order. The question of the evolutionary relationship between anthocyanin and betalain families remains moot. Stafford (1994) addresses the issue with attention to the possible metabolic and control functions that might be played by betalain pigments. She poses the logical questions: (1) Did loss of "anthocyanidin synthase" activity occur first with subsequent development of the capacity to make betalains? or, (2) Did the betalain pathway evolve in parallel with the anthocyanidin pathway and survive through selection? Detailed information at the enzyme level and using the tools of molecular genetics is needed before these questions can be answered.

Some steps in that direction have been taken by Downie and Palmer (1994) who described a phylogeny of Caryophyllales based on structural and inverted repeat variation in the chloroplast genome as determined by restriction site analysis. The only conclusion of immediate relevance to us is that Caryophyllaceae and Molluginaceae, the only two anthocyanin-producing families in the order, were not basal to the group, although they did emerge in the same part of the resultant trees. Additional molecular phylogenetic studies are clearly needed before the significance of the positioning of the anthocyanin-producing families can be assessed.

Affinities of Julianaceae

Julianaceae is a small family of two genera, *Amphipterygium* with four species native to Central America, and the monotypic Peruvian *Orthopterygium*. The affinities of the family had been debated for at least a hundred years, one school placing it near Juglandaceae (walnuts) in Amentiferae on the basis of its pollination biology (lack of petals; wind pollinated; male flowers in loose catkins). Others (Cronquist, 1981) have placed Julianaceae close to, or included within, Anacardiaceae. This latter position is strongly supported by studies of wood anatomy (Stern, 1952; Youngs, 1955). An examination of the flavonoid chemistry of leaves and heartwood of *A. adstringens* (Young, 1976) provided additional support for alignment with or near Anacardiaceae. Sixteen compounds were identified among which were seven 5-deoxyflavonoids. All seven had been reported from members of Anacardiaceae and, moreover, some of these, e.g., sulfuretin, fisetin, and fustin, had been cited as characteristic of that family by Hegnauer (1964). Additional support for the alignment of these families came from identification of the unusual 6,3′,4′-trihydroxy-4-methoxyaurone (rengasin) [3-15] as a component of *Amphipterygium*.

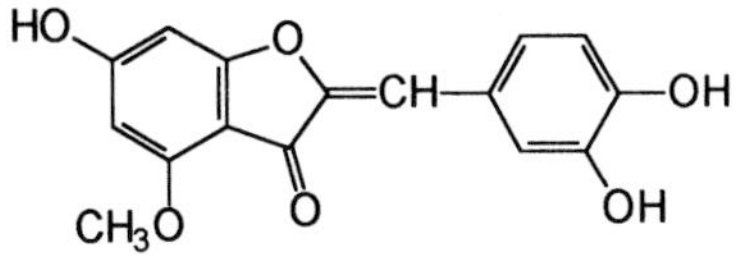

HO, O, =CH—, —OH, CH_3O, O, OH

3-15

This compound had until that time been found only in members of Anacardiaceae. An examination of representatives of Juglandaceae (Young, 1979) failed to find any of the flavonoids isolated from *Amphipterygium* or Anacardiaceae. Based upon the close chemical and morphological similarity between *Amphipterygium* and members of tribe Rhoeae, Young (1976) suggested that the relationships of Julianaceae to Anacardiaceae would be better represented by placing *Amphipterygium* and *Orthopterygium* within that family as subtribe Julianiinae within tribe Rhoeae. Serological data reported by Petersen and Fairbrothers (1979) support this position.

A Family Home for *Polypremum*

Polypremum procumbens, the sole member of the genus, occurs in southeastern United States. Its familial alignment has been a subject of disagreement for many years. de Candolle (1830, 1845) first placed *Polypremum* in Rubiaceae but shifted it to Loganiaceae in a later treatment. Opinion was, and continues to be, divided on this issue with several workers in favor of Loganiaceae (Leeuwenberg, 1980; Tahktajan, 1986; Mabberley, 1987; Thorne, as cited by Scogin and Romo-Contreras, 1992) while others tend toward an association with *Buddleja*. In this latter group are Bentham (1846), who assigned the genus to tribe Buddleieae of Scrophulariaceae, and Cronquist (1981) who "reluctantly" included it in Scrophulariales. Cronquist's (1981) view was that *Polypremum* clearly belongs in Loganiaceae if that family is defined in

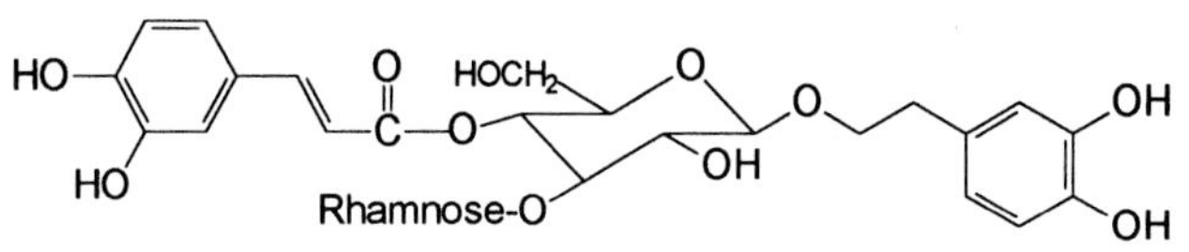

3-16

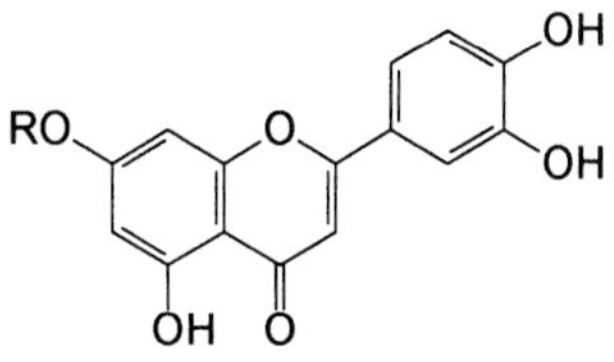

3-17 R = Glucose
3-18 R = Rutinose

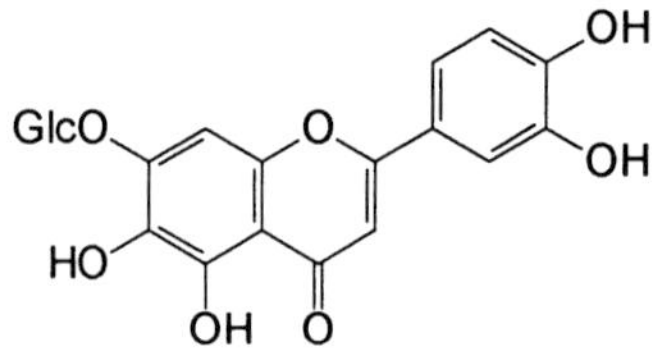

3-19

the traditional broad sense to include *Buddleja*, but not when Buddlejaceae is allowed to stand on its own. Cronquist (1981) noted that anatomical and chemical features had not been determined. Scogin and Romo-Contreras (1992) addressed the lack of chemical information with convincing success. They identified four phenolic compounds, the phenylpropanoid glycoside acetoside [3-16], and the flavones, luteolin 7-*O*-glucoside [3-17] and 7-*O*-rutinoside [3-18] and 6-hydroxyluteolin 7-*O*-glucoside [3-19]. Acetoside is characteristic of members of Scrophulariales (Scogin, 1992), including *Buddleja* (Ahmad, 1986). 6-Hydroxyluteolin is considered to be characteristic of Scrophulariales by some workers (Tomas-Barberan *et al.*, 1988). Neither acetoside nor 6-hydroxyluteolin has been reported from Loganiaceae.

Gesneriaceae

Gesneriaceae is a large family consisting of 146 genera with about 2400 species with its highest concentration in the Old World tropics (Mabberley, 1987). The genera have been placed in two subfamilies, Gesnerioideae and Cyrtandroideae, depending upon two major structural features. Gesnerioideae have equal cotyledons (isocotyly) and the ovary is partly or entirely inferior, while Cyrtandroideae have unequal cotyledons (anisocotyly) and the ovary is always superior. In addition, Gesnerioideae are almost entirely found in the New World whereas Cyrtandroideae are almost completely Old World taxa. A third subfamily, Coronantheroideae, has been suggested to accommodate a few Old World genera that lack the specialized characters of the other two subfamilies. A significant difference in flavonoid chemistry between Gesnerioideae and Cyrtandroideae was first noted by Harborne (1966). 3-Deoxyanthocyanins were present in 18 species representing 11 genera of Gesnerioideae but were absent from 25 species belonging to eight genera of Cyrtandroideae. An additional difference in the flavonoid profiles of these two taxa was noted, namely the presence of chalcones or aurones in members of Cyrtandroideae and their absence from members of Gesnerioideae. In the following year, Harborne (1967) reported 3-deoxyanthocyanins in a further 11 species of Gesnerioideae representing seven additional genera, and anthochlors in a further six species of Cyrtandroideae. J. B. Lowry (1972) examined members of five genera of Malaysian Cyrtandroideae and reported the presence of normal anthocyanins as would be expected from the subfamily. He made no comment on anthochlors. In a subsequent paper, he (Lowry, 1973) reported the presence of cyanidin and pelargonidin 3-*O*-glucosides from *Rhabdothamnus solandri*, which is monotypic and the sole member of Gesneriaceae in the New Zealand flora. He also noted the absence of anthochlors, flavonols, and "leucoanthocyanidins" in this species.

Harborne (1966) commented on other occurrences of 3-deoxyanthocyanins in plants including their presence in the moss *Bryum cryophyllum* (Bendz *et al.*, 1962). Other reports from flowering plants included several gesneriads (Robinson *et al.*, 1934), *Camellia sinensis* (Roberts and Williams, 1958), *Sorghum vulgare* (Poaceae) (Stafford, 1965), *Chiranthodendron pentadactylon* (Sterculiaceae) (Pallares and Garza, 1949), and *Arrabidaea chica* (Bignoniaceae) (Chapman *et al.*, 1927). The report from Bignoniaceae was of particular interest since the family is considered to be closely related to Gesneriaceae. Presence of 3-deoxyanthocyanins in both families would provide a significant chemical link between them. However, Harborne (1967) did not detect 3-deoxyanthocyanins in 16 species representing 11 genera (3 tribes) from Bignoniaceae; normal anthocyanins were present.

The 1966 paper by Harborne also included the first identification of 3-deoxyanthocyanins as the pigments responsible for coloration in eight species of ferns representing the following genera: *Adiantum, Blechnum, Dryopteris, Osmunda*, and *Pteris*. The 3-deoxyanthocyanins identified were all present as glycosides. Treatment of plant material with strong acid resulted in the appearance of normal anthocyanidins, i.e., cyanidin, pelargonidin, and delphinidin. Under the conditions employed it is likely that these compounds arose from proanthocyanidin oligomers.

Although it is not relevant to the 3-deoxyanthocyanin story, it is interesting to point out a study by Kvist and Pedersen (1986) in which they examined 590 specimens representing 91 genera of Gesneriaceae using electron paramagnetic resonance (EPR) spectroscopy. This method allows study of dihydric phenols as their semiquinone radicals. Signals observed in spectra of plant extracts can be compared with those generated from known compounds. The occurrence patterns of some 15 different compounds was reported. One of the conclusions of some taxonomic significance in the present context was that the EPR data for members of Coronantheroideae show the presence of many compounds seen in one or both of the other two subfamilies, thus arguing against separate taxonomic recognition of the subfamily.

Datiscaceae

The small family Datiscaceae are considered by some to consist of three genera, *Datisca*, *Tetrameles*, and *Octameles* (Cronquist, 1981; Mabberley, 1987). *Datisca* consists of two species, *D. cannabina*, whose range extends from Asia Minor to India (Mabberley, 1987), and *D. glomerata*, which occurs in western North America from northern Baja California to northwestern California. *Tetrameles* and *Octameles* are monotypic tree genera of southeast Asia. *Datisca cannabina* has a long history as a source of yellow pigment used for dying silk. Attention to its flavonoids was drawn by Grisebach and Grambow (1968) who were interested in the biosynthesis of 3,5,7-trihydroxyflavone (galangin) [3-20] and 3,5,7,2′-tetrahydroxyflavone (datis-

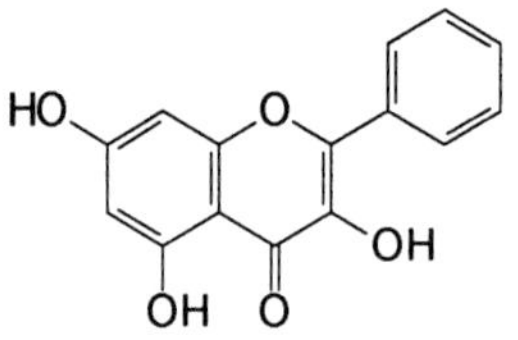

HO O OH OH O
3-20

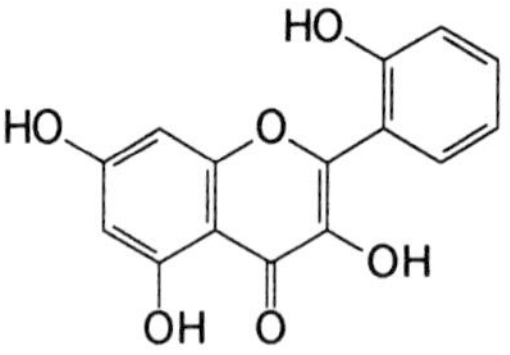

HO HO O OH OH O
3-21

cetin) [3-21] known to be constituents of this species. The availability of *D. glomerata* from several sites in California prompted me to examine that species for galangin, datiscetin, and related compounds (7-methyl ethers) (Bohm, 1988). In addition to that material, specimens of the Asian taxa were obtained so that all members of the family could be examined. The flavonoid profiles of both *T. nudiflora and O. sumatrana* were limited to common glycosides of kaempferol and quercetin, in which respect they resembled *Datisca*, but the absence of the unusual B-ring flavonoids clearly serves as a distinguishing feature.

Arguments have been advanced suggesting that the two Asian genera should comprise Tetramelaceae leaving *Datisca* as the sole genus of Datiscaceae (Airy Shaw, 1964). Studies on vegetative features and wood anatomy led Davidson (1973, 1976) to suggest that the differences were best accommodated by recognizing two tribes, Datisceae and Tetrameleae, the differences being insufficient in his view to support separate family status. Boesewinkel (1984) found only minor differences between ovule and seed structure which did not support splitting the family. Whatever the taxonomic status decided upon, tribe or family, clear distinction between the two groups of taxa is supported by the flavonoid chemistry. A recent study of the nucleotide sequence of the *rbc*L gene (Swensen *et al*., 1994) does not support the monophyly of Datiscaceae *s.l.* with the further outcome that *Datisca* appears to lie closer to members of Cucurbitaceae and Begoniaceae than it does to either *Tetrameles* or *Octameles*. The relationship with Begoniaceae is not a new idea; Brown (1938) had suggested such an alignment on the basis of nectary structure. Flavonoids have little further to add insofar as both Cucurbitaceae and Begoniaceae have pigment profiles based upon common flavonols; neither has shown any tendency to make the B-ring compounds that typify *Datisca*.

Applications Involving Ferns

A substantial literature concerning flavonoids of ferns and fern allies (e.g., Psilophytales above) has accumulated, a good deal of which was reviewed by Markham (1988). Here we will examine two examples where flavonoids have served well to,

(1) support definition of a new genus, and (2) suggest the need for further study toward recognition of additional species within a genus. The first example comes from studies of the western North American genus *Pityrogramma*, which includes the silverback and goldenback ferns. These plants gained their common names through the existence of deposits of white to orange material on the undersides of leaves. These exudates arise from glandular structures and consist of lipid-soluble substances including a wide selection of flavonoid derivatives. By way of introduction to *Pityrogramma* flavonoid chemistry, it is interesting to look at the first compound obtained from a member of the genus. Nilsson (1959) reinvestigated a compound that had originally been isolated from leaves of *P.* (*Ceropteris*) *triangularis* by Blasdale (1903) many years before. Nilsson determined that the compound, which was called "ceroptene," was a modified chalcone, modified in the sense that the *gem*-dimethylation prevents aromatization of the A-ring [3-22]. Other compounds identified over the years from members of the genus include chalcones, dihydrochalcones, flavanones, and a variety of flavone derivatives. *Pityrogramma triangularis* had also attracted attention because of the ready recognition of several extreme forms, which Weatherby (1920) recognized as varieties. Further variation

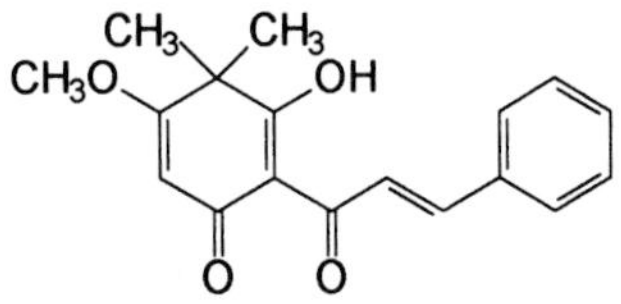

3-22

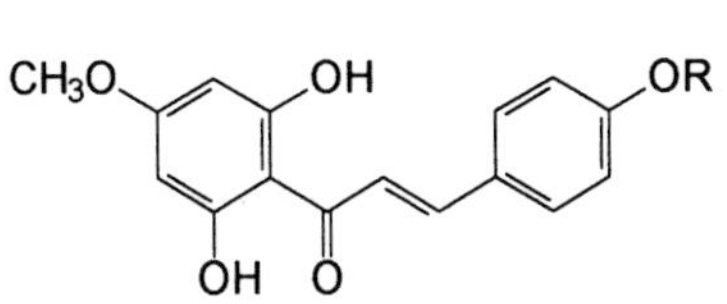

3-23 R = H
3-24 R = CH_3

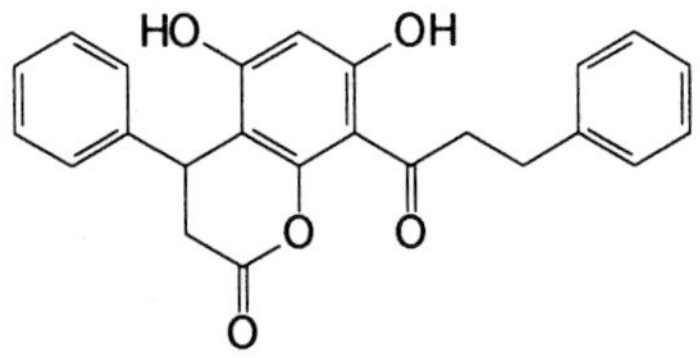

3-25

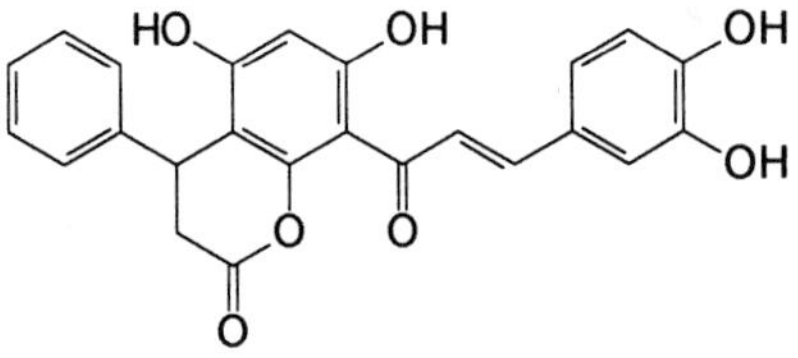

3-26

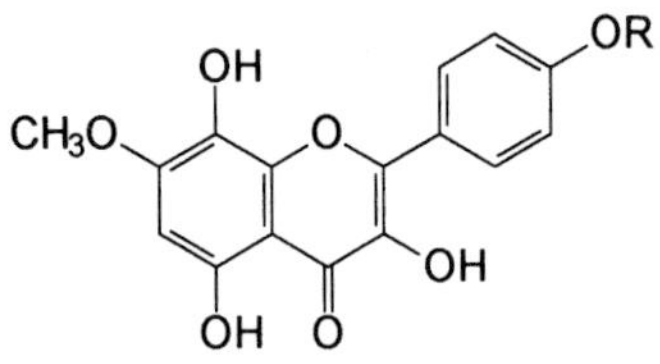

3-27 R = H
3-28 R = CH_3

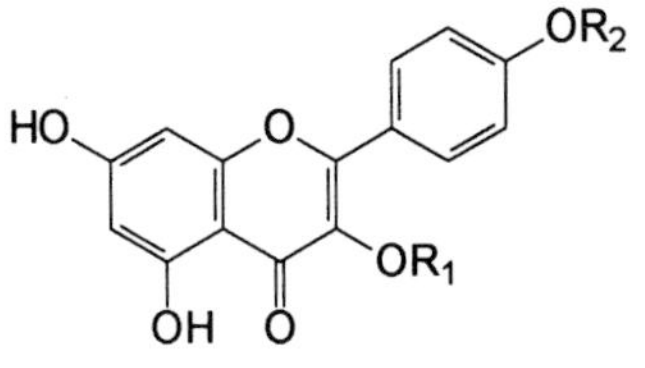

3-29 $R_1 = R_2 = H$
3-30 $R_1 = H_3$, $R_2 = H$
3-31 $R_1 = H$, $R_2 = CH_3$

within the species, in terms of both chromosome number and flavonoid chemistry, has been documented by Dale Smith and his colleagues at University of California, Santa Barbara (Smith *et al.*, 1971; Smith, 1980).

Although considered for many years as a species with several varieties, *Pityrogramma triangularis* departs from other members of the genus in a number of features. Tryon (1962) pointed out that *P. triangularis* differs from members of the "central group" of species in the genus and suggested that it might warrant recognition at the genus level. However, his treatment of the genus was conservative with emphasis on the similarities rather than the differences among the species. More recently, Tryon and Tryon (1982) again suggested that the southwestern North American complex might warrant reconsideration but were reluctant to do so without further information. It seems that sufficient information has now been accumulated for serious reappraisal of the situation (Yatskievych *et al.*, 1990). Based upon differences in rhizome scale features, stipe structure, laminar shape, shape of pinnae, venation, spore structure, and base chromosome number, *P. triangularis* has been segregated from the rest of the genus and accorded generic status under the name *Pentagramma*. The flavonoid content of *Pentagramma* is fully in accord with this decision. Wollenweber and Dietz (1980) have surveyed all of the taxa comprising *Pityrogramma* (original version) and found that members of the *Pentagramma* complex share none of the compounds by which the remaining species of *Pityrogramma* can be characterized, 2′,6′,4-trihydroxy-4′-methoxychalcone [3-23], 2′,6′-dihydroxy-4′,4-dimethoxychalcone [3-24], their respective dihydrochalcones, cinnamoylneo-flavonoids of the sort represented by [3-25] and [3-26], and a few other related compounds. The *Pentagramma triangularis* profile consists of ceroptene [3-22], *O*-methylated flavonols, and a *C*-methyldihydrochalcone, although not all of its varieties contain the same profiles (Smith, 1980; Wollenweber *et al.*, 1994).

The second example of flavonoids as aids in unraveling taxonomic problems, or perhaps only helping to point them out, comes from work on the genus *Notholaena*. *Notholaena standleyi* is a fern of dry regions of southwestern United States and northern Mexico. It has been recognized for some time that plants from different places within the species' range possess somewhat different colored exudates on the undersides of their leaves. Three color forms are known, a gold race, a yellow race, and a yellow-green race. The gold race occurs in Arizona and far northwestern Mexico (but is known from Baja California) where it occurs predominantly on rocks of igneous origin. The yellow race is more widespread and scattered in its distribution occurring in southeastern Arizona, New Mexico, extreme southeastern Colorado, Oklahoma, Texas, and in north central Mexico. Members of this race also occur on igneous rocks but are occasionally found on sandstones as well. The yellow-green race occurs in north central Mexico and adjacent Texas and is found growing on limestones. There are no morphological features by which individuals of these races can clearly be distinguished although there is a tendency of yellow and yellow-green colored plants to be, on the average, taller than gold race plants. Seigler and Wollenweber (1983) studied the flavonoid profiles of 59 specimens representing all three color forms collected from the entire range of the species. Their results show that the three forms can be clearly distinguished on the basis of their exudate flavonoids. Both qualitative and quantitative differences were observed among color forms and profiles showed minimal variation within individual populations. No plants having intermediate flavonoid profiles were found. The flavonoid profiles were based upon kaempferol and herbacetin (8-hydroxykaempferol) methyl ethers. The most clear-cut difference observed is the absence of herbacetin derivatives from yellow race plants. Herbacetin 7-methyl ether [3-27] and 7,4′-dimethyl ether [3-28] occurred consistently but in lesser amounts in gold race plants and both were seen sporadically in yellow-green race plants. The major components of yellow race plants were kaempferol [3-29], kaempferol 3-methyl ether [3-30], and kaempferol 4′-methyl ether [3-31]. Kaempferol 3-methyl ether was not seen in the gold race and occurred only sporadically in yellow-green plants. The two most obvious differences are the absence of herbacetin derivatives in the yellow race which, of course, suggests a major difference in hydroxylation enzymes, and the absence of 3-methyl ethers in the gold race. The major differences involve the *O*-methyltransferase enzymes, both in terms of presence/absence as well as in terms of amount of substrate converted to the respective products. This system would be an interesting one in which to probe the methyltransferases in detail.

Taxonomic conclusions are not so straightforward, at least in terms of the available data. So far as is known, *N. standleyi* reproduces without undergoing meiosis, as judged by spore number, and that there is little gene flow between individuals or populations. There was no evidence of sexuality in any of the collections used in the flavonoid study (Seigler and Wollenweber, 1983). The flora in this region of southwestern North America and northern Mexico is thought to be a very old one that developed in response to increasing aridity of the area (Tryon, 1969, 1972; Tryon and Tryon, 1973). The consistent flavonoid profiles within the three pigment types, the differences among them, and the apparent lack of gene flow among the races, suggests that these forms may be remnants of formerly sexually reproducing populations that have become ecologically isolated and are currently surviving in a more or less stable environment. An examination of levels of genetic variation using electrophoretic analyses of allozymes would assist in assessing the degree of isolation of the species.

Generic Status in Podocarpaceae

Podocarpaceae is a family of evergreen, usually dioecious, trees primarily of the Southern Hemisphere but having representatives as far north as Japan. Flavonoids of the New Zealand members of the family have been scrutinized in detail by K. R. Markham and his associates in an ongoing attempt to help define generic limits. Two examples will serve to demonstrate the usefulness of flavonoid profiles in this regard, one involving *Podocarpus* itself and one involving *Dacrydium s.l.* We will look at them in that order.

Podocarpus, taken in the broad sense, has long been considered a large and heterogeneous assemblage that has been assorted in a variety of sections and subsections (Buchholz and Gray, 1948). With the accumulation of additional information on members of the genus, it became evident that segregation into additional genera was reasonable. This was done by de Laubenfels (1969, 1978) who recognized five genera. The New Zealand members of *Podocarpus s.l.* [seven species in Allan (1961)] were reallocated in the following manner: *Podocarpus* in the strict sense consisting of four species, *P. acutifolius, P. hallii, P. nivalis*, and *P. totara; Dacrycarpus dacrydioides*; and *Prumnopitys* with *P. ferruginea* and *P. taxifolia*. Markham and coworkers (1985a) identified the flavonoids of all of these species. Major flavonoids met in the survey were *C*-glycosylflavones, flavonol 3-*O*-glycosides, flavonol 3-methyl ether glycosides, and dihydroflavonol glycosides. Naringenin and its 7-*O*-glucoside were present in lesser amounts in two species of *Podocarpus*. Although there was some qualitative overlap in compound distribution, the occurrence pattern of the major compound types clearly shows correlation with the newer taxonomy. For example, flavonol 3-*O*-diglycosides occur uniquely in the three species of *Prumnopitys* studied (the two New Zealand plus one non-New Zealand species). Although present as minor constituents in two species of New Zealand *Podocarpus*, flavonol 3-methyl ethers and compounds with B-ring trioxygenation are the predominant compounds in the three species of *Dacrycarpus* studied (two non-New Zealand species). A further distinguishing feature of *Dacrycarpus* and *Podocarpus* is the presence of *C*-glycosylflavones as major components while such compounds occur only as trace constituents in a non-New Zealand species of *Prumnopitys* (*P. andina*).

The second example comes from a study of *Dacrydium s.l.* by Markham and coworkers (1989). The situation with regard to this genus is akin to that we just saw with *Podocarpus s.l.* The seven species of *Dacrydium* listed in Allan (1961) are considered by Quinn (1982) to constitute *Dacrydium s.s.* with one species, *Halocarpus* with three, *Lepidothamnus* with two, and *Lagarostrobus* with one. Flavonoid data strongly support segregation of *Lepidothamnus*, whose flavonoid profile is dominated by flavone *O*-glycosides, and *Halocarpus*, whose profile is dominated by *C*-glycosylflavones. Support for the other two genera is less convincing especially in view of the differences seen between *Dacrydium cupressinum*, the sole New Zealand member of the genus, and members of the genus from elsewhere. *Lagarostrobus* and *D. cupressinum* profiles are dominated by flavonol 3-*O*-glycosides but *D. cupressinum* can be distinguished by the presence of flavones and flavonols having trioxygenated B-rings. It can be noted in passing that all four genera reported in this study possess biflavonoid derivatives.

Peperomia on the Juan Fernandez Islands and Tristan da Cunha

Peperomia (Piperaceae) is a very large genus, perhaps 1000 species, distributed widely in the tropics and warmer regions of the earth. Of interest to us here are the species of *Peperomia* endemic to the Juan Fernandez Islands, Chile, and their relationship with *Peperomia* on the Tristan da Cunha Islands. The Juan Fernandez Islands lie about 600 km off the coast of Chile at 80°W, 33°S. The archipelago consists of two large islands, Masatierra (Isla Robinson Crusoe) and Masafuera (Isla Alejandro Selkirk), and a small one, Santa Clara, which lies off the southwestern corner of Masatierra and probably was separated therefrom by erosion. Masatierra is the older island (3.7–4.4 million years) and lies farther to the west. Masafuera has been dated at 1–2 million years. These islands are of volcanic origin (Stuessy *et al.*, 1984). The Tristan da Cunha group lies in the southern Atlantic Ocean (ca. 12°W, 37°S), and is associated with the Mid-Atlantic Ridge. Three major islands comprise the group, Tristan (0.5 million years old), Inaccessible (2.9 million years old), and Gough (18 million years old). We will be concerned only with Inaccessible in this example.

Three species of *Peperomia* are endemic to Juan Fernandez Islands, *P. margaritifera*, restricted to Masatierra, *P. skottsbergii*, restricted to Masafuera, and *P. berteroana*, which occurs on both islands. A fourth species, *P. fernandeziana*, occurs on both islands and on the Chilean mainland. At the time the flavonoid study of the Juan Fernandez Island species was undertaken (Valdebenito *et al.*, 1990a,b, 1992) it was known that a single representative of the genus, *P. tristanensis*, occurs on Inaccessible Island. During the course of the study herbarium specimens of *P. tristanensis* were examined and found to be more closely related to *P. berteroana* of the Juan Fernandez Islands than previously recognized and that both taxa are best considered as subspecies of *P. berteroana*: *P. berteroana* subsp. *berteroana* of the Juan Fernandez Islands and *P. berteroana* subsp. *tristanensis* of Inaccessible Island. Flavonoid analysis of the relevant taxa, shown in Table 3.7, confirmed the close association of the two. A further outcome of the flavonoid analysis arises from the fact that the pigment profile of *P. berteroana* subsp. *berteroana* from Masatierra and Masafuera are different and that the profile of the flavonoids from Inaccessible is identical to that of plants from Masafuera. The authors concluded that the presence of *Peperomia* on Inaccessible Island was due to long distance dispersal from Masafuera and that a likely agent of dispersal was the petrel. Dispersal in the reverse direction was considered unlikely owing to the rare status of the taxon on Inaccessible and its comparative abundance on Masafuera. It is important to appreciate that resolution of the extremely interesting relationships within this group of plants was made possible only after those workers, having examined plant material from all sources, recognized the significance of the morphological similarities they observed. This is a particularly pointed example of the importance of correct identification and painstaking study of research material.

Table 3.7 Flavonoids of *Peperomia berteroana* subspecies from the Juan Fernandez Islands[1] and Tristan da Cunha (Valdebenito *et al.*, 1990)

	Flavonoids[2]				
Taxon	1–7	8	9	10	11–14
Peperomia berteroana ssp. *berteroana* (MT)	+	+	+	+	–
Peperomia berteroana ssp. *berteroana* (MF)	+	–	–	–	+
Peperomia berteroana ssp. *tristanensis*	+	–	–	–	+

[1]MT = Masatierra; MF = Masafuera

[2]1 = Acacetin; 2 = Acacetin 7-sulfate; 3 = Diosmetin 7-*O*-Glc; 4 = Diosmetin 7-*O*-Glc-Rhm; 5 = Luteolin 7 = *O*-Glc; 6 = Luteolin 7-*O*-Ara-Ara; 7 = Luteolin 7-sulfate; 8 = Apigenin 7-*O*-Glc-sulfate; 9 = Luteolin 7-*O*-Glc-Rhm; 10 = Diosmetin 7-*O*-Glc; 11 = Acacetin 7-*O*-Glc-disulfate; 12 = Diosmetin *C*-Glc-Ara; 13 = Diosmetin 7-*O*-Glc-Ara; 14 = Diosmetin 7-sulfate; Ara = arabinose; Glc = glucose; Rhm = rhamnose

Ranunculus on the Kerguelen Islands

The Kerguelen Islands lie in the southern Indian Ocean at ca. 49°S, 69°E. Owing to the extreme southerly location the Kerguelen Islands have a rather depauperate flora but it is not without interesting facets. A study of three species of *Ranunculus* (Ranunculaceae) was described by Hennion and coworkers (1994). The species in question are *R. biternatus* and *R. pseudotrullifolius*, which occur on the islands as well as in the Magellanic region of South America, and *R. moseleyi*, which is endemic to the island. The disjunct species are thought to have inhabited the island for about 8000 years.

Quercetin was the only flavonoid aglycone detected in the three species but the array of glycosidic derivatives was comparatively complex involving glucose, arabinose, and xylose. The simplest derivatives were the 3-*O*-diglucoside and 3-*O*-xylosylglucoside. A 3-*O*-diglucoside-7-*O*-glucoside and a 3-*O*-xylosylglucoside-7-*O*-glucoside were also identified. The remaining seven compounds were all 3-*O*-diglycoside-7-*O*-glucosides with varying levels of acylation involving *p*-coumaric, caffeic, and ferulic acids. Although there was a certain level of pigment variation evident among individuals of all populations of the three species, there were several clear-cut differences between the profile of *R. biternatus* and the summed profile of *R. moseleyi* and *R. pseudotrullifolius*. The combination of karyological, morphological, and flavonoid data led the authors to conclude that *R. moseleyi* was derived from *R. pseudotrullifolius* comparatively recently, although no time estimate is realistic in situations of this sort.

Arctostaphylos

Arctostaphylos (Ericaceae) is a genus of perhaps 50 species centered in western North America with two species enjoying a circumpolar distribution (Mabberley, 1987). In the introduction to his study of the flavonoids of the genus, Denford (1981) noted that as many as 100 taxa have been defined at one time or another in the genus but that many of these are based on trivial features. An analysis of the flavonoids of 16 of the North American species revealed the existence of a rich array of flavonol glycosides based on kaempferol, quercetin, and myricetin. Quercetin 7-*O*-glucoside was identified from four species (see below) while all other glycosides were linked through the 3-OH group. A numerical analysis based on a single linkage method yielded two groups, each of which consisted of eight species. Group 1, characterized by quercetin and myricetin glycosides, encompassed *elegans, hookeri, manzanita, nevadensis, pacifica, patula, pungens*, and *stanfordiana*. Group 2, whose pigment profiles were based upon kaempferol and quercetin glycosides, consisted of *andersonii, canescens, crastacea, edmundsii, glandulosa, myrtifolia, obispoensis*, and *viscida*. An interesting finding involved the existence of a pair of species in each group, *A. nevadensis* and *A. pacifica* from Group 1, and *A. crastacea* and *A. myrtifolia* from Group 2, that had quercetin 7-*O*-glucoside in addition to their normal arrays of 3-*O*-glycosides. No explanation was offered for this apparent parallelism. It would be of interest to compare the properties of the *O*-glycosyltransferases and the encoding DNA sequences in these species. Is the appearance of the 7-*O*-glycosylation system the result of convergence or are the species that possess this enzyme activity more closely related than the overall flavonoid data suggest?

Chenopodium fremontii

The study of the flavonoids of *Chenopodium fremontii* by Crawford and Mabry (1978) was one of the first to document infraspecific variation. Most studies published before that time had either been based upon massed samples of plant material, which obscures any differences that might exist among populations, or had simply not detected infraspecific differences. Studies of 22 populations of *C. fremontii* representing the species in Wyoming (5 populations), Nebraska (2 populations), Colorado (2 populations), New Mexico (4 populations), Arizona (6 populations), and California (3 populations) revealed flavonoid profiles based on kaempferol, kaempferol 7-methyl ether, quercetin, quercetin 7-methyl ether, and isorhamnetin. Kaempferol, quercetin, and isorhamnetin derivatives were present in all populations while kaempferol and quercetin 7-methyl ethers were present only in the two northern populations. Whereas the 7-methyl ethers set the northern and southern races apart, four chemical races could be defined on the basis of the respective glycoside profiles. There are clear-cut similarities between the two northern races and between the two southern races. The glycosidic profile of the northern races differ only in the presence of robinobiosides in the northwestern population, whereas the southern populations can be distinguished on the basis of rutinosides being present only in the southeastern populations. The chemistry of the four races is summarized in Table 3.8. As the authors pointed out, the different profiles manifested in the races are based upon only a small number of enzyme/gene differences. However, it is this sort of apparent stepwise alteration of flavonoid profiles that one might expect in a species with widely separated populations. Differentiation could well be expected to occur

Table 3.8 Flavonoid races in *Chenopodium fremontii* (from Crawford and Mabry, 1978)

	Aglycones[1]				Glycosides[2]					
	K	Q	I	R	G/G	Ara	Rut	Rob	Sof	2RG
Northwestern pop'n. (Wyoming & No. Colorado)	+	+	+	+	+	−	−	+	+	−
Northeastern pop'n. (Nebraska)	+	+	+	+	+	−	−	−	+	−
Southwestern pop'n. (California)	+	+	+	−	+	−	+	+	+	+
Southeastern pop'n. (Arizona, New Mexico, So. Colorado)	+	+	+	−	+	−	+	+	+	+

[1]K = Kaempferol; Q = Quercetin; I = Isorhamnetin; R = K and Q 7-methyl ethers
[2]G/G = Glucosides and galactosides; Ara = Arabinosides; Rut = Rutinosides; Rob = Robinobiosides; Sof = Sophorosides; 2RG = 2-Rhamnosylgalactosides

through utilization of different sugars in the formation of diglycosides, or through formation of a new diglycoside through change of position of attachment of the outer sugar. These changes can be seen to have occurred in *C. fremontii*. The position of *O*-methylation raises an interesting, and, on the basis of available information, unresolvable question. Does the presence of 7-methyl ethers represent a derived property of the northern two races, or does it represent the remnants of an ancestral type that possessed the capacity to methylate flavonols at both the C-3′ and C-7 positions?

Additional insight into questions of this sort may be gleaned by looking at relationships among *C. fremontii* and related species as was done in the Crawford and Mabry (1978) paper. On morphological grounds *C. fremontii* is considered to be the "least modified taxon" among the species considered in their paper (*C. incanum, C. desiccatum, C. atrovirens, C. hians, C. leptophyllum*, and *C. pratericola*). The close morphological similarity between *C. fremontii* and *C. incanum* is borne out in the flavonoid profile of the latter species, which consists only of quercetin and isorhamnetin glycosides, the capacity to accumulate kaempferol derivatives having been lost. The loss of kaempferol derivatives could argue for a change in the efficiency of utilization of its precursor, dihydrokaempferol, through conversion to dihydroquercetin. Resolution to problems of this sort require in-depth studies of enzyme concentration and activity and are, unfortunately, rarely done. *Chenopodium desiccatum*, with only isorhamnetin glycosides, has undergone apparent simplification. The other suite of species (*C. atrovirens, C. hians, C. leptophyllum*, and *C. pratericola*) exhibit a profile based solely on quercetin, which represents another simplification relative to *C. fremontii*. These four species are of more limited occurrence and exhibit leaf structural features of species adapted to drier habitats. What adaptive value the simpler flavonoid profile provides is not known but the reduction in complexity seems to be a feature often observed in species that show more limited ranges or specializations for soil type [see Mabry in Bendz and Santesson (1974) for general discussion, and the work of Mears cited in the *Parthenium* example below].

To round out the western *Chenopodium* work, the interested reader might care to look into the systematic study of the narrow-leaved species (Crawford, 1975), the variation in seed protein profiles of *C. fremontii* (Crawford, 1976), or a study of character compatibility, using both morphological and chemical characters, and phyletic relationships within the genus (La Duke and Crawford, 1979).

Lasthenia californica

Lasthenia californica (Asteraceae) is the most widespread and morphologically variable of the 17 species that constitute this mostly Californian genus (Ornduff, 1966). This particular species occurs from southern Oregon throughout California into northern Baja California and into central Arizona. The initial study of the flavonoids of the entire genus showed a reasonable level of congruence with the sectional taxonomy of the genus (Bohm *et al*., 1974). That study was based on a limited sampling within each species, however. To counter the criticism that variation might have been missed owing to sampling error, a broad program of sampling was undertaken which eventually showed that not only did different populations of species differ, but individuals within a single population could exhibit different flavonoid profiles as well. One of the populations involved in the search for variation lies in the Jasper Ridge Biological Preserve of Stanford University. The population of *L. californica* in the research area grows on an outcrop of serpentine-derived soil along with other serpentine-tolerant or serpentine endemic species. Individual plants collected alone several transects, analyzed by one-dimensional thin layer chromatography, showed the existence of four major, easily discernible flavonoid profiles (Bohm *et al.*, 1989). All four chemotypes exhibited an array of aurone and chalcone glycosides and flavonol glucuronides, which constituted the C-profile. The B-profile added luteolin 7-*O*-glucoside to the anthochlors and glucuronides. The A-profile consisted of anthochlors, glucuronides, eriodictyol 7-*O*-glucoside, and flavonol diglycoside sulfates. The D-profile consisted of anthochlors, glucuronides, the eriodictyol glycoside, and flavonol diglycosides that lacked sulfation. Only a few plants (out of several thousand tested) had the D-profile and were not considered further.

Several permanent transects, from 40 to 65 m long, were established along which individual plants in flower were sampled at 1 m intervals every year for six years. A definite pattern emerged in which the A-profile plants and the B/C-profile plants maintained their relative positions over that time period. The transects were revisited off and on for several more years so that it is possible to state in 1996 that the pattern of occurrence of these chemotypes has remained essentially unchanged for 15 years. The existence of the chemotypes within one population led to an examination of the flavonoid profiles of *L. californica* populations throughout its entire range (northern Baja California to southwestern Oregon and into central Arizona). With few exceptions, populations from the southern part of the range, essentially San Francisco Bay area and southward, were characterized by high frequencies of the A-profile, while the two populations from northern California and two from Oregon consisted exclusively of the B/C-profile (Desrochers and Bohm, 1993). More recent studies, including additional populations from southern Oregon, have corroborated those findings (Rajakaruna and Bohm, unpubl.). A single population in the southern part of the range, near Santa Margarita in San Luis Obispo County, was atypical in having only B/C-profile plants. A biosystematic study of *L. californica* (Desrochers and Bohm, 1995) offered further support for the existence of two distinct biological entities within the broader context of *L. californica*; differences in both achene structure and electrophoretic banding patterns showed strong north–south differences among populations and correlated nearly perfectly with the flavonoid profiles. Furthermore, the two chemotypes appear not to interbreed, despite their apparent opportunity to do so in mixed populations. This behavior may be accounted for by their different flowering times; the B/C chemotype has an earlier flowering time than does the A type. Although the morphological differences and flowering time disparity between the two chemotypes are insufficient to support formal taxonomic recognition of distinct taxa, it is abundantly clear that differentiation among populations of *L. californica* has occurred and is continuing to occur.

Some Flavonoid Studies Involving Saxifragaceae and Related Groups

In the first example in this section we will look at the genus *Itea* which consists of 10 (Mabberley, 1987) or 20 (Cronquist, 1981) species. The difference in opinion regarding species number does not alter the fact that the genus is represented in North America by only one, the rest being eastern and southeastern Asian. The North American species, *I. virginica*, known locally as "Virginia willow" occurs on the Atlantic coastal plain from southern New Jersey southward to Florida and eastward as far as southern Illinois, and in eastern parts of Missouri, Oklahoma, and Texas. The Asian species occur from Japan southward to the Philippines, tropical India and Java, and into the northwestern Himalayas (Spongberg, 1972). *Itea* and the monotypic African genus *Choristylis* have been considered by some to constitute the family Iteaceae but by others to belong in Grossulariaceae (the current family). Earlier taxonomic opinion had *Itea* within subfamily Escallonioideae in a very broadly based Saxifragaceae (Engler, 1891), but in a later view (Engler, 1928), it constituted its own subfamily, Iteoideae, also in Saxifragaceae. Hutchinson (1959, 1967) placed *Itea* in its own family, Iteaceae. Cronquist (1981) placed *Itea* in his broadly defined Grossulariaceae. A considerable amount of information has been gathered in an effort to determine a comfortable resting place for *Itea* including morphological, anatomical, pollen structure, development, and chromosome numbers. Some data suggest a close relationship with Saxifragaceae, although probably not as one of the genera therein, while other data suggest that any close relationship with Saxifragaceae is unlikely (see Bohm *et al.*, 1988 for citations).

Flavonoids were isolated from *I. japonica*, *I. parviflora*, and *I. virginica* and examined chromatographically from an additional 12 species (Bohm *et al.*, 1988). The major components identified from the three bulk samples, and detected chromatographically in the rest, were the *C*-glucosylflavones orientin, isoörientin, vitexin, and isovitexin. In addition to the free phenols, X″-*O*-glucosides and X″-*O*-xylosides of each were identified. No flavonols were detected in any of the specimens examined. The lack of flavonols is interesting in light of the flavonoid data that have been accumulated over the years for members of Saxifragaceae and Hydrangeaceae (Bohm *et al.*, 1985a,b; Nicholls *et al.*, 1986). Species representing some 30 genera from those two families consistently yielded kaempferol, quercetin, and, in most cases, myricetin 3-*O*-mono-, di-, and triglycosides. The occasional flavone or dihydroflavonol was observed but in none of these was there any indication of *C*-glycosylflavones. These observations are rewarding in that there is such a clear-cut difference between *Itea* and members of two groups with which *Itea* was supposed to be related. However, all available non-flavonoid data taken together, including recent DNA sequences (D. Soltis and P. Soltis, unpubl. data), clearly point to a close relationship of *Itea* with other groups that include Saxifragaceae *s.s.* Those workers go on to define Saxifragales as encompassing Saxifragaceae, Crassulaceae, *Itea*, and *Pterostemon* (a Mexican genus with two species) as a close-knit group, as well as a number of other genera some of which are not usually seen as closely associated with Saxifragaceae. The postulated relationship has a moderately high likelihood of reflecting the evolutionary relationships between and among these various groups.

Both species of *Pterostemon* have recently been examined for flavonoids using a combination of TLC and HPLC analyses (Bohm, unpubl.). Their flavonoid profiles were essentially identical and consisted of the same *C*-glycosylflavones that had been reported from *Itea* (*I. virginica* was used as a standard) plus quercetin 3-*O*-glycosides typical of Saxifragaceae. The presence of *C*-glycosylflavones in *Itea* and *Pterostemon* adds additional weight to the suggestion that they are closely related taxa. The absence of flavonol 3-*O*-glycosides from the pigment profile of *Itea* and their presence along with *C*-glycosylflavones in *Pterostemon* suggests that we may be observing an evolutionary replacement of one type of flavonoid with another with the sequence being flavonol 3-*O*-glucosides only → flavonol 3-*O*-glycosides plus *C*-glycosylflavones → *C*-glycosylflavones only. An apparent loss-replacement sequence of this sort deserves detailed scrutiny. Since it is rather unlikely that the entire suite of flavonol biosynthetic genes has been lost in this process, it would be of some interest to learn how they have been silenced. Of equal interest would be an examination of the *C*-glycosylation process. Since *C*-glycosylation does not appear to occur in taxa other than *Itea* and *Pterostemon*, the appearance of the necessary enzyme(s) raises profound questions concerning their origin.

Chromosomes and Flavonoids

In this section we will examine a number of cases where there has been some alteration in the genome of the species. This may involve simply doubling of the chromosome number (autopolyploidy), it may involve doubling through hybridization (allopolyploidy), or it may involve some structural change in a chromosome (e.g., inversion or translocation). Although we do not know the mechanism or mechanisms that result in changes of flavonoid profile *t* accompanying chromosome changes, it is clear that these changes can be significant. Our failure to understand the mechanisms by which chromosomal changes affect flavonoid biosynthesis hinges on our ignorance on how the flavonoid biosynthetic pathway is controlled. We will begin with a straightforward example of autopolyploidy.

Tolmiea menziesii, the sole member of this genus in Saxifragaceae, occurs from Alaska southward to northern California along the coast and inland as far as the Cascade Mountains. The flavonoid profile of *Tolmiea* had been described as part of the survey of a larger survey of Saxifragaceae (Bohm, 1979) before Soltis (1980b) discovered that two cytotypes of the species existed, a diploid with $2n=14$ and a tetraploid with $2n=28$. Populations occurring from central Oregon south to northern California were all diploid, while all populations counted from northern Oregon north to Alaska were tetraploids. It was argued by Soltis (1984) that the *Tolmiea* situation represented one of the best documented examples of autopolyploidization. Since chromosome numbers of the specimens studied in the initial flavonoid survey (Bohm, 1979) had not been determined, although inclusion of both cytotypes in that study was likely, a reanalysis of the species was undertaken using material from the species' entire range (29 populations) (Soltis and Bohm, 1986). The results of the expanded study agreed very well with those from the earlier one; the flavonoid profile of the species consisted of mono-, di-, and triglycosides of kaempferol, quercetin, isorhamnetin, and myricetin. What little intrapopulational variation was seen did not correlate with ploidy level. An examination of the anthocyanin content of material from 16 populations, again representing the entire range of the species, showed no appreciable differences (P. Soltis and D. Soltis, 1986). Flavonoid data are in agreement with gross morphology, karyology, ribosomal genes, and electrophoresis, all of which suggest that the tetraploid cytotype has arisen by autopolyploidization. This is in line with conclusions drawn on other suspected autopolyploidization systems, e.g., *Galax* (Diapensiaceae) in the Appalachian Mountains (Soltis *et al.*, 1983) where overall flavonoid profiles were not significantly altered by changes in chromosome number.

Recent work on flavonoid composition of diploid and tetraploid lines of *Chamomilla recutita* (Asteraceae) showed an increase in apigenin content, as well as increases in size of various organs, as generally associated with polyploidization, but no apparent qualitative differences in flavonoid profile (Repcak and Martonfi, 1995).

Such simple situations are not seen in all instances of apparent autopolyploidization. In a detailed study of *Phlox* species (Polemoniaceae) evidence was presented that showing that flavonoid patterns can indeed be altered by chromosome doubling (Levy and Levin, 1971, 1974, 1975; Levy, 1976). Changes in the arrays of compounds were classified into three groups: (1) appearance of novel *C*-glycosylflavones; (2) disappearance of *C*-glycosylflavones typical of the diploid parent; and (3) loss of tissue specificity with regard to *C*-glycosylflavone production. In the latter category, several possibilities were seen: (1) compounds present in leaves and flowers of the diploid occur only in (usually) flowers of the tetraploid; (2) loss of compound from one or both tissue types in tetraploids; and (3) compounds present in one tissue of the diploid appear in the other tissue of the tetraploid. These are complex problems whose resolution will likely only come from understanding the factors that control flavonoid biosynthesis. Several examples of the use of flavonoids to study cases of natural hybridization will be presented below. It will be noted in some of these that flavonoid profiles in hybrid plants do not agree with the expected additivity. The types of profile changes seen in the *Phlox* hybrid studies should be kept in mind while reading that section.

Two examples from the flavonoid literature that try to relate chromosomal changes and flavonoid profiles are instructive. The first concerns the flavonoid chemistry of *Briza media* (Poaceae) (Williams and Murray, 1972; Murray and Williams, 1973, 1976). Three flavonoid profiles were observed: (1) a typical diploid type; (2) a typical tetraploid type, and (3) a modified tetraploid type that lacked an acylated derivative of 8-*C*-galactosylluteolin. Two different sets of $3n$ aneuploid individuals were examined, one of which was trisomic for both acrocentric chromosomes and showed a typical tetraploid flavonoid profile. Other $3n$ plants were trisomic for only one of the acrocentrics and showed the modified tetraploid (acyl group lacking) flavonoid profile. The authors suggested that control of the acylation step resides on one of the acrocentric chromosomes. As Giannasi and Crawford (1986) point out in their review of the topic what is needed in this sort of work is a system with a known chromosomal marker with which flavonoid changes could be correlated and thus followed throughout the changes.

The second example comes from a study of the flavonoids of *Gibasis schiedeana* (Commelinaceae) by Martinez and Swain (1977) and their attempts to rationalize the pigment profiles in terms of a major chromosomal translocation. Baseline was established using *G. consorbrina* and *G. karwinskyana* both of which are known to form tetraploids ($2n=20$; $x=5$) whose flavonoid profiles closely parallel the diploids ($2n=10$; $x=5$). In the case of *G. schiedeana*, where the diploid number is still $2n=10$, the tetraploid is $2n=16$ ($x=4$). The reduction in chromosome number occurred through centric fusion (Robertsonian fusion) of two of the acrocentric chromosomes. Small amounts of chromosome material is left over from this process and can be observed as B-chromosomes. The $2n=16$ tetraploids lacked a few of the compounds that would have been seen in normal tetraploids. The authors suggested that elements essential for forming the full flavonoid complement had been lost. A problem, however, was that flavonoid profiles did not correspond to presence or absence of B-chromosomes, which presumably would contain the missing elements. Although interesting, and certainly suggestive, a great deal of additional work is required before any relationship between chromosomal structure and flavonoid biosynthesis can be established.

A comparatively straightforward interaction between ploidy level and flavonoid content was described by Griesbach and Kamo (1996) using *Petunia* 'Mitchell.' The flavonoid profiles of haploid, diploid, and tetraploid individuals were identical consisting of quercetin glycosides: 3-*O*-glucoside, 3-*O*-sophoroside, 7-*O*-glucoside, 3,7-di-*O*-glucoside, 3-*O*-sophoroside-7-*O*-glucoside, and a caffeoyl ester of the latter. Quercetin 3-*O*-sophoroside and the 3,7-di-*O*-glucoside were the major components. Artificial polyploidization was accomplished using colchicine. HPLC analysis of the flavonoid profiles showed that although the total flavonoid content was not altered by the change in chromosome number, the amount of the major flavonoid, the sophoroside, increased while the concentration of quercetin 3,7-di-*O*-glucoside fell. The changes in these two compounds were statistically significant at the 99.9% probability level. There were increases in the amounts of quercetin 3-*O*-glucoside and 7-*O*-glucoside in the polyploids but the differences were apparently not statistically significant. Quantitative changes of this sort, especially in view of qualitative constancy, suggests that some regulatory factor(s) has been affected by the chromosome doubling. With the ready availability of plant material and a simple array of compounds, this system would seem to be an ideal one in which to pursue the underlying biochemistry of glycosylation. The main issue is the apparent change in activities of the two enzymes involved, 7-*O*-glucosyltransferase and 2″-*O*-glucosyltransferase. It would be highly instructive to learn if the change in glycoside concentrations was caused by a change in enzyme levels. Such information is not available and is critically needed if we are to understand, (1) how ploidy changes affect pigment profiles, and (2) more generally, how glycosylation reactions are controlled.

Fossil Flavonoids

In the section above on criteria for usefulness, it was stated that flavonoids were among the most stable compounds used for chemotaxonomic studies. This is usually taken to mean that little or no decomposition is to be expected in normal extraction and analytical procedures and that flavonoids can be extracted from well cared for herbarium specimens up to perhaps 100 years of age. The stability feature, however, has been taken to far greater heights through the work of David Giannasi, now at the University of Georgia, and Karl Niklas, now at Cornell University, in their study of flavonoids from fossils found in the Succor Creek (Miocene) Flora of eastern Oregon and the Miocene Clarkia Lake deposits of northern Idaho (Niklas and Giannasi, 1977a,b, 1978; Giannasi and Niklas, 1977, 1981). The fossilized plants from these locations had not undergone mineralization, the result of which is loss of all organic matter, but had been preserved in ash falls under conditions where decomposition was minimal. The age of the Succor Creek Flora was taken to be 16.7 to 25×10^6 K/A years (potassium argon years indicating the method of dating).

Several examples will demonstrate the part flavonoids played in helping to establish affinities of the fossil taxa. *Acer oregonianum* (Aceraceae, maple), one of the Succor Creek fossil species, has been mostly closely matched with the extant *A. macrophyllum*, the big leaf maple, common on the coast from Alaska south to southern California (Graham, 1965). The fossil material exhibited a flavonoid profile consisting of kaempferol and quercetin 3-*O*-glycosides and two compounds provisionally identified as "tannins" by Niklas and Giannasi (1978). The presence of flavonols in *A. oregonianum* is in agreement with other species of maple, but the presence of the two assumed tannins suggests affinities with Asian maples in which group the tannins were seen consistently.

Fossil *Quercus* leaves are a major component of the Succor Creek Flora, but assignment was not straightforward owing to similarities with other members of Fagaceae, in particular, *Castanea, Castaneopsis*, and *Lithocarpus*. One of the oak fossils that was identified, however, as *Q. consimilis*, whose nearest relative was thought to be either of the east Asian species, *Q. stenophylla* or *Q. myrsinaefolia*. The flavonoid profile of *Q. consimilis* consisted of kaempferol and quercetin 3-*O*-glycosides, an apigenin 6-*C*-glycoside, an apigenin 8-*C*-glycoside, and ellagic acid. *Quercus stenophylla* and *Q. myrsinaefolia* had the flavonol 3-*O*-glycosides but lacked *C*-glycosylflavones. In addition, they had several flavonoids not seen in the fossil species. These observations led to a survey of 30 species of oaks, from North America, Europe, and Asia, with the interesting outcome that 28 had only flavonols while two, *Q. chenii* and *Q. acutissima*, had *C*-glycosylflavones as well as flavonols. Flavonoid data were clearly instrumental in arriving at both relationships and biogeographical connections of the fossil species with extant oaks in eastern Asia.

The third example of the use of palaeochemical data involves the genus *Pseudofagus* (Fagaceae), first described by Smiley and Huggins (1981). The flavonoid profile of *P. idahoensis* was determined to be quercetin 3-*O*-glucoside and isorhamnetin 3-*O*-glucoside plus two unidentified compounds. Morphological information and steroid chemistry clearly established the fossil material as belonging to Fagaceae, and further suggested that the closest related genus is *Fagus*. The absence of ellagic acid from the phenolic profiles of *Pseudofagus* and all species of *Fagus* further supports the view that these two genera are closely related. The presence of isorhamnetin (quercetin 3′-methyl ether) in the fossil, along with onocerane and 5α-cholestane, provides a species-specific chemical profile for *P. idahoensis* (Giannasi and Niklas, 1981).

Well preserved specimens of *Platanus* were also reported from the Miocene Clarkia Flora (Smiley and Huggins, 1981). Sufficient material was obtained to allow Rieseberg and Soltis (1987) to identify kaempferol and its 3-*O*-galactoside and 3-*O*-glucoside, and quercetin and its 3-*O*-arabinoside, 3-*O*-galactoside, and 3-*O*-glucoside. These workers also examined seven of the nine extant members of the genus all of which exhibited profiles consisting of kaempferol and quercetin 3-*O*-mono- and 3-*O*-diglycosides. The absence of diglycosides from the fossil species cannot be taken as significant in view of the possible loss of an outer sugar. The closest flavonoid match between the fossil profile and an extant species involved the Asian species *P. orientalis*. This conclusion was couched in the most cautious terms as one would expect. Nonetheless, a closer relation of the fossil's profile with that of an Asian species is in line with observations with the other fossil flavonoid studies mentioned above.

Looking at Lupines (*Lupinus*)

Lupinus is a genus of legumes consisting of some 200 species, many of which are widespread and conspicuous members of the North American flora. A study of the flavonoid chemistry of western North American lupines documented a rich array of pigments (Nicholls and Bohm, 1982). In most instances where plant material was available from more than one population some quantitative variation was apparent when thin layer chromatograms were compared. Significant qualitative differences were noted with a few species but invariably these involved groups whose taxonomy was problematic. In one or two cases rather large quantitative differences were evident and it is one of these systems that we will look at here.

Lupinus sericeus occurs from southeastern British Columbia and southwestern Alberta southwards to the Kaibab Plateau in northern Arizona covering extensive areas in Alberta, British Columbia, eastern Washington, Idaho, western Montana, northeastern Oregon, Utah, and northern Arizona. Taxonomic difficulties with this species can be better appreciated by reflecting on the fact that no fewer than 23 synonyms for *L. sericeus* are recorded in *Vascular Plants of the Pacific Northwest* (Hitchcock *et al.*, 1961) and that Fleak (1971) studied 61 taxa thought (by some) to belong to the *L. sericeus* complex. Comparisons of two-dimensional thin layer chromatograms representing 181 individual plants from 32 populations representing the entire range of the species showed major differences in size of the orientin (8-*C*-glucosylluteolin) spot. High performance liquid chromatography (HPLC) confirmed the qualitative comparisons. The mass of information generated by the HPLC analyses was subjected to multivariate statistical analysis. Principal components analysis (PCA) showed a large degree of dispersion in the data with orientin dominating axis I (72% of total variation accounted for), isoörientin dominating axis II (12%), and orientin 3′-*O*-glucoside dominating axis III (10%). Despite the large contribution from orientin, no clear-cut grouping of populations appear to exist in most of the northern and central parts of the range. The southernmost populations are more discernible than those throughout the rest of the range, but the obvious result of these analyses is the existence of clinal variation in a north–south axis, with the high-orientin plants originating from the southernmost populations. This agrees well with an observation made by Fleak (1971) on plants from the southern populations. He noted that they consistently displayed heavy anthocyanin pigmentation in the lower part of the stems. Stem pigmentation was sufficiently variable in plants from the rest of the range of the species that Fleak and Dunn (1971) chose to recognize the southern plants as *L. sericeus* subsp. *huffmanii*. Accumulation of high concentrations of orientin in these individuals provides support for their view. Flavonoid data provided no support, however, for three other subspecies defined by Fleak (1971).

An inherent problem with any study of such wide ranging organisms is that one has no idea of the effect that local variables might have on the flavonoid chemistry (or any other trait for that matter). In order to overcome this problem (and obviously attendant criticism) it is necessary to study plants that have been grown in a common garden setting. This was done in the *L. sericeus* study using seeds collected from 20 individuals representing six populations, two from southern Utah, one from northern Utah, and one each from Idaho, eastern Washington, and western Montana. HPLC analysis showed no significant differences in concentration of any of the compounds, which led to the conclusion that differences in the field-collected samples were caused by some environmental factor that was eliminated when cultivated plants were used. It is interesting to note, however, that high levels of stem anthocyanins were still evident in the cultivated plants. This study clearly demonstrates the influence that environmental factors can have on the significance of any data set. Unfortunately many quantitative studies using flavonoids have not taken this problem into consideration.

Sereno Watson (1873) established two sections within *Lupinus* to accommodate annual lupines, sect. *Platycarpos* for those species with terminal, racemose inflorescences (10 species including *L. kingii*), and sect. *Lupinellus* to accommodate the single species *L. uncialis* with its solitary, axillary flowers. The species that comprised sect. *Platycarpos* were further arranged into two informal groupings by Smith (1944): "Pusilli" with scattered flowers with glabrous keels, and "Microcarpi" with whorled flowers having ciliate keels. The pattern of occurrence of *C*-glycosylflavones and flavone *O*-glycosides in all members of both sections has been determined (Nicholls and Bohm, 1987). The profiles of the two sections are distinct although they have luteolin, chrysoeriol 7-*O*-diglucoside, vitexin, and cytisoside (8-*C*-glucosylacacetin) in common. *Lupinus uncialis* uniquely has isovitexin and isoörientin, but lacks many of the apigenin and luteolin derivatives that characterize section *Platycarpos*. Significant differences in flavonoid profiles also support Smith's (1944) view that two groups of species can be discerned within sect. *Platycarpos*. Both qualitative and quantitative differences were observed with the only profile not clearly fitting in being that of *L. kingii*. The profile of *L. kingii* is very simple with a significantly greater resemblance to members of the Microcarpi group while its floral features are more akin to those of the Pusilli group. It was suggested that *L. kingii* deserves a more detailed biosystematic examination (Nicholls and Bohm, 1987).

Red Oak in the Southern Appalachians

The study of *Lupinus sericeus* described above demonstrates the importance of common garden experiments in determining the influence of environmental factors on the expression of flavonoids. A study of red oak in the southern Appalachians provides an excellent example of another approach to documenting environmental effects, the reciprocal transplant experiment. *Quercus rubra* occurs in eastern North America from Ontario southward to the Gulf Coast states and can be found throughout a fairly wide elevational range. The species is quite variable in leaf and acorn morphology over its range which has led to recognition of varietal forms. A study of flavonoids of red oak showed the existence of two chemotypes that intergrade between 600 and 1050 m (McDougal and Parks, 1984). The chemotypes consisted of a lower elevation form that produced kaempferol and quercetin 3-*O*-glycosides and a higher form that produced kaempferol 3-*O*-glycosides, quercetin 3-*O*-glycosides, and myricetin 3-*O*-glycosides.

In order to assess the impact of environmental factors on flavonoid expression, acorns collected from a range of trees spanning 1455 m were planted at two sites, one at 75 m and one at 1140 m elevation. Leaves of the resulting young trees were harvested after five years and subjected to flavonoid analysis. Myricetin-rich high elevation trees retain their capacity to make myricetin glycosides even when grown at the lower elevation and the quercetin-rich low elevation trees retain their original flavonoid synthetic capacities when grown at the higher elevation site. It is interesting to note that the high and low elevation chemotypes correspond to the varieties *Q. rubra* var. *borealis* and *Q. rubra* var. *rubra*, although there is a good deal of overlap in the mid elevations.

Silene pratensis Chemotypes

In my opinion, the most impressive display of flavonoid genetic detective work has come from the group at the University of Utrecht who undertook a study of flavonoid variation in *Silene pratensis* (Caryophyllaceae). Several glycosylated derivatives of isovitexin (6-*C*-glucosylapigenin) are known throughout the species' range in Europe. The derivatives involve one or more of arabinose, xylose, glucose, and rhamnose, and can involve either an A-ring hydroxyl group or an hydroxyl group on the *C*-bound sugar. The various compounds identified do not occur randomly but fall into fairly distinct geographical units. Genetic studies established the existence of three unlinked flavone glycosylating loci, *fg*, *g*, and *gl* (van Brederode and Nigtevecht, 1975). Let's look at each of these in turn. Two alleles were found at the *fg* locus with the dominant one (*Fg*) responsible for glucosylation of the C-2″ hydroxyl group. Three alleles were identified for the *g* locus, *g*, g^G, and g^X. The alleles g^G and g^X are responsible, respectively, for placing glucose and xylose at the 7-OH of the flavonoid nucleus. Three alleles are also known for the *gl* locus, *gl*, gl^R, and gl^A. The alleles gl^R and gl^A, respectively, control placement of rhamnose and arabinose at the 2″-OH group of the *C*-bound sugar. Alleles g^X and gl^A are rare in *S. pratensis*, hence the compounds whose synthesis they control, 7-*O*-xylosylisovitexin and 2″-*O*-arabinosylisovitexin, respectively, are likewise rare.

The more common alleles show distinct geographical patterning. For example, g^G and fg are present in most populations throughout Europe with a frequency of 1.0, except in the central regions where the frequency may be less than 1.0 and may be as low as zero (Mastenbroek *et al*., 1982). A detailed examination of 285 populations of *S. pratensis* provided evidence for the existence of at least three genetically different flavonoid races: (1) western and southern Europe with g^G, gl, and fg; (2) Hungary and Romania with g, gl^R, fg or Fg; and (3) USSR (sic) and Fennoscandia with g^G, gl^R, and fg (Mastenbroek *et al*., 1983a). Table 3.9 summarizes information on the genotypes and flavonoid profiles.

The data presented by Mastenbroek *et al.* (1983a) were subjected to clustering analysis (Mastenbroek *et al*., 1983b) one outcome of which shows the power of the analytical methods used. One of the clusterings showed an accumulation of populations that had a relatively high frequency (ca. 0.30) of *Fg* which, as pointed out above, is normally rare in *S. pratensis*. The presence of this gene could be taken as evidence of some early hybridization event involving the closely related species *S. dioica* in which the frequency of *Fg* is always high. Populations from Russia have other alleles, e.g., V_g *and* V_x, so the overall genetic situation is not a straightforward one involving only the genes listed in the table. Despite the uncertainty about some aspects of the flavonoid genetics of *S. pratensis*, the geographical mapping of the glycosylation reactions in this taxon represent one of the most impressive attempts to understand pigment variation yet attempted. It should stand as a model for this type of research.

Table 3.9 Genetic control of isovitexin glycosylation in *Silene* (from Mastenbroek *et al*., 1983)

Locus	Allele[1]	Sugar transferred	Product formed
fg	Fg	Glucose	Isovitexin 2″-*O*-glucoside
g	g^G	Glucose	Isovitexin 7-*O*-gucoside
	g^X	Xylose	Isovitexin 7-*O*-xyloside
gl	gl^R	Rhamnose	Isovitexin 2″-*O*-rhamnoside
	gl^A	Arabinose	Isovitexin 2″-*O*-arabinoside

[1]The null alleles, fg, g, and gl, have not been included since they do not code for any glycosyltransferase

Barley

One of the major steps in the evolution of Human society has been the development of agriculture. Successful agriculture was based upon cultivation of plant species whose roots, fruits, leaves, or seeds could be relied upon consistently to provide dietary requirements. Among the earliest plants to be domesticated were the cereal grasses, barley, oats, wheat, and rye. One of the most interesting aspects of ethnobotany is the attempt to trace the history of the development of these important crops. A major difficulty encountered in this kind of research, however, is the length of time that the plant has been under cultivation. The longer the time since domestication the more difficult it becomes to trace the lineage. In some cases, hybridization between two natural species has occurred; in others the progenitor species may no longer exist. Archeological evidence suggests that cereal harvesting was practiced in the Jordan Valley between 9000 and 8000 B.C. (Sauer, 1994). The history of the domestication of barley has been well described (Harlan, 1976; Zohary and Hopf, 1988; Bar-Yosef and Kislev, 1989; Sauer, 1994).

Barley belongs to the genus *Hordeum*, which consists of some two dozen or so taxa, only a few of which are relevant to the development of the domesticated plant. As related by Sauer (1994), *H. distichum* (*H. vulgare* subsp. *distichum*) was the first domesticated barley and has served as the progenitor for all other varieties. *Hordeum vulgare* subsp. *vulgare* originated by human selection for six-rowed varieties (native barley has two rows of grains). *Hordeum distichum*, although still cultivated in parts of Europe, was displaced by *H. vulgare* as the main crop variety. *Hordeum spontaneum*, which occurs naturally in the so-called Fertile Crescent, is frequently encountered as an agricultural weed in southwestern Asia. Mabberley (1987) suggests that *H. distichum* may be a hybrid between *H. vulgare* and *H. spontaneum*.

Fröst and coworkers (1975), in an introduction to their study of the use of flavonoids in unraveling the phylogeny of barley, state that three closely related diploid, interbreeding species can be distinguished within *Hordeum* section *Cerealia*, *H. vulgare, H. spontaneum*, and *H. agriocrithon*. *Hordeum agriocrithon*, a six-rowed, brittle-eared species from Tibet, was included in their study because some workers had argued that forms with two-rowed heads had evolved from a six-rowed progenitor (this idea is not accorded much weight by current workers). Some of the relationships among this group of taxa become clearer in light of detailed flavonoid surveys.

An examination of 1424 local varieties of *H. vulgare* was undertaken, along with much smaller samples of *H. spontaneum* (two populations) and *H. agriocrithon* (one population), using thin layer chromatography (Fröst *et al*., 1975). Three distinctly different profiles were observed for the *H. vulgare* samples, designated A, B, and C, with lesser differences within B identified as B_1 and B_2. The *H. spontaneum* profile, designated chemotype S, was markedly different from those of *H. vulgare*, while the specimen of *H. agriocrithon* matched the *H. vulgare* A profile. Chemotypes A and B occur throughout the cultivated area of *H. vulgare* whereas the C chemotype occurs predominantly in populations sampled in Ethiopia. In an effort to get more reliable data for *H. spontaneum* and *H. agriocrithon*, Fröst and Holm (1975) investigated a further 279 collections of the former and 10 of the latter. Fifty percent of the *H. spontaneum* collections gave chemotype A, 32% gave chemotype B, 3% gave chemotype C, and the remaining 15% were chemotype S. Despite the occurrence of only 15% S chemotypes in the *H. spontaneum* samples, Fröst and Holm (1975) consider this to be the ancestral profile. This conclusion was based upon controlled crossing experiments that showed the dominance relationships in the F1 generation to be $S > B > C > A$. The high frequency of A and B chemotypes in this species was attributed to repeated hybridization with selection of features from *H. vulgare*, including the flavonoid profile, which allowed survival in disturbed sites. The samples of *H. agriocrithon* gave only chemotypes A and B, which led these workers to conclude that this species is of no importance in the phylogeny of barley in agreement with others who do not consider it to be a genuine wild species.

Although the chromatographic spot patterns provided reliable markers for establishing relationships among this set of taxa, full structural information was desired. The flavonoid profiles of all five chemical races, A, B_1, B_2, C, and S were established by Fröst *et al.*, (1977). The five races had the following compounds is common: luteolin 7-*O*-glucoside, tricin 7-*O*-glucoside, isoörientin 7-*O*-glucoside, plus a lucenin isomer and its glucoside. Chemotype A is distinguished from the others in its capacity to accumulate apigenin 7-*O*-glucoside, chrysoeriol 7-*O*-glucoside and galactoside, vitexin 7-*O*-rhamnosylglucoside, and isovitexin 7-*O*-diglucoside. The other chemotypes are distinguishable on the basis of various combinations of vitexin 7-*O*-glycosides, isovitexin glycosides, and isoörientin glycosides. The differences among the five chemotypes was discussed in terms of the structural modifications exhibited by the compounds in each.). The least modified flavonoids, e.g., those with simplest glycosylation patterns, are those seen in chemotype S, the most modified are seen in chemotype A. The ascending evolutionary order, according to those workers, is thus: $S \rightarrow C \rightarrow B_1 \rightarrow B_2 \rightarrow A$.

Roses

It could probably be argued that the rose (cultivars of *Rosa*) is not only the most popular garden plant but also the most readily recognizable one. The popularity arises from the existence of an almost unlimited number of color varieties as well as the availability of miniature, shrub, and climber forms. Maintaining some accounting of the number of cultivars and authenticating new forms are daunting tasks. One of the problems confronting rose specialists is being able to confirm parentage of new hybrids and to trace the ancestry of older ones. The history of the rose of commerce is a tangled and complex one. Essentially, however, hybridization has brought together the diploid genomes of the "tea" roses of the Orient ($2n=14$) and the tetraploid "cabbage" roses of the west (Mabberley, 1987). Through the years, classification methods have relied on morphological features and statistical indices (Roberts, 1977), but those techniques are now being supplemented with data from other sources: meiotic figures, isozymes, DNA restriction fragments (see Raymond *et al.*, 1995 for leading references), and flavonoids (e.g., Asen, 1982). The utilization of HPLC methods in the analysis of floral pigments has made a more rigorous chemical classification of roses possible (Asen, 1982; Van Sumere *et al.*, 1993). The successful application of these methods to problems in identifying rose cultivars and hybrids has been amply demonstrated (Biolley *et al.*, 1992, 1993). A recent HPLC study of ancient rose varieties has proven equally useful (Raymond *et al.*, 1995). Their analytical system effectively distinguished 15 petal flavonols based on kaempferol and quercetin, with two pairs (kaempferol 3-*O*-glucoside and 3-*O*-glucuronide; kaempferol 4′-*O*-glucoside and 3-*O*-glucuronide) left unresolved. Statistical analysis of the HPLC fingerprints of 30 varieties, selected for their historical significance, showed clear-cut clustering of groups of varieties along with their respective founder species, *R. chinensis*, *R. gallica*, or *R. moschata*.

FLAVONOIDS IN THE STUDY OF HYBRIDS

One of the most useful applications of flavonoids has involved their potential as indicators of natural hybridization. Since flavonoids are known to be codominantly inherited (each parent contributes equally to the offspring) one needs only determine the flavonoid chemistry of putative parents and compare those flavonoid profiles with the hybrid. Of course, in order for this technique to work the parent species must obviously exhibit different profiles and the degree of intraspecific variation must be comparatively small. The reader will notice that in most of the examples given below, one or more "other" compounds appear in hybrids suggesting that more than just additivity is at work. Despite the fact that so little is known about the control of flavonoid biosynthesis, workers studying natural hybrids nonetheless tend to suggest that the hybridization process somehow has affected it.

Baptisia, Where it all Began

The power of flavonoids as indicators of hybridization was elegantly demonstrated within selected species of *Baptisia*, a legume genus consisting of 17 species native to southeastern United States. Pioneering work in the study of hybridization in this genus was carried out at the University of Texas by R. E. Alston and B. L. Turner (Alston and Turner, 1959, 1962, 1963b; Turner and Alston, 1959). Among the early studies was an examination of *B. leucophaea* and *B. sphaerocarpa*, which undergo extensive hybridization near the Texas Gulf coast. These species are easily distinguishable morphologically so that it was possible to identify the hybrids with certainty and assess the contribution of each parent to the hybrid's pigment profile. Six species-specific components were thus identified, three from each species. Although chromatographic analysis was not necessary for identification of hybrid individuals, the indications were that the analytical approach might prove to be very useful in situations involving more cryptic specimens. This was proven to be the case in a study of a very complex population composed of *B. leucophaea*, *B. sphaerocarpa*, *B. leucantha*, and different, potentially hybrid individuals. Using individuals of known specific purity, 20 morphological characters were measured and the data used to estimate a tri-hybrid index. The index values allowed the workers to estimate the percentage contribution from each pure species for any hybrid individual. A convincing example of the usefulness of this method was given by Baetke and Alston (1968) who examined every (!) individual in a hybridizing population of *B. sphaerocarpa* and *B. leucophaea*. Chromatographic and morphological analysis of some 1100 plants (in an area measuring 240 ft × 300 ft) resulted in the identification of 470 *B. leucophaea*, 561 *B. sphaerocarpa*, 83 F_1 hybrids, 37 individuals backcrossed to *B. leucophaea*, and 19 individuals backcrossed to *B. sphaerocarpa*.

Chromatographic analyses of this sort, and others, allowed the components (presumptive flavonoids) to be assigned to one of four classes:

(1) components common to all parental species involved in the hybridizing system.
(2) components that are species-specific and reliable because of consistent presence.
(3) species-specific components that are not consistently present (they may be present but below the limits of detection).
(4) hybrid-specific components.

In the following cases we will meet situations where flavonoids that appear in hybrids were not seen in either parent. These usually involve substitutions in the sugars, e.g., glucose for galactose, or changes in sugar combinations, and are frequently rationalized on the basis of some change in the control of flavonoid biosynthesis caused by bringing together two different genomes. Occasionally compounds seem to vanish completely in the process of hybridization, which may also be due to a disturbed glycosylation system. An additional factor that should be borne in mind involves alteration of organ specificity of certain compounds. For example, they may appear in leaves in hybrids whereas in parent plants they were floral components.

Lathyrus chrysanthus × *L. chloranthus*

As part of an effort to introduce yellow pigmentation into the cultivated sweet pea (*Lathyrus odoratus*), Markham and coworkers (1992) studied the flavonoid chemistry of two yellow-flowered species, *L. chrysanthus* and *L. chloranthus*, along with a hybrid between the two. *Lathyrus chloranthus* has the simpler flavonoid profile with kaempferol, quercetin, and isorhamnetin 3-*O*-glycosides, including 2-*O*-xylosylglucosides. *Lathyrus chrysanthus* has some of the same 3-*O*-glycosides but lacks the xylosylglucosides. It also has quercetin 7-*O*-rhamnoside and a series of 3,7-di-*O*-glycosides involving the same aglycones. The flavonoid profile of the hybrid consists of most (possibly all) of the 3,7-diglycosides, kaempferol and isorhamnetin 3-*O*-glycosides, and two hybrid compounds. The latter were identified as quercetin and isorhamnetin 3-*O*-xylosyl(1 → 2)glucoside-7-*O*-rhamnosides. These compounds occur in neither parent and result from the combination of the 3,7-diglycosylation capacity of *L. chrysanthus* with the capacity of *L. chloranthus* to make the xylosyl(1 → 2)glucosides. This is one of the most convincing instances of the formation of a true hybrid compound.

An Example from *Sideritis*

Sideritis is a genus in Lamiaceae (the mint family) that comprises some 100 species native to the northern temperate Old World and Macaronesia. Much of the taxonomic difficulty in the genus in Southwestern Europe is thought to result from morphological similarities among species and hybridization (Heywood, 1972). Flavonoid data have been successfully applied to the system involving *S. serrata, S. bourgaeana*, and their putative hybrids in Spain. Ferreres and coworkers (1989) studied vacuolar and exudate flavonoids of both parents, their hybrid offspring, and individuals thought to be backcross hybrids. Vacuolar flavonoids consisted of glycosides of 5,7,8,4′-tetrahydroxyflavone (isoscutellarein), 5,7,8,3′,4′-pentahydroxyflavone (hypolaetin), and 5,7,8,4′-tetrahydroxy-3′-methoxyflavone (hypolaetin 3′-methyl ether). The exudate flavonoid pool consisted of a series of 6- and 6,8-oxygenated flavones having two or three *O*-methyl groups. In most instances, suspected hybrid individuals showed the additivity expected of first generation hybrids. One of the sets of compounds by which the two parents can be distinguished involves the B-ring methyl ethers of 5,3′,4′-trihydroxy-6,7,8-trimethoxyflavone. *Sideritis serrata* produces only the 4′-methyl ether 5,3′-dihydroxy-6,7,8,4′-tetramethoxyflavone [3-32]; *S. bourgaeana* only the 3′-isomer 5,4′-dihydroxy-6,7,8,3′-tetramethoxyflavone [3-33]. Hybrid plants could be identified by the presence of both isomers. The 3′,4′-dimethyl ether [3-34], which would have been a hybrid compound, was *not* observed in any of the plants. This is in agreement with the idea that neither of the monomethyl ethers would be accepted as substrate by either *O*-methyltransferase (De Luca and Ibrahim, 1985a,b). Suspected backcross individuals had proportionately more of the compound(s) characteristic of the recurrent parent.

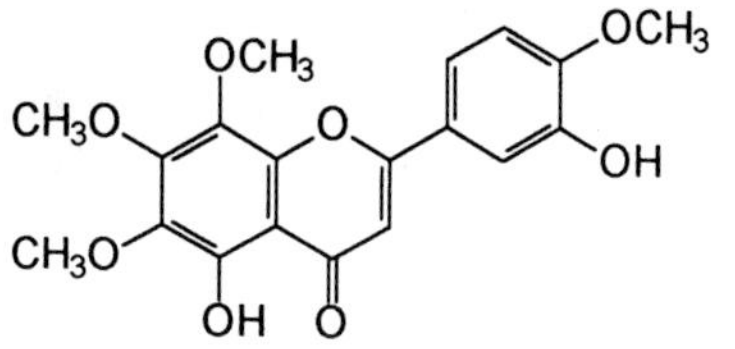

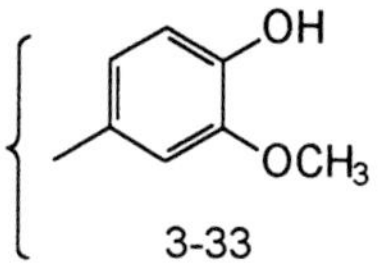

Is *Arnica gracilis* a Hybrid Species?

The flavonoids of North American members of *Arnica* attracted the attention of Keith Denford, formerly of the University of Alberta, and his students, who have made excellent use of them as indicators of evolutionary relationships among several troublesome taxa in western North America. Part of their work involved a study of the flavonoid profiles of *A. cordifolia, A. gracilis*, and *A. latifolia* to test the hypothesis that *A. gracilis* is of hybrid origin and that the other two species served as parents. *Arnica gracilis* occurs on exposed, rocky slopes in alpine regions of the Rocky Mountains from Alberta south to Wyoming and in the Cascades of British Columbia and Washington. The question as to whether *A. gracilis* is a "good" species no doubt arose from its close morphological similarity to *A. latifolia*. (Wolf and Denford, 1984). The flavonoid differences between *A. cordifolia* and *A. latifolia* were determined to be sufficiently different and their profiles sufficiently invariant that flavonoid profiles were considered ideal tools for examining the putative hybrid nature of *A. gracilis*. Table 3.10 lists the compounds identified in the three taxa. Flavonoids identified in *A. gracilis* show additivity with three exceptions. Five compounds occur in all three taxa, four are shared by *A. cordifolia* and *A. gracilis*, and three are shared by *A. latifolia* and *A. gracilis*. One of the exceptions to additivity, the presence of apigenin 7-methyl ether (genkwanin) in *A. cordifolia*, can be ignored because the compound was seen in only one of 18 populations of the species examined. The other exceptions involved matters of more substance. Luteolin 4′-methyl ether (diosmetin) and apigenin 7-*O*-glucoside occur in *A. gracilis* but not in either putative parental species. What can be inferred about biosynthetic processes in the hybrid based on these anomalies? The accumulation of luteolin 4′-methyl ether (diosmetin) by *A. gracilis* suggests that control of *O*-methylation has been disturbed in the hybridization process such that a new substrate, luteolin in this case, can be

Table 3.10 Flavonoids of *Arnica cordifolia, A. gracilis*, and *A. latifolia* (from Wolf and Denford, 1984)

	A. cord.	*A. grac.*	*A. lati.*
Apigenin 7-*O*-glucoside	−	+	−
Apigenin 7-methyl ether (one pop'n)	+	−	−
6-Methoxyapigenin	+	+	+
Luteolin 7-*O*-glucoside	+	+	−
Luteolin 4′-methyl ether	−	+	−
6-Methoxyluteolin 7-*O*-glucoside	+	+	−
Kaempferol 3-*O*-glucoside	+	+	+
Kaempferol 3-*O*-galactoside	−	+	+
6-Methoxykaempferol 3-*O*-glucoside	+	+	−
Quercetin 3-*O*-glucoside	+	+	+
Quercetin 3-*O*-diglucoside	+	+	+
Quercetin 3-*O*-gentiobioside	+	+	+
Quercetin 3-methyl ether	+	+	−
Patuletin (6-methoxyquercetin)	−	+	+
Patuletin 3-*O*-glucoside	−	+	+

tolerated. *O*-Methylation occurs in both of the putative parent species using 6-hydroxyapigenin, 6-hydroxyluteolin, quercetin, and 6-hydroxykaempferol as substrates in *A. cordifolia*, and quercetagetin as substrate in *A. latifolia*. The other anomalous compound found was apigenin 7-*O*-glucoside, which was observed only in *A. gracilis*. An explanation of this finding might be found in the lowered efficiency of use of the apigenin precursor (naringenin) in *A. gracilis* so that the flavone might be formed and accumulated, whereas in the two putative parent species naringenin is completely consumed in forming the variety of compounds seen in those two taxa. Resolution to problems of this sort can be sought in a better understanding of quantitative differences in these compounds among the species, and the nature and efficiency of the enzyme reactions by which the various compounds are made. A numerical analysis of *Arnica* is also enlightening (Wolf and Whitkus, 1987).

An Example from *Parthenium*

Parthenium (Asteraceae) consists of 16 species (some writers say 20) with representatives in North and Central America, northern South America, and the West Indies (Karis and Ryding in Bremer, 1994). Two of its species have gained levels of fame (notoriety in the case of the latter) for quite different reasons. *Parthenium argentatum* is a desert species successfully cultivated as a source of rubber, while *P. hysteropherus*, which has become a global nuisance, presents a serious problem owing to its capacity to cause severe contact dermatitis in humans.

The flavonoid chemistry of the North American members of the genus have been extensively examined by Mears (1980a,b) who found, among other things, that flavonoid diversity varied markedly among certain species depending upon the soil type upon which they were growing. This phenomenon came into play in part of the examination of putative hybrids between *P. incanum* and *P. argentatum*. These two species occur in northern Mexico and western Texas and are known to hybridize in that area. Although the two species coexist geographically, they occur on soils of different origin. *Parthenium argentatum* occurs on desert limestone whereas *P. incanum* occurs in more mesic areas. Using plants collected in areas where hybridization had presumably not occurred, Mears was able to establish the flavonoid profiles of the pure parental species. *Parthenium incanum* was thus shown to possess a flavonoid profile that consisted of a single free flavonoid, kaempferol 3-methyl ether, and a total of eight glycosides of kaempferol, quercetin, and quercetagetin. Similarly, *P. argentatum* was shown to accumulate an array of free flavonoids consisting of *O*-methylated derivatives of kaempferol, quercetin, 6-hydroxykaempferol, and quercetagetin. Mears' flavonoid occurrence data for the two species and the hybrid are summarized in Table 3.11 except that quercetin 3,3′-dimethyl ether is marked as present in *P. incanum* whereas it was marked as absent in his original table (Table 3.1). (This modification was based upon his reporting of one of the quercetin 3,3′-dimethyl ether 7-*O*-glycosides (marked #2 here) was present in some of the specimens of *P. incanum*. The existence of this glycoside obviously requires the presence of the aglycone.) An examination of the table shows that the hybrid individuals, even with the variation taken into consideration, reflect input from the two species suggesting that they both have contributed genetic material. Thus, the hybrid exhibits the higher frequency of 6-hydroxylation and the higher level of *O*-methylation characteristic of *P. argentatum* plus the capacity to make the number-2 glycoside of kaempferol seen only in *P. incanum*.

An interesting situation exists with respect to the presence of quercetin 3,3′-dimethyl ether 7-*O*-glycoside number-1 in hybrid individuals and its apparent absence from the two parent species. The hybrids could have inherited the capacity to make the aglycone from either parent although *P. argentatum* is the more likely contributor on the basis of this aglycone's presence as a major component of that

Table 3.11 Flavonoids of *Parthenium incanum* X *argenteum* (Mears, 1980)

	I. incan.	hybrid	*I. argen.*
Quercetagetin 3,7-dimethyl ether	–	+	+
Quercetagetin 3,6-dimethyl ether	+	+	+
6-Hydroxykaempferol 3,7-dimethyl ether	–	+	+
6-Hydroxykaempferol 3,6,7-trimethyl ether	–	+	+
Kaempferol 3-methyl ether	+	+	+
Kaempferol 7-methyl ether	–	+	+
Quercetin 3,3′-dimethyl ether	(+)	+	+
Quercetin 3,3′-dimethyl ether 7-*O*-Gly #1	–	+	–
Quercetin 3,3′-dimethyl ether 7-*O*-Gly #2	+	+	+
Kaempferol 3-*O*-Gly #1	+	+	+
Kaempferol 3-*O*-Gly #2	+	+	–
Kaempferol 3-*O*-Gly #3	–	+	–
Quercetin 3-*O*-Gly #1	+	+	+
Quercetin 3-*O*-Gly #2	–	+	+
Quercetin 3-*O*-Gly #3	+	+	+
Unknown #1	+	+	–
Unknown #2	+	+	–
Unknown #3	–	+	–
Unknown #4	–	+	–

species and its apparent transitory presence in *P. incanum* (the presence of the glycoside requires the aglycone to have been present). That this compound exists as the 7-*O*-glycoside is of interest because this position of glycosylation was not seen in any other flavonoid in this set of taxa. Is this a case of control of glycosylation being upset by the hybridization process, or has there been an alteration in the nature of the glycosyltransferase itself that allowed the site of glycosylation to be altered from position C-3 to position C-7? This is a particularly interesting point insofar as data available to date suggest that position specificity, i.e., C-3 vs. C-7, is a characteristic of the enzymes responsible for glycosylation (Chapter Six). Detailed studies at the level of enzyme specificity, enzyme structure, and the relatedness (homology) of the genes coding for the enzymes in question are necessary to answer these questions.

An Example from *Lasthenia*

An interesting example of suspected hybridization in Nature involves members of the genus *Lasthenia* growing in California. A detailed monographic study of *Lasthenia* by Ornduff (1966) resulted in his recognition of six sections three of which had species that produced anthochlors (aurones and chalcones), *Baeria, Burrielia*, and *Hologymne* (Bohm *et al*., 1974). All members of the first two sections have flavonoid profiles dominated by anthochlors. Section *Hologymne*, on the other hand, consists of three species (four taxa), *L. chrysantha, L. glabrata* subsp. *glabrata, L. glabrata* subsp. *coulteri*, and *L. ferrisiae*. On the basis of morphological and chromosomal information, Ornduff (1966) suggested that *L. ferrisiae* arose through hybridization between *L. glabrata* subsp. *coulteri* and *L. chrysantha*. Flavonoid data support this view. The pigment profile of *L. glabrata* subsp. *coulteri* consists of quercetin 3-*O*-galactoside, quercetin 3-*O*-glucuronide, and a quercetin 3-*O*-rhamnosylgalactoside but lacked anthochlors. The profile of *L. chrysantha* consisted of the same three quercetin glycosides plus anthochlors. The flavonoid profile of *L. ferrisiae* was shown to consist of anthochlors, the three quercetin glycosides named above as well as quercetin 3-*O*-glucoside. The additivity of flavonoid profiles can be taken as support for the hybrid origin of this species. The presence of a new quercetin glycoside is not unusual for hybrid systems. Its appearance may have resulted from some disruption in the control of glycosyltransferase system(s) with regard to substrate selectivity, viz., glucose vs. galactose. Another possibility is that quercetin 3-*O*-glucoside is normally present in the plant but lies below the levels of detection and that the process of hybridization in some way has changed the concentrations of sugars available or glycosides produced. Again, only detailed enzymological studies can resolve this question.

Polyploid Species of *Viguiera*

Viguiera (Asteraceae) is a large genus with possibly as many as 180 species and a range that includes much of North and South America (Karis and Ryding in Bremer, 1994). Most of the flavonoid information available on this genus comes from the work of Edward Schilling at the University of Tennessee and his colleagues. One of the systems that attracted that group's attention was *V. triangularis* and its possible progenitors (Schilling, 1989). This species is interesting because it belongs to the "Baja California group," which contains three polyploid species with flavonoid profiles having a higher degree of similarity than was seen among the diploid species. Since polyploids are themselves potentially of mixed parentage, a study of hybridization within the group takes on more of a detective-work aspect than in cases involving interaction between species of lower ploidy. The polyploids in the Baja group are not well differentiated morphologically and have often been lumped with the diploid *V. parishii*. The flavonoids obtained from leaf resins of the polyploid species, however, are sufficiently different to suggest that they would be useful indicators of possible ancestry. The morphological differences between *V. triangularis* and *V. parishii* lie mainly in size differences of leaves, pales, and achenes. Their flavonoid differences, on the other hand, are quite distinct. Major compounds from *V. parishii* are 6-hydroxyapigenin 7-methyl ether and the chalcone isoliquiritigenin 4′-methyl ether, with smaller amounts of apigenin, luteolin, and 6-methoxyapigenin. The flavonoid profile of *V. triangularis* is more complex the major components being luteolin, 6-hydroxyapigenin 7-methyl ether, 6-methoxyapigenin, the 6,4′- and 7,4′-dimethyl ethers of 6-hydroxyapigenin, the 6- and 7-methyl ethers of 6-hydroxyluteolin, and isoliquiritigenin 4′-methyl ether. Present in lesser amounts are apigenin 7-methyl ether (wogonin), and the 6,7-dimethyl ether of 6-hydroxyapigenin. The close morphological similarity between *V. parishii* and *V. triangularis* suggest that the former may have supplied part of the genome of the polyploid species. Flavonoid chemistry supports this view since the pigment profile of *V. parishii* consists of a subset of the compounds identified in *V. triangularis*.

Flavonoids can assist in helping to find other possible parental species. *Viguiera microphylla* has morphological and geographic features that, along with a similar flavonoid profile (methyl ethers of 6-hydroxyapigenin and 6-hydroxyluteolin), suggest a possible relationship with *V. triangularis*. It was not possible to rule out *V. tomentosa*, however, since it has the capacity to make luteolin and the 7-methyl ethers of 6-hydroxyapigenin and 6-hydroxyluteolin. Flavonoids helped to eliminate other species from consideration. Thus, *V. laciniata* could be eliminated on the basis of its capacity to accumulate flavonols, naringenin derivatives, and several anthochlors not seen in any other species in the group. *Viguiera subincisa* was removed from contention because of its unique capacity to make 6,3′-dimethyl ether of 6-hydroxyluteolin, and *V. reticulata* was eliminated on the basis of its capacity to make apigenin 4′-methyl ether.

In Vino Veritas (with Apologies)

One of the problems facing maintenance of wine quality (adherence to the rules of "Appelations controllées") involves detecting wine made from hybrid grapes. A sophisticated, yet simple to execute, method for detecting adulteration arising from use of hybrid grapes was developed by Ribéreau-Gayon (reviewed by Ribéreau-Gayon, 1982). The method was based upon the finding that anthocyanins produced by *Vitis vinifera* (the "legal" species) are monoglycosides whereas those produced by the adulterating species, *V. riparia* and *V. rupestris*, are diglycosides, and that the diglycoside character is dominant over the monoglycoside. Crosses between *V. vinifera* and either *V. riparia* or *V. rupestris* produce an F_1 generation all members of which contain anthocyanidin diglycosides. Crossing F_1 hybrids should show a 3 : 1 ratio of diglycoside bearing offspring to monoglycoside-bearing ones. The results are in general agreement with genetic theory. Further genetic tests involved crosses between: (1) *V. vinifera* and a homozygous dominant hybrid that was diglycoside dominant (Seyve-Villard 2318) which produced progeny all of which contained diglycosides; (2) *V. vinifera* and a recessive homozygote (Seyve-Villard 18402) gave progeny that had no diglycosides; and (3) *V. vinifera* and a heterozygous hybrid from which came offspring half of which produced monoglycosides and half produced diglycosides. A drawback of the method is that some grape varieties, as illustrated here with the Seyve-Villard cultivars, can confuse the issue. However, involvement of the major adulterants, *V. riparia* and *V. rupestris* can be readily detected.

Flavonoids of Elm Hybrids

One of the most devastating plant diseases in North American history was (is?) the Dutch elm disease caused by *Ophiostoma ulmi*. Some of the Asian species of elm, *Ulmus pumila* (Ulmaceae), the Siberian elm, for example, are resistant to the fungus and have been used in breeding experiments aimed at producing a hybrid tree that might be reintroduced into areas stricken by the disease. Such a breeding program was described by Heimler and coworkers (1993) who worked with *U. carpinifolia, U. pumila, U. parvifolia*, and *U. japonica*. They developed a high performance liquid chromatographic (HPLC) system that allowed them to analyze the flavonoid fraction of parent, hybrid, and putative hybrid plants. Only two compounds were identified, quercetin 3-*O*-rutinoside (rutin) and quercetin 3-*O*-glucoside (isoquercitrin), but the HPLC results subjected to multivariate discriminant analysis provided a reliable means of identifying the parentage of suspected hybrids (often obtained from arboreta from open pollinated stands of elms). Some information on the suspected parental species was necessary, of course, but information on the exact parental trees was not required.

Populus Hybrids

Greenaway and coworkers have made an extensive study of the bud exudate chemistry of *Populus* species using gas chromatography-mass spectrometry (GC-MS) (1988, 1990a, 1992a,b). Their method has proved of great utility in determining the hybrid status of certain species. An excellent example of the utility of their approach can be seen in a study that clearly helped establish the parentage of *P.* x *jackii*, which was thought to have been derived from crosses between *P. balsamifera* and *P. deltoides* (Greenaway *et al*., 1990b). Three qualitative features were particularly useful. Dihydrochalcones, present as major components of

P. balsamifera but absent from the profile of *P. deltoides*, were present in the putative hybrid species. Similarly, but reversing the situation, flavanones methylated or esterified at C-3 are major components of the profile of *P. deltoides* but were not observed in *P. balsamifera*. Both specimens of *P.* x *jackii* examined had members of this group of flavonoids. The third qualitative difference involved several terpenoids that were not seen in *P. deltoides*, but were present in both *P. balsamifera* and *P.* x *jackii*. Flavanone and flavone profiles also showed additivity in the buds of *P.* x *jackii*.

Possible Hybrid in *Coprosma*

Wilson (1979) studied the flavonoid profiles of 47 named and unnamed species of New Zealand *Coprosma* (Rubiacecae). The flavonoid profiles of most species were based on one or more of the common flavonols, kaempferol, quercetin, or myricetin, glycosylated at C-3 with some also glycosylated at C-7. A few species had profiles based on apigenin and luteolin glycosides. The important feature to come out of this study was that each species exhibited a unique combination of flavonoid glycosides. Two species are of significance in the present context. *Coprosma colensoi* and *C. banksii* differ only in leaf shape, and when the two leaf forms are found growing together intermediate forms are also present. The *Flora of New Zealand* (Allan, 1961) suggests that hybridization has been involved. Eagle (1982), however, *sees C. banksii* as only a form of *C. colensoi*. Wilson studied three specimens, one each *colensoi*-like, *banksii*-like, and an intermediate form. The flavonoid profiles of the three specimens were identical consisting of a set of apigenin and luteolin 7-*O*-, 4′-*O*-, and 7,4′-*O*-glycosides. Although the sample size is small, the results certainly support the view that the three specimens belong to the same species, *C. colensoi*, and that hybridization has not occurred.

HOW DID FLAVONOIDS ORIGINATE IN THE FIRST PLACE?

This simply worded question is at once the most fundamental and the most difficult to answer. What was the driving force behind selection of the first organism that had the capacity to produce a flavonoid? The development of the flavonoid biosynthetic pathway is often linked with the emergence of terrestrial plant life with the concomitant need for protection from mutagenic radiation if an individual line were to be successful. The known capacity of flavonoids to absorb ultraviolet radiation in the range most detrimental to nucleic acids, and possibly other systems, has often been put forward as the driving force behind selection of those organisms best adapted to that task (Lowry *et al.*, 1980). The capacity of flavonoids to function in plant-microbial interactions has also been mentioned as a reasonable driving force (Graham, 1993; Shirley, 1996 and citations in both). Helen Stafford (1991), however, pointed out an extremely important fact: the initial appearance of flavonoids would not have involved large enough amounts of these compounds to serve effectively as UV shields. It is her contention that their initial function would likely have been as internal regulatory factors and that that function provided the producing organisms a selective advantage. Evidence for this view comes from the long-held view that polyphenolic compounds can have significant effects on growth through their capacity to interact with the hormone indoleacetic acid (IAA) oxidase (Furuya *et al.*, 1962; Galston, 1969). Monohydroxy B-ring flavonoids (e.g., kaempferol derivatives) were implicated as cofactors of the peroxidase that destroys IAA, whereas dihydroxy B-ring flavonoids (e.g., quercetin derivatives) inhibit the destructive enzyme. Thus, a balance between active forms of these compounds (whatever they might be) would modulate hormone activity through selective destruction or inactivation of the growth hormone. Recent studies have implicated both mono- and dihydroxy flavonoids as inhibitors of IAA transport across the plasma membrane (Jacobs and Rubery, 1988). Regulation of transport of growth hormones, as well as their function, could provide the fine tuning necessary for control of growth in highly competitive situations. A further indication of possible hormone-associated behavior of flavonoids comes from the observation that "lunularic acid," a stilbene ($C_6C_2C_6$) derivative, acts in mosses in a fashion similar to the action of the hormone abscissic acid (ABA) in higher plants (Gorham, 1990). With the selection of flavonoid-bearing organisms it would be only a matter of time before some variant(s) developed the capacity to make larger amounts of these compounds whereupon their other physical characteristics, specifically their UV absorbing capacity, would be brought into play. Their potential toxicity to other organisms could have entered the scene at any time since competing organisms certainly existed at all stages during which phenolic metabolism was developing, and compounds can have antibiotic activity at comparatively low concentrations. Only much later did oxidation of a colorless flavonoid (flavanone or dihydroflavonol) occur to produce the anthocyanins which, in addition to absorbing potentially harmful UV radiation, can reflect other wavelengths to produce the colors, some visible to vertebrates, some to insects, that would serve to attract pollinators.

It has also been observed recently that certain flavonoids appear necessary for the successful growth of pollen tubes, without which the pollinators' efforts would be futile (Mo *et al.*, 1992; Taylor and Jorgensen, 1992). The discovery that flavonoids are involved in this critical stage of plant reproduction provided the germ for the development of a most ingenious hypothesis involving the impact of flavonoids on plant evolution (Jorgensen, 1993). We will look at his idea next.

Jorgensen's (1993) idea builds on the hypothesis that land plants might have originated by an endocellular union between a green alga and a fungus, rather than simply as an evolutionary extension of a green alga (Atsatt, 1988, 1991). A green alga would contribute to the union the capacity for photosynthesis while the fungus would have certain properties not possessed by its algal associate, namely the capacity for invasive intercellular growth (such as seen in haustorial growth) and the capacity for absorptive nutrition. The parallel between the fungal capacity for tip growth, such as seen in invasion of a host organism by haustoria, and the characteristic growth of a pollen tube through stylar tissue is the key connection in this synthesis. Jorgensen sees the symbiotic association beginning with an unspecialized green alga, which had evolved the capacity to make UV-shielding flavonoids, and a fungus (or other tip-growing organism) that were living in close proximity in shallow water. The fungus then developed the capacity to recognize the flavonoids as an indicator of the presence of the UV-protective alga. In time, the recognition would develop into a dependence upon the compounds. Eventually, the association between the two organisms would culminate in the fungus becoming an endosymbiont living within the cells of the alga. As a final stage, there would be some level of fusion between the genomes of the two organisms with some genetic information lost and some maintained. The capacity for tip growth and its control by flavonoids would have been part of the information maintained by the "new" organism. This dependence of the tip growth capacity of the new line of organisms would eventually manifest itself in such phenomena as the growth of the pollen tube. Jorgensen (1993) describes each of these steps in considerable detail and goes on to suggest several ways by which his hypothesis might be tested. Most of these tests would involve determining gene phylogenies in appropriate organisms and are beyond the scope of this book. A thorough study of Jorgensen's paper will repay the reader's efforts!

CHAPTER FOUR

EXTRACTION, PURIFICATION, AND IDENTIFICATION OF FLAVONOIDS

INTRODUCTION

In this chapter we will look at how flavonoids are extracted from plant material, how flavonoid mixtures can be resolved into individual components, how these components can be purified, and how structures of the individual components can be determined. Because the literature on all of these aspects of flavonoid science is large, and for the most part, easily accessible, we will not review the entire subject in detail. Rather, sufficient information will be presented to acquaint the reader with the history of the development of various techniques, particularly chromatographic methods and the application of spectroscopic methods for the determination of flavonoid structures. In order to appreciate the power of current spectroscopic methods, the reader will also be introduced to classical methods used for identification of natural products. Several case studies will be described in some detail to demonstrate the usefulness of these techniques.

It is a general phenomenon in science that major advances are made as a result of the development of new techniques. This is no less true in the study of flavonoids where the development of paper chromatography opened up the field to many who otherwise would have found classical chemical methods too daunting. Similarly, the application of standardized ultraviolet (UV) spectroscopic analysis of flavonoids provided a straightforward means of determining simple structures. Eventually, the wedding of chromatographic and UV spectroscopic technology produced an extremely power tool, namely, a high performance liquid chromatographic system (HPLC) monitored by diode array detectors. The versatility of this analytical tool will be demonstrated below. Another event of major significance was the advent of nuclear magnetic resonance spectroscopy (NMR), which began with 60 MHz proton magnetic resonance (PMR) and has evolved to include studies of both ^{1}H and ^{13}C spectra, including sophisticated applications of two-dimensional coupling analyses involving both nuclei. Examples of the applications of these methods will also be found below. Application of mass spectroscopy (MS) has been as important to flavonoid structural work as it has been to other areas of natural product chemistry. The application of MS methods to flavonoid structural analysis, particularly with regard to the analysis of complex glycosides, will be seen in case studies below.

Before we look more closely at the methods, and the history of their development, it is useful to identify some of the major sources of information available on the subject. One of the earliest treatments of the subject, entitled *Methods in Polyphenol Chemistry* (Pridham, 1964), arose from a meeting of the Plant Phenolics Group at Oxford University in 1963. It consists of chapters contributed by experts in the various fields including UV, infrared spectroscopy (IR), NMR, paper, thin layer and column chromatography, electrophoresis, and quantitative methods. In some cases, as in the chapter on paper chromatography of phenolic compounds by E. C. Bate-Smith, the presentations represented original work. The treatment of UV spectroscopy of phenolic compounds by J. B. Harborne remains, in this author's opinion, one of the best introductions to the subject available to this day. Further applications of UV and visible light spectroscopy in flavonoid identification appear in *The Comparative Biochemistry of Flavonoids* (Harborne, 1967). Largely through the efforts of workers at the University of Texas in Austin, led by T. J. Mabry, a set of standardized techniques for the use of UV and proton NMR spectroscopy became available (Mabry *et al*., 1970). This was followed a few years later by Markham's (1982) *Techniques of Flavonoid Identification*. This small but very important contribution to the field includes many detailed procedures and is *the* source to consult for anyone interested in learning the subject. Various aspects of flavonoid analysis can also be found in *The Chemistry of Flavonoid Compounds* (Geissman, 1962) as well in volumes of *The Flavonoids* series (Harborne *et al*., 1975; Harborne and Mabry, 1982; Harborne, 1988, 1994). Volume 1 of the comprehensive series *Methods in Plant Biochemistry*, edited by P. M. Dey and J. B. Harborne, deals specifically with plant phenolic compounds (Harborne, 1989). In addition to chapters on various flavonoid classes, the volume includes treatments of simple phenols and phenolic acids, lignins, tannins, quinonoid compounds, and lichen substances. The chapters are concerned primarily with reviews of the relevant literature and do not dwell on description of procedures. A recent book by Waterman and Mole (1994) entitled *Analysis of Phenolic Plant Metabolites* offers a very current description of most of the methods available for the study of phenolic compounds in plants. The book is one in the series *Methods in Ecology* (edited by J. H. Lawton and G. E. Likens) and also includes a concise overview of chemical ecology of plants.

THE EXTRACTION PROCESS

The initial process in flavonoid analysis obviously involves extracting the compound, or more likely, compounds from the plant. Factors to be taken into consideration include the form in which the plant material exists, that is, whether it is fresh or dried, and the solubility properties of the flavonoids. Solubility of flavonoids depend to a large degree on whether they occur bound to one or more sugar residues, which renders them highly polar, or whether they occur in the free form in which case they are much less polar, and in the case of highly alkylated flavonoids, quite lipid soluble. If one is dealing with the vacuolar components of the plants, then the compounds will likely exist as glycosides and a polar solvent will be required for efficient extraction. Three solvents that find frequent use are acetone, ethanol, and methanol. These can be used as such or in varying mixtures with water. Experience in our laboratory has shown that two or three extractions with 80% aqueous methanol serves to remove the bulk of soluble materials. Since fresh plant material has a significant amount of water in the tissues, the actual concentration of water in the extracting mixture will be less than 80% methanol. One can modify the proportions to maximize extraction depending upon the situation at hand; there is no *best* system. This is well documented by Waterman and Mole (1994, pp. 46–47) who list 33 different extraction regimes that have been reported in recent plant biochemical literature. The variation in these procedures, as pointed out by those authors, involved extraction times that ranged from 30 seconds to 96 hours and ratios of solvent volume to sample weight from 2 to 200! In the present author's experience, and as was clearly pointed out by Waterman and Mole (1994), each laboratory tends to work out the method that best serves their particular requirements. For additional reviews the reader should consult works by Seshadri (1962a), Markham (1975), Hostettmann and Hostettmann (1982), and Harborne (1990).

Flavonoids also occur as components of glandular exudates of some plants (Wollenweber, 1982, 1984, 1986; Wollenweber and Dietz, 1981). These compounds usually exist as aglycones and frequently are characterized by moderate to high levels of *O*-methylation. *C*-Alkylated flavonoids or flavonoids esterified to some aliphatic or aromatic acid are also encountered from time to time. Combinations of these features are also seen. These highly non-polar compounds can be removed from the leaf surface by brief rinsing in an appropriate solvent such as acetone or dichloromethane. In most cases, the extraction can be accomplished in a matter of a few seconds. The longer leaf material remains in contact with the solvent the greater is the amount of leaf waxes and related compounds that are also extracted. Since these compounds can complicate subsequent chromatographic steps, it is best to avoid them at the outset. It has been our experience that a two or three second rinse with acetone is sufficient to remove the bulk of flavonoid material from the leaves of the goldback and silverback ferns (*Pentagramma*) or from leaves of *Balsamorhiza* and *Wyethia* (Asteraceae). Incidentally, extraction of exudate flavonoids from these species followed by thin layer chromatography provides a useful introduction to these techniques in the teaching laboratory.

CHEMICAL DEGRADATION

General Introduction

In the early days of natural product chemistry, before spectroscopic methods became available, the structure of a newly isolated compound was determined by careful degradation into successively smaller molecules until all pieces matched known compounds. By working backwards from the identified products it was possible to piece together a likely structure for the unknown. Confirmation of the structure of the unknown compound then required total synthesis using starting materials of known structure and well established reactions. Both the degradation and synthetic phases of this approach are labor intensive and require an extensive knowledge of organic chemistry. Since flavonoid aglycones are comparatively simple compounds, the classical approach is not a particularly arduous one, except in the case of biflavonoids and higher oligomers. Somewhat more troublesome was the establishment of structures of glycosidic derivatives, which entailed, in addition to the steps necessary to establish the nature of the aglycone, an additional exercise in carbohydrate chemistry. In the case of di- or triglycosides, or flavonoids linked to sugars at several sites, the problems could be quite challenging. An additional factor was the need for sufficient quantities of purified compound to allow the worker some margin of error. As we will see below, sophisticated mass spectral and nuclear magnetic resonance techniques now allow a worker to determine complex flavonoid structures, including glycosidic derivatives, with, at most, only a few milligrams of material. Instead of a detailed knowledge of degradation methods, the modern worker must understand the scope (and limitations) of atom–atom interactions as detected by nuclear magnetic methods, and the application of highly specialized mass spectral techniques to establish the identity of the component monosaccharides and the linkages within complex glycosides. But before we get to these methods it is useful to examine some of the specific chemical reactions that have been used in establishing flavonoid structures in the past.

Flavonoid Degradation

Once an unknown compound has been identified as belonging to a particular class of flavonoids, e.g., flavanone, flavone, or flavonol, the next goal is to determine the nature of its substitution. This involves determining the nature of any glycosidic involvement and establishing the number and positions of attachments of hydroxyl, methoxyl, or other substituents on the main skeleton. We will defer discussion of the glycosidic condition until later. Suffice it to say here that any sugar residues will be removed by acid hydrolysis so that we can turn our attention to the aglycone. A first step in assessing the substitution pattern of a flavonoid often involved splitting the molecule into two fragments, one arising from the A-ring and the other from the B-ring. A frequently used approach is to degrade the unknown compound by treatment with base, examples of which from the earlier literature demonstrate its widespread use.

In the first example, the dihydrochalcone phloretin [4-1] was fused with alkali to yield phloroglucinol [4-2] and 4-hydroxydihydrocinnamic acid (3-(4-hydroxyphenyl)propionic acid) [4-3]. The original assignment of structure to the C_6C_3 fragment was incorrect (substituted at C-2 rather than C-3). The structure of phloretin was proved by synthesis from phloroglucinol and 3-(4-acetoxyphenyl)propionitrile (the Hoesch reaction; see Chapter Five) (Fischer and Nouri, 1917). In a study of the structure of the flavanone homoeriodictyol [4-4], which had been shown by elemental analysis to contain a methoxy group, two different alkaline degradations were employed (Power and Tutin, 1907; cited by Venkataraman, 1962). In the first, homoeriodictyol was heated with 30% NaOH which yielded phloroglucinol and ferulic acid [4-5]. In the second, fusion of the flavanone with alkali yielded phloroglucinol and vanillic acid [4-6]. Since phloroglucinol and both acids were known compounds, it was possible to confirm the structures of the degradation products by comparison of their physical properties with data from the literature.

A somewhat more complicated situation existed in the case of two compounds obtained from the bark of Ponderosa pine (*Pinus ponderosa*) by Puri and Seshadri

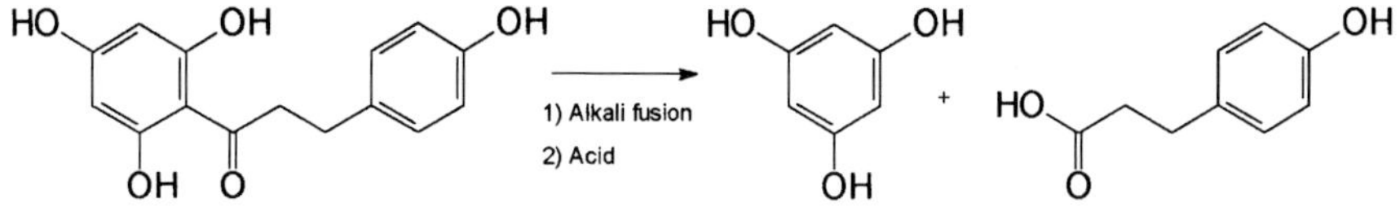

1) Alkali fusion
2) Acid

4-1 4-2 4-3

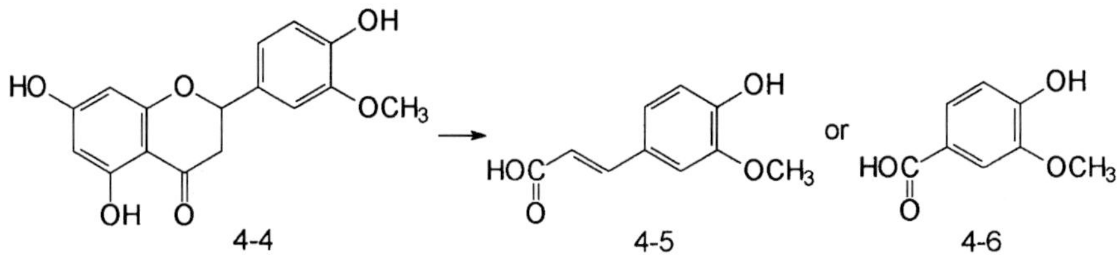

4-4 4-5 or 4-6

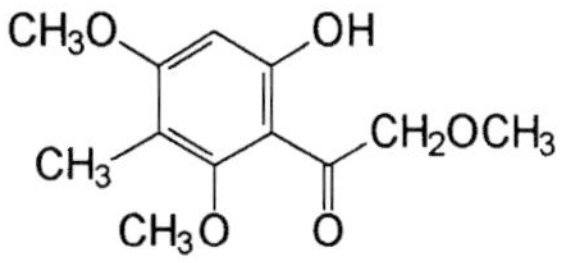

4-7 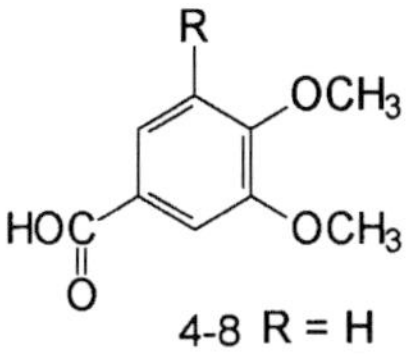 4-8 R = H
4-9 R = OCH_3

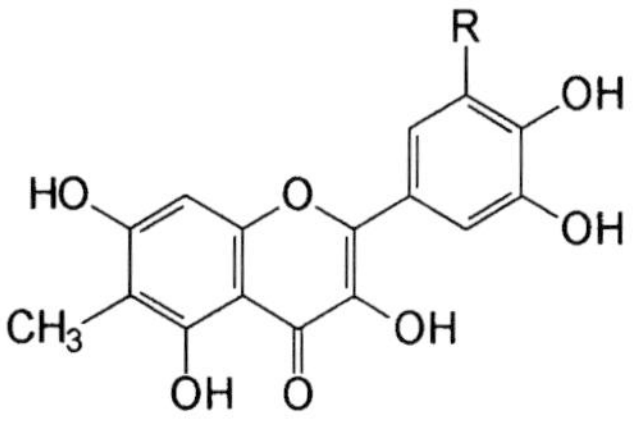

4-10 R = H
4-11 R = OH

(1955). The compounds were shown to be flavonols, each of which possessed a *C*-methyl group; one had a total of five hydroxyl groups, the other had six. Owing to the sensitivity of flavonols with B-ring hydroxyl groups to severe oxidative damage in the presence of base, further work on the compounds was done on their permethyl derivatives. Upon treatment with 8% alcoholic KOH both permethylated compounds gave ketone [4-7]. In addition, one gave 3,4-dimethoxybenzoic acid (veratric acid) [4-8], while the other gave 3,4,5-trimethoxybenzoic acid (trimethyl gallic acid) [4-9]. Since both compounds gave the ketone [4-7], they must have identical A-ring substitution patterns. They differ in the nature of their B-rings; pinoquercetin is the 6-*C*-methyl derivative of quercetin [4-10, R=H], while pinomyricetin is the

6-*C*-methyl derivative of myricetin [4-11, R=OH]. Further examples of this sort can be found in the discussion of flavonoid structural analysis by Venkataraman (1964). The interested reader will find a wealth of additional information in F. M. Dean's (1963) masterful *Naturally Occurring Oxygen Ring Compounds*.

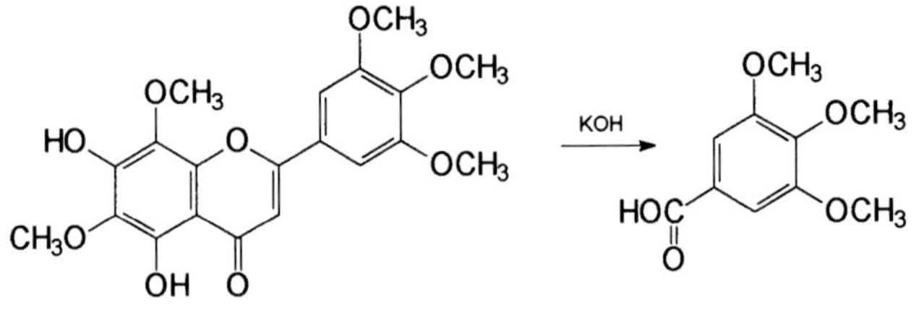

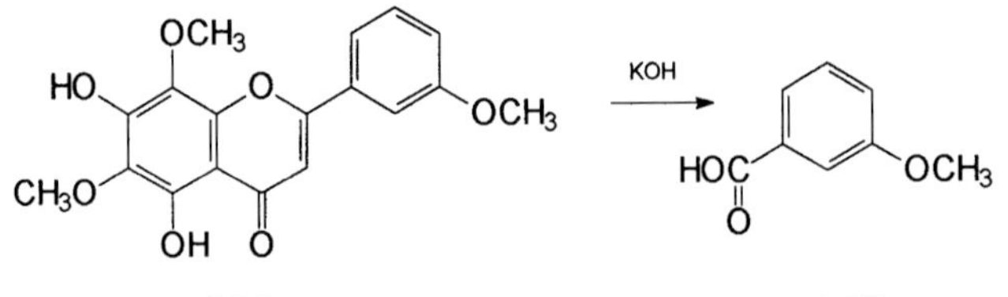

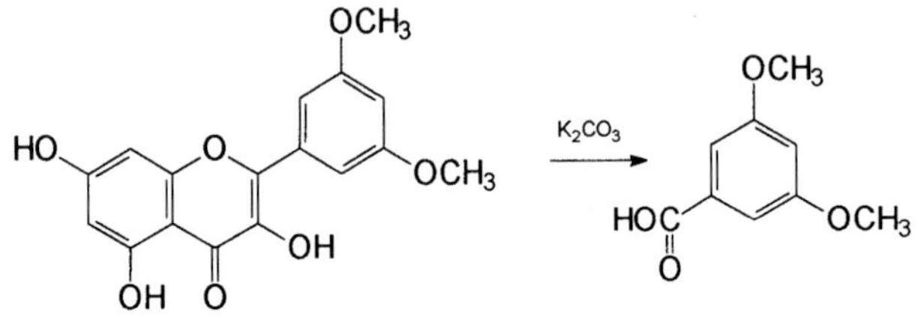

Examples from the more recent literature demonstrate that the general procedure is still highly useful. Dominguez and Torre (1974) described the isolation and structural determination of two compounds from *Gymnosperma glutinosum* (Asteraceae). A combination of chemical degradation results along with UV, MS, and NMR data led these workers to assign one of these compounds the structure 5,7-dihydroxy-6,8,3′,4′,5′-pentamethoxyflavone [4-12]. It is interesting to note that these workers reported a "Zeisel number" of five for these compounds. The Zeisel number arises from a quantitative analysis for *O*-methyl groups based upon hydrolysis of an ether with boiling HI in glacial acetic acid (Shriner *et al*., 1956). (The advantage of NMR analysis should be obvious in this situation!) Fusion of the compound with KOH afforded 3,4,5-trimethoxybenzoic acid [4-13] in agreement with the spectral data. The second compound was also shown to have five methoxy groups, but MS fragmentation suggested that only one of these was on the B-ring. In addition to peaks arising from the methyl groups, the NMR spectrum showed a four proton multiplet centered at 7.5δ representing protons on the B-ring. The absence of splitting characteristic of a 4′-substituted flavonoid suggested that some other arrangement existed. Only two possibilities exist, either a 2′-methoxy- or 3′-methoxyflavonoid. Microfusion of the unknown with KOH afforded 3-methoxybenzoic acid [4-15] which was identified by comparison with known compound using TLC and PC. The structure of the flavonoid is, therefore, 5-hydroxy-6,7,8,3′-pentamethoxyflavone [4-14]. A similar situation was encountered by Dominguez and coworkers (1976) in a study of *Baileya multiradiata* (Asteraceae). Spectral information suggested that one of the compounds isolated had the structure 3,5,7-trihydroxy-3′,5′-dimethoxyflavone [4-16]. In this instance, microfusion with potassium carbonate at 200°C afforded 3,5-dimethoxybenzoic [4-17] acid fully in accord with the suggested structure. An obvious problem with alkaline degradations arises when any part of the molecule under investigation is particularly sensitive to oxidation in the presence of alkali. This was overcome to a significant degree by Saxena and coworkers (1988) who carried out degradations of aurones and chalcones, both of which are notoriously sensitive to base, with alkaline hydrogen peroxide in the presence of triethylammonium chloride (TEBA). The reaction was successfully conducted at the milligram level with the resulting arylcarboxylic acid products compared to standards using TLC.

In order to appreciate fully the power (and convenience) of currently available spectroscopic techniques (as well as the intellectual accomplishments of earlier flavonoid chemists!) it is instructive to look in some detail at one of the more challenging problems that faced workers in this field, namely, the biflavones. Several biflavones exist which, when treated with boiling hydriotic acid (HI), are converted to amentoflavone, which we know today to consist of two apigenin units joined *via* the 8- and 3′-positions in the respective units. The following presentation is taken, with some modification, from Kawano (1962). One of these *O*-methylated amentoflavones, "sciadopitysin," isolated from a species of *Sciadopitys* (Taxodiaceae), had been shown to have the elemental formula $C_{33}H_{24}O_{10}$, which corresponds to an amentoflavone trimethyl ether [4-18]. At the time of the original work, three problems had to be addressed: (1) the nature of the flavone unit(s); (2) the sites of *O*-methylation; and (3) the points of linkage between the two units. A series of alkaline degradations of sciadopitysin and its trimethyl ether (amentoflavone hexamethyl ether) provided information on the fundamental flavone units and the points of attachment, while the location of the *O*-methyl groups required further study. Treatment of sciadopitysin with aqueous KOH afforded *p*-methoxybenzoic acid (anisic acid) [4-19], *p*-methoxyacetophenone [4-20], 2,6-dihydroxy-4-methoxyacetophenone [4-21], a ketoflavone, and a carboxyflavone. These results allowed the workers to conclude that at least one of the B-rings of the natural product has a 4′-methoxy group, that the A-ring is based on the phloroglucinol oxygenation pattern, and that there is C-7 methoxy group.

Subsequent work was done with sciadopitysin trimethyl ether (amentoflavone hexamethyl ether). Boiling ethanolic KOH treatment of the totally methylated unknown gave anisic acid, *p*-methoxyacetophenone, 2-hydroxy-4,6-dimethoxyacetophenone [4-22], 2-hydroxy-4,6-dimethoxybenzoic acid [4-23], a phenolic ketone analyzing for $C_{19}H_{20}O_6$ ("substance A"), and an acidic phenolic compound analyzing for $C_{18}H_{18}O_7$ ("substance B"). It was suggested that substance-A had structure

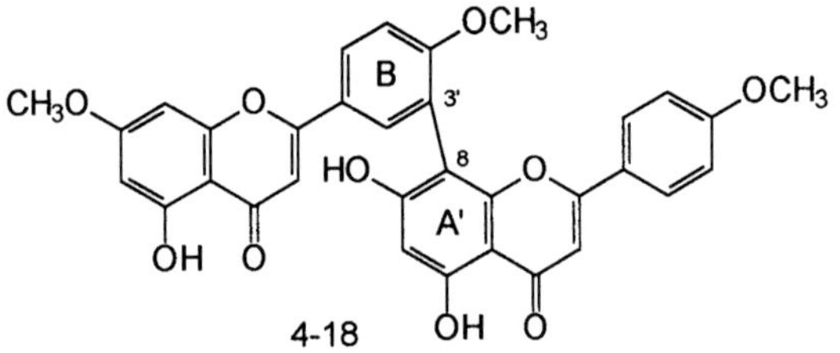

4-18

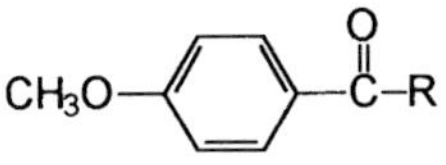

4-19 R = OH
4-20 R = CH_3

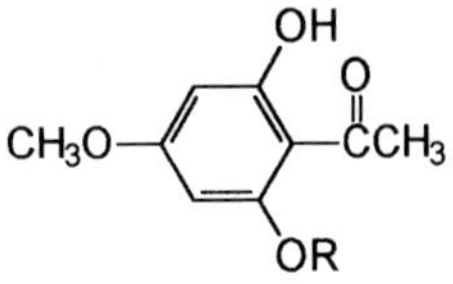

4-21 R = H
4-22 R = CH_3

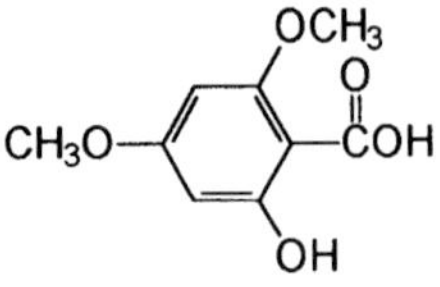

4-23

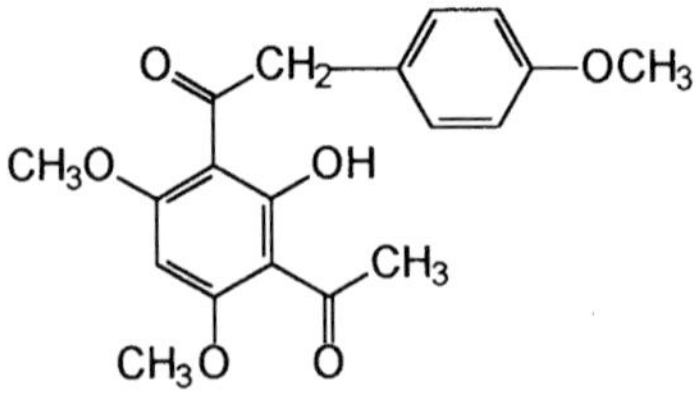

4-24

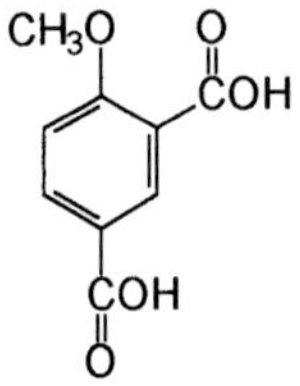

4-25

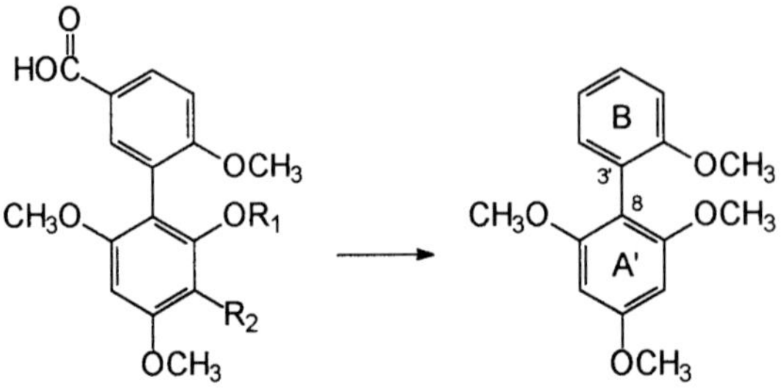

4-26 R_1 = CH_3, R_2 = CCOH (with two C=O)
4-27 R_1 = CH_3, R_2 = H

4-28

[4-24] but this structure did not agree with authentic material. A third degradation protocol, this time using methanolic $Ba(OH)_2$, worked well proving good yields of only three products, anisic acid, the dimethoxyacetophenone, and substance B. Oxidation of substance B with alkaline $KMnO_4$ lead to the formation of 4-methoxyisophthalic acid [4-25]. Substance B was then subjected to *O*-methylation followed by alkaline permanganate oxidation of the product. This resulted in the formation of the acidic biphenyl derivative [4-26], which yielded the acid [4-27] on continued heating. Decarboxylation of [4-27] by heating with copper powder, in turn, produced 2,2′,4,6-tetramethoxybiphenyl [4-28], which was shown to be identical to material synthesized by an unequivocal route (the Ullmann reaction). Two further observations are relevant: (1) substance B gave a positive Gibbs reaction indicating that there was an unsubstituted position *para* to a phenolic group; and (2) it could be *O*-methylated only with difficulty as would be expected for a phenolic group that is hydrogen-bonded to a carbonyl function.

Independently, a study of the structure of "ginkgetin," an amentoflavone dimethyl ether from *Ginkgo biloba*, was being undertaken by Baker and coworkers (1959). Ginkgetin tetramethyl ether (thus, equivalent to sciadopitysin trimethyl ether and amentoflavone hexamethyl ether) was subjected to alkaline H_2O_2 oxidation, which resulted in the formation of anisic acid, 2-hydroxy-4,6-dimethoxybenzoic acid, and a phenolic acid with the general structure $C_{12}H_4(OMe)_3(OH)(COOH)_2$ thought to be a diphenyl derivative. Oxidation of either ginkgetin or sciadopitysin with H_2O_2 afforded 4-methoxyisophthalic acid. Evidence from UV spectroscopy and biosynthetic concerns suggested that the biflavone was based upon two apigenin derivatives joined most likely *via* a $3' \rightarrow 8$ linkage.

Structures of Flavonoid Glycosides

We return now to a consideration of how the structures of flavonoid glycosides are determined. Three things must be established: (1) whether the compound in question is an *O*-glycoside or a *C*-glycoside (or possibly both); (2) the identity of the sugar or sugars; and (3) the site or sites of attachment. In the case of *O*-glycosides, it is comparatively easy to remove the sugar or sugars from the glycoside by acid hydrolysis. A more or less standard procedure for hydrolysis of a flavonoid glycoside into its components was described by Markham (1982, pp. 52–53). The specimen is heated with 2N HCl in methanol (1:1) under reflux for 60 minutes. After evaporating the reaction mixture to dryness under reduced pressure (water pump vacuum is usually sufficient), the residue is dissolved completely in methanol: water (1:1) and a small portion chromatographed on cellulose (e.g., paper or Avicel tlc) using 15% acetic acid along with a sample of the original glycoside. Successful hydrolysis is indicated by the disappearance of the starting material and the appearance of a spot with a lower R_f value. If there is no change in chromatographic behavior the original compound may be a *C*-glycoside or an *O*-glucuronide. Glucuronides can be hydrolyzed using a glucuronidase preparation whereas *C*-glycosides are resistant to enzyme as well as acid hydrolysis.

Differences in the rate of hydrolysis of glycosides depends upon both the nature of the sugar and the position of attachment of the sugar on the flavonoid skeleton. Using the hydrolytic scheme described above, except that ethanol was used instead of methanol, Harborne (1965) categorized glycosides as easily hydrolyzed, slowly hydrolyzed, and acid resistant. In the first category one finds the 3-*O*-glucosides, galactosides, and rhamnosides of kaempferol and quercetin along with kaempferol 7-*O*-glucoside. These glycosides can be completely hydrolyzed in from three to six minutes, the 3-*O*-rhamnosides being the most labile. In the slowly hydrolyzed category one finds a wide array of compounds including quercetin 4′-*O*-glucoside, which can be totally hydrolyzed in 10 minutes; apigenin 7-*O*-glucoside (15 min.); quercetin 7-*O*-glucoside, kaempferol, dihydrokaempferol, and naringenin 7-*O*-glucosides (25 min.); peonidin and cyanidin 3-*O*-glucosides, luteolin 7-*O*-glucoside, and kaempferol 3-*O*-glucuronide (45 min.); apigeninidin (3-deoxypeonidin) 5-*O*-glucoside and quercetin 3-*O*-glucuronide (60 min.). In the acid resistant category were kaempferol and quercetin 7-*O*-glucuronides and apigenin 4′-*O*-glucuronide (180 min.); and, slowest of all, apigenin 7-*O*-glucuronide (250 min.).

Flavonoid 5-*O*-glycosides are especially labile to acid hydrolysis. This sensitivity to acid presents a problem in that the absence of flavonoid 5-*O*-glycosides where they might otherwise be expected, e.g., in citrus products or species of *Prunus*, may be due to their destruction during isolation. Glennie and Harborne (1971) showed that the hydrolysis of 5-*O*-glucosides of luteolin, tricin, and several flavonols occurred much more rapidly than the corresponding 7-*O*-glucosides. For example, the time required for 50% hydrolysis of luteolin 5-*O*-glucoside was 10 seconds compared to 20 minutes for luteolin 7-*O*-glucoside. More recently, studies by Geibel and Feucht (1991) have shown that hydrolysis of 5-*O*-glucosides can also occur in the presence of acetic acid, oxalic acid, or malic acid. Isoflavone and flavone 5-*O*-glucosides were the most labile in 1 M malic acid at 60°C with flavanone 5-*O*-glucoside more stable under these conditions. Flavone 7-*O*-glucosides were not affected under these conditions.

Comparative rates of hydrolysis of flavonol glucosides using a β-glucosidase preparation were also reported by Harborne (1965). Rapid hydrolysis (under 1 hour for complete hydrolysis) was recorded for 3-*O*-glucoside, 7-*O*-glucoside, 4′-*O*-glucoside, 3,7-di-*O*-glucoside, 3,4′-di-*O*-glucoside, 3-*O*-galactoside, and 7-*O*-glucoside-3-*O*-sophoroside. In the case of the last compound hydrolysis produced the flavonol 3-*O*-sophoroside as sole product. Flavonol 3-*O*-gentiobioside was only hydrolyzed very slowly (up to 24 hours for completion). Sophorosides, rhamnosides, and a *p*-coumaroylglucoside resisted hydrolysis.

Although the use of HCl for hydrolysis of glycosides is very convenient, there are some attendant problems. In the first place, prolonged heating in the presence of strong acid may result in serious degradation of certain compounds, or, of lesser magnitude but still perplexing, the formation of rearranged products that do not reflect the true flavonoid synthetic potential of the plant. A purely technical problem associated with use of HCl is the need to eliminate excess acid after hydrolysis is complete. Procedures have been developed for accomplishing this (neutralization of excess acid with NH_4OH for example), but a simpler solution involves use of an acid that brings about the desired hydrolysis and is easy to remove from the system. It was pointed out by Markham (1982) that trifluoroacetic acid (b.p. 72.4°C) serves well. It is easily removed by evaporating the reaction mixture under reduced pressure.

As we saw above, the sugars in *C*-glycosides cannot be removed by acid hydrolysis. However, such a compound's behavior in the presence of strong acid can give additional clues about its structure. Treatment of a *C*-glycoside with strong acid results in the Wessely-Moser rearrangement, which involves opening of the heterocyclic ring, rotation of the A-ring around the A-ring-carbonyl single bond with formation of the isomeric forms, followed by re-establishment of the heterocyclic ring. Chromatography of the reaction mixture shows the presence of two compounds. As an example, if vitexin, which is 8-*C*-glucosylapigenin, is heated with acid, the result will be a mixture of the starting compound and its 6-*C*-glucosyl isomer, isovitexin. The formation of a new compound also occurs if a di-*C*-glycosylflavone has different sugars at C-6 and C-8, whereas there is no change in chromatographic behavior if the same sugar occurs at both positions. (The rearrangement occurs but the product is the same as the starting material.)

Several methods are available for selective degradation of the *C*-bound sugars in these glycosides including periodate oxidation (see Chopin and Bouillant, 1975, pp. 671–672). The resulting fragments can be readily identified and the position of attachment to the flavonoid skeleton is marked by the presence of a formyl group. Alternatively, removal of the sugar can be accomplished by oxidation of the glycoside with ferric chloride, but this procedure results in the total destruction of the flavonoid part of the molecule. *C*-Glucosylflavones yield a mixture of glucose and arabinose while *C*-rhamnosides yield only rhamnose. As we shall see below, the structures of *C*-glycosylflavones can be determined much more readily by using common spectroscopic techniques.

The classical approach for determining the position of attachment of *O*-linked glycosides involves exhaustive *O*-methylation of the glycoside followed by removal of the sugar(s) by acid hydrolysis. The resulting *O*-methylated aglycone is then identified by comparison with authentic specimens, either by comparing its physical properties (e.g., its melting point or the melting point of a derivative) with published data, or by direct comparison with a compound of known structure (e.g., mixed melting point, chromatographic properties). If neither known compounds nor published information is available, identity of the aglycone can be established by controlled degradation reactions of the sort described earlier in this chapter. Unsubstituted phenolic groups in the unknown flavonoid glycoside would appear as *O*-methyl groups in the aglycone liberated by hydrolysis. The approach is demonstrated in the simple model compound, luteolin-3′-*O*-glucoside [4-29]. Preliminary

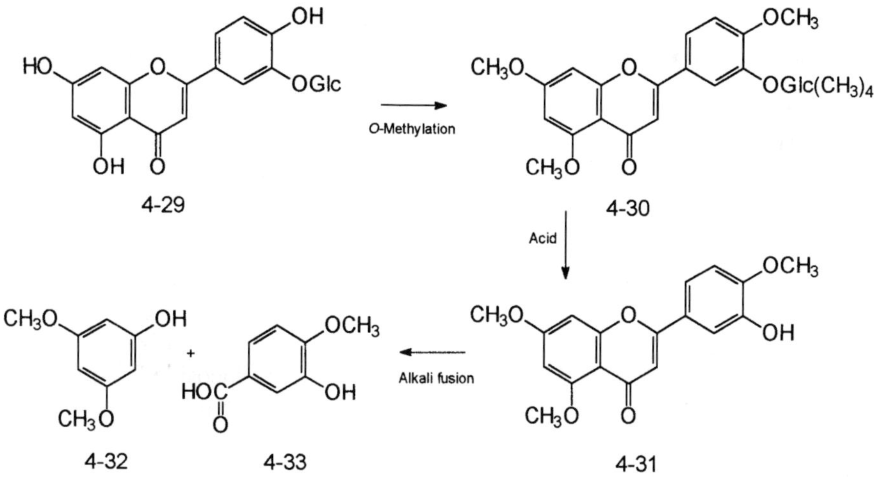

investigations, e.g., color tests, solubility, chromatographic properties, and ultraviolet spectroscopy would have revealed that the compound is a flavone monoglycoside with a sugar attached through a B-ring hydroxyl group. Phenolic *O*-methylation would afford a trimethyl ether [4-30], acid hydrolysis of which would yield the aglycone [4-31] and the sugar moiety. If the flavone trimethyl ether could not be identified at this stage, it would be possible to subject it to alkaline degradation, which, in this case, would yield phloroglucinol dimethyl ether [4-32] and 3-hydroxy-4-methoxybenzoic acid (isovanillic acid) [4-33]. The formation of phloroglucinol dimethyl ether means that the A-ring oxygens were unsubstituted in the original glycoside, and were, therefore, available for *O*-methylation. The free phenolic group arose from the heterocyclic oxygen. Isolation of isovanillic acid indicates that the C-4′ hydroxyl was unsubstituted in the original. This leaves the hydroxyl at C-3′ as the only remaining position to which the sugar could have been attached. This approach will be seen in Case Studies No. 1 and 2 at the end of this chapter.

Many flavonoid glycosides, of course, are known that involve more than one sugar. Although most flavonols occur as 3-*O*-glycosides and most flavones and flavanones as 7-*O*-glycosides, glycosidic linkages through any available phenolic position, or positions, are possible. Flavonoid diglycosides, for example, may have both sugars linked together, as in the case of the commonly occurring diglycoside rutin, which is a quercetin 3-*O*-rhamnosylglucoside isomer. Rutin has the rhamnose group linked through the C-6 hydroxyl of glucose, but linkage through hydroxyls on C-4, C-3, or C-2 are equally possible (and known). One must also be aware of the stereochemistry of the glycosidic linkages and the nature of the sugar units themselves, i.e., whether they are in the furanose (five-membered ring) or pyranose (six-membered ring) forms. The two sugars may be attached at two different positions, e.g., quercetin 3-*O*-glucoside-7-*O*-rhamnoside. Although this dimonoside consists of the same components as rutin, i.e., one equivalent each of quercetin, glucose, and rhamnose, its physical properties are different. These isomers would have somewhat different chromatographic properties and their UV absorption behavior, using spectral shift reagents (see below), would allow a clear distinction to be drawn between them. It would also be possible to distinguish between them by studying the products of hydrolysis. In this specific case, use of β-glucosidase would yield either a 3-*O*-rhamnoside or a 7-*O*-rhamnoside, which are easily distinguishable from each other. The dimonoside could, of course be the isomeric 3-*O*-rhamnoside-7-*O*-glucoside. This compound would also be distinguishable from rutin by means of chromatographic properties and UV behavior, but further work would have to be done in order to determine which of the two dimonosides was involved.

Problems mount when one encounters triglycosides. In the simplest case, where only one sugar is involved, the position or positions of the outer glycosidic linkages, and their stereochemistries, offer real challenges if one wishes to be more specific than merely reporting the presence of "a flavonol 3-*O*-triglucoside." Further complicating the picture would be a triglycoside involving more than one sugar. Position of linkages and stereochemistry are still problems, but one also has to deal with the sequence of sugars. If two glucoses and a rhamnose are detected, for example, is the sequence, reading from the aglycone outward, glucose-glucose-rhamnose, glucose-rhamnose-glucose, or rhamnose-glucose-glucose? The reader can imagine the difficulty of determining the structure of a flavonol tetraglycoside that involved substitution through hydroxyls at C-3 and C-7 and three different sugars! Classical methods of analysis can handle problems of this sort but the efforts required are considerable. As we will see below, the application of NMR and MS methods can reduce the effort required by orders of magnitude.

Interconversion of Flavonoids

Comparison of degradation fragments with known compounds played an important part in the various studies described above. It is sometimes possible, however, to circumvent the degradation route and identify an unknown flavonoid through conversion to a known flavonoid. The subject of interconversions of flavonoids was reviewed by Seshadri (1962b) who described many examples of this approach from the earlier literature. Incidentally, many of these interconversion reactions have found extensive use in flavonoid synthesis as well. Although flavonoid interconversions are used less frequently for identification at the present time, owing to the power of spectroscopic methods, a few examples to show the utility of the general approach will be given. Chalcones [4-34] and flavanones [4-35] are easily interconvertible so that an unknown compound belonging to either of these classes can be converted to the other for which characteristic information may already be available. A flavanone can be converted to a flavone [4-36] or a flavonol [4-37] using a variety of reagents that selectively remove

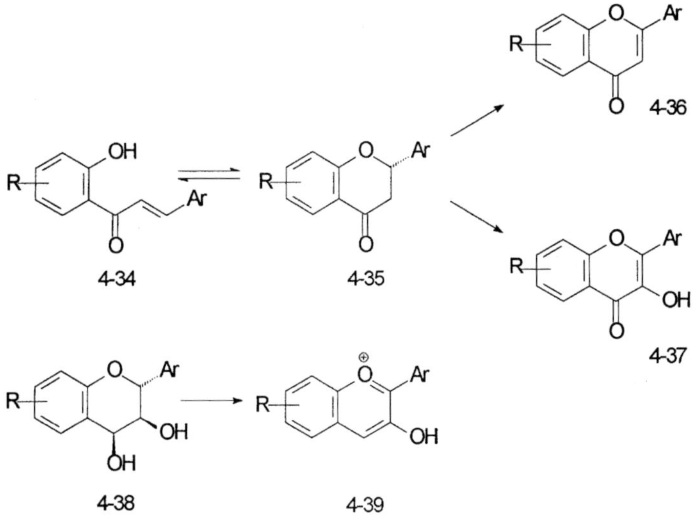

4-34 4-35 4-36 4-37 4-38 4-39

hydrogens from C-2 and C-3 (Seshadri, 1962b) and, in the case of the formation of a flavonol, insert oxygen at C-3. Again, the resulting flavone or flavonol may be a known compound, which would allow one to establish the structure of the flavanone (or chalcone) without further degradation. An interconversion that has been used extensively in chemotaxonomic surveys involves the treatment of flavan-3,4-diol derivatives with acid, resulting in their conversion to the corresponding anthocyanidins, e.g. [4-38] to [4-39]. The oxygenation pattern of the B-ring of the starting compounds can often be determined directly by observing the color of the resulting pigment: 4′-hydroxylation leading to pelargonidin; 3′,4′-dihydroxylation leading to cyanidin; and 3′,4′,5′-trihydroxylation leading to delphinidin. Each of these products exhibits characteristic absorption maxima in the ultraviolet and visible spectra which make establishment of the oxygenation pattern of the unknown flavan-3,4-diol comparatively easy.

CHROMATOGRAPHY

Paper Chromatography (PC)

No single separation technique has had as much impact on the application of chemical data to problems in plant systematics as has paper chromatography (PC). The technique has a modest history. In 1901 Goppelsroeder used strips of filter paper dipped into solutions to achieve separation of anthocyanins in what was essentially an application of ascending paper chromatography. Much more recently, Bate-Smith (1964), addressing the Plant Phenolics Group Symposium at Oxford in 1963, suggested that were it not for paper chromatography the organization would not have come into being! Going a step further, I think that it is safe to say that the combination of PC with ultraviolet absorption spectroscopy (UV) allowed workers, most of whom had comparatively little chemical background, to participate in one of the most intense data gathering exercises in the history of plant systematics, although we are witnessing a similar level of enthusiasm at the present time with regard to DNA sequence studies. With a glass chamber capable of being tightly sealed, a supply of chromatography paper (sheets of Whatman Nos. 1 and 3 were commonly used), a few easily available solvents, a bottle of ammonium hydroxide, a source of UV light (366 nm), and a dark room, it became possible to determine spot patterns of whatever plant was under investigation. With two chambers, it was possible to do two-dimensional chromatography, which significantly increased the information value of the pattern of spots obtained. By judicious choice of solvent mixtures it was possible to achieve separations based upon both the nature of the flavonoid, i.e., class of compound along with an idea of its substitution pattern, and the level of glycosylation.

The most useful pair of solvent systems for general flavonoid analysis consisted of a mixture of *n*-butanol, water, and acetic acid in the proportions (4 : 1 : 5). It is interesting to note that this solvent mixture, originally developed by Partridge (1947) as a means of separating sugars, became one of the standard systems in the flavonoid field. A disadvantage is that this mixture consists of two-phases so that equilibration and separation in a separatory funnel was required. The top phase was used to irrigate the paper while the lower phase was used to saturate the atmosphere in the chromatography chamber. A modification of this procedure involved substituting *t*-butanol for the *n*-butanol. This so-called TBA system had the advantage of being a monophasic system but the running times were longer. The second direction for the two-dimensional development was essentially an aqueous system. Various concentrations of acetic acid in water have been used. In his surveys of the plant kingdom for flavonoids Bate-Smith (1962) employed 6% acetic acid, while our experience has been that 15% acetic acid gives better resolution. The combination of TBA and 15% acetic acid was used in the standardized approach advocated by Mabry and coworkers' (1970) in *The Systematic Identification of Flavonoids*. Markham (1982) followed their lead (he was one of the coworkers, after all!) in his methods book. In his description of chromatographic methods, he presents a map of a two-dimensional paper chromatogram showing regions where various classes of flavonoids would be expected to appear (see his Fig. 2.2).

To be sure, other solvent systems have been developed and used quite successfully. The Forestal solvent, for example, has been used by many workers. It consists of acetic acid, conc. HCl, and water in the proportions (30 : 3 : 10). It was originally developed for separation of anthocyanidins (Bate-Smith, 1954) for which purpose it works very well with R_f values for the common anthocyanidins of 0.30 (delphinidin), 0.50 (cyanidin), and 0.68 (pelargonidin). Kaempferol, quercetin, and myricetin move well in this system as well with R_f values of 0.58, 0.41, and 0.27, respectively. A mixture of phenol and water (4 : 1) was also commonly used but had the disadvantages of long drying times, presence of residual phenol which made use of phenol-detecting reagents useless, and, not the least consideration, its toxicity. A mixture of benzene, acetic acid, and water (125 : 72 : 3) was frequently used in the resolution of mixtures of benzoic and cinnamic acids and mixtures of flavonoid aglycones. Owing to its toxicity, we have replaced benzene with 1,2-dichloroethane in our laboratory with reasonable success. Most authors describe a variety of solvent systems that have proven to be useful (Markham, 1982, Table 2.2; Harborne, 1990, Table 1.9). Many of these solvents have continued to be successfully used in thin layer chromatographic applications using cellulose-based coatings (e.g., Avicel microcrystalline cellulose).

Since most flavonoid derivatives are not detectable in visible light (exceptions are anthocyanins, aurones, and chalcones) some means for locating compounds on the completed chromatogram are necessary. The simplest of these is to capitalize on flavonoids' capacity to absorb UV radiation. Depending on their structure, flavonoids may appear as a variety of variously colored spots, perhaps the most common being as absorbing (purple or black) spots against a reflective background. Fuming the chromatogram with ammonia vapors, which causes certain phenolic groups to ionize, intensifies the reaction. Several chromogenic reagents have also been widely used. Among these are such common reagents as a solution of sodium carbonate, which has much the same effect as ammonia fumes but the colors last longer, and alcoholic aluminum chloride, which forms brightly colored complexes with most flavonoids.

An interesting and simple solution to a particular problem of differentiating between two isomeric flavonol methyl ethers has been described. Quercetin 3′-methyl ether (isorhamnetin) and quercetin 4′-methyl ether (tamarixetin) and their respective glycosides are difficult to separate by normal chromatographic techniques. Saleh (1976) noted that the two compounds respond differently to sodium ethoxide (0.05 M in ethanol); although both yield a yellow color with the reagent, the color with tamarixetin is stable for as much as 24 hours while that obtained with isorhamnetin fades after about 2 hours. This is due to the inherent stability of flavonols with unsubstituted hydroxyl groups at C-3 and C-4′. The colors are different under UV as well but are not as pronounced as in visible light. It would have been interesting to see what the color response of a mixed sample of the two compounds might have been. Despite this niggling criticism, this simple test shows the extent to which routinely used reagents can be taken in the hands of an experienced flavonoid chemist.

One of the routinely reported characteristics of flavonoids studied by PC methods is the R_f value. This value is determined by dividing the distance traveled by the compound by the distance traveled by the solvent front. Comparison of R_f values of an unknown run in several solvent systems with published values is considered to be a useful means, if not of establishing the compound's identity, then at least a way of limiting the number of structural possibilities. Since R_f values can vary depending upon a number of factors (e.g., temperature, differences in solvent composition, amount of material spotted) it is useful to include a sample of a compound of similar structure (if possible) to serve as chromatographic standard. After satisfactory purification of the new compound has been achieved, co-chromatography with an authentic specimen provides additional evidence of its structure.

It is often possible to cut out the flavonoid spot, elute the material from the paper with methanol, filter to remove cellulose fibers, and determine the UV absorption spectrum of the compound without further purification. This can be hazardous, however, as other compounds may have chromatographic properties similar to those of the target flavonoid but different spectroscopic properties giving rise to misleading results. It is generally wise to purify the unknown compound by running in other chromatographic systems. If insufficient material is available, one can resort to preparative paper chromatography in which the crude extract is banded on thick sheets of paper and developed one-dimensionally in an appropriate solvent. The desired band is located (non-destructively) and eluted from the paper. Rechromatography in other systems is almost always necessary. Although this method is straightforward, it is not as convenient as column chromatography nor as efficient as high performance liquid chromatography.

Thin Layer Chromatography (TLC)

Whereas paper chromatography (PC) works especially well with polar compounds, such as flavonoid glycosides, it has serious limitations for studies of non-polar compounds. The adsorbent of choice for these compounds is often silica gel. The availability of thin layer plates coated with high quality silica gel made the study of a wide variety of compounds much more accessible. Thin layer chromatography has several additional advantages among which are much reduced space for chambers, smaller volumes of solvents needed, the need for only very small amounts of plant material, and, with the availability of commercially manufactured plates, a highly reproducible method capable of providing semi-quantitative results. An additional feature of TLC is the tendency of spots to remain compact and not spread out the way they often do on paper chromatograms. Time is also a factor; TLC plates can be developed in a fraction of the time required by PC.

For those workers who are more comfortable with PC, the commercial availability of cellulose TLC plates provides an opportunity to couple the use of familiar solvent systems with the several advantages of the latter. Some of the disadvantages of PC solvents are carried over into the TLC system, however, principally the tendency of spots to diffuse. The availability of polyamide as a chromatographic medium has overcome most, if not all, of the objections of converting from PC to TLC (see Hörhammer, 1962, for a review of early work). The polyamides are a family of polymers characterized by repeating amide linkages and a variety of different chain lengthening units. Perhaps the most familiar of these is polyvinylpolypyrrolidone, known familiarly as PVP. PVP attracted a considerable amount of attention when its capacity to eliminate phenolic compounds from enzyme preparations was described some years ago. It is the strong capacity of PVP, and related polymers, to bind phenolic compounds that makes them so valuable for the study of the phenols themselves. Use of these for resolving mixtures depends upon the degree to which phenolic compounds are hydrogen-bonded to the medium. Compounds with many phenolic groups are more tightly bound than are compounds with fewer such groups. By using eluting solvents of increasing polarity, methanol or acetone with increasing amounts of water, for example, it is possible to resolve quite complex mixtures of phenolic compounds efficiently. Some of the products that have found widespread use in flavonoid analysis are Polyamid DC-6.6, which consists of alternating units of 1,6-diaminohexane and butane-1,4-dicarboxylic acid; Polyamid DC-6, which consists of repeating units of 5-aminopentane carboxylic acid, and Polyamide DC-11, whose backbone is built on the 11-carbon omega-amino carboxylic acid homologous to the Polyamid 6 monomer (all from Macherey-Nagel). A further adaptation of these materials is seen in the availability of acetylated products, e.g., Polyamid-DC 6-Ac, which eliminate any basicity of the polymer at the same time providing one additional amide group.

We have used both Polyamid DC-6 and DC-6.6 in our laboratory with equal success in resolving flavonoid glycoside mixtures. Development in the first direction is done with a mixture of water, *n*-butanol, acetone, and dioxane (70 : 15 : 10 : 5) that very successfully separates the common flavonol glycosides on the basis of their level of glycosylation with the higher glycosides moving progressively further on the plate. Development in the second direction is done with a mixture of 1,2-dichloroethane, methanol, methylethyl ketone (butanone), and water (50 : 25 : 21 : 4), which resolves compounds according to the number of free hydroxyl groups available. Typically, within each glycoside class, using the common flavonols as examples, kaempferol 3-*O*-glucoside would migrate farther than quercetin 3-*O*-glucoside, which in turn migrates farther than myricetin 3-*O*-glucoside. Rhamnosides migrate faster than glucosides so that kaempferol 3-*O*-glucoside and quercetin 3-*O*-rhamnoside overlap to some extent. The capacity of this system to allow differentiation between glucosides and rhamnosides also manifests itself in the behavior of the flavonol 3-*O*-diglycosides. In our experience, quercetin 3-*O*-rutinoside migrates farther than does the corresponding 3-*O*-diglucoside. A graphical representation of a thin-layer separation of kaempferol and quercetin 3-*O*-monoglucosides and 3-*O*-rutinosides is given in Figure 4.1. The “Aq” and “Org” terms refer, respectively, to the aqueous and organic solvent systems described above. This system is useful for judging the overall complexity of an extract. There may be some overlapping of compounds in actual applications, but, with experience, and the judicious use of standards, this does not detract from the overall usefulness of the system.

Polyamide TLC has proven to be exceptionally useful in the study of non-polar flavonoids. That this is the case is borne out no better than in the work of E. Wollenweber and his associates on the nature of exudate flavonoids from a wide variety of plants. Representative of his work is the analysis of exudate compounds from leaves of *Acacia neovernicosa* (Wollenweber and Seigler, 1982)

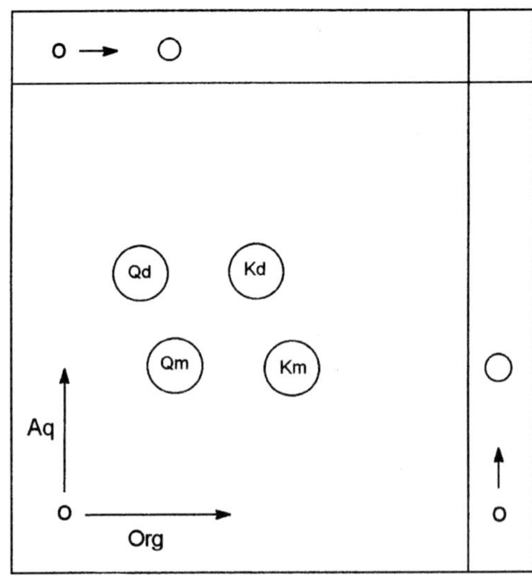

Figure 4.1 2-D TLC showing resolution of a mixture of kaempferol (K) and quercetin (Q) 3-*O*-mono- and di-glycosides. Quercetin 3-*O*-mono glucoside is the standard. See text for solvent mixtures.

where compounds were run in four TLC systems. The first two systems employ Polyamid DC-11 and their solvents A (toluene-petrol, b.p. 100–140°-butanone-methanol, 12:6:2:1) and B (toluene-dioxane-methanol, 8:1:1), while solvents C (toluene-butanone, 9:1) and D (toluene-dioxane-acetic acid, 90:25:4) were run on silica gel plates (SILGUR-25, UV_{254}; Macherey-Nagel). The latter product contains a dye whose fluorescence is quenched in the presence of the UV-absorbing flavonoids. Work in our laboratory on exudate flavonoids has routinely used Polyamid DC-6.6 plates and a mixture of ethyl formate, cyclohexane, *n*-butyl acetate, and formic acid (50:25:23:2) as the irrigating solvent. Detection of compounds is routinely done first by examining plates under UV with and without ammonia fuming, and then by spraying with Naturstoffreagenz A (0.5% in methanol–water, 1:1).

Naturstoffreagenz A, a complex of diphenylboric acid and ethanolamine, is widely used as a spray reagent for flavonoids (and other natural products) (Neu, 1957). This reagent is particularly useful in flavonoid studies owing to its capacity to form complexes that exhibit different colors under UV depending upon the hydroxylation pattern of the flavonoid involved. In many cases it is possible to arrive at a reasonable idea of a compound's structure based upon a combination of its chromatographic characteristics and its behavior under UV before and after spraying with this reagent. For example, kaempferol 3-*O*-glycosides give a green color, quercetin 3-*O*-glycosides give a yellow-orange color, and myricetin 3-*O*-glycosides give a darker orange color. Quercetagetin 3-*O*-glycosides, which have the quercetin-type B-ring and an extra hydroxyl group at C-6, give a brick red color. Fuming sprayed chromatograms with ammonia enhances these colors in most instances. We have found that while naringenin derivatives (flavanones with an unsubstituted 4′-hydroxyl group) give a dark (absorbing) reaction to the spray, eriodictyol derivatives (flavanones with a 3′,4′-dihydroxy B-ring) develop a red color upon standing. Fuming of the spot with ammonia causes the color to disappear temporarily. After dissipation of the ammonia, the red spot color returns. One final note on this phenomenon: use of eriodictyol 7-*O*-glucoside as an internal chromatographic standard provides a useful marker.

It has been our experience that the colors formed on polyamide thin layer chromatograms are much more striking and structurally informative than those produced with Naturstoffreagenz A on cellulose media. Additional information of TLC methods can be found in Markham (1982) and in Harborne (1990). The latter source lists TLC systems that have been found useful for all major classes of phenolic compounds (his Table 1.10).

Column Chromatography (CC)

Column chromatography has long been an important method of isolating natural products [see Thompson and coworkers (1959) for an extensive review of early applications]. Applications to flavonoid isolation employ many of the same adsorbents that have been used in PC and TLC procedures, namely, cellulose, silica gel, and polyamide. The procedure is a straightforward one that consists simply of loading a crude extract of the plant material on top of a column of adsorbent and eluting compounds using solvent mixtures of changing polarity. Mixtures of non-polar compounds can be resolved using polyamide columns and conditions essentially scaled up from the TLC system. Adsorbents for CC generally have larger particle size so that flow rates are reasonable, but resolution suffers. Additional purification steps using TLC will likely be required. Scaling up from a successful silica gel TLC procedure is also common. Cellulose column chromatography is a rather crude method that does not, at least in our experience, yield very good resolution. This problem has been elegantly circumvented by the development of the Sephadex line of chromatographic products. Of particular value in flavonoid studies is Sephadex LH-20. Sephadex is a polysaccharide with a narrow molecular weight range. The LH-20 modification involves treating the polysaccharide with propylene oxide which replaces each hydroxyl hydrogen in the polymer with the 2,3-dihydroxypropyl group. The outcome of this procedure is a polymer with increased hydrophobicity. We have used Sephadex LH-20 as a column packing for resolution of complex mixtures of flavonoid glycosides with a considerable degree of success. Although the packing itself is comparatively expensive it can be recycled almost indefinitely. Our routine use of this material involves packing a column using a slurry of LH-20 in degassed 30% methanol in water. The plant material is dissolved in the same concentration of methanol-water, filtered, and added to the top of the column. Elution with mixtures of methanol and water, changing by increments of 10%, proceeds with fractions being taken as needed. Progress of development can easily be monitored using a 366 nm UV lamp. This procedure has the distinct advantage of removing the usually very large amounts of blue fluorescent material (phenolic acid derivatives) in early fractions. Higher level glycosides elute next followed by diglycosides. Monoglycosides elute next when the solvent mixture is about 70–80% methanol. Aglycones are the last to be removed and usually require 100% methanol. The enriched mono- and diglycoside fractions, essentially free of interfering cinnamic and benzoic acid derivatives, can then be fractionated using TLC, as desired.

We have also used a column procedure for the isolation of bulk quantities of non-polar flavonoids. This procedure is based upon an industrial grade of PVP and uses increasing amounts of methanol in dichloromethane as eluting solvent. The column is packed in a mixture of dichloromethane-methanol (3 : 1), the extract dissolved in the same mixture is added and the column is developed with the 3 : 1 mixture. Progress of development can be monitored with a 366 nm UV lamp. Flavonoids with four or five *O*-methyl groups elute first, followed by those with three, and so forth. It is possible to alter the solvent mixtures, 3 : 1, 2 : 1, 1 : 1, 1 : 2, etc., as development progresses. With appropriate sized columns this procedure can easily accommodate 10 grams of plant extract.

High Performance Liquid Chromatography (HPLC)

High-performance liquid chromatography (or high pressure liquid chromatography) (HPLC) represents an extremely valuable addition to the methods available to the natural product chemist. HPLC relies on either a column of densely packed medium coated with an appropriate absorbent or a capillary column with the inner wall coated with the absorbent. In either case, high pressure is required in order to achieve solvent flow. Detection of emerging components can be done in a variety of ways perhaps the most useful of which involves use of a diode array. In this way, characteristic absorption peaks of emerging components can be recorded which facilitates identification. The main requirement for HPLC, other than an appropriate detector, is a means of producing and maintaining a constant flow of solvent. Solvents can be delivered as a uniform mixture or the concentrations of solvent components may be changed through the course of development using a gradient generator. An important feature of HPLC methods, of course, is that they are readily quantified. Since HPLC is a widely used technique further technical details are hardly necessary. To illustrate the use of this extremely powerful methods, several examples from the flavonoid literature are presented below. It ought to be borne in mind that, as in most other analytical systems, each worker tends to develop a system unique to the problems at hand.

An early application of reverse phase HPLC to the problem of resolving complex mixtures of flavonoids comes from the study of Vande Casteele and coworkers (1982). Using a LiChrosorb RP-18 column and 5% aqueous formic acid as the developing solvent, those authors accumulated data for 141 flavonoids ranging from aglycones to triglycosides. Grayer and colleagues (1996) investigated the external leaf flavonoids of *Ocium basilicum* (Lamiaceae) as an aid in judging the specific status of *O. minimum*, a taxon frequently considered by some as a variety of *O. basilicum*. Leaf flavonoids were resolved using a reverse-phase 100RP-18 column (4.0 mm i.d. × 25 cm) and a linear gradient of solvent-A (2% acetic acid in water) and solvent-B (methanol-acetic acid-water; 18 : 1 : 1). Development was initiated using 60% A and 40% B going to 100% B in 20 minutes, with development continuing with 100% B

Table 4.1 HPLC Analysis of Flavones from *Ocium* (from Grayer *et al*., 1996)

Flavone		
Common name	Structure[1]	R_{time} Min[2]
Cirsileol	53′4′/67	15.8
Apigenin	574′/-	16.3
Cirsimaritin	54′/67	17.7
Cirsileneol	54′/673′	17.8
Eupatorin	53′/674′	17.9
Ladanein	56/74′	18.7
Nevadensin	57/684′	19.8
Acacetin	57/4′	20.5
Genkwanin	54′/7	20.7
Salvigenin	5/674′	21.6
Gardenin B	5/6784′	23.0
Apig-7,4′diMe	5/74′	24.8

[1]Hydroxyl positions/methoxyl positions
[2]See text for HPLC conditions

for an additional five minutes. The flow rate was maintained at 1.0 ml min^{-1} at 25° with detection at 335 nm. The 12 compounds resolved using this system are listed in Table 4.1 in the order of their increasing elution times along with their structures. Attempts to correlate retention time (relative polarity) with structures in a set of closely related compounds is generally not reliable so that identification of an unknown compound is best done by combining retention times with other analytical data.

Anthocyanidin and flavonol glycosides of genetically transformed lisianthus (*Eustonia grandiflorum*, Gentianaceae) were isolated by Markham (1996) using, first, an analytical HPLC system consisting of an Econosphere C18 column using a 1 : 1 mixture of solvent-A (1.5% H_3PO_4 in water) and solvent-B (water-acetonitrile-acetic acid-H_3PO_4, 107 : 50 : 40 : 3) at a flow rate of 0.8 ml min^{-1} with detection at 352 and 530 nm. A second, preparative, separation was run using a Whatman Partisil 10, ODS-2 column with a solvent mixture consisting of 43% A and 57% B at 2.0 ml min^{-1} flow rate and detection at 440 nm. Retention data were not provided in that paper.

Saito and coworkers (1996) employed Waters C18 columns to resolve mixtures of complex acylated anthocyanidin glycosides in their study of pigments of the Japanese morning glory (*Pharbitis nil*). The solvent system used was a linear gradient of solvent B from 40–85% in solvent A over a period of 30 min. Solvent A was a 1.5% solution of H_3PO_4 in water. Solvent B consisted of 1.5% H_3PO_4, 20% acetic acid, and 25% acetonitrile in water. Elution was monitored at 550 nm. A later paper (Tatsuzawa *et al*., 1994), detailing a study of pigments from the orchid cultivar *x Laeliocattleya* cv Mini Purple, described resolution of 18 anthocyanin peaks using the same solvent systems.

Ducrey and coworkers (1995) employed thermospray liquid chromatography-mass spectrometry (TSP/LC-MS) and HPLC coupled with diode array UV detection to study the flavonoids of several species of *Epilobium*. Data from mass spectral and UV analyses, before and after post-column addition of shift reagents, were combined to give structural information on 19 compounds. The advantage of this combination of methods lies in the capacity to obtain information on crude extracts using small amounts of plant material. This paper is a very useful source of background information and technical details and is highly recommended to the interested reader.

Tomás-Barberán and coworkers (1990) employed HPLC, TLC, and UV analytical techniques to examine a large number of B-ring deoxyflavonoids. These compounds often pose problems in survey studies when only limited amounts of plant material are available. The standardized methods described by those workers included HPLC separations using a LiChrospher 100 RP-18 column and a 65:35 mixture of methanol and water delivered isocratically at a flow rate of 1.0 ml min^{-1}. Elution of compounds was monitored using a diode-array detector. TLC was done on commercial silica gel plates with two solvent systems: (1) chloroform; and (2) 1:1 light petroleum and ethyl acetate. Standard UV methods were employed for identification purposes. The authors pointed out, in essential agreement with Grayer *et al.* (1996), whose data were presented in Table 4.1, that there was no correlation between the number of methoxy groups and retention time, at least under the conditions employed.

A discussion of the application of HPLC methods to the separation and identification of lipophilic flavonoids (other than the B-deoxy flavonoids), was described recently by Greenham and coworkers (1995). The variety of compounds tested included *O*-methylflavonoids, *C*-methylflavonoids, *C*-prenylflavonoids, and compounds with combinations of these features (45 compounds in all). HPLC was carried out using a Bondapak phenyl C_{18} reverse phase column. Isocratic elution employed a 2:3 mixture of acetic acid and water (1:50) (solvent-A) and methanol-acetic acid-water (18:1:1) (solvent-B). Gradient elution employed a 2:3 mixture of solvent-A and solvent-B changing to 100% solvent B over a period of 20 minutes.

An application from our laboratory shows the usefulness of HPLC techniques for the analysis of very small amounts of plant material. Some years ago we (Bohm *et al.*, 1988) found that the flavonoid profile of species of *Itea* consisted of *C*-glycosylflavones and not the array of flavonol 3-*O*-glycosides normally seen in plants belonging to Saxifragaceae (considered in the broadest sense). Recent studies by Douglas Soltis and (personal communication) at Washington State University, using gene sequence data, have concluded that *Itea*, flavonoids notwithstanding, fits well into the saxifrage group and shares a clade with the Mexican genus *Pterostemon*, which consists of two species. We were able to acquire two small leaves of each species of *Pterostemon* but the amount of material was far short of the amount normally required to isolate and identify an array of flavonoids. By a combination of TLC and HPLC (using diode array detection), however, we have been able to identify the pigments present in the Mexican species. Comparison of 2D TLC of the *Pterostemon* extracts with those of cultivated and herbarium specimens of *Itea* showed the presence of *C*-glycosylflavones in *Pterostemon*. Comparison of the HPLC traces of all three species showed that the *C*-glycosylflavones were the same in all of them. Further, flavonol 3-*O*-glycosides were clearly present in both *Pterostemon* specimens.

Thermospray Liquid Chromatography-Mass Spectrometry (TSP LC-MS)

Thermospray liquid chromatography coupled with mass spectrometry offers a very efficient means of monitoring the output of HPLC systems with mass spectrometry. As pointed out by Wolfender and colleagues (1994) in a comprehensive review of the technique, the incompatibility of the working conditions of HPLC and MS prevented wedding of the techniques. This incompatibility prevented workers from using an extremely efficient detection system, mass spectrometry, with one of the most efficient separation systems, namely, HPLC. These problems have been overcome by the use of the thermospray approach in which direct electrical heating of a capillary, which acts as the vaporizer, converts the liquid eluate into a vapor, part of which is led into the mass spectrometer. Four different ionization modes are available for any given analytical system. Greater technical detail, including ionization modes, and references to the development literature, can be found in the review by Wolfender and his colleagues.

The technique has been applied to a wide variety of natural products including flavonoids of various sorts. Fourteen examples are identified in the Wolfender review article a few of which should offer the interested reader an opportunity to judge whether the technique would be a likely solution to problems at hand: poly-*O*-methylated flavones in oranges (Hadj-Mahammed and Meklati, 1987); procyanidins in cider apples (Mellon *et al.*, 1987); isoflavone aglycones as phytoestrogens in soybean products (Setchell *et al.*, 1987); apigenin 7-*O*-glucosides and derivatives in *Chamomilla recutita* (Carle *et al.*, 1993); flavonoid glycosides in *Epilobium* (Ducrey *et al.*, 1994), and anthocyanins in grape hybrids (Tamara *et al.*, 1994). In most of the cases cited the analyses of the compounds in question were at least as efficient and as seen with "conventional" methods of HPLC. Since the techniques involves "soft" ionization of eluate constituents, it is routinely possible to detect the presence of conjugated forms (e.g., glycosides) of the compounds of interest. Other applications are described in Wolfender and Hostettmann (1995).

Gas Chromatography (GLC)

Gas chromatography has had only comparatively limited application in flavonoid studies despite the availability of a large selection of stationary phases and relatively inexpensive instruments. One of the drawbacks may be the perception that GLC methods are suitable only for volatile plant components such as the simpler terpenes or esters. This need not be a deterrent in view of the availability of highly efficient means of rendering most compounds volatile enough to be analyzed by GLC, e.g., *tert*-butyl ethers, methyl ethers, methyl esters, and trimethylsilyl (TMS) ethers (Mole and Waterman, 1994, p. 159). Vande Casteele and coworkers (1976) used N,N-*bis*-(trimethylsilyl)trifluoroacetamide as silylating agent in a GLC study of naturally occurring non-volatile phenolic compounds. The columns used were SE-30 and SE-32 coated on Chromosorb and run from 80–210° and from 210–300°. Mixtures with up to 36 components could be satisfactorily resolved.

"Classical" means of detecting compounds in a GLC output often involved such methods as flame ionization (FID), which is sufficiently sensitive to provide reliable results but is totally destructive. A major advance occurred when a gas chromatograph was coupled with a mass spectrometer (GC-MS). Output from a GLC analysis could then be read in terms of the molecular ion and major fragments of the mixture as they emerged from the column. If the mass spectrometer were additionally equipped with a data base of molecular ions and fragment ions it became possible to read the output directly in terms of likely structures. Excellent examples of the application of GC-MS to resolution of complex mixtures of flavonoids and related phenolic compounds come from the work of Greenaway and colleagues on the bud exudates of *Populus* species (Greenaway *et al*., 1988, 1990, 1992a,b). The number of compounds resolved in these studies is quite impressive, 26 compounds in the case of *P. violascens* (Greenaway *et al*., 1992a), 30 in the case of *P. laurifolia* (Greenaway *et al*., 1992b), and 40 in the study of *P. deltoides* (Greenaway *et al*., 1990). The total number of peaks observed, as opposed to identified, in the GC-MS traces from these species is often much larger.

Some technical aspects of the use of GC-MS have been addressed by Creaser and colleagues (Creaser *et al*., 1989, 1991a,b, 1992). For example, they compared the use of methyl vs. trimethylsilyl derivatives, different types of column, and column heating using a set of simple flavones (Creaser *et al*., 1989). In a subsequent study they investigated the origin and control of multi-peak formation in the GC-MS analysis of flavanone trimethylsilyl ethers using capillary column GC (Creaser *et al*., 1991a). It was found that derivatization temperature, time, and capillary column injection technique all had an effect on the interconversion of flavanones and chalcones.

Gas chromatography has also been applied to the analysis of the sugar components obtained from flavonoid glycosides. Analyses can be done by comparing retention times of the trimethylsilyl (TMS) ethers obtained from hydrolysis of a flavonoid with those of standard sugars (Markham, 1982, p. 55). An alternative approach involves reduction of the sugars to their respective sugar alcohols, preparation of the acetates, and analysis by GLC. Owing to the widespread availability of NMR instruments, GLC-based sugar analyses of flavonoid glycosides are rarely found in the literature.

Capillary Electrophoresis (CE)

Tomas-Barberan and coworkers (1996c) asked the question "Is capillary electrophoresis the HPLC of the 90s in polyphenol analysis?" The technique has been applied to a number of flavonoid systems with some degree of success. McGhie and Markham (1994) point out that the separatory efficiency of CE is superior to that of HPLC and can be enhanced for flavonoid analysis by using borate-containing buffers. In that paper McGhie and Markham studied the effect of flavonoid structure on electrophoretic mobility and concluded that the following factors contribute to separations: molecular size, number and location of unsubstituted hydroxyl groups, nature of glycosidic components, and presence of ionizable functions on the individual flavonoids. Some of these factors can act in concert as in the case of higher levels of glycosylation which not only increase the size of the molecule but also decrease the number of ionizable phenolic groups with the result being a decrease in mobility. Applications of the general technique can be found in the work on medicinally interesting flavonoids (Seitz *et al.*, 1991) and in the analysis of the flavonoids of sugarcane (McGhie, 1993). Morin and coworkers (1993) demonstrated that borate complexation of flavonoid 3-*O*-glycosides aided in their resolution. The order of movement in the column was as follows: Q-Glc-Ara > Q-Glc > Q-Gal > Q-Rhm > Q-Ara.

A study of isotachophoretic analysis of flavonoids and phenolcarboxylic acids of possible relevance to the pharmaceutical industry has been described by Seitz and colleagues (1991). Details were given for leading and terminating electrolytes and for the addition of hydroxypropylmethylcellulose (HPMC) to both electrolytes to reduce electroendosmosis. The detection limit of the system was 3.8 nmol for quercetin 3-*O*-rutinoside (rutin).

Centrifugal Partition Chromatography

Centrifugal partition chromatography (CPC), also called centrifugal countercurrent chromatography (CCCC), is a liquid/liquid chromatographic system that uses two immiscible phases with the liquid stationary phase being retained by a centrifugal field. The technical background and relevant literature on this high-tech methodology has been reviewed in some depth by Marston and coworkers (1990).

SPECTROSCOPIC METHODS

Infrared Spectroscopy (IR)

Infrared spectroscopy has been a major source of information for natural product chemists for many years. The characteristic absorption of energy in the infrared region of the electromagnetic spectrum by different functional groups (causing bending, stretching, rocking of bonds, etc.), gives the chemist a powerful means of determining what the essential features of any given molecule are. The method is an extremely useful one that has the capacity to determine the environment in which a given functional group is located; e.g., whether a carbonyl group is free, hydrogen bonded, or conjugated with an α,β-double bond. If no other spectroscopic information is available, the IR spectrum will at least tell the chemist the class of compound to which the unknown belongs. This information is vital and serves as a key in deciding the direction of subsequent steps. Since flavonoids manifest themselves readily by means of their informative UV spectra, IR is rarely used. To be sure, there are structural features of flavonoids that are revealed by IR analysis (Wagner, 1964), but, for the most part, these are more of theoretical than practical interest. Nonetheless, detailed descriptions of flavonoid structural analyses often include IR information, as we will see in some of examples below, e.g., structural work on "chrysograyanone" by Arisawa and coworkers (1992) in the section on X-ray diffraction, and in some of the case studies.

Ultraviolet Spectroscopy (UV)

It might almost seem that ultraviolet absorption spectroscopy (UV) was designed specifically with flavonoids in mind! Few techniques provide such a wealth of information about a group of naturally occurring compounds as does UV when applied to flavonoids. In the case of most of the simpler flavone and flavonol glycosides, which are the most widely distributed flavonoids, a reasonable idea of a compound's structure can be arrived at by observing its chromatographic behavior, its color reaction with selected chromogenic reagents, and its UV spectrum. It is not possible to identify the sugar components of a glycoside, of course, but it is possible to determine chromatographically what level of glycosylation exists and, on the basis of its UV characteristics, where the sugar or sugars are attached. These techniques have been standardized and described in detail (Mabry *et al.*, 1970; Markham, 1982) and are available to anyone who has an interest in these data and has access to the necessary apparatus. In addition to those sources, discussions by the following workers provide valuable examples and tables of wavelength data: Jurd (1962); Mabry and Markham (1975); Harborne, 1964, 1990). Many of the chapters in *The Flavonoids* series also contain spectral data for individual classes of flavonoids.

If one were to record the UV spectra of flavonoids representing all structural classes in the wavelength range 220–600 nm, the most obvious features would be the existence of two principal maxima. The longer wavelength absorption is referred to as Band I, the shorter as Band II. In certain cases, flavonoid spectra consist of a single major band along with one of much lesser intensity. The exact wavelengths and magnitude of the maxima depend upon a number of factors: (1) the nature of the C-ring; (2) the point of attachment of the B-ring; and (3) the nature of the substituents on the A- and B-rings. Risking over simplification, it is convenient to consider Band I as reflecting substitution on the B-ring, and Band II as reflecting A-ring characteristics. Flavones exhibit Band I absorption in the range 310–350 nm with Band II absorption in the range 250–280 nm. Flavonols (3-OH) have Band I absorption in the range 350–385 nm with Band II much the same as in flavones. Flavonols that are substituted at position-3 (*O*-alkyl or *O*-sugar) have Band I absorption shifted to shorter wavelengths, 330–360 nm; Band II is again found in the range 250–280 nm. Flavones and flavonols are characterized by two chromophores, the cinnamoyl function that consists of the B-ring and the three carbons of the C-ring; and the benzoyl function that consists of the A-ring and the C-4 carbonyl carbon. Hydroxyl groups at the 4′-position in the former and at the 7-position in the later act as auxochromes (they contribute electrons to the aromatic system) the result of which is a shift of the respective absorption maxima to longer wavelengths (bathochromic shifts). Differences in the absorption characteristics of the two ring systems can be appreciated by comparing the length of their respective conjugated systems. This can be seen in the resonance diagrams of the A-ring [4-40 and 4-41] and B-ring [4-42 and 4-43] systems, respectively.

It is instructive to examine the contribution to the UV spectrum made by isolated hydroxyl groups. This can be seen in the position of Band I in a series of hydroxy flavones [see Harborne (1964) for complete data]. Flavone itself has a maximum at 297 nm. Placing an hydroxyl at C-7 increases the maximum to 307 nm, while 4′-hydroxyflavone has a maximum at 327 nm. Although the contribution is not as large as that brought about by the 4′-hydroxy group, Band I is affected by addition of an hydroxyl group to the A-ring. Hydroxyl groups at C-5 and C-3 have a significant effect giving bathochromic shifts of 40 and 47 nm, respectively. These shifts are due to hydrogen bonding between the phenolic hydrogens on those groups and the carbonyl oxygen at position-4. In general, if those positions lack hydroxyl groups or if they are occupied by *O*-alkyl or *O*-glycosyl groups, the UV maxima tend to be shifted somewhat to shorter wavelengths (hypsochromic shift). It should also be noted that the pyrone ring system is involved in the electronic activities of flavones and flavonols as is shown in resonance structures [4-42 and 4-43].

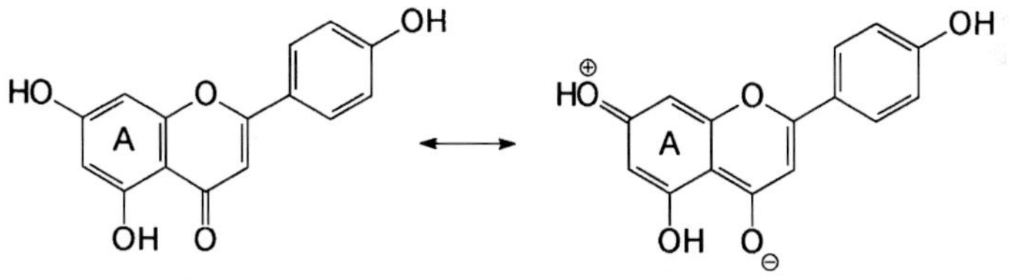

4-40 4-41

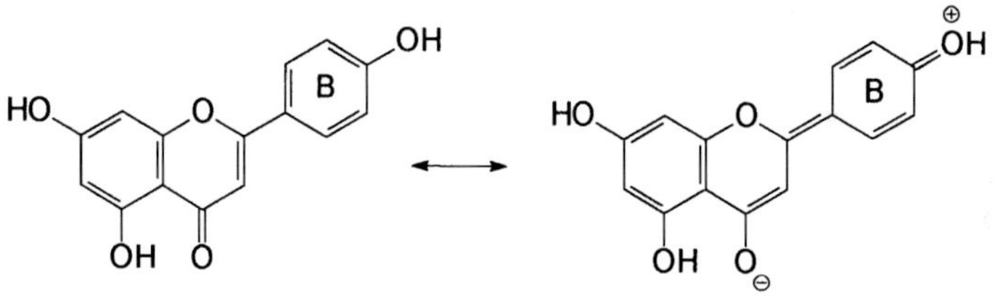

4-42 4-43

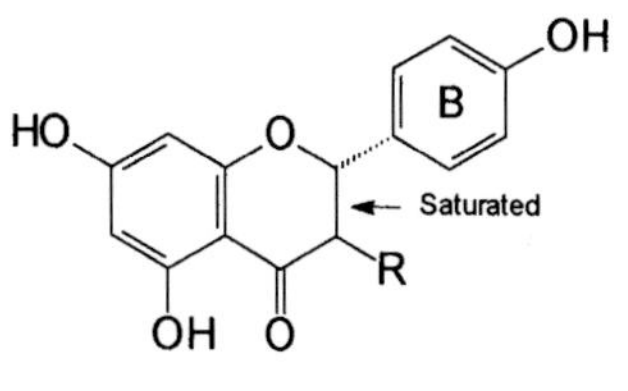

4-44

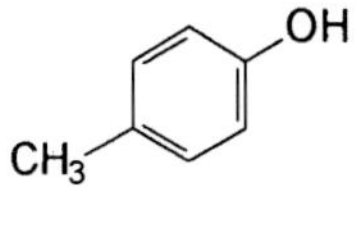

4-45

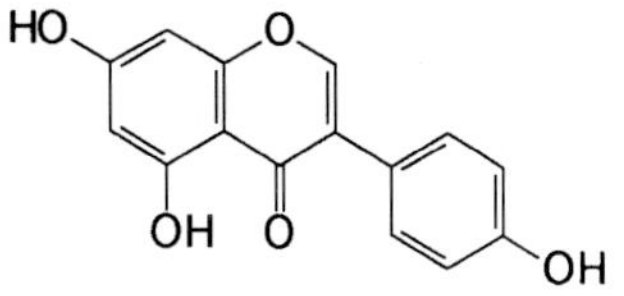

4-46

The contribution of the C2-C3 double bond to the UV absorption spectrum can be appreciated by comparing flavone and flavonol spectra with those of flavanones and dihydroflavonols [4-44 where R = hydrogen or hydroxyl, respectively]. In these latter compounds, Band I is reduced to little more than a shoulder in the 300–330 nm range, with Band II as the main peak in the range 277–295. Reduction of the C2-C3 double bond eliminates the cinnamoyl chromophore leaving only the A-ring benzoyl system intact. Contribution from the B-ring to the spectrum of flavanone derivatives is reduced to that of a simple phenol [4-45]. Some unusual absorption spectra for flavanones have been observed, however. Jurd (1962) noted that three distinct bands had been recorded for 6,7,3′,4′-tetramethoxyflavanone. Although spectra of this sort are very difficult to rationalize in terms of the simple relationships noted above, their appearance generally indicates that the compounds under study have unusual features.

Although isoflavones do have the C2-C3 double bond [4-46], the location of the B-ring at C-3 prevents conjugation of the phenyl group with the pyrone carbonyl group. This results in a significantly diminished contribution of the B-ring to the spectrum. Isoflavone spectra, therefore, consist of a prominent band in the 250–270 nm range with a peak of very low intensity in the 300–330 nm range. It is not uncommon for the longer wavelength absorption to consist merely of an inflection in that wavelength range.

The UV spectra of chalcones and aurones are characterized by very intense Band I absorptions in the 365–390 nm range for the former and 390–430 nm for the latter. Band II absorptions for both of these classes of flavonoids are of much lower intensity (ca. 30%) and occur in the range 240–270 nm. The presence of a chalcone or an aurone in a flavonoid mixture is usually evident from their chromatographic behavior and is easily confirmed by their UV spectra.

The most easily recognized of flavonoids are the anthocyanins. These compounds have two absorption bands, one of medium intensity in the range 270–280 nm, and one of higher intensity in the visible region, 465–560 nm (all measurements done in methanolic HCl). We recognize absorptions in this region, of course, as visible color. There are three main groups of anthocyanins depending upon B-ring oxygenation. Pelargonidin derivatives, those with 4′-hydroxylation, have a Band I ranging from 498–513 nm. Cyanidin derivatives, anthocyanins with 3′,4′-dioxygenation, have a Band I maximum in the vicinity of 520 nm, while those with 3′,4′,5′-trioxygenation have a maximum at about 532 nm. A special class of anthocyanidins comprises the small group of pigments that lack the C-3 hydroxyl group, hence the name 3-deoxyanthocyanins. Members of this group have Band I maxima somewhat lower than those of the true anthocyanins, e.g., apigeninidin 5-*O*-glucoside has a maximum at 477 nm while the maximum for luteolinidin 5-*O*-glucoside is observed at 495 nm (Harborne, 1967). Harborne (1967, p. 17) points out that the long wavelength maximum for anthocyanins is dependent upon the solvent. For example, cyanidin 3-*O*-rutinoside has a maximum of 533 nm in ethanolic HCl, 523 nm in methanolic HCl, and 507 nm in aqueous HCl. It is obvious that, in comparative studies, uniform conditions must be maintained.

Although absorption maxima for an unknown pigment can be of considerable help in assigning it to the appropriate flavonoid class, we can go much further in determining its structure by taking advantage of some of the chemical properties of phenols. Two important features of the phenolic group are its acidity, i.e., its capacity to become ionized in the presence of base to yield the phenoxide ion, and, in certain circumstances, its capacity to form complexes with metal ions or other charged groups. First, we will examine a flavonoid's response to alkali. The phenolic hydrogen is a weak acid (pK of the tyrosine phenolic function is ca. 10.0) and can be easily ionized in the presence of strong bases such as sodium hydroxide or sodium methoxide. Treatment of a flavonoid with strong base will result in ionization of all of its hydroxyl groups with resulting large bathochromic shifts of both absorption maxima. Judicious use of sodium methoxide has proven to be a useful diagnostic tool, however. A bathochromic shift of the longer wavelength maximum indicates an unsubstituted C-4′ hydroxyl group. If this peak decreases in magnitude upon standing, indicating decomposition of the flavonoid skeleton, several possibilities exist: (1) unsubstituted 3- and 4′-hydroxyls; (2) *o*-dihydroxy groups in the A-ring; (3) 3′,4′,5′-trihydroxylation (or equivalent). However, if the Band I peak in strong base is stable, and there has been no decrease in intensity, it can be concluded that the 4′-OH-3-OR combination exists, where R may be either an alkyl group or a sugar. If the shifted Band I peak is stable in base but shows a decreased intensity the 3-OH group is free and the 4′-position is substituted (again, either alkyl or sugar).

The use of a weaker, and hence more selective, base has been found to be very useful. The base of choice is the acetate ion, which has the capacity to ionize only the most acidic phenolic hydroxyl groups. Addition of anhydrous sodium acetate to a methanolic solution of an unknown flavonoid will bring about ionization of the C-7 hydroxyl group with a resulting bathochromic shift of the Band II maximum of up to 20 nm. Absence of a shift indicates that the oxygen at C-7 is not free. Acetate also brings about a bathochromic shift in the Band I maximum caused by ionization of the 4′-hydroxyl group. A very useful combination of reagents involving sodium acetate and boric acid allows one to establish the presence of *o*-dihydroxy groups owing to the capacity of the borate ion to complex with *o*-dihydroxy groups [4-47]. After measuring the acetate-derived shift of Band I, solid sodium borate is added to the test solution with shaking. Complexation with *o*-dihydroxy groups causes the Band I maximum to shift back towards the methanol maximum to give an overall shift relative to methanol of 12–36 nm [values from Markham (1982)]. If *o*-dihydroxyl groups are absent the acidity of the boric acid will neutralize the acetate ions with concomitant return of the absorption maximum to its original methanol value. *o*-Dihydroxy groups on the A-ring of flavanones and dihydroflavonols give similar results.

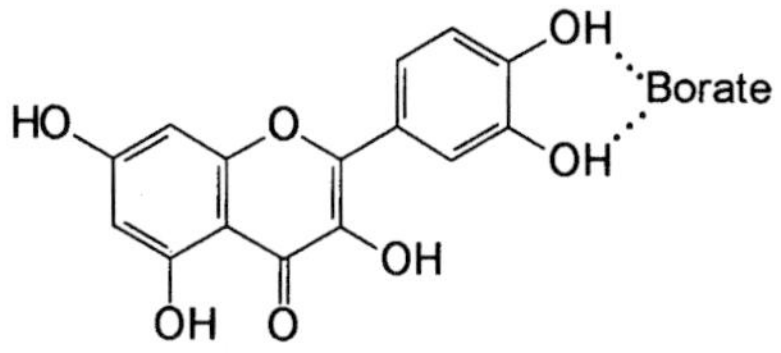

Borate

4-47

Al^{+3} (Unstable in acid)

Al^{+3} (Stable in acid)

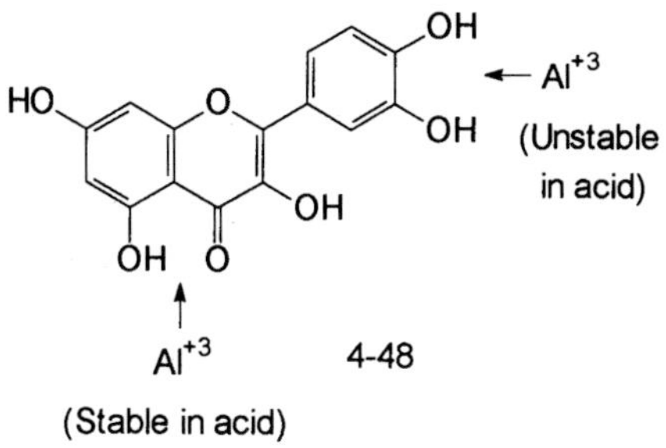

4-48

A solution of aluminum chloride, usually in some alcohol, has been used as chromogenic spray for locating compounds on chromatograms for many years. Brightly fluorescent spots are an indication that some compound capable of forming a complex with the aluminum ion is present. This reaction forms the basis of an important UV spectral test. Addition of aluminum chloride to a test solution of the flavonoid results in complexation of aluminum ions with the 5-hydroxy-4-keto system as well as with *o*-dihydroxy groups [4-48] leading to a large (50–60 nm) bathochromic shift of Band I. Advantage is taken of the acid stability of the keto-aluminum complex and the acid lability of the complex with the *o*-dihydroxy system. Addition of aqueous HCl results in disruption of the complex with the *o*-dihydroxy system and a bathochromic shift, relative to the maximum in methanol, of about 40 nm. This test is sensitive to the presence of 6-oxygenation with which one sees an aluminum/HCl shift of only about 20 nm relative to the methanol maximum. This general introduction to the use of standard shift reagents comes, with slight modifications, from Markham (1982). The interested reader is strongly urged to consult that source for technical details and additional examples.

A considerable amount of effort on the part of Prof. Ragai Ibrahim and his colleagues, at Concordia University in Montreal, has resulted in a much better understanding of the nature, biochemistry, and distribution of flavonoid sulfates (Barron *et al*., 1988). One of the outcomes of that research was the realization that certain spectroscopic methods routinely used for flavonoids do not necessarily work with flavonoid sulfates. This matter was resolved by the development of UV-spectral methods that use HCl and aryl-sulfatase preparations to detect 3- and 4′-sulfated flavonoids (Barron and Ibrahim, 1988). An extensive list of flavonoid sulfates with spectral characteristics is presented in that paper.

Nuclear Magnetic Resonance Spectroscopy (NMR)

Nuclear magnetic resonance spectroscopy is based upon the capacity of certain atomic nuclei to absorb energy in the radio frequency range under such conditions that the electronic environment of the nuclei can be assessed. Of particular interest to us, of course, is the fact that both hydrogen (^{1}H) and carbon (^{13}C) nuclei are susceptible to this type of probing. Since the effect involves a magnetic interaction, the essential requirement of the nuclide of interest is that its spin number (I) must be other than zero. Hydrogen (protium) presents no problems because it has $I=1/2$. The problem lies with carbon whose principal nuclide, ^{12}C, belongs to the group of nuclides with $I=0$, making it undetectable by this method. We are saved from a dilemma, however, by the existence in Nature of a small population of carbon atoms that have an extra neutron, i.e., ^{13}C, which, fortunately for the organic chemist, has $I=1/2$. However, its natural abundance is only 1.108%, which, along with a lower sensitivity compared with protons, and other problems, places certain restraints on the use of ^{13}C NMR. After a brief mention of useful literature on the theory and application of NMR methods, we will examine the use of ^{1}H NMR. ^{13}C NMR will be discussed later in this section.

Very good, practical, and "user friendly" treatments of all aspects of NMR spectrometry of organic compounds (as well as IR, UV, and MS) can be found in the books by Silverstein *et al.*, the 4th (1981) and 5th (1991) editions of which were consulted for the present discussions. The second edition of Günther's *NMR Spectroscopy* (1995) provides a thorough and highly technical coverage of principles and practice. A technically uncompromising treatment of one- and two-dimensional NMR methods can be found in a recent book by Friebolin (1993). A long-standing source of information on ^{13}C NMR is the work by Stothers (1972). The various treatments of flavonoid analysis referred to earlier in this chapter (Mabry *et al.*, 1970; Markham, 1982) also contain specialized treatments of NMR methods. A detailed treatment of ^{13}C NMR of flavonoids, including 125 flavonoid spectra and a useful index, can be found in Markham *et al.* (1982).The most recent member of *The Flavonoids* series contains a chapter by Markham and Geiger (1994) that deals with a variety of topics including two-dimensional NMR as well as the use of hexadeuterodimethylsulfoxide as a solvent for analyses of flavonoids and flavonoid glycosides. Extensive tables and many sample spectra are included in this valuable work.

The analytical use of NMR methods involves determining the "chemical shift" of the protons (or carbon atoms) in a molecule. This is a complicated business and best approached for our purposes from an empirical perspective. In the simplest terms, the chemical shift of an atom is an indication of the effect of its electronic environment on how the nucleus responds to the applied energy. The absorption of energy by certain protons, those associated with benzene rings, for example, will be different from those in an *O*-methyl group. This is frequently discussed in terms of one nucleus being more or less "shielded" from the applied magnetic field compared to some standard compound. A useful correlation of chemical shifts with electronegativity exists (although several important exceptions are known) (Silverstein *et al.*, 1981). For example, an aliphatic carbon to which a *C*-methyl group is attached is considerably less electronegative than is the oxygen atom in an *O*-methyl group. This is indicated in their respective chemical shifts, δ of less than 1.0 for the former and in the range of δ 3–4 in the latter. At the other extreme is the flavonoid 5-OH group, which has δ 12–14 [when measured in hexadeuterodimethylsulfoxide (DMSO-d_6) (Markham, 1982, p. 73)]. In this situation the phenolic hydrogen is under the strong influence of both the 5-oxygen and the C-4 carbonyl oxygen. It should be noted here that, since δ units are expressed in parts per million, the notation "ppm" is often used for chemical shift values.

Since the methyl groups in tetramethylsilane absorb at higher fields than almost any other organic protons, TMS, as it is commonly known, has been widely used as a standard. The chemical shift (δ) values cited in this discussion are all measured relative to TMS, which is taken as having $\delta = 0$. Most protons fall within the range of $\delta = 0$ to 10, with certain exceptions, one of which was noted above. Protons will have δ values that are generally characteristic of the functional group of which they are a part, although some overlap does exist. The reader should also be aware that literature values for chemical shifts may vary somewhat. For example, an aromatic proton range of δ 6.5–8.5 is cited by Friebolin (1993) whereas in Markham's Table 6.1 (1982) the range is given as δ 6.0–8.0. As in all other analytical techniques, experience and a sound understanding of the fundamentals of the subject are better than strict reliance on a tabulated value! Table 4.2 lists proton chemical shift values that are useful in flavonoid studies as you will see in several of the Case Studies below.

If all that an NMR spectrum provided was a view of the kinds of hydrogens present in a molecule (it also tells us how many of each are present) the method would be a valuable one. However, there is more. Because each hydrogen nucleus generates its own magnetic field, it can affect other nearby nuclei. This phenomenon, called "coupling" or "spin–spin coupling," results in the splitting of principle resonance peaks into multiplets (doublets, triplets, etc.). For example, hydrogens on adjacent carbons of a benzene ring undergo "*ortho*-coupling" which results in a splitting of each resonance into a doublet. The magnitude of the coupling effect is characteristic of the relative locations of the atoms involved. For *ortho*-coupled hydrogens, for example, we see a "coupling constant" of about 9 Hz, which would be written as $J = 9$ Hz. For protons further apart, such as those on the A-ring of apigenin, we see *meta*-coupling, whose J value is about 2.5 Hz. *para*-Coupling is also possible but the J value, which is quite small, would not have a visible effect on the spectrum unless a high-powered instrument is used. We can illustrate coupling using acacetin [4-49] as a model. The proton NMR of acacetin (run as its trimethylsilylated derivative) is represented in Figure 4.2. (The diagram is intended to show the main features of the spectrum and has not been drawn to scale.) We can start our analysis by looking at the resonance near δ 4.0, which arises from the methoxy group protons. *O*-Methyl protons appear in this general region and occur as singlets (they are equivalent). The next most prominent resonance arises from H-3, which is isolated from other protons in the molecule. The H-3 proton appears as a singlet at ca. δ 6.3. Since there is only one such hydrogen, and there are no others in the immediate vicinity with which it can couple, its resonance appears as a singlet. The H-6 and H-8 resonances appear at ca. δ 6.2 and δ 6.5, respectively, and show *meta*-coupling ($J = 2.5$ Hz). The remaining resonances arise from the four hydrogens on the B-ring. The H-2′ and H-6′ pair occur in identical environments and are centered at ca. δ 7.7 while the H-3′ and H-5′ pair, also in identical environments,

Table 4.2 ^{1}H NMR Chemical shifts for flavonoids (Markham and Mabry, 1975)

Position	Range of δ values
H-5	7.7–8.2
H-6	5.7–6.4
H-8	5.9–6.5
H-2 (isoflavones)	7.6–7.9
Chalcone α-proton	6.7–7.4
Chalcone β-proton	7.3–7.7
Aurone benzylic proton	6.5–6.7
H-3 (flavones)	ca. 6.3
H-2 (flavones)	5.0–5.5
H-2 (dihydroflavonols)	4.8–5.0
H-3 (dihydroflavonols)	4.1–4.3
Methoxyl protons	3.5–4.1
H-3 (flavonones)	ca. 2.8

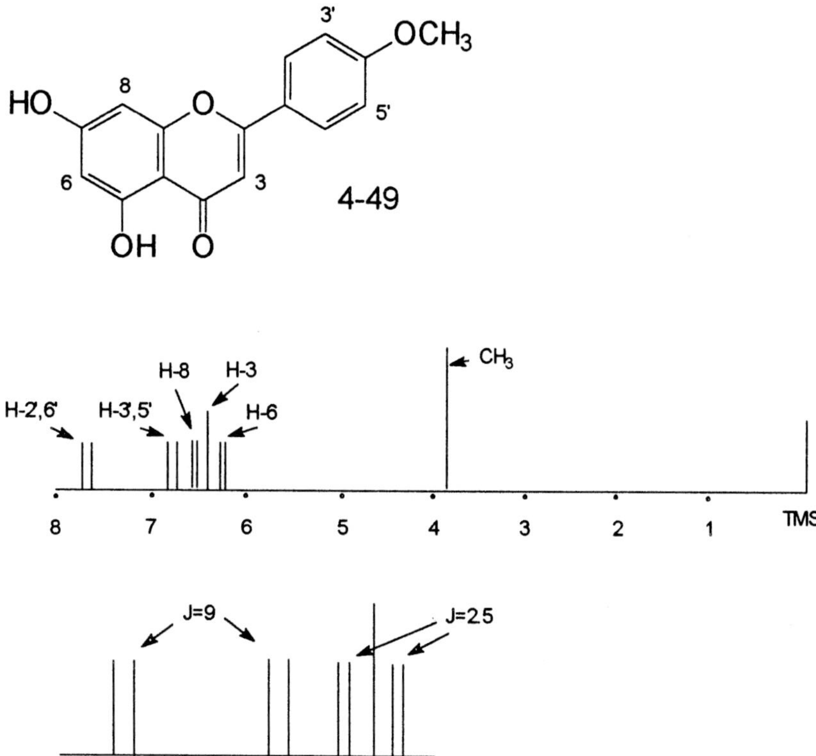

Figure 4.2 Graphic representation of a ^{1}H NMR spectrum using acacetin as a model compound. Bottom panel shows expansion of the spectrum in the range of 6–8. Neither vertical scale nor couplings drawn to scale.

appear as a resonance centered at ca. δ 6.8. Notice that these resonances are described as being "centered." Their observed positions in the spectrum are the result of *ortho*-coupling ($J = 9$ Hz). The aromatic and aromatic-like proton spectrum of acacetin, as is the case in many flavonoids, is crowded into a comparatively small range of δ values, thus making precise measurements difficult. Instruments have the capacity to produce "expanded" spectra, which make interpretation simpler; the expansion of a segment of the acacetin spectrum also appears in Figure 4.2.

Proton NMR is also very useful in helping to determine the structures of flavonoid glycosides. Several diagnostic pieces of information can be gleaned from a flavonoid glycoside spectrum. For example, the anomeric hydrogen of glucose has a chemical shift (ca. δ 5.85) that is clearly distinct from that of rhamnose (ca. δ 4.2). In addition, the configuration at C-1 of the sugar can often be determined on the basis of the coupling of H-1 with H-2 of the sugar. α-Glucosides [4-50] and β-glucosides [4-51] can be readily distinguished. β-Glucosides have H-1 (the anomeric proton) and H-2 oriented axially which gives rise to a moderately large coupling constant ($J = 6$), whereas in α-glucosides the H-1 hydrogen is oriented equatorially and has a smaller effect on H-2, the result of which would be a smaller J value. Another useful feature is the appearance of the rhamnose *C*-methyl proton resonance at ca. δ 0.85, which would appear as a singlet and integrate for three protons.

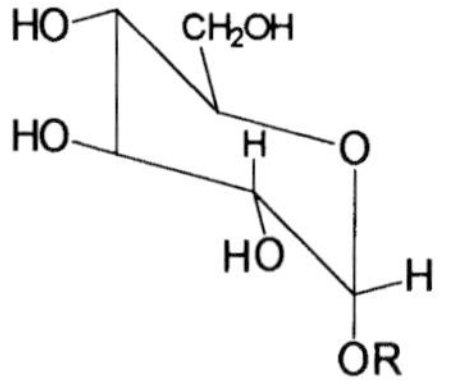

alpha-Glucoside
anomeric and C-2 Hs are cis

4-50

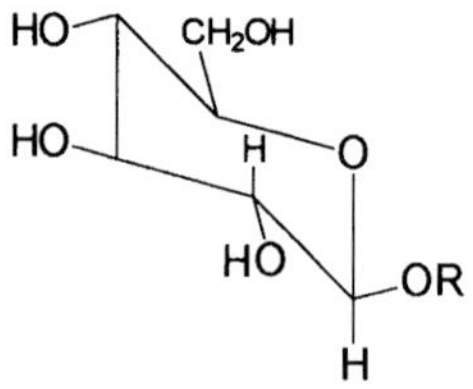

beta-Glucoside
anomeric and C-2 Hs are trans

4-51

Other resonances of importance in flavonoid studies include the C-2 proton in isoflavones [4-52]. This isolated proton appears as a singlet at about δ 7.5–8.0, which is sufficiently downfield relative to the 3-hydrogen of flavones (ca. δ 6.3) that the existence of one type over the other is immediately obvious. The NMR spectra of flavanones are also easily interpreted. In addition to their aromatic proton spectra, flavanones have characteristic resonances arising from interactions between H-2 and H-3_{cis} and H-3_{trans} [4-53]. (The *cis* and *trans* notations in the structure refer to the protons only.) The H-2 proton occurs downfield in the vicinity of δ 5–6 and, because it is influenced by both C-3 protons, appears as a double doublet. The protons on C-3 appear as a multiplet in the range of δ 2.5–3.0 the complex nature of which arises from interaction with the C-2 proton. The situation with dihydroflavonols is simpler because there are only two hydrogens to contend with, one at C-2 and one at C-3 [4-54]. Each appears as a doublet, the C-2 proton at ca. δ 5.2 (J=ca. 5 Hz), and the C-3 proton at ca. δ 4.2 (J=11 Hz). The stereochemistry of [4-54] is (2*R*,3*R*), but three other arrangements of the substituents are possible. Needless to say, NMR is very helpful in establishing the stereochemistry of these compounds.

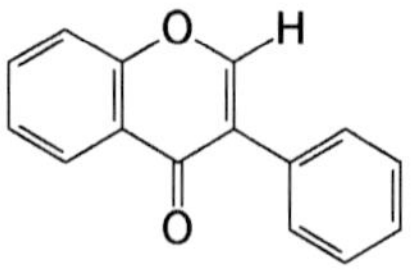

4-52

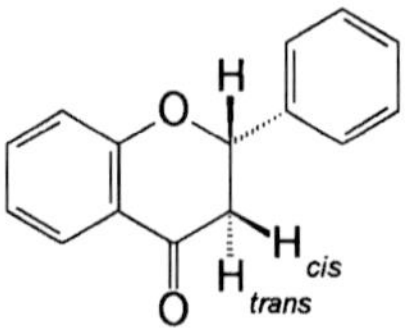

4-53

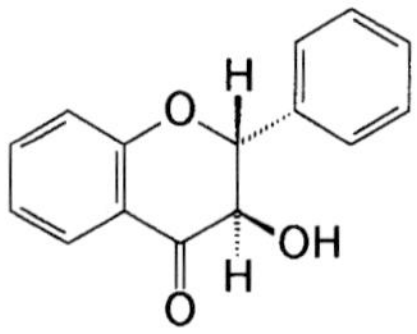

4-54 (2*R*,3*R*)

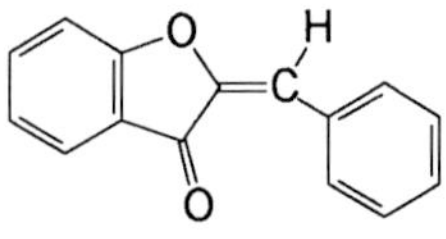

4-55

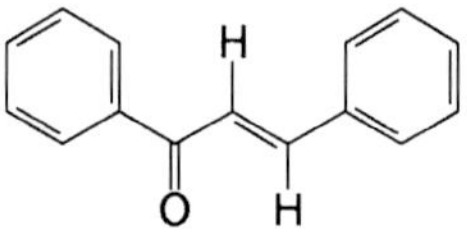

4-56

Aurone A-ring and B-ring protons have chemical shifts in the same general regions seen for other flavonoid structural types. The unique resonance in aurones [4-55] arises from the =CH- proton which occurs in the range δ 6.37–6.94 (Huke and Gorlitzer, 1969). (Note in structure [4-55] that the proton and phenyl group are interchangeable so that two isomers are possible.) In NMR spectra of chalcones [4-56] H_α and H_β (the protons on the three-carbon bridge) appear as doublets at δ 6.7–7.4 and δ 7.3–7.7, respectively [data from Markham and Mabry (1975)] although a few examples of compounds with slightly different values are known. As was pointed out in an earlier treatment of chalcone NMR (Bohm, 1989), the cases where these unusual values occur are frequently associated with compounds having unusual B-ring oxygenations. Usually, the chalcone α- and β-protons appear as two doublets with J values in the range 14–17 Hz, again with some extension of the range in unusual cases. These high J values arise from the *trans*-orientation of the two protons. *cis*-Chalcones exhibit much smaller couplings. A prominent singlet in the δ range 12.0–14.6 indicates the existence of a 2′-hydroxyl group (strong chelation with the carbonyl oxygen). Most other hydroxyl protons in chalcones occur at δ values of ca. 10.

The number of *O*-methyl groups present in a flavonoid can be determined easily by integrating the methoxyl proton resonances, which occur as a set of singlets near δ 4.0. If two *O*-methyl groups happen to fall at the same value the integral curve will reveal that six protons are present. The protons of the methylenedioxy function (—O—CH_2—O—) appear as a singlet near δ 6.0. More complex substituents, such as the prenyl group, give correspondingly complex NMR spectra. An example of a prenylated flavonoid can be found in Case Study No. 6. Phenolic proton resonances appear in a very wide range and may overlap with other, non-phenolic hydrogens. Phenolic protons are labile, however, and will undergo exchange with the solvent if the opportunity is available. This capacity can be exploited by treating the flavonoid with D_2O. The phenolic proton will be replaced by deuterium with the result that the phenolic proton resonance disappears.

The development of ^{13}C NMR methods has been of incalculable importance to natural product chemists. Despite the power of the newer technique, however, ^1H NMR still remains an important tool in its own right. The situation is very well summarized by Markham *et al.*, (1982, p. 19), which we quote here in full. “It should be stressed, however, that ^{13}C-NMR is in no sense superseding ^1H-NMR, but is complementary to it. Information gained relates to the carbon ‘backbone’ of the molecule while ^1H-NMR gives information about the structural environment of each proton. Further, it is possible to obtain a good ^1H-NMR spectrum on a sample that is much too small for carbon-13 analysis.”

As in the case of protons, the use of ^{13}C NMR in determining the structure of a molecule is based upon interpretation of the chemical shifts of carbon atoms relative to some standard substance. Chemical shifts of carbon atoms appear in a range of about 240 ppm downfield relative to TMS (a few appear above the TMS signal but these do not concern us). This is approximately 20 times the range for protons, thus providing highly resolved spectra of quite complex molecules. The chemical shift of a carbon is primarily dependent on electron density around that atom. Aliphatic methyl groups, for example, appear at about δ 17, unsubstituted aromatic carbons appear in the range δ 90–135, while the C-4 atom in flavonoids (the carbonyl carbon) occurs much farther downfield, at about δ 182 in apigenin. Details of ^{13}C chemical shifts will be presented below.

As the reader will recall, proton–proton coupling is an extremely important source of structural information for the study of flavonoids. It helps to establish substitution patterns of the aromatic rings, the nature of the heterocyclic ring, and the nature of more complex substituents such as prenyl or other *C*-alkyl groups. In contrast, ^{13}C–^{13}C coupling does not occur under normal analytical conditions. Owing to the very low frequency of occurrence of ^{13}C-atoms (ca. 1.1%), the probability of two of them lying adjacent to one another in a molecule is very small (ca. 1/10,000). (The use of ^{13}C-enriched precursors in biosynthetic studies, however, does exploit the capacity of carbon atoms to couple with one another.) Coupling of ^{13}C-atoms with protons also occurs. In certain circumstances this "heteronuclear" interaction is desirable; in others it is not. Let us examine both scenarios. In complex molecules, splitting of a carbon resonance by nearby protons can result in a spectrum with a good deal of overlap that is difficult to interpret. In this situation it is desirable to eliminate carbon–proton coupling and produce a "pure" carbon spectrum. One of the ways by which this can be accomplished is "noise-decoupling," which involves irradiating all ^{1}H frequencies (broad band) while the ^{13}C-spectrum is being recorded in what is generally referred to as a "double resonance" experiment. The result is a completely decoupled spectrum from which chemical shift values can be read directly. If one uses a lower energy to irradiate the protons, called "^{1}H off-resonance decoupling," it is possible to reduce proton–carbon interactions to those that involve only directly bonded C—H pairs, i.e., CH_3, CH_2, and CH. Associated with broad band decoupling is an enhancement of ^{13}C signal intensity called the "nuclear Overhauser effect" (nOe). This response is an important tool for studying interactions (coupling, also referred to as connectivity) between selected atoms.

One of the disadvantages of ^{13}C compared to ^{1}H NMR arises from the fact that relaxation times (T_1) for carbon atoms tend to be quite different compared to relaxation times for protons. (Relaxation time is a measure of the lifetime of the excited nucleus.) This means, in effect, that the height of a carbon resonance peak may not necessarily represent the entire population of that particular carbon. The similarity of T_1 values for protons, on the other hand, means that peak dimension is proportional to the number of protons in any particular environment. Special conditions must be met if ^{13}C resonances are to be used for quantitative purposes. Routine analyses do not provide this information.

Another apparent drawback of ^{13}C NMR spectroscopy is the relative insensitivity of the carbon atom compared to that of the proton. This requires a much larger sample of unknown compound or, if the sample size is limiting, application of more sophisticated techniques such as Fourier Transform NMR (FT-NMR). Although this techniques allows for spectra of small amounts of material to be obtained, it does require longer times for signal acquisition. Adapting a familiar saying, instrument time is money.

Many tables of ^{13}C chemical shifts have been published; the most useful for our purposes are those compiled by Markham (1982), Markham *et al.* (1982), and by several workers that appear in Harborne (1990). Much of the following discussion

Table 4.3 ^{13}C NMR Chemical shifts for common flavones (compiled from Markham *et al.*, 1982)

Position	Range of δ values	Position	Range of δ values
C-2	163.6–164.6	C-1′	120.9–123.5
C-3	102.8–104.0	C-2′	106.0–128.8
C-4	181.6–182.3	C-3′	114.8–150.8
C-5	161.5–162.2	C-4′	137.9–162.8
C-6	98.2–99.0	C-5′	112.1–146.5
C-7	163.7–165.6	C-6′	118.7–128.8
C-8	92.9–94.2		
C-9	157.3–157.9		
C-10	103.3–105.0		

will use data from the tables in Markham *et al.* (1982). Table 4.3 presents the range of chemical shifts reported for seven flavones: apigenin, apigenin 4′-methyl ether (acacetin), apigenin 7-methyl ether (wogonin), luteolin, luteolin 3′-methyl ether (chrysoeriol), luteolin 4′-methyl ether (diosmetin), and 5,7,3′,4′,5′-pentahydroxyflavone (tricetin). It should be immediately apparent that the ranges for A-ring and heterocyclic ring carbons are narrow enough to allow assignment of structural elements with reasonable certainty. Note that C-9 (attached to the heterocyclic oxygen) and C-10 (attached to carbonyl carbon), normally not involved in flavonoid nomenclature, are identified for the purpose of ^{13}C NMR. It is often helpful to see chemical shifts presented graphically. Relevant information is presented in Figure 4.3: chemical shift values for the A-ring of apigenin; B-ring chemical shifts (A to F) as they relate to the different flavones, (from Markham *et al.*, 1982); and comparisons of the heterocyclic ring carbons of flavones, flavonols, isoflavones, flavanones, and dihydroflavonols, G to K, respectively.

The dramatic effect of hydroxylation of the aromatic nucleus can be seen in a comparison of the δ values for B-ring carbons in flavone itself compared to those for apigenin (all comparisons based on data in Table 7.2 in Markham *et al.*, 1982). Adding an hydroxyl function to C-4′ changes the δ value from 131.4 to 161.1, a down-field shift of nearly 30 ppm. The δ value for C-3′ and C-5′ are altered from δ 128.7 to 116.0, a change of just over 12 ppm. The *meta* carbons, C-2′ and C-6′, experience a change of δ value from 126.0 to 128.4, while the value for C-1′ moves from 130.8 in flavone to 121.3 in apigenin. Shifts of similar magnitude are seen in A-ring carbons as well: δ 133.8 for C-7 in flavone to δ 163.7 in apigenin; δ 125.1 for C-5 in flavone to δ 161.5 in apigenin. A diagrammatic representation of the ^{13}C spectrum of acacetin (modified from spectrum No. 20 in Markham *et al.*, 1982) is shown in Figure 4.4.

Glycosylation has a significant effect on the carbon to which the sugar is attached as well as on other carbons in the same ring (in the case of 5-*O*-glycosides the effect can be seen in the C-ring as well). In the case of *O*-glucosides, the δ value for the ring carbon is shifted upfield by as much as 2 ppm while the δ values for the carbons adjacent (the *ortho*-carbons) are shifted downfield from 1–4 ppm, the larger of the two shifts generally being associated with the carbon *para* to the point of glycosylation. *C*-Glycosylation causes a downfield shift of about 10 ppm of the aromatic ring

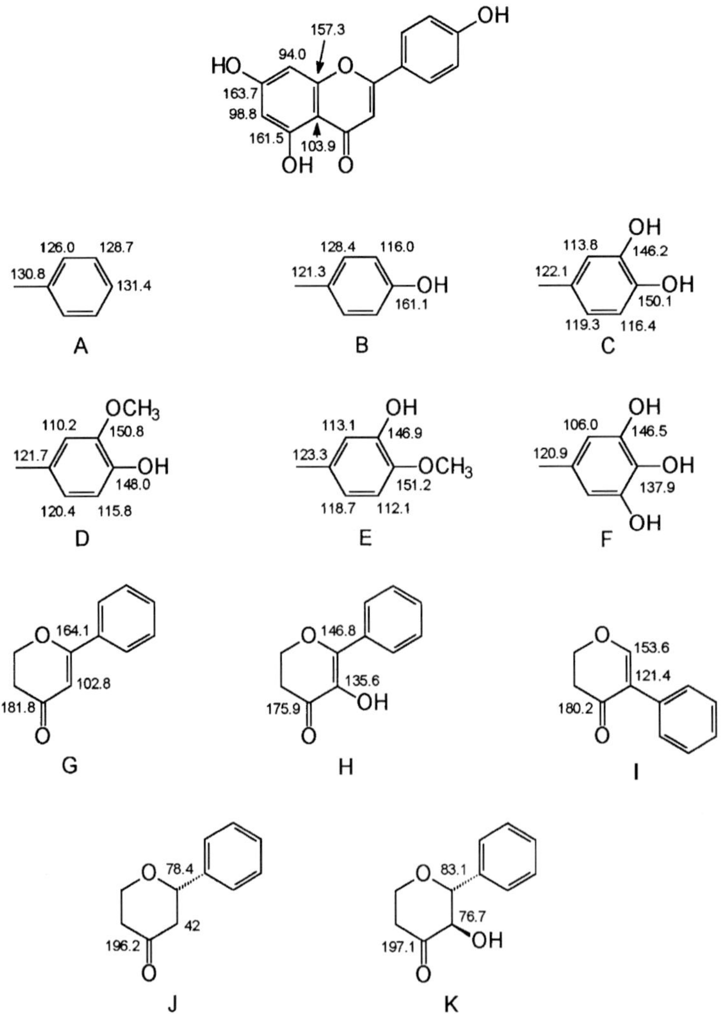

Figure 4.3 Chemical shift values for flavonoid B-ring and C-ring carbons for usual substitution patterns and C-ring oxidation levels. Values for the A-ring carbons of apigenin are also given.

carbon to which the sugar is attached with only a slight change in adjacent carbons. Acetylation has a pronounced effect on the δ values of the carbon to which the acetylated hydroxyl group is attached. Markham *et al.* (1982) state that an upfield shift of 6.6–15.6 ppm can be expected with *ortho-* and *para*-carbons shifting downfield by 4.1–12.1 and 2–7.9 ppm, respectively.

Many flavonoids are known that have some kind of substitution at C-6 or C-8 (or both). ^{13}C NMR provides a very efficient means of determining whether one or both of these positions are occupied in an unknown compound. Using 5,7-dihydroxyflavanone as a starting point, we see that C-6 and C-8 have δ values of 96.1 and 95.1,

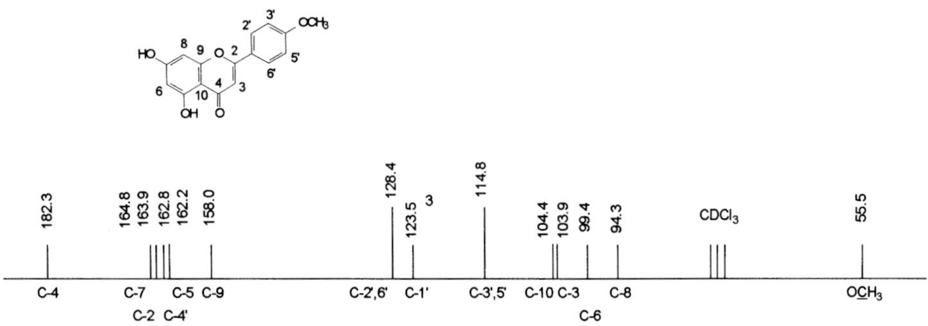

Figure 4.4 Diagrammatic representation of the ^{13}C NMR spectrum of acacetin. Modified from Markham and Chari (with Mabry) (1982). The spectrum was run in DMSO-d_6 + $CDCl_3$ at 25.2 MHz at ambient temperature.

respectively (data from Markham *et al.*, 1982, Table 2.6). A 6-*C*-methyl group shifts the C-6 resonance to δ 102.1 and reduces the C-8 δ value by 0.4 ppm. *C*-Methylation at C-8 has a comparable effect. 6-Hydroxylation has a more dramatic effect; 5,6,7,3′,4′-pentahydroxyflavone has δ 140.4 for C-6 and δ 93.6 for C-8.

Nearly as revolutionary as the advent of NMR in the first place, the development of an array of two-dimensional NMR techniques has had an immense impact on natural product chemistry. With only a few milligrams of compound it is now possible to obtain information on essentially all connectivities (another name for couplings) in any molecule. The techniques are based on a variety of pulse sequences and can involve proton–proton or carbon–proton interactions. The seemingly limitless variations include such techniques as COSY (correlated spectroscopy) and HOHAHA (Homonuclear Hartmann-Hahn spectroscopy), from which one can obtain information on almost all proton–proton connectivities, HETCOR (heteronuclear chemical shift correlation) and the related HMQC (heteronuclear multiple quantum coherence) which provide information on carbon–proton connectivities, HMBC (heteronuclear shift correlations *via* multiple bond connectivities) and NOESY (Nuclear Overhauser and exchange spectroscopy), an enhancement of the NOE approach particularly useful for large molecules with conformational and configurational problems. Many other techniques exist, some of which are particularly useful for some structural problems, while others are more suited to different problems (Silverstein *et al.*, 1991, Chapter Six; Markham and Geiger, 1994; Günther, 1995). Interpretation of reported spectra requires practice, and hands-on experience is vital for a full appreciation of these techniques. Results from studies that utilized some of these techniques appear in the Case Studies at the end of this chapter. In those instances the reader will find a simple statement to the effect that carbon-X and proton-Y (say) have been shown to be coupled using a particular technique. An understanding of the fundamental principles upon which that analysis is based is not, strictly speaking, necessary for our purposes.

Two recent examples of flavonoid structural elucidation using some of these techniques will be presented here. Both involve anthocyanin derivatives but applications can, obviously, be applied to any flavonoid. In the first of these, homonuclear

Hartmann-Hahn (HOHAHA) spectroscopy was used, in conjunction with mass spectral measurements, to determine the structure of acylated anthocyanins in grapes (Tamura *et al*., 1994). One of the compounds investigated by these workers is structure [4-57], malvidin 3-(6-*O*-*p*-coumaroylglucoside)-5-*O*-glucoside. Those workers

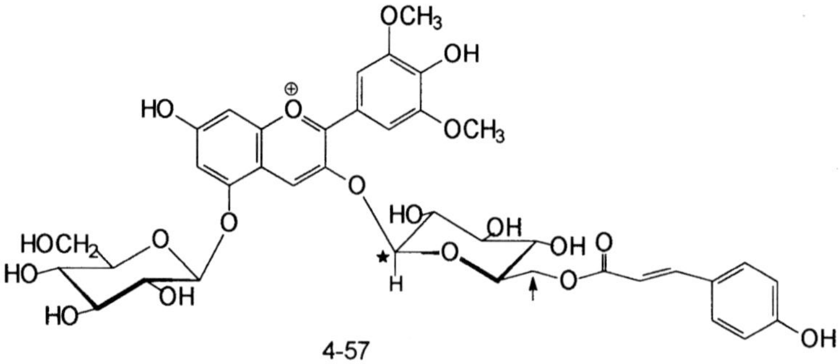

4-57

found that ^{1}H NMR pulse techniques, particularly HOHAHA spectroscopy, gave reliable information on the position of acylation and the stereochemistry of the sugars involved. Irradiation of the anomeric proton of the 3-*O*-glucoside group (starred in the structure) showed connectivity with the protons of the methylene group (arrow) that was attached to the *p*-coumaroyl function. Further, all C-H signals showed coupling constants of more than 7 Hz, which places them all in axial (*trans*) orientation. It follows that the sugar is glucose and that it exists in the pyranose form. The second example involves a more difficult situation since the compound possesses features that clearly mark it as unusual. Bakker and coworkers (1997) identified a compound from some red wines that was clearly an anthocyanin but differed from known pigments in having an additional three carbon unit attached at C-4. Extensive HMBC and NOE experiments indicated that the correct structure for the compound is [4-58]. The usual presentation of results of this sort involve structures with arrows indicating the various couplings. Such illustrations are often so "busy" with arrows that the underlying structure is difficult to find! For example, the structure drawn in the Bakker *et al*. paper has 23 arrows. In order to simplify the picture somewhat, I have prepared a table of connectivities to accompany the structure. Readers wishing to examine the original paper are, of course, encouraged to do so. The structure and the connectivity data appear together in Figure 4.5.

One of the more difficult problems in flavonoid structure analysis involves determining the sequence of units in an oligosaccharide when different sugars are involved. The classical method involved exhaustive methylation, treatment with acid, and identification of the hydrolysis products. Alternatively, partial acid hydrolysis of the unknown glycoside followed by chromatographic comparison of hydrolysis products with standards can be done. Both of these approaches require a reasonable amount of unknown and present numerous technical obstacles. A solution to this problem was described by Guinaudeau *et al*. (1981) in a study of two triglycosides obtained from leaf extracts of *Colubrina faralaotra* (Rhamnaceae). Both compounds yielded glucose, galactose, and rhamnose upon acid hydrolysis. Based upon a method used in studies of cardiac glycosides, these workers measured the T_1 values

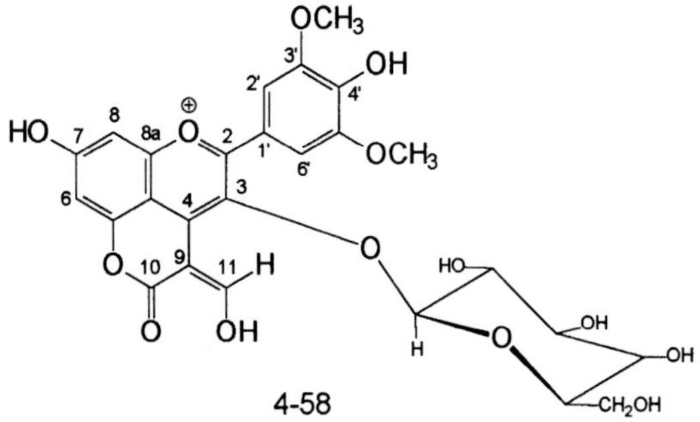

4-58

NOEs: Anomeric H and H-11,
H-11 and anomeric H,
anomeric H and H-6'

HMBCs: C-10 and H-11,
C-5 and H-6,
C-6 and H-8,
C-7 and H-6,
C-7 and H-8,
C-8 and H-6,
C-8a and H-8,
C-2 and H-2',
C-1' and H-2',
C-6' and H-2',
C-4' and H-2',
C-3' and H-2',
C-4' and Methyl Hs,
C-3 and anomeric H,
C-4 and H-11

Figure 4.5 Examples of anthocyanins whose structures were established using ^{1}H and ^{13}C NMR coupling analysis.

(relaxation times) for the *Colubrina* compounds and found that the carbons of the distal sugar residue had the longest values, the proximal unit carbons had the shortest, and the central sugar unit had intermediate values. This technique has likely not been used to maximum advantage in the field of flavonoid oligosaccharides.

Mass Spectroscopy (MS)

Unlike other spectrometric methods, where absorption of electromagnetic energy at various wavelengths brings about detectable changes in the orientation of nuclei (NMR) or π-electrons (UV), or induces vibration, bending, or stretching of bonds (IR), mass spectroscopy involves assaulting a molecule with sufficient energy to break it up into smaller pieces. The disrupting force is a beam of electrons and the detection apparatus is such that each fragment is recorded in terms of its mass to charge ratio, m/z. The practical significance of this is that the pieces are identifiable and are formed by more or less well known mechanisms. With a group of identified fragments in hand, the natural product chemist then tries to reassemble them into an intact molecule. Flavonoids seem particularly amenable to this process.

The first step in the degradation process involves formation of the "molecular ion" which results from the removal of an electron to produce a "fragment" having the molecular weight of the target flavonoid aglycone. (Glycosides will be dealt with below.) It is thus possible to establish the molecular weight of the parent compound directly, although this is often done using high resolution mass spectroscopy. Wollenweber and Dietz (1979) prepared a table of mass spectral parent ions from which it is possible to determine possible structures of unknown flavonoids. For example, $m/z = 314$ indicates $C_{17}H_{14}O_6$ which could accommodate an aurone, a flavone, or a flavonol with two hydroxyl and two methoxyl groups. Similarly, $m/z = 316$ indicates $C_{17}H_{16}O_6$ which could have originated from either a chalcone with three hydroxyl and two methoxyl groups or a flavanone or dihydroflavonol with two hydroxyl and two methoxyl groups.

Structural information about the target molecule can be gleaned two ways, either through the elimination of characteristic little pieces, e.g., methyl groups, or through major fission of the molecule and identification of the resulting pieces. Both sources of information are important in flavonoid studies. Removal of small pieces of a flavonoid often provides very important insights into its structure. For example, flavonoids with methoxy groups at C-6 [4-59, written as the molecular ion] or C-8 give a very prominent peak at an *m/z* value 15 mass units less than the molecular ion, i.e., $[M\text{-}15]^+$. If the methoxy group is at C-6, the peak arises from fragment [4-60], which, although prominent, is smaller than the molecular ion. If the methoxy group is at C-8, however, the peak is of greater intensity than the molecular ion. Much smaller [M-15] peaks indicate loss of methyl groups from some methoxyl groups at other positions. Loss of 17 mass units results from loss of hydroxyl (OH) and generally indicates the presence of a 2′-hydroxy group [4-62], while loss of 31 mass units can be attributed to loss of a 2′-methoxy group. Loss of 28 or 29 mass units can arise from extrusion of the carbonyl group (loss of CO, sometimes HCO) from the C-ring of flavonols and dihydroflavonols. 3-Methoxyflavones will often lose the elements of CH_3 and CO resulting in a peak equal to the molecular ion minus 43 mass units [4-61]. Loss of the prenyl group from *C*-prenyl flavonoids results in the formation of an $[M\text{-}55]^+$ or $[M\text{-}56]^+$ ion. The fragmentation leaves a methylene group attached to the original point of prenylation which takes the form of a benzyl carbocation [4-63]. *C*-Glycosylflavones will undergo loss of all their sugar carbons except the one attached to the flavonoid nucleus, which will be retained as a benzylic carbocation [4-63], which, as we just saw, also originates from a *C*-prenyl group. In the case of the *C*-glycosylflavone, the result of this scissions is a molecular ion equal to the parent flavone plus 13 mass units (CH). 6-*C*-Glycosylflavones can be distinguished from 8-*C*-glycosides by the appearance of an intense peak at [M-31] in spectra of the former.

The molecular ion of flavonoids, e.g., [4-64], can undergo fission in different ways, one by the reverse Diels-Alder route to yield the A_1 [4-65] and B_1 [4-67] fragments, while the second route gives the B_2 fragment [4-68] from the B-ring and fragments arising from the A- and C-rings. Other fragments that may be appear include $[A_1+H]^+$ [4-66] and [4-69], which is the ethylenic analogue of B_1. Fragments [4-65], [4-66], and [4-68] can lose carbon monoxide to produce ions 28 mass units lighter. As Markham (1982, 1989) points out in his discussions of flavonoid mass spectroscopy, the dominant route of decomposition depends to a certain extent on the nature of the aglycone: flavones and isoflavones tend to give $A_1^+ = (A_1+H)^+$ plus

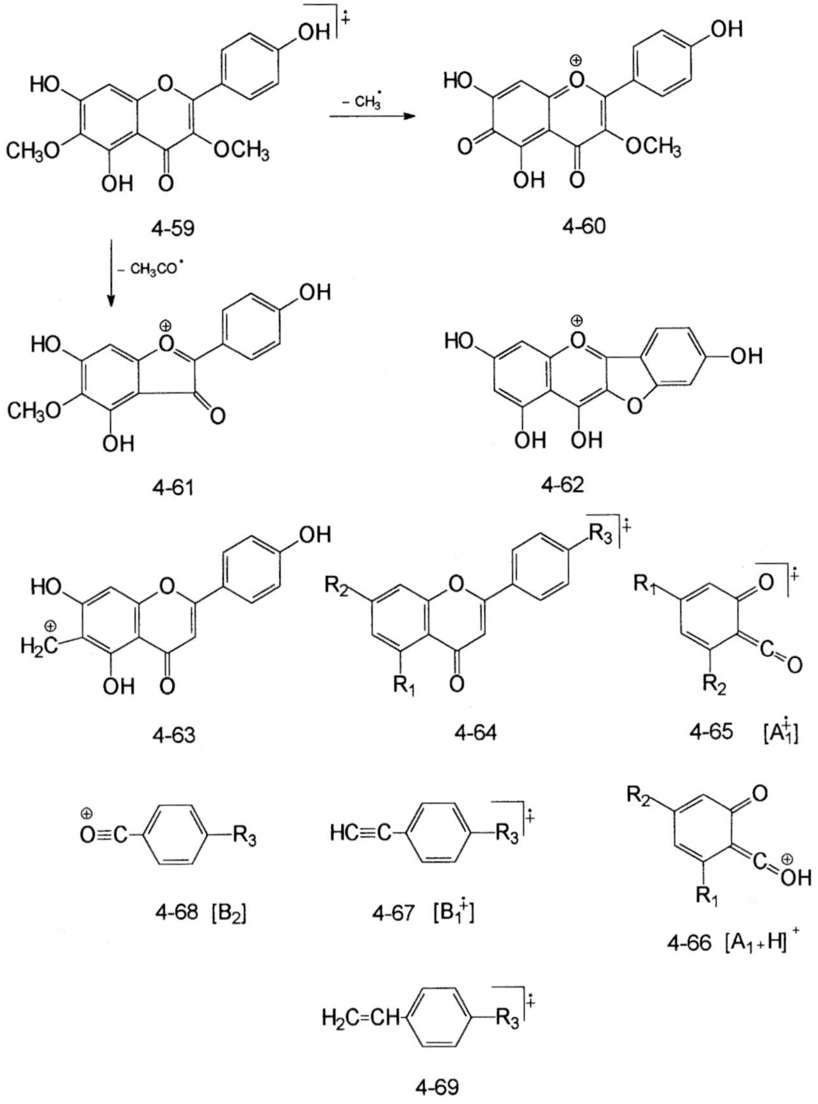

B_1^+ fragments; flavonols A_1^+ plus B_2^+ fragments; flavanones A_1^+ plus $(B_1+2H)^+$ fragments; and dihydroflavonols A_1^+ plus $(B_1+H_2O)^+$ fragments. The reader should consult the major compilation of electron impact analyses of flavonoids described by Hedin and Phillips (1992). It is also important to note Markham's (1982) warning that none of these pathways necessarily has to give any detectable fragments. (Unlike NMR methods, MS tends to be a bit fickle!) Chalcones give fragments arising from fission on either side of the carbonyl group although those that have a 2′-hydroxyl group tend to isomerize to the flavanone and follow the fragmentation path of that type of compound. Biflavonyls linked through oxygen give fragments arising from the two flavone units, while *C*-linked biflavonoids tend to give more complex fragmentations which include biphenyl derivatives.

Flavonoid glycosides offered a particular challenge since the use of normal 70 eV electron beams, although giving excellent data on the aglycone part of the molecule, destroy the sugar unit(s), usually without leaving a useful trace. One solution to this problem is to prepare the permethylated derivatives which behave in a more restrained manner under electron bombardment and yield a useful molecular ion. Markham (1982) provides a table listing values for 6- or 8-mono-*C*-pentosylapigenin, 6- or 8-mono-*C*-rhamnoxylapigenin, up to 6,8-di-*C*-hexosylapigenin. By adding 30 mass units for each hydroxyl group or associated methinyl group, the formula can be applied to any *C*-glycosylflavone. Markham's (1982) book should be consulted for technical details and relevant literature citations.

Although use of permethylated and trimethylsilylated derivatives overcame the main problem, there are some attendant difficulties with using these derivatives including preparation time, incompleteness of the reactions, and possible rearrangements (Hedin and Phillips, 1991). A major advance in mass spectroscopic analysis of carbohydrate derivatives, of which flavonoid glycosides are but one example, has involved development of chemical ionization mass spectrometry (CI-MS). In this technique, the target molecule is bombarded in the presence of a carrier gas, such as ammonia, methane, or isobutane. Strong molecular ions are generally observed in the form of a "quasimolecular" ion $[M+1]^+$. Hedin and Phillips (1991) cited a simple example of the use of CI-MS in their study of apigenin 7-*O*-glucoside. In the presence of methane, fragments with $m/z=163$, 145, 127, and 108 mass units were observed that agreed with the pattern obtained from fragmentation of glucose itself. Along with a large number of mono-, di-, trisaccharides, and other types of sugar derivatives, their list of examples contained detailed information on CI-MS of esculetin 6-*O*-glucoside (a coumarin), apigenin 7-*O*-glucoside, hesperetin 7-*O*-rutinoside, and quercetin 3-*O*-rhamnoside. An example from the recent flavonoid literature involves the work of Liu and colleagues (1994) on flavonol glycosides from *Cephalocereus senilis*. CI-MS of one of the glycosides gave $[M+H]^+$ at m/z 741 (1%); $[M+H\text{-rhamnose}]^+$ at m/z 595 (5%); $[M+H\text{-rhamnose-galactose}]^+$ at m/z 433 100%); and the aglycone at m/z 287 $[M+H]^+$ (72%). A second glycoside consisting of three sugars, galactose, glucose, and rhamnose, gave, in the FAB-MS, a peak with m/z 902 for $[M]^+$ (35%) and one at m/z 901 $[M-H]^+$ (65%).

The term FAB, which stands for fast-atom bombardment, entered the discussion without prelude. This technique involves bombarding the target molecule with high energy atoms in the presence of a viscous medium such as glycerol. Xenon or argon are typically used as the projectile. This technique provides a softer disrupting beam and pseudomolecular ions are commonly produced. This procedure is discussed in detail by Watson (1985). A comparatively simple example of the use of FAB-MS comes from the study of flavone derivatives from *Perilla ocimoides* (Lamiaceae) (Yoshida *et al*., 1993). A pseudomolecular ion with $m/z=623$ was observed for a 7-*O*-diglucuronide of apigenin. A more complex example from the recent flavonoid literature comes from the work of Kijima and colleagues (1995) on flavonol glycosides of *Alangium premnifolium*. "Alangiflavoside" gave galactose, glucose, and two units of rhamnose on acid hydrolysis. Negative-ion high resolution fast-atom bombardment MS gave *m/z* 901.2580 for $[M-H]^-$; $C_{39}H_{49}O_{24}$ requires 901.2614. Alangiflavoside was acetylated and subjected to FAB-MS in the presence of *m*-nitrobenzylalcohol (same function as glycerol) to give *m/z* 1491 $[M+H]^+$ as expected for the tetradecaacetate.

Our last example of the use of FAB-MS involves a unique flavonoid derivative from *Pteridium aquilinum* (Imperato, 1996a). Acid hydrolysis afforded kaempferol, *p*-coumaric acid, apiose, the amino acid glycine, and some intermediates that gave additional glycine upon treatment with stronger acid. Alkaline hydrolysis gave similar results except that kaempferol 3-*O*-apioside was also obtained. The positive ion FAB-MS gave a pseudomolecular ion $[M+H]^+$ at *m/z* 964, along with a series of ions resulting from the sequential loss of six glycine units: 906, 849, 792, 735, 678, and 621. Other ions observed were those for kaempferol 3-*O*-(*p*-coumaroyl)apioside at *m/z* = 565; kaempferol 3-*O*-apioside at *m/z* = 419; heptaglycine at *m/z* = 401; *p*-coumaroylapioside-glycine-glycine at *m/z* = 393; kaempferol at *m/z* = 287; and apiose-glycine-glycine at *m/z* = 247.

OTHER ANALYTICAL METHODS

X-Ray Diffraction

Although X-ray diffraction analysis is not routinely used in flavonoid studies, it is a powerful method that has found application in a number of instances. The technique essentially allows the worker to "see" the molecule in question in 3-dimensional space and to glean from the observation a clear picture of the arrangement of its component atoms. In the case of flavonoids, X-ray diffraction can reveal such features as C-5 hydroxyl–carbonyl H-bonding, C-3 hydroxyl–carbonyl H-bonding, the conformation of the C-ring in flavanones and dihydroflavonols, and the orientation of the 2-phenyl group (or 3-phenyl in the case of isoflavones) in relation to the benzopyran system. A detailed picture of the molecule in space often allows an X-ray crystallographer to choose between alternative structures for an unknown flavonoid. An example of this application is shown next.

It has long been known that cryptostrobin and strobopinin constituted the isomeric pair 5,7-dihydroxy-8-*C*-methylflavanone [4-70] and 5,7-dihdyroxy-6-*C*-methylflavanone [4-72]. There has been less agreement, however, on which is which. Another possibility is that the presence of one as a natural product could lead to the other *via* a Wessely-Moser rearrangement. The problem was solved by the use of X-ray diffraction analysis of (-)-cryptostrobin, which had been isolated from *Agonis spathulata* and assigned the structure [4-70] by Cannon and Martin (1977). Byrne and coworkers (1982) prepared the 6-bromo derivative of (-)-cryptostrobin [4-71] and subjected a crystal of the product to X-ray analysis. The results clearly indicated that the *C*-methyl group was correctly positioned at C-8. Likewise, the stereochemistry at C-2 is *S* as originally suggested.

Ceratiola ericoides (Empetraceae), a member of the Florida shrub vegetation, produces an exudate that contains a dihydrochalcone derivative, "ceratiolin," as a major component. Tanrisever and coworkers (1987) determined that ceratiolin is a

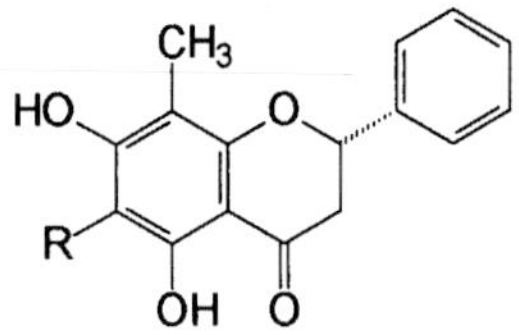

4-70 R = H
4-71 R = Br

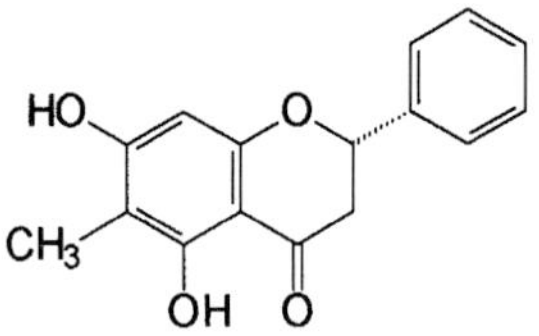

4-72

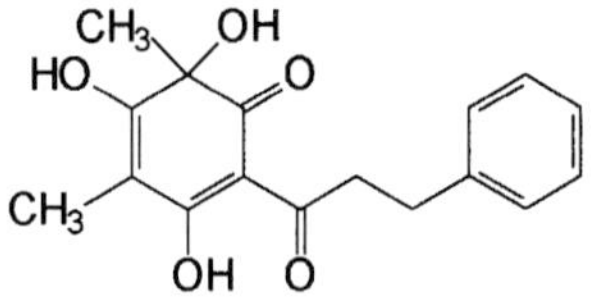

4-73

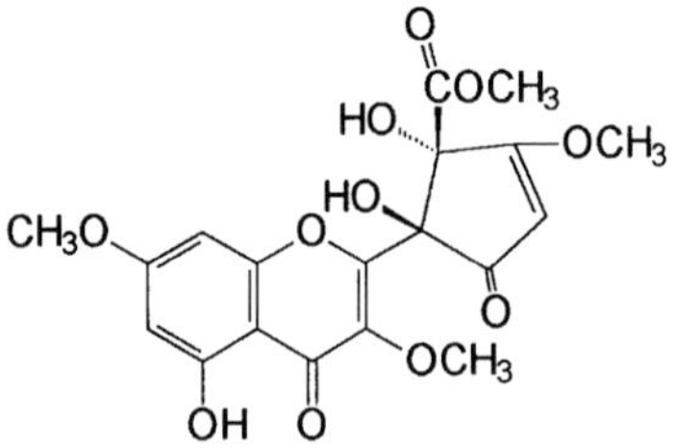

4-74

dihydrochalcone with a modified A-ring, 3-phenyl-1-(2,4,5-trihydroxy-3,5-dimethyl-6-oxo-1,5-cyclohexadienyl)-1-propanone [4-73]. Confirmation of this structure came from two sources, a total synthesis by Obara and coworkers (1989), and an X-ray diffraction analysis by Tak and coworkers (1993). From the X-ray data, it was confirmed that the A-ring was fully substituted and that it exists in a half-chair conformation with the carbonyl C atom 0.203(1) Angstrom unit out of the diene plane and the tetrahedral C atom 0.184(1) Angstrom unit to the opposite side of the plane. The hydroxyl group on the tetrahedral carbon forms an intramolecular hydrogen bond with the A-ring carbonyl oxygen while the hydrogen of the *o*-hydroxyl group forms an intramolecular hydrogen bond with the β-phenylpropionyl group.

A flavonoid with a highly modified B-ring was isolated from *Chrysosplenium grayanum* by Arisawa and coworkers (1992). The combination of ^{13}C and ^{1}H NMR, UV, IR, MS, and X-ray analysis led to the decision that the B-ring consisted of a cyclopentenyl system substituted with a carbonyl group, two hydroxy groups, a methoxy group, and a methoxycarbonyl function as shown in [4-74]. X-Ray data established the β-configuration of both the hydroxy group at C-1′ and the methoxycarbonyl function at C-5′. The X-ray diffraction data also indicated that "chrysograyanon" is a planar molecule in agreement with the conclusion based on 2D NMR data. X-Ray data have been useful in helping to establish the structures of other complex flavonoid derivatives among which are silidianin, one of the antihepatotoxic components of *Silybum marianum* (Abraham *et al*., 1970), and a proanthocyanidin derivative that exhibited two linkages between the component units (Crespo-Irizar *et al*., 1992).

X-Ray diffraction studies of simple model compounds have provided a good deal of information on the orientation of flavonoid molecules in space. A study of 3-chloroflavanone (Tomlin and Cantrell, 1990) showed that the pyran ring is distorted into a C(2)-sofa configuration and that the 2-phenyl group is oriented equatorially and the chlorine atom at C-3 is oriented axially. Owing to steric hindrance between the phenyl and chloro groups, the B-ring is twisted out of the plane of the pyrone ring by 37°. In 5-hydroxyflavone (Shoja, 1990), 6-hydroxyflavone and 2′-hydroxyflavone the benzopyran ring systems and the B-rings are nearly coplanar (Seetharaman and Rajan, 1992, 1995). Wallet and coworkers (1989, 1990) have shown that while 2′-methoxyflavone is very nearly planar, 2′,6′-dimethoxyflavone is not, as might have been expected. The crystal structures and hydrogen-bonding properties of 3-hydroxyflavone, 2′-methyl-3-hydroxyflavone, and 3′-methyl-3-hydroxyflavone were studied by Etter and colleagues (1986) who showed that all three compounds had five-membered rings arising from the C-3 hydroxyl and the carbonyl oxygen. 3-Hydroxyflavone showed the greatest planarity of the three while the other two compounds showed significant deviations from planarity.

Conformational analyses of 3,5,7,2′,4′-pentahydroxyflavone (morin) and 3,5,7,3′,4′,5′-hexahydroxyflavone (myricetin), as their triphenylphosphine oxide complexes, was carried out by Cody and Luft (1994) in an effort to gain insight into the hydrogen-bonding capacities of bioactive flavonoids. The interested reader will find citations in that paper to work involving interactions of morin and myricetin with a number of enzyme systems, including HIV-1 protease, lipoxygenase, and iodothyronine deodinase. Another application of X-ray analysis of flavonoid interactions with biological molecules can be found in a paper in which Wallet and coworkers (1993) describe a study of the interaction between 2′,6′-dimethoxyflavone and *o*-phosphoric acid as a model for flavone-nucleotide interactions.

Resonance Raman Spectroscopy (RR)

Resonance Raman spectroscopy provides a means by which molecules can be examined *in situ* with only minimal harm being done to the cell. In a Raman experiment, a laser is employed to generate a monochromatic beam of light which is focused on the sample. The resulting scattered light is collected and focused on a monochromator where individual spectral lines are detected and an appropriate spectrum produced. The result is a vibrational spectrum that reflects the characteristic absorption of energy by different functional groups within the target molecule. The method has had comparatively little use in the study of flavonoids except in the case of anthocyanins whose chemical structure in the cell has been the subject of intense scrutiny for many years as exemplified by two works by Merlin and associates (Merlin *et al*., 1985, 1994). Initial stages in the study involved analysis of known compounds in aqueous solution under a variety of conditions in order to establish the spectral characters of the pigments. The more recent report described the groups' results with *Vitis vinifera* cv. Pinot noir and *Malva sylvestris* (Merlin *et al*., 1994). Results from this study showed that the main pigment inside Pinot noir berries is malvidin 3-*O*-glucoside whereas in the skin cells it was present in the quinonoidal base form. In the upper epidermis of the common mallow (the *Malva* sp.) only malvidin 3,5-di-*O*-glucoside could be detected and that it existed solely in the flavylium ion form.

CASE STUDIES

The case studies below have been selected from the literature to exemplify methods that have been successfully used for the elucidation of structures of members of several classes of flavonoids. Some of the examples involve primarily degradation and use spectroscopic analysis minimally, such as seen with the *Citrus* flavanone glycosides. Others make extensive use of spectroscopic methods with only minimal chemical manipulation, as seen with sigmoidin I. In the case of brickellin, new data allowed an older structural hypothesis to be rejected. Chemical syntheses have been described in some cases that confirm the structure of compounds. Many hundreds of papers have been published in which most of the methods described here have been used, and it is to that literature that a worker in the field must ultimately go for specific details. Leading references to that literature have been cited in the pages above.

Case Study No. 1. *Citrus* Flavanones

The work featured in this study represents a comparatively straightforward, yet excellent, example of the use of classical degradation reactions to establish the structure of the aglycone and the sugar components of a well known flavonoid glycoside. The information presented below is taken from two excellent reviews to which the reader is referred for specific information and detailed references to the literature (Horowitz, 1964; Horowitz and Gentili, 1969). The compound in question is hesperidin [4-75], known from several *Citrus* species, e.g., *C. sinensis* (the sweet orange), *C. aurantium* (the Seville or bitter orange), and *C. limon* (the lemon). Acid hydrolysis afforded one equivalent each of D-glucose, L-rhamnose, and the aglycone, hesperetin [4-76]. Spectral data and alkaline cleavage that yielded phloroglucinol derivatives [4-77] from the A-ring and isovanillic acid (3-hydroxy-4-methoxybenzoic acid) [4-78] from the B-ring led to the structure 5,7,3′-trihydroxy-4′-methoxyflavanone for hesperetin. This structure was confirmed by synthesis by several groups. Carefully controlled acid hydrolysis of hesperidin, in order to avoid ring opening, afforded optically active (-)-hesperetin. This in turn was subjected to ozonolysis which yielded L-malic acid indicating that the aglycone must have the 2*S*-configuration. Total methylation of hesperidin followed by acid hydrolysis gave the chalcone [4-79] which lacks *O*-methylation at the C-4′ position. This position is equivalent to the 7-position in the flavanone which indicates that hesperidin consists of the aglycone bound to a disaccharide through the 7-oxygen. Examination of the diglycoside obtained by careful hydrolysis of hesperidin showed that it was identical to rutinose

At the time of the hesperidin work, it was known that some of the diglycosides from *Citrus* species were rutinosides, 6-*O*-α-L-rhamnosyl-β-D-glucosylpyranosides. It was assumed that naringin, the intensely bitter glycoside from *C. paradisi* (grapefruit), was the naringenin 7-*O*-rutinoside (naringenin had been identified as 5,7,4′-trihydroxyflavanone). Other intensely bitter glycosides were also known from *Citrus*, namely poncirin and neohesperidin. The aglycone of poncirin had been

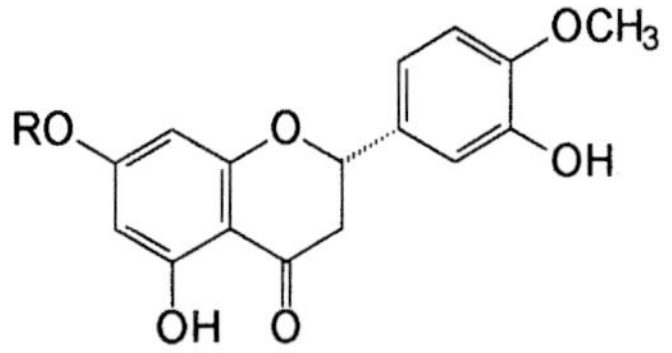

4-75 R = Neohesperidose (Neo)
4-76 R = H

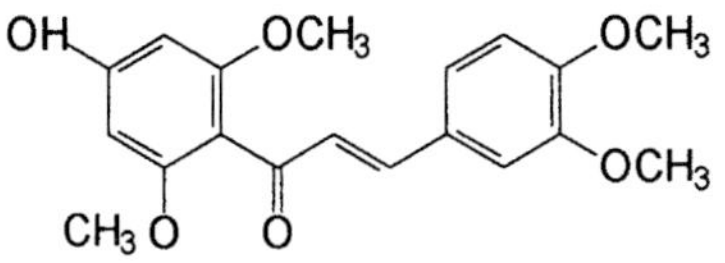

4-79

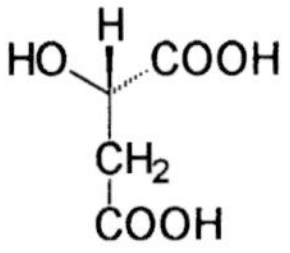

L-Malic acid

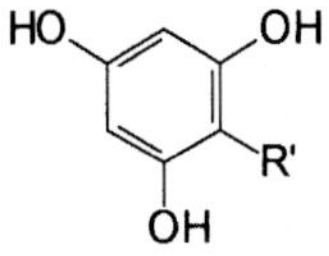

4-77

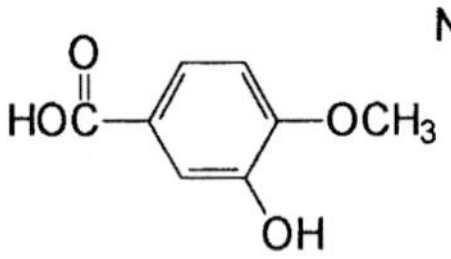

4-78

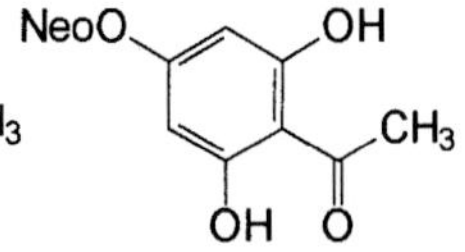

4-80

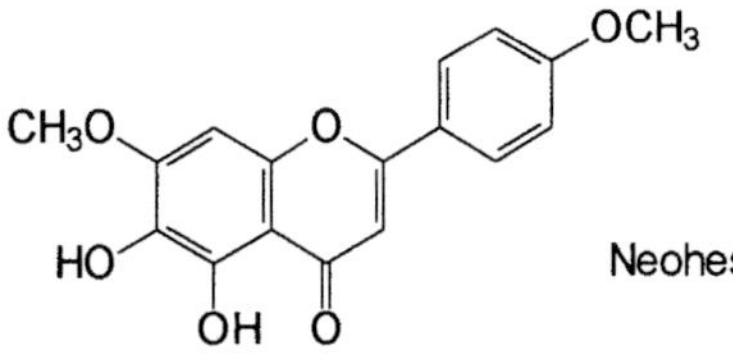

4-81

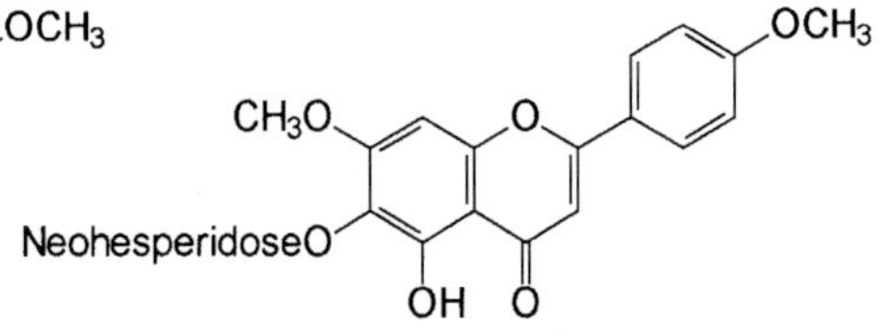

4-84

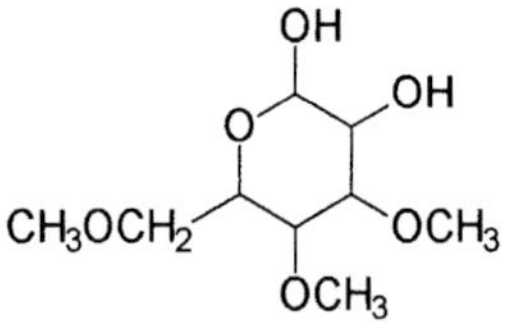

4-82

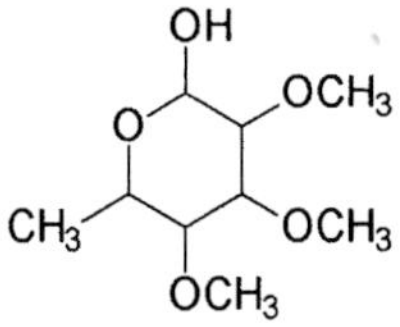

4-83

identified as naringenin 7-methyl ether (isosakuranetin), while that of neohesperidin was the same as hesperidin. It was also known that the disaccharide obtained from these compounds, called neohesperidose, was an isomer of rutinose. Work by Horowitz and Gentili (1969) showed that naringin, poncirin, and neohesperidin all yielded the same A-ring fragment, phloroacetophenone 4′-*O*-neohesperidoside [4-80] upon alkaline cleavage. This meant that all of the bitter glycosides of *Citrus*, including neoeriocitrin, an eriodictyol derivative, were characterized by the presence of neohesperidose at the 7-position. This result was of immediate significance because bitterness correlated with the presence of neohesperidose, while the rutinosides of the same flavanones were tasteless. The problem of accounting for the bitterness of some *Citrus* compounds and the tastelessness of others had been recognized for some time. Determination of the structure of neohesperidose was the obvious next step. Exhaustive methylation of naringin followed by acid hydrolysis yielded 2,3,4-tri-*O*-methyl-L-rhamnopyranose [4-83] and 3,4,6-tri-*O*-methyl-*β*-D-glucopyranose [4-82]. This means that rhamnose is connected to glucose *via* a (1 → 2) linkage. Therefore, neohesperidose is 2-*O*-α-L-rhamnopyranosyl-D-glucose. The stereochemistry was confirmed through synthesis from 1,3,4,6-tetra-*O*-acetyl-α-D-glucose and 2,3,4-tri-*O*-benzoyl-α-L-rhamnopyranosyl bromide, followed by ester exchange to yield neohesperidose *β*-heptaacetate. The heptaacetate was identical in all respects to the acetylated disaccharide obtained from naringin by ozonolysis and acetylation. The bitter flavanone glycosides of *Citrus* species are thus the 7-*O*-neohesperidosides of the corresponding aglycones. 2(*S*)-Stereochemistry has been established for all of them. The effect of disaccharide structure on the taste of these compounds and their rutinoside analogues will be discussed in Chapter Eight.

Case Study No. 2. A Flavone Neohesperidoside

Case Study No. 1 provided an example of the steps followed to determine glycoside structures, notably neohesperidosides, before the availability of proton magnetic resonance spectroscopy. This next study gives the reader a chance to compare the classical approach with currently available methods. The target compound in this example is a flavone neohesperidoside from *Gelonium multiflorum* (Euphorbiaceae) (Das and Chakravarty, 1993). The UV spectrum of the unknown, with maxima at 278 nm and 330 nm, suggested that the compound was a flavone. Acid hydrolysis yielded glucose, rhamnose, and an aglycone that gave $[M]^+$ $m/z = 314$, which corresponds to $C_{17}H_{14}O_6$. The UV spectrum of the aglycone in the presence of $AlCl_3$ and HCl indicated a 5,6-dihydroxyflavone. This was confirmed by preparing a diacetate using acetic anhydride-pyridine and a monomethyl ether using diazomethane (the 5-OH group is not methylated with diazomethane because of hydrogen bonding with the carbonyl group). These data led to the conclusion that the aglycone was 5,6-dihydroxy-7,4′-dimethoxyflavone (scutellarein 7,4′-dimethyl ether) [4-81].

Since the unknown glycoside gave a shift with $AlCl_3$ indicating the presence of a 5-OH group, the glycosidic linkage must be through the oxygen at position-6. This was confirmed by examination of the ^{13}C NMR spectrum of the glycoside which showed an up-field shift of 1.9 ppm for the C-6 resonance and down-field shifts of 6.4, 4.3, and 2.3 ppm for the C-5, C-7, and C-9 carbons, respectively. This is in full agreement with the expected down-field shifts of carbons located *ortho* and *para* to the position of attachment of the sugar. The ^{13}C chemical shift for the glucose anomeric carbon was shielded by about 4 ppm while the C-2 carbon of glucose was deshielded by about 3.5 ppm compared to methyl-*β*-D-glucopyranoside, which indicated that the rhamnose is linked through the glucose C-2 oxygen. The α-linkage of the rhamnose and *β*-linkage of the glucose were indicated by the multiplicities of the appropriate proton signals in the ^1H NMR spectrum of the diglycoside.

The ^{13}C NMR spectrum of the diglycoside agreed well with published data for α-L-rhamnopyranosyl(1 → 2)-β-D-glucopyranoside (neohesperidose). Final confirmation of the inter-saccharide linkage was obtained by permethylation of the flavone glycoside, hydrolysis of the product, and identification of the fragments as 3,4,6-tri-*O*-methyl-D-glucose [4-82] and 2,3,4-tri-*O*-methyl-L-rhamnose [4-83], which is in complete agreement with the observations described above for the *Citrus* glycosides. The structure of the diglycoside is, thus, [4-84].

Case Study No. 3. 8,2′-Dihydroxyflavone

This simple flavone is included among the case studies because of its highly unusual substitution pattern. Although flavonoids with oxygenation at C-8 are comparatively common in some plant groups (certain groups within Asteraceae), as are those with a C-2′ hydroxyl group (common in Lamiaceae), the presence of these as the only substituents on the flavonoid skeleton is highly unusual [4-85]. Discovery of this compound in a species of *Primula*, however, is not surprising in light of earlier studies of the genus. Previous studies of *P. pulverulenta* had revealed the presence of several unusual flavonoids: flavone itself; 5-hydroxyflavone; 2′-hydroxyflavone; 2′-methoxyflavone; 5,8-dihydroxyflavone; 5,2′-dihydroxyflavone; 5-hydroxy-2′-methoxyflavone; 2′,5′-dihydroxyflavone and its natural 5′-acetate; 2′,4-dihydroxychalcone; and 5,8,2′-trihydroxyflavone (see Wollenweber *et al.*, 1988 for citations). The new compound emerged from a more recent study of leaf exudate of this species by those workers. Data upon which the structure was deduced are described below.

The electron impact mass spectrum of the new compound gave $m/z = 254$ suggesting a flavone with two hydroxyl groups. *O*-Methylation gave a compound with $m/z = 282$ in agreement with the presence of two available hydroxyls. Mass spectral fragmentation of the flavone gave an $[A_1^+]$ fragment with $m/z = 136$ [4-86], an $[A_1 + H]^+$ fragment with $m/z = 137$ [4-86 plus proton], and a $[B_1^+]$ Fragment with $m/z = 118$ [4-87]. These fragments indicate that each ring has one of the hydroxyl groups. This conclusion was supported by mass spectral analysis of the dimethyl ether derivative as well. The UV spectrum of the flavone had maxima at 264 and 330 nm, which is within the range of a flavone, but not definitive with regard to placing both hydroxyl groups. It was clear, however, that C-5 lacked substitution as indicated by the lack of any change of the original spectrum in the presence of aluminum chloride. The spectrum of the original compound was not affected by the addition of sodium acetate, which rules out C-7 as the A-ring hydroxyl site. Addition of sodium methoxide solution caused a bathochromic shift of 63 nm in band 1, which is considered evidence for hydroxylation at C-2′ rather than at C-4′ (Tanaka *et al.*, 1986). The NMR spectrum, determined at 270 MHz using a mixture of deuterated dimethylsulfoxide and deuterated acetone (2 : 1) as solvent, exhibited a broad doublet at 7.51 ppm characteristic of a C-5 proton *ortho* coupled to a proton at C-6 ($J =$ 7.69 Hz). Further splitting indicated additional, *meta*, coupling. These observations can only be accommodated by placing the A-ring hydroxyl group at C-8.

Recognizing the lack of model compounds against which their spectral data could be compared, these workers undertook a synthesis of 8,2′-dihydroxyflavone.

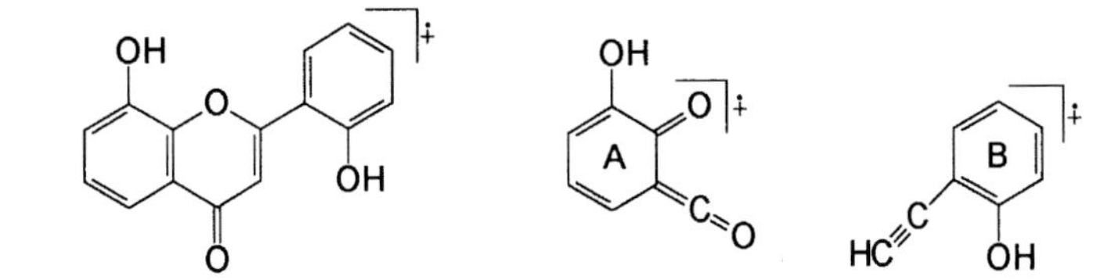

4-85 m/z 254 $C_{15}H_{10}O_4$ 4-86 4-87

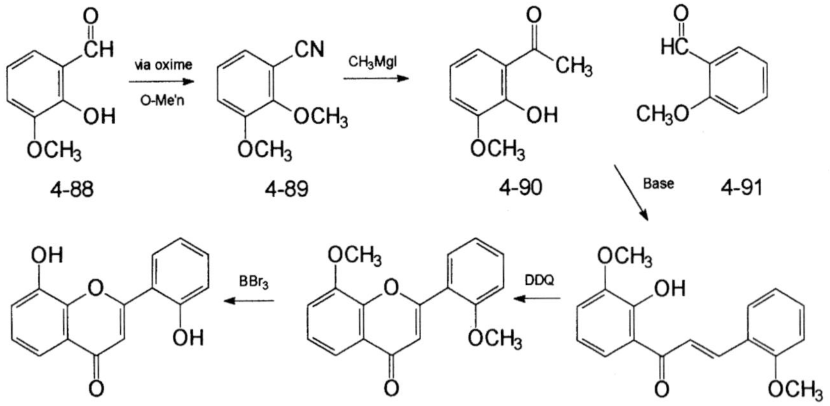

4-88 4-89 4-90 4-91

4-94 4-93 4-92

This was accomplished by conversion of 2-hydroxy-3-methoxybenzaldehyde (*o*-vanillin) [4-88] to 2-hydroxy-3-methoxycyanobenzene through the action of hydroxylamine hydrochloride and sodium formate. The product was then methylated to give 2,3-dimethoxycyanobenzene [4-89]. Conversion of the nitrile to 2-hydroxy-3-methoxyacetophenone [4-90] was accomplished using the methyl Grignard reagent (from CH_3I and Mg). Condensation of the acetophenone with 2-methoxybenzaldehyde [4-91] in the presence of base afforded 2′-hydroxy-4′,4-dimethoxychalcone [4-92] which, in turn, was oxidatively cyclized to 8,2′-dimethoxyflavone [4-93] through the action of DDQ. Demethylation was accomplished with boron tribromide (BBr_3) to yield a product that was identical in all respects to the natural product [4-94].

Case Study No. 4. A Chalcone from *Clerodendron*

This example is based upon the work of Roy and Pandey (1994) who described two known flavones, 7-hydroxyflavone and 5,7-dihydroxy-6,4′-dimethoxyflavone (pectolinarigenin), a known flavanone, 7-hydroxyflavanone 7-*O*-glucoside, and a new chalcone glycoside from aerial parts of *Clerodendron phlomis* (Verbenaceae). Although the example is a very straightforward one, it nevertheless shows the usefulness of integrated spectroscopic analyses. It also represents a situation wherein the aglycone of the original unknown is converted to an isomeric form and it is with the latter that some of the analytical work is done.

The unknown glycoside, which crystallized as yellow granules, gave the typical UV spectrum of a chalcone with Band I at 364 nm with a lesser peak at 245 nm. The IR spectrum had, among others, a band indicating the presence of a hydrogen bonded carbonyl group. FAB-MS of the glycoside gave a molecular ion with $m/z = 610$, as well as significant fragments with $m/z = 166$ and 120, to which we will return in due course. The $m/z = 610$ ion requires the presence of two sugar units and since only glucose was recovered after acid hydrolysis, the unknown must be reckoned to be a diglucoside. Acetylation of the glycoside afforded a nona-acetate. Since a diglucoside would account for eight acetate groups there must be an additional hydroxyl group available on the phenolic portion of the glycoside. ^{13}C NMR of the glycoside showed signals at δ 101.1 and 101.7 assignable to two anomeric carbons of the glucose groups. ^{1}H NMR also showed two anomeric protons, at δ 5.1 and δ 5.3, each of which appeared as doublets with $J = 3$ Hz. This splitting was taken as evidence of α-glycosidic linkages (Note. Roy and Pandey show β-linkages in their structures 1 and 4, however).

Hydrolysis of the chalcone glucoside afforded an aglycone whose UV spectrum was typical for a flavanone. The aglycone had an elemental formula represented by $C_{16}H_{14}O_5$, exhibited a methoxyl group in the ^{1}H NMR (δ 3.72), and formed a diacetate. Other resonances in the ^{1}H NMR spectrum were typical of a 4′-substituted B-ring (δ 7.25 and 6.80, d, $J = 9$ Hz), a 5,7-disubstituted A-ring (δ 5.83 and 6.02, d, $J = 2.2$ Hz), and the arrangement of the C-2 and C-3 protons expected for a flavanone (δ 2.70 mult., integrating for two protons and one proton at δ 5.25, dd, $J = 5$ Hz). The melting points of the flavanone and its diacetate were in agreement with published values for naringenin 5-methyl ether [4-98]. Mass spectral analysis of the flavanone gave fragments of $m/z = 166$ [4-96] and $m/z = 120$ [4-97] in agreement with a flavanone having one hydroxyl group and one methoxyl group on the A-ring and one hydroxyl group on the B-ring. Taking this structure back to the chalcone form, we see that 2′,4′,4-trihydroxy-6′-methoxychalcone accounts for all observations [the aglycone of 4-95].

The final step in the analysis involves placing the glucose units. UV data allowed placing one glucose on each ring as indicated by two observation. First, addition of sodium methoxide resulted in a bathochromic shift of Band I with no increase in intensity indicating that the 4-hydroxyl was not free. The second observation was the lack of a bathochromic shift with sodium acetate showing that the 4′-position was also occupied. The authors concluded that the new compound was 2′,4′,4-trihydroxy-6′-methoxychalcone 4′,4-di-*O*-α-glucoside [4-95]. The presence of two α-linked glucoside units in this compound represents a comparatively rare finding in plants. Owing to this unusual feature, it would seem prudent to look at this aspect again.

Case Study No. 5. A *C*-methylaurone from *Cyperus*

The unknown compound in this study, from underground organs of *Cyperus capitatus* (Seabra *et al.*, 1995) had UV maxima at 270 and 398 nm, the latter shifting to 446 nm in the presence of sodium methoxide and to 432 nm with $AlCl_3$. Addition of HCl to the $AlCl_3$ mixture yielded an absorption identical to that of the original

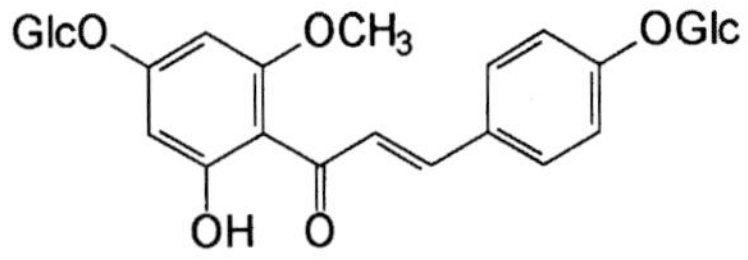

m/z = 286 $C_{16}H_{14}O_5$ (aglycone)

4-95

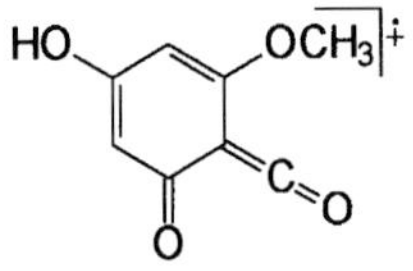

4-96

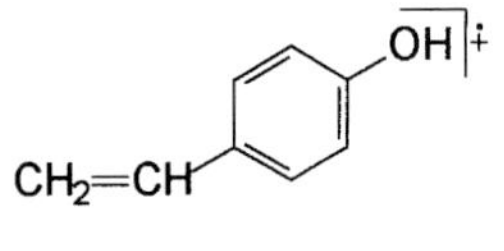

4-97

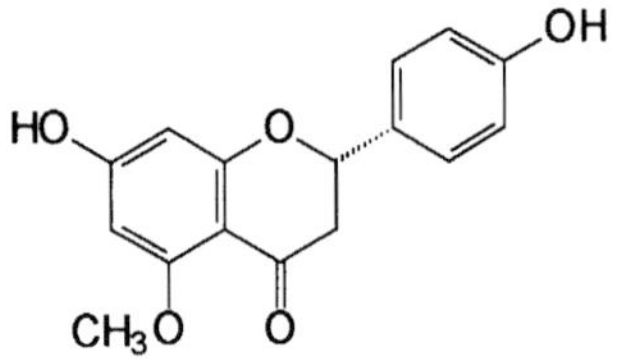

4-98

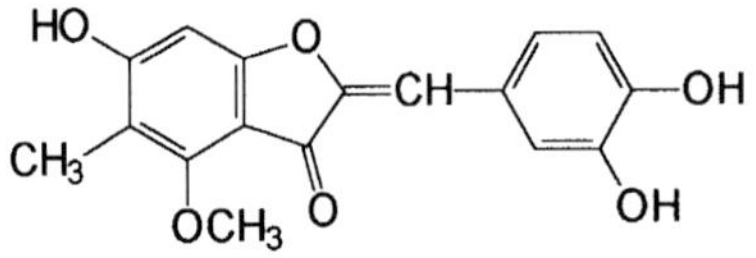

m/z = 314 $C_{17}H_{14}O_6$

4-99

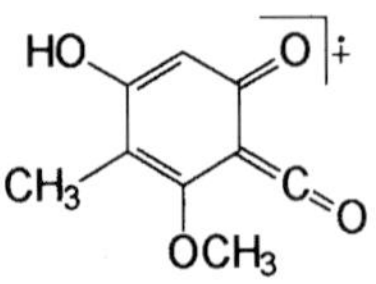

4-100

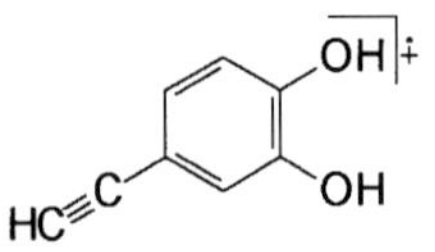

4-101

methanol solution. The compound gave a bathochromic shift of both maxima with sodium acetate and the reaction with added boric acid indicated the presence of a 3′,4′-dihydroxy B-ring. The spectral data, along with color on chromatograms and reaction with Naturstoffreagenz A, suggested that the compound was an aurone with a substituted oxygen at position-4. These suggestions were supported by NMR data.

The ^{1}H NMR spectrum showed the typical B-ring catechol-type resonance pattern with one proton at δ 7.42 (d, J=2 Hz), one proton at δ 7.19 (dd, J=6.9 Hz and 2 Hz), and one proton at δ 6.82 (d, J=6.9 Hz) representing the protons at C-2′, C-6′, and C-5′, respectively. A resonance at δ 4.03 (3H) was attributed to the methoxyl methyl group, and a resonance at δ 1.9 (3H) was attributed to an aromatic *C*-methyl group. The benzylic and C-7 protons appeared together as a singlet at δ 6.54. ^{13}C NMR data confirmed the existence of the 4,6-dioxygenation pattern in the A-ring. The *C*-methyl group was located at position-5 on the basis of comparison with literature values for 4-*O*-methyl-5-*C*-methyl sulfuretin rather than for the 4-*O*-methyl-7-*C*-methyl isomer. The δ 61.43 resonance for OCH_3 is characteristic of a sterically crowded methoxyl near a methyl group which, in the present case, is only possible if the methoxy group is at C-4 and the *C*-methyl group is at C-5. The unknown was assigned structure [4-99]. Electron impact MS gave $[M]^+$ m/z=314 (equivalent to $C_{17}H_{14}O_6$) and fragments with m/z=180 and 181 representing $[A_1]^+$ [4-100] and $[A_1+H]^+$, respectively, and m/z=134 attributable to $[B_1]^+$ [4-101].

Case Study No. 6. Sigmoidin-I

Sigmoidin-I is one of four compounds isolated from the roots of *Erythrina sigmoidea* by Nkengfack and coworkers (1994c). Three of the compounds were identified as the known isoflavonoids neoisobavaflavone, and corylin, both isoflavones, and the pterocarpan phaseollidin. High resolution MS of sigmoidin I gave a value of 354.1463 indicating $C_{21}H_{22}O_5$ (requires 354.1467). The unknown was optically active, $[\alpha]_D$ at $22° = -2.9°$, and gave positive color tests with $FeCl_3$, indicating the presence of a free hydroxyl group, and with Mg-HCl indicating the possibility of a flavanone. The UV spectrum, with a maximum at 276 nm and a lesser peak at 310 nm, further supported the possibility that the unknown had a saturated C-ring. Acetylation yielded a diacetate with $m/z = 438$ that did not give a $FeCl_3$ color test. Treatment of the unknown with acid brought about the formation of a new compound having the same elemental analysis. Both the ^1H NMR and MS analyses provided vital information. The proton NMR showed the AMX system typical of the C-2 and C-3 protons of an isoflavanone. Heterocyclic ring resonances were noted at δ 4.30 (dd, $J = 11.0$, 5.4 Hz) and δ 4.47 (dd, $J = 11.2$ Hz) for the C-2 protons and δ 4.02 (dd, $J = 11.5$, 5.4 Hz) for the C-3 proton. HETCOR experiments showed correlation of these protons with the C-2 and C-3 carbons at δ 70.9 and δ 47.7, respectively, confirming the existence of the isoflavanone structure. The carbonyl carbon appeared at δ 193.1.

A three-proton singlet at δ 3.58 indicated the presence of a methoxy group, while a prenyl group was indicated by the typical array of nine protons. The two *C*-methyl protons appeared at δ 1.55 and δ 1.60 (both singlets integrating for three protons), the methylene carbon appeared at δ 3.10 as a two proton doublet ($J = 7.2$ Hz) and the methine proton appeared as a triplet at δ 5.15 ($J = 7.2$ Hz). The conversion of the *C*-prenyl group to a *gem*-dimethylchromene in the presence of acid, as mentioned above, is in line with the prenyl function lying *ortho* to one of the phenolic hydroxyl groups.

The next step in the elucidation of this compound's structure involved the assignment of the aromatic protons. A doublet at δ 7.70 ($J = 8.7$ Hz) was assigned to a proton at C-5. The *J* value was typical of an *ortho*-coupled proton so position-6 must also bear a proton. The C-6 proton at δ 6.40 was a double doublet that showed both *ortho*-coupling ($J = 8.7$ Hz) and *meta*-coupling ($J = 2.2$ Hz). *meta*-Coupling is only possible if there is a proton at C-8. The resonance at δ 6.27, which appears as a doublet, shows the expected *meta*-coupling ($J = 2.3$ Hz). These data account for protons at C-5, C-6, and C-8. A bathochromic shift of 12 nm of Band II of sigmoidin-I in the presence of sodium acetate serves to locate one of the compound's hydroxyl groups at C-7, as shown in the partial structure [4-102].

With the A-ring substitution pattern solved, attention can now be turned to the B-ring, which we know must accommodate the methoxyl group, the prenyl group, and the remaining hydroxyl group. We also know that the latter two functions must be placed on adjacent carbons owing to the compound's facile cyclization to the chromene. Two proton resonances assignable to B-ring protons appeared as singlets at δ 6.32 and δ 6.64. The absence of splitting indicates that these protons lie *para* to

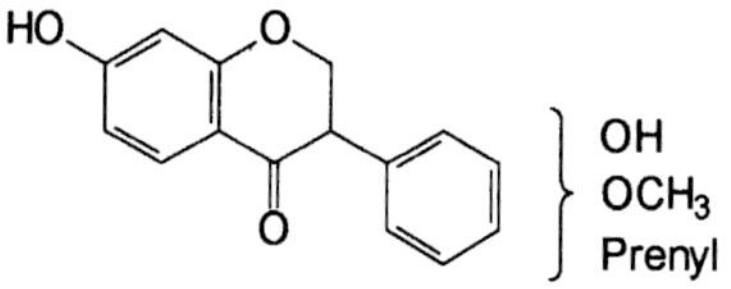

4-102

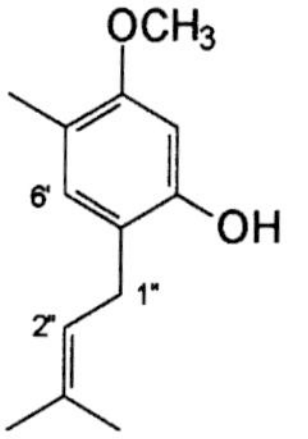

4-103

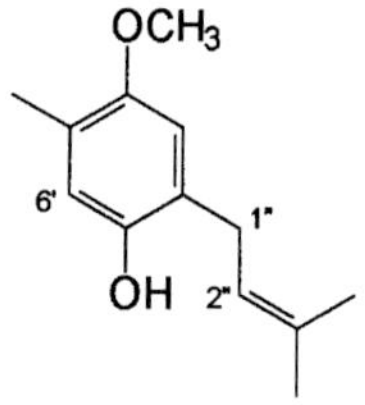

4-104

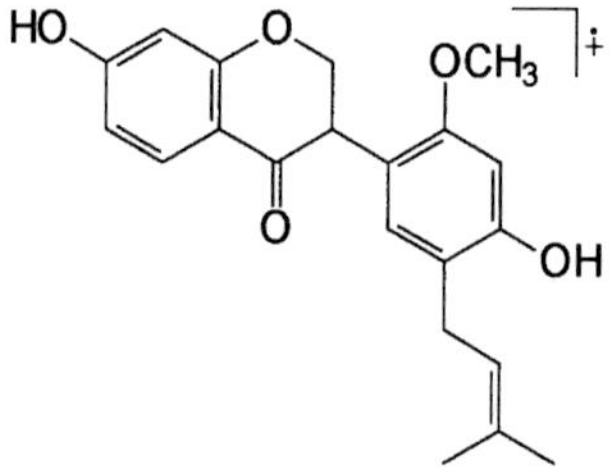

4-105 m/z = 354

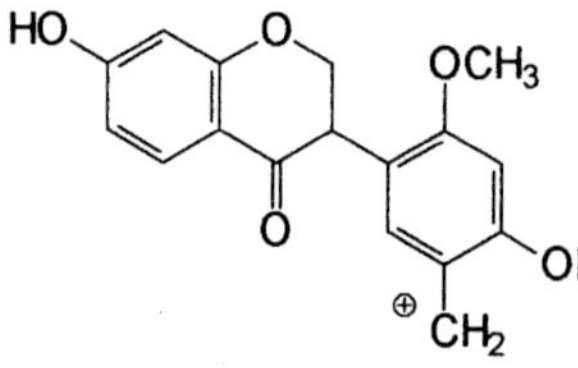

4-106 m/z = 299

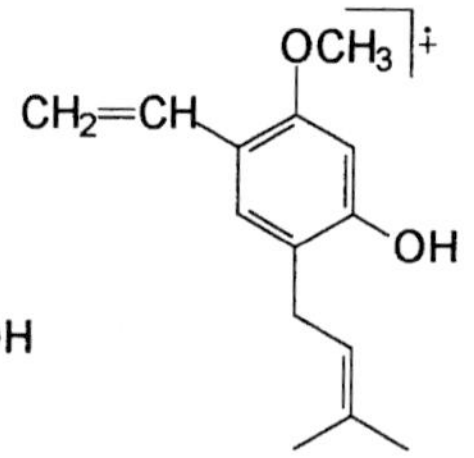

4-107 m/z = 218

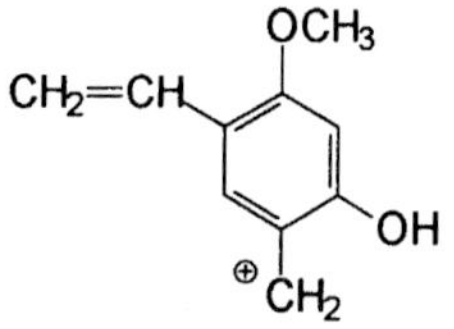

4-108 m/z = 163

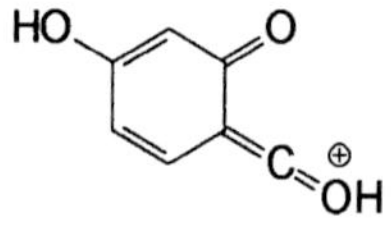

4-109 m/z = 137

one another, which can only be accommodated by protons at positions-2′ and 5′ (or the equivalent 3′ and 6′ positions following the numbering in the original paper). Since the hydroxyl and prenyl groups are on adjacent carbons, the methoxyl group must be located at C-6′. The two possible arrangements are shown below as [4-103] and [4-104]. On biosynthetic grounds, structure [4-103], with normal 4′-oxygenation, would be preferred but enough flavonoids with 2′,5′-dioxygenation are known that more rigorous chemical evidence is required to establish the correct substitution pattern. Such evidence came from studies of coupling between C-1″ and C-2″ of the prenyl function and the C-6′ proton. Interactions between those carbons and the C-6′ proton are possible only if the prenyl function is located adjacent to C-2′ as is the case in [4-103], hence the unknown must have structure [4-105, but without the charge!].

Data from electron impact MS were in agreement with the structure arrived at above. Major fragments observed, along with intensity values, were the molecular ion at m/z 354 (52%), m/z 299 (6%) [4-106], m/z 218 (91%) [4-107], m/z 203 (10%) [4-107 minus methyl?], m/z 163 (100%) [4-108], and m/z 137 (17%) [4-109]. The m/z 137 fragment agrees with the location of one of the hydroxyl groups on the A-ring, while the fragment with m/z 299 represents the loss of the C_4H_7 characteristic of

C-prenyl flavonoids. Fission of sigmoidin-I into A-ring fragment, $m/z = 137$, and the B-ring unit, $m/z = 218$, occurs by the normal retro Diels-Alder route. Note that the *m/z* 218 and 163 fragments are incorrectly drawn in the original paper.

Case Study No. 7. 5-Deoxyflavonoids from *Parkia*

In pursuit of the compound or compounds responsible for the antimolluscicidal activity of *Parkia clappertoniana* (Fabaceae), Lemmich and coworkers (1996) identified two 5-deoxyflavonoids. One of them, 7-hydroxy-3,8,4′-trimethoxyflavone [4-110], offered little challenge and will not be discussed here other than to say that the some of its structural features, in particular the absence of oxygen at C-5 and the presence of the 8-methoxy group suggested the possibility of similar features in the unknown. The UV spectrum of the unknown compound had maxima at 249 and 346 nm suggestive of a flavonol. Addition of $AlCl_3$ resulted in no change in maxima from which it was possible to conclude that the compound did indeed lack a free hydroxyl group at C-3 and/or C-5. A free hydroxyl group in the B-ring was indicated by a bathochromic shift in Band I in the presence of sodium methoxide. Sodium acetate had no effect on the spectrum.

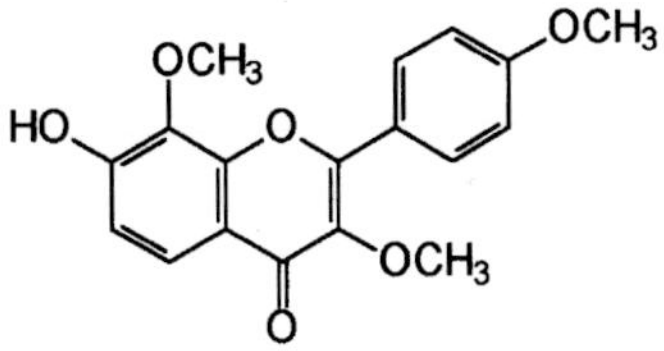

4-110

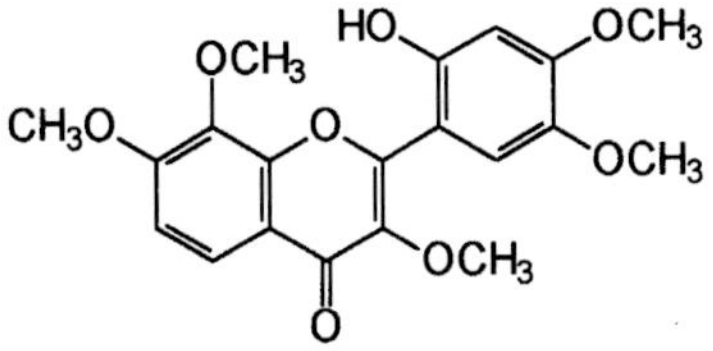

4-111

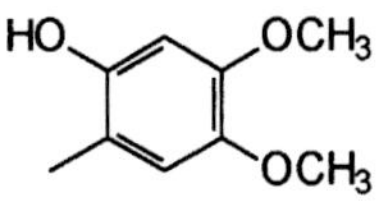

4-112

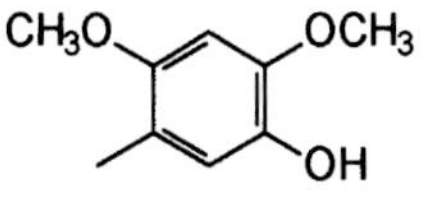

4-113

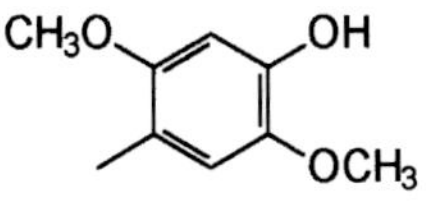

4-114

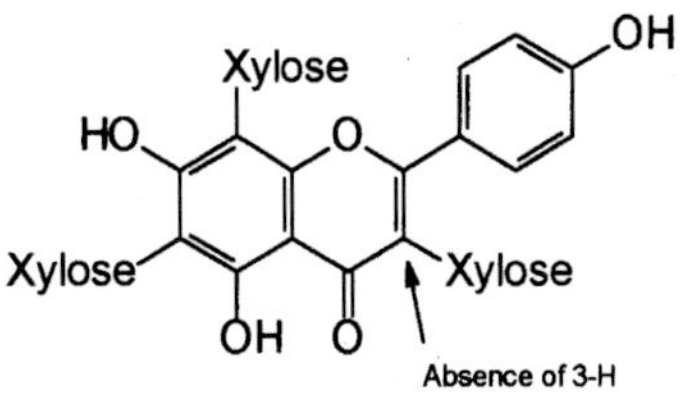

4-115

Mass spectral fragmentation gave $[M]^+$ with $m/z=388$ and $[M\text{-}15]^+$ (100%) at $m/z=375$. The 388 mass would arise from a flavonoid with the elemental formula $C_{20}H_{20}O_8$, which could be accommodated by a flavone with one hydroxyl and five methoxyl groups. Both the ^1H and ^{13}C NMR spectra showed resonances for five *O*-methyl groups. The presence of a free phenolic group was confirmed by the observed disappearance of a resonance at δ 8.20 in the presence of D_2O. The remaining signals in the proton spectrum consisted of a pair of doublets at δ 7.08 and δ 8.01 ($J=8.0\,Hz$), and a pair of singlets at δ 6.64 and δ 7.32 each of which integrated for a single proton. The first pair shows *ortho*-coupling while the second pair shows no coupling, which is characteristic of protons that lie *para* to one another. Returning to the MS study we can begin to get some idea of the disposition of the methoxy groups by noting that the base peak, $m/z=373$, represents the molecular ion minus a methyl group. Since its intensity is greater than the molecular ion itself it is reasonable to place a methoxy group at C-8. (Recall that a methoxy group at C-6 would also gives an intense peak but one that is generally not as intense as the molecular ion.) Loss of 43 mass units ($m/z=345$, 17%) placed a methoxy group at C-3. Major scission of the molecule afforded a fragment with $m/z=181$, which could have been either the $[A_1+H]^+$ or B_2^+ fragment. In either case, the parent molecule must have had two methoxy groups in the A-ring, and with one at C-3, we are left with locating the points of attachment of the other two on the B-ring. The presence of *para* protons can be accommodated by three arrangements of one hydroxyl and two methoxyl groups, [4-112, 4-113, and 4-114]. Structure [4-114] was eliminated on UV spectral data. Two pieces of evidence point to [4-112] as the correct orientation. Loss of 17 mass units (m/z 371, 34%) is characteristic of 2′-hydroxyflavonoids. Irradiation of the phenolic proton at δ 8.02 led to a greater effect on the signal at δ 6.64 than on the signal at δ 7.32. The δ 6.64 signal arises from the proton on C-3′, hence the unsubstituted phenolic function must lie at C-2′. This accounts for both B-ring protons and, with *ortho*-coupling observed for the A-ring protons, the structure of the compound must be 2′-hydroxy-3,7,8,4′,5′-pentamethoxyflavone [4-111].

Case Study No. 8. A C-glycosylflavone

C-Glycosylflavones offer an interesting challenge in that both the identity of the constituent sugar(s) and the position(s) of attachment must be determined for a molecule that is essentially acid resistant. Some rather drastic methods have been used for the analysis of *C*-glycosylflavones the cost of which is often the loss of large parts of the molecule. In view of such drastic measures, the usefulness of NMR methods can hardly be exaggerated. Both approaches were used in the present study. Imperato (1993) obtained a compound from the fern *Asplenium viviparum* that had the UV spectral characteristics of a flavone with unsubstituted hydroxyl groups at C-5, C-7, and C-4′. Despite the compound's mobility in aqueous chromatographic media, suggesting that it was at least a diglycoside, acid hydrolysis yielded no sugar. Neither did acid hydrolysis yield any isomerized product. Therefore, the compound was judged to be a 6,8-diglycosylflavone with the same sugar at both

C-6 and C-8 (lack of discernible Wessely-Moser rearrangement products). Treatment of the compound with hot hydriotic acid (HI) yielded apigenin (sugars were destroyed) while oxidation with $FeCl_3$ gave xylose as the only sugar (aglycone was destroyed).

Mass spectral analysis at this point should have confirmed the existence of 6,8-di-*C*-xylosylapigenin, but this was not to be. The FAB spectrum gave $[M+H]^+$ at $m/z=667$ indicating that the molecular weight of the unknown is 666, equivalent to $C_{30}H_{34}O_{17}$, or one xylose unit larger than expected for a di-*C*-xyloside. The ^{1}H NMR spectrum showed the presence of three anomeric protons, at δ 4.98, δ 4.82, and δ 4.63, along with 15 sugar protons confirming the presence of three xylose units. There were no A-ring proton signals in the spectrum, but there was a typical pattern for a 4′-substituted B-ring. Significantly, the spectrum lacked a signal for a C-3 proton indicating that the third xylose unit was linked at that position. ^{13}C NMR spectral data confirmed this in that the signals for C-3, C-6, and C-8 all showed down-field shifts typical of flavonoid ring carbons involved in *C*-glycosyl linkages. Although di-*C*-glycosylflavones are very common in plants, this represented the first report of a naturally occurring tri-*C*-glycoside [4-115].

Case Study No. 9. *O*-Methylated Flavonols from *Blumea*, Corrections

Four flavonols were described as new natural products from *Blumea malcomii*, a weedy species of India, by Kulkarni and coworkers (1987). The compounds were described as 6-hydroxy-3,5,7,4′-tetramethoxyflavone [4-116]; 6,2′,5′-trihydroxy-3,5,7-trimethoxyflavone [4-117]; 6,5′-dihydroxy-3,5,7,2′-tetramethoxyflavone [4-118]; and 6-hydroxy-3,5,7,2′,5′-pentamethoxyflavone [4-119] on the basis of spectroscopic measurements. These new structures attracted the particular attention of Markham (1989) who recognized two unusual features of the compounds, namely, the presence of 5-*O*-methylation in all of them and the presence of the highly unusual 2′,5′-oxygenation pattern in three of them. His re-examination of the original data showed several problems that led to reassignment of structures for all four compounds. Permethylation of [4-116] had produced a penta-*O*-methyl flavonoid identical to totally *O*-methylated penduletin, 3,5,6,7,4′-pentamethoxyflavone [4-120], confirming the oxygenation pattern of the original material. The product of permethylation exhibited a blue fluorescence characteristic of 3,5-di-*O*-methyl flavonoids, whereas the original compound appeared dark brown under UV. This latter behavior is characteristic of flavonoids with an hydroxyl group at C-5. An unsubstituted 5-OH is also indicated by the UV absorption maximum in the presence of $AlCl_3$, i.e., a bathochromic shift of band-I of ca. 15 nm. In addition to indicating that the 5-OH is unsubstituted, this shift is characteristic of unsubstituted 5-OH groups that have an adjacent 6-OCH_3 function. The presence of a significant $[M-15]^+$ fragment in the EIMS spectrum also argues for a methoxy function at C-6. The corrected structure is, therefore, 5-hydroxy-3,6,7,4′-tetramethoxyflavone [4-120 without the methyl group at C-5].

The remaining three compounds were convertible to the same permethyl ether [4-121], which exhibited blue fluorescence under UV. As pointed out above, this

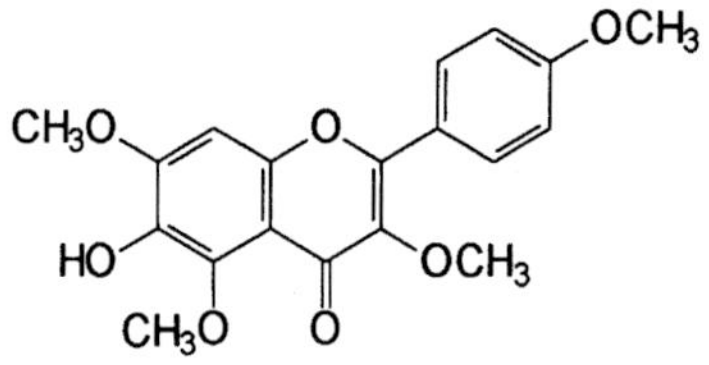

4-116

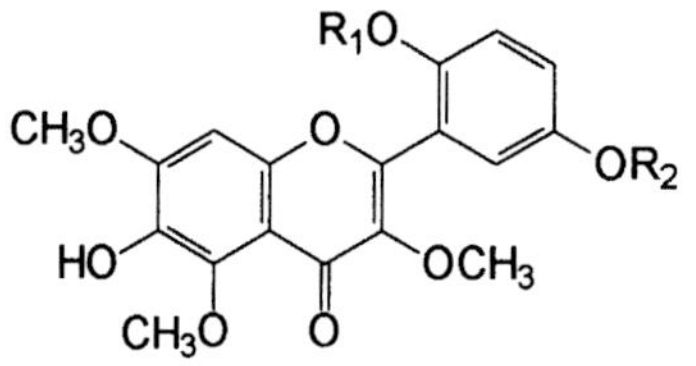

4-117 $R_1 = R_2 = H$

4-118 $R_1 = CH_3$, $R_2 = H$

4-119 $R_1 = R_2 = CH_3$

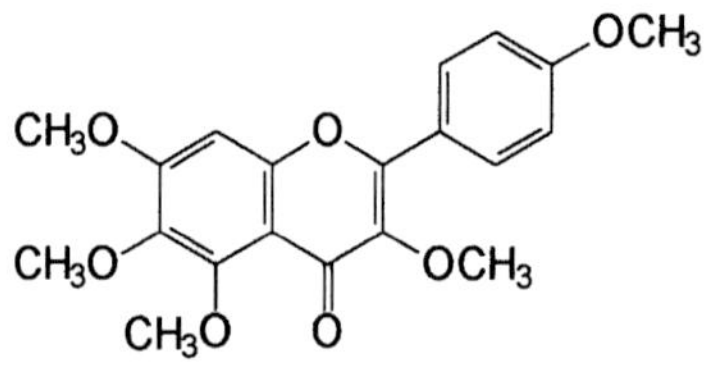

4-120

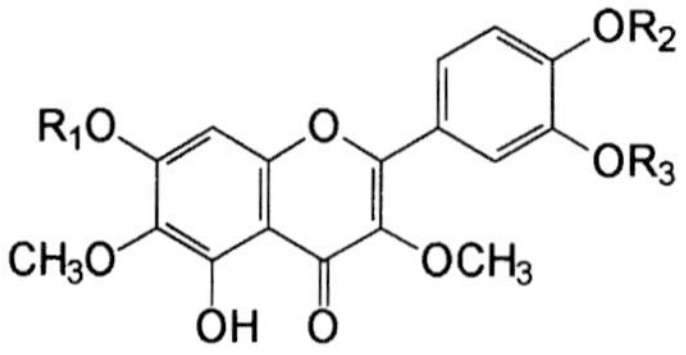

4-121

4-122 $R_1 = CH_3$, $R_2 = R_3 = H$

4-123 $R_1 = Glc$, $R_2 = H$, $R_3 = CH_3$

4-124 $R_1 = R_2 = R_3 = CH_3$

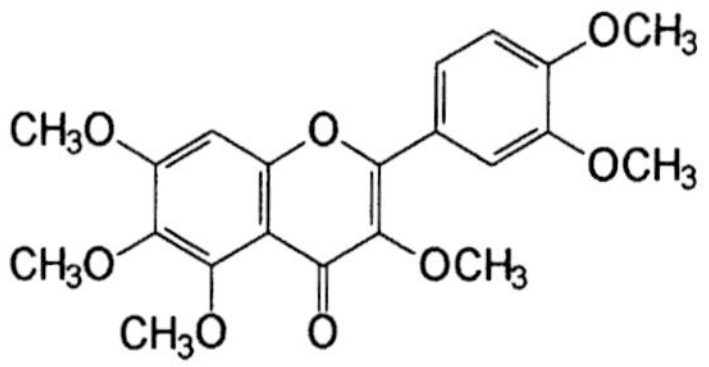

behavior is characteristic of 3,5-di-*O*-methyl flavonoids. Once again, the three compounds were described as being dark under UV indicative of the presence of the 5-OH function. The methanolic UV spectra of all three compounds responded to $AlCl_3$ as would be expected for compounds bearing the 5-hydroxy-6-methoxy arrangement, not the reverse as claimed by the original workers. Other spectral information indicates that all three compounds have 5-hydroxy-6,7-dimethoxy A-rings. The UV spectra of the three compounds were very similar to those of three well known flavonol derivatives, quercetagetin-3,6,7-trimethylether [4-122], quercetagetin-3,6,3′-trimethylether 7-*O*-glucoside [4-123], and quercetagetin-3,6,7,3′,4′-pentamethyl ether [4-124], respectively. Further information came from the ^1H NMR spectral data. Quercetagetin derivatives exhibit two proton signals for H-2′ and H-6′ at lower and one for H-3′ at a higher field position, which is the situation evident in the spectra of the supposed 2′,5′-dioxygenated flavonols. However, known 2′,5′-dioxygenated flavonoids exhibit the reverse of this, one proton (H-6′) resonating at a lower field position and two (H-3′ and H-4′) resonating at higher positions. Comparison of the spectra of the permethyl ethers of the three compounds with quercetagetin hexamethyl ether showed nearly perfect matches in all cases. In addition, if the unknowns had had either a 2′-hydroxy or a 2′-methoxy function, the EIMS spectrum would have shown significant fragments at $[M\text{-}17]^+$ or $[M\text{-}31]^+$, respectively. Neither of these fragments was observed.

Case Study No. 10. Brickellin

This case study involves reinvestigation of the structure of a flavonoid that originally drew attention because of its unusual 2′,3′,4′,6′-tetrasubstituted B-ring. The molecule in question, named "brickellin," was obtained originally from two species of *Brickellia*, *B. veronicaefolia* and *B. chlorolepis*. The background surrounding this problem was briefly reviewed by Roberts and coworkers (1984) who also put forward, admittedly tentatively, 5,6′-dihydroxy-6,7,2′,3′,4′-pentamethoxyflavone [4-125], as the compound's structure. Unlike the case with the *Blumea* compounds, where ample data

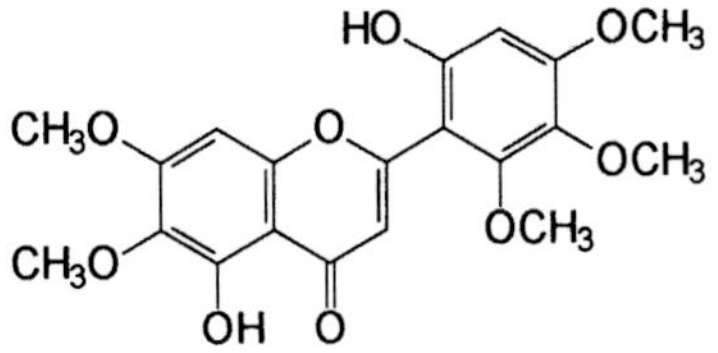

4-125

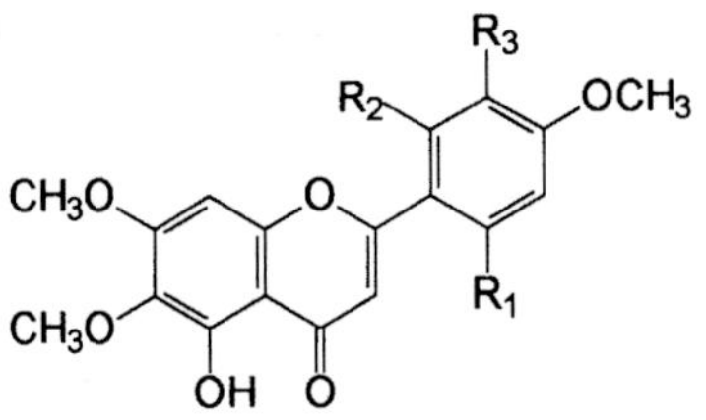

4-126 $R_1 = R_3 = CH_3$, $R_2 = OH$

4-127 $R_1 = R_2 = CH_3$, $R_3 = OH$

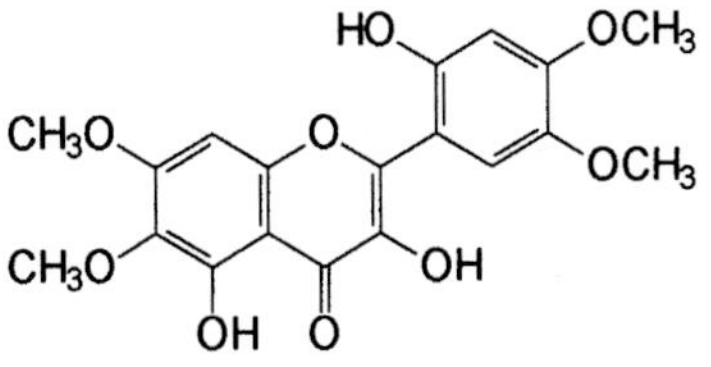

4-128

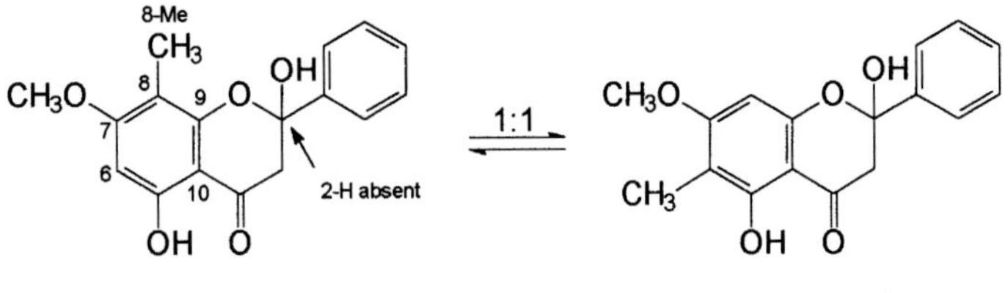

4-129

4-130

were simply misinterpreted, the situation with respect to brickellin was the apparent scarcity of data. One of the difficulties surrounding this molecule was the very simple ^{1}H NMR spectrum consisting of three uncoupled skeletal proton signals, five OCH_3 signals, and two phenolic hydroxyl proton signals. This arrangement was consistent with the $[M]^+$ at *m*/*z* 404, corresponding to $C_{20}H_{20}O_9$. Color under UV and the presence of a signal in the ^{1}H NMR spectrum at δ 13.05 for a phenolic proton hydrogen bonded to a carbonyl placed one of the hydroxyl groups at C-5. A prominent $[M\text{-}15]^{+}$ peak was taken as evidence for a 6-methoxy function with a second methoxy function located at C-7 as indicated by the lack of any change in the UV spectrum in the presence of sodium acetate. An aromatic signal at δ 6.45 was assigned to H-8 rounding out the A-ring. This left one hydroxyl group, three methoxyl groups and two protons to be accounted for. Fragments in the EIMS at $[M\text{-}17]^{+}$ and $[M\text{-}31]^{+}$ pointed to the existence of both 2′-OH and 6′-OCH_3. Irradiation of the 4′-methoxyl group resulted in an NOE enhancement of the H-5′ signal, irradiation of the 7-methoxyl group gave an enhancement of the H-8 signal, and radiation of the 2′-methoxyl caused an enhancement of the 3-H signal. An anomalous UV spectrum, where Band-I was only about one-third the intensity of Band-II was interpreted as arising from the substituents at C-2′ and C-6′ interfering with coplanarity of the flavone system. The $AlCl_3$/HCl spectrum, with its 32 nm bathochromic shift of Band-I, was also unusual; the shift for 6-methoxyflavonoids is usually in the range of 18 nm.

Owing to the scarcity of flavonoids with four B-ring substituents and some of the puzzling spectroscopic features of brickellin, it was considered desirable to synthesize some B-ring tetrasubstituted flavonoids by unequivocal means so that their spectral properties could be examined (Iinuma *et al.*, 1985). The three isomeric compounds [4-125], [4-126], and [4-127] were synthesized and their UV, ^{1}H NMR, ^{13}C NMR, and MS spectral characteristics measured and shown *not* to match those of brickellin. An alternative structure [4-128] was suggested that was consistent with the spectroscopic data. Since the A-ring was not in question, it was necessary to see what effect moving the 6′-methoxyl group to C-3 would have. The two B-ring protons lie *para* to one another in [4-128] and would show no coupling, and the 2′-OH group would still give the observed $[M\text{-}17]^{+}$ fragment. This compound was synthesized and shown to be identical in all respects to brickellin from natural sources. Details of this synthetic route will be discussed in the following chapter.

Case Study No. 11. A 2-hydroxyflavanone

Although only occasionally reported as natural products (see Chapter Two), 2-hydroxyflavanones occupy an interesting position in flavonoid chemistry because of the suggestion that they may be involved in flavone biosynthesis in a fashion akin to the involvement of 2-hydroxyisoflavanones in the formation of isoflavones (Chapter Six). It is instructive to examine a recent report of a new 2-hydroxyflavanone isolated from *Friesodielsia enghiana* (Moraceae) by Fleischer and coworkers (1997). The new compound was isolated in a nearly 1:1 mixture of a 6/8-*C*-methyl isomer pair [4-129 and 4-130]. Mass spectral analysis provided a molecular weight value of 300 which accords to $C_{17}H_{16}O_5$ as the elemental formula. The UV spectrum was that of a flavanone with the IR spectrum indicating the presence of a hydrogen bonded carbonyl function. The presence of a 5-hydroxyl group was confirmed by the appearance of a signal at δ 12.64 in the ^{1}H NMR. Three protons at δ 7.43 and two at δ 7.68 were interpreted as arising from an unsubstituted

B-ring. A peak in the mass spectrum at $m/z = 77$ lent support to this feature. A single signal at δ 6.02 indicated that only one position on the A-ring lacked substitution. A signal at δ 3.16 indicated the presence of a methoxyl function and a singlet at δ 2.28 that integrated for three protons pointed toward an aryl *C*-methyl group. These groups were conclusively located by studying long range correlations (HMBC) that showed coupling of H-6 with C-8 and C-10, and 8-CH_3 with C-7 and C-9. The substitution pattern of the A-ring was thus established as 5-hydroxy-7-methoxy-8-*C*-methyl. Flavanones normally exhibit a double doublet in the vicinity of δ 5.20 arising from coupling of the C-2 proton with the axial and equatorial protons on C-3. The absence of a C-2 proton signal indicated that that position was occupied instead by an hydroxyl function. A signal at δ 101.8 in the ^{13}C NMR spectrum, indicative of a doubly oxygenated sp^3 system, confirmed the assignment [4-129]. Additional support for locating a carbon of that sort at C-2 came from its interaction with protons on C-2′ and C-6′. The isomeric compound was shown to have the *C*-methyl group at C-6 through study of long range couplings.

Case Study No. 12. Acylated Anthocyanin

One of the more difficult flavonoid structural problems to deal with using classical methods would be a molecule glycosylated in two or more positions that also had an acyl function on one of the sugar units. The situation is even more challenging when different sugars are involved. Now add an additional sugar attached to the acylating acid and the problems are of such magnitude that one must ask whether the efforts might outweigh the significance of the results. No such defeatist attitudes are necessary with the availability of current NMR methods, however! As the reader may recall from our discussions of anthocyanidin glycosides in Chapter Two, such compounds do indeed exist. The example chosen for this study is a comparatively straightforward one involving a cyanidin derivative from the Japanese morning glory *Pharbitis nil* whose structure was determined by Saito and coworkers (1993). These workers combined results from total and partial acid hydrolyses with data from a variety of NMR methods to arrive at the complete structure with only 10 mg. of purified unknown.

Total hydrolysis yielded cyanidin, glucose, and caffeic acid. Partial hydrolysis yielded four products: cyanidin 3-*O*-glucoside [4-131], cyanidin 3-*O*-sophoroside (2-glucosylglucoside) [4-132], cyanidin 3,5-di-*O*-glucoside [4-133], and cyanidin 3-*O*-sophoroside-5-*O*-glucoside [4-134]. Deacylation of the unknown gave caffeic acid glucoside and cyanidin 3-*O*-sophoroside-5-*O*-glucoside. The FAB-MS of the unknown gave a value of 1098 for the molecular weight, corresponding to $C_{48}H_{56}O_{29}$, which translates to one equivalent of cyanidin, four of glucose, and one of caffeic acid. Four anomeric proton signals in the ^1H NMR spectrum indicated that the four glucoside bonds have the β-orientation. The structure of the molecule determined by these workers is given as [4-135] with the glucose units designated as A for the 3-proximal glucose, B for the 5-*O*-glucose unit, C for the distal glucose at position-3, and D for the glucose attached to the caffeic acid residue. The C-6 protons of three of glucose units gave signals in the normal, or unsubstituted, range

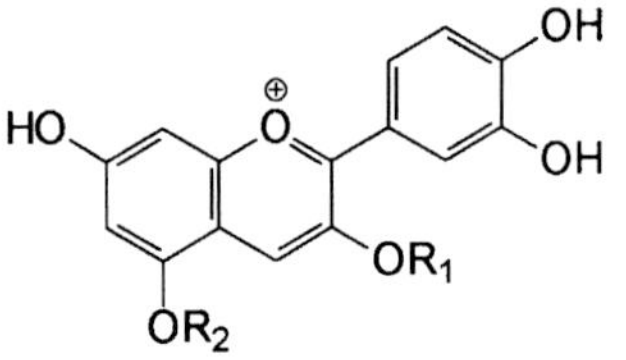

4-131 R_1 = Glc, R_2 = H
4-132 R_1 = Soph, R_2 = H
4-133 R_1 = R_2 = Glc
4-134 R_1 = Soph, R_2 = Glc

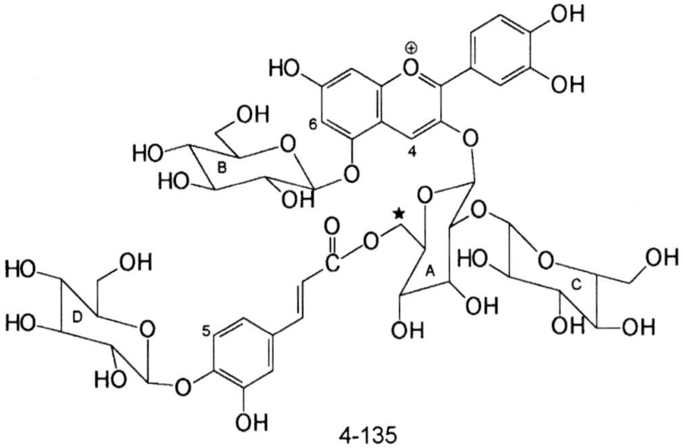

4-135

δ 3.01-3.78, while one gave signals of δ 4.32 and δ 4.39 indicating that the latter was the point of attachment of the caffeic acid unit. A ^1H-^1H COSY spectrum indicated an interaction of the anomeric proton of glucose A with the protons on the acylated sugar methylene group, which is only possible if the caffeic acid group is attached at the C-6 position of glucose A , marked with a star in [4-135]. Other interactions of importance include one between the glucose A anomeric proton and the cyanidin 4-proton, which supports locating that sugar at C-3. An interaction between the glucose B anomeric proton and the cyanidin H-6 proton showed that glucose unit B is linked through the 5-oxygen. Glucose C was shown to be linked through the C-2 oxygen of glucose A. Finally, the glucose D anomeric proton was shown to interact with the C-5 proton of caffeic acid leading to the conclusion that glucose D was linked to the caffeic acid unit via position-4. The structure of the pigment was thus established as cyanidin 3-*O*-(2-*O*-(β-D-glucopyranosyl)-6-*O*-(*trans*-4-*O*-(β-D-glucopyranosyl)caffeyl)-β-D-glucopyranosyl)-5-*O*-(β-D-glucopyranoside). Despite the apparent ease of this analysis, it should be borne in mind that the workers had considerable experience with anthocyanins of this sort and, moreover, had a library of standard compounds that allowed them to define the nature of the problem at the outset.

Case Study No. 13. A Biflavonoid

Three biflavonoids were isolated from *Dysoxylum lenticellare* (Meliaceae) by He and coworkers (1996). Two of them were identified as the known bilobetin [4-136] and

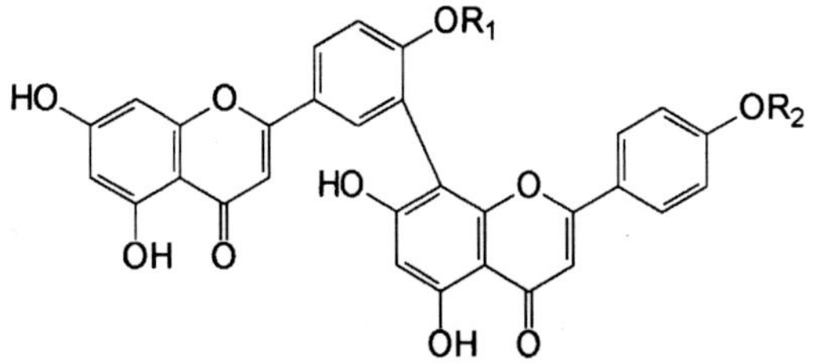

4-136 $R_1 = CH_3$, $R_2 = H$

4-137 $R_1 = R_2 = CH_3$

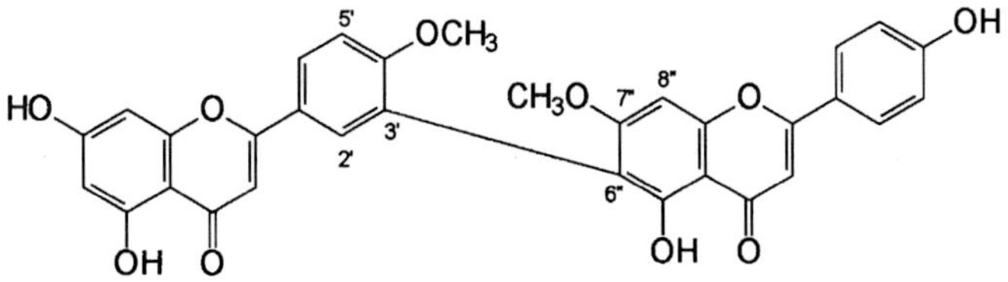

4-138

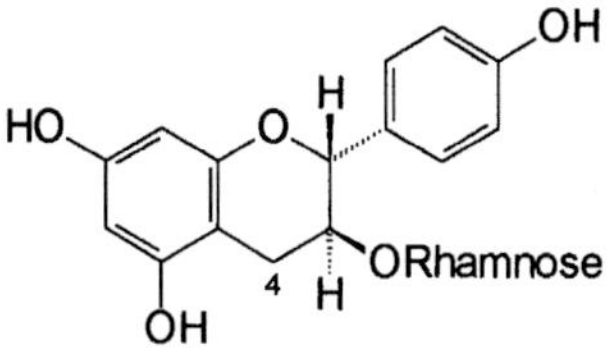

4-139

isoginkgetin [4-137], which are the 4′-methyl ether and 4′,4‴-dimethyl ether, respectively, of amentoflavone [apigenin(3′ → 8″)apigenin]. The structure of the third compound was determined in the following way. The size of the unknown was determined from its high resolution MS value of 566.1221 which indicated an elemental composition of $C_{32}H_{22}O_{10}$ (calc. 566.1206). This corresponds to a biflavonoid with two *O*-methyl groups. The UV spectrum exhibited maxima at 331 nm and 265 nm as would be expected for flavone derivatives. Sodium acetate brought about a 6 nm bathochromic shift of band II suggesting a free 7 or 7″ hydroxyl group. Methoxide caused a bathochromic shift of 51 nm suggesting that the 4 or 4‴-positions were unsubstituted. 3′,4′-Dihydroxylation was discounted since there was no reaction toward borate. The $AlCl_3$ spectrum was characteristic of a flavone with an unsubstituted hydroxyl group at C-5.

The ^1H NMR spectrum of the unknown showed signals at δ 7.83 (d, $J = 2.5$ Hz), δ 8.06 (dd, $J = 8.8$, 2.5 Hz), and δ 7.22 (d, $J = 8.8$ Hz), which were attributed to the protons on C-2′ (*meta*-coupled), C-6′ (*ortho*- and *meta*-coupled), and C-5′ (*ortho*-coupled), respectively. The appearance of only three B-ring protons in this set pointed to C-3′ as one of the points of linkage between the two units. *meta*-Coupled protons with δ 6.18 and δ 6.47 were assigned to positions C-6 and C-8. A second set of B-ring protons was evident with signals at δ 7.99 (d, J=8.9 Hz) and δ 6.94 (d, J=8.9 Hz), which were assigned to the protons on C-2‴ and C-6‴ and C-3‴ and C-5‴, respectively. These results exclude C-6, C-8, C-3‴, C-5‴, C-2‴, and C-6‴ as possible linkage points between the rings. The resonance at δ 6.94, arising from a single and uncoupled proton, was assigned to C-8″. The NOESY (NOE correlation) spectrum showed a cross peak at δ 3.80 indicating an interaction between H-8″ and the 7″-OCH_3 group. The interflavone linkage was, thus, shown to be between C-3′ and C-6″. Additional evidence to support this conclusion came from the ^{13}C NMR spectrum where the C-6″ resonance appeared at δ 109.1 indicative of substitution at that position. Further interactions were observed in long-range carbon–proton couplings between H-2′ and C-6″, H-8″ and C-7″, and 5″-OH and C-6″ all of which are in line with a C-3′-C-6″ linkage. These interactions can be appreciated better by referring to structure [4-138].

Other relevant information from the ^{13}C NMR spectrum included the observation of two carbonyl resonances in the vicinity of δ 182, which excluded C-5 and C-5′ as positions for methylation. (The presence of 5-*O*-methylation would have shifted the carbonyl resonance to about δ 177). Other positions excluded for locating the methoxyl groups were 6, 8, 2‴, 6‴, 3‴, and 5‴ since the δ values for all of these agreed closely with the corresponding carbons in robustaflavone [apigenin(3′→6″) apigenin] which would be the parent (unmethylated) compound. This leaves only positions 4′ and 7″ available for the two *O*-methyl groups. Further support for locating methoxy groups those positions came from the observation that only C-5′ and C-8″ exhibited up-field shifts (5.0 and 2.8 ppm, respectively) as expected for carbons adjacent to carbons bearing methoxy groups. The NOESY spectrum exhibited cross peaks between H-8″ at δ 6.94 and 7″-OCH_3 at δ 3.80, as well as between H-5′ at δ 7.22 and 4′-OCH_3 at δ 3.78. This study represents an excellent example of the application of NMR techniques to an unknown compound whose structure determination by classical methods of degradation would have been very demanding of time and effort. The interested reader would do well to study the original paper in detail. A recent addition to the study of biflavonoids employed proton-detected C-H correlation (Geiger *et al.*, 1993).

Case Study No. 14. A Flavan-3-ol Glycoside

This case study involves a straightforward analysis of a rare flavonoid. Whereas flavan-3-ols are commonly occurring constituents of plants, e.g., the catechin family of compounds, and are the building blocks of proanthocyanidins (condensed tannins), a report of one as a naturally occurring glycoside is noteworthy. Such was the case in the description of (+)-afzelechin 3-*O*-α-L-rhamnopyranoside [4-139] as a constituent of the bark of *Cassipourea gerrardii* (Rhizophoraceae) (Drewes *et al.*, 1992). The compound exhibited a major peak in the UV at 276 nm and was optically active. Enzymatic hydrolysis with hesperidinase yielded afzelechin, which was also optically active, and rhamnose, identified by TLC using an authentic specimen. The substitution patterns of the aromatic rings were clearly established from the ^{1}H NMR data: the A-ring spectrum consisted of two *meta*-coupled protons (δ 5.94 and δ 6.09, $J=2.29$ Hz), while the B-ring pattern consisted of a pair of doublets (δ 7.28 and δ 6.88, $J=8.57$ Hz). The protons on C-2, C-3, and C-4 were characteristic of *trans*-oriented substituents. Thus, H-2 appeared as a doublet at δ 4.74 ($J=7.85$ Hz), H-3 appeared as a multiplet (triple doublet) at δ 4.02 ($J=7.85$, 5.67, and 8.03 Hz), with the C-4 equatorial proton at δ 2.67 (d, $J=16.27$, 8.33 Hz) and the C-4 axial proton at δ 2.92 (dd, $J=16.27$, 5.75 Hz).

The identification of rhamnose was confirmed through its typical anomeric signal at δ 4.32 (d, $J=1.56$ Hz), its methyl signal at δ =1.26 (d, $J=6.2$ Hz), and the ^{13}C NMR signals for the remaining ring carbons. The α-stereochemistry of the glycosidic link was established by noting that the C-3 and C-5 signals of rhamnose appeared at δ 72.5 and δ 69.4, respectively, whereas signals at δ 75.4 and 73.5 would have been expected had the β-anomer been involved.

Although this example might be considered analytical over-kill by some, several interesting and potentially very useful observations were made from a delayed homonuclear COSY study of the compound (using $J=5$ Hz): (1) There was weak but distinct coupling between the phenolic protons of C-5 and C-7 while that at C-4′ was unaffected. (2) Distinct coupling between the C-4 hydroxyl proton and the signals arising from the C-2′ and C-6′ protons (but not with the protons on C-3′ and C-5′) and between the C-7 hydroxyl proton and the protons on C-6 and C-8. (3) Very strong coupling was noted between protons on C-2′ and C-6′ and the proton on C-2, and somewhat weaker coupling between protons on C-3′ and C-5′ and the C-2 proton. (4) Strong coupling was also observed between the signals for both the C-6 and C-8 protons and the two protons on C-4. In addition to providing further confirmation of the compound's structure, these observations provide an excellent example of a potentially useful technique for studying long-range proton–proton coupling.

CHAPTER FIVE

SYNTHESIS AND INTERCONVERSIONS OF FLAVONOIDS

INTRODUCTION

It is generally acknowledged that the structure of a naturally occurring compound has not been established with certainty until it has been synthesized from known starting materials using reagents and methods that give unequivocal results. This "rule" applies with no less rigor to flavonoids. There are several cases in the literature in which structures, deduced from limited spectroscopic data, have been shown to be in error when the compounds in question were synthesized. There are, of course, other reasons for synthesizing these compounds in the laboratory. Synthesis of flavonoids, and flavonoid precursors, with isotopically labeled atoms in defined positions have enabled workers to elucidate biosynthetic pathways. In addition, various derivatives of flavonoids, naturally occurring or not, have been prepared for study of their usefulness as flavoring or coloring additives for foods, for their potential use as drugs, or for their usefulness as agents against plant pests.

The flavonoid synthetic literature is vast and deserving of a book in its own right. It will only be possible here to present an overview of the subject with a selection of examples to demonstrate the major synthetic strategies. Wagner and Farkas (1975) presented a detailed account of flavonoid synthesis in the first volume in *The Flavonoids* series. Fundamental ideas of flavonoid synthesis were discussed by Gripenberg (1962) and in the same volume there was a lengthy chapter by Seshadri (1962a) on interconversions of flavonoid structural types. Several of the chapters in the later volumes of *The Flavonoids* series include summaries of recent synthetic work dealing with particular flavonoid structural classes.

Before we examine specific reactions, it is instructive to look very briefly at the challenges to be met in flavonoid synthesis. One of the major challenges, of course, is building the flavonoid skeleton. This entails establishing the three-carbon bridge (or the two-carbon bridge in the case of isoflavonoids) that links the two aromatic rings. Since the actual condensation reactions by which the flavonoid skeleton is formed are comparatively straightforward, the real challenge may lie in construction of the appropriately substituted A-ring and B-ring synthons. There is some latitude in solving this problem since the substitution patterns can be established at the level of the A-ring and B-ring synthons or the substitution pattern can be modified to a certain extent after the flavonoid skeleton has been made. The more complicated the substitution pattern exhibited by the target compound, the more challenging is the task for the synthetic chemist! Another problem in flavonoid synthesis involves establishing the desired oxidation level of the heterocyclic ring. Some syntheses yield the desired heterocyclic ring directly while in other cases some interconversion is necessary. Finally, methods must be sought for the preparation of the conjugated forms in which many flavonoids exist, e.g., glycosides, sulfates, or esters.

The information in this chapter appears in a series of more or less self-contained discussions, although there is a good deal of inter-relatedness among the various topics. We will start with a brief description of some of the classical flavonoid synthetic routes that are known by the names of their developers, viz., Claisen–Schmidt, Baker–Venkataraman, Allan–Robinson, and Algar–Flynn–Oyamada. Next there is a section on acylation of phenols (sometimes called Friedel-Crafts acylation). Immediately following these fundamental synthetic reactions there is a discussion about reactions that can be used to convert flavonoids, mostly chalcones and flavanones, into other flavonoids. Following this general section will be a series of discussions of individual flavonoid types and selected special topics. The aim of this chapter is to present a mix of classical synthetic work and more recent innovations to give the reader as broad an overview of the subject as possible within obvious space limitations. Sufficient references to the literature, both classical and current, have been given to provide access to general aspects of the subject as well as to experimental details.

Claisen–Schmidt Condensation

Owing to the availability of starting material, and comparative ease with which the reaction can be run, one of the most frequently used means of establishing the flavonoid nucleus involves the Claisen–Schmidt reaction between substituted 2-hydroxyacetophenones [5-1] and benzaldehyde derivatives [5-2]. The product of this reaction is a 2′-hydroxychalcone [5-3] (or the isomeric flavanone) bearing A-ring substituents provided by the acetophenone (indicated as R_1) and B-ring substituents provided by the benzaldehyde (indicated at R_2). The major challenge in using this route involves establishing the desired substitution pattern in the starting materials. Wagner and Farkas (1975, p. 132) list over two dozen acetophenone derivatives that had been used in flavonoid syntheses at that time; many more could be added to the list today. It should be noted that ω-methoxy [5-4] and ω-benzyloxyacetophenones [5-5] have been very successfully used in the Claisen–Schmidt reaction. Use of these starting materials establishes oxygenation at C-3 which is important for subsequent conversion of the flavanones to flavonols. The ω-methoxy function becomes the 3-methoxy group, of course, while the ω-benzyloxy function allows for the synthesis of flavonols *per se* by removal of the benzyl protecting group.

The Claisen–Schmidt reaction is routinely run using aqueous sodium or potassium hydroxide or ethanolic sodium ethoxide at about 50°C over a period of several hours. The benzaldehyde is often used in slightly more than equivalent amounts. Optimization is necessary in all situations. An example of the classical synthesis of a chalcone by the Claisen–Schmidt reaction is the preparation of coreopsin, 2′,4′,3,4-tetrahydroxychalcone 4′-*O*-glucoside [5-8], one of the pigments responsible for the flower color of *Cosmos sulfureus* and *Coreopsis* species. Coreopsin was synthesized by the base catalyzed condensation of resacetophenone (2′,4′-dihydroxyacetophenone) 4′-*O*-tetraacetylglucoside [5-6] with 3,4-dihydroxybenzaldehyde [5-7] in the presence

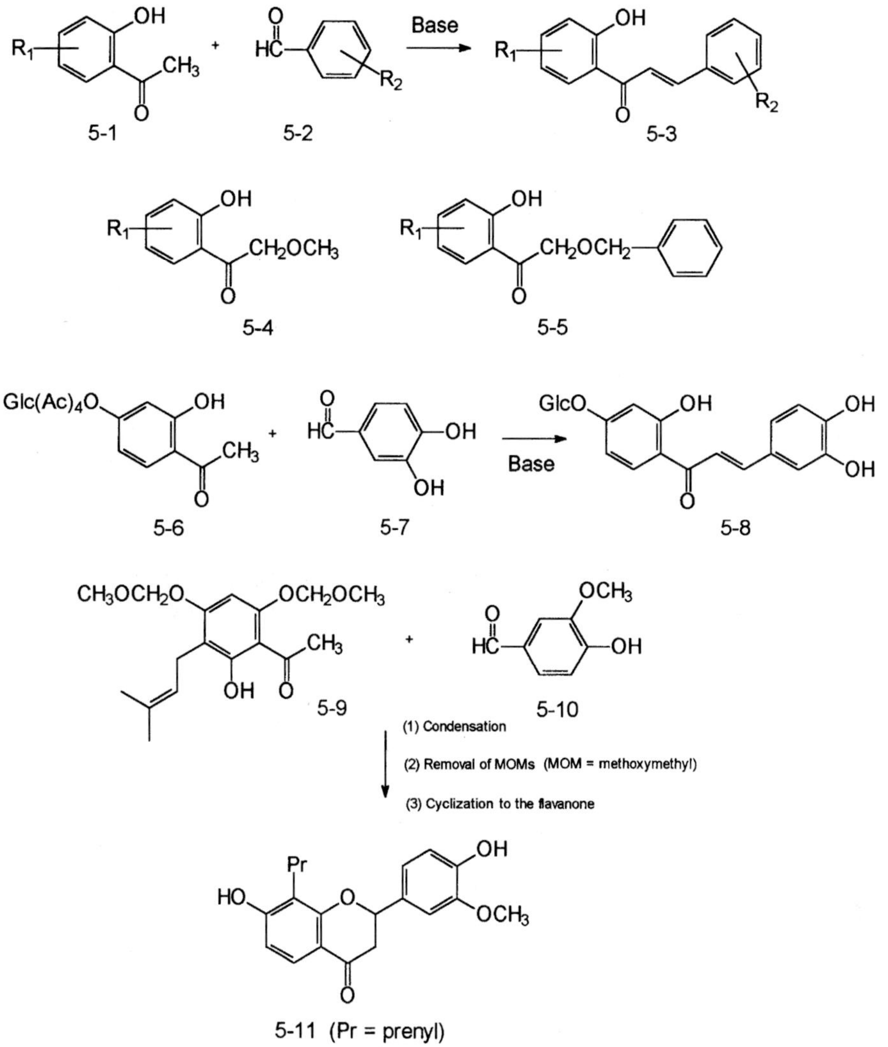

Scheme 5.1 The Claisen–Schmidt reaction.

of base (an inert atmosphere prevents oxidation of the *ortho*-dihydroxy system). The alkali serves two functions in this process; in addition to catalyzing the condensation, it removes the acetate groups from the sugar moiety. As an indication that this synthetic approach is still in wide use, we can look briefly at a recently described synthesis. Bu and Li (1996) were interested in preparing exiguaflavanone-K and (±)-leachianone-G, two flavanones from *Sophora* with suspected biological activity. The acetophenone required for the syntheses involved *C*-prenylation of phloroacetophenone with prenyl bromide in tetrahydrofuran followed by protection of two of the hydroxyl groups using chloromethyl methyl ether.

2′-Hydroxy-4′,6′-dimethoxymethoxy-3′-*C*-prenylacetophenone [5-9] was then condensed with either vanillin [5-10], leading to exiguaflavanone-K [5-11] or 2,4-dihydroxybenzaldehyde, leading to leachianone-G. Condensations were achieved by stirring a solution of the acetophenone and a 10% excess of the aldehyde in an aqueous ethanolic solution of KOH for 36 h at room temperature under argon. Following the usual workup, the crude chalcones were purified by silica gel column chromatography. Both chalcones were prepared in 84% yield. The authors prepared the desired final products by cyclization of the chalcones to the flavanones with ethanolic sodium acetate and removal of the methoxymethyl protecting groups with dilute HCl in methanol.

Other basic catalysts that have been used successfully include magnesium *t*-butoxide (Guthrie and Rabjohm, 1957), potassium carbonates (Rochus and Kickuth, 1957), and various forms of barium hydroxide (Sathyanarayana and Krishnamurty, 1988; Aguilera *et al.*, 1987; Fuentes *et al.*, 1987). The condensation will also occur under acid catalysis. Climent and coworkers (1995) list HCl, $AlCl_3$ (Calloway and Green, 1937), BF_3 (Breslow and Hauser, 1940), $POCl_3$ (Kaval and Shah, 1962); borax (Jadhav and Kulkarni, 1951), Al_2O_3 (Nondek and Malek, 1980), $TiCl_4$ (Mazza and Guarna, 1980), and zeolites (Corma *et al.*, 1989) as acid catalysts that have been used. It is in the realm of the zeolites that Climent and coworkers (1995) have made a significant contribution to the development of industrial-level syntheses of chalcones and flavanones. These workers prepared a series of hydrotalcites by mixing a solution of aluminum nitrate and magnesium nitrate with a solution of sodium carbonate (containing sufficient sodium hydroxide to maintain the desired pH). The gels were aged and calcined in order to obtain Mg–Al mixed oxides. Using the synthesis of 2′,4′,4-trimethoxychalcone as a test system (the product is of pharmaceutical interest), it was established that the best yields (85% with a reaction time of 20 h at 170°C) were obtained with a catalyst with Al/(Al + Mg) molar ratios of 0.25–0.30. In addition to catalyst composition, surface area and reaction temperature also influenced yield.

In a synthetic procedure that is mechanistically akin to the Claisen–Schmidt reaction, an acetophenone and an ester of an aromatic acid are condensed in refluxing xylene in the presence of powdered sodium. This reaction was used by Jain and Prasad (1989) to prepare β-hydroxy-2′,4′-dimethoxy-5′-*C*-prenylchalcone (pongagallone A) and its 3,4-methylenedioxy analogue (pongagallone B) two "dibenzoylmethane" derivatives known from *Pongamia glabra* (Gandhisadan *et al.*, 1987). The syntheses involved condensation of 2,4-dimethoxy-*C*-prenylacetophenone [5-12] with either methyl benzoate [5-13] or methyl 3,4-methylenedioxybenzoate [5-14] to yield the corresponding "dibenzoylmethane" derivatives [5-15] or [5-16], respectively. Since there is no 2′-hydroxyl group in either compound cyclization to the 2-hydroxyflavanone and elimination of water to form the flavones cannot occur. The compounds are, therefore, locked in the "dibenzoylmethane" form. ^{1}H NMR spectra of the products showed the presence of an enolic OH in both compounds (δ 16.3 in pongagallone A and δ 16.2 in pongagallone B). Furthermore, signals for single α-protons appeared in the spectrum of each compound, δ 7.86 for pongagallone A, δ 7.83 for pongagallone B. Therefore, the correct names for these compounds must indicate that they are β-hydroxychalcones and not dibenzoylmethanes. These reactions are summarized in Scheme 5.2.

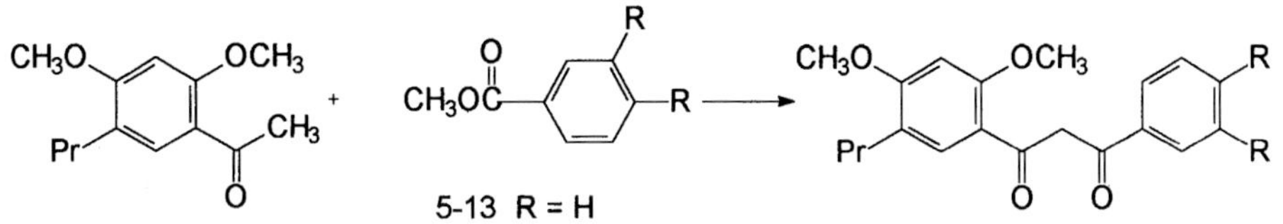

5-12

5-13 R = H

5-14 R = OCH_2O

5-15 R = H

5-16 R = OCH_2O

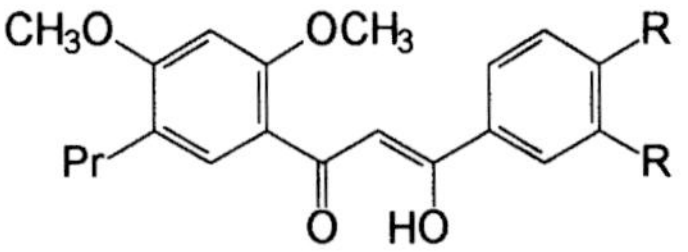

5-17 R = H

5-18 R = OCH_2O

Scheme 5.2 Synthesis of pongagallones A and B.

Baker–Venkataraman Rearrangement

The Baker–Venkataraman rearrangement involves acylation of a 2′-hydroxyacetophenone with an aromatic acid chloride at oil bath temperature in the presence of a base such as potassium carbonate or pyridine. The resulting ester is treated with a strong base, potassium hydroxide or sodium hydride are frequently used, which brings about reaction between the acetophenone methyl group (its carbanion) and the carbonyl carbon of the ester. The initial product of this reaction is thought to be the 2-hydroxy-2-aryl flavanone which opens to yield the β-diketone. Treatment of the β-diketone with acid results in recyclization to the hemiketal followed by elimination of water to form the flavone. Wagner and Farkas (1975) listed 34 compounds whose syntheses were based on the Baker–Venkataraman method. Additional historical comments on this and closely related reactions can be found in Gripenberg (1962, pp. 409–413). A typical example of this method involves the conversion of the 3-methoxy-4-benzyloxybenzoyl ester of 2,5-dihydroxy-4,6-dimethoxyacetophenone [5-19] to 5,6,4′-trihydroxy-7,3′-dimethoxyflavone [5-20] (Wagner *et al.*, 1973). Removal of the benzyl group from the 4′-position and the methyl group from the 5-position afforded the desired compound [Scheme 5.3].

Allan–Robinson Synthesis

The Allan–Robinson synthesis is a variation of the Baker–Venkataraman route in which a 2′-hydroxyacetophenone derivative is heated with the anhydride of an aromatic acid and the sodium salt of that acid, or in the presence of pyridine or

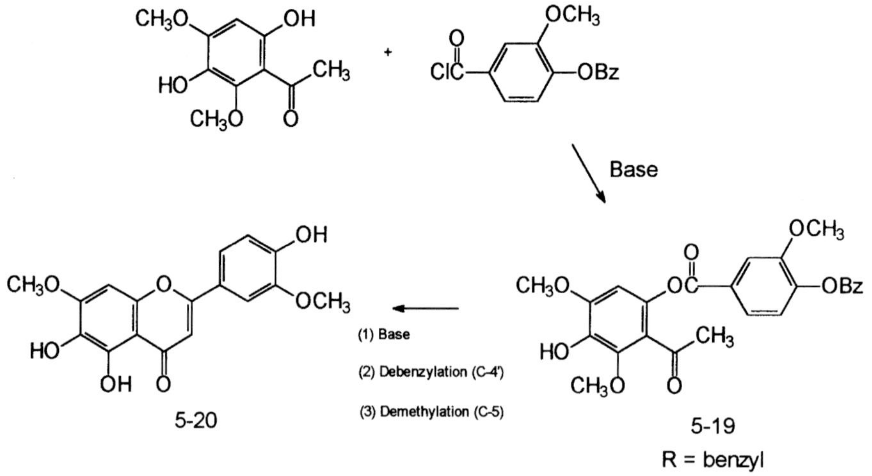

Scheme 5.3 Baker–Venkataraman rearrangement in the synthesis of 5,6,4′-trihydroxy-7,3′-dimethoxyflavone.

triethylamine. The intermediate esters and dibenzoylmethanes are not isolated. An example of the Allan–Robinson synthesis comes from the work of Fukui and colleagues (1968) who were interested in preparation of 5,7,4′-trihydroxy-3,6,3′-trimethoxyflavone (jaceidin) [5-25]. They prepared the required acetophenone by means of the Hoesch reaction, which involves treatment of a phenol with an acetonitrile derivative in the presence of HCl. In this instance the phenol used was 2-methoxyphloroglucinol [5-21] and the nitrile derivative was methoxyacetonitrile [5-22]. Note that the methoxy function is destined to become the 3-methoxy function of the finished flavonoid. After purification, acetophenone [5-23] was heated with *O*-benzylvanillic anhydride [5-24] and triethylamine to yield a product removal of the protecting group from which afforded jaceidin. Gripenberg (1962) noted that this method gives best results when the acetophenone has an oxygen function at the ω-position, thus making it a useful route for the preparation of flavonols as just seen. The ω-substitution can be either a methoxy or an aroyloxy group. Wagner and Farkas (1975) cited the synthesis of digicitrin [5-28], one of the most highly substituted flavonoids known at the time, from 2′-hydroxy-3′,4′,5′,6′,ω-pentamethoxyacetophenone [5-26] and the anhydride of 3,4-dimethoxy-5-benzyloxybenzoic acid [5-27], as an example of the Allan–Robinson method.

Additional comments on the Hoesch reaction would seem to be in order since the reaction is an important one in the synthesis of phenolic ketones, e.g., acetophenones, which can be of particular value in flavonoid syntheses. As mentioned above, the Hoesch reaction involves condensing an appropriately substituted phenol with acetonitrile (CH_3CN), or an acetonitrile derivative, in the presence of dry HCl and anhydrous zinc chloride although other catalysts have been used, e.g., trifluoromethyl sulfonic acid (CF_3SO_3H). The Hoesch reaction has also been run using benzylnitrile derivatives (benzylnitrile is $C_6H_5CH_2CN$) to synthesize deoxybenzoins, which are important starting materials for the synthesis of isoflavones (see below).

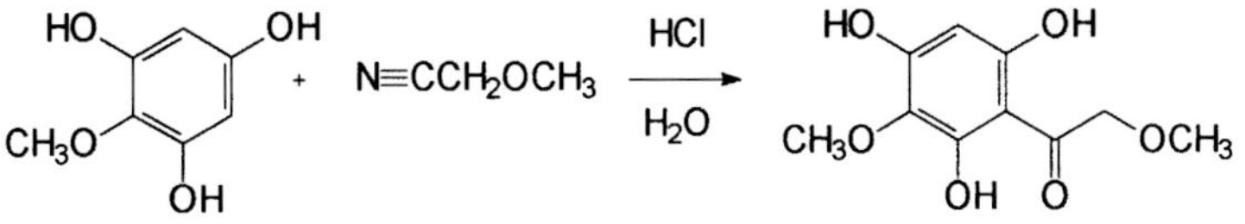

HO OH CH3O OH 5-21 + N≡CCH2OCH3 5-22 HCl H2O → HO OH CH3O OH O OCH3 5-23

N(CH2CH3)3

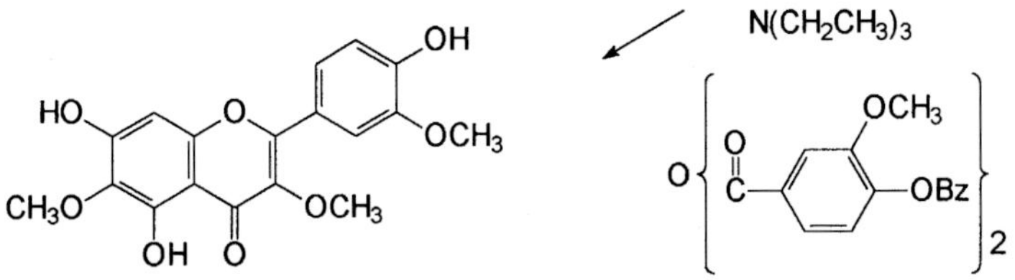

5-25 ← 5-24

Scheme 5.4 Allan–Robinson synthesis of jaceidin, including the Hoesch synthesis of the starting acetophenone.

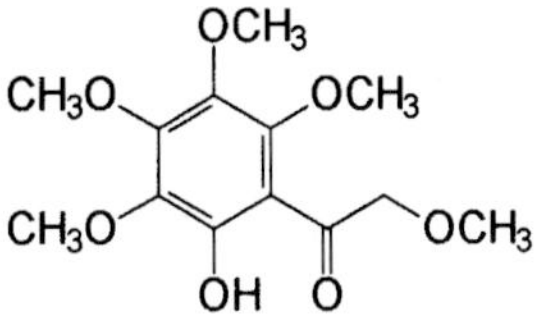

5-26 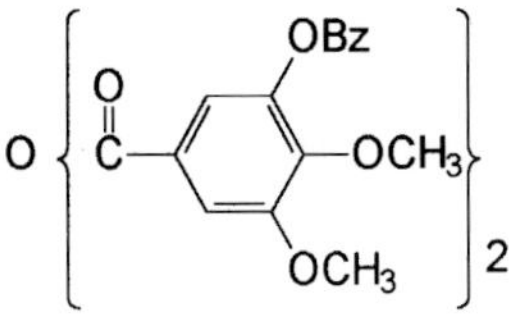 5-27

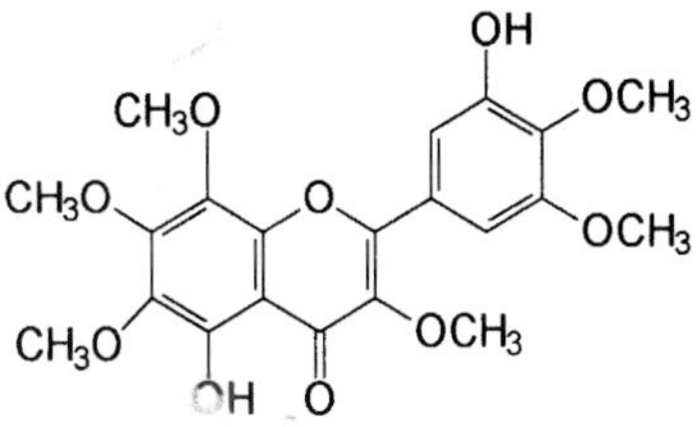

5-28

A typical example using acetonitrile comes from the work of Raju and coworkers (1992) on the synthesis of methoxychalcones from *Mimosa tenuifolia*. The required starting compound, 2,4-dihydroxy-3-methoxyacetophenone, was synthesized by condensation of pyrogallol 2-methyl ether (1,3-dihydroxy-2-methoxybenzene) with acetonitrile in the presence of trifluoromethyl sulfonic acid. An application of historical interest involves the synthesis of phloretin (2′,4′,6′,4-tetrahydroxydihydrochalcone) and a description of the preparation of the nitriles of phenylcarboxylic acids (Fischer and Nouri, 1917).

Algar–Flynn–Oyamada Reaction

At about the same time in 1934 Algar and Flynn and Oyamada independently described a method for converting chalcones to flavonols or flavonol 3-methyl ethers in a single step. The method, which came to be known as the AFO reaction, involves oxidizing a chalcone with H_2O_2 in an alkaline medium. The chalcone epoxide formed in the first stage of the reaction is attacked by the chalcone 2′-phenoxide group, represented in [5-29], establishing the six-membered heterocyclic ring with the epoxy oxygen becoming the flavanone 3-hydroxyl group. Further oxidation yields the flavonol. Two comparatively recent examples of the application of the AFO reaction are the syntheses of 3,5,6,7,8,2′,4′,5′-octaoxygenated flavones, some of which were useful in confirming the structure of flavonols isolated from *Gutierrezia microcephala* (Iinuma *et al.*, 1986a,b), and the synthesis of 6-*C*-methylmyricetin 3-methyl ether [5-31] (Islam and Hossain, 1993). In the case of the latter synthesis, 2′-hydroxy-4′,6′,3,4,5-pentamethoxymethyl-5′-*C*-methylchalcone [5-30] was converted in about 53% yield to the *O*-methylated flavonol. Removal of the *O*-methyl groups with HCl afforded several products that were separated by TLC yielding the desired compound in about 34% yield. A recent detailed report by Bennett and coworkers (1996) re-examines the mechanism of the AFO reaction particularly with regard to the existence of chalcone epoxides under certain circumstances. Mechanistic details notwithstanding, the AFO reaction is a useful means of synthesizing 3-hydroxyflavanones (dihydroflavonols) and flavonols.

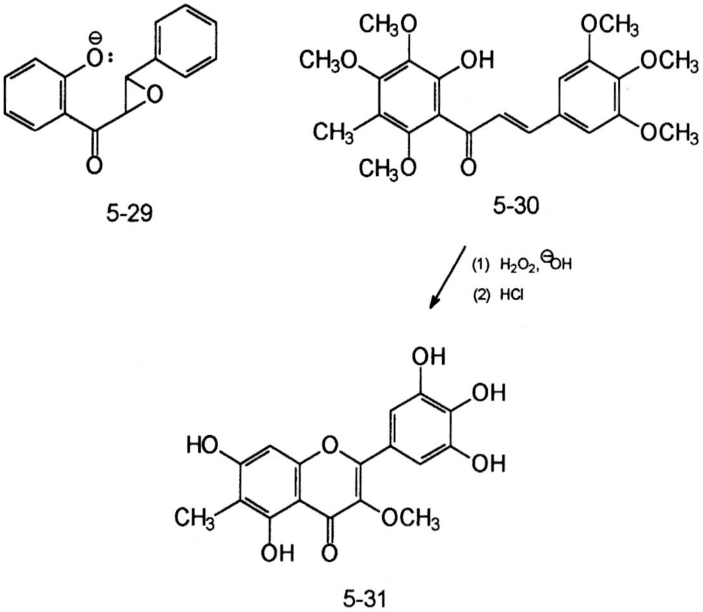

5-29

5-30

(1) H_2O_2, $^{\ominus}OH$
(2) HCl

5-31

Scheme 5.5 Algar–Flynn–Oyamada oxidation of a chalcone.

Acylation of Phenols

Chalcones can be synthesized by the direct acylation of a phenol. In this approach the phenol becomes the chalcone A-ring while the acylating agent provides both the B-ring carbons and the three-carbon bridge. Several different catalysts have been used in this reaction two of which are seen in the examples below [Scheme 5.6]. A very simple example of phenolic acylation comes from the work of Talapatra *et al.* (1986) who synthesized 7-hydroxyflavanone [5-34] in 60% yield by condensing resorcinol [5-32] with cinnamic acid [5-33] in the presence of polyphosphoric acid. Examples with more highly substituted phenols come from the work of Ichino and colleagues (1988) involving their syntheses of 2′-hydroxy-4′,5′,6′-trimethoxychalcone (helilandin) [5-37], 2′,6′-dihydroxy-3′,4′-dimethoxychalcone (pashanone), and several of their isomers. Helilandin was prepared by the Friedel-Crafts reaction between 3,4,5-trimethoxyphenol [5-35] and cinnamoyl chloride [5-36] using $AlCl_3$ in nitrobenzene at room temperature. For the synthesis of pashanone, these workers first prepared the isopropyl ether of [5-35] followed by the condensation with

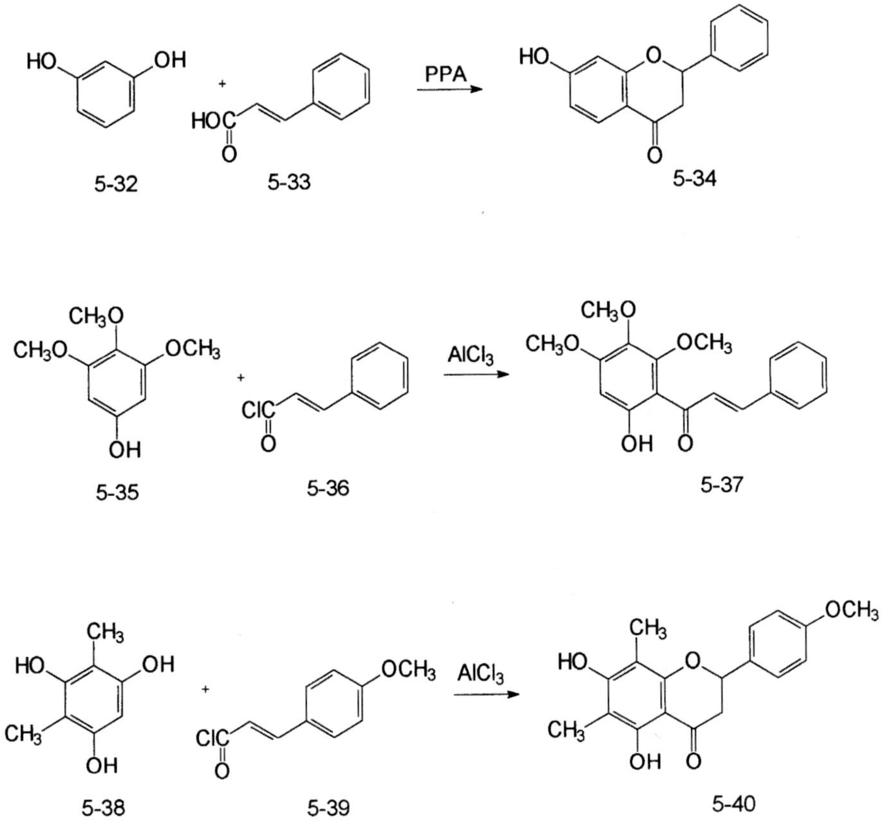

HO OH + HOC(=O) PPA → HO O O

5-32 5-33 5-34

CH3O CH3O OCH3 OH + ClC(=O) AlCl3 → CH3O CH3O OCH3 OH O

5-35 5-36 5-37

CH3 HO OH CH3 OH + ClC(=O) OCH3 AlCl3 → CH3 HO O CH3 OH O OCH3

5-38 5-39 5-40

Scheme 5.6 Phenol acylation as a route to the flavonoid skeleton.

cinnamoyl chloride as above to yield 2′-hydroxy-4′,5′-dimethoxy-6′-isopropoxychalcone. Removal of the isopropyl function with boron trichloride in dichloromethane at −78°C yielded a product identical in all respects to the natural chalcone. These reactions also work when the C_6C_3 moiety, the B-ring to be, is substituted as seen in the next example. Xie and coworkers (1986) condensed di-*C*-methylphloroglucinol [5-38] with *p*-methoxycinnamic acid chloride [5-39] in the presence of $AlCl_3$ to yield 5,7-dihydroxy-4′-methoxy-6,8-di-*C*-methylflavanone [5-40] whose common name "matteucinol" reflects its natural occurrence in the fern genus *Matteuccia*. These workers also synthesized matteucinol by Friedel Crafts acylation of di-*C*-methylphloroglucinol to give the corresponding acetophenone derivative followed by condensation with anisaldehyde (*p*-methoxybenzaldehyde) in the presence of base (the Claisen–Schmidt reaction). A useful synthesis based on direct acylation of a phenol is described in the dihydrochalcone section below (work of Sato *et al*., 1995).

Flavonoid Conversions

Because chalcones and their isomeric flavanones are so readily available by the Claisen–Schmidt reaction and by direct alkylation of phenols, it is not surprising that a considerable amount of work has gone into studying their conversion to other types of flavonoids. Major discussions of flavonoid interconversions are those of Seshadri (1962) and Wagner and Farkas (1975). Comments on individual flavonoid reactions can be found scattered in all members of *The Flavonoids* series.

It seems appropriate to start this discussion with reactions of chalcones since they are so readily available, as described above. The most commonly seen interconversion involves isomerization of a chalcone to the corresponding flavanone. This can be accomplished using acidic or alkaline conditions with the choice often predicated upon the potential sensitivity of other functional groups that might be present on the molecule. For example, cyclization of a chalcone glycoside would require minimal exposure to acid conditions to avoid cleavage of the glycosidic bond(s). Some of the more gentle treatments include simply allowing the chalcone to stand in aqueous solution, treatment with sodium acetate for a short time (e.g., 5 min), or treatment with pyridine at room temperature for several hours (Wagner and Farkas, 1975, p. 134). Catalytic hydrogenation readily converts chalcones to dihydrochalcones details of which reaction are presented in the section on dihydrochalcones below.

Many reagents have been used for the conversion of chalcones and/or flavanones to flavones. Several of these have involved halogens in one form or another as reviewed by Wagner and Farkas (1975, pp. 135–136). Chalcones can be converted to α,β-dibromodihydrochalcones [5-41] (often incorrectly called dibromochalcones) which upon treatment with methanol yield α-bromo-β-methoxydihydrochalcones [5-42]. These compounds undergo thermal cyclization to flavones although there is some risk of partial demethylation (Farkas *et al*., 1969). Treatment of a flavanone [5-43] with *N*-bromosuccinimide [5-44] in the presence of benzoylperoxide yields the 3-bromoflavanone [5-45], which readily losses hydrogen bromide to give the corresponding flavone (Aft, 1965). Dehydrogenation of flavanone glycosides has been accomplished using iodine in acetic acid with sodium acetate and acetic anhydride (Mahesh and Seshadri, 1955). The flavone glycosides were released following treatment of the acetates with sodium methoxide (Wagner *et al.*, 1968). Iodine in dimethylsulfoxide (DMSO) has been used by Doshi and coworkers (1986). A recently described reagent for converting 2′-hydroxychalcones to the corresponding flavones is hypervalent iodine. For example, Litkei and coworkers (1995) used iodosobenzene diacetate [5-46] in a study of the cyclodehydrogenation of a number of 2′-hydroxychalcones. Fieser and Fieser (1961, p. 789), who write this compound as $(C_6H_5IOCOCH_3)^+CH_3COO^-$, point out that it is comparable to lead tetraacetate as an oxidizing agent.

Selenium dioxide (SeO_2) has been used successfully to dehydrogenate a variety of flavanones to the corresponding flavones. The reaction is usually run in refluxing isoamyl alcohol. Examples of the use of this reagent include the synthesis of 5,7,4′-trihydroxy-6,3′-dimethoxyflavone (jacein) (Wagner *et al.*, 1973); 5,3′,4′-trihydroxy-6,7-dimethoxyflavone (cirsiliol) and 5,4′-dihydroxy-6,7,3′-trimethoxyflavone (cirsilineol) (Matsuura *et al.*, 1973); 5,4′-dihydroxy-7,8,3′,5′-tetramethoxyflavone and isomers (Chandra and Babber, 1987); and 5,6,7,4′-tetramethoxyflavone and 5,6,7-trimethoxy-3′,4′-methylenedioxyflavone (Bidhendi and Bannerjee, 1989). Selenium dioxide in DMSO has also been used (Makrandi and Seema, 1989).

A very useful reagent for conversion of either chalcones or flavanones into the corresponding flavones is 2,3-dichloro-5,6-dicyano-*p*-benzoquinone (DDQ) [5-47]. Typical examples of the use of this reagent include synthesis of hepta- and octamethoxyflavones related to the agehoustins (Iinuma *et al.*, 1984a), flavones related to compounds from *Brickellia* (Iinuma *et al.*, 1984a,b,c), assorted flavones with 2′,3′,4′,6′- and 2′,4′,5′-methoxylated B-rings (Iinuma *et al.*, 1984a,b,c, 1986a,b,c, 1987), 6,7-dimethoxy-3′,4′-methylenedioxyflavone (Parmar *et al.*, 1989), and a group of *C*-prenylated flavones related to the ovalichalcones (Hossain and Islam, 1993). Imafuku and coworkers (1987) subjected several 2′-hydroxychalcones to DDQ oxidation in an effort to determine how different arrangements of substituents affected the nature of the product, whether flavanone, flavone, or aurone, and yield.

Correa and coworkers (1971) employed the unusual reagent trityl perchlorate (trityl is triphenylmethyl) [5-48] in acetic acid and acetic anhydride to convert 3′-hydroxy-7,4′-dimethoxyflavanone to the corresponding flavone acetate which yielded tithonine, an unusual flavone from *Tithonia tubaeformis*, upon treatment with base. Treatment of 2′-hydroxychalcones with sodium periodate ($NaIO_4$) in DMSO has been shown to be an efficient means of converting chalcones having a variety of hydroxyl and methoxyl substitution patterns (Hans and Grover, 1993). Nickel peroxide was examined as a reagent in this reaction by Mallik and coworkers (1989), and heating chalcones with Pd/C in vacuum has been investigated (Bose *et al.*, 1971). The final example in this section involves the photosensitized dehydrogenation of flavanones to flavones using 2,4,6-triphenylpyrylium tetrafluoroborate (TPT) [5-49] as a photosensitizer (Climent *et al.*, 1989). Moderate yields were realized in most cases, but the reaction was affected by the electronic nature of the substituents on the rings. Scheme 5.7 presents structures of a number of the more complex reagents.

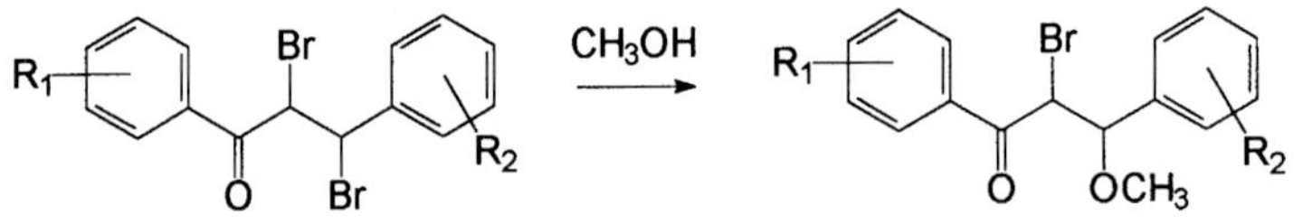

5-41 5-42

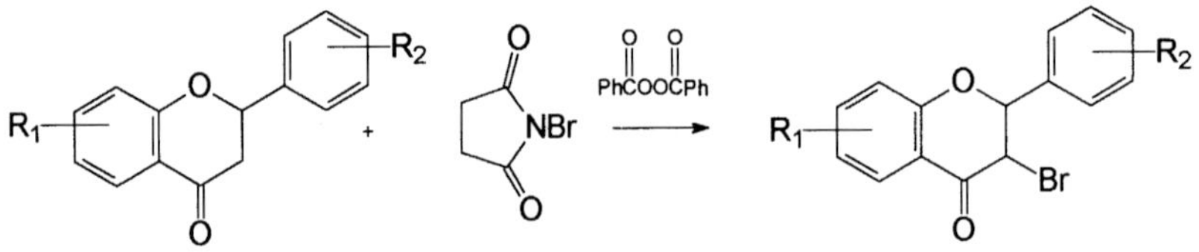

5-43 5-44 5-45

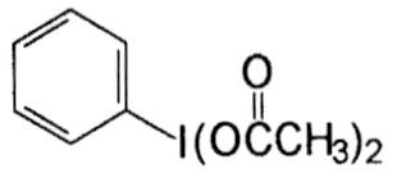

5-46

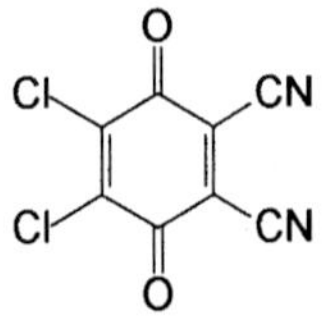

5-47

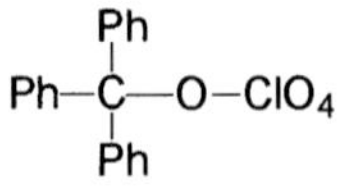

5-48

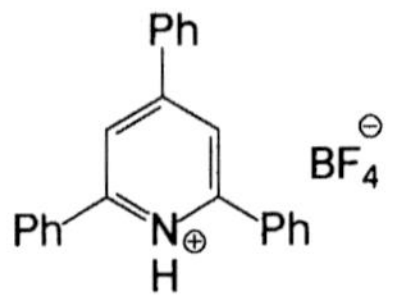

5-49

Scheme 5.7 Reagents used for selected flavonoid conversions. Ph = phenyl.

Anthocyanidins

The chemistry of anthocyanidins and anthocyanins was reviewed by Seshadri (1962, pp. 166–169) and by Hayashi (1962, pp. 262–268), both of which are chapters in the Geissman volume. Wagner and Farkas (1975) refer the reader to those and a few earlier works, e.g., a review by Jurd (1972). Since there are comparatively few anthocyanidins known in Nature, one might expect only a few recent synthetic studies. Because anthocyanidin derivatives were early recognized as one of the main sources of floral color, it is not surprising that the history of these compounds goes back some years. The first total synthesis of an anthocyanidin, in fact, was the work of Willstätter and Zechmeister in 1914. It is interesting to look at their synthesis, which consisted of three fundamental stages: (1) preparing the A-ring synthon; (2) constructing the coumarin system (the C-ring); and (3) introducing the B-ring. The A-ring was provided by phloroglucinol which was converted, using HCN and HCl, to phloroglucinaldehyde [5-50], thus providing a reaction site for condensation with methoxyacetic anhydride (in the presence of sodium methoxyacetate; note similarity with Allan–Robinson condensation). The resulting coumarin derivative [5-51] was saponified and then methylated to give 3,5,7-trimethoxycoumarin [5-52]. The B-ring is introduced by using *p*-methoxyphenyl magnesium bromide Grignard reagent [5-53], followed by treatment with HCl. Removal of the methyl groups yielded pelargonidin chloride. The reactions are shown in Scheme 5.8.

While the above synthesis was presented for historical interest, Hayashi (1962) pointed out that a more convenient and direct synthesis of anthocyanins was developed that has been used extensively since its introduction (Pratt and Robinson, 1922; Robertson and Robinson, 1928; Robertson *et al.*, 1928; Murakami and Robinson, 1928; Bradley and Robinson, 1928). As shown in Scheme 5.9, the synthesis involves condensation of an appropriately substituted *o*-hydroxybenzaldehyde [5-54] and an acetophenone derivative (providing the B-ring and two of bridge carbons in this synthesis) [5-55]. The condensation is accomplished by treatment with dry HCl in an appropriate solvent. Removal of the protecting groups affords the anthocyanidin, cyanidin chloride [5-56] in this example. Further work in the laboratory of Robert Robinson (eventually, Sir Robert) led to the synthesis of anthocyanins as well (Reynolds *et al.*, 1934).

3-Deoxyanthocyanins are restricted to one subfamily of the tropical family Gesneriaceae and to a smattering of species elsewhere in the plant kingdom. They can be synthesized by reduction of the corresponding flavone using sodium amalgam in methanol containing HCl. A possible mechanistic route to these compounds was discussed by Hayashi (1962, p. 265).

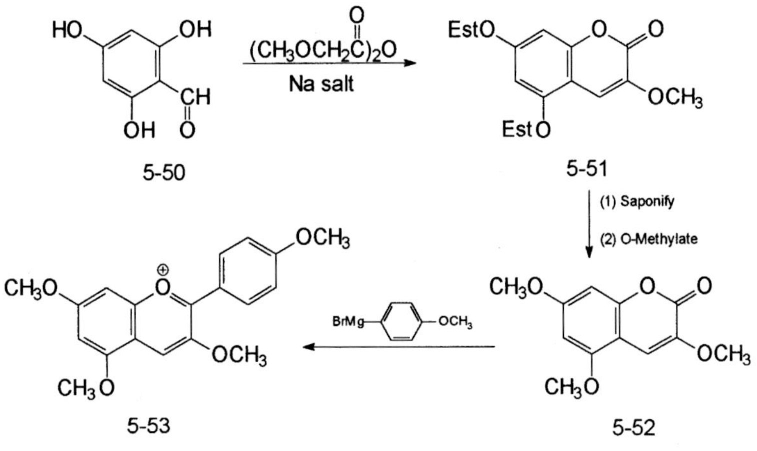

HO OH CH O OH 5-50 $(CH_3OCH_2C(=O))_2O$ Na salt EstO O O OCH_3 EstO 5-51 (1) Saponify (2) O-Methylate CH_3O O O OCH_3 CH_3O 5-52 BrMg— —OCH_3 OCH_3 CH_3O $\oplus$ O OCH_3 CH_3O 5-53

Scheme 5.8 Anthocyanidin synthesis by Willstätter and Zechmeister (1914).

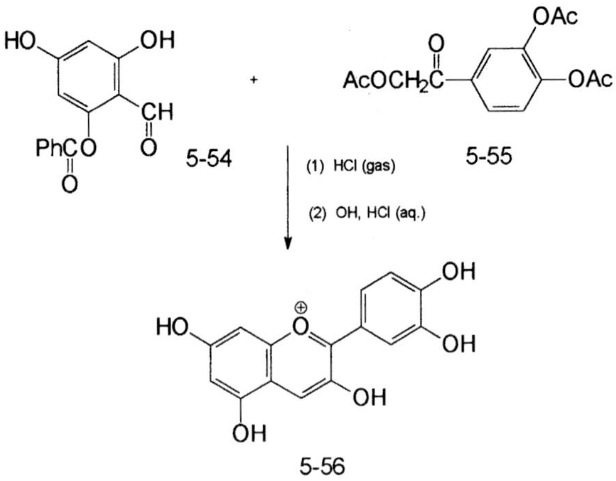

HO OH CH O PhCO O 5-54 + AcOCH$_2$C(O) OAc OAc 5-55 (1) HCl (gas) (2) OH, HCl (aq.) HO O⊕ OH OH OH OH 5-56

Scheme 5.9 Anthocyanidin synthesis after Robinson *et al.*

Aurones

Although aurones have been synthesized through conversions of other flavonoid derivatives, mainly chalcones, the only practicable route involves condensation of suitably substituted coumaranones and benzaldehydes. With the appropriate choice of starting materials and conditions it is possible to synthesize aurone glycosides with little problem. The reaction will occur under either acidic or basic conditions depending upon the nature of the substitution pattern of the coumaranone (4,6-dihydroxycoumaranone doesn't work) and the sensitivity of other functional groups in the starting materials. A generally useful alternative involves refluxing the starting materials in acetic anhydride; no other catalyst is required. This is a particularly useful method when aurone glycosides are the desired end product. Two syntheses of sulfurein will show the versatility of this synthetic approach. Soon after the discovery of aurones as a new class of flavonoids, Geissman and Moje (1951) accomplished a synthesis of sulfuretin by condensing 6-hydroxycoumaranone [5-57] with 3,4-dibenzyloxybenzaldehyde to provide 3′,4′-dibenzylsulfuretin [5-58]. That product was glycosylated by treatment with α-bromoacetoglucose followed by the removal of the protecting groups to yield sulfurein [5-61] identical in all respects with the naturally occurring compound. A few years later, Farkas and Pallos (1959) accomplished the same synthesis by condensing the coumaranone tetraacetylglucoside [5-59] with 3,4-dihydroxybenzaldehyde in boiling acetic anhydride to afford sulfurein hexaacetate [5-60]. Sulfurein was recovered after removal of the acetate groups. This can generally be accomplished with ammonium hydroxide in methanol. Both reaction sequences can be seen in Scheme 5.10. The condensation of coumaranones and benzaldehydes has been discussed in detail by Farkas and Pallos (1967).

Other routes to aurones are not of practical value owing to competing processes. At best, yields are comparatively low and there is always the need to separate the

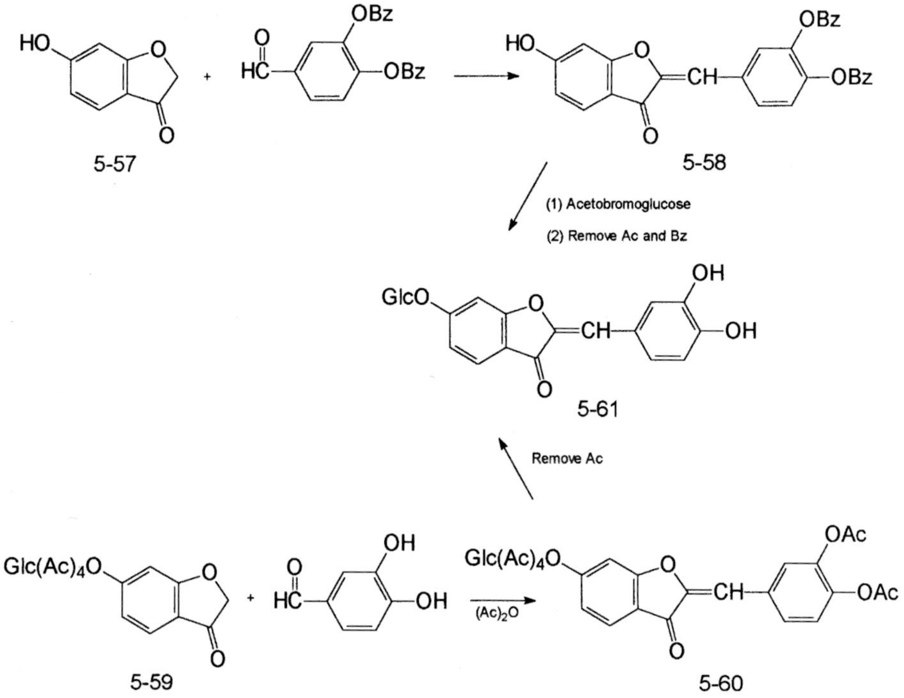

Scheme 5.10 Synthesis of sulfurein by different routes. Ac = acetyl, Bz = benzyl.

desired compound from other flavonoid products. Depending upon the substitution patterns of the starting compound and the specific reaction conditions, one may get some aurone from the based catalyzed ring closure of dihydrochalcone α,β-dibromides [5-41] although the normally expected product from this reaction is the flavone. The "AFO" reaction, described in the introductory comments above, is primarily used to prepare flavonols but aurones can occur following α-attack of the 2′-phenoxy ion on the chalcone epoxide (for a review see Bennett *et al.*, 1996). Imafuku and coworkers (1987) studied the cyclodehydration of 2′-hydroxychalcones using 2,3-dichloro-5,6-dicyano-p-benzoquinone (DDQ). When chalcones, with various B-ring substitutions, had hydrogen at C-4′, mixtures of flavones (32–42% yields) and aurones (9–17%) were formed. When there was a methoxyl function at C-4′, flavones were the major products with only traces of aurones.

Biflavonoids

Biflavonoids fall into two groups, the biphenyl type in which the units are linked through a carbon–carbon bond, and the biphenyl ether type in which the linkage is through oxygen. Within each of these subclasses there is a great deal of opportunity for variation owing to differences in the positions of interunit linkage and whether the two units are identical (symmetrical biflavonoids) or different (unsymmetrical biflavonoids). Two general approaches are available for synthesizing biflavonoids: (1) construction of the two subunits independently followed by a coupling reaction, and (2) establishment of the interflavonoid link before the remainder of the flavonoid skeleton is constructed. As in the case of all other flavonoid syntheses, it is possible to modify a biflavonoid after the inter-unit link has been established. An excellent review of earlier work can be found in the chapter by Wagner and Farkas (1975). The first biflavonoid to be synthesized was ginkgetin hexamethyl ether [5-64]. Nakazawa (1959) accomplished the synthesis by condensing 3′-iodo-5,7,4′-trimethoxyflavone [5-62] with 8-iodo-5,7,4′-trimethoxyflavone [5-63] using the Ullmann reaction with dimethylformamide (DMF) as solvent. The Ullmann reaction involves treatment of aryl iodides with activated copper powder and results in bond formation involving the carbons at which the iodine atoms were attached. Loss of some methyl groups under these reaction conditions was avoided in a subsequent synthesis of ginkgetin in which the reactants were heated at 225–230°C for 40 minutes without solvent (Nakazawa and Ito, 1963). A recent modification of the Ullmann reaction was used by Muller and Fleury (1991) to synthesize amentoflavone derivatives. The modification used flavone 8-boronic acids, e.g., [5-65], and iodo-flavones and gave higher yields than with the classical Ullmann approach.

A recent example of the application of the Ullmann reaction involves the synthesis of 5,5″-dihydroxy-4′,4‴,7,7″-tetramethoxy-8,8″-biflavone [5-66], a compound known from species of *Araucaria*. The compound is noteworthy because it was the first optically active biflavone isolated from Nature. Although the compound lacks chiral centers, diastereomers (tropoisomers) exist because of restricted rotation around the 8-8″ bond. The synthesis (Zhang *et al*., 1995) involved treatment of 8-iodo-5,7,4′-trimethoxyflavone with copper powder in dimethylformamide (DMF) at reflux for 20 h, which produced the (±)-hexamethyl ether in 40% yield. Selective demethylation at positions 5 and 5″ yielded the desired compound. Resolution of the isomers was accomplished by treating the mixture with (+)-camphorsulfonyl chloride and triethylamine, followed by crystallization.

Biflavonoids linked through oxygen can be prepared by reaction between an iodoflavone and the sodium or potassium salt of a second flavone. This method is exemplified by the synthesis of hinokiflavone pentamethyl ether [5-67] by condensation of 8-iodo-5,7,4′-trimethoxyflavone and 5,7-dimethoxy-4′-hydroxyflavone under modified Ullmann reaction conditions (Krishnan *et al*., 1966). Cupressuflavone hexaacetate, formed by the coupling of two of the iodoflavone units, was the anticipated byproduct of this reaction and could be separated from the desired product chromatographically. These reactions are shown in Scheme 5.11.

Several methods have been developed for oxidative coupling of two flavone subunits. Ali and Ilyas (1986) treated 2′-hydroxy-4′,6′-dimethoxychalcone [5-68] at room temperature with iodine in methanolic potassium hydroxide solution in a successful attempt to form the bichalcone [5-69]. The bichalcone was readily isomerized to the biflavanone derivative [5-70]. Kikuchi and coworkers (1991) used silver oxide in refluxing benzene to bring about the coupling at C-8 of two units of 7-hydroxy-5,6-dimethoxyflavone [5-71] to yield, after demethylation, 5,5″,6,6″,7,7″-hexahydroxy-8,8″-biflavone (8,8″-bibaicalein) [5-72]. These reactions are shown in Scheme 5.12.

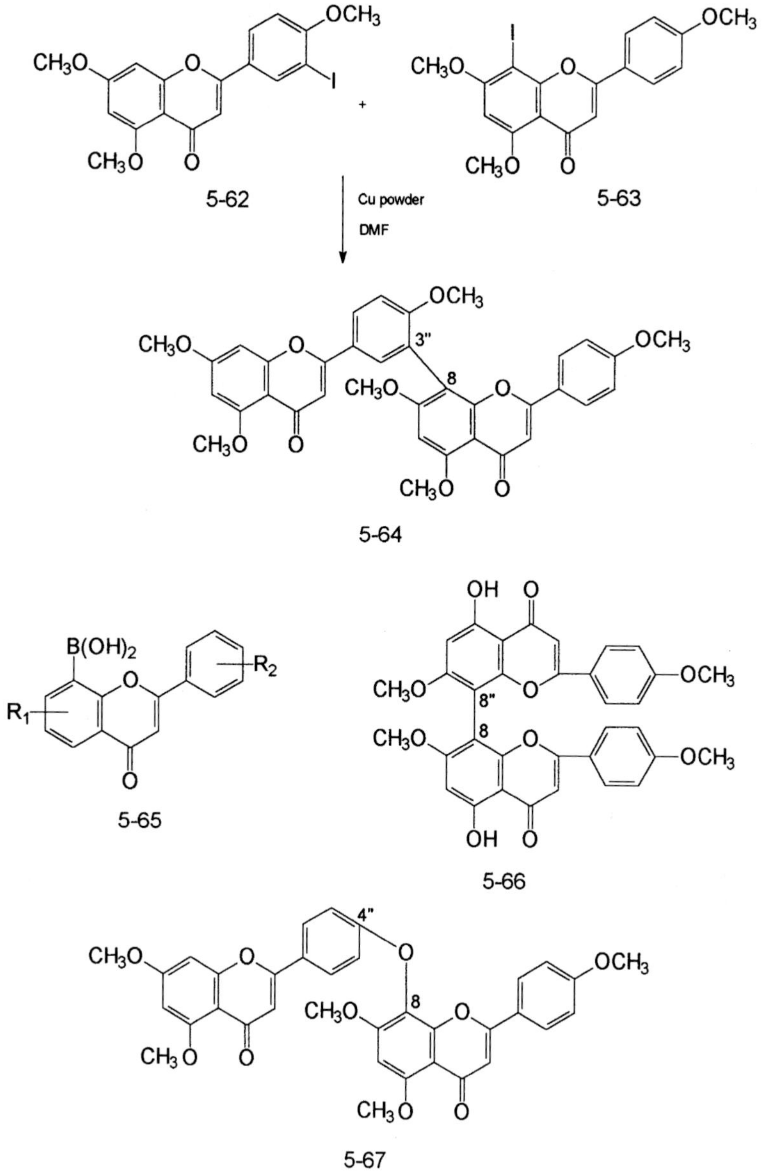

5-62 + 5-63 → (Cu powder, DMF) → 5-64

5-65

5-66

5-67

Scheme 5.11 A synthesis and some examples of biflavonoids.

Efforts have also been directed toward coupling reactions at the acetophenone level. Ahmad and Razaq (1976), for example, prepared the phloroacetophenone dimethyl ether dimer [5-73], condensed it with *p*-hydroxybenzaldehyde using Claisen–Schmidt conditions, and subjected the resulting chalcone [5-74] to selenium oxide to yield the 7,7″-dimethyl biflavone [5-75] as shown in Scheme 5.13. The same

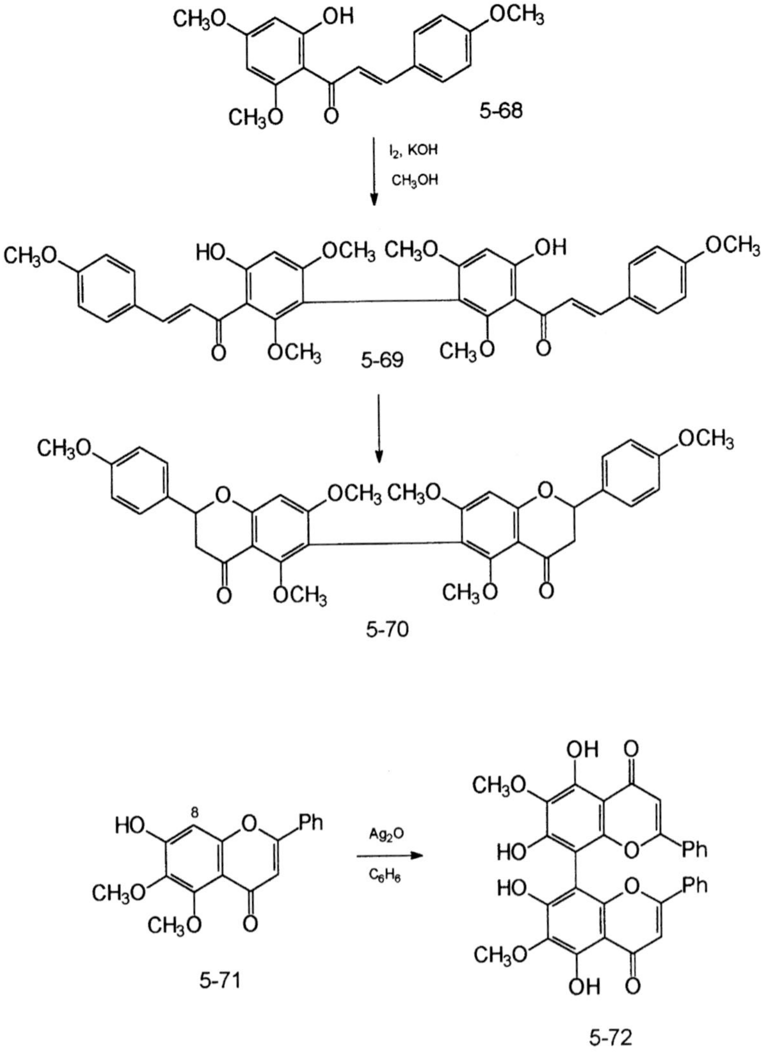

5-68

I_2, KOH
CH_3OH

5-69

5-70

5-71

Ag_2O
C_6H_6

5-72

Scheme 5.12 Oxidative coupling of flavonoid subunits.

dimeric acetophenone was used to prepare B-ring deoxybiflavones (Khan and Aqil, 1995). Phloroacetophenone dimethyl ether had been successfully dimerized using ferric chloride in dimethylformamide, written as $[Fe(DMF)_3Cl_2][FeCl_4]$ (Parthasarathy *et al.*, 1977). Those workers then condensed the dimeric acetophenone with *p*-methoxybenzaldehyde (anisaldehyde) to form the bichalcone which was subsequently converted to the corresponding biflavanone and biflavone. The ferric chloride-DMF system was also used by Khan and coworkers (1985a,b) to synthesize a series of *O*-methylated biflavonoids based on 6,6″-biflavone, 6,6″-bisisoflavone, 6,6″-bidihydroflavonol, and 5,5″-bisaurone. The oxidative dimerization of phloroacetophenone dimethyl ether, resacetophenone (2,4-dihydroxyacetophenone), and resacetophenone monomethyl ether (2-hydroxy-4-methoxyacetophenone) using silica-bound ferric chloride was examined by Parthasarathy and Gupta (1984). Coupling of phloroacetophenone dimethyl ether yielded only one product whereas coupling of the two resorcinols gave all three possible dimers from each. Thus, with the appropriate structural elaborations, biflavones with 6,6″, 6,8″, and 8,8″ linkages are possible. In the case of the resorcinols, however, separation of the individual dimers would be necessary.

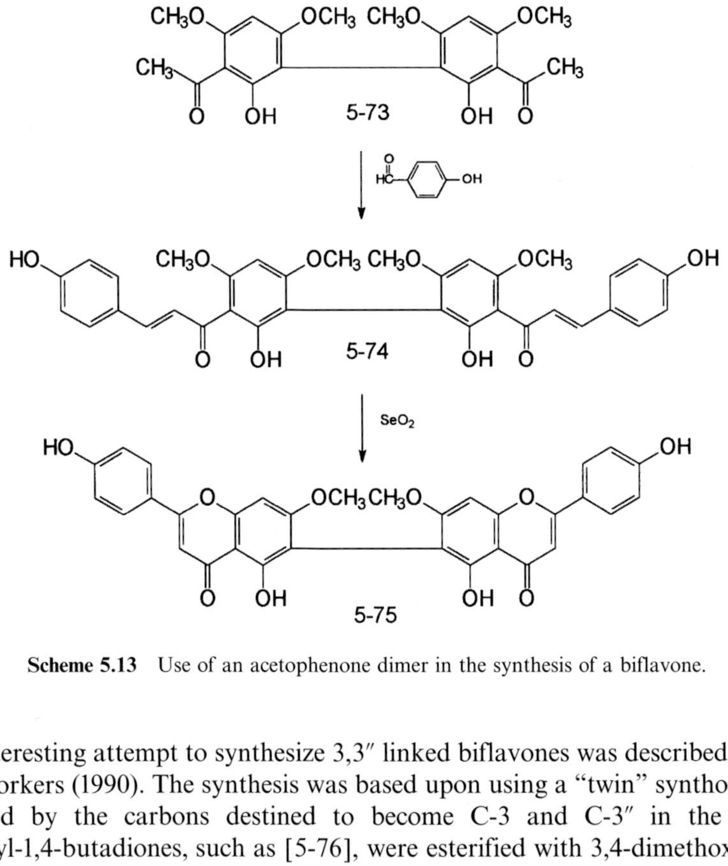

Scheme 5.13 Use of an acetophenone dimer in the synthesis of a biflavone.

An interesting attempt to synthesize 3,3″ linked biflavones was described by Khan and coworkers (1990). The synthesis was based upon using a "twin" synthon already connected by the carbons destined to become C-3 and C-3″ in the product. 1,4-Diaryl-1,4-butadiones, such as [5-76], were esterified with 3,4-dimethoxybenzoyl chloride (veratroyl chloride) to yield intermediates of the sort [5-77], which were then subjected to conditions that normally lead to formation of the flavone ring system. The product of the reaction [5-78], however, was not the desired biflavone. Regardless of the conditions tried, the second condensation/cyclization could not be accomplished (Scheme 5.14).

The synthesis of biflavonoids having different subunits, whether they involve the same flavonoid class with different *O*-methylations or completely different flavonoid classes, requires a good measure of synthetic ingenuity. Fundamentally, one has to devise a synthetic strategy where one of the flavonoid units is built onto another one.

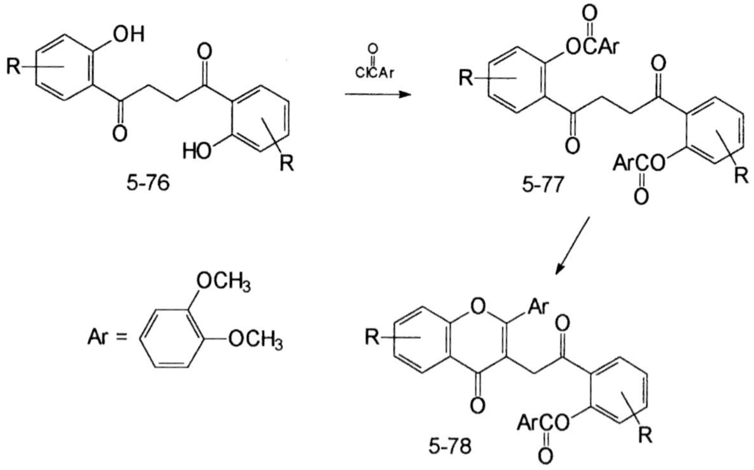

Scheme 5.14 Unsuccessful attempt to synthesize biflavonoids using 1,4-aryl-1,4-butadiones.

Part of Muller and Fleury's (1991) study of arylboronic acids involved a condensation of 3′-iodo-5,7,4′-trimethoxyflavone [5-79] with 2,4,6-trimethoxyphenylboronic acid [5-80] to yield 3′-(2,4,6-trimethoxyphenyl)-5,7,4′-trimethoxyflavone [5-81]. Acylation of [5-81] with *p*-methoxycinnamic acid [5-82] in the presence of boron trichloride afforded the chalcone [5-83], cyclization of which gave biflavone [5-84]. Although the biflavone synthesized by those workers consists of two identical subunits, compounds with different subunits could be constructed using this method by substituting some other cinnamic acid derivative (changing the R group in 5-82). This synthesis is pictured in Scheme 5.15.

A different strategy was employed by Ikeshiro and Konoshima (1972) in their synthesis of fukugetin heptamethyl ether [5-89], which consists of different subunits, apigenin trimethyl ether linked *via* C-3′ and luteolin tetramethyl ether linked *via* C-8. The C-8 linkage was established at the outset by synthesis of 8-chloromethyl-5,7,3′,4′-tetramethoxyflavone (not shown) which was converted in several steps to the acid chloride [5-85]. Reaction of the acid chloride with phloroglucinol dimethyl ether yielded ester [5-86] which, upon treatment with titanium tetrachloride, underwent a Fries rearrangement to yield the acetophenone derivative [5-87]. The newly established acetophenone provides the A-ring and two of the bridge carbons of the second flavone unit. The remainder of the second unit was established at this point by condensation with of *p*-methoxybenzaldehyde (anisaldehyde). Conversion of the flavonochalcone [5-88] to the biflavone [5-89] yielded a product identical in all respects to the per-*O*-methylated derivative of fukugetin. The steps of this synthesis can be found in Scheme 5.16.

Attempts to synthesize biflavonoid derivatives (proanthocyanidin dimers) that contain the 7,8,4′-trihydroxyflavan nucleus were described recently by Malan and coworkers (1997). The target compound, which was isolated from *Acacia caffra*, was formed by acid catalyzed condensation of two units of 4,7,8,4′-tetrahydroxyflavan, which was prepared by reduction of the corresponding flavanone.

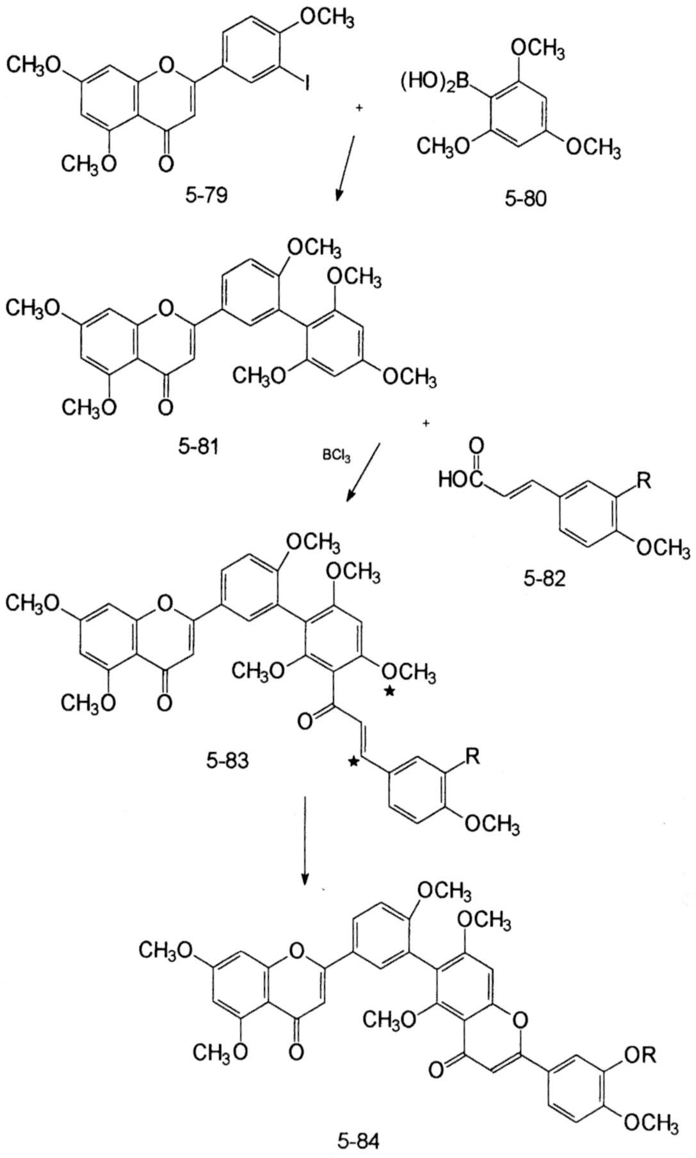

Scheme 5.15 Constructing a second flavonoid unit piece by piece.

Chalcones

We have already seen the two major routes by which chalcones are prepared, the Claisen–Schmidt condensation of a 2′-hydroxyacetophenone with a benzaldehyde derivative, and the direct acylation of a phenol. Many chalcones occur as natural products, of course, and their synthesis by unequivocal means helps to establish their

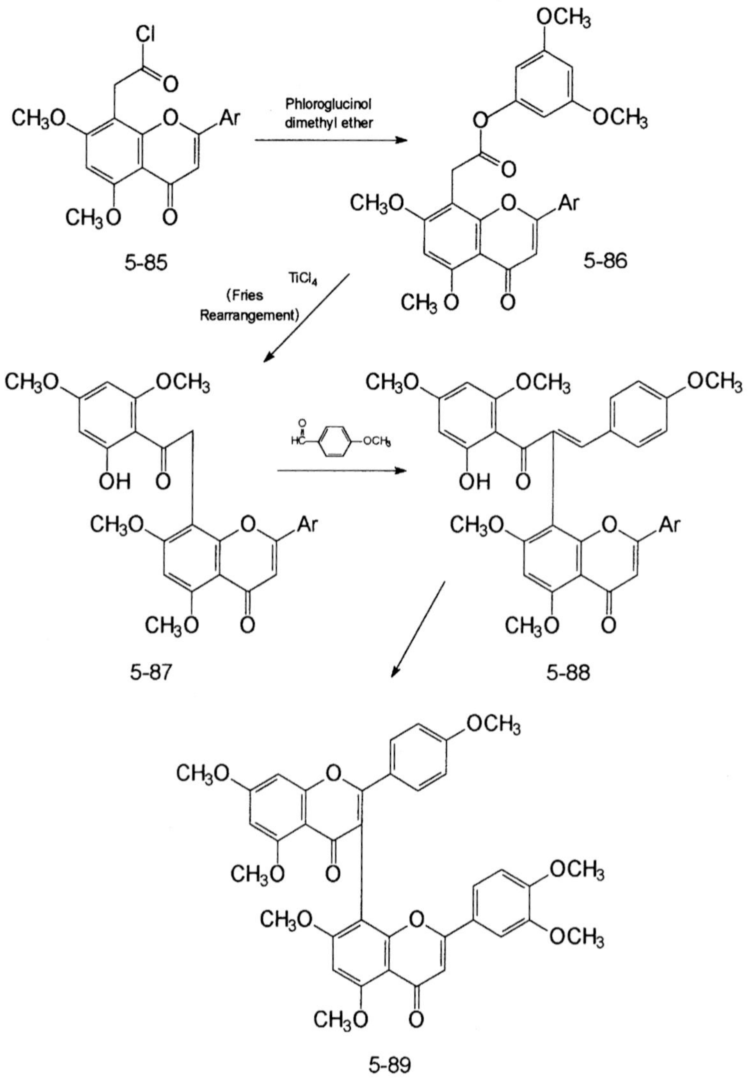

Scheme 5.16 Synthesis of fukugetin heptamethyl ether.

structures. Other chalcones are synthesized as means to an end; they are, in other words, convenient synthons. Table 9.16 in Bohm (1994) lists some of the chalcone syntheses (dihydrochalcones and flavanones as well) that were published during the time period 1986–1991.

Some highly original synthetic approaches to chalcones, and from them to other flavonoid types, have been examined in recent years. The full potential of many of these has not been established. One of these is the "isoxazoline" route (Thomsen and Torssell, 1988; Almtorp *et al.*, 1991). Their synthesis of 2′,4′,4-trihydroxychalcone (isoliquiritigenin) from 2,4-dihydroxybenzaldehyde serves well as an example. The benzaldehyde hydroxyl groups were first protected by dimethyl *t*-butyl silyl ether groups to yield [5-90]. The aldehyde was then converted to the oxime [5-91]. The oxime was chlorinated using *N*-chlorosuccinimide to give compound [5-92]. Reaction of [5-92] with 4-benzoyloxystyrene [5-93] led to the isoxazoline derivative [5-94]. Desilylation and catalytic hydrogenation afforded the β-hydroxydihydrochalcone [5-95] which, upon acid catalyzed elimination of water, gave isoliquiritigenin. The reaction sequence is presented in Scheme 5.17. Attempts to apply this synthetic approach to isoflavones were described by these workers in the earlier paper (Thomsen and Torssell, 1988).

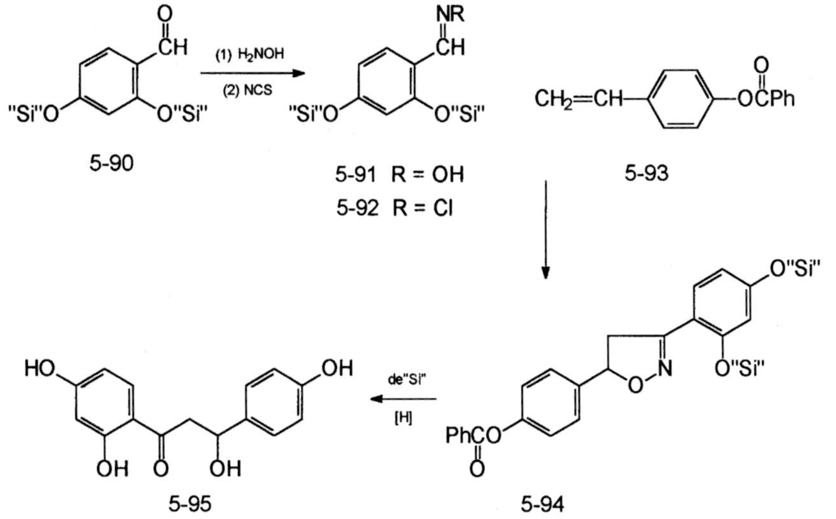

Scheme 5.17 Synthesis of a β-hydroxydihydrochalcone *via* an isoxazoline. Ph = phenyl, "Si" = dimethyl-*t*-butylsilyl.

Dihydrochalcones

Simple dihydrochalcones, those that lack complex or sensitive side chain substituents, can be prepared readily by catalytic reduction of the corresponding chalcone. Palladium on carbon is the usual catalyst for this reduction and high pressure is not necessary. For example, Uyar and coworkers (1978), studying the chemistry of *Myrica gale*, synthesized 2′,6′-dihydroxy-4′-methoxy-3′,5′-di-*C*-methyldihydrochalcone [5-97] by reduction of the corresponding flavanone [5-96] using 10% Pd/C. Similarly, Jain and coworkers (1986) reduced several *C*-benzylated chalcones to the corresponding dihydrochalcones using 20% Pd/C in ethyl acetate at room temperature. The tendency of 2′,4′,6′-trihydroxychalcones to cyclize to flavanones, which are more difficult to reduce to dihydrochalcones than are the chalcones themselves, led Krishnamurty and Satyanarayana (1989) to examine alternative methods for preparing dihydrochalcones. The method chosen for their synthetic studies, catalytic transfer hydrogenation, involves the use of sodium formate as the source of hydrogen and Pd/C as catalyst. Examination of a number of experimental protocols led to the use of a mixture of sodium formate, formic acid, isopropanol, and Pd/C, with which it was possible to obtain yields of up to 65% of 2′,4′,6′,4-tetrahydroxydihydrochalcone from naringenin.

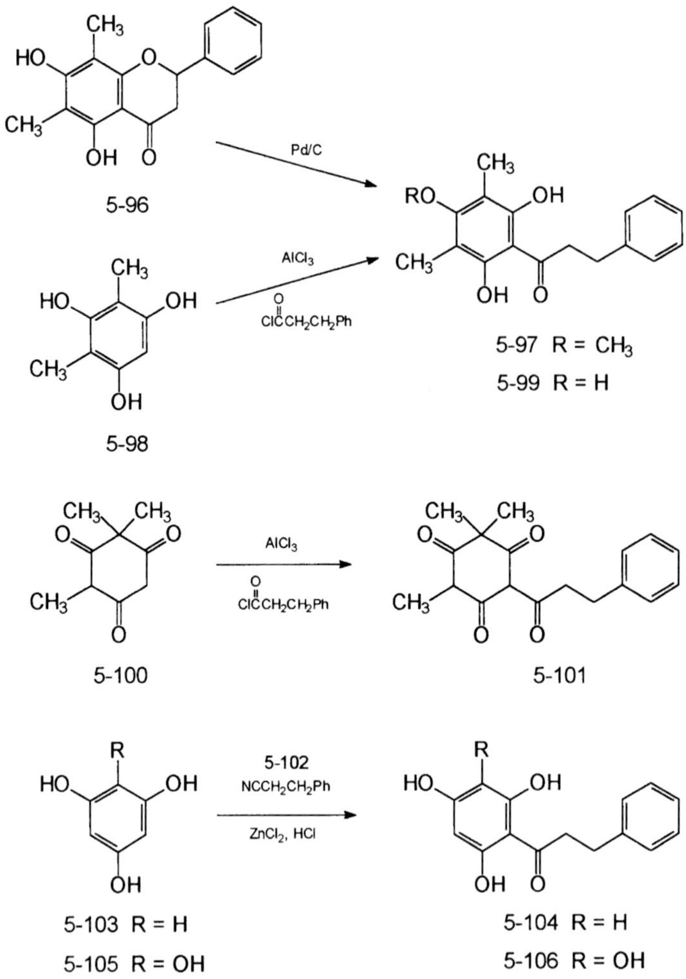

Scheme 5.18 Synthetic routes to dihydrochalcones.

Acylation of the di-*C*-methylphloroglucinol [5-98] with 3-phenylpropionyl chloride by the Friedel-Crafts route was used by Misirlioglu and coworkers (1978) to synthesize 2′,4′,6′-trihydroxy-3′,5′-di-*C*-methyldihydrochalcone [5-99]. The same workers also synthesized 2,2,6-trimethyl-4-(3-phenylpropionyl)cyclohexane-1,3,5-trione [5-101] by Friedel-Crafts condensation of 3-phenylpropionyl chloride with

2,2,4-trimethylhexane-1,3,5-trione [5-100]. A direct and efficient synthesis of dihydrochalcones using 3-phenylpropionitrile [5-102] was described by Sato and coworkers (1995). A mixture of the phenol, 3-phenylpropionitrile, and zinc chloride was saturated with HCl in the cold to yield dihydrochalcones directly. Phloroglucinol [5-103] afforded an 88% yield of 2′,4′,6′-trihydroxydihydrochalcone [5-104]; when 1,2,3,5-tetrahydroxybenzene [5-105] was used as the starting phenol the yield of 2′,3′,4′,6′-tetrahydroxydihydrochalcone [5-106] was 94%.

The major challenge in synthesizing flavonoids with only hydroxyl or methoxyl substituents lies in arriving at starting materials that have the desired substitution patterns. There is usually little need to worry about the fate of these groups during the synthetic manipulations, with the exception of selective dealkylation in some instances (this is covered below in the section on protecting groups). Synthesis of flavonoids that bear sensitive substituents, however, require judicious choice of starting materials, reagents, and reaction conditions. The presence of an acid-labile function, for example, obviously would require working as near neutrality as possible. Similarly, to avoid loses through oxidation of sensitive phenolic compounds, e.g., those with *o*-dihydroxyl groups, care must be taken to prevent contact with air and avoid steps involving strong oxidizing agents. The specific example described below involves the synthesis of a dihydrochalcone that has an unsaturated side chain, in this case the geranyl function. Since the geranyl function contains two double bonds, use of catalytic hydrogenation, normally part of dihydrochalcone synthesis, is precluded. The compound in question is 2′,4′,3,4-tetrahydroxy-5-*C*-geranyldihydrochalcone [5-115], a component of *Artocarpus communis* (Fujimoto *et al*., 1987) that was shown to inhibit 5-lipoxygenase (Koshihara *et al*., 1987). The synthesis of this compound, accomplished by Nakano and coworkers (1989), is an excellent example of synthetic inventiveness. Construction of the B-ring started with 3,4-dimethoxybenzoic acid [5-107], which was converted into 2-(3,4-dimethoxyphenyl)-4,4-dimethyloxazoline [5-108] using 2-amino-2-methylpropanol. The oxazoline was converted to the 2-lithio [5-109] derivative by reaction with *n*-butyllithium in dry tetrahydrofuran at −45°C under argon and subsequently treated with geranyl bromide to yield the *C*-geranyl derivative [5-110]. Compound [5-110] was then converted stepwise to the benzaldehyde, the benzyl alcohol, and the benzyl bromide [5-111]. That series of reactions completed construction of the B-ring synthon. Attention was then turned to constructing the A-ring synthon. 2,4-Dimethoxymethoxyacetophenone [5-112] was condensed with diethyl carbonate using sodium hydride in refluxing toluene to give the doubly protected and activated β-ketoester [5-113] which was then condensed with the 2-*C*-geranyl-3,4-dimethoxybenzyl bromide [5-111] using sodium hydride in dimethylformamide to yield the protected dihydrochalcone [5-114]. The final steps involved removal of the methoxymethyl groups with HCl and the *O*-methyl groups with boron tribromide. The resulting dihydrochalcone [5-115] was identical in all respects to the natural product. The synthetic sequences are presented in Scheme 5.19.

α-Hydroxydihydrochalcones represent a small but interesting subgroup of flavonoids characterized by the chirality of the α-carbon. Typical members of this group are α,2′,4′,4-tetrahydroxydihydrochalcone and its 4-methyl ether from the timber tree *Pericopsis elata* (Fabaceae) and α,2′-dihydroxy-4′,4-dimethoxydihydrochalcone from

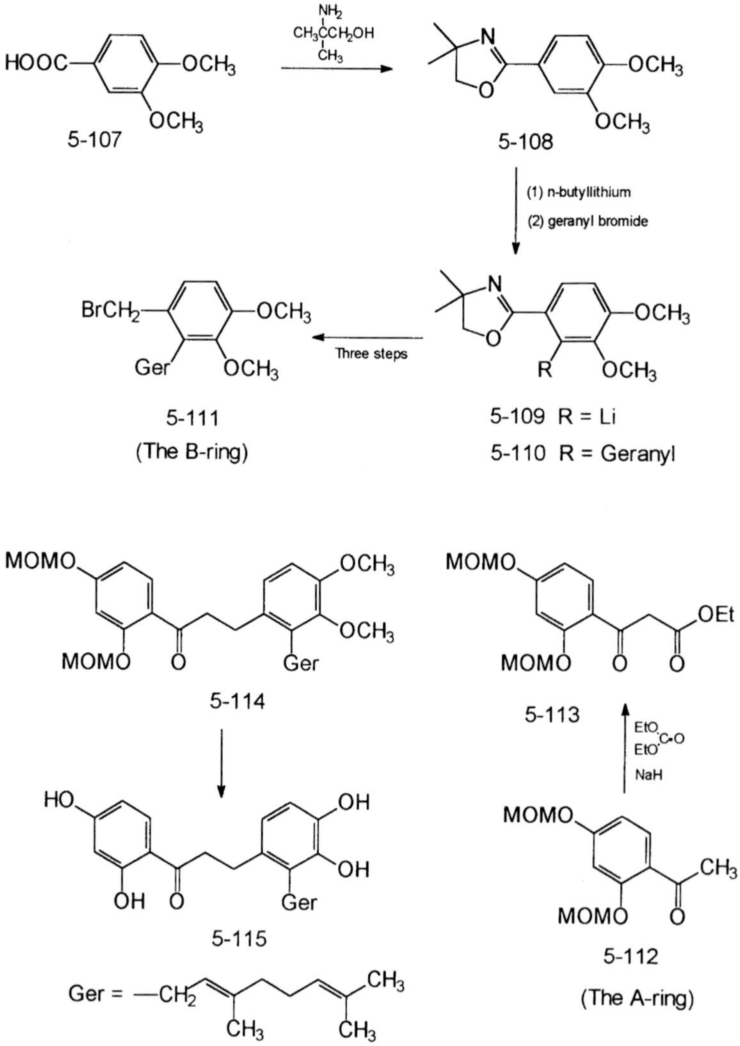

Scheme 5.19 Synthesis of a dihydrochalcone containing a functional group sensitive to reduction. MOM = methoxymethyl.

Pterocarpus angolensis. Augustyn and coworkers (1990) undertook enantioselective synthesis of certain members of this group in order to assess their absolute configuration. The synthetic route involved treatment of (*E*)-chalcones with a triphasic mixture consisting of sodium hydroxide in 30% hydrogen peroxide, carbon tetrachloride, and either poly-L-alanine, which gave (−)-($\alpha R,\beta S$)-chalcone epoxides [5-116], or poly-D-alanine which gave (+)-($\alpha S,\beta R$)-chalcone epoxides [5-118]. Catalytic reduction of [5-16] gave (+)-(αR)-α-hydroxydihydrochalcones [5-17], while [5-118] gave (−)-(αS)-α-hydroxydihydrochalcones [5-119]. See Scheme 5.20 for structures.

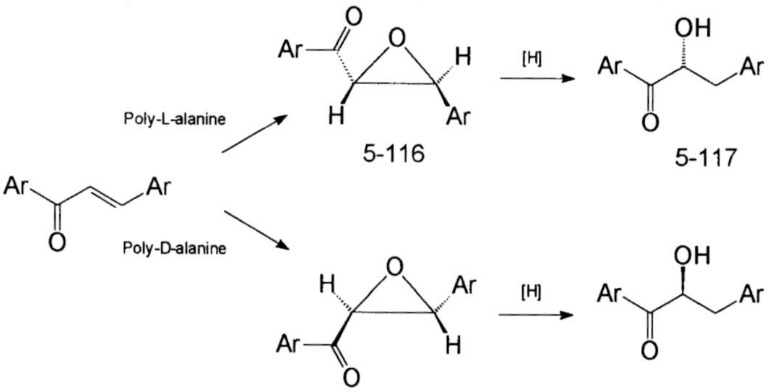

Poly-L-alanine
Poly-D-alanine
[H]
5-116
5-117
5-118
5-119

Scheme 5.20 Synthesis of enantiomeric alpha-hydroxydihydrochalcones.

Flavones and Flavonols

It may seem odd not to have a lengthy section specifically dedicated to the synthesis of flavones and flavonols. Their importance and widespread occurrence would certainly seem to qualify them for special attention! However, several of the more important ways of preparing these compounds have been described above, in the sections on the name reactions, in the discussion of the conversions of chalcones and/or flavanones, and in a number of papers describing selective *O*-alkylation and dealkylation. Extensive reviews of the natural occurrence of flavones and flavonols have appeared regularly in *The Flavonoids* volumes (Wollenweber, 1982, 1994; Wollenweber and Jay 1988) wherein the interested reader will find citations leading to identification of specific flavonoids many of which also describe synthetic work on the compounds.

Flavonolignans

As the reader may recall from Chapter Two, flavonolignans are formed from a flavonoid and 3-methoxy-4-hydroxycinnamyl alcohol (coniferyl alcohol) by radical coupling and are of particular interest, as we shall see, because of their anti-hepatotoxic properties (Chapter Eight). Among the many studies of these unusual compounds have been numerous attempts at their synthesis. We will examine the total syntheses of two of these compounds here, silymarin (silybin) and silychristin. These compounds have the 5,7-dihydroxydihydroflavonol structure in common but differ significantly in the nature of their B-rings.

In silymarin (silybin) the C_6C_3 unit is attached to the B-ring by way of a *p*-dioxane-like arrangement as seen in [5-120]. Note that a stereochemical representation of this compound is not presented here. Because the stereochemistry of silymarin as drawn by Mishima and coworkers (1971) (their structure 1) is opposite that given in a recent paper by Wagner (1986), we will focus on the chemistry of the synthesis. Since the A- and C-rings would be provided by readily available phloroacetophenone, attention was turned to construction of the B-ring. The target compound was 6-formyl-2-(4-hydroxy-3-methoxyphenyl)-3-hydroxymethylbenzo-1,4-dioxane [5-121], which is the benzaldehyde derivative needed for Claisen–Schmidt condensation with phloroacetophenone. This compound was prepared by two different routes. In the first, the catechol derivative [5-122] was condensed with the bromoester [5-123] using potassium *t*-butoxide to produce the ether [5-124]. Both the carbonyl and ester groups were reduced using sodium borohydride. Treatment of the diol with acid resulted in elimination of the methoxymethyl protecting group [5-125] and brought about ring closure to form the *p*-dioxane structure [5-126]. Susceptible groups were reprotected for subsequent steps. Oxidation of the aromatic *C*-methyl generated the aldehyde function needed for the final condensation step. Condensation of phloroacetophenone trimethoxymethyl ether with the aldehyde using sodium hydride in benzene afforded the intermediate chalcone. Oxidation of the chalcone with alkaline hydrogen peroxide produced the chalcone epoxide which upon treatment with acid gave the desired product [5-121]. The reactions can be followed by referring to Scheme 5.21.

The second route involved condensation of 2-methoxymethoxy-5-methylphenoxyacetic acid [5-127] with benzylvanillin [5-128], in the presence of sodium hydride and lithium diisopropylamide to form the hydroxyacid [5-129]. The remaining steps involved cleavage of the methoxymethoxy function, ring closure to the *p*-dioxane, reduction of the acid group to hydroxymethyl, and generation of the aldehyde function at C-5. Condensation of the aldehyde with the acetophenone and appropriate manipulation of protecting groups resulted in desired product. It is interesting to note that, in addition to accomplishing the synthesis of silymarin by two routes, these workers also accomplished the first chemical synthesis of dl-taxifolin (dihydroquercetin)! In order to arrive at suitable conditions for the condensation of the benzodioxane-aldehyde with phloroacetophenone, a model system consisting of phloroacetophenone trimethoxymethyl ether and 3,4-dimethoxymethoxybenzaldehyde was studied. These reactants were condensed in refluxing in benzene in the presence of sodium hydride to give the chalcone. The purified chalcone was oxidized with alkaline hydrogen peroxide, followed by removal of the protecting groups, to yield a dihydroflavonol identical in all respects to the natural product. The work of Tanaka and colleagues (1989) can be consulted if the reader is interested in studying the total synthesis of silychristin, a flavanolignan characterized by a furano ring rather than by the *p*-dioxane system seen in the above study.

Isoflavonoids

Early work in the synthesis of isoflavonoids has been extensively reviewed (Warburton, 1954; Venkataraman, 1959; Baker and Ollis, 1956). More concise updates have occurred from time to time as parts of chapters in the Geissman volume (Ollis, 1962, pp. 385–389), and in *The Flavonoids* series (Wagner and Farkas, 1975, pp. 184–199;

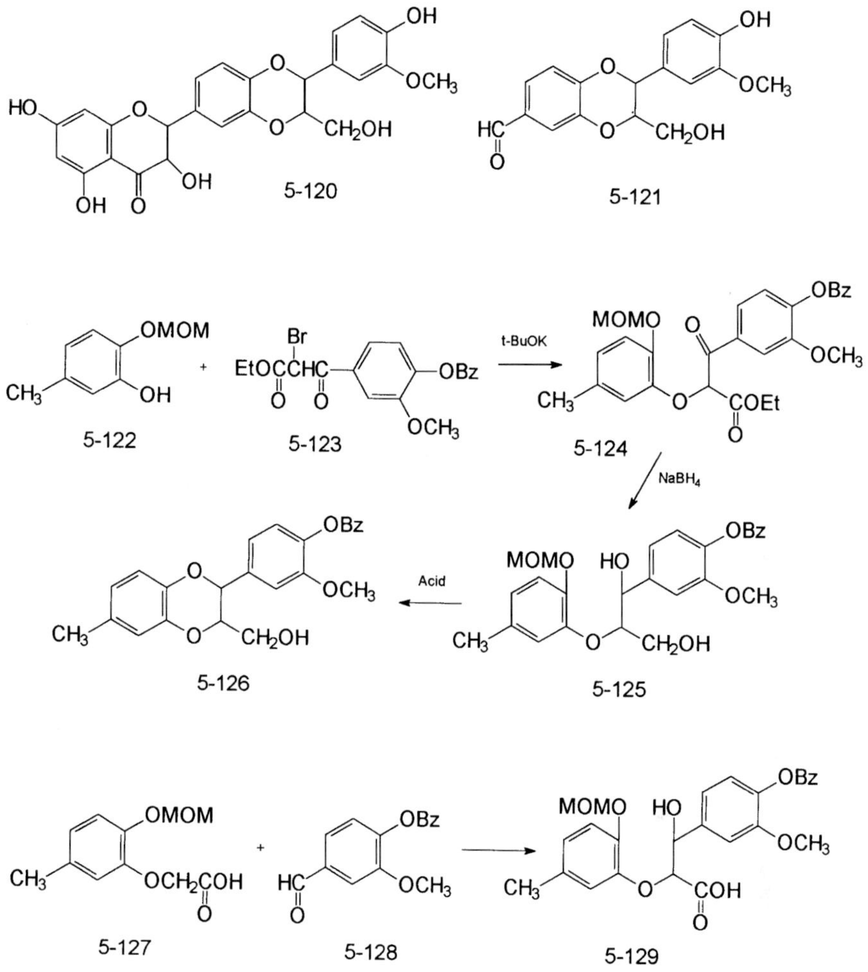

5-120 5-121

5-122 + 5-123 → (t-BuOK) 5-124 → (NaBH$_4$) 5-125 → (Acid) 5-126

5-127 + 5-128 → 5-129

Scheme 5.21 Two synthetic approaches to the flavanolignan system as exemplified by silymarin (silybin). Bz = benzyl, MOM = methoxymethyl.

scattered in Dewick, 1982, 1988, 1994). Donnelley and Bolton (1995) comment on several recent innovations.

Synthesis of isoflavonoids presents a fundamentally different problem than is encountered in the synthesis of flavonoids with the C_6—C_3—C_6 arrangement. Three main synthetic approaches have been followed for the formation of isoflavonoid derivatives: (1) the C_6—C_2—C_6 plus C_1 route, which involves formylation of a desoxybenzoin; (2) C_6—C_3—C_6 to C_6—C_2—C_6—C_1 rearrangement; and (3) the C_6C_3 plus C_6 route involving arylation of a preformed chromanone system. The historical development of these routes is extensive and well beyond the scope of this book. Although some early examples of syntheses will be given here, no effort has been made to cover all aspects of the subject.

The addition of a one-carbon unit to a preformed fourteen-carbon molecule is one of the earliest methods to be used. It is based upon the formylation of a 2-hydroxy-benzoin using a variety of reagents to provide the formyl group. The hydroxyl oxygen becomes the heterocyclic oxygen in C-ring through hemiacetal formation and elimination of water as shown in the following general reaction (Scheme 5.22). Several different reagents have been used as the source of the one-carbon unit including ethyl formate [5-133] and sodium or sodium *t*-butoxide (Mahal *et al.*, 1934), triethyl orthoformate [5-134] and base (Kagal *et al.*, 1956; Levai and Sebok, 1992), ethoxalyl chloride [5-135] (Baker *et al.*, 1953a,b), zinc cyanide and HCl (Farkas *et al.*, 1958), dimethylformamide [5-136] and $POCl_3$, *N,N*-dimethylformamide dimethylacetal [5-137] (Pelter and Foot, 1976; Pelter *et al.*, 1978; Dominguez *et al.*, 1991), methanesulfonyl chloride [5-138] (Bass, 1976), *N*-formylimidazole

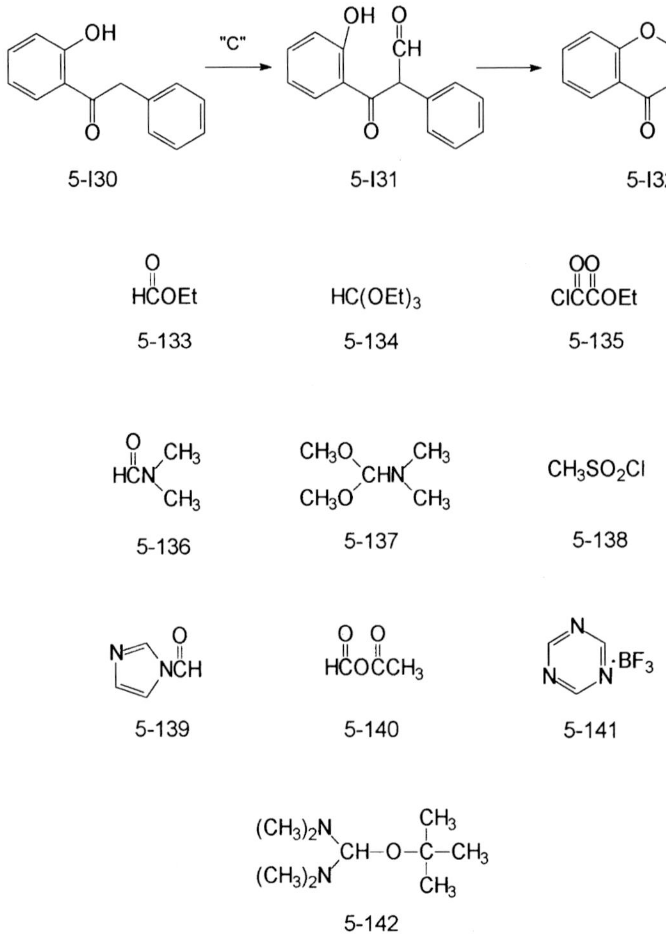

OH
"C"
OH O
CH
O
5-130
5-131
5-132
HCOEt
5-133
HC(OEt)3
5-134
ClCCOEt
5-135
HCN
CH3
5-136
CH3O
CHN
5-137
CH3SO2Cl
5-138
N
NCH
5-139
HCOCCH3
5-140
·BF3
5-141
(CH3)2N
CH-O-C-CH3
5-142

Scheme 5.22 General reaction for the synthesis of isoflavones, and an assortment of reagents used to provide the "C" unit.

[5-139] (Krishnamurty and Prasad, 1977), acetoformic anhydride [5-140], 1,3,5-triazine-boron trifluoride-ether-acetic anhydride [5-141] (Jha *et al*., 1981), and *bis*-(dimethylamino)-*t*-butoxymethane (Bredereck's reagent) [5-142] (Schuda and Price, 1987). Some of the restrictions encountered with the general method include the need to protect all phenolic hydroxyls (other than the one at C-2), the failure of desoxybenzoins with certain oxygenation patterns to react, and difficulties in obtaining good yields of the desired desoxybenzoins in the first place. Some problems are unique to the use of a particular reagent, as in the case of ethoxalyl chloride where decarboxylation of the intermediate carboxylic acid derivative may cause problems.

The Hoesch reaction, which we met earlier in this chapter, is an important source of deoxybenzoins. The reaction involves treatment of a mixture of a phenol [5-143], an appropriate benzonitrile [5-144], and $ZnCl_2$ with dry HCl. The initial product is the ketimine hydrochloride [5-145], which is readily hydrolyzed to the corresponding ketone [5-146] with aqueous acid. An alternative method to using the Hoesch reaction involves Friedel-Crafts alkylation of a phenol (the A-ring to be) with a phenylacetic acid derivative (the B-ring and two-carbon bridge). An elegant example of this approach is the condensation of the phenol [5-147] with 3,4-methylene-dioxyphenylacetyl chloride [5-148] in the presence of titanium tetrachloride in dichloromethane at $-78°C$ to yield the desoxybenzoin [5-149]. Treatment of the desoxybenzoin with bis(dimethylamino)-*t*-butoxymethane (Bredereck's reagent) [5-142] led to the desired product [5-150], calopogonium isoflavone (Iyer and Iyer, 1989). These reactions are summarized in Scheme 5.23.

Many of the problems attendant upon use of the desoxybenzoin route to isoflavones are overcome through the oxidative rearrangement of chalcones. Boron trifluoride-catalyzed rearrangement of chalcone epoxides was used widely for the synthesis of a variety of isoflavones but poor yields in many cases led to a search for a more reliable reagent. Thallium (III) nitrate was found to be very effective in bringing about the desired rearrangement and worked well under a variety of conditions (Farkas *et al*., 1974; Taylor *et al*., 1980). Two examples are summarized below (Scheme 5.24). The first is a synthesis of erythrinin-A in which the chromanochalcone [5-151] was treated with thallium (III) nitrate and methyl orthoformate [$HC(OCH_3)_3$] to yield the dimethylacetal [5-152], which was readily cyclized to the isoflavone [5-153] by treatment with acid. This treatment also removed the methoxymethyl protecting group. The synthesis was completed by treatment of [5-153] with DDQ to introduce the double bond into the chromane system to give erythrinin-A [5-154] (Suresh *et al*., 1985). Our second example comes from a synthesis of parvisoflavone-B [5-157] (Tsukayama *et al*., 1991). The step of interest to us involves the conversion of chalcone [5-155] to the isoflavone derivative [5-156] using thallium (III) nitrate and methanol in 1,2-dichloroethane, and then acid; the intermediate was not isolated. In this synthesis the chromene double bond was introduced in a sequence of reactions starting from the chromone carbonyl group.

Dimethylacetal [5-158] was prepared by treatment of the corresponding chalcone with thallium (III) nitrate and trimethyl orthoformate (Camarda *et al*., 1982). The product of that condensation was converted to isoflavone [5-159] upon acid treatment.

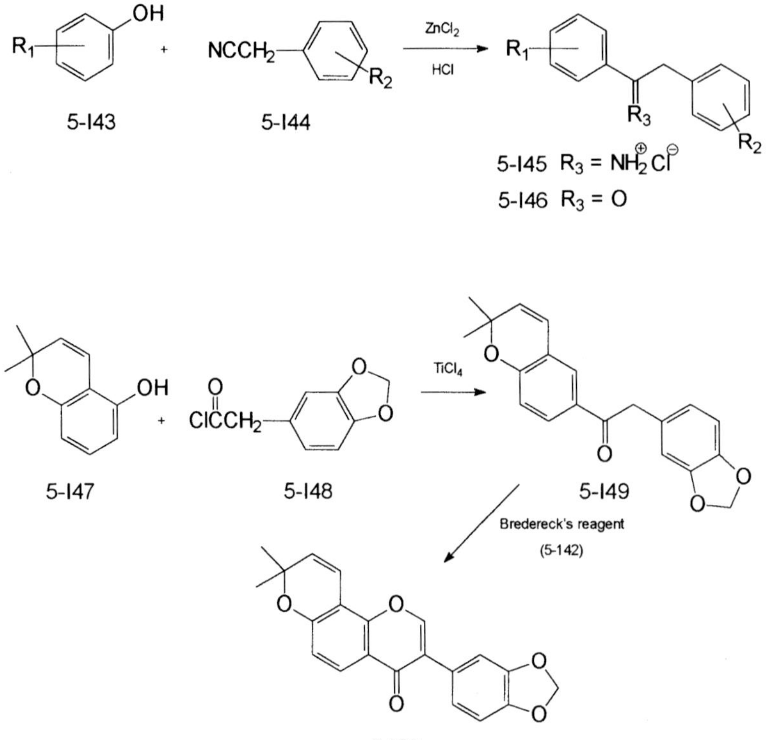

Scheme 5.23 Classical Hoesch synthesis of desoxybenzoins (top), and the synthesis of calopogonium isoflavone using the Friedel-Crafts reaction.

Reduction of [5-159] with 10% palladium on barium sulfate afforded three products, the debenzylated isoflavone [5-160], 6,7,3′-trihydroxy-2′,4′-dimethoxyisoflavanone [5-161], and the corresponding isoflavan [5-162]. The isoflavan was identical to the natural pigment ($\pm$)-bryaflavan from the West Indian legume *Brya ebenus.* Reduction of the isoflavanone with lithium aluminum hydride afforded the isoflav-3-ene [5-163], which was shown to be identical with the pigment from the tropical African timber tree *Baphia nitida* (Fabaceae). Structures of these products can be found in Scheme 5.25. Isoflav-3-enes can be useful intermediates in the synthesis of other isoflavonoid types, e.g., 6a-hydroxypterocarpans. Additional examples of isoflavone syntheses are discussed by Dewick (1988, 1994) and by Donnelly and Boland (1995).

Several variants in the use of thallium have been developed and used to excellent effect. For example, Khanna and coworkers (1992) used thallium (III) *p*-toluenesulfonate (TTS) in propionitrile to prepare isoflavones, in some cases in nearly quantitative yields. The presence of strong electron-withdrawing groups were a problem with this reagent, however, causing the formation of flavones as well as the desired isoflavones. Sensitivity to electronic effects was overcome to a large extent by the use of thallium (III) perchlorate (TTPC) (Singh and Kapil, 1993). The active

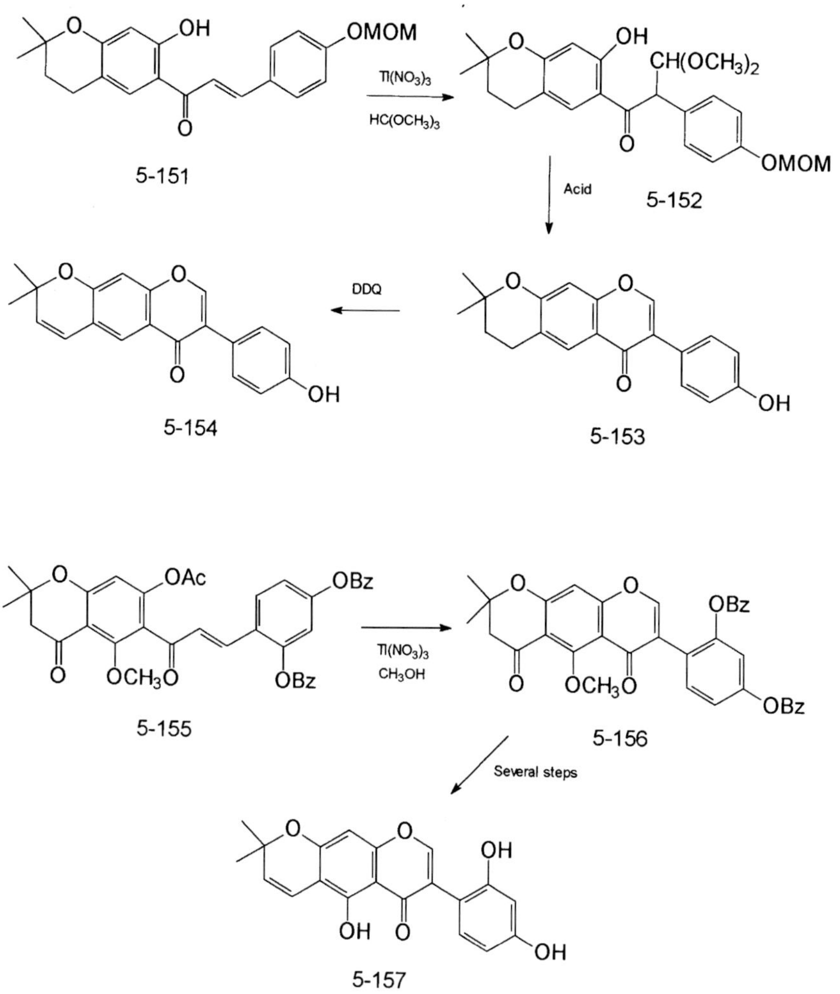

Scheme 5.24 Synthesis of two isoflavones using thallium (III) nitrate. Ac = acetyl, Bz = benzyl, MOM = methoxymethyl.

TTPC is generated *in situ* by exchange of perchlorate ion with either thallium nitrate (TTN) or acetate (TTA). Donnelly and Boland (1995) point out that one of the serious disadvantages in the use of thallium salts lies in their extreme toxicity and the need to use stoiciometric quantities. They go on to describe several alternative routes to isoflavones that require less dangerous reagents.

One way to avoid using thallium-based reagents involves direct alkylation of chromanones. Several very cleverly devised reactions using organometallic reagents have been described that take advantage of the reactivity of the 3-C position, two of which are summarized in Scheme 5.26. 3-Phenylsulfonylchromanone-4 derivatives [5-164] were treated with potassium hydride and triphenylbismuth carbonate to yield the 3-phenyl-3-phenylsulfonyl derivative [5-165]. When treated with zinc in acetic

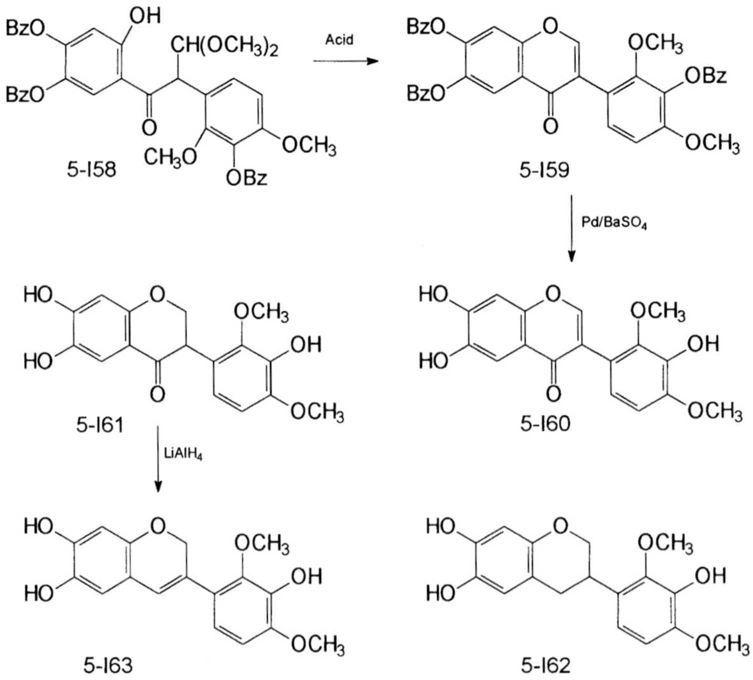

Acid

5-158

5-159

$Pd/BaSO_4$

5-160

5-161

$LiAlH_4$

5-163

5-162

Scheme 5.25 Palladium on barium sulfate reduction of an *O*-benzylated isoflavone and lithium aluminum hydride reduction of an isoflavanone.

acid [5-165] was reduced to the corresponding isoflavanone, whereas treatment with aluminum chloride for a few minutes resulted in formation of the isoflavone (Santhosh and Balasubramanian, 1992). Donnelly and Boland (1995) pointed out that, despite the good yields from this reaction, the difficulty of preparing bismuth derivatives precluded its wide use. They then described syntheses using aryllead (IV) reagents, which are more easily prepared than their bismuth counterparts. Work from Donnelly's laboratory (Donnelly *et al.*, 1993a) exemplifies this alternative approach. Starting with chromanones having a variety of substitution patterns, 3-allyloxycarbonylchroman-4-ones [5-166] were prepared using lithium bis(trimethylsilyl)amide (LHMDS) and allyl cyanoformate [5-167]. Treatment of the esters with aryllead (IV) triacetate in the presence of pyridine resulted in the 3-aryl-3-allyloxycarbonylchromanones [5-168], which could then be converted either into isoflavanones or isoflavones depending upon choice of conditions. Further studies from that group (Donnelly *et al.*, 1993b) dealt with arylation of 3-phenylthiochroman-4-ones with aryllead (IV) triacetates. Again, choice of oxidizing or reducing conditions lead, respectively, to either isoflavones or isoflavanones (nickel boride was used as reducing agent for the latter).

We have just seen two very sophisticated ways to synthesize isoflavanones. More "ordinary" ways to make these compounds also exist. Isoflavones are efficiently and selectively reduced to isoflavanones with diisobutyl aluminum hydride (DIBAL)

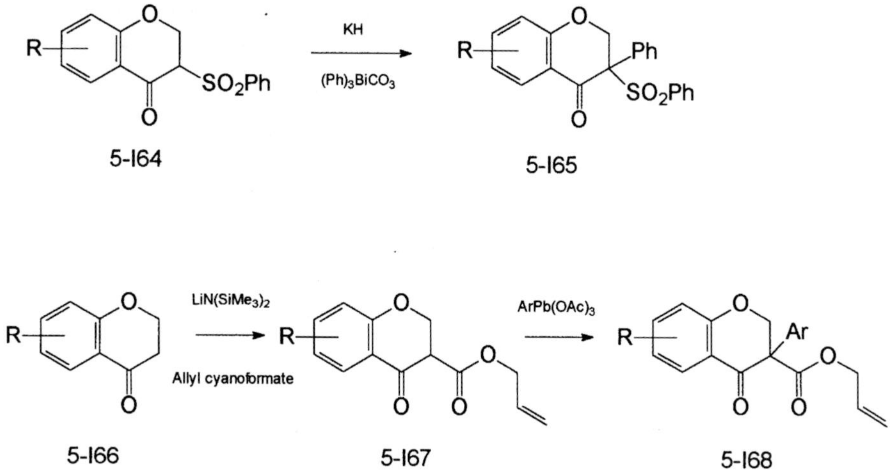

Scheme 5.26 Use of organometallic reagents based on bismuth (top), and lead to effect *C*-alkylation in the synthesis of isoflavonoids.

(Major *et al.*, 1988). The reduction can also be effected with sodium hydrogen telluride (Jain *et al.*, 1991) or using catalytic transfer hydrogenation with ammonium formate as the hydrogen donor and Pd/C as the catalyst (Krishnamurty and Satyanarayana, 1986; Wähälä and Hase, 1989). In a synthetic procedure that parallels the route used for isoflavones, a methylene group is introduced directly into a suitable desoxybenzoin [5-169] (Scheme 5.27). Ethoxymethyl chloride supplied the methylene group in a synthesis of the hydroxymethyl desoxybenzoin derivative [5-170] which was subsequently converted to sativanone [5-171] (Jain and Nayyar, 1987).

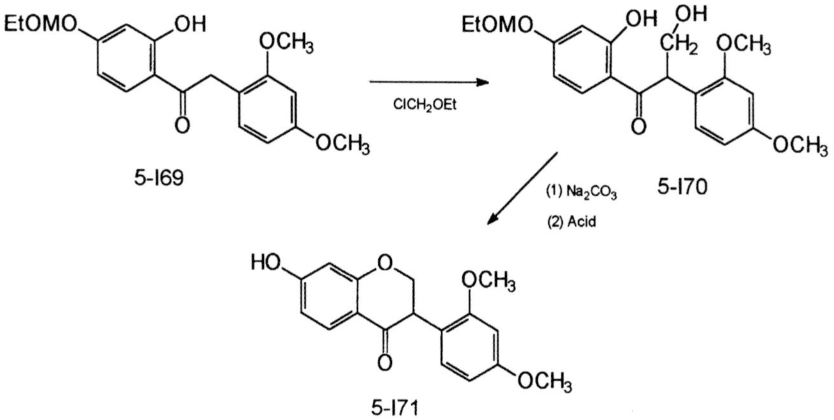

Scheme 5.27 Synthesis of an isoflavanone directly from a desoxybenzoin. EtOMO = ethoxymethoxy.

A considerable amount of work has been done on the synthesis of tetracyclic isoflavonoids no doubt motivated by the important roles some of these compounds play in plant defense against microorganisms, e.g., the phytoalexins. The reader is well advised to consult the comprehensive chapters by Dewick (1982, 1988, 1994) and the excellent recent review by Donnelly and Bolton (1995) for many examples and lists of relevant citations. A few samples of these syntheses demonstrating the challenges and the inventiveness in meeting them follow.

One of the larger groups of multiring isoflavonoids are the pterocarpans. These compounds are widely distributed in members of Fabaceae and have attracted a good deal of attention including, of course, efforts to synthesize them in the laboratory. As Dewick (1994) points out, about the most straightforward route to these compounds is simply the sodium borohydride reduction of a 2′-hydroxyisoflavone that has the appropriate substitution pattern (Krishna Prasad *et al.*, 1986). The general reaction is shown in Scheme 5.28 (top reaction). At the other extreme, are syntheses such as those based upon the Heck arylation of a chromene system using lithium tetrachloropalladate as catalyst. This reaction is shown in Scheme 5.28 (bottom reaction) and treats the synthesis of leiocarpin [5-176] which involves reaction between the dichromene system [5-174] and the mercury derivative 2-chloromurcurio-4,5-methylenedioxyphenol [5-175] (Narkhede *et al.*, 1989, 1990). The reaction runs at room temperature in acetone and has the distinct advantage of selectively operating on the simpler of the two chromene systems. A further example of this reaction is discussed by Donnelly and Boland (1995, p. 332).

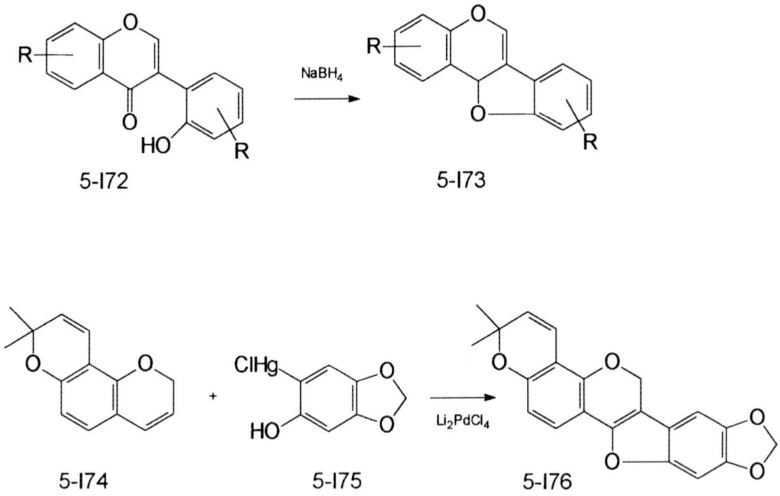

R O O HO R $NaBH_4$ R O O R

5-172 5-173

O O ClHg O + HO O Li_2PdCl_4 O O O O O

5-174 5-175 5-176

Scheme 5.28 Comparison of simple reduction (top), and Heck acylation in the synthesis of the pterocarpan ring system.

Several 6a-hydroxypterocarpans have figured significantly in the study of phytoalexins, a prominent member of which is pisatin, whose role in the phytoalexin story is discussed in Chapter Seven. It was the first phytoalexin to be identified so it seems appropriate to examine its synthesis here. The problems to be contended with in this synthesis involve establishment of the A-ring and B-ring substitution patterns, preparation of an intermediate compound that could be used for 6a-hydroxylation, the hydroxylation process itself, and establishment of the pyran ring. Mori and Kisida (1988, 1989) accomplished these stages in the following fashion. The A-ring and B-ring substitution patterns were established through the synthesis of 2′-hydroxy-7-methoxy-4′,5′-methylenedioxyisoflavone [5-177, R=H], the dimethyl *t*-butyl silyl ether of which was reduced to the isoflavan-4-ol [5-178] with sodium borohydride. Elimination of water yielded the isoflav-3-ene [5-179], which was converted to the diol [5-180] with osmium tetroxide. The isomers of [5-180] were resolved by HPLC of their (+)-camphor-10-sulfonates followed by removal of the silyl function with tetrabutylammonium fluoride. Cyclization completed the synthesis of (+)-pisatin [5-181].

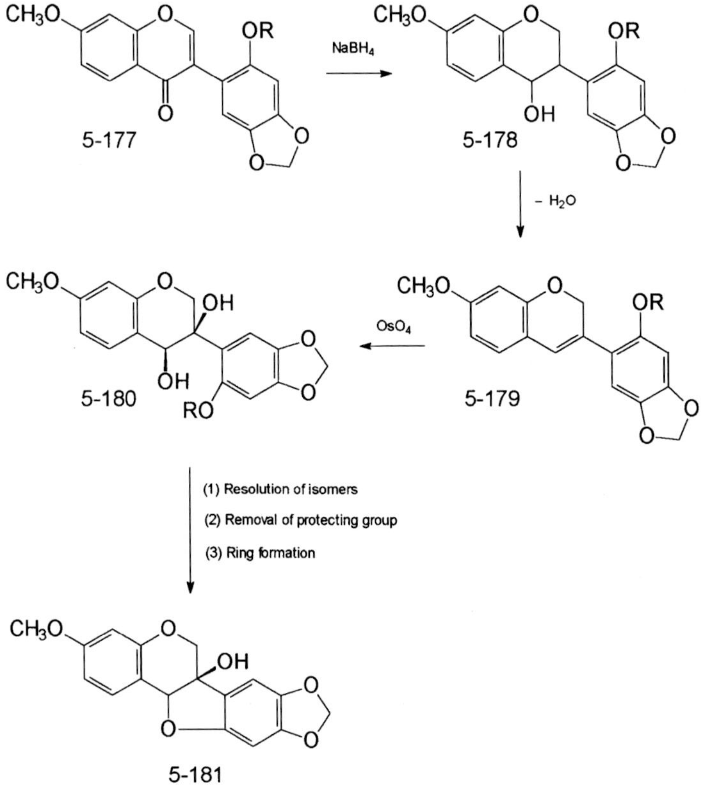

CH_3O OR O 5-177 —$NaBH_4$→ CH_3O OR OH 5-178

— H_2O →

CH_3O OR 5-179 —OsO_4→ CH_3O OH OH RO 5-180

(1) Resolution of isomers
(2) Removal of protecting group
(3) Ring formation

CH_3O OH 5-181

Scheme 5.29 Synthesis of (+)-pisatin, a 6a-hydroxypterocarpan. The protecting group (R) is dimethyl *t*-butylsilyl.

Rotenoids are distinguished from other multiring isoflavonoids by the presence of an additional carbon atom attached to the C-2 position of the isoflavone skeleton. This “extra” carbon, starred in [5-183], arises in Nature from the methyl carbon of a 2′-methoxy function. The classical chemical synthesis actually parallels the biosynthetic process through construction of a functionalized ether at that position (Robertson, 1933). The carbethoxymethyl ether [5-182] was converted to the dehydrorotenone [5-183] with sodium acetate and acetic anhydride. Reduction of the B-ring has been accomplished in a number of ways including by catalytic hydrogenation or by reduction by sodium borohydride and reoxidation of the hydroxyl group at C-12 to a carbonyl (Miyano and Matsui, 1958). Several very clever, and quite complicated, syntheses of rotenoids have been described in the recent literature a sampling of which appear in Dewick (1994, pp. 164–165) and in Donnelly and Boland (1995, pp. 335–336). An interesting example of inventive chemistry involves the use of aryl propargyl ethers [5-184]. The reaction progresses by way of a Claisen rearrangement of the ether to intermediate [5-185] and subsequent cyclization to the desired product [5-186] (Scheme 5.30, middle series of reactions). Our final example of a multiring isoflavonoid synthesis comes from the work of Lai and colleagues (1989) who employed a condensation of benzoyl chloride derivatives, here represented by [5-187], with 4-phenylsulfonyl chroman [5-188] as the source of the B- and D-rings (Scheme 5.30, bottom set). The products of this condensation were compounds of the sort seen as [5-189]. The presence of the sulfonyl group aids in activating the chroman ring carbon for coupling with the acyl chloride. Removal of the sulfur-containing function using Raney-Ni afforded [5-190] dehydrogenation and selective demethylation of which led to ring closure to form the final rotenoid derivatives [5-191].

Protecting Groups in Flavonoid Synthesis

The synthesis of flavonoids provides an excellent opportunity to study the art of using protecting groups. The critical role that protecting groups play in allowing manipulations in one part of a molecule without disrupting things in another part can hardly be exaggerated. Dewick (1988, p. 147) specifically addresses the subject. A study of many of the syntheses featured above shows the frequent involvement of a de-*O*-methylation step, such as seen in some of the biflavone syntheses. Horie and his many coworkers at the University of Tokushima continue to add new applications of selective *O*-alkylation and dealkylation reactions in flavonoid synthesis. Examples from their series, effectively titled "Studies of the selective *O*-alkylation and dealkylation of flavonoids," include a synthesis of 5,6-dihydroxy-3,8,4′-trimethoxyflavone (Horie *et al.*, 1982, No. 5 in their series), a synthesis of 5,6,7-trioxygenated flavones related to pectolinarigenin (Horie *et al.*, 1985, No. 7), studies of 7-hydroxy-3,5,8-trimethoxyflavones (Horie *et al.*, 1987, No. 10), studies of 5,8-dihydroxy-6,7-dimethoxyflavones (Horie *et al.*, 1995, No. 17), a synthesis of 5,6,7-trihydroxy- and 5,6-dihydroxy-7-methoxyisoflavones (Horie *et al.*, 1996a, No. 20), and a synthesis of 3,5,7-trihydroxy-6,8-dimethoxy- and 5,7-dihydroxy-3,6,8-trimethoxyflavones (Horie *et al.*, 1996b, No. 21). Other workers have added to the list of compounds synthesized using selective demethylation as we see in a paper describing selective demethylation of 5-*O*-methyl flavonoids (Khan *et al.*, 1994). Another very commonly used protecting device is the benzyl group excellent use of

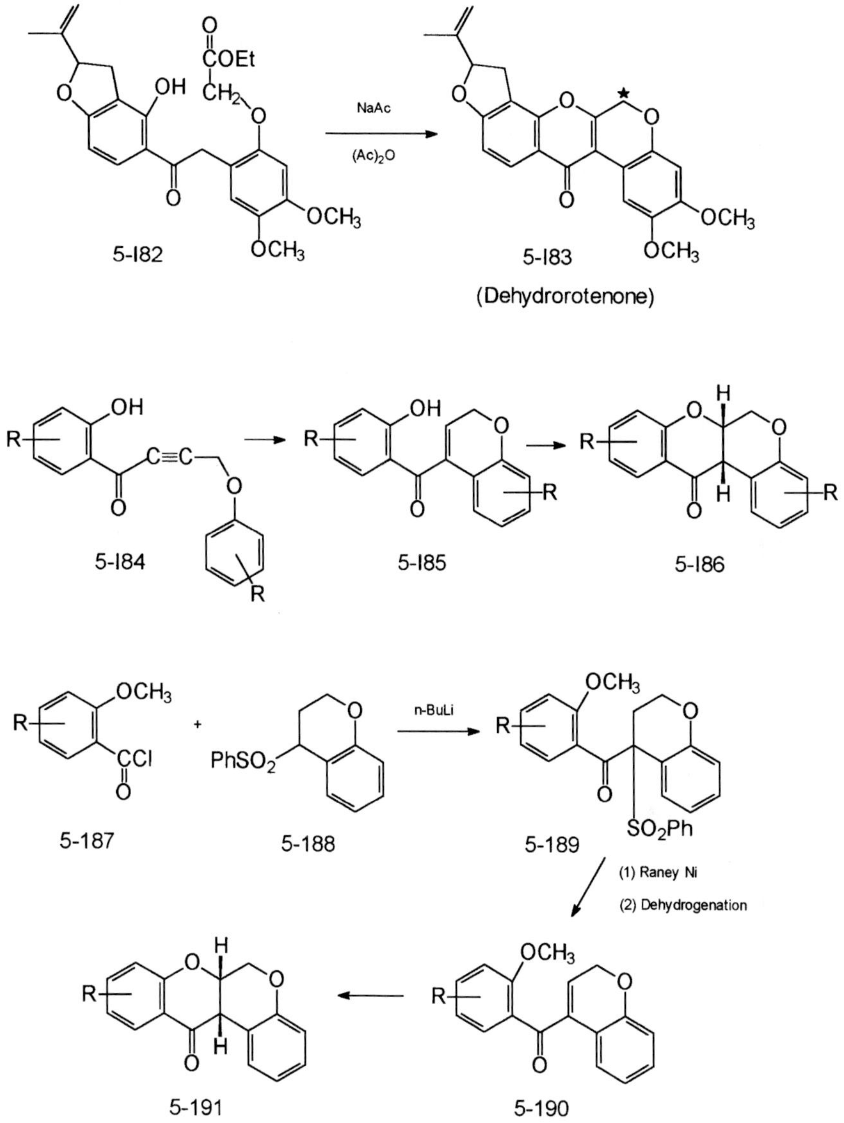

Scheme 5.30 Synthetic routes to rotenoids. Ph = phenyl.

which we saw above in the flavonolignan syntheses. Removal of the benzyl group is done by hydrogenation under mild conditions, which is important if acid- or base-sensitive functional groups are present in the compound. Straightforward applications of debenzylation can be seen in the synthesis of 5,3′,4′-trihydroxy-3,6,7,8-tetramethoxyflavone (Parmar *et al.*, 1987), where the B-ring entered the scene as the 3,4-dibenzyloxybenzoic acid anhydride (and potassium salt), and a synthesis of 3,5,6,7,8,2′,4′,5′-octaoxygenated flavones by Iinuma and coworkers (1986b) in which removal of the benzyl groups from 3-hydroxy-5,6,8,4′,5′-pentamethoxy-7,2′-dibenzyloxyflavone was accomplished using 10% Pd/C in ethyl acetate in an atmosphere of hydrogen.

A typical example of selective demethylation can be seen in the synthesis of a series of isoflavones by Horie and coworkers (1996a). 5-Methoxy groups were selectively cleaved with 5% (w/v) anhydrous aluminum bromide in acetonitrile. 7-Methoxy groups were cleaved with 30% anhydrous aluminum chloride in acetonitrile at 70°C for periods up to 48 h.

Whereas the methyl group can be comparatively difficult to remove, the methoxymethyl group is readily removed using moderately gentle acid hydrolysis. Several examples of the use of this protecting group appeared in the examples above including the flavonolignan syntheses. A related protecting agent is the methoxyethoxymethyl group which was successfully employed in a synthesis of 8-(1,1-dimethylallyl)-apigenin [5-198], one of the components of *Platanus acerifolia* buds. Raguenet and coworkers (1996) employed benzyl chloride and potassium carbonate in dimethylformamide to prepare 2′,4′-dibenzylphloroacetophenone [5-192] which was condensed with 4-methoxyethoxymethoxybenzaldehyde [5-193] to yield chalcone [5-194]. The chalcone underwent cyclization in the presence of DDQ to yield flavone [5-195]. Removal of the benzyl groups was accomplished by allowing the flavone to stand for several hours with Pd/C and ammonium formate. Prenylation at the C-7 hydroxyl group to give [5-197] was accomplished using prenyl bromide in the presence of potassium carbonate and tetrabutylammonium iodide. Claisen rearrangement of the 7-prenyl ether was then effected using acetic anhydride and anhydrous sodium acetate to produce the desired product [5-198]. Under these conditions the protecting group at C-4′ was also removed. The sequence of reactions is shown in Scheme 5.31.

Side chain unsaturations, such as the double bond in dimethylchromeno flavonoids, can be "protected" by delaying their establishment until late in a synthesis. Two examples of this type of manipulation were seen in examples above. In one of them, a chromone system was maintained until the remainder of the molecule had been completed, whereupon it was reduced to the benzyl alcohol which was then induced to undergo elimination of water to form the desired double bond. In the other example, the chromano system was chlorinated on the benzylic carbon using N-chlorosuccinimide. Elimination of hydrogen chloride established the double bond.

Enzyme-Mediated Acylation

In this short section we will examine some recent research that deals with the use of enzymes as tools in organic synthesis. The availability of a few enzymes that are soluble in non-protic solvents and still maintain their catalytic properties has stimulated study of the applicability of these methods to certain problematic situations (Riva *et al*., 1988, and citations therein). This group developed methods for acylating flavonoid glycosides using 2,2,2,-trifluroethyl butanoate in pyridine with the enzyme subtilisin (Danieli *et al*., 1989, 1990). In the first of these papers the

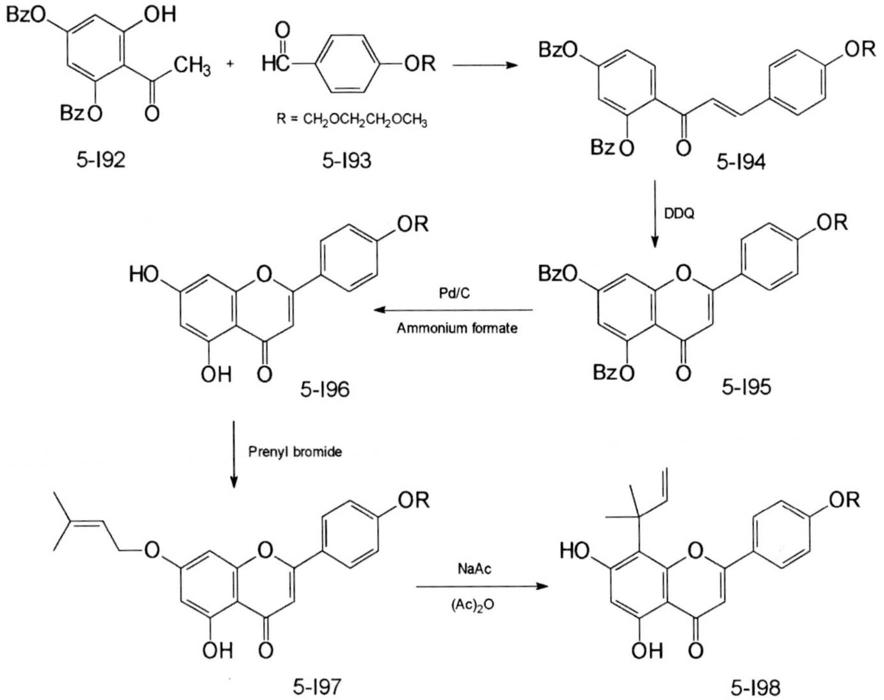

Scheme 5.31 An example of the use of protecting groups in flavonoid synthesis. Ph = phenyl.

workers found that quercetin 3-*O*-glucoside and luteolin 7-*O*-glucoside yielded mixtures of 3″-*O*-mono-, 6″-*O*-mono-, and 3″,6″-*O*-diacyl derivatives. Quercetin 3-*O*-rhamnoside was not acylated under these conditions. In their second paper, these workers applied the reaction to flavonol 3-*O*-diglycosides. Rutin and hesperidin, which have 6-*O*-(α-L-rhamnosyl)-D-glucose in common, were acylated at the 3-position of the glucose moiety. Naringin, which has the 2-*O*-(α-L-rhamnosyl)-D-glucose unit was butanoylated at C-6 of the glucose. When a glucosylarabinoside was used, esterification produced a complex mixture of products including esterification of the C-3 position of the glucose moiety and the C-4 position of the arabinose.

The preceding examples show the utility of non-aqueous enzyme-assisted syntheses but were not used to produce naturally occurring acylated flavonoid glycosides. More recent work by these investigators have addressed this issue. Quercetin 3-*O*-(6″-*O*-malonyl-β-D-glucoside) was prepared by a two step synthesis first involving introduction of a methyl malonate group at the 6″-position using 2-chloroethyl methyl malonate and subtilisin. A second enzyme, "biophine esterase," was then employed to aid in the removal of the methyl group (Danieli *et al.*, 1993). Although the desired product was obtained, the procedures were complicated and the yields low. A significantly more efficient synthesis was developed that is also easier to perform (Riva *et al.*, 1996). The first step involved esterification of quercetin 3-*O*-glucoside at the 6″-hydroxyl group with dibenzylmalonate using a lipase from

Candida antarctica as catalyst. The reaction was run for 12 days at 45°C in an acetone–pyridine mixture and afforded excellent yields (74% with quercetin 3-*O*-glucoside, 79% with rutin). In the case of rutin, esterification again occurred at the 3-hydroxyl group of the glucose unit. Treatment of naringenin 7-*O*-(2-*O*-α-L-rhamnosyl)-*β*-D-glucoside (naringin) under the same conditions resulted in a 69% yield of the benzylmalonate attached at the glucose C-6 position. All of the benzylmalonates synthesized in this study were debenzylated in high yield using Pd/C. Additional information on enzyme-assisted acylation of natural products (not restricted to flavonoids) can be found in recent papers by Danieli and Riva (1994, 1996) and by Danieli *et al*. (1997).

CHAPTER SIX

BIOSYNTHESIS AND GENETICS

INTRODUCTION

Few subjects in the chemistry and biochemistry of plant natural products have been discussed as often, or in as much detail, as has flavonoid biosynthesis. From the beginnings, when the individual steps in the pathway were being elucidated, through the "middle years" when the enzymes active in the pathway were being studied, to the present molecular genetic phase, the subject has been the focus of almost countless re-examinations. A few of the steps in the overall pathway have not yet revealed their innermost secrets (e.g., formation of aurones and dihydrochalcones), but the mainstream of metabolic activity is very well understood indeed. Taking the lead from the Geissman volume, where Birch (1962) reviewed the biosynthesis of flavonoids and anthocyanins (in seven pages with 16 citations), each volume of *The Flavonoids* series has contained a chapter updating current knowledge in the field. In the first volume, Hahlbrock and Grisebach (1975) reviewed the field (in 49 pages with over 160 citations) including one of the first extensive discussions of enzymology. Ebel and Hahlbrock (1982) updated the subject using 38 pages (ca. 180 citations) focusing on enzymology and an increasing understanding of the mechanism of some of the reactions. Regulation of the pathway began to require more space and genetic information became a regular subject as well. The 1988 volume contained a chapter (26 double-column pages with small print!) by Heller and Forkmann that utilized information from over 250 citations on subjects that included the origin of flavonoid precursors (the phenylpropanoid pathway), details of individual reactions in the flavonoid pathway, and new genetic information. In the most recent volume, Heller and Forkmann (1994) again reviewed the subject (36 pages with well over 200 citations) providing very detailed accounts of most aspects of the subject. They did not include extensive references to flavonoid genetics, however. This subject received special attention in this volume in the form of its own chapter (28 pages with over 300 citations) prepared by Forkmann (1994). Other major treatments of the subject include the chemically and biochemically detailed treatment of *Flavonoid Metabolism* by Stafford (1990). Of somewhat broader scope, is the erudite (what else?) treatment of *The Shikimate Pathway* by Haslam (1974).

The challenge facing the present writer, then, is to present an overview of flavonoid biosynthesis, the genes controlling the various steps, and recent attempts to understand the various process at the level of the gene. A strict historical approach offers an interesting look at the individual accomplishments of the various workers who have contributed to our current understanding of the subject, but is difficult to follow in terms of the individual steps in the pathway. Because we know a good deal about most of the steps in the formation of flavonoids, it is easier to let the narrative follow the pathway with occasional excursions into related material. Although the first step in the formation of flavonoids technically involves condensation of a phenylpropanoid derivative with malonyl coenzyme-A, current discussions of the pathway often start with the first committed step in the phenylpropanoid pathway, i.e., the formation of cinnamic acid from phenylalanine. We will take that approach here, but first we have to attend to some other matters.

Until radiolabelled precursors became available in the middle 1950s, it was not possible to say with any degree of certainty that the formation of a natural product followed one particular pathway and not another. With the availability of labeled precursors, however, this picture changed dramatically. We will see below just how effectively pioneering workers in the flavonoid biosynthetic field used ^{14}C- and ^{3}H-labelled precursors, and, more recently, how important the use of ^{13}C-labelled compounds has become. But these early workers were pioneers only in the sense that they had new techniques available to apply. Others had been pondering the origin of flavonoids, and other natural phenolic compounds as well, for many years. Some of the very early ideas, such as the origin of flower pigments (we know them now as anthocyanins) from the breakdown of chlorophyll, which we saw in Chapter One, were quite imaginative but doomed because of the total ignorance in those days of the molecular nature of living things. In more enlightened times, when knowledge of both cell structure (and hence an appreciation of natural complexity) and chemical structure was expanding rapidly, more sophisticated hypotheses concerning the routes of formation of biochemical molecules emerged. Early thinking on the subject of flavonoid biosynthesis was reviewed by Geissman and Hinreiner (1952a,b), who presented an overview of structures known at the time and discussed several of the major ideas then current about how these compounds came about. I have chosen a few of the more important issues for discussion here.

Although many of the ideas of the time were useful, at the very least for the purpose of stimulating discussion, many others were compromised by two factors. Theories of biosynthesis (or biogenesis as it was often called then) were often based on natural products whose structures were subsequently found to be wrong. Secondly, reactions suggested to explain formation of a given natural compound often involved conditions that do not occur in plant cells. Strong acid or base catalysis, high pressure, and elevated temperatures, all or some of which are required for some of the reactions invoked, simply are not available in living organisms, which as we know, require reactions to proceed at ambient temperature in an aqueous, buffered medium. Much, if not most, of biochemical speculation was done by scientists who were primarily organic chemists unfamiliar for the most part with the intricacies of the living cell. It should also be remembered that the first enzyme was not obtained in crystalline form until 1926 when James Sumner, working at Cornell University, obtained crystals of urease from jack beans. Acceptance of the idea that water soluble, high molecular weight proteins, i.e., enzymes, catalyzed all biological reactions, was not immediate, especially in view of the strong opposition to such "heresy" from German biochemists such as Richard Willstätter. It was not until the 1930s that the idea of enzymes became widely accepted.

On a more positive note, some important relationships between simple compounds and more complex natural products were beginning to emerge. It had been recognized for many years that there were similarities between certain amino acids and parts of the larger, more complex alkaloids. For example, the tyrosine molecule [6-1], or the *p*-hydroxyphenylethylamine part at least, could be recognized as part of several alkaloids, for example in isoquinoline derivative [6-2]. Similar observations had been made with regard to tryptophan [6-3] and the indole alkaloids, e.g., [6-4]. Subsequent work, of course, has shown that both of these amino acids are important precursors for certain alkaloids. The same sort of relationship was noted between simple C_6C_3 compounds, such as phenylalanine, tyrosine, and cinnamic

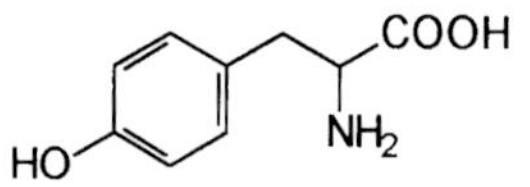

COOH
HO
NH_2
6-1
CH_3O
CH_3O
N
CH_3
6-2 CH_3

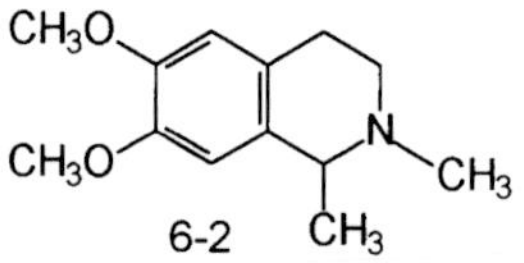

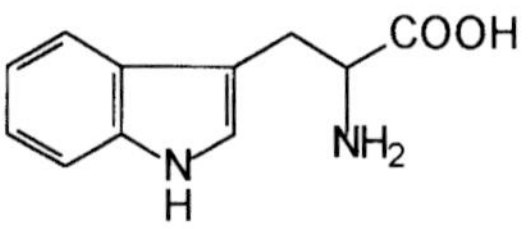

COOH
N H
NH_2
6-3
N H
N
CH_3
CH_3 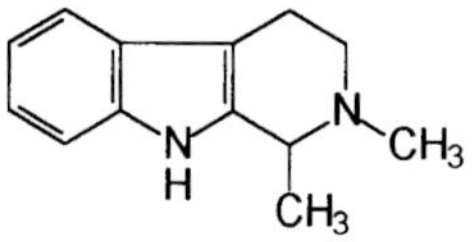
6-4

acid, and flavonoids whose fundamental $C_6C_3C_6$ skeleton had been established. Another key observation was that chemical conversion of a member of one flavonoid structural class to another was possible as had been demonstrated in the conversion of certain colorless flavonoids to anthocyanidins (Blank, 1947). T. R. Seshadri, the illustrious Indian natural product chemist, suggested that one fully formed flavonoid could serve as a precursor for another. In model systems he and his students had successfully hydroxylated quercetin to gossypetin using potassium persulfate ($K_2S_2O_8$) (Scheme 6.1). Continuing in the same vein, Seshadri suggested that reduction of a flavonoid might also be possible. This is represented in Scheme 6.1 by the second reaction where gossypetin is seen to lose one of its B-ring hydroxyls to become herbacetin. We now know that establishment of hydroxylation patterns occurs at an early state of flavonoid biosynthesis and that removal of an hydroxyl group does not occur with flavonoids under normal circumstances.

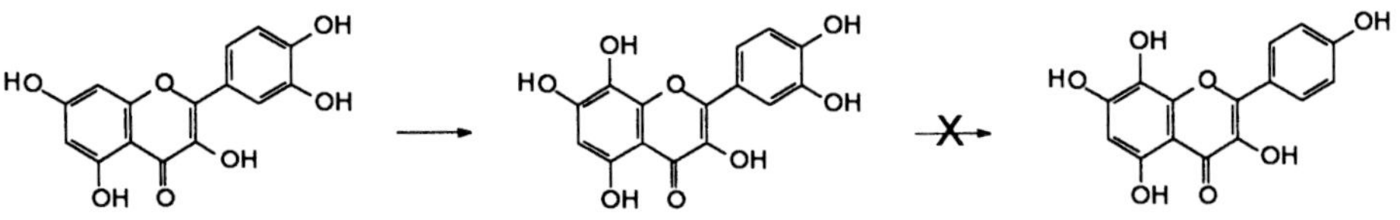

HO O OH OH OH OH O
Quercetin
HO OH O OH OH OH OH O
Gossypetin
HO OH O OH OH OH O
Herbacetin

Scheme 6.1 Hypothetical conversion of quercetin to herbacetin via gossypetin.

The suggestion by Frey-Wyssling (1938) that the origin of secondary plant products involved the degradation of amino acids to olefins with subsequent combinations in various ways is an interesting one. It is wrong of course, but it hints at the importance of phenylalanine, although the way in which the amino acid is metabolized does not reflect biochemical reality. In his scheme alanine gave rise to ethylene, leucine to isoprene (2-methylbuta-1,3-diene), and phenylalanine to phenylethylene (styrene) (Scheme 6.2). These alkenes could then recombine in different ways. The aromatic ring, for example, could arise from condensation of two ethylenes with one isoprene, or it could come directly from phenylalanine. Condensation of an equivalent each of phenylethylene, isoprene, and ethylene was thought

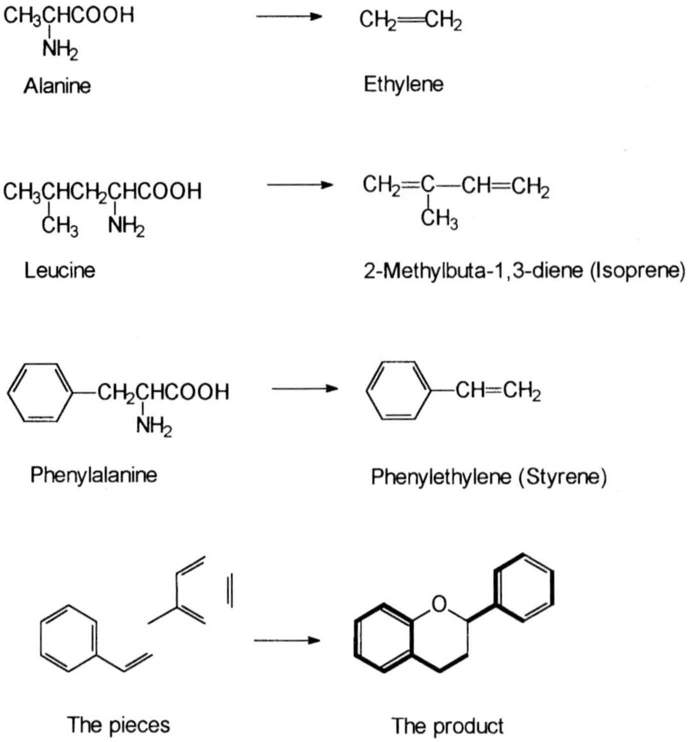

$CH_3CH(NH_2)COOH \longrightarrow CH_2{=}CH_2$

Alanine — Ethylene

$CH_3CH(CH_3)CH_2CH(NH_2)COOH \longrightarrow CH_2{=}C(CH_3){-}CH{=}CH_2$

Leucine — 2-Methylbuta-1,3-diene (Isoprene)

$C_6H_5{-}CH_2CH(NH_2)COOH \longrightarrow C_6H_5{-}CH{=}CH_2$

Phenylalanine — Phenylethylene (Styrene)

The pieces — The product

Scheme 6.2 The Frey-Wyssling olefins.

a possible route to the flavan structure as illustrated in the lower portion of the Scheme. The possibility that amino acids might undergo deamination leading to substances capable of producing flower pigments was seriously argued by Onslow in 1931. We know today, of course, that loss of ammonia from phenylalanine leads to *trans*-cinnamic acid (cinnamate) and that all of its nine carbons are used to make the B-ring and three-carbon bridge of flavonoids. This occurs by mechanisms unknown to Frey-Wyssling and Onslow but their vision of the involvement of phenylalanine was a correct one.

Robert Robinson (1936), the renowned English organic chemist, pointed out that the oxidation level of some anthocyanins is the same as that of normal carbohydrates and that the latter might well serve as precursors for the pigments. The reactions involved combining, in effect, two hexoses and a triose to yield a fifteen carbon intermediate, which, after elimination of water, is converted to the anthocyanidin. The existence of polyhydroxy compounds in Nature, cyclitols such as inositol [6-5], carbocyclic acids such as quinic acid [6-6] and shikimic acid [6-7], gallic acid [6-8],

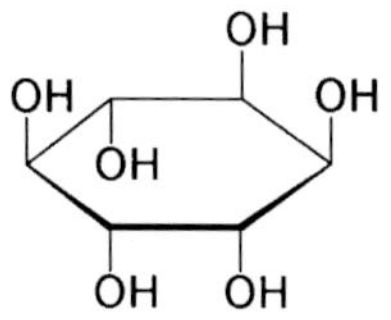

6-5

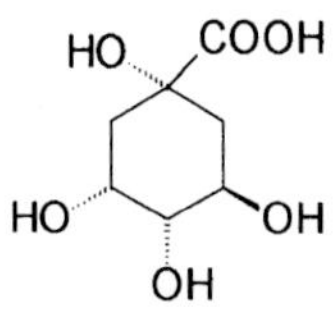

6-6

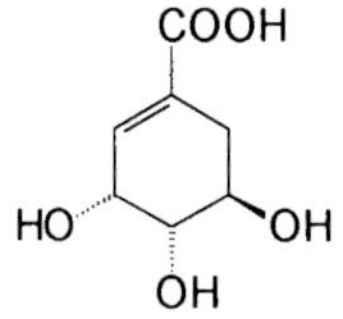

6-7

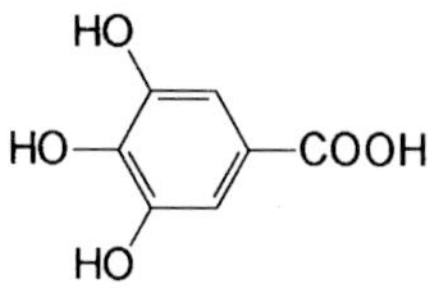

6-8

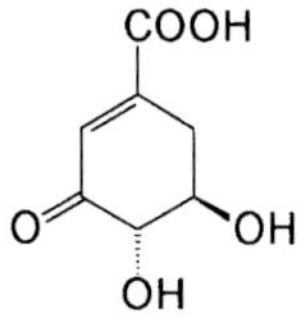

6-9

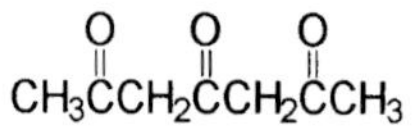

6-10

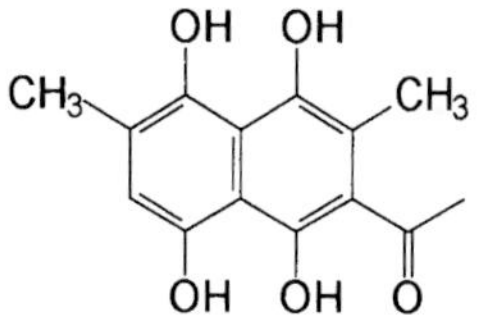

6-11

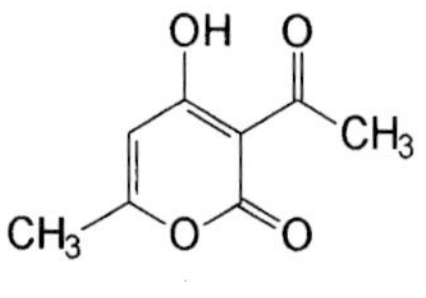

6-12

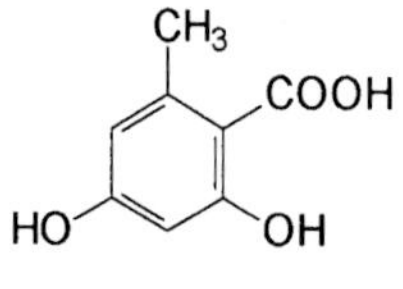

6-13

and, obviously, the common sugars, was used in support of the argument that a polyhydroxylated intermediate could exist. Flavonoids are not formed in this way, but certain aspects of the hypothesis and supporting arguments have a hint of prophecy and are worth noting. Sugars are indeed involved in the pathway leading to the aromatic amino acids. The initial reaction involves condensation of erythrose 4-phosphate and phosphoenol pyruvate to form the seven-carbon sugar

3-deoxyarabinoheptulonic acid 7-phosphate. Cyclization of the heptose to a carbocyclic acid, 5-dehydroquinic acid [6-9] establishes the ring system that is destined to become aromatic. The pathway is, of course, the shikimic acid pathway. (Although the full carboxylic acid name is used in this discussion, e.g., shikimic acid, the acid exists in the cell in the ionized form, shikimate in this instance.)

We have to go back in history now to pick up another thread of the story. Around the turn of the century, the English chemist J. N. Collie and his students (Collie, 1893, 1907; Collie and Myers, 1893; Collie and Chrystall, 1907; Collie and Hildrich, 1907) at the University of London were studying the chemical properties of polyacetyl compounds. They noted among other things, the ease with which compounds such as diacetylacetone (2,4,6-heptatrione) [6-10] dimerized to the naphthalene derivative [6-11], and how readily dehydroacetic acid [6-12] (formed by pyrolysis of acetoacetic ester) could be converted into 2,4-dihydroxy-6-methylbenzoic acid [6-13]. Collie recognized the similarity of [6-11] to naturally occurring naphthalene derivatives and was aware that [6-13], orsellinic acid, was a known naturally occurring compound. He postulated that such compounds might be made in Nature from polyacetate derivatives emphasizing the comparative ease (dilute base, room temperature) with which the condensation/cyclizations occurred. A discussion of the chemical conversion of polyacetate compounds into phenolic compounds can be found in the review by Money (1970).

So far as can be determined, Collie's ideas were seminal in providing the conceptual foundation upon which the polyacetate hypothesis was built (Richards and Hendrickson, 1964). It was another half century, however, before the idea was taken up again. A. J. Birch (Birch and Donovan, 1953; Birch, 1957) saw the value in Collie's initial idea, realized its potential, and developed the "acetate hypothesis" essentially as we know it today. The fundamental idea is that linear polyacetate chains have the capacity, through aldol or Claisen-type condensations, to form a variety of 1,3-di- or 1,3,5-trioxygenated carbocycles, which, upon undergoing keto–enol tautomerization, are converted into aromatic ring systems. The reactivity of the polyacetate chain lay, of course, in the alternating carbonyl carbons and active methylene groups, from which a nucleophile can be generated.

Precedent for the existence of linear carbon chains involving alternating carbonyl and methylene groups came from the field of fatty acid biosynthesis (Little and Bloch, 1950). It is now well known that the acetate unit, activated as acetyl coenzyme-A, is used to initiate a process whereby the chain is lengthened through condensation with malonyl coenzyme-A (with concomitant loss of CO_2) followed by a series of reduction reactions that, in effect, replace the carbonyl oxygen with two hydrogens. The four-carbon acid (still linked to coenzyme-A) undergoes another condensation with malonyl coenzyme-A followed by the reduction reactions and so on until a particular chain length is achieved. The same type of condensation reaction, in which an acetyl coenzyme-A unit condenses with malonyl coenzyme-A, initiates the polyacetate pathway, but subsequent reduction does not occur. (In a few special cases limited reduction does occur.) The process can be visualized by assuming that the growing polyacetate chain is held on the enzyme surface through hydrogen bond interactions involving carbonyl groups on the growing chain and functional groups on the protein. Specific protein topologies are also necessary in order to bring reactive positions on the polyacetate chain(s) close enough for bond formation. Polarization of a carbonyl function and an active methylene group, which would be necessary for aldol-type condensations, would also be facilitated by functional groups on the enzyme.

A critical development in the acetate hypothesis, of particular interest to us, was the realization that acids other than acetic could function as the "starting unit." The A-ring of flavonoids possesses the 1,3,5-trioxygenation pattern characteristic of a polyacetate-based compound, with the B-ring arising from some aromatic moiety. A C_6C_3, or phenylpropanoid, precursor activated in such a way that condensation with malonyl coenzyme-A would occur was considered as one possibility. There is an alternative route, however, which involves an appropriately substituted and activated benzoic acid derivative (C_6C_1) as starting group. Condensation with four malonyl coenzyme-A units, as opposed to three in the phenylpropanoid case, would provide the necessary carbon skeleton with oxygens, or their equivalents, in the required places. The cinnamoyl and benzoyl routes are illustrated in Scheme 6.3. Note that the product of the benzoyl condensation would be a dibenzoyl methane

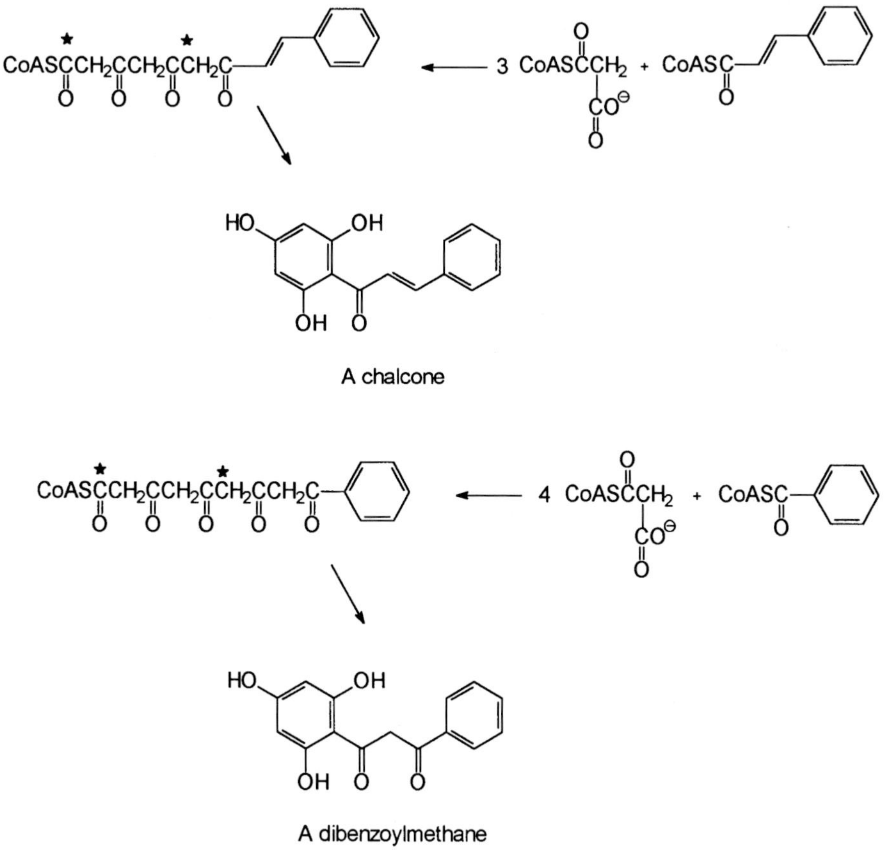

CoASCCH2CCH2CCH2C ← 3 CoASCCH2 + CoASC

CO⊖

HO OH

OH O

A chalcone

CoASCCH2CCH2CCH2CCH2C ← 4 CoASCCH2 + CoASC

CO⊖

HO OH

OH O O

A dibenzoylmethane

Scheme 6.3 Cyclizations of cinnamoylpolyacetyl chains.

derivative of the sort we saw in Chapter Two. (Recall that these compounds can exist in the β-hydroxychalcone form as well.) One of the features of polyacetate chains is the possibility of alternative foldings. In the case of the fifteen-carbon flavonoid precursor, two foldings are possible, the one shown in forming the normal flavonoid C_{15} system, and one that results in two aromatic rings linked by a two-carbon bridge. This pathway is known and will be met again below.

Biosynthetic research concerning naturally occurring compounds can be broken into several different stages, each building on, and often overlapping with, to a greater or lesser extent, the preceding ones. First, there is generally a theoretical stage, which for flavonoids was discussed briefly above. Next comes the first experimental work designed to elucidate the metabolic pathway(s) involved. Enzyme studies usually come next and often involve studies of early steps in the pathway before later steps have been worked out. Work can diverge at this point with some researchers attempting to put the newly established biosynthetic pathways into physiological perspective, while others attempt to establish the genetic background that controls the pathway. In the case of flavonoids, significant genetic information was already available from classical studies of flower color inheritance patterns. In fact, some of these studies were very useful in helping to establish the sequence of individual steps in the pathway. Once the enzymology and mode of inheritance have been established, attention turns to detailed analysis of the molecular genetics of the individual genes. Sequence determination allows one to establish relatedness of genes and from that infer evolutionary directions. It is also possible to probe the control systems to see how they react to signals from the environment. We will see all of these stages in our look at flavonoid biosynthesis. We will also examine efforts to alter the genetic composition of plants in an effort to achieve novel effects.

EXPERIMENTAL BEGINNINGS

In 1957 independent reports from four laboratories described experimental studies of flavonoid biosynthesis using ^{14}C-labelled precursors: Geissman and Swain (1957) used *Fagopyrum esculentum* (buckwheat), Grisebach (1957) used red cabbage seedlings, Shibata and Yamazuki (1957) used *Fagopyrum cymosum*, and Underhill and coworkers (1957), and Watkin and coworkers (1957) used *Fagopyrum tataricum*. Grisebach studied the biosynthesis of the anthocyanidin cyanin, whereas the other groups targeted quercetin in their studies. The Canadian workers (Underhill *et al*., 1957; Watkin *et al*., 1957) found that phenylalanine and cinnamic acid were excellent precursors of the B-ring carbons and the three-carbon bridge, while acetate labelled the A-ring carbons. When β-^{14}C- or ring-^{14}C-labelled cinnamic acid were fed and the isolated quercetin, after *O*-methylation, subjected to alkaline degradation, it was found that all of the label resided in the veratric acid fragment with none in the acetophenone fragment Underhill *et al*., 1957). When α-^{14}C- or carboxy-^{14}C-labelled cinnamic were administered no label was detected in the veratric fragment but the acetophenone fragment was labelled. In an experiment with U.L.-^{14}C-phenylalanine (U.L. = uniformly labeled), the amount of label present in the acetophenone fragment was equivalent to 2/9 of the label in the quercetin while that in the veratric acid fragment was equivalent to 7/9. In a study of the origin of the A-ring carbons

(Watkin *et al.*, 1957), it was shown that $^{14}CO_2$ or ^{14}C-sucrose labelled both rings of quercetin equally, but with acetate, labeled in either C-1 or C-2, label was incorporated significantly in the A-ring. Underhill and coworkers (1957) also observed incorporation of shikimic acid into quercetin. ^{14}C-Labeled *p*-hydroxybenzoic, 3,4-dihydroxybenzoic, and caffeic acids were also fed to buckwheat seedlings but were not incorporated to any significant degree. They also reported that phloroglucinol was not incorporated into quercetin. Two years later, a dihydrochalcone derivative, phloretin from apple leaves, was added to the list of flavonoids whose biosynthesis was being studied (Hutchinson *et al.*, 1959). Experiments with ^{14}C-labelled acetate clearly showed incorporation of label into the flavonoid A-rings while ^{14}C-phenylalanine was incorporated into the B-rings in complete agreement with the cabbage and buckwheat studies.

The possibility of direct incorporation of phloroglucinol into flavonoids was re-examined by Ali and Kagan (1974) who fed ^{14}C-labelled phloroglucinol to buckwheat seedlings under a variety of conditions and for varying periods of time. Short term (3 hr) exposure to the precursor afforded rutin labeled in the sugar moieties whereas longer term experiments (<95 hr) resulted in label distributed about equally in both rings of the quercetin molecule as well as in the sugars. These results are explainable in terms of extensive degradation of the labeled phloroglucinol and re-incorporation of carbon therefrom in the form of smaller molecules, e.g., acetate or even carbon dioxide. These authors also examined the possibility of a bio-Fries rearrangement of phloroglucinyl cinnamate into cinnamoyl phloroglucinol (2′,4′,6′-trihydroxychalcone). Label from this precursor was not significantly incorporated into quercetin (compared to incorporation of cinnamate). The sum of all of the above observations clearly indicates that: (1) the A-ring of flavonoids is not derived from an intact phloroglucinol unit (or biological equivalent); (2) flavonoid biosynthesis does not involve C_6C_1 precursors; and (3) because *p*-coumaric acid was incorporated while caffeic acid was not, the final B-ring oxygenation pattern is probably not established at the cinnamic acid level (this topic will reappear later). The literature on pathways to aromatic compounds was reviewed by Neish in 1960.

There followed in the next few years a wave of papers from many laboratories involving a wider array of plant species, use of cultured cells, and more sophisticated labelling experiments. Studies at the enzyme level were not far off and the importance of understanding the genetics of flavonoid biosynthesis began to attract a great deal of attention. The subject has been reviewed frequently details of which appeared in this chapter's introductory paragraph. Before we begin an examination of the biosynthesis of individual flavonoid types themselves, and control of their formation, it is necessary to attend to the origin of the C_6C_3 precursors.

THE GENERAL PHENYLPROPANOID PATHWAY

Phenylalanine Ammonia Lyase (PAL)

The finding that phenylalanine and certain cinnamic acid derivatives were readily incorporated into flavonoids naturally focused attention on the exact nature of the C_6C_3 precursor involved. Loss of nitrogen from phenylalanine was obviously indicated since cinnamic and *p*-coumaric acids were well incorporated. The connection between L-phenylalanine and the cinnamic acids was established when Koukol and Conn (1961) isolated and purified phenylalanine ammonia lyase (PAL, EC 4.3.1.5). Detailed studies have shown that PAL catalyzes the *anti*-elimination of ammonia from phenylalanine along with the pro-3*S* proton (starred) as shown in the following reaction sequence [Scheme 6.4]. Most preparations of PAL also showed activity with tyrosine (R=OH) converting it into the well-known plant metabolite, *p*-coumarate, albeit at much lower efficiency than the conversion of phenylalanine to cinnamate. All efforts to determine the existence of a tyrosine ammonia lyase (TAL) as an enzyme in its own right have failed. No "TAL" preparation is known that doesn't function more efficiently with phenylalanine. In addition, several workers were unable to detect any TAL activity in preparations from a variety of plant species (Hanson and Havir, 1981, p. 580).

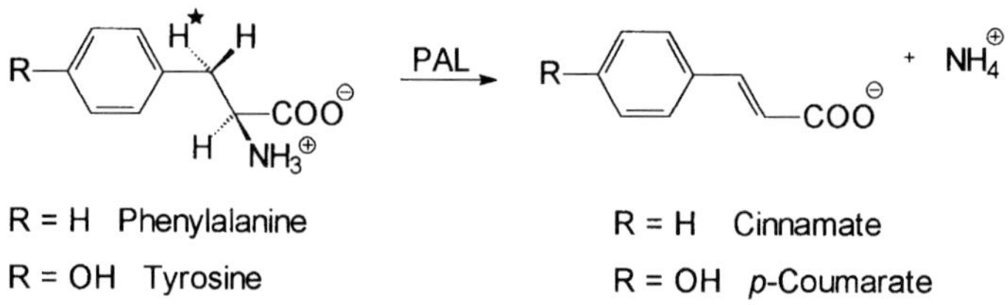

R = H Phenylalanine → (PAL) R = H Cinnamate + $NH_4^{\oplus}$

R = OH Tyrosine → (PAL) R = OH *p*-Coumarate

Scheme 6.4 The reaction catalyzed by phenylalanine ammonia lyase. The star marks the *pro*-3s-hydrogen that is lost in the stereospecific elimination of the ammonium ion.

A difficulty inherent in studying enzyme activity with plant extracts, no matter how pure they may appear, is the problem of how many polypeptide chains are actually present. In the case of PAL and TAL, it had never been demonstrated unequivocally that only one polypeptide was involved. This issue was addressed by Rösler and coworkers (1997) who isolated a PAL coding region from *Zea mays* and, after necessary manipulations, obtained protein using a cooperative strain of *Escherichia coli*. The homogeneous protein was shown to possess both PAL and TAL activities. Although the two activities had different Michaelis constants (K_m) and different turnover numbers, their catalytic efficiencies (k_{cat}/K_m) were shown to be quite similar. The homogeneous enzyme showed similar pH optima for the two substrates in the range 8.0–8.5, which is not in agreement with earlier reports of 8.7 for phenylalanine and 7.7 for tyrosine (Havir *et al*., 1971). Rösler and his colleagues (1997) suggested that the discrepancy in pH optima may have arisen from use of different isozymes by the earlier workers. These data suggest that maize can produce *p*-coumaric acid from either phenylalanine or tyrosine. The authors speculated about the physiological significance of the enzyme's capacity to deaminate both amino acids. They point out that with variable ratios of concentration of the two amino acids, the relative contribution of the two pathways may change. Several possible reasons were given for expecting the ratio of the amino acids to vary including different rates of their formation as well as differences in their transport across plastid membranes. Further experimentation is clearly needed to resolve the issue.

The intense interest that was generated in all aspects of PAL resulted in an extensive literature that has been reviewed by, among others, Hanson and Havir (1972a,b, 1981), Camm and Towers (1973), Jones (1984), Cramer *et al.* (1989), Hahlbrock and Scheel (1989), Lois and Hahlbrock (1992), and van der Meer *et al.* (1992). A paper by Howles and coworkers (1996), in which they discuss over-expression of the enzyme in transgenic tobacco plants, should be consulted for recent information on control points for flow of carbon into phenylpropanoid biosynthesis.

The enzyme is ubiquitous in higher plants and occurs in a number of fungi as well. It serves to shunt carbon out of primary and into secondary metabolism. While not the key enzyme in flavonoid biosynthesis, PAL does serve as the gateway for entry of carbon into the various pathways that rely on a supply of phenylpropanoid derivatives, *viz.*, cinnamic acids, benzoic acids, lignin, flavonoids, and stilbenes. PAL has been cloned from a number of species [Forkmann (1994) listed nine] where it occurs in small multigene families. Individual genes are differentially expressed during plant development or in response to a variety of environmental stimuli, e.g., light, wounding, and microbial elicitors (Jones, 1984).

Four isomeric forms of PAL have been resolved from preparations of *Phaseolus* through chromatofocusing (Bolwell *et al.*, 1985). Each individual isomeric form displayed normal Michaelis-Menten kinetics. Three isoforms of PAL were reported from cell cultures of *Medicago sativa* (Fabaceae) that had been treated with an elicitor preparation. A recent study of structure and evolutionary relationships of a PAL gene in *Nicotiana* species, designated *gPal-1*, provides detailed information on its intron/exon structure and its differential expression in different plant tissues (Fukasawa-Akada *et al.*, 1996). It was also shown that the concentration of mRNA coded by the gene increased rapidly after wounding of the plant.

Cinnamate 4-hydroxylase (C4H)

The second reaction in this series is the synthesis of *p*-hydroxycinnamate from cinnamate and involves cinnamate 4-hydroxylase (C4H, EC 1.14.13.11). A preparation from spinach that could hydroxylate cinnamic acid was reported by Nair and Vining in 1965. Enzyme activity from pea seedlings was reported a few years later by Russell and Conn (1967) from microsomal preparations of peas seedlings. Properties of the enzyme and possible metabolic and developmental controlling factors were described by Russell (1971). The enzyme requires NADPH, oxygen and acts as a typical cytochrome P450 system requiring a second enzyme system, NADPH:cytochrome P450 reductase (Potts *et al.*, 1974). It has also been shown that the hydroxylation reaction proceeds by the "NIH shift" mechanism (Russel *et al.*, 1968). Cloning and the study of expression of cytochrome P450 genes involved in controlling flower color have been described by Holton *et al.* (1993a). Reference to plant cytochrome P450 appears often below. The range of reactions catalyzed by the P450s, many of which are involved in polyphenolic biosynthesis, has been reviewed in depth by Bolwell and associates (1994).

Although there was some debate at the outset of studies with C4H, it is now known that the enzyme, with a molecular weight of about 57,000, is associated with endoplasmic membrane. Enzyme activity of C4H can be reconstituted from purified NADPH-cytochrome P-450 (cytochrome c) reductase and partially purified P-450 (from *Helianthus tuberosus*) in the presence of dilauroyl phosphatidylcholine as liquid phase (Benveniste *et al*., 1986). At the time of preparation of his review for the 1994 volume of *The Flavonoids* series, Forkmann stated that C4H had not been cloned owing presumably to the difficulty of isolating it from the membrane.

The importance of the phenylpropanoid pathway in providing starting materials for pathways other than that leading to the flavonoids, lignification for example, has led to a good deal of experimentation. It is important to know something about these other pathways and appreciate the mechanisms by which carbon flow is regulated among them. A few examples will provide the interested reader with access to the relevant literature. An interesting earlier work on wheat leaf response to infection by *Botrytis cinerea* showed that activation of both cinnamate 4-hydroxylase and hydroxycinnamate:coenzyme-A ligase occurred before the onset of lignification in wounded leaves (Maule and Ride, 1983). A recent study of the effect of stress factors in the formation of compression wood in *Pinus taeda* (loblolly pine) provides an excellent example involving a growth response (Zhang and Chiang, 1997). For a recent view of phenylpropanoid metabolism as it applies to lignification, the reader should consult a timely review by Douglas (1996), although his elevation of *Arabidopsis* to the status of "model tree" will undoubtedly cause the odd raised eyebrow.

4-Coumarate:CoA Ligase (4CL)

The third enzyme in this sequence, 4-coumarate:CoA ligase (4CL, EC 6.2.1.12), converts *p*-coumarate to its coenzyme-A ester, thus activating it for reaction with malonyl CoA. The enzyme has a strict requirement for ATP and Mg^{+2}. The reaction proceeds *via* the cinnamoyl-AMP intermediate. Enzyme activity was first reported from preparations of *Beta vulgaris* by Walton and Butt (1970, 1971) with subsequent reports by many workers from a diverse array of plant species. Molecular weight measurements have given values of 55,000 and ca. 60,000 (Knobloch and Hahlbrock, 1975, 1977). The chemical synthesis and properties of hydroxycinnamoyl coenzyme-A derivatives have been described by Stöckigt and Zenk, 1975). Isozymes have been detected in several species as well, e.g., two each in *Glycine max* and *Pisum sativum* and three in *Petunia hybrida*. Results from specificity studies of these systems are particularly interesting because they demonstrate the capacity of plants to direct metabolites to specific pathways with a high degree of selectivity. In the *Glycine max* system, for example, isoenzyme 1 was shown to activate caffeate (3,4,-dihydroxycinnamate), ferulate (3-methoxy-4-hydroxycinnamate), and sinapate (3,5-dimethoxy-4-hydroxycinnamate) at levels approximately one half that observed for *p*-coumarate, while isoenzyme 2 excludes sinapate but activates *p*-coumarate, caffeate, and ferulate about equally (Knobloch and Hahlbrock, 1975). The *Petunia* situation is somewhat more complex. Ranjeva and coworkers (1975a,b, 1976, 1979) separated the three forms of 4CL from *P. hybrida* using DEAE-cellulose and hydroxyapatite chromatography and found that all three forms activated *p*-coumarate, but differed significantly in their capacity to activate other cinnamic acids. Isoenzyme 1a activated caffeate even more efficiently than it did *p*-coumarate, but showed no activity toward either ferulate or sinapate. Isoenzyme 1b activated sinapate at about 80% of *p*-coumarate activation, while isoenzyme 2 activated ferulate but not sinapate. The roles of the three ligases were rationalized in terms of the three pathways into which cinnamate derivatives flow, flavonoid biosynthesis (isoenzyme 1a), formation of cinnamate esters (isoenzyme 1b), and lignin biosynthesis (isoenzyme 2). Supporting these divisions of labor were observations that the flavonoid pathway isoenzyme was sensitive to feedback inhibition by naringenin while the others were sensitive to cinnamate esters. The combination of pronounced substrate specificities, the existence of isoenzymes, and the regulatory effect of a number of product molecules is fully in accord with an enzyme system intimately involved in metabolic control (Hahlbrock and Grisebach, 1979).

A recent study of tobacco (*Nicotiana tabacum*) 4-coumarate:coenzyme-A ligase revealed some interesting features of this particular 4CL system (Lee and Douglas, 1996). Several cDNA clones encoding 4CL were obtained from young shoot tips and subjected to restriction analysis and DNA sequencing allowing these workers to distinguish groups of similar cDNAs. Two of these, referred to as 4CL1 and 4CL2, were selected for detailed study. Cloning of the DNAs and expression of the enzyme activity were done using standard procedures (see their paper for details). Substrate specificity of the resulting enzymes were then compared with the enzyme obtained directly from plant tissue using *p*-coumarate, cinnamate, caffeate, ferulate, and sinapate. Taking the activity with *p*-coumarate as 100, activities of the other cinnamate derivatives with 4CL1 were, respectively, 21, 17, 73, and 0. The values for 4CL2 were 29, 25, 62, and 0. The enzyme from plant tissue showed no activity with cinnamate, 17 and 60 for caffeate and ferulate, respectively, and a small level (4%) with sinapate. The reactivity with cinnamate is puzzling insofar as B-ring deoxyflavonoids are not known from tobacco. No explanation for this unexpected behavior was offered.

One of the long-standing questions in flavonoid biosynthesis concerns the point at which the B-ring oxidation pattern is established. Early feeding experiments with labeled cinnamic acid derivatives gave enough incorporation under some circumstances to suggest that the substitution level might be established before the actual formation of the flavonoid skeleton. Although there are a few cases where this may be the case, the general understanding is that the 4′-hydroxyl group enters *via* *p*-coumarate with further substitution occurring at the flavonoid level. Evidence to support this view also comes from studies of utilization of 4-coumaroyl CoA by chalcone synthase (see below). Experiments have shown that the 4-coumarate:CoA ligase from cultured parsley cells has comparatively low K_m values and high rates of conversion for *p*-coumarate, caffeate, and ferulate, while the chalcone synthase from these cells only utilizes 4-coumaroyl-CoA. Further, *O*-methyltransferases exist in these cells that function to produce *O*-methylated flavonoids indicating that utilization of feruloyl CoA for flavonoid formation is not necessary (Kreuzaler and Hahlbrock, 1973). This enzyme failed to activate 4-hydroxybenzoic acid, several phenylacetic acids, phenylalanine, tyrosine, and a selection of aliphatic mono- and dicarboxylic acids (Knobloch and Hahlbrock, 1975, 1977).

An interesting exception exists with regard to establishment of the B-ring oxygenation pattern. Data from genetic and biochemical investigations led Kamsteeg and coworkers (1980, 1981) to conclude that 3-hydroxylation of 4-coumaroyl CoA (thus yielding caffeoyl CoA) accounts for the 3′-hydroxyl group in anthocyanins in petals of *Silene dioica*. The *Silene* situation may be unique, but the lack of any extensive comparative study of this phenomenon limits our capacity to generalize. Considering the size of the plant kingdom and the very limited number of species that have been examined in detail in this regard, it would be foolhardy to make any pronouncements.

Over the past few years, a great deal of attention has been paid to factors that control expression of 4CL. In addition to the role that 4CL plays in activating *p*-coumarate for flavonoid biosynthesis, the enzyme has attracted the attention of many workers interested in the part this enzyme plays in plant growth and response to stress of varying sorts. Much of the detailed information currently available on 4CL, in fact, comes to us *via* studies of lignification, floral development, and pathogen attack. For example, the structures of two *4CL* genes in parsley were studied by Douglas and coworkers (1987) who showed that both were activated by UV irradiation and by an elicitor from *Phytophthora megasperma* f. sp. *glycinea*. Representative of contemporary studies of regulation of the *4CL* gene are the studies of Reinold and coworkers (1993) on temporal and spatial regulation in a cell type-specific manner during floral development in tobacco. These workers employed tobacco plants that had been engineered to contain a second *4CL* gene obtained from parsley. Probes were used such that the gene's activity in carpels, anthers, petals, and sepals could be monitored throughout floral development. Cell-specific expression in anthers was observed in endothecial cells during deposition of secondary wall material. They also discussed the activation of *4CL* in maturing pollen grains where, and when, it is known that phenylpropanoid and flavonoid biosyntheses are occurring.

In a recent report by Lee and coworkers (1995), 4CL was shown to be encoded by a single copy gene in *Arabidopsis thaliana*. The gene was activated in the early stages of seedling development at a time of rapid cell wall growth with its concomitant high demand for lignin precursors. In mature plants 4CL activity was associated with bolting stems, which is known to be a time of rapid lignification. The authors also discussed the effect of wounding on *4CL* gene expression as well as the plant's response to *Pseudomonas syringae* pv. *maculicola* infection. The effect of *Phytophthora infestans* (*Pi*) infection of potato on both PAL and 4CL induction was described by Fritzmeier and coworkers (1987). These workers observed almost immediate, transient, and coordinate increases in the rates of transcription and concentrations of the respective mRNAs upon treatment of potato cell cultures with culture filtrate from *Pi*.

Uhlmann and Ebel (1993) determined that 4CL is likely encoded in *Glycine max* (soybean) by a small gene family and that the members of this set of genes are differentially expressed in response to treatment with an elicitor from *P. megasperma*. These workers discussed their observations in terms of a mechanism by which the plant can regulate the flow of phenylpropanoid metabolites to different pathways. Lee and Douglas (1996) have documented the existence of a *4CL* gene family in tobacco. Two classes were distinguished which shared about 80% nucleotide and amino acid sequence identity. The genes are inducible by wounding and by methyl jasmonate (see below). The relationship of *N. tabacum* to its ancestors, *N. sylvestris* and *N. tomentosiformis*, was discussed in terms of inheritance patterns of the two *4CL* classes.

THE FLAVONOID PATHWAY

There are several points in the flavonoid pathway where the product of one reaction provides the substrate for two or more subsequent steps. These "branch points" are extremely important since factors controlling the on-going reactions can direct precursors exclusively in one direction rather than in any of the others, depending upon the physiological state of the plant at that instant. Understanding the workings of these control points is crucial to understanding the overall economics of the pathway. For example, an important branch point compound is naringenin, which can be converted to flavones, dihydroflavonols, or isoflavones. Dihydroflavonols can, in turn, be converted to flavonols, flavan-3-ol (catechin) (Singh *et al.*, 1997) derivatives, or they can be channeled into the formation of anthocyanidins *via* flavan-3,4-diols. Isoflavones can be converted into a wide variety of isoflavonoid derivatives including phytoalexins. Flavanones probably also play a significant role in the biosynthesis of *C*-glycosylflavones. We will embark on our tour of flavonoid biosynthesis now with an examination of the first committed step in the process, the formation of chalcones.

Chalcone Synthase (CHS)

The debate as to whether a chalcone or a flavanone was the first product of flavonoid biosynthesis was conclusively decided with the observation of an enzyme that catalyzed the reaction between *p*-coumaroyl CoA and malonyl CoA to yield the 2′,4′,6′,4-tetrahydroxychalcone. The first report of chalcone synthase (CHS, EC 2.3.1.74) activity came from studies using irradiated cultures of parsley cells (Kreuzaler and Hahlbrock, 1972). The presence of chalcone isomerase accounted for the formation of naringenin, but subsequent studies showed that the true product is the chalcone (Heller and Hahlbrock, 1980; Light and Hahlbrock, 1980). Support for this conclusion came from studies of the enzyme in other tissues, tulip anthers and *Cosmos* floral tissue, by Sütfeld and Wiermann (1980, 1981). Since the initial work was published, CHS has been reported from and studied in depth in many more systems. That the enzyme is represented in non-vascular plants was demonstrated by Fischer and coworkers (1995) who studied CHS activity in the liverwort *Marchantia polymorpha*. The liverwort enzyme was detected by reactivity against antibodies raised against CHS from three flowering plant species. Apparent molecular weights of 46,000 and 77,000 were observed on chromatograms, both of which numbers are in line with the following. Most recent determinations have shown the molecular weight for CHS to be in the range 75,000–88,000 for the native enzyme.

Denaturing gel electrophoresis gave values between 40,000 and 44,000. No cofactors are required.

The substrate specificity of CHS has attracted a great deal of attention over the years. Evidence from early studies suggested that appropriately substituted cinnamoyl CoA derivatives could provide the flavonoid B-ring substitution pattern. Hess (1964) observed that ferulic and sinapic acids with ^{14}C-labeled *O*-methyl groups were incorporated into the correspondingly substituted anthocyanins in *Petunia hybrida*. However, only 12–65% of the incorporated label was found in the anthocyanins thus suggesting extensive demethylation of the precursor acids. Similar results were obtained by Ebel and coworkers (1970) in a study of 5,7-dihydroxy-4′-methoxyflavone (acacetin) biosynthesis in *Robinia pseudoacacia* (Fabaceae). They observed that with older leaves and a feeding/metabolism period of 64 hours, the ratio of tritium to ^{14}carbon ($^{3}H/^{14}C$) was reduced to 23% when multiply labeled *p*-methoxycinnamic acid was administered. The acid was labeled with tritium in the *O*-methyl group and had ^{14}C in the *O*-methyl group and at the β-carbon. When younger leaves and shorter metabolism times were used, however, the ratio did not change, suggesting incorporation of the intact acid. Heller and Forkmann (1994, p. 509) discussed experiments with *Glycine max* (soybean) that showed that the enzyme has a much higher preference for *p*-coumaroyl CoA than for caffeoyl CoA or feruloyl CoA. In general, these other acids were never accepted with anything approaching the activity with *p*-coumaroyl CoA, and in some preparations they were not accepted by the enzyme at all. In extracts of *Pinus sylvestris* seedlings, however, both *p*-coumaroyl CoA and cinnamoyl CoA were readily accepted as substrates for CHS and incorporated into naringenin and pinocembrin 5,7-dihydroxyflavanone (pinocembrin), respectively (Rosemann *et al*., 1991; Fliegmann *et al*., 1992). Formation of the flavanones was accounted for by the presence of CHI (chalcone isomerase) in the crude extracts. The finding that cinnamoyl CoA is used in the biosynthesis of pinocembrin (the corresponding chalcone, actually) is important in explaining the existence of B-ring deoxyflavonoids. If *p*-coumaroyl CoA were the precursor to pinocembrin, and other B-ring deoxyflavonoids, one would expect labeling experiments to reflect the removal of the hydroxyl group. This appears not to happen. Experimental evidence supporting the requirement for the appropriately substituted cinnamic acid first became available from a study of the formation of 3,5,7,2′-tetrahydroxyflavone (datiscetin) [6-14]. The most efficient precursor for this unusually substituted flavonoid turned out to be 2′,4′,6′,2-tetrahydroxychalcone [6-15] (Grisebach and Grambow, 1968). Since B-ring deoxyflavonoids occur so widely in the plant kingdom (Bohm and Chan, 1992), it would be of interest to see if CHS from a wide selection of species is equally able to accommodate cinnamoyl CoA as well as *p*-coumaroyl CoA.

To this point in our discussion, we have paid attention only to those compounds characterized by having the phloroglucinol-type A-ring substitution, 2′,4′,6′- in the case of chalcones, 5,7- in the case of flavanones, flavones, etc. There exists, however, a large group of flavonoids that lack 5-oxygenation (or 6′-oxygenation in chalcones). 5-Deoxyflavonoids are most abundant in members of Fabaceae, but occur in many non-legumes as well. The first information available on the biosynthesis of this family of compounds in cell free systems came from two Japanese groups.

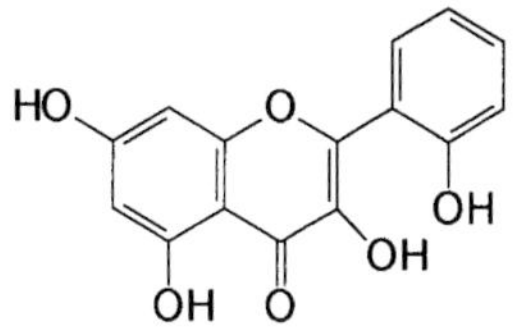

6-14

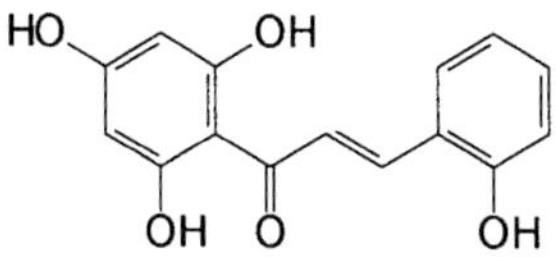

6-15

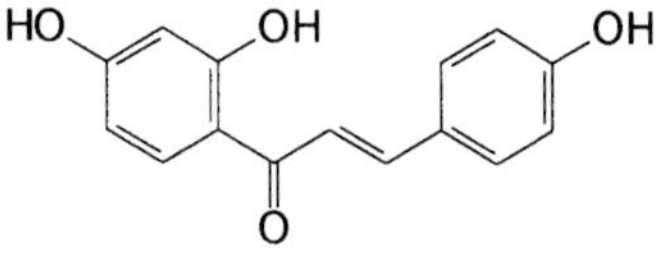

6-16

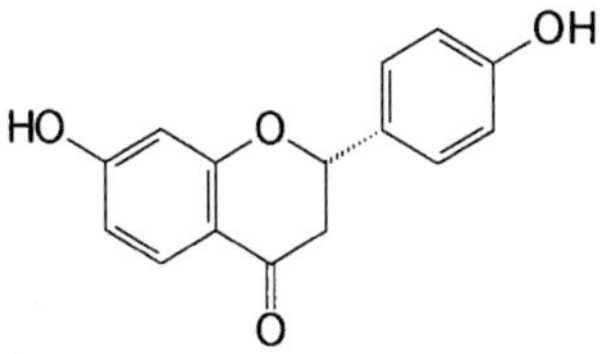

6-17

Ayabe and coworkers (1988a,b) observed the formation of 6′-deoxychalcone [6-16] and 5′-deoxyflavanone [6-17] from *p*-coumaroyl CoA and malonyl CoA in extracts of *Glycyrrhiza echinata* protoplasts and cells in the presence of high concentrations of NADPH. Hakamatsuka and colleagues (1988) obtained similar results from cell cultures of *Pueraria lobata*, again in the presence of high concentrations of NADPH. The latter workers noted that two enzymes were likely involved in the reaction, one of which appeared to degrade over time. Welle and Grisebach (1988), using CHS preparations from either soybean or parsley, concluded that a deoxychalcone synthase is *not* involved but that there is a second enzyme that coacts with CHS in the formation of the deoxychalcone. The enzyme has marked similarities with other oxido-reductases (Welle *et al.*, 1991) and probably acts at the polyketide level to reduce one of the carbonyl groups, qualifying it as a polyketide reductase (PKR). It is interesting to note that the purified PKR from soybean functions in the presence of CHS preparations from plants that do not accumulate 5-deoxyflavonoids.

Chalcone synthase catalyzes the condensation of a polyketide in the fashion shown in Scheme 6.5 with the two carbons involved in the reaction, C-1 and C-6, marked with stars. The carbanion generated at the star-marked methylene carbon would react with the terminal carbonyl group following the "star route" to the chalcone with subsequent cyclization to the flavanone and ultimate formation of other flavonoids. An alternative possibility for cyclization exists, however. If a carbanion is generated on the C-2 methylene group followed by interaction with the cinnamoyl carbonyl group, the "double arrow route" in Scheme 6.5, the product is a 2-carboxydiphenylethylene derivative. Decarboxylation results in the formation of a stilbene, resveratrol, in this case. The formation of different products from the same polyketide precursor because of different foldings on the enzyme surface is well known. As an exercise, readers might wish to convince themselves of the generality of this phenomenon by seeing how the tetraketide precursor 3,5,7-triketobutanoyl coenzyme-A can be converted into

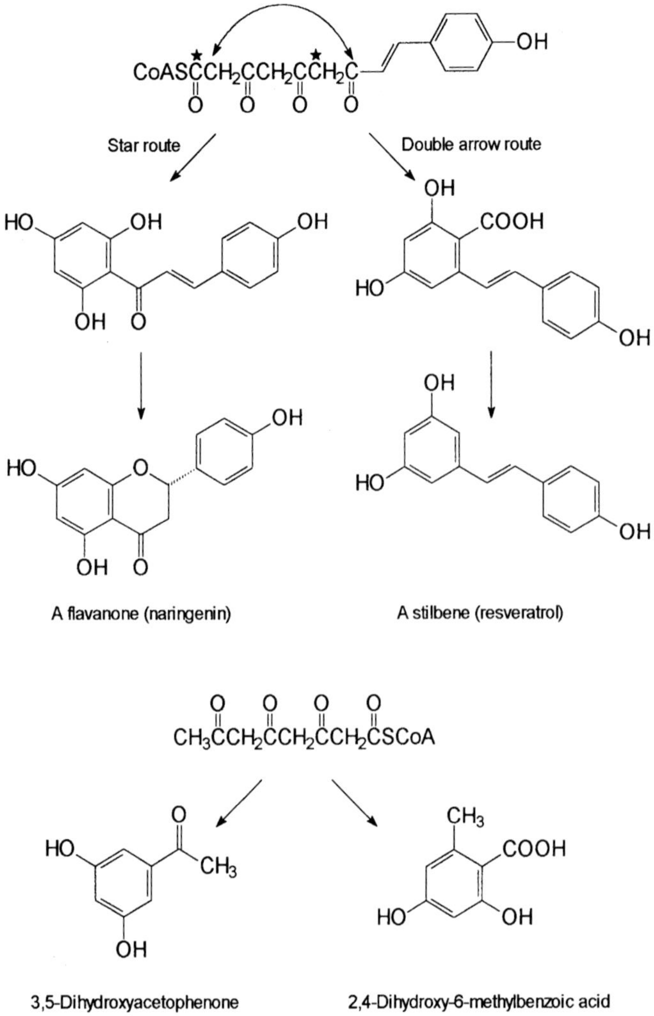

Star route

Double arrow route

A flavanone (naringenin)

A stilbene (resveratrol)

3,5-Dihydroxyacetophenone

2,4-Dihydroxy-6-methylbenzoic acid

Scheme 6.5 Alternative routes of cyclization to form either a chalcone or a stibene (above), or an acetophenone or a benzoic acid derivative (below).

either 3,5-dihydroxyacetophenone or 2,4-dihydroxy-6-methylbenzoic acid by altering the way in which it is folded.

The existence of stilbene synthase in cell cultures of *Picea excelsa* has been demonstrated by Rolfs and Kindl (1984). Although the enzymes accept the same substrate and have similar physical properties, they can be distinguished by means of gel electrophoresis and reaction to monospecific antibodies. It seems reasonable to suggest that the two enzymes are closely related and differ functionally only in the manner in which the polyketide substrate is folded on their respective surfaces. In an effort to gain a better understanding of the function and control of stilbene synthase (STS) genes, Fischer and coworkers (1997) employed a line of transgenic tobacco that carried the *STS* gene from grape. It was found that the appearance of stilbenes in tobacco, which does not normally make these compounds, interfered with flower color, presumably by competing with chalcone synthase for precursors, and, surprisingly, brought about male sterility. No mechanism was determined for the latter effect but it was suggested that this sort of construct might be useful in the development of novel flower colors.

Flavonoid and stilbene syntheses in grape were studied in detail by Sparvoli and coworkers (1994) who found that the genes involved, with two exceptions, were expressed in response to light. The exceptions were the genes coding for PAL and STS (StSy in the notation in the Sparvoli *et al.* paper) which appeared to be constitutive. These two genes also appeared to be members of large gene families whereas the other genes in the pathway consisted of smaller numbers, one to four per haploid genome. A recent paper by Jeandet and colleagues (1995) demonstrated a potential relationship between stilbene (resveratrol) synthesis in grape species and anthocyanin accumulation in fruits. The reduction in the UV-induced biosynthesis of resveratrol accompanied the increase in anthocyanin pigments in the fruit. The phenomenon was discussed in terms of chalcone synthase out-competing stilbene synthase for substrate. The intimate connection between the formation of these two polyphenols is clearly indicated by these observations.

The genetic control of the *CHS* gene has been extensively studied in many organisms and has been reviewed in a number of publications (van Tunen *et al.*, 1990; Dangl, 1992; van der Meer *et al.*, 1992). A few examples will show the level of complexity with which geneticists must cope in order to understand how this key enzyme is controlled. In *Petunia*, for example, CHS is encoded by a family of genes, as many as 12 according to Koes and coworkers (1987, 1989). Not all of these appear to be expressed, however, with about 90% of the total CHS mRNA arising from *CHSA*, whereas *CHSJ* contributes only about 10% to floral organs. These two genes can also be induced in young seedlings by irradiation with UV. Two other genes, *CHSB* and *CHSG*, are also UV-inducible in young seedling tissues but only at low levels. Conditional mutants of the chalcone synthase gene were examined by Mol and coworkers (1983) in their study of gene expression in *Petunia*. The situation in *Pisum* involves three genes, *CHS1*, *CHS2*, and *CHS3*, all of which are expressed, and two control loci termed *A* and *A2* (Harker *et al.*, 1990). *CHS1* and *CHS3* are expressed in both root and petal tissues while *CHS2* is expressed only in root tissue. The products encoded by *A* and *A2* are required for the expression of *CHS1* and *CHS3* in petals, whereas in root tissue all three genes can be induced by $CuCl_2$ regardless of the genotype of the two regulatory genes. Two forms of chalcone synthase, both induced by light, were obtained from spinach in a state of apparent homogeneity (Beerhues and Wiermann, 1988; Beerhues *et al.*, 1988). The two forms exhibit very similar properties and were not distinguished by antibodies raised from both. The enzymes are cytosolic and appear not to be associated with either tonoplast or endoplasmic reticulum. A small fraction of the enzyme was localized in the chloroplast stroma. The second paper (Beerhues *et al.*, 1988) contains a valuable summary of opinion as to the cellular localization of flavonoid synthesis.

In *Matthiola*, chalcone synthase is controlled by gene *f*, in *Antirrhinum* (snapdragon) by *niv* (or *nivea*), and in *Zea mays* (maize) by *2c*. The recessive condition in each of these results in the lack of enzyme and, therefore, the absence of anthocyanins (actually, all flavonoids) (Spribille and Forkmann, 1981, 1982; Dooner, 1983). In the case of wild-type genotypes of *Matthiola*, one sees a dramatic increase in CHS activity that parallels floral development. In the *ff* genotype, however, there is a reduced and nearly constant amount of inactive enzyme protein present in floral tissues (Rall and Hemleben, 1984).

Studies of promoters associated with the flavonoid pathway have provided important insights into developmental and environmental factors that influence the expression of structural genes. An excellent example comes from the work of Schmid *et al*. (1990) on regulation of a bean *CHS* promoter in transgenic tobacco. A 1.4 kilobase fragment associated with the *CHS8* gene from bean (*Phaseolus*) was engineered into tobacco plants using a *CHS8-GUS* gene fusion. (*GUS* serves as the reporter gene whose location in tissue can be established either biochemically or histologically.) The promoter was highly active in root apical meristem and in petals, exclusively in those cells that accumulate anthocyanins. Only minor activity was observed in other floral structures, mature leaves, or stems. Wounding of mature leaves activated the promoter in a well-defined area closely associated with the wounding site. Since the source plant is a legume, it has the natural capacity to form nitrogen fixing nodules in the roots. The observation in these studies that the legume *CHS* promoter is expressed in tobacco root cortical tissue represents an excellent example of site specialization. In the source legume, of course, flavonoid biosynthesis is necessary to produce the *nod* gene inducers necessary for establishing the symbiotic association.

Not all localizations of the *Phaseolus* promoter were as easily explained. For example, the *CHS8* promoter was strongly expressed in the root apical meristem and at the point of lateral root initiation, but not in the shoot apex. This is in contrast to the situation with regard to the promoter in bean that controls phenylalanine ammonia-lyase (PAL), which has been shown to be very active in the shoot meristem (Liang *et al*., 1989). Schmid and his coworkers (1990) speculated that the high root tip *CHS* promoter activity might be explained by the synthesis of flavonoids involved in modulation of polar auxin transport. This is in accord with evidence supporting the idea that auxins are critical for the formation of lateral roots (Wightman *et al*., 1980). This example demonstrates nicely how a single reaction in a pathway can be under the control of several different factors: (1) wounding; (2) the synthesis of messenger compounds; and (3) the synthesis of compounds required for modulation of plant growth hormones.

Fritze and coworkers (1991) were successful in linking the 1.1-kb snapdragon chalcone synthase promoter to the *GUS* reporter gene and incorporating the construct into the tobacco genome. Defined deletions were used to determine if certain segments of the promoter could be identified as controlling factors for *CHS* expression in particular tissues of the plant. A sampling of their results will show the detailed nature of the system (as well as the power of this technique). A promoter fragment truncated to −39 bp activates transcription in 4-week old roots, a longer fragment, extending to −197 bp, directs *CHS* expression in petals and seeds, and a regulatory element located between −661 and −566 bp is active in all tissues except petals.

To conclude this section, it is useful to direct the reader's attention to an application that uses chalcone synthase as a tool to address broader issues. A number of different genes or gene products have been used over the years as markers of evolutionary relationships among plants, *viz*., cytochrome c, various ribosomal sequences, *rbc*L, ITS, *matK*, and others (Crawford, 1990; Soltis *et al*., 1992). Niesbach-Klösgen and coworkers (1987) compared *CHS* sequences of *Hordeum vulgare* (1477 bp), *Magnolia liliiflora* (1359 bp), *Petunia hybrida* (1477 bp), *Ranunculus acer* (1334 bp), and *Zea mays* (1461 bp) with sequences of *Antirrhinum majus* and *Petroselinum hortense* from the work of others. The sampling of species has been too small to date to allow construction of any phylogeny but some interesting points did emerge from that study. Chalcone synthase is G/C rich in monocotyledons (65.7–69.3%) as compared to dicotyledons (45.5–53.9%) and monocotyledons show a strong bias toward G or C in the third codon position. Evidence was also presented indicating that the first exon of *Petroselinum* has evolved differently from the same exon in the other tested species. More extensive surveys of this sort would be most welcome.

Chalcone Isomerase (CHI)

Much of the early confusion surrounding the nature of the first product of flavonoid biosynthesis was caused by the subject of this section, chalcone isomerase. Chalcone isomerase (CHI, EC 5.5.1.6) is the second enzyme in the flavonoid pathway acting to catalyze the cyclization of a chalcone to the corresponding flavanone. Chalcone isomerase was the first enzyme in the flavonoid pathway to be studied (Moustafa and Wong, 1967). Working with a preparation from soybean, they showed that 2′,4′,4-trihydroxychalcone [6-18] was converted to (-)-(2*S*)-7,4′-dihydroxyflavanone [6-19]. Further information on the involvement of chalcones and flavanones in flavonoid biosynthesis soon appeared (Wong, 1968; Wong and Grisebach, 1969). Additional evidence for a specific isomerase were obtained by Wiermann (1972) from a study of the isomerase reaction in anthers of *Lilium caudidum* and *Tulipa* cv. "Apeldoorn." Enzyme preparations from these plants acted upon 2′,4′,6′,4-tetrahydroxychalcone but were not active with 2′,4′,4-trihydroxychalcone (a 6′-deoxychalcone). Wiermann observed that the level of the isomerase was highest when the concentration of chalcone was decreasing rapidly and the concentration of flavonols was increasing, which clearly marked the enzyme as playing a key role in the flavonoid biosynthetic pathway. Subsequent studies by others have supported and extended these findings. For example, in several plants devoid of 5-deoxyflavonoids, Forkmann and Dangelmayr (1980) found that the chalcone isomerases accept only 2′,4′,6′-trihydroxylated substrates. However, plants that accumulate both 5-oxy- and 5-deoxyflavonoids possess chalcone isomerases that accommodate both 2′,4′-dihydroxy- and 2′,4′,6′-trihydroxychalcones, but show a higher specificity toward the latter (Chmiel *et al*., 1983).

The cyclization reaction involves an overall *syn*-addition and yields the flavanone (2*S*)-isomers exclusively, regardless of the substitution pattern of the chalcones. However, cyclization of 2′,4′,6′-trihydroxychalcones [6-20] can occur spontaneously

6-18 CHI 6-19

6-20 Spontaneous 50% (2S) 6-21; 50% (2R) 6-22

to produce a racemic mixture of (2*S*), [6-21] and (2*R*) isomers [6-22]. Isolation of an optically active flavanone, then, is taken as clear indication that the compound is a true natural product. Conversely, isolation of an optically inactive flavanone usually indicates that isomerization of a naturally occurring chalcone has occurred during extraction or purification.

Mutants in the *CHI* structural gene in *Dianthus* (Forkmann and Danglemayr, 1980) and in *Callistephus* (Kuhn *et al*., 1978) result in flowers that lack anthocyanins but have a yellow-color owing to the accumulated 2′,4′,6′,4-tetrahydroxychalcone. A small amount of anthocyanin is occasionally noted in *chi* genotypes owing to the spontaneous isomerization of the chalcone to naringenin, which is then converted to anthocyanin by the remaining enzymes of the pathway (assuming no further lesions exist). Tissue specificity in control of CHI activity has also been observed, as in the case of the *Po* mutants in *Petunia* that only affects pollen (van Tunen *et al*., 1988). Further observations of these tissue specific expression of CHI have been detailed by van Tunen and coworkers (1989, 1990) using *Petunia hybrida* inbred line V30. Two isomerase genes, *CHIA* and *CHIB*, were observed, each of which was shown to be associated with promoters, P_{A1} and P_{A2} in the case of *CHIA*, and P_B in the case of *CHIB*. The P_{A1} promoter is active in corolla and tube tissue whereas the P_{A2} promoter, which yields a larger transcript, is active in pollen grains during late stages of anther development. The *CHIB* gene appears to have only the one promoter, which is active in early stages of anther development. Other plant genes that are under tandem promoter regulation were mentioned briefly in those workers' 1990 paper.

Flavanone 3-Hydroxylase (F3H)

Flavanone 3-hydroxylase (F3H in Martin and Gerats, 1993; FHT in Heller and Forkmann, 1994; EC 1.14.11.9; F3H will be used in this treatment) catalyzes the stereospecific 3β-hydroxylation of (2*S*)-flavanones [6-23] to the (2*R*,3*R*)-3-hydroxy-flavanones (dihydroflavonols) [6-24]. Hydroxylation at the 3-position of both naringenin and eriodictyol *in vitro* was observed by Forkmann and coworkers (1980) using preparations of *Matthiola incana* flowers. The enzyme is soluble and requires 2-oxoglutarate, Fe^{+2}, ascorbate, and molecular oxygen. Subsequent studies by several groups documented the occurrence of F3H in other genera including *Antirrhinum*, *Dahlia*, *Petroselinum*, *Petunia*, *Streptocarpus*, *Verbena*, and *Zinnia*. The native enzyme, obtained from a line of *Petunia*, thought to be a dimer with molecular weight about 74,000, accepts only (2*S*)-flavanones as substrates (Britsch and Grisebach, 1986). 5,7,3′,4′,5′-Pentahydroxyflavanone was not accepted by the enzyme from *Petunia* but was converted efficiently into dihydromyricetin by an enzyme from *Verbena* (Forkmann and Stotz, 1984).

R_1 and R_2 = H or OH depending upon the organism.

The genetic control of F3H has been demonstrated in several systems including white flowered variants of *Dahlia variabilis*, *Streptocarpus hybrida*, *Verbena hybrida*, and *Zinnia elegans* (Forkmann and Stotz, 1984). Other examples include the recessive allele at the *An3* locus in a white-flowered line of *Petunia*, known to be involved in flavanone 3-hydroxylation (Froemel *et al*., 1985). The use of white flowered mutants of normally pigmented species has proved to be a very useful means of studying anthocyanin biosynthesis. The lack of color in these mutants is due to the absence of one (or more) of the enzymes necessary for the biosynthesis of anthocyanins. However, the lack of any given enzyme does not necessarily mean that all other enzymes of the pathway are also missing. This phenomenon can be demonstrated simply by administration of a missing precursor to colorless tissues and observing the formation of anthocyanidin by constitutive enzymes of the pathway. For example, in a mutant deficient in flavanone 3-hydroxylase, application of the dihydroflavonol provides substrate for the downstream enzymes that results, ultimately, in formation of the anthocyanin, which is easily observed visually.

Flavanone 3-hydroxylase is encoded by a single gene in *Arabidopsis thaliana* (Pelletier and Shirley, 1996). The enzyme has 72–94% amino acid sequence homology with enzymes prepared from other plant species. Studies with etiolated seedlings exposed to white light showed that F3H expression was coordinated with expression of chalcone synthase (CHS) and chalcone isomerase (CHI) but not to expression of dihydroflavonol reductase (DFR). These workers suggested that F3H may be of pivotal importance in regulation of flavonoid biosynthesis since its expression is coordinated with different subsets of flavonoid biosynthetic enzymes in different plant species. Charrier and coworkers (1996) also consider flavanone 3-hydroxylase to be in a position of central importance in controlling flavonoid biosynthesis. They introduced an alfalfa *F3H-promoter-GUS* fusion into *Nicotiana benthamiana* to examine the sites of flavonoid synthesis. They observed activity to be widely expressed in flowers, stems, leaves, and roots and determined that there was a high correlation between these sites and the presence of flavonoids.

Biosynthesis of Anthocyanins (DFR, ANS, and LDOX)

Anthocyanins are the most conspicuous products of flavonoid biosynthesis and, as such, have attracted a great deal of interest. Studies range from classical genetic analysis of flower color inheritance patterns, through establishment of their chemical structures, to efforts to understand the factors involved in their biosynthesis. Contemporary efforts to understand the metabolic pathway at the level of the gene have added immeasurably to our knowledge of this particular group of pigments, as well as to an understanding of how plant metabolic processes are controlled in general. The result of these efforts is a vast literature. What follows is only a sampling with sufficient references to the original and review literature to provide the interested reader with leads for further study. Though long out of print, no work treats the history of the subject better than Muriel Onslow's *The Anthocyanin Pigments of Plants*, the second edition of which was published in 1925. The work represents an outstanding synthesis of information including citations to publications from the mid 17th century. A recent work that deserves serious reading is Helen Stafford's (1990) *Flavonoid Metabolism*. Other timely reviews include the work of Weiring and de Vlaming (1984) on *Petunia*, a series of papers in a volume edited by Styles and colleagues (1989), and general reviews by Martin and Gerats (1993) and Holton and Cornish (1995). Each of the volumes of *The Flavonoids* series contains relevant information on chemical and biochemical aspects of anthocyanins, of course.

One of the first studies of flavonoid biosynthesis was that of Hans Grisebach (1957) who demonstrated that the carbon skeleton of cyanidin was formed from the same precursors as other flavonoid types, the A-ring from acetate and the B-ring and carbon bridge from a phenylpropanoid derivative. Over the next several years a number of workers demonstrated that chalcones, flavanones, and dihydroflavonols were all readily incorporated into anthocyanins. Genetic information was also accumulating that pointed toward involvement of these other flavonoid types (Harrison and Stickland, 1974, 1978; Stickland and Harrison, 1974, 1977). Often it was possible to induce the formation of pigment in white flowers by administering one of the pathway intermediates that lay beyond the genetic block. Other major contributors at the time were Kho and coworkers (1975, 1977) working with *Petunia hybrida*, Forkmann (1977, 1980) working with *Matthiola incana*, and McCormick (1978) working with corn. Another source of information involved the use of inhibitors that inactivated PAL, thus preventing any carbon from flowing into the phenylpropanoid and flavonoid pathways (Amrhein, 1979). Addition of a suspected intermediate on the anthocyanin pathway to the "phenylpropanoid-starved" tissue resulted in formation of pigment. With the starting and end points established it then remained to establish the detailed pathway. That part of the story comes next.

It seemed reasonable to focus attention on dihydroflavonols as likely late intermediates on the anthocyanin pathway simply because they are such prominent compounds in many plants and they served very well in precursor studies. A major step in answering the question as to what the next compound in the pathway beyond the dihydroflavonol came from studies of polyphenolic biosynthesis in cultured cells of Douglas fir (*Pseudotsuga menziesii*). Stafford and Lester (1982) were the first to detect dihydroflavonol reductase (DFR) activity; they also identified the product as a flavan-3,4-diol. The reaction is represented in the reduction of (2*R*,3*R*)-dihydrokaempferol [6-25] to (2*R*,3*S*,4*S*)-2,3-*trans*-5,7,4′-trihydroxyflavan-3,4-*cis*-diol [6-26]. The enzyme has now been purified from, or detected in, a wide spectrum of plant species including monocots, dicots, and gymnosperms; all DFR preparations require NADPH. Specificity of the enzyme varies a good deal both in what substrates it will accept and how efficiently any given dihydroflavonol will be converted to the flavandiol. For example, DFR from *Callistephus* and *Dianthus* will

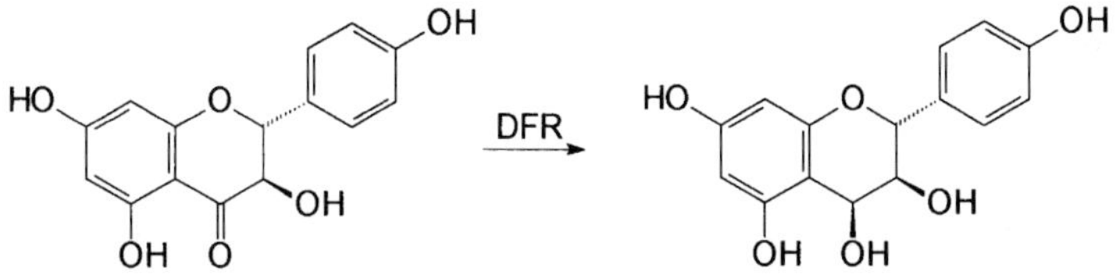

6-25 6-26

accept dihydrokaempferol, dihydroquercetin, and dihydromyricetin but handles dihydrokaempferol least well of the three. *Matthiola*, *Dahlia*, and *Dianthus* DFRs readily accept dihydromyricetin despite the absence of delphinidin glycosides in any of their species. The enzymes from *Petunia*, *Nicotiana*, and *Lycopersicon* (tomato) work best with dihydromyricetin, somewhat less efficiently with dihydroquercetin, and not at all with dihydrokaempferol. In these species dihydrokaempferol will accumulate if the hydroxylation system, which normally converts it to dihydroquercetin, is inactive. Incidentally, the lack of natural capacity to convert dihydrokaempferol to anthocyanins in *Petunia* has been "rectified" by genetic engineering using a DFR construct from corn (see Chapter Eight for details).

Mutations in the gene(s) coding for DFR result in white or ivory-colored flowers and may affect pigmentation in leaves as well. Lesions have been studied in a number of plants, notably in *Petunia* (*an6*) (Beld *et al*., 1989), *Antirrhinum* (*Pallida*) (Coen *et al*., 1986), *Callistephus* (*f*), *Dianthus* (*a*), and *Matthiola* (*e*) (Forkmann, 1989), and maize (*A1*) (O'Reilly *et al*., 1985; Reddy *et al*., 1987). Forkmann (1994) lists 10 species, including the ones just named, that have yielded some information on genetic control of DFR activity. These mutants tend to accumulate the appropriate dihydroflavonol, unless other processes intercede. The situation with regard to the *Pallida* locus in *Antirrhinum*, which includes transposable elements affecting promoter sequences in some individuals and structural sequences in others, offers an excellent example of the level of complexity that can confront workers in the field (Martin and Gerats, 1993). A detailed description of *Petunia* flavonoid genetics by Wiering and de Vlaming (1984) appeared in a monograph on the genus (Sink, 1984). In their treatment, they describe experiments demonstrating that the formation and accumulation of dihydroflavonol glycosides is a natural process in this species and that they are not involved (as precursors) in anthocyanin biosynthesis. It appears, however, that they can serve as a source for dihydroflavonols through removal of the sugar (Schram *et al*., 1981, 1982). At the time of preparation of their article (Wiering and de Vlaming, 1984), four genes were known that controlled formation of anthocyanins in *Petunia*, *An1*, *An2*, *An6*, and *An9*. The first two are involved in controlling the glycosylation of the newly formed anthocyanidin at the 3-OH position. Evidence also suggests that *An2* is a regulatory gene that controls not only the glucosylation reaction, but also rhamnosylation (*Rt*) and *O*-methylation of anthocyanins at 3′-OH and 5′-OH (Gerats *et al*., 1984). The other two genes, *An6* and *An9*, are thought to be involved in some way with chemical modification of the dihydroflavonols. Some comments on these reactions follow.

Flavan-3,4-diols, which are also called leucoanthocyanidins, are very unstable molecules and are likely converted very soon after formation into the corresponding anthocyanidin. Adequate demonstration of this conversion in a cell-free system has not been accomplished, although Stafford, and others, use the designation ANS, standing for anthocyanin synthase, to provide a name that can identify the overall reaction (Stafford, 1990). Other workers refer to the enzyme as leucoanthocyanidin dioxygenase (LDOX) (Sparvoli *et al*., 1994, for example). Whatever the mechanism is, the conversion of leucoanthocyanidin to anthocyanin is likely both very fast and closely associated with a glucosyltransferase that converts the newly formed anthocyanidin into the corresponding 3-*O*-glucoside (the anthocyanin). A mechanism has been postulated for this overall conversion, however (Heller and Forkmann, 1988, p. 410). The reaction as viewed by those workers involves hydroxylation at C-2, loss of the hydroxy group at C-4 through dehydration, loss of the hydroxyl group from C-2, and *O*-glucosylation. One of the possible routes is shown in Scheme 6.6. It is also possible that *O*-glucosylation might occur before loss of the hydroxyl group. That tritium is retained at C-4 throughout the reactions was shown by Heller and coworkers (1985) in the incorporation of 4-tritio-leucopelargonidin into pelargonidin derivatives in *Matthiola*. Heller and Forkmann (1988, p. 409) suggest that the glycosylation reaction is actually part of the anthocyanin biosynthetic process and not just a modification reaction as it is usually considered for other flavonoids types.

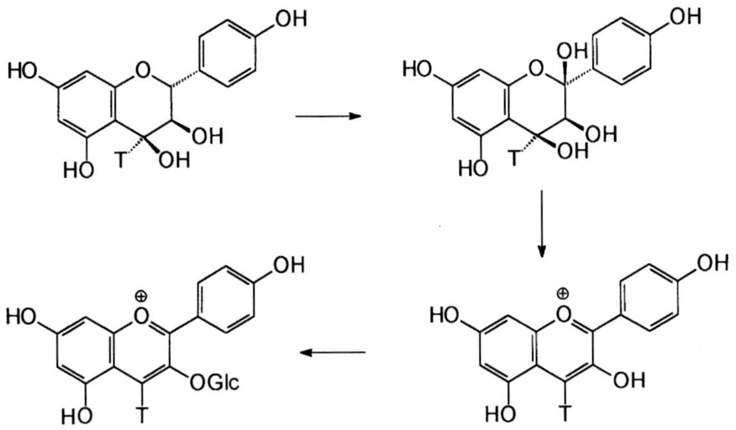

Scheme 6.6 A possible mechanism for conversion of a flavan-3,4-diol to an anthocyanin.

The anthocyanidin is probably unstable under the conditions in which it is formed, whereas formation of the glucoside stabilizes the product. 3-*O*-Glucosylation of anthocyanidins has been studied by several groups including Teusch and coworkers (1986b) who used preparations from flowers of *Matthiola incana*. The enzyme accepted either anthocyanidins or flavonols and is thus categorized as a UDP-glucose:anthocyanidin/flavonol 3-*O*-glucosyltransferase. Many anthocyanins consist of more than one sugar group, which is typical of the pigments in *M. incana* where 5-*O*-glucosylation of anthocyanins has also been studied (Teusch *et al.*, 1986a). These workers found that the best substrate for 5-*O*-glucosylation was the *p*-coumaroyl derivative of 3-*O*-xylosylglucoside. In descending order of effectiveness thereafter were 3-*O*-xylosylglucoside, 3-*O*-glucoside (acylated with *p*-coumaric acid), and the 3-*O*-glucoside itself, which was a very poor substrate.

Flavonoid pseudobases were discussed in Chapter Two as one of the forms in which anthocyanins can exist under certain conditions. They were also invoked as intermediates in one of the possible biosynthetic routes to anthocyanins of Heller and Forkmann (1988) mentioned above. Support for the existence of pseudobase structures of the sort shown as [6-27] came from NMR studies of anthocyanins under various pH conditions (Chapter Two), as well as the finding that the conversion of pseudobase to anthocyanin in *Pisum sativum* is under genetic control (Crowden, 1982). The pseudobase of malvidin 3-*O*-rhamnoside-5-*O*-glucoside was shown to accumulate in white flowers of *am* mutants of pea. Brief exposure of the petals to HCl brought about rapid formation of the anthocyanin, which was isolated and identified.

Recent molecular genetic information has provided additional insights as to the nature of the so called "late step" in anthocyanin formation. Clones of genes involved with the conversion of leucoanthocyanidin to anthocyanin, *A2* from maize, *candica* from *Antirrhinum*, and "late step" genes from *Callistephus* (*A*), *Matthiola* (not named), and rice (not named), appear to code for the same type of enzyme. Striking sequence homology to genes coding for flavanone 3-hydroxylase suggest that a 2-oxoglutarate-dependent dioxygenase is involved in the final conversion to

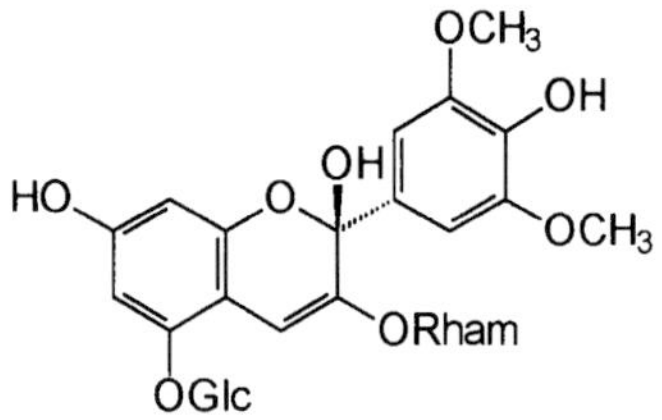

6-27

anthocyanin (Menssen *et al.*, 1991; Britsch, 1992; Martin and Gerats, 1993). Observations of this sort underlie the importance of combining biochemical and genetic information, neither source alone being capable of solving the problem.

Biosynthesis of 3-Deoxyanthocyanidins: Flavanone 4-Reductase (FNR)

A small group of anthocyanin-like pigments exists that are characterized by the absence of oxygenation at C-3. These 3-deoxyanthocyanins occur in a few fern and moss species and in scattered flowering plant species but are primarily known from New World Gesneriaceae (subfamily Gesnerioideae). Since the pathway to an anthocyanidin involves the sequence flavanone to dihydroflavonol to flavan-3,4-diol, the formation of a 3-deoxyanthocyanidin might be expected to follow a similar path without the intervening 3-hydroxylation step. Reduction of a flavanone to a flavan-4-ol would be the expected first step in the biosynthesis of a 3-deoxyantho-cyanins. The sequences are illustrated in Scheme 6.7. That reduction of the carbonyl function of flavanones occurs is suggested by the existence of a small number of flavan-4-ols in Nature, e.g., 4,5,7,4′-tetrahydroxyflavan (apiforol) [6-28] and 4,5,7,3′,4′-pentahydroxyflavan (luteoforol) [6-29]. Oxidation of the two flavan-4-ols would yield, respectively, apigeninidin [6-30] and luteolinidin [6-31]. Soluble enzyme preparations of *Sinningia cardinalis* (Gesneriaceae) catalyzed an NADPH-dependent reduction of (2*S*)-naringenin and (2*S*)-eriodictyol to the respective flavan-4-ols (Stich and Forkmann, 1988a). 5,7,3′,4′,5′-Trihydroxyflavanone was also reduced to the corresponding alcohol by this enzyme preparation although tricetinidin derivatives do not appear to occur in this species. The close similarities between the properties of the enzyme from *Sinningia* and the dihydroflavonol reductase from *Matthiola* led those authors to speculate that the two enzymes are closely related. It is particularly interesting that the *Sinningia* enzyme will also reduce dihydroflavonols to flavan-3,4-diols, which is an obligatory step on the pathway to anthocyanidins. The absence of dihydroflavonols in *Sinningia*, and presumably other members of this unusual subfamily, is likely due to the absence of the required flavanone 3-hydroxylase. Selection of an enzyme capable of reducing flavanones instead of dihydroflavonols has allowed these plants to produce floral pigments with properties akin to those of anthocyanins.

Subsequent studies of flavanone 4-reductase (FNR) have shown that the enzyme is also present in flowers of a *Columnea* species (also Gesneriaceae) (Stich and

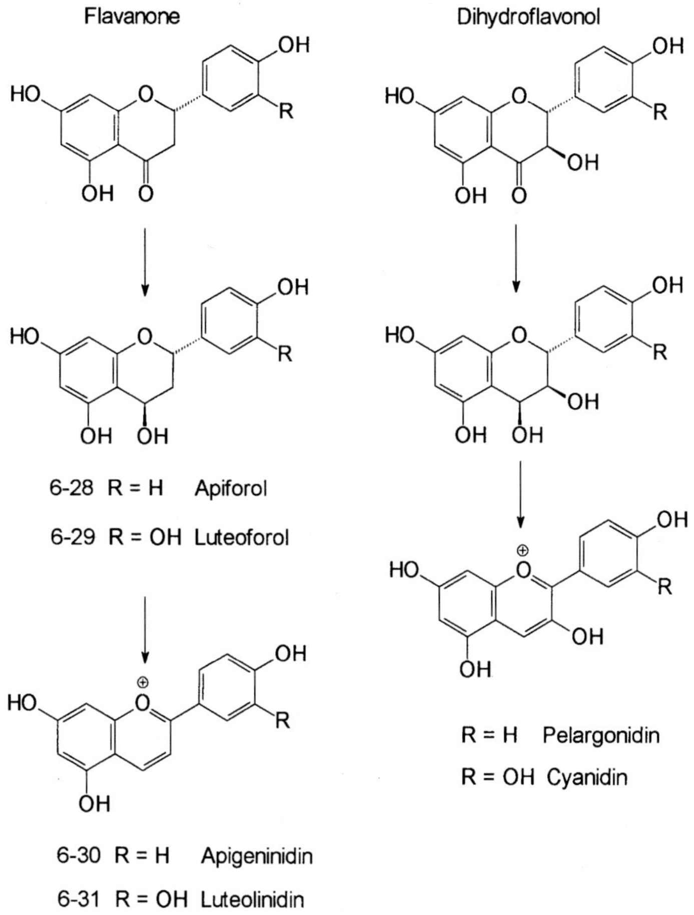

Flavanone

Dihydroflavonol

6-28 R = H Apiforol

6-29 R = OH Luteoforol

6-30 R = H Apigeninidin

6-31 R = OH Luteolinidin

R = H Pelargonidin

R = OH Cyanidin

Scheme 6.7 Biosynthesis of 3-deoxyanthocyanidins and anthocyanidins compared.

Forkmann, 1988b). FNR activity also has been detected in preparations from corn (*Zea mays*), which is known to accumulate apigeninidin and luteolinidin glucosides. A comparative study of the enzyme from a members of Gesneriaceae, corn, and the moss and fern species would provide a very interesting clue as to the evolution of pigment biosynthesis across a wide range of plant species.

Flavone Synthases (FNS)

The conversion of a flavanone to a flavone involves removal of hydrogen atoms from C-2 and C-3 to yield a double bond between those carbons. There have been suggestions that the reaction proceeds *via* a 2-hydroxyflavanone intermediate but no convincing evidence to support this idea has been forthcoming even though a few

2-hydroxyflavanones are known to occur naturally. Two flavone synthase enzymes have been described, one apparently quite restricted in its distribution, the other much more widespread. Flavone synthase I (FNS I) was first observed in young parsley leaves (*Petroselinum crispum*, Apiaceae) (Sutter *et al*., 1975). Subsequent work using cultured cells of parsley demonstrated that the enzyme required oxygen, 2-oxoglutarate, Fe^{+2}, and ascorbate as cofactors (Britsch *et al*., 1981). The enzyme has been purified to near homogeneity and shown to have a molecular weight of ca. 48,000 and to consist of two subunits with molecular weights between 24,000 and 25,000 (Britsch, 1990). The cofactor requirements were confirmed with the purified enzyme and experiments were performed to determine the sensitivity of the enzyme to changes in substrate structure. The enzyme converted (2*S*)-naringenin [6-32] and (2*S*)-eriodictyol [6-33] to apigenin [6-34] and luteolin [6-35], respectively, refused (2*R*)-naringenin, and was competitively inhibited by (2*S*)-naringenin 7-*O*-*β*-D-glucoside (prunin). 5,7,3′,4′,5′-Trihydroxyflavanone and 3,5,7,4′-tetrahydroxyflavanone (dihydrokaempferol) were also refused by the enzyme. These experiments demonstrated that FNS I is sensitive to stereochemistry at C-2, the presence of an hydroxyl group at C-3, and the nature of the B-ring substitution. The refusal of FNS I to accept a 3-hydroxyflavanone as substrate is interesting because, as we shall see presently, the enzyme (FLS) that converts 3-hydroxyflavanones to flavonols bears a number of striking similarities to FNS I including discriminating substrates on the basis of the presence or absence of the 3-hydroxyl group. Apparently, FNS I activity has been detected in other members of Apiaceae (Britsch, cited by Heller and Forkmann, 1994) but apparently not in plants from other families.

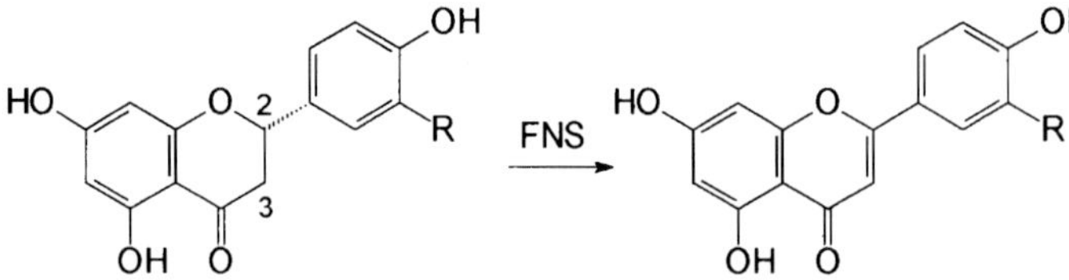

6-32 R = H

6-33 R = OH

6-34 R = H

6-35 R = OH

Flavone synthase II (FNS II) seems to be the more widespread flavone synthesizing enzyme in plants having been reported to date from members of Asteraceae, Fabaceae, and Gesneriaceae (Heller and Forkmann, 1994) in addition to Scrophulariaceae. The first report of this enzyme activity came from work of Stotz and Forkmann (1981) using flowers of *Antirrhinum majus*, the snapdragon (Scrophulariaceae). The enzyme has an absolute requirement for NADPH and molecular oxygen and is inhibited by compounds that typically inhibit cytochrome-P450-dependent mono-oxygenases. The substrate requirements for FNS II are very similar to those already mentioned above for FNS I.

Flavonol Synthase (FLS)

Flavonol synthase (FLS, no EC number as per Heller and Forkmann, 1994) catalyzes the dehydrogenation of 3-hydroxyflavanones to the corresponding flavonols as illustrated in the conversion of [6-36] to [6-37]. The earliest demonstration of flavonol synthase activity came from the same study that yielded flavone synthase I (Britsch *et al.*, 1981). The enzyme is soluble and requires oxygen, 2-oxoglutarate, Fe^{+2}, and ascorbate. The similarities of flavone synthase I and the newly observed flavonol synthase gave rise to speculation as to whether there were in reality two enzymes. However, clear differences between the two enzymes were subsequently demonstrated (Kochs and Grisebach, 1987). The same cofactor requirements as seen above were also noted for the enzyme from *Matthiola incana*

6-36 → (FLS) → 6-37

R_1 and R_2 = H or OH depending upon the organism.

(Spribille and Forkmann, 1984). Those authors also showed that the enzyme was equally capable of dehydrogenating dihydrokaempferol and dihydroquercetin to the corresponding flavonols. FLS preparations from *Petunia* likewise converted dihydrokaempferol and dihydroquercetin to the flavonols but showed poor conversion of dihydromyricetin to myricetin (Forkmann *et al.*, 1986). Genetic control of the reaction was established by the same workers when they showed that *Petunia* mutants with the *fl1fl1* genotype failed to form flavonols whereas lines that contained the wild-type *Fl1*__ genotype were normal. More recent studies of FLS have added *Tulipa* anthers (Beerhues *et al.*, 1989) and *Dianthus* (carnation) flowers (Eidenberger *et al.*, 1992). Manipulation of the *FLS* gene has been successful in transforming a line of petunias so that they exhibit brick red flowers, which, until that time, was not possible because of the inability of this plant's FLS to accept dihydrokaempferol as substrate while being quite capable of converting dihydroquercetin and dihydromyricetin to their respective flavonols. Meyer and coworkers (1987) generated the new petunia genotype by "transplanting" the *FLS* gene from maize, which does accommodate dihydrokaempferol.

It is also important to note that the highest FLS activity is associated with early stages of growth. In flowers it is most active in small, unpigmented buds, with activity dropping off dramatically as the anthocyanin content begins to increase. This is, of course, explained by the early synthesis of dihydroflavonols followed by their subsequent conversion, *via* the appropriate flavan-3,4-diols, to anthocyanidins.

Stich and coworkers (1992) studied this reaction in detail during early flower development in *Dianthus caryophyllus* (carnation). A recent study of the control of flavonol biosynthesis during anther and pistil development and during pollen tube growth using *Solanum tuberosum* (potato) was described by van Eldik and coworkers (1997).

Formation of Isoflavonoids

Not long after it was determined that the flavonoid skeleton is built from a C_6C_3 unit plus three C_2 units, experiments were undertaken to determine if isoflavonoids arose from the same precursors. It had been suggested by Robinson (1955) that the branched isoflavonoid structure might be the result of an acid catalyzed Wagner-Meerwein rearrangement of a 3-hydroxyflavanone (a dihydroflavonol). Feeding experiments using clover plants showed that the carboxyl carbon of phenylalanine is retained in the biosynthesis of 7-hydroxy-4′-methoxyisoflavone (formononetin) [6-38], as shown below, and that there is a migration of the aryl group from C-2 to C-3 during the process (Grisebach and Dörr, 1959, 1960). Subsequent studies showed that dihydroflavonols are not involved (Wong, 1965; Barz and Grisebach, 1966). Experiments did not resolve the question as to whether a chalcone or a flavanone was the natural precursor for isoflavonoid formation at that time, however. The situation was altered significantly when Hagmann and Grisebach (1984) observed isoflavone synthase activity in a preparation of cell cultured soybeans (*Glycine max*) that had been challenged by an elicitor from *Phytophthora*. The enzyme activity required NADPH and oxygen and converted (2*S*)-7,4′-dihydroxyflavanone (liquiritigenin) [6-39] and (2*S*)-5,7,4′-trihydroxyflavanone (naringenin) [6-40] into 7,4′-dihydroxyisoflavone (daidzein) [6-41] and 5,7,4′-isoflavone (genistein) [6-42], respectively. Studies of the isoflavone synthase system (Kochs and Grisebach, 1986; Hakamatsuka *et al.*, 1994) indicate that the enzyme has all of the characteristics of a cytochrome P450-dependent mono-oxygenase. Of critical importance in resolving the question of mechanism was the isolation and identification of 2,5,7,4′-tetrahydroxyisoflavanone [6-43]. This new compound could be dehydrated to the corresponding isoflavone by an enzyme preparation that did *not* require either NADPH or oxygen. Supporting data were reported by Hashim and coworkers (1990) who identified a mixture of 2,3-*trans* diastereoisomers of 2,7,4′-trihydroxyisoflavanone [6-43 without the 5-OH] from elicitor-treated cells of *Pueraria lobata* and showed that dehydration to the corresponding isoflavone (daidzein) occurred. Studies with ^{18}O showed that the hydroxyl oxygen came from atmospheric oxygen and that the carbonyl group of isoliquiritigenin was retained in the isoflavone.

One of the early attempts to rationalize the formation of isoflavones was that of Pelter and coworkers (1971) who suggested that phenolic oxidation of a chalcone might lead to the formation of a spirodienone intermediate. Intermediates of this sort had been postulated for non-enzymatic reactions involving aryl migration. Heller and Forkmann (1988) presented a mechanism credited to L. Crombie that involved formation of a flavan-3,4-epoxide intermediate which was subsequently converted to a spirodienone on the pathway to a 2-hydroxyisoflavanone. Both the epoxy and

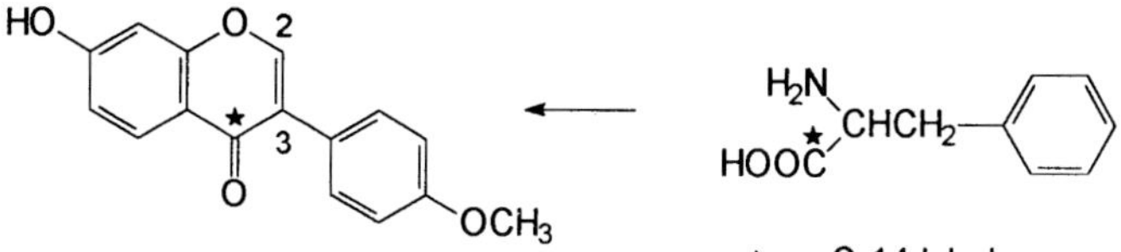

HO O 2 3 O OCH3
6-38

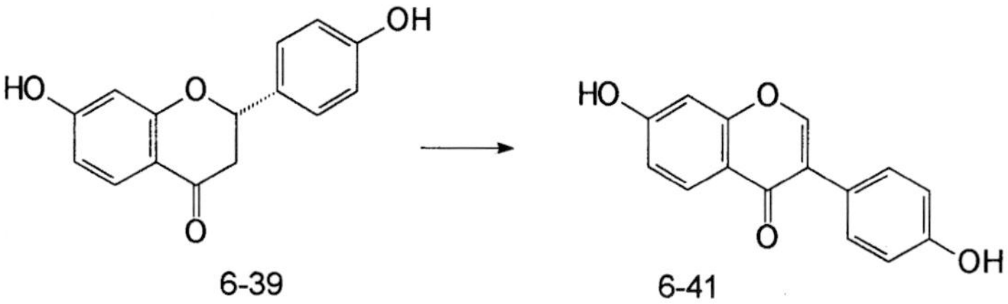

H2N CHCH2 HOOC

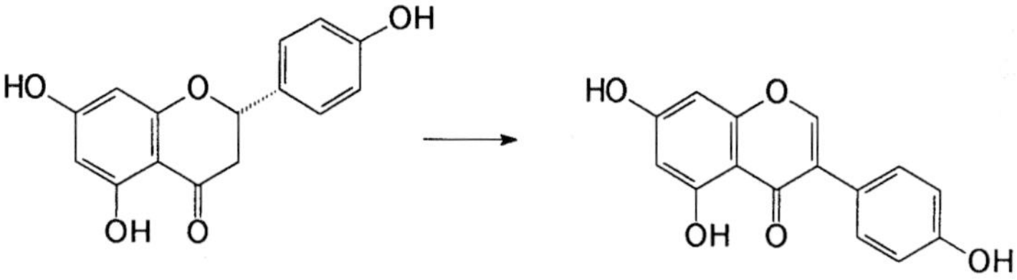

★ = C-14 label

HO O OH O
6-39

HO O O OH
6-41

HO O OH OH O
6-40

HO O OH O OH
6-42

HO O OH OH O OH
6-43

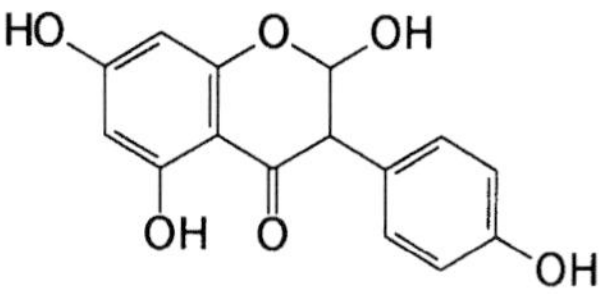

spirodienone intermediates are shown in Scheme 6.8. A problem with mechanisms that involved the 4′-oxygen lies with the widespread occurrence of isoflavonoids that lack substitution at that position. A mechanism that does not involve the 4′-oxygen was proposed by Hashim and coworkers (1990). In this series of steps, shown in Scheme 6.9, a hydrogen atom is abstracted from C-3 followed by rapid rearrangement of the aryl group from C-2 to C-3. Reaction of the newly established free radical at C-2 with water yields the 2-hydroxyisoflavanone which then undergoes dehydration catalyzed by 2-hydroxyisoflavanoane dehydratase (IFD). Dewick (1994) points out that isoflavone synthases/dehydratases may exist that have unusual substrate specificities. It is possible, of course, that isoflavones may be formed by way of different mechanisms in different organisms.

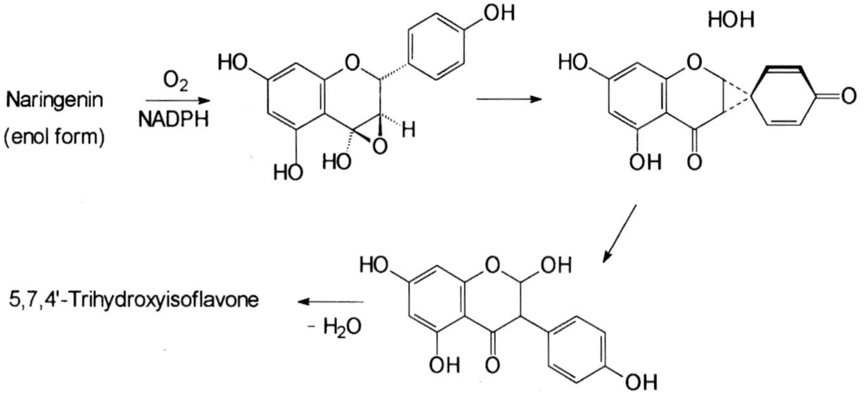

Naringenin (enol form) $\xrightarrow[\text{NADPH}]{O_2}$ → → 5,7,4'-Trihydroxyisoflavone ($-H_2O$)

Scheme 6.8 Isoflavone biosynthesis via epoxide and spirodienone.

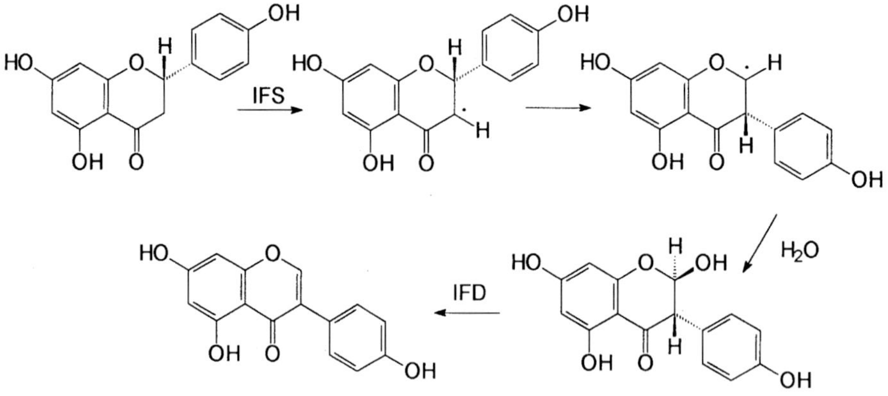

IFS → → H_2O → IFD

Scheme 6.9 Isoflavone biosynthesis via free radical mechanism.

There has been a good deal of interest in establishing the biosynthetic interrelationships among the various classes of isoflavonoids, particularly in light of the biological activity of many members of the group (Dewick, 1994). Although there is a large amount of information available on the general subject of isoflavonoid biosynthesis and interconversions, we will focus our attention on the formation of two principal types, pterocarpans and rotenoids. These isoflavonoid types are characterized by the presence of additional ring systems, excluding the cyclized prenyl system, and provide opportunities to comment upon some of the intermediate structures involved. Our first example comes from the work of Dewick and Steele (1982) involving the biosynthesis of the pterocarpan phaseolin [6-49] from $CuCl_2$-treated *Phaseolus vulgaris* seedlings. Good incorporation of carbonyl-^{14}C-2′,4′,4-trihydroxychalcone [6-44], 4-^{14}C-7,4′-dihydroxyisoflavone (daidzein) [6-45], 2-^3H-7,2′,4′-trihydroxyisoflavone [6-46], and the pterocarpans [6-47] and [6-48] led to the pathway shown below. Pterocarpan [6-47] was labeled at C-2 with tritium and compound [6-48] was labeled at C-6, C-6a, C-11 and the carbons of ring-D with carbon-14. The sequence of steps in the biosynthesis are shown in Scheme 6.10. Note that the labeling patterns of the intermediates are cumulative, which is to say that compound [6-46] would have been labeled with carbon-14 at C-4 from its precursor, [6-45], but [6-46] was prepared with tritium at C-2 in order to study formation to [6-47]. Had doubly labeled compounds been used, the cumulative patterns would have been observed in reality.

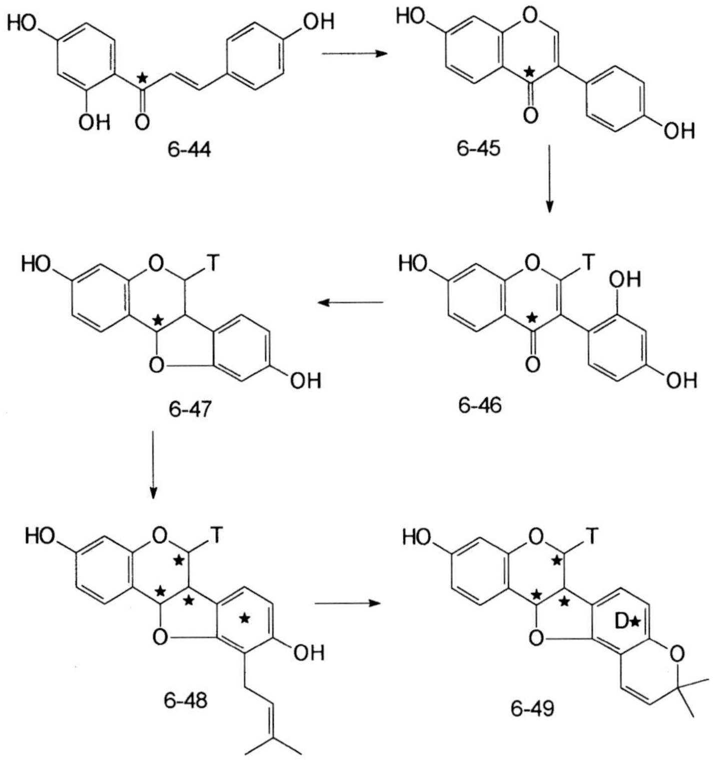

HO OH OH O 6-44 → HO O O OH 6-45

HO O T OH O OH 6-46 → HO O T O OH 6-47

HO O T O OH 6-48 → HO O T O D★ O 6-49

Scheme 6.10 Biosynthetic pathway to phaseollin. T = tritium and the star represents C-14. The labeling in phaseollidin (6-48) and phaseollin is cumulative.

The biosynthesis of two other pterocarpans has been studied using enzyme preparations from *Cicer arietinum* cell suspension cultures (Bless and Barz, 1988). These preparations were capable of converting the isoflavanone vestitone [6-50] to medicarpin [6-51] and maackian [6-52]. The possible participation of a 4-hydroxyisoflavan [6-53], which would seem a logical intermediate, was mentioned, but there appear to be no data to support its involvement. Perhaps it does exist transiently but is never free of the enzyme surface. These compounds are illustrated in Scheme 6.11. A recent paper by Fischer *et al.* (1990) described a pterocarpan

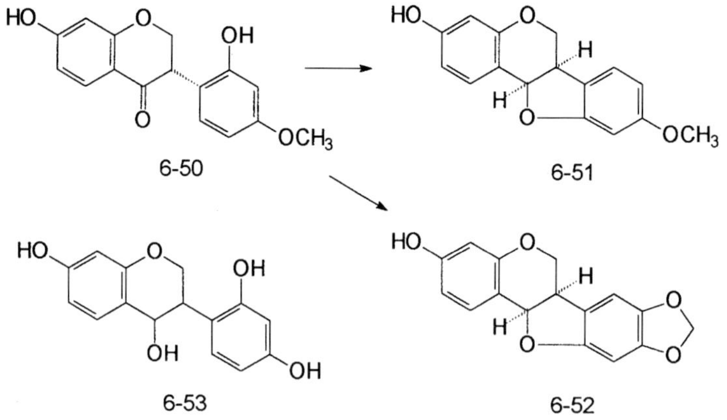

Scheme 6.11 Biosynthesis of the pterocarpans medicarpin [6-51] and maackian [6-52] from the isoflavanone [6-50].

synthase from elicitor-challenged soybean cultures. Additional information on biosyntheses of derived isoflavonoids can be found in an excellent paper by Martin and Dewick (1980).

Rotenoids are isoflavonoid derivatives that have a fourth ring that involves the C-2 carbon (using the isoflavonoid numbering system) and what appears to be the carbon of a B-ring methoxy group. Early literature on the chemistry and biosynthesis of natural rotenoids has been reviewed by Crombie (1984). An example, rotenone itself, is shown as [6-54], along with the labeling pattern that was expected, and observed, from 3-^{14}C-phenylalanine (Crombie *et al.*, 1968, 1973). In those papers, Crombie and his coworkers characterized the biosynthesis of rotenoids as occurring in four stages: chalcone–flavanone stage, the flavanone–isoflavone stage, the hydroxylation-*O*-methylation stage, and the rotenoid stage. The crux in the biosynthesis involves establishment of the final ring system. Along with the observation that C-6a (C-2 of the isoflavone) arose from 3-^{14}C-phenylalanine, in agreement with the occurrence of a phenyl shift, Crombie and coworkers (1968) established that methyl ^{14}C-methionine fed to *Derris elliptica* and *Amorpha fruticosa* seedlings yielded label exclusively in *O*-methyl groups and C-6 in the rotenoids. The final step involves formation of the characteristic rotenone ring, most likely, by addition of a free-radical derived from the 2′-methoxy function to the α, β-unsaturated carbonyl system of the isoflavone nucleus. [1,2-^{13}C]-Acetate was shown to label the carbons that are equivalent to the A-ring of the parent isoflavone (Bhandari *et al.*, 1992). Results of these labeling experiments are summarized in Scheme 6.12.

Details of an *O*-methyltransferase involved in the biosynthesis of (+)-pisatin [6-56] from (+)-6a-hydroxymaackiain [6-55] have been reported by Sweigard and coworkers (1986). The bulk of enzyme activity appeared after pea seedlings had been induced by microbial infection or treatment with $CuCl_2$, although there also appeared to be a low level of constitutive activity present. Thus, the terminal step in (+)-pisatin biosynthesis is accomplished by a dedicated enzyme, marked as

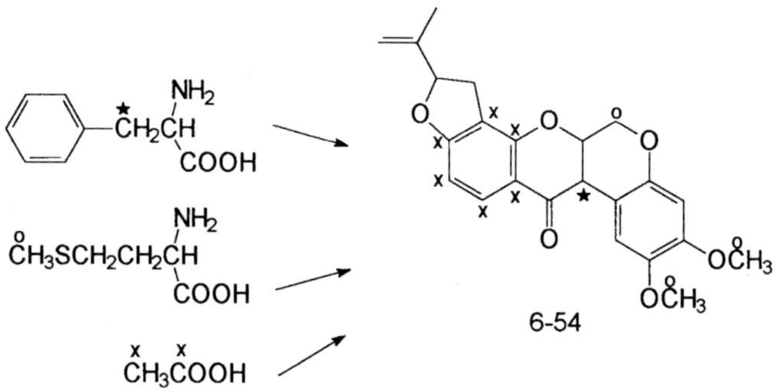

Scheme 6.12 Biosynthesis of rotenone from phenylalanine, methionine, and acetate. Carbon-14 locations indicated in each precursor and in the product with star or small letter.

OMT, and not by a general purpose *O*-methyltransferase suggesting that the entire biosynthetic pathway to this phytoalexin is under close genetic control.

Many phytoalexins and related isoflavonoids synthesized by legumes feature one or more *C*-prenyl groups, which may undergo cyclization with an adjacent phenolic hydroxyl group to yield a chromene derivative. Zähringer and coworkers (1981) studied the *C*-prenylation of a pterocarpan in elicitor-treated soybean cell cultures and cotyledons and noted that prenylation occurred at two positions in glycinol, equivalent to C-6 and C-8 in the isoflavonoid numbering system (starred carbons in 6-57). The product of prenylation at C-6 is glyceollidin II [6-58]. The *C*-prenyl group of glyceollidin II is subsequently cyclized to produce glyceollin II [6-59] and glyceollin III [6-60]. Glyceollidin I is the 8-*C*-prenyl isomer of [6-58] and glyceollin I is the corresponding isomer of [6-59]. The mechanism and stereochemistry of the enzymatic conversion of the *C*-prenyl function to the cyclic form has been studied in detail (Bhandari *et al*., 1989). Other investigators have addressed the mechanism of origin of the 2,2-dimethylchromen ring system (Crombie *et al*., 1982; Begley *et al*., 1986). For example, Crombie and coworkers (1986) showed that a cyclase enzyme from *Tephrosia vogellii* brought about the reaction apparently without either an epoxy or hydroxy intermediate contrary to what might be expected based on similar reactions in the laboratory.

C-Prenylation of isoflavones using enzyme preparations from *Lupinus albus* (white lupine) has been described by Laflamme and coworkers (1993). 5,7,4′-Trihydroxyisoflavone (genistein) and 5,7,2′,4′-tetrahydroxyisoflavone (2′-hydroxygenistein) were converted to their 6-, 8-, and 3′-*C*-prenyl derivatives by an enzyme that appeared to be associated with microsomal membrane fraction (a detergent was required for solubilization of the active protein). The reaction required γ,γ-dimethylallyl pyrophosphate (DMPP) and manganese ions; isopentenyl pyrophosphate (IPP) did not yield any flavonoid products but it did inhibit the reaction competitively. Differences in prenyltransferase activities in relation to source of enzyme, type of detergent used, and different stabilities were taken to indicate the existence of a family of enzymes each of which is likely to exhibit position specificity (or at least selectivity).

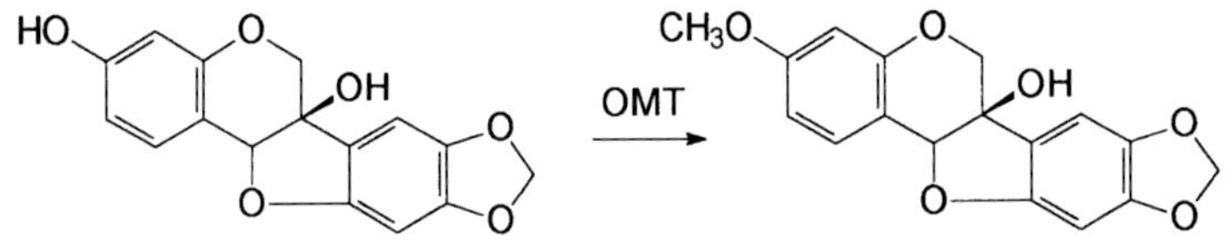

OMT

6-55 6-56

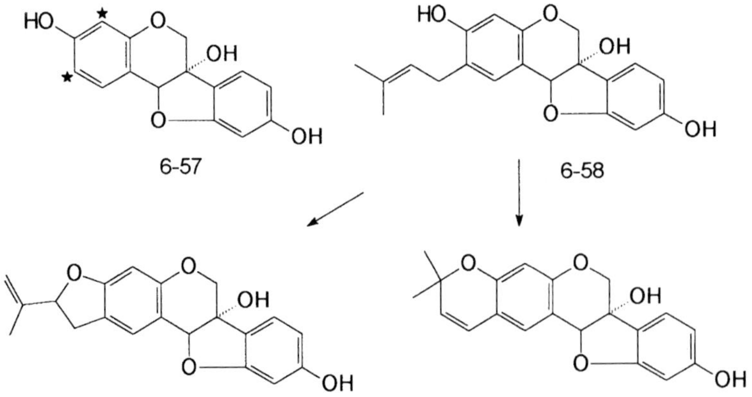

6-57 6-58

6-60 6-59

C-Glycosylflavones

Whereas *O*-glycosylation is a terminal or near terminal step in flavonoid biosynthesis, the formation of *C*-glycosylflavones follows a different path. The first indication that this was the case came from feeding studies in which *Spirodela polyrhiza*, *S. oligorhiza*, and *Lemna minor* (all Lemnaceae) were administered ^{14}C-apigenin and ^{14}C-luteolin. These plant species were chosen for study because they accumulate both *C*-glycosyl and *O*-glycosyl derivatives of those flavones. After an appropriate period of metabolism, the flavonoids of the test species were isolated and examined for radiolabel. The only compounds that had become labelled were flavone *O*-glucosides; no label was detected in any of the *C*-glycosylflavones (Wallace *et al.*, 1969). Subsequent experiments using 2-^{14}C-naringenin, however, led to the production of apigenin and luteolin 7-*O*-glucosides as above, but labeled vitexin (8-*C*-glucosylapigenin) and orientin (8-*C*-glucosylluteolin) were also observed (Wallace and Grisebach, 1973). These observations strongly pointed toward an early establishment of the flavonoid *C*-sugar linkage.

It was not until comparatively recently, however, that formation of a *C*-glycosylflavone was observed in a cell-free preparation, although the cell-free formation of a *C*-glycosyl anthroquinone had been described some years ago (Grün and Franz, 1981). Kerscher and Franz (1987, 1988) showed that extracts from buckwheat could catalyze the formation of vitexin and isovitexin. Flavones were not accommodated as substrate but, quite surprisingly, neither were flavanones! They found that *C*-glycosylation required a 2-hydroxyflavanone and that the enzyme accepted either 2,5,7,4′-tetrahydroxyflavanone or 2,5,7-trihydroxyflavanone as substrate; neither 2,5-dihydroxyflavanone nor 2,4-dihydroxydibenzoylmethane was glycosylated. 2,4-Dihydroxybenzoylmethane was shown, however, to act as a competitive inhibitor. UDP-Glucose was the preferred sugar nucleotide, but UDP-galactose and UDP-xylose were also accepted by the enzyme; UDP-glucuronic acid was not. The enzyme worked most efficiently at about pH 9.8 suggesting the involvement of a carbanion in the reaction, as would be expected for a *C*-alkylation. The mechanism of the reaction, other than the suggestion that some dibenzoylmethane derivative is involved in the enzyme-substrate interaction (Kerscher and Franz, 1987), remains to be determined.

Finishing Touches: Hydroxylation, *O*-Methylation, *O*-Glycosylation, *O*-Acylation, etc.

Although the most elementary of flavonoid compounds, i.e., 2′,4′,6′,4-tetrahydroxychalcone, naringenin, apigenin, dihydrokaempferol, and kaempferol, do exist in Nature, they usually do not exist as such, and almost never alone. The vast majority of naturally occurring flavonoids are characterized by some additional structural feature or features such as "extra" hydroxyl groups, methyl ether groups, glycosidically linked sugars, and very often, some combination of these. In addition to these common adornments, one finds *C*-alkyl groups (other than *C*-glycosylflavones), acylated glycosides, and sulfates. Owing to their intimate involvement in the biosynthesis of anthocyanins, hydroxylation, *O*-methylation, and *O*-glycosylation have attracted the lion's share of attention. This is reflected in the very extensive literatures that have accumulated that deal with these processes, particularly in regard to methylation and glycosylation. We will start, however, with a short discussion of hydroxylation of flavonoid A- and B-rings.

The first evidence for the enzymatic hydroxylation of a flavonoid was that of Fritsch and Grisebach (1975) who showed that microsomal preparations of *Haplopappus* cell cultures had the capacity to convert naringenin to eriodictyol and dihydrokaempferol to dihydroquercetin. The reaction required NADPH and molecular oxygen. There followed papers from a number of laboratories describing the enzyme from a wide array of plant species. A more detailed examination using parsley (Hagmann *et al*., 1983) confirmed that the enzyme was a cytochrome P-450-dependent mono-oxygenase. This was later confirmed using enzyme preparations from several additional species in the presence of specific inhibitors (Stich *et al*., 1988). Hagmann and coworkers (1983) observed that the enzyme would accept naringenin, dihydrokaempferol, apigenin, or kaempferol as substrate and in each case produce the corresponding 3′-hydroxy derivative, eriodictyol, dihydroquercetin, luteolin, or quercetin, respectively, but would not convert *p*-coumaric acid to caffeic acid. The flavonoid 3′-hydroxylase from *Petunia* did not accept leucopelargonidin as substrate (Stotz *et al*., 1985).

The failure of flavonoid 3′-hydroxylase to function with *p*-coumaric acid supports the long held view that the flavonoid B-ring oxygenation pattern is established at the 15-carbon stage. An exception to this generalization involves *Silene* wherein Kamsteeg and coworkers (1981) have identified gene *P* that controls hydroxylation of *p*-coumaroyl CoA to caffeoyl CoA. Although the majority of plant species examined fall into the former category, sampling has not been extensive. Examination of a wider representation of species would clearly help to put the generalization on a firmer ground. It would also be of interest to find out how widely occurring the capacity to effect hydroxylation of *p*-coumaroyl CoA is in Caryophyllaceae (the family to which *Silene* belongs) and related families in Caryophyllales.

The first observation that an enzyme exists that is capable of hydroxylating flavonoids at both C-3′ and C-5′ came from the work of Stotz and Forkmann (1982) who were studying enzymes in delphinidin-containing flowers of *Verbena.* This enzyme is also found in microsomal preparations and requires NADPH and molecular oxygen. When naringenin [6-61] is the substrate, the product is 5,7,3′,4′,5′-pentahydroxyflavanone [6-62] and when dihydrokaempferol [6-63] is used dihydromyricetin [6-64] is obtained. Substrates that already have the 3′,4′-dihydroxy pattern, such as eriodictyol or dihydroquercetin, are converted to the corresponding 3′,4′,5′-trihydroxyflavonoid. Genetic studies had determined that a single gene controls the existence of delphinidin derivatives in *Verbena* (Beale, 1940; Beale *et al.*, 1940). The presence of the 3′,5′-hydroxylase is strictly correlated with the presence of delphinidin derivatives. The enzyme is controlled by gene *R* in *Callistephus* (Forkmann, 1977) and *Hf1* in *Petunia* (Tabak *et al.*, 1981).

Little is known about hydroxylation of the flavonoid A-ring despite the very extensive occurrence of 6-, 8-, and 6,8-oxygenated flavonoids. A-Ring hydroxylation probably occurs at the flavonoid skeleton level, but the processes involved have not been studied in any detail. We know from such studies as those of Ibrahim and coworkers on sequential *O*-methylation (see below), however, that hydroxylation of quercetin derivatives to give the corresponding quercetagetin derivatives occurs after about half of the *O*-methyl groups have been put in place. Much work is needed in this area.

Compounds bearing methyl groups bound to oxygen or nitrogen atoms are extremely common in Nature. The biochemistry of methylation, and demethylation, has been intensively studied in both animal and plant systems. Much of the background information on methylation in plants has been expertly reviewed by Poulton (1981). As we saw in Chapter Two, flavonoids bearing one or more *O*-methyl groups are extremely common components of a variety of plant species. Among these are *O*-methylated anthocyanins whose biosynthesis and genetic background were part of the overall analysis of the formation of these plant pigments. The conversion of cyanidin [6-65] to peonidin [6-66] and delphinidin [6-67] to petunidin [6-68] and/or malvidin [6-69] in *Petunia* were shown by Jonsson and coworkers (1982) to involve enzymes that used anthocyanidin 3-*O*-(*p*-coumaroyl)-rutinoside-5-*O*-glucosides as substrates. The functioning of these *O*-methyltransferases depended, therefore, upon the availability of anthocyanin derivatives that were already at a level of some structural complexity. Further studies from the same group (Jonsson *et al.*, 1983a,b) showed that four genes were involved in the *O*-methylation process, *Mt1* and *Mf1* and *Mt2* and *Mf2*. All four code for enzymes that can methylate both

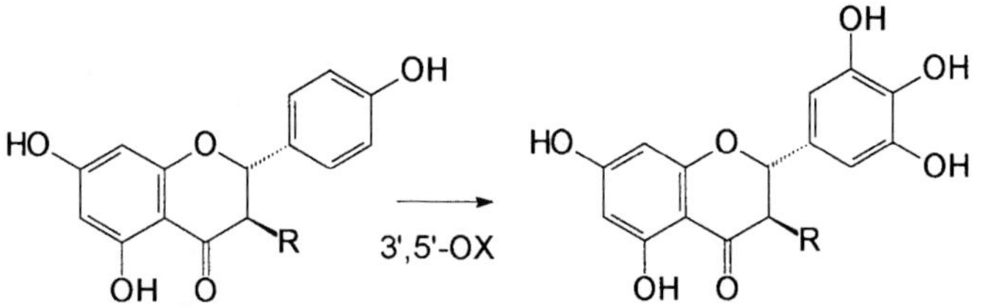

3′,5′-OX

6-61 R = H

6-63 R = OH

6-62 R = H

6-64 R = OH

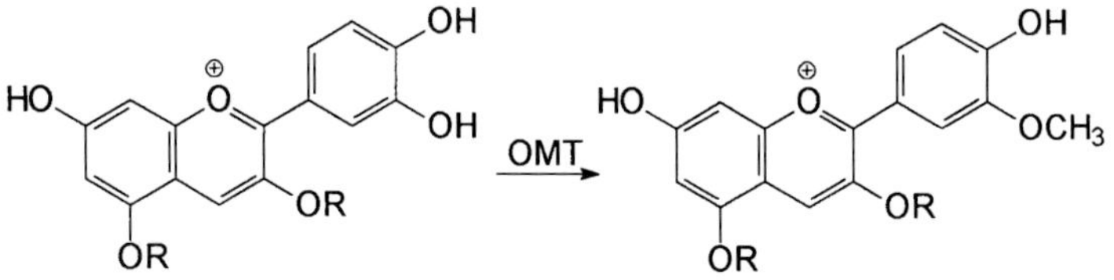

OMT

6-65

6-66

OMT

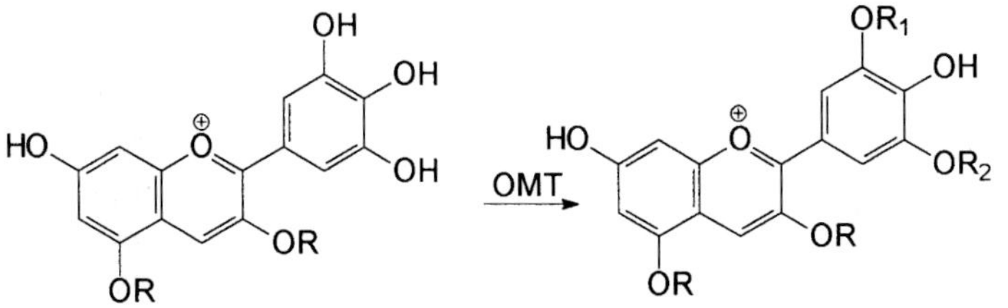

6-67

R = Sugars

6-68 $R_1 = CH_3$, $R_2 = H$

6-69 $R_1 = R_2 = CH_3$

3′- and 5′-hydroxyl groups but double methylation occurs best with the enzymes coded for by the *Mf* genes.

O-Methylation systems have been described involving a variety of plant species and flavonoid types. Among these are the 4′-*O*-methylation of isoflavones in *Cicer arietinum* (Wengenmayer *et al.*, 1974), the conversion of apigenin to its 4′-methyl ether (acacetin) by an enzyme from *Robinia pseudoacacia* (Kuroki and Poulton, 1981), *O*-methylation of a variety of flavones and flavonols by preparations from *Lotus corniculatus* (Jay *et al.*, 1983), 5-*O*-methylation of isoflavones in *Lupinus luteus* (Khouri *et al.*, 1988a), flavonoid 3′-*O*-methylation in maize (Larson, 1989), and 2′-*O*-methylation of isoliquiritigenin in alfalfa (Maxwell *et al.*, 1993). Methylation of *C*-glycosylflavones has been studied in *Silene pratensis* and *Avena sativa* by Knogge and Weissenböck (1984). In *Silene*, isoörientin [6-70] and isoörientin 2″-*O*-rhamnoside [6-71] were converted to isoscoparin [6-72] and its 2″-rhamnoside

[6-73]. The efficiency of methylation of isoörientin 2″-*O*-rhamnside was such to suggest that this may be the normal substrate in the plant. In *Avena*, *O*-methylation of vitexin 2″-*O*-rhamnoside [6-74] produced the 7-methyl ether [6-75].

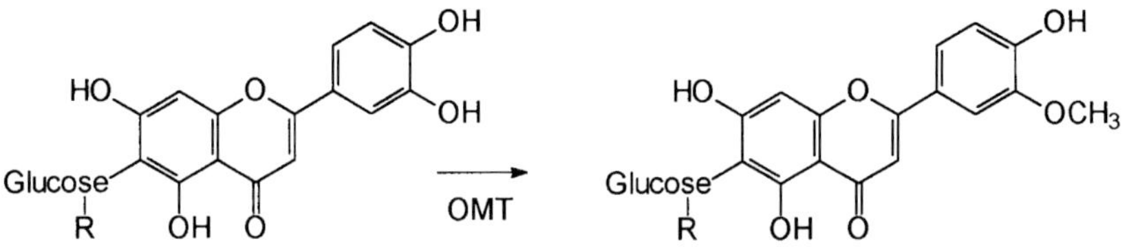

OMT

6-70 R = H

6-71 R = 2"-Rhamnose

6-72 R = H

6-73 R = 2"-Rhamnose

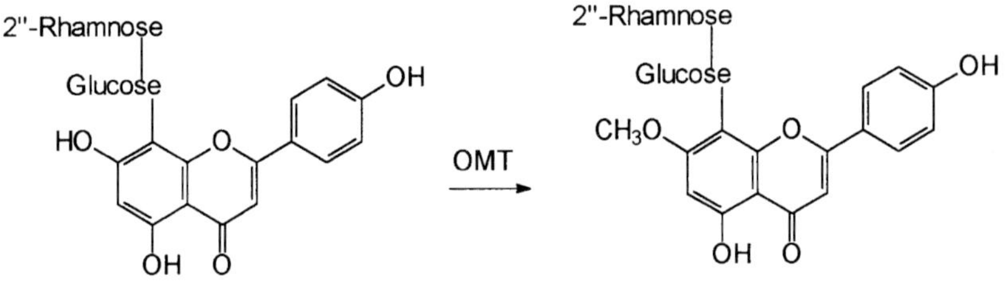

OMT

6-74

6-75

One of the more complex systems in plants about which we have detailed information involves synthesis of the family of *O*-methylated flavonols by *Chrysosplenium americanum*. This species elaborates an array of flavonols based upon quercetin and quercetagetin some of which also have an hydroxyl group at C-2′ (Collins *et al*., 1981). Studies of the enzymology of *O*-methylation in *Chrysosplenium* have shown the existence of O-methyltransferases that exhibit a high degree of position specificity (De Luca and Ibrahim, 1982, 1985a,b; De Luca *et al*., 1982; Ibrahim *et al*., 1987, 1989; Khouri and Ibrahim, 1987; Khouri *et al*., 1988b; Ibrahim, 1992). The entire set of reactions (Gauthier *et al*., 1996) is too complex to reproduce in its entirety, but a shortened version is useful to demonstrate the sequential formation of products. With quercetin as the starting compound, a sequence of *O*-methylations occurs each of which involves a specific hydroxyl group. The first three steps involve methylation of quercetin [6-76] to give quercetin 3-methyl ether [6-77], quercetin 3,7-dimethyl ether [6-78], and quercetin 3,7,4′-trimethyl ether [6-79]. At this point, quercetin 3,7,4′-trimethyl ether [6-79] can serve as substrate for three different reactions: (1) hydroxylation at C-6 to yield quercetagetin trimethyl ether [6-80]; (2) methylation of the C-3′-OH to yield [6-81]; or (3) hydroxylation at C-2′. We will look only at the 2′-hydroxylation branch. The product of this hydroxylation reaction is 5,2′,5′-trihydroxy-3,7,4′-trimethoxyflavone [6-82].

This compound can undergo two different reactions: (1) methylation at the 5′-hydroxyl group by an enzyme (3′/5′-OMT) specific for *meta*-hydroxy groups (Gauthier *et al*., 1996) to yield the tetramethyl compound [6-83]; and (2) glucosylation of the 2′-hydroxyl group to yield [6-84]. This trimethyl glucoside serves as substrate for a 5′-*O*-methyltransferase to give compound [6-85]. Compound [6-85] does not appear to undergo any further modifications. Since *O*-glycosylation is often referred to as the terminal step in flavonoid biosynthesis, it is interesting to note this exceptional case involving *O*-methylation of the glucoside [6-84]. The reactions are summarized in Scheme 6.13.

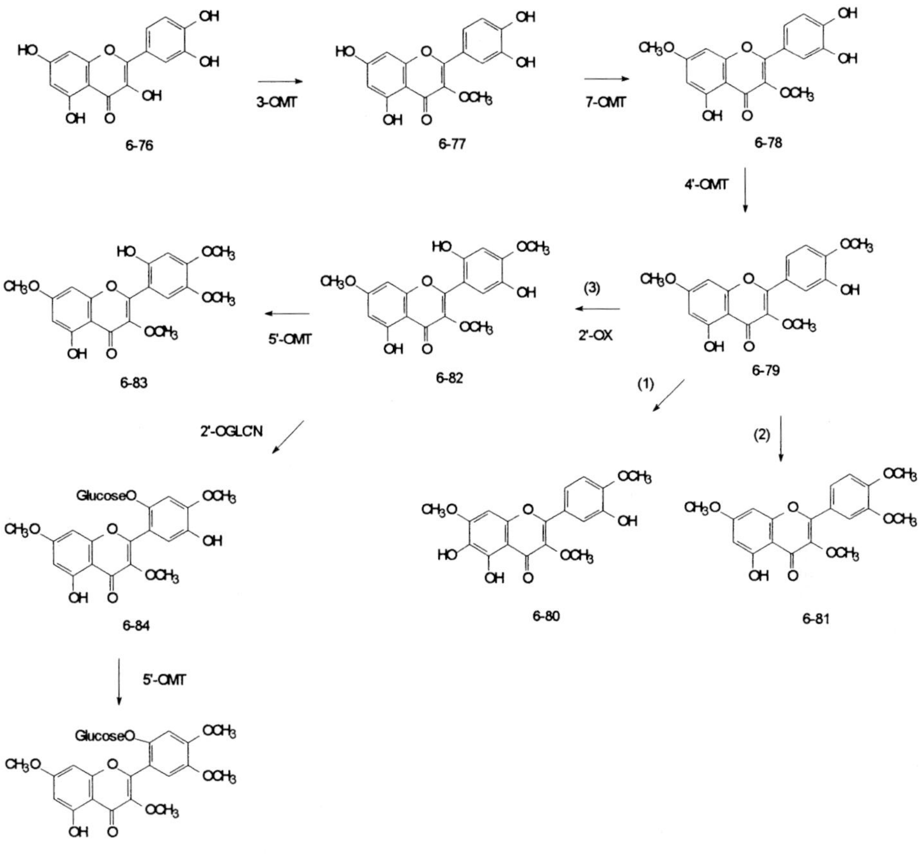

Scheme 6.13 A partial sequence of reactions reported from *Chrysosplenium americanum*.

Gauthier and coworkers (1996) have studied the 3′/5′-OMT system in detail including sequence comparisons, which showed about 67–85% similarity to other plant *O*-methyltransferases. The enzyme shows strict specificity for the 3′/5′ (*meta*) position in partially *O*-methylated flavonols exemplified in the conversion of [6-82] to [6-83]. We are told that sequence studies on other enzymes in the pathway are underway.

We have looked at this system in some detail because it is the most complex one for which data are available and because it demonstrates so clearly the highly selective nature of the enzymes involved. It will be of considerable interest to learn about the evolutionary relationships of the *O*-methyltransferases involved. When those relationships have been established, it will also be possible to address the question of relationships within the genus *Chrysosplenium*. Although all members of the genus accumulate *O*-methylated flavonols, only the opposite-leaved species, as opposed to the alternate-leaved species, appear to have the capacity to make 2′-oxygenated flavonols (Bohm and Collins, 1979). A comparison of the nucleic acid sequences of the *O*-methyltransferase genes of representatives of the two groups could provide useful insights into the relationships between the two sets of species.

Most flavonoids occur in Nature bound to one or more sugars, commonly glucose, galactose, rhamnose, arabinose, xylose, glucuronic acid, and galacturonic acid. Hahlbrock and Grisebach (1975) reviewed early studies of glycosylation of flavonoids using cell-free preparations. The UDP and TDP-activated forms of the sugars were the only forms accepted by the enzymes. All of the examples listed in their table involved flavonols as substrates with 3-*O*-glycosides as the sole products. Detailed studies of flavonoid biosynthesis in cultured parsley cells revealed, among many other things, that flavone 7-*O*-glycosides, flavonol 7-*O*-glycosides, and flavonol 3,7-di-*O*-glycosides were formed. The corresponding glucosyltransferases were isolated from the cells and shown to possess position specificity (Sutter *et al.*, 1972; Sutter and Grisebach, 1973). UDP-Glucose:flavone/flavonol 7-*O*-glucosyltransferase utilizes either UDP- or TDP-glucose and can convert either flavones or flavonols to the corresponding 7-*O*-glucoside. The enzyme accepts neither flavonol 3-*O*-glycosides nor flavanones as substrates. The second enzyme, UDP-glucose:flavonol 3-*O*-glucosyltransferase is specific for the 3-OH group including quercetin 7-*O*-glucoside. It does not accept dihydroflavonol, which is a further indication that the glucosylation step occurs late in pathway. Parsley also accumulates flavone diglycosides that consist of glucose and the branched sugar apiose [6-86]. Formation of the diglycoside is catalyzed by UDP-D-apiose:flavone 7-*O*-β-D-glucoside apiosyltransferase. The reaction is specific for the 2-OH group of glucose to yield flavone 7-*O*-β-D-apiofuranosyl($1 \rightarrow 2$)-β-D-glucoside (Ortmann *et al.*, 1970). The apigenin diglycoside is commonly known as "apiin."

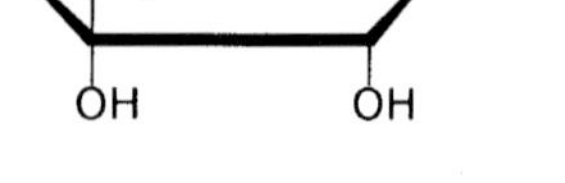

O CH$_2$OH H,OH OH OH

6-86

There followed in the next several years many reports of glycoside biosynthesis involving a variety of plant species and increasingly more complex glycosidic combinations. A sampling of these follows. *Glycine max* (soybean) cells in suspension culture afforded a flavonol 3-*O*-glucosyltransferase that accepted kaempferol, quercetin, isorhamnetin, and fisetin as substrates but was inactive against 2′,4′,4-trihydroxychalcone, naringenin, dihydroquercetin, the isoflavone daidzein, cinnamic acids, and simple phenols. A low level of vanillic acid glucosyltransferase was detected but was separable from the flavonoid glucosyltransferase (Poulton and Kauer, 1977). The specific activity of the enzyme increased with age of the culture reaching a maximum late in the growth cycle. Protoplasts of *Hordeum vulgare* (barley) contain an enzyme capable of forming vitexin 7-*O*-glucoside (saponarin) from the flavone and UDP-glucose (Blume *et al.*, 1979). Position specificity and a high degree of overall substrate specificity characterized an enzyme from *Cicer arietinum* (chick pea) that transfers glucose (from UDP-glucose) to the 7-hydroxyl group of some isoflavones (Köster and Barz, 1981). The enzyme worked well with the 4′-methoxyisoflavones biochanin A and formononetin but less well with genistein and daidzein, both of which have a 4′-hydroxyl group. The enzyme did not accept isoflavanones, flavones, flavanones, flavonols, coumarins, or benzoic and cinnamic acids. Ultraviolet light-induced flavonoid formation in *Anethum graveolens* (dill, Apiaceae) cell cultures including an enzyme that catalyzed the formation of quercetin 3-*O*-*β*-glucuronide (Möhle *et al.*, 1985). Activity of other enzymes earlier in the pathway could also be detected. Teusch (1986) described an enzyme from *Matthiola incana* petals that has the capacity to transfer a xylose residue to the glucose of anthocyanidin 3-*O*-glucosides. Whereas other anthocyanins are accepted by the enzyme, no activity was observed with the following sugar derivatives: ADP-, CDP-, or GDP-glucose, UDP-mannose, UDP-galactose, and UDP-glucuronate.

Position specificity was reported for *O*-glucosyltransferases in young leaves of *Euonymus alatus* forma *ciliato-dentatus* (Celastraceae) (Ishikura and Yang, 1994). Two *O*-glucosyltransferases were obtained, F3GT and F7GT, which catalyzed transfer of glucose specifically to the 3- and 7-positions, respectively. F3GT was itself resolved into two isoforms, F3GT1 and F3GT2. These two enzymes were specific in their requirement for kaempferol and a few other flavonols but would not accept flavonol 7-*O*-glucosides or flavonol 3-*O*-glucosides. F7GT appeared to exist as a single form, was specific for the 7-position, but would accept kaempferol and quercetin 3-*O*-glucosides as substrate although at lowered efficiency. There was also evidence presented for the existence of a UDP-D-xylose:flavonol 3-*O*-xylosyltransferase in this plant.

In the section above on *O*-methyltransferases we saw the intermix of *O*-methylation and *O*-glucosylation required to build some of the complex flavonol derivatives that characterize *Chrysosplenium americanum*. In addition to analyses of *O*-methyltransferases in that species, the Concordia group has also studied a B-ring glucosyltransferase in detail. The enzyme appears to be specific for hydroxyl groups at C-2′ or C-5′and requires the presence of the *para*-dihydroxy arrangement for activity. The enzyme has similar K_m values for the two positions and the cofactor, UDP-glucose (Bajaj *et al.*, 1983). Subsequent studies of this system dwelt on establishment of a monoclonal antibody specific to the 2′-*O*-glucosyltransferase (Latchinian and Ibrahim, 1989).

Some of the earlier attempts to study the enzymatic formation of flavonoid diglycosides came from the work of Barber (1962), who demonstrated the formation of quercetin 3-O-rutinoside by transfer of rhamnose to quercetin 3-*O*-glucoside, and the Freiburg group who investigated the formation of diglucosides containing apiose, as described above. The first report of the formation of a linear triglycoside came from the work of Shute and colleagues (1979) who investigated the synthesis of kaempferol and quercetin 3-*O*-triglucosides in *Pisum sativum* seedlings. Incubation of either flavonol and UDP-glucose with a cell-free preparation of seedlings yielded the corresponding 3-*O*-triglucoside. The reaction sequence proceeds *via* the monoglucoside and diglucoside, either of which could be converted to the triglucoside under the reaction conditions employed. The number of proteins involved was not determined.

The existence of linear triglycosides in the anthers of *Tulipa* cv. Apeldoorn presented a somewhat more complex problem in that they consist of three different sugars, glucose, rhamnose, and xylose. Three enzymes were obtained that accounted for the formation of the flavonol 3-*O*-xylosylrhamnosylglucosides in this tissue (Kleinehollenhorst *et al.*, 1982). The enzymes identified were UDP-glucose:flavonol 3-*O*-glucosyltransferase (GT-I), UDP-rhamnose:flavonol 3-*O*-glucoside rhamnosyltransferase (GT-II), and UDP-xylose:flavonol 3-*O*-glycoside xylosyltransferase (GT-III). The sequence of reactions is summarized in Scheme 6.14. Differences in physical properties and biochemical requirements distinguished the three enzymes. For example, GT-I and GT-II had approximate molecular weights of 40,000 while GT-III showed a molecular weight of 30,000. GT-III required ammonium or calcium ions; the other two enzymes had no such requirement. GT-I and GT-III were sensitive to the presence of sulfhydryl reagents while GT-II was not. GT-I accepted flavonol aglycones as substrate but did not yield product when offered flavonol 3-*O*-glycosides, flavones, flavanones, dihydroflavonols, anthocyanins, or simple phenols, including *p*-coumaric acid. GT-II and GT-III had no affinity for aglycones but both could convert diglycosides to triglycosides. The highest activity of all three enzymes appeared in the tapetum fraction of anthers, whereas the pollen fraction showed only minor activity.

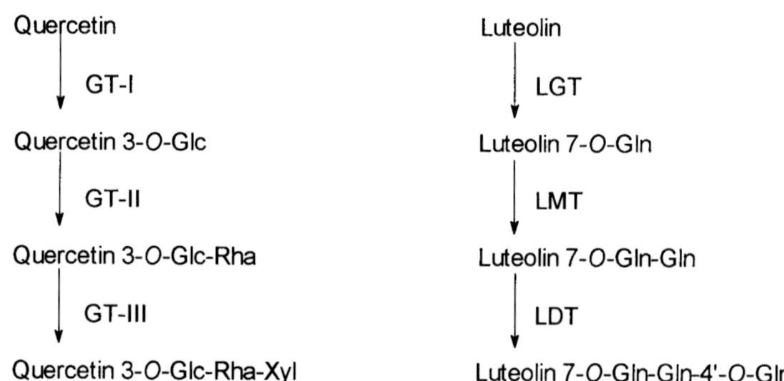

Scheme 6.14 Order of attachment of monosaccharides in the biosynthesis of a quercetin 3-*O*-triglycoside and a luteolin 7-*O*-diglycoside 4′-*O*-monoglycoside.

Primary leaves of *Secale cereale* (rye) accumulate flavone triglycosides as components of their flavonoid profile, the major one of which is luteolin 7-*O*-[β-D-glucuronosyl(1 $\rightarrow$ 2)β-D-glucuronide]-4′-*O*-β-D-glucuronide. Schulz and Weissenböck

(1988) obtained from young rye leaves three enzymes that are responsible for a strictly sequential transfer of glucuronic acid units to flavone. UDP-Glucuronate: luteolin 7-*O*-glucuronosyltransferase (LGT) gets the process started by catalyzing the formation of luteolin 7-*O*-glucuronate. The second enzyme, UDP-glucuronate: luteolin 7-*O*-glucuronide-glucuronosyltransferase (LMT), specifically attaches the second glucuronate residue at the C-2″ position of luteolin 7-*O*-glucuronide. The third enzyme (LDT) requires luteolin 7-*O*-diglucuronide as substrate and acts to put the third glucuronate unit at C-4′. This sequence of reactions is also summarized in Scheme 6.14. The enzymes are highly selective for these substrates; much reduced activities were observed with apigenin and chrysoeriol glucuronides. Similarly, UDP-xylose, UDP-galactose, and UDP-glucose were used poorly by the enzymes. Estimated molecular weights were 34,000 for LGT, 37,000 for LMT, and 29,000 for LDT. Comparable numbers were observed with SDS electrophoresis indicating that the enzymes are monomeric. Under the conditions of the analyses, none of the enzymes was reversible, whereas some reactions involving UDP-glucose: flavonol glucosyltransferases are freely reversible.

The genetics of flavonoid biosynthesis in maize has been studied in great detail including the control of anthocyanin glycosylation. Much of the work has been done with the mutant "bronze" so called because its kernels have a bronze color rather than the purple normally associated with anthocyanins. The *Bz1* locus controls the synthesis of UDP glucose-flavonoid 3-*O*-glucosyl transferase (UFGT) which catalyzes glucosylation of the 3-OH group of flavonols and anthocyanidins (Larson and Coe, 1977). In *bz1* mutants the anthocyanidin, cyanidin in this case, accumulates and undergoes polymerization leading to the formation of a bronze colored product. Other glucosyltransferase genetics studies have involved the *An4* locus in petunia (Gerats *et al.*, 1985), which converts anthocyanidins to the corresponding 3-*O*-glucosides, and an enzyme from *Silene* that transfers glucose to the 5-OH of anthocyanidin 3-*O*-glycosides (Kamsteeg *et al.*, 1978). Many flavonoids occur as diglycosides common among which are rutinosides. Kamsteeg and coworkers (1979) have also examined the genetics of the rhamnosyltransferase reaction in *Silene* which leads to the formation of anthocyanidin 3-*O*-rutinosides. The *Rt* locus in petunia has been identified as the factor that codes for the rhamnosyltransferase in that species (Wiering and de Vlaming, 1984). This enzyme is specific for anthocyanidin 3-*O*-glucosides and for rhamnose but is not specific for the B-ring oxygenation pattern of the anthocyanin. For information on cloning of glycosyltransferase genes, the reader is referred to Forkmann (1994, Table 12.7).

Many flavone and flavonol glycosides, and anthocyanins, occur naturally as acylated derivatives. The acylation can involve aliphatic or aromatic acids bound to sugar hydroxyls. (Sulfates are some times included here, but we will look at these separately below.) As we saw in Chapter Two, aliphatic acyl groups arising from acetic or malonic acids are common with occasional appearance of others, e.g., succinic acid. Benzoic and cinnamic acid derivatives are commonly seen as well. In a few cases glycosides are known that contain a mixture of aliphatic and aromatic acyl groups.

One of the reactions examined during extensive studies of flavonoid biosynthesis and control in parsley was the enzymatic formation of malonyl esters of apigenin 7-*O*-glucoside (cosmosiin), apiin, and chrysoeriol 7-*O*-apiosyglucoside (gravebioside B).

The acylation reaction is the terminal step in the biosyntheses of these compounds (Hahlbrock, 1972; Kreuzaler and Hahlbrock, 1973). Subsequent detailed studies of the malonyltransferases of parsley led to the detection, separation, and purification of two distinct proteins (Matern *et al.*, 1981, 1983a), one of which accepted either flavone or flavonol 7-*O*-glucosides as substrate while the other only functioned with flavonol 3-*O*-glucosides. Succinoyl anthocyanins have been found to occur in a number of plants species including the corn flower (*Centaurea cyanus*), in which the enzymatic esterification reaction has been described (Yamaguchi *et al.*, 1995). The enzyme preparation from the flowers used succinoyl-coenzyme A and the 3-*O*-glucosides of cyanidin and pelargonidin but would not acylate the 3,5-diglucosides. The enzyme was also capable of using malonyl coenzyme A at about the same rate. It was not determined whether separate enzymes are involved or whether the acyltransferase can function with different acids. An enzyme that catalyzes malonylation of cyanidin 3-*O*-glucoside has recently been isolated from *Lactuca sativa* (lettuce) leaves, also by Yamaguchi and coworkers (1996).

An often-asked question concerning flavonoids involves the need for some of the structural modifications that we see. It is generally believed that one of the reasons for the development of phenolic glycosylation is that the process converts the poorly soluble aglycones into water-soluble glycosides enabling them to be transported from the site of synthesis to other places within the plant. The same question can be asked about the need for acylation. At least one reasonable answer to that question comes from additional studies of malonylglucosides in parsley by Matern and his coworkers (1983b, 1986). The first of those two studies examined the conformational changes that apigenin 7-*O*-(6-*O*-malonylglucoside) undergoes as one changes the concentration of dissolved proteins and the composition of the solvent in which the components are dissolved. These changes take on particular significance in light of the finding that the acylated apigenin glucoside is taken up rapidly by vacuoles whereas apigenin 7-*O*-glucoside *per se* is not taken up at all. It would be of interest to learn whether the malonyl group is the "password" by which a transport protein recognizes the compound to be carried across the membrane. A similar function was suggested for malonylated anthocyanidin 3-*O*-glucosides in *Callistephus chinensis* from which Teusch and Forkmann (1987) obtained an enzyme that catalyzed their formation from anthocyanidin 3-*O*-glucosides. Those workers noted that the 3,5-di-*O*-glucosides were significantly poorer substrates.

An example of genetic control of acylation can be seen in *Matthiola incana* where the gene *u* governs the presence of acylated anthocyanins (Teusch *et al.*, 1986a). In the dominant form (u^+) the enzyme transfers a *p*-coumaroyl unit to anthocyanidin 3-*O*-glucosides and 3-*O*-xylosylglucosides but gives no product with anthocyanidin 5-*O*-glucosides. No acylated compounds are present in homozygous recessive plants.

An unusual situation was observed by Glässgen and Seitz (1992) in a study of acylation in cell cultures of carrot (*Daucus carota*). In these tissues, the cofactor involved is an hydroxycinnamoyl-1-*O*-glucose derivative, e.g., [6-87]. The enzyme transfers the cinnamoyl function specifically to the C-6 hydroxyl group of glucose in flavonol 3-*O*-(2″-*O*-xylosyl-6″-*O*-glucosylgalactoside). *p*-Coumaroyl, feruloyl, and sinapoyl-1-*O*-glucose esters were accepted by the enzyme.

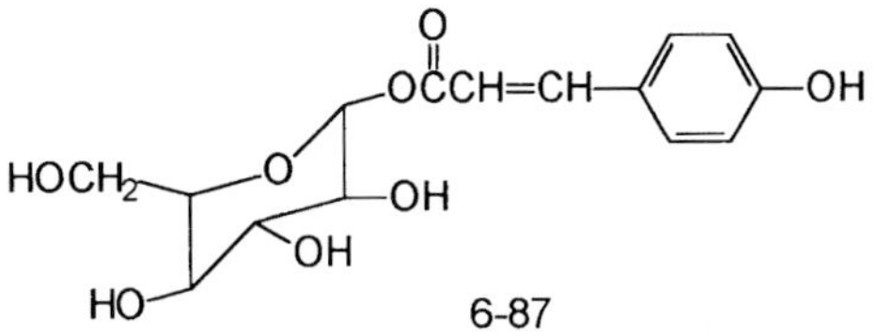

6-87

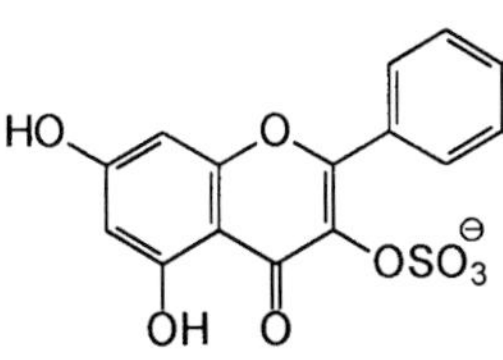

6-88

Several plant species have the capacity to synthesize flavonoid sulfates with linkages either through a sugar hydroxyl of a flavonoid glycoside or one or more phenolic oxygens (Barron *et al.*, 1988). A typical example of a directly sulfated flavonoid is quercetin 3-sulfate [6-88]. Studies of the enzymology of the sulfation process have been described (Varin *et al.*, 1987; Varin and Ibrahim, 1989, 1991, 1992). Enzyme preparations have been obtained from two species of *Flaveria*, *F. chloraefolia* and *F. bidentis*. The enzymes accept a variety of flavonols and use 3′-phosphoadenosine 5′-phosphosulfate (PAPS) as cofactor. Some differences were noted in the relative synthetic capacity of the enzyme(s) from the two species. The most efficient substrate for the enzyme from *F. chloraefolia* was kaempferol which was converted to the 3-*O*-sulfate. The *F. bidentis* system, however, had the capacity to produce quercetin tetrasulfates. Other studies showed that there were enzymes with position specificity similar to that observed for *O*-glycosylation and *O*-methylation. In addition to showing specificity toward the 3-OH (FL3OS), the 3′-OH (FL3′OS), or the 4′-OH (FL4′OS), there was a distinct ordering of substitution with the sulfation of quercetin at 3-OH occurring before sulfation of the 3′-OH. There was also evidence that a *F. chloraefolia* enzyme has the capacity to make 7-*O*-sulfates. Varin and coworkers (1992) resolved the 3-sulfotransferase from the 4′-sulfotransferase and studied their amino acid sequences (following cloning). The 3-sulfotransferase had 311 amino acids and a molecular weight of 36,442, while the 4′-sulfotransferase consisted of 320 amino acids and a molecular weight of 37,212. An estimate of 35,000 had been determined by chromatographic methods.

Coordination and Control

We will conclude this chapter with a few observations on the general subject of coordination and control of the flavonoid biosynthetic pathway. The subject has been mentioned in different contexts so far and will occur again in Chapter Seven when we deal with biological roles of flavonoids. Intense research activity during the 1960s and early 1970s informed us that the flavonoid pathway consists of a sequence of reactions each step of which builds on the product of the previous one. This was, of course, what one would have expected in light of general biosynthetic experience. Once a reaction sequence has been dissected into its component steps and the enzymes responsible for each step identified, a key question can be asked, namely, what controls the system? This question was attacked most enthusiastically by Klaus

Hahlbrock and his colleagues in Freiburg who published an important series of papers establishing the critical part played by light in the process. Descriptions of some of their work using cultured parsley cells (*Petroselinum*) can be found in papers by Hahlbrock and Wellmann (1970), Hahlbrock *et al.*, 1971; and by Hahlbrock and Ragg (1975). A concise review of much of the earlier work on this system can be found in Hahlbrock (1976). The studies showed that two groups of enzymes are involved, group I consisting of the three enzymes of the phenylpropanoid pathway (PAL, C4H, and 4CL), and group II, consisting of the enzymes of the flavonoid pathway starting with chalcone synthase (CHS). Treatment of cultured parsley cells with light brought about formation of both groups. Group I enzymes reached their maximum activity after about 15 hours, while group II enzyme activity peaked at about 24 hours. Group I enzyme activity tends to drop to pre-initiation levels quite precipitously, while group II enzyme activity drops off more slowly. Flavonoid glycoside levels begin to rise soon after the appearance of detectable enzyme activity.

The coordination of enzyme activity and development in plants has been studied in a number of organisms indicating that a close relationship between these processes is likely a general property of plants. Some of the earlier studies employed only a few of the enzymes involved, often PAL, CHS, and another one farther along the pathway. More recently, with the availability of significantly improved enzyme assay procedures, all or nearly all of the enzymes can be studied, which allows a much more detailed picture to be drawn of the overall process. An example of this approach comes from the detailed examination of the anthocyanin pathway in developing berries of *Vitus vinifera* L. cv. Shiraz by Boss and coworkers (1996). In flowers and berry skins, expression of PAL, CHS, CHI, F3H, DFR, and LDOX, but not UFGT, was detected until about four weeks postflowering with a reduction in expression from six to eight weeks postflowering. Activity of CHS, CHI, F3H, DFR, LDOX, and UFGT increased 10 weeks postflowering coinciding with the onset of anthocyanin formation. In grape berry flesh, neither PAL nor UFGT was detected at any time during development, but activity of CHS, CHI, F3H, DFR, and FDOX was detected up to four weeks postflowering. UFGT appears to be regulated independently of the other genes which prompted those authors to conclude that anthocyanin biosynthesis in *Vitus* is controlled at a later stage than so far reported for maize, petunia, or snapdragon.

A perfectly efficient biosynthetic pathway should, theoretically at least, produce end product with little or no accumulation of intermediates. We know from experience, however, that most pathways in plants are not perfectly tuned, and that the accumulation of intermediates is the rule rather than the exception. It is not difficult to imagine circumstances where an accumulated intermediate might be "put to use" by the organism for some task other than simply being converted to the next intermediate. Perhaps it can serve as a herbivore feeding deterrent, for example, or maybe it can enhance flower color leading to more visits by pollinators. Following this line of thought, it seems reasonable to speculate that the set of metabolic conditions leading to the accumulation will likely be maintained since they clearly benefit the organism. In order to assure a constant supply of the intermediate, some control of the overall process would also be advantageous. That such situations exist is indicated by recent studies of the control of flavonoid biosynthesis in *Antirrhinum* flowers (Moyano *et al.*, 1996). These workers observed that the apparent redundancy in *myb* gene function in snapdragon appears to be involved with modularization of pigment biosynthesis. In other words, an early stage and a late stage in the pathway appear to be controlled independently. Specifically, this system allows flavanones to be formed (and accumulated!) independently of anthocyanins the advantage of which is that both compound types are available for pollinator attraction.

The study of plant photoreceptors has been an area of active research throughout the history of plant biology. Even a cursory review of the subject is well beyond the scope of this book, but it is useful to point out an example of current research in the area that deals with the flavonoid biosynthetic pathway. The example chosen involves a study of light-regulated expression of the chalcone synthase gene in *Arabidopsis* (Batschauer *et al.*, 1996). Those workers discussed the difficulty in sorting out which photoreceptor might be involved in any given process listing three, phytochrome, blue/UV-A, and UV-B. The difficulties are magnified by the fact that phytochromes absorb in both long wavelength (red and far-red) and short wavelength regions. The *Arabidopsis* work employed mutants deficient in the phytochromes which, of course, eliminated them from the experimental system. In the absence of these complicating factors, these workers established that receptors in the blue/UV-A region are involved in triggering enzyme activity.

Additional insights into the complexity of the interaction of a gene with light can be gleaned from a detailed study of developmental and UV-light regulation of the chalcone synthase promoter in snapdragon (Fritze *et al.*, 1991). We saw some of the results of these workers' activities earlier in this chapter in a discussion of promoter regions and tissue specific transcription of the enzyme. Included in that work, and of particular relevance to the topic of light activation of enzymes, was the finding "... that UV light induction and sequences important in directing expression in leaf tissue work cooperatively in the intact promoter." One of the possible mechanisms "... might be that factors important in expression in leaf tissue cooperate with UV light-induced factors that recognize separate promoter elements." Details of how the light receptor communicates the appropriate signal or signals to the promoter region are not available for this system but will certainly be the subject of continued research. The interested reader should consult the papers by the Fritze and Batschauer groups for leading references to the relevant literature.

One of the subjects in the study of light-induced flavonoid biosynthesis involves the formation of apigenin glycosides and related compounds in parsley (papers by Hahlbrock and his colleagues cited above). It has also been demonstrated that the formation of apiin (apigenin-7-*O*-apiosylglucoside) and apiin 6″-*O*-malonate can be increased by a factor of about two by subjecting parsley leaves to 0.2 ppm ozone for a period of 10 hours. This is in accord with other reports of enhanced synthesis of phenolic compounds in response to low levels of ozone, stilbenes in Scots pine (Rosemann *et al.*, 1991) and isoflavonoids in some legumes (Rubin *et al.*, 1983). Also of interest is the induction of apiin biosynthesis by 12-oxo-phytodienoic acid, a precursor of jasmonic acid (Dittrich *et al.*, 1992). Although known as naturally occurring for some years, jasmonic acid and related compounds have only comparatively recently been shown to function as signal molecules in plants. The structure of jasmonic acid is shown as [6-89]. Information on these compounds, along with extensive references to the literature, can be found in recent research papers by Blechert *et al*. (1995) and Ellard-Ivey and Douglas (1996) and in a review by Creelman and Mullet (1997).

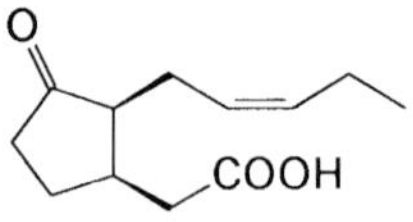

O

COOH

(3*R*,7*S*)-Jasmonic acid

6-89

In addition to the effect of light and air pollutants such as ozone, other environmental factors can have significant effects on flavonoid synthesis. A recent study by Dedaldechamp and colleagues (1995), using grape cell cultures, revealed the sensitivity of flavonoid synthesis to changes in mineral nutrients. Under conditions of phosphate deprivation there was a large enhancement of dihydroflavonol reductase (DFR) activity. This resulted in a significant increase in anthocyanin accumulation as well as an increase in the production of acylated peonidin 3-*O*-glucoside. Current investigations in our laboratory (Rajakaruna and Bohm, unpubl. results) have revealed significant differences in soil mineral content and soil water content between different flavonoid chemotypes of *Lasthenia californica*. Although closely correlated, soil mineral chemistry and flavonoid profiles do not appear to be related in a cause and effect manner, however. Soil chemistry and flavonoid variation have been observed in other systems including *Mimulus cardinalis* (Scrophulariaceae) (Pollock *et al*., 1967), *Pulsatilla alpina* (Ranunculaceae) (Horovitz, 1976), *Richea* (Epacridaceae) (Menadue and Crowden, 1983), and in *Cassinia vauvilliersii* (Asteraceae) (Reid and Bohm, 1994). In none of these situations does there appear to be a straightforward cause and effect relationship between pigment chemistry and soil composition. And, in most instances, plants cultivated in a common garden retain their particular flavonoid features.

We saw in one or two instances above that certain enzymes in the flavonoid pathway are active in specific plant tissues but remain essentially inactive in others. In such situations, it is logical to speculate about what specific functions the compounds perform. Where flavonoid synthesis is initiated in response to invasion by a pathogenic organism, the role is not difficult to imagine. In other cases, however, in tapetal walls for example, the ultimate function of the product is not immediately obvious. In other cases still the reports of flavonoids in only one of the phases of a sexually dimorphic organism has led to speculation that they may have a significant role to play in reproduction. Examples of this include the report of the aurone bractein (4,6,3′,4′,5′-pentahydroxyaurone) [6-90] in sporophytes of the moss *Funaria hygrometrica* (Weitz and Ikan, 1977). Markham and Porter (1978) identified another aurone derivative, aureusidin 6-*O*-glucuronide [6-91], from antheridiophores of the following liverworts: *Marchantia berteroana, M. polymorpha*, and *Conocephalum supradecompositum*. The compound could not be detected in the leafy, vegetative part of the plants. The presence of this compound for a period of time during which reproductive structures are functional strongly suggests that it performs a significant function. What that function is remains a mystery. There has been speculation, however, that the compound may function in a chemotactic role in "attracting" gametes. An interesting situation has been described in *Matteuccia struthiopteris*, the ostrich fern, where the haploid and diploid phases exhibit significantly different phenolic profiles (Araki and Cooper-Driver, 1993). After spore germination, for a period of some weeks, the haploid gametophytic phase invests its phenolic metabolic energy in the production of a variety of benzoic acid and cinnamic acid derivatives including chlorogenic acid, 1-caffeoylglucose 3-sulfate and 1-*p*-coumaroyl glucose 6-sulfate [6-87 with sulfate on the glucose]. Apparently, no flavonoids are produced. After fertilization, however, initiation of flavonoid biosynthesis occurs resulting in the formation of kaempferol and quercetin 3-*O*-glucosides, kaempferol 7-*O*-glucoside, the *C*-methylflavanone matteucinol 7-*O*-glucoside [6-92], and the *C*-glucosylflavones vitexin [6-93] and isovitexin [6-94]. There appears to be no information available on the mechanism by which flavonoid biosynthesis is switched on when the diploid generation is regenerated. It would be

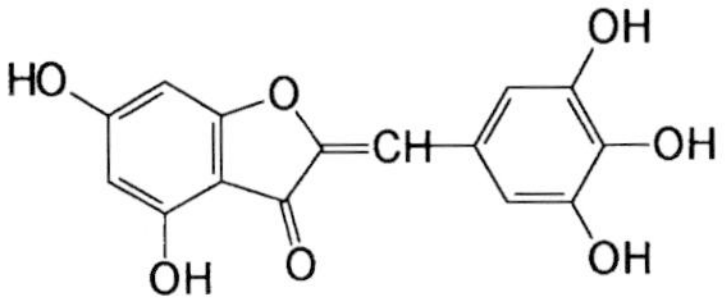

6-90

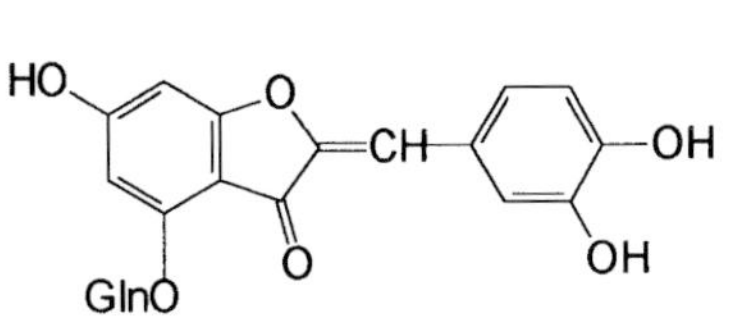

6-91

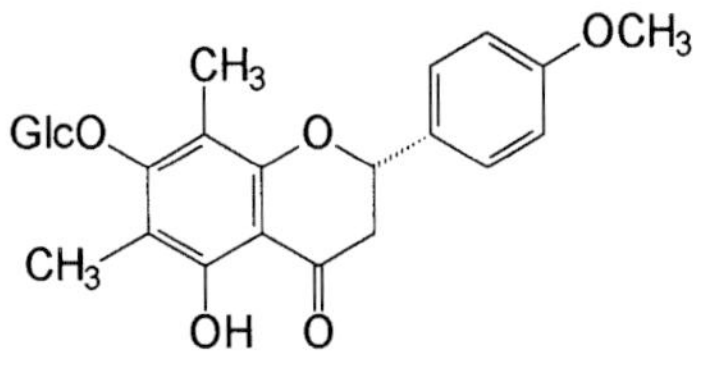

6-92

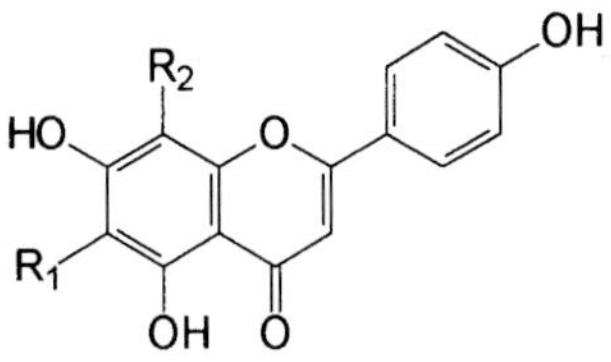

6-93 R_1 = H, R_2 = Glc

6-94 R_1 = Glc, R_2 = H

of interest to learn if only the entry enzyme, chalcone synthase, needs to be "awakened," or whether all flavonoid synthetic enzymes are absent, or if present, are silenced in the haploid phase. Detailed studies of haploid tissue at the macromolecular level would be of interest in this system. It would also be of interest to see how widely distributed this phenomenon is in pteridophytes.

CHAPTER SEVEN

FLAVONOID FUNCTIONS IN NATURE

INTRODUCTION

There was a time in the early days of this century when the compounds we now call secondary metabolites were considered to be waste products of plant metabolism. Since the plant had no obvious means of excreting these waste materials, it was nonetheless necessary to get rid of them in some way. The vacuoles were considered a likely dumping ground since no other use for them had been established. Trees and woody shrubs solved the problem by disposing of unwanted material by shipping it off to the heart wood. One presumes that aquatic plants disposed of their wastes by using the surrounding water to wash them away. It is difficult to establish the exact time when it occurred to someone that this waste disposal concept was wrong and that some, perhaps even most, of these compounds actually served a useful purpose for the producing plants. The long-known bitter or toxic properties of some natural products, alkaloids are a good example here, could well have suggested that they might be useful to the plant in warding off herbivores. The tendency of certain insects to assemble around sweet-smelling flowers might have led someone to suspect that the plant had some "strategy" for attracting the insects. These ideas may appear naïve to us today, inundated as we are with sophisticated concepts of population dynamics, species-specific pollination systems, chemical communication, and the general ideas of co-evolution, but the path to our present level of understanding has been a comparatively slow one. It is exceedingly difficult to answer the question, "What is compound X doing in plant Y?" The opposite approach, where one looks for the cause of an observed phenomenon in plant Y, has a far greater likelihood of being productive. That approach is the physiological equivalent of biological activity-directed fractionation of a drug plant aimed at isolating the active component. The awareness of the near universality of secondary metabolites in plants by physiologists and ecologists has resulted in an increasing number of natural functions being discovered. This trend will only accelerate in the future.

It is difficult to see any underlying structural feature of flavonoids that leads to a specific biological function, although it is entirely possible that there is some subtle feature or features of which we are unaware. In more technical terms, no obvious structure-function relationship has emerged that underlies why certain flavonoids can function as messengers to nodulating bacteria, while others function as deterrents to herbivory (I dislike the term antifeedant!), while still others stimulate insects to lay eggs. Adding to the picture are the observations that the same compound, or compound type, can perform different roles in different organisms, and that different compounds or compound types can perform the same general task. This apparent redundancy in ecological interactions has been addressed in some detail by contributors to a volume edited by Romeo *et al.* (1996). Thus, with no obvious underlying structure-function relationship evident, I have opted simply to take the case-study approach to the subject. What follows, is a series of short discussions intended to acquaint the reader with the variety of flavonoid functions that have been identified.

An important contribution to the chemical ecology literature is *Introduction to Ecological Biochemistry* by J. B. Harborne (1990, 3rd ed.). The topics covered include plant–plant interactions, plant–animal interactions, plant–micro-organism interactions, and an interesting chapter on plant chemistry and food choice by higher animals (us!).

Flavonoids as Ultraviolet Shields

One of the most frequently cited functions for flavonoids is to serve as ultraviolet filters. Some have suggested that this may have been the driving force for their selection and, in turn, for the early evolution of land plants. There seems to be little doubt that flavonoids function in this capacity at the present time but their original functions may have been quite different (Stafford, 1991). We will examine some of these ideas later. Ultraviolet radiation is generally classified into three wavelength ranges: UV-A (320–390 nm), UV-B (280–320 nm), and UV-C (<280 nm). Radiation reaching the earth is high in the UV-A region, drops sharply in the UV-B region, and drops to nearly zero at 290 nm (Robberecht, 1989). The absorption spectrum of DNA lies in the range 240–310 nm, within the lower range of wavelengths that reach the surface of the planet. The epidermal layer of plants has been shown to absorb 90–99% of the incident ultraviolet radiation (Robberecht and Caldwell, 1978; Caldwell *et al.*, 1983). The localization of flavonoids in the epidermal layers of plants and their known ultraviolet absorption properties leads to the logical suggestion that they can serve as shields to potentially harmful radiation. There is a growing body of evidence that this suggestion is well founded.

Recent studies strongly implicate flavonoids as radiation shields in a variety of flowering plants and also indicate that they are not alone in serving this function. Significant contributions have been made using mutants of *Arabidopsis thaliana* (Brassicaceae). For example, Li and colleagues (1993) showed that exposure of plants to high UV-B brings about an increase in the synthesis of ultraviolet absorbing compounds arising from phenylalanine. The *transparent testa-4* (*tt4*) mutant, which has reduced capacity to synthesize flavonoids but normal sinapate levels [7-1; sinapic acid], is significantly more sensitive to UV-B than plants with normal levels of both compound types. Two other mutants, *tt5* and *tt6*, with reduced levels of both flavonoids and sinapate derivatives, are highly sensitive to ultraviolet radiation damage. Shirley and coworkers (1995) described 11 loci, collectively identified as *transparent testa*, that control flavonoid biosynthesis in *Arabidopsis*. Mutations at these loci disrupt the formation of the normal brown seed coat. Lois (1994) showed that irradiation of *Arabidopsis* with ultraviolet of different wavelengths and intensities brought about increased levels of flavonoids, in agreement with the observations of Li and coworkers (1993). The amount of pigment produced appeared to depend on the developmental stage of the leaf at the time of irradiation. It was also noted that the response was limited to the area immediately surrounding the site of irradiation. In a second paper, Lois and Buchanan (1994) described a chemically induced mutant of *Arabidopsis* that showed an increased sensitivity to ultraviolet radiation and that the mutation involves a single gene whose blockage prevented the formation of ultraviolet absorbing flavonoids. One of the compounds was shown to be a kaempferol rhamnoside.

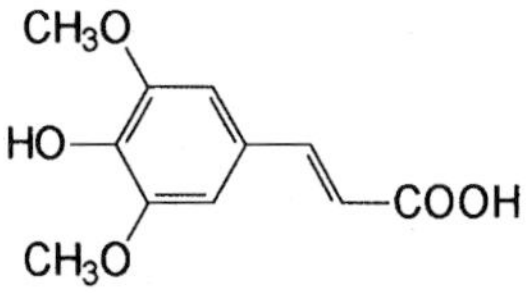

CH_3O
HO
COOH
CH_3O

7-1

Flavonoids have been implicated in providing protection from solar UV-B radiation in other species. For example, Robberecht and coworkers (1983) reported that 95% of incident UV-B radiation was attenuated by the epidermis of *Oenothera stricta* (Onagraceae). Stapleton and Walbot (1994) used analyses of cyclobutane pyrimidine dimer level and pyrimidine(6,4)pyrimidone damage in DNA to measure the effect of ultraviolet radiation on maize (*Zea mays*). Radiation damage to plants was more severe in the case of lines that have reduced levels of flavonoids (anthocyanins) than in those with normal pigment complement. UV-B radiation induced the formation of a complex array of acylated flavonol glycosides, including kaempferol and quercetin 3-*O*-glucoside-3″,6″-di-*O*-*p*-coumarates, in Scots pine (*Pinus sylvestris*) (Jungblut *et al.*, 1995a,b). The quercetin conjugate was found in cotyledons while the kaempferol conjugate was found in primary leaves. Both were present in the epidermal cells of their respective leaves. In the case of *Quercus ilex*, the holly oak, ultraviolet absorbing compounds are located in non-glandular leaf hairs (Skaltsa *et al.*, 1994). The compounds responsible were shown to be acylated derivatives of kaempferol glycosides. Removal of trichomes resulted in a significant reduction of photosystem II photochemical efficiency. Other plant species with leaf hairs have been shown to contain ultraviolet-absorbing pigments and that this may be a major function of plant hairs. (Karabourniotis *et al.*, 1992, 1993).

In an earlier citation, involving the work of Li and coworkers (1993), we saw that sinapic acid derivatives as well as flavonoids were involved as ultraviolet screening compounds. Further work on this subject was recently described by Sheahan (1996) who demonstrated that sinapate esters in *Arabidopsis thaliana* are capable of greater attenuation of ultraviolet radiation than flavonoids. This work suggests that other hydroxylated cinnamic acid derivatives may also be effective in screening plants from harmful ultraviolet radiation.

Flavonoids and the Attraction of Pollinators

We will see in the section immediately following this one that certain flavonoids are required for germination of pollen grains and for successful pollen tube growth.

But an important event must happen first: pollen must be deposited on the stigma in the first place! There are a number of ways in which this can happen: it can arrive by wind, or through visitation by an insect, or by a bird, or occasionally by other animals, bats and mice among others. Wind pollination does not concern us here, but the means by which a flower attracts the other pollinators does. The subject has attracted a great deal of attention in the scientific literature for at least 200 years, involving a range of interests running the gamut from natural historical accounts of "pollination sightings" to the realm of carefully controlled experiments in animal behavior. Several books and articles are available that chronicle most aspects of the subject, particularly *The Principles of Pollination Ecology* by Faegri and van der Pijl (1971), the profusely illustrated *The Pollination of Flowers* by Proctor and Yeo (1972), *The Pollination of Flowers by Insects* (Richards, 1978), and a study of floral colors in the high arctic (Kevan, 1972). *Pollination Biology*, a collection of articles by specialists in the field, edited by L. Real (1981), provides an excellent and thoroughly referenced introduction to problems confronting pollination biologists and contains, pertinent to our immediate interests, a chapter by N. Waser on the adaptive nature of floral traits. Aimed at a more general audience, Barrett (1987) describes mimicry in plants using excellent photographs to illustrate two impressive examples. The first involves color patterning in an orchid flower that mimics a female bee resulting in visits by male bees in a process referred to as pseudocopulation that leads to pollen transfer. The second example shows the flower of *Stapelia nobilis* (Asclepiadaceae) whose purple and brown spot patterning resembles spoiled flesh and is attractive to carrion flies. Adding to the effect, *Stapelia* also has a strong aroma of amines reminiscent of rotting flesh.

Extensive surveys of floral anthocyanins have shown that the pigment arrays are more highly correlated with pollinator class than with taxonomic relationships within the respective groups. Eighty-seven species (ca. 250 known) of *Pentstemon* (Scrophulariaceae) were studied by Scogin and Freeman (1987) who found that whereas anthocyanidins (the aglycones) are under strong selection as pollinator attractants, the anthocyanins (the glycosides) are only weakly selected at the interspecific level and are thus useful as taxonomic indicators. A similar conclusion concerning correlation between flower color, anthocyanidin type, and pollinator, was drawn by Saito and Harborne (1992) who studied the floral pigmentation of 49 species and cultivars representing 12 genera in Lamiaceae. In a comparison of two closely related species of *Mimulus*, Wilbert and coworkers (1997) observed quite different arrays of floral pigments. *Mimulus lewisii* has pink flowers and is regularly pollinated by bumblebees, while *M. cardinalis* has red flowers and is pollinated by hummingbirds. The total anthocyanin concentration in *M. cardinalis* is about two times that in *M. lewisii*, while the cyanidin bioside concentration of *M. lewisii* is about ten times that present in *M. cardinalis*. Qualitative differences were also observed with the presence of two pelargonidin biosides restricted to *M. cardinalis*.

The classic view of pollination biology states that flower visitors have relatively fixed color preferences. This idea leads to the hypothesis that there will be specific color associations for each type of flower visitor, e.g., hummingbirds prefer red flowers, bees prefer yellow or blue. In turn, this might suggest that these vectors have limited visual sensitivities. In view of several pieces of work, this appears not to be the case. Goldsmith (1980) has shown that hummingbirds are sensitive to a spectrum that ranges from deep red to the near ultraviolet. Similarly, both honey bees and bumblebees are sensitive from the red-orange or red to near UV wavelengths (Goldsmith and Bernard, 1974; Kevan, 1978). That the visual ranges of these animals, both bees and hummingbirds, is broader than is often cited is demonstrated by the fact that both can be rapidly trained to respond to any wavelength within the range as long as a food reward is involved. A natural test of the breadth of the hummingbird spectrum was documented by Waser (1983, Table 1) who reported some of his observations of flowers visited by hummingbirds in the vicinity of the Rocky Mountain Biological Laboratory in Colorado. Red-flowered species visited by hummingbirds included *Ipomopsis aggregata* and *Castilleja miniata*, which were described as being visited commonly and moderately, respectively, by the birds. In both cases, experimental verification of pollination was obtained. Both *Epilobium angustifolium* and *Hydrophyllum fendleri*, with purple and pale purple flowers, respectively, were classified as moderately visited species. *Delphinium nelsonii*, with blue flowers, was visited by hummingbirds and was successfully pollinated in the process. The pale blue flowers of two *Mertensia* species were visited moderately often, but *Iris missouriensis*, which has similarly colored but structurally more complex flowers, was only rarely visited. *Pedicularis bracteosa*, with yellow flowers, and *Frasera speciosa*, with green flowers, were both visited moderately often by hummingbirds.

Earlier in this chapter we saw how flavonoids protect plants by shielding them from potentially harmful radiation. The absorption of ultraviolet radiation by flavonoids also finds a use in attracting potential pollinators. General discussions of flower color are often concerned with the basic flavonoid chemistry involved, the anthocyanins themselves, non-anthocyanic flavonoids, the possible involvement of co-pigments, etc. On closer examination of a random selection of flowers, however, one can see that many exhibit patterns of pigmentation. There is often an overall background color, either anthocyanin or carotenoid, underlying markings of different hue. Arrangements of the markings may take many forms, among which are lines or arrays of spots. Other patterns may be very complex as in the case of some orchids whose lower lip pigmentation mimics a female bee (see Barrett, 1987 for illustration). Examples of pigment patterns in some common household or garden plants, *viz*., lily, Cape primrose (*Streptocarpus*), violet, and "daisy," are shown in Figure 7.1a–d, respectively.

The study of floral pigment patterns attracted a good deal of attention during the late 1960s and into the 1980s, but the earliest awareness of the phenomenon, at least in the scientific literature, may have been the observation by Sprengel in 1793 that certain arrangements of color patterns provided nectar guides by which insect visitors were led to their reward. One of the earliest experimental approaches to the subject was the use of a television camera to emulate the insect eye (Eisner *et al*., 1969). Using a lens that absorbed visible wavelengths but allowed transmission in the ultraviolet, those workers observed that bird and bat-pollinated flowers lacked nectar guides while insect-pollinated flowers had distinctive markings. That study was followed by one in which black and white photography was employed to register ultraviolet reflectance patterns (Eisner *et al*., 1973). Both of these methods emphasized

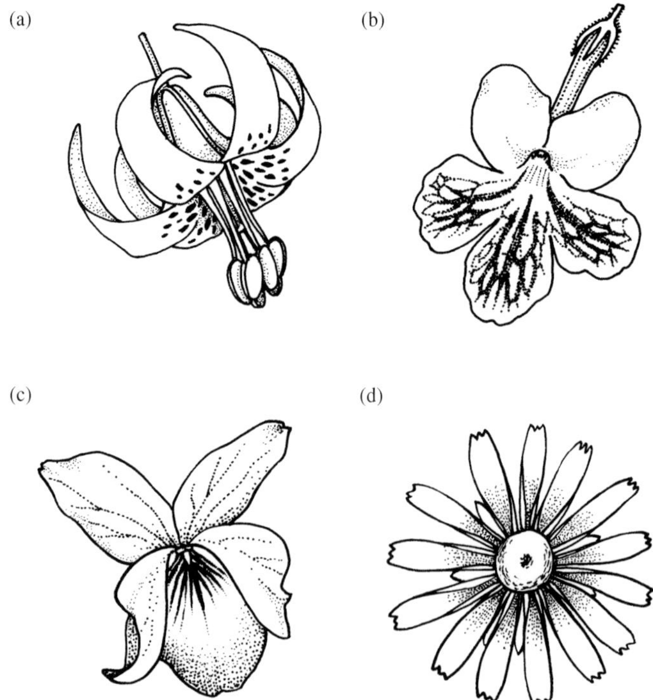

Figure 7.1 Illustrations of typical flavonoid deposition patterns in flowers. Upper left, Columbia lily (*Lilium columbianum*); upper right, Cape primrose (*Streptocarpus* sp.); lower left, violet (*Viola* sp.); and lower right, a "daisy".

the behavior in ultraviolet without recording patterns that can be seen in visible light. McCrea and Levy (1983) resolved this by developing a method that employs daylight-balanced color film with illumination from a blacklight fluorescent source, which provided near ultraviolet, and a filtered daylight fluorescent source. This combination approximates the sensitivity of the honeybee eye. Striking photographs of several flowers are presented in that paper that compare the human visual system with the insect visual system.

Horovitz and Cohen (1972) surveyed 29 species of Brassicaceae (mustard family) for ultraviolet reflectance patterns and a survey of a large number of species of Asteraceae was described by King and Kranz (1975). Complex patterns were observed in both families suggesting that floral patterning is widespread in the plant kingdom. Detailed studies of several taxa have shown localized deposition of a variety of flavonoids in members of Asteraceae: *Bidens* (Scogin and Zakar, 1976; Scogin, 1983); *Coreopsis* (Scogin, 1976, 1983; Scogin *et al.*, 1977), and *Viguiera* (Rieseberg and Schilling, 1985). Dement and Raven (1974) determined that flavonoids were responsible for the nectar guides in flowers of *Oenothera* (Onagraceae, evening primrose) as well. Chalcones and aurones play major roles in *Bidens*, *Coreopsis*, and *Viguiera*, but flavones and flavonols may also be involved.

Thus, 6-hydroxyquercetin (quercetagetin) [7-2] and its methyl ether (patuletin) [7-3] were identified as the ultraviolet absorbing compounds in flowers of *Rudbeckia hirta* (Asteraceae) (Thompson *et al.*, 1972). Additional examples from Asteraceae include species of *Coreopsis*, *Eriophyllum*, and *Helianthus* (Harborne and Smith, 1978) and other genera (Scogin, 1978). Compounds involved in these genera include quercetin 3-*O*- and 7-*O*-glucosides (*H. annuus*), the chalcone coreopsin [7-4] and the aurone sulfurein [7-5] (*Helianthus* species other than *H. annuus*), and quercetagetin and patuletin 7-*O*-glucosides (*Eriophyllum* and *Geraea*).

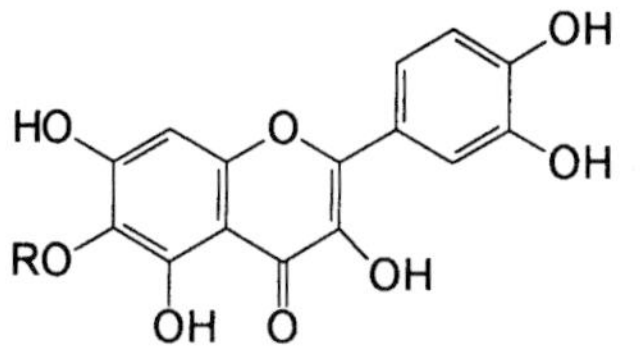

7-2 R = H

7-3 R = CH_3

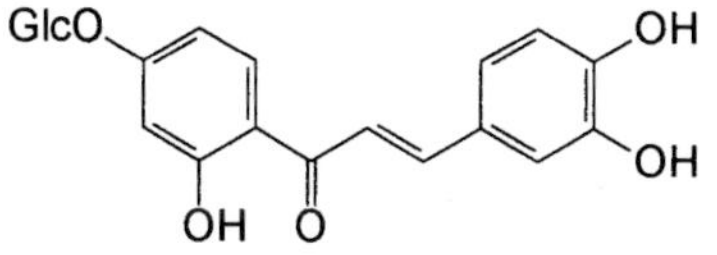

7-4

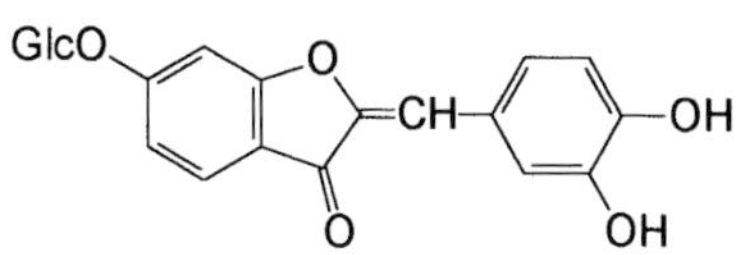

7-5

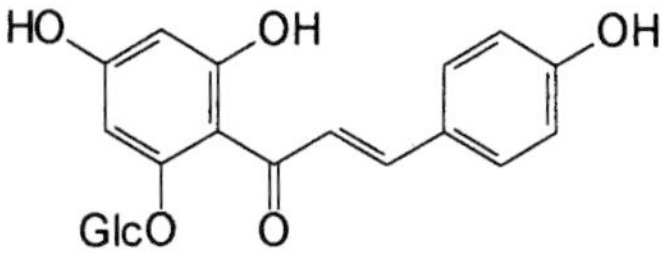

7-6

In flowers of *Mimulus luteus*, which are characterized by reddish spots on a uniform yellow background, cyanidin 3-*O*-glucoside and quercetin 3-*O*-glucoside are localized in spots on an otherwise uniform background of carotenoids and herbacetin 7-*O*-glucoside (Bloom and Vickery, 1975). Ultraviolet radiation is absorbed by the flavonoids but is reflected by the carotenoid-bearing tissue resulting in the appearance of dark spots on a white or pale field. In the case of the asteraceous species listed above, the absorbing pigments are generally concentrated in the proximal part of the ray florets. The outer portion of these florets reflect radiation giving a "bulls-eye" effect. The ultraviolet absorbing pigment is the same in the flowers of *Oenothera hookeri* where the chalcone isosalipurposide [7-6] occurs in the basal quarter of the petals (Dement and Raven, 1974). In the case of *Torenia baillonii* (Scrophulariaceae) the same sort of patterning is seen except that the flavonoids responsible are anthocyanins (Lang and Potrykus, 1971). Petal spotting in *Clarkia* species is caused by local accumulation of anthocyanins against a background of the same anthocyanins present in lower concentrations (Dorn and Bloom, 1984).

Studies of floral tissue of *Balsamorhiza careyana*, *Lasthenia* (*chrysostoma*) *californica*, *Rudbeckia hirta* (Brehm and Krell, 1975) and *Viguiera* (Rieseberg and Schilling, 1985) demonstrated that the ultraviolet absorbing pigments were located in epidermal papillae whose structure and arrangement were such that incoming radiation would be maximally focused on absorbing chromoplasts throughout the day. Scanning electron microscopic examination of cross-sections of ray florets of *Viguiera dentata* showed the presence of ultraviolet-absorbing flavonoids in the adaxial epidermal cells (Rieseberg and Schilling, 1985). Cells in the distal parts of the ray florets had loosely arranged pale yellow chromoplasts while those in the proximal parts had densely packed yellow–orange chromoplasts. They observed the characteristic anthochlor color test (orange–red with base) with sections cut from the proximal part of the florets and no test with sections from the distal part, thus confirming the localization of these compounds in the ultraviolet absorbing part of the florets. The epidermal localization of absorbing pigments was confirmed in a survey of 201 species representing 60 families of angiosperms (Kay and Daoud, 1981). Results from a study of *Zinnia angustifolia*, *Z. elegans*, and their hybrids (Boyle and Stimart, 1989) agreed in general with the above in that floral pigments are present mainly in the upper epidermis with a lower concentration in the lower epidermis and none in the internal parenchyma. Differences in color among the varieties was due to the composition of the pigment mixture and not to differences in cellular location of the pigments. This is not true for all plant species, however.

Extensive surveys of orchid species for pigment disposition, both anthocyanins and carotenoids, have revealed a good deal of variation, however. Studies of *Vanda* and related genera (Matsui *et al.*, 1984) revealed that the presence of anthocyanins in epidermal cells is relatively rare. Similarly, in a study of floral color of *Cattleya* and related genera Matsui and Nakamura (1988) found that the purplish red flowers of most *Cattleya* species contained anthocyanins in parenchymatous cells whereas in a few exceptional cases, *C. intermedia* var. *aquinii* and *C. leopoldii*, anthocyanins were localized in the epidermal cells of splashed or spotted petals. In the case of *Laelia* milleri these workers found that carotenoids were present in both epidermal and parenchymatous cells with anthocyanins present only in the epidermis, while in *L. tenebrosa*, anthocyanins were observed in parenchymatous cells only. In a related study, Matsui (1988) noted that in most *Sophronitis* hybrids anthocyanins were present in both epidermal and parenchymatous cells.

In order to appreciate the significance of floral pigment patterning, it is instructive to look for a moment at some classic experiments in which insect behavior was monitored in controlled environments. Kugler (1943) performed many experiments with artificial feeders including several different types of color patterns. The response of bumblebees was dramatic in that feeders with markings elicited extension of the proboscis and feeding far more often than did feeders that lacked markings. Additional studies were done by Daumer (1958) who arranged ray florets from different asteraceous species in different ways and then recorded bee behavior. When florets with normal proximal pigmentation (Figure 7.2a) were presented to bees they responded as one would expect in a natural setting by extending their proboscises and walking toward the nectar reward. When the ray florets were reversed in orientation (Figure 7.2b), however, with the pigmented parts distal to the disk, bees would land on the outer part of the floret, extend their proboscises and walk around but fail to receive any reward. A test of pattern sensitivity in a more natural setting was undertaken by Scora (1964) who used unspotted mutants of *Monarda punctata* (Lamiaceae) as test organisms within a population of normally spotted plants.

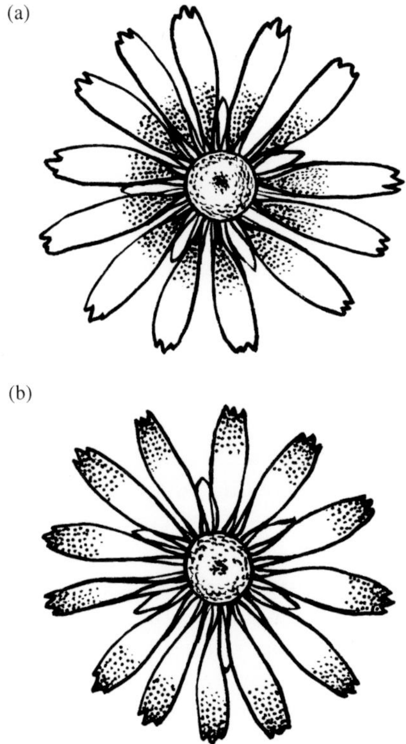

Figure 7.2 An illustration of normal deposition of flavonoid pigments in the proximal portion of ray florets (7.2a), and experimentally adjusted in the distal portions (7.2b) of a daisy-like flower head.

Bees and wasps would approach the unspotted flowers in a mixed population but would ignore them. In a population consisting mainly of unspotted plants the insects would approach the flowers and land on them but did not behave as they normally would and soon left. In behavioral terms, the insects failed to receive the stimulus normally provided by the pigment spot which released the next behavioral activity.

Once a flower has been pollinated and pollen tube growth has begun, deposition of additional pollen represents not only a waste of genetic material but also a waste of the pollinator's time and energy. In order to prevent this sort of unrewarding duplication, a signaling mechanism has evolved that informs potential pollinators which flowers have already been successfully visited and should be avoided. Although color change during floral maturation and development has been recognized for over two centuries (e.g., Sprengel, 1793), it was not until very recently that the subject was thoroughly reviewed (Weiss, 1995). That this phenomenon has undoubtedly evolved many times is indicated by its occurrence in at least 77 flowering plant families. Weiss has also recorded visitors from 15 families of insects and four families of birds. Not all members of a genus, nor all genera in a family, nor all families in an order show color change, or if members of related taxa do undergo changes, they need not involve the same colors or even the same organ.

Color change could involve entire flowers or parts of flowers; Weiss (1995) showed 10 different categories. The changes involved at least seven different physiological processes, and included all three of the major plant pigment families, carotenoids, anthocyanins, and betalains. Color changes result from gain or loss of anthocyanins, gain or loss of carotenoids, gain of betalains (loss of betalains not encountered), movement of a floral part resulting in the appearance of color change (unmasking), or change in pH.

Several examples will illustrate the variety of processes that are involved. Three species of *Lantana* (Verbenaceae) utilize different types of color change to yield three different patterns. In the first, *L. camara*, flowers start out yellow but change to red, whereas in *L. hirta* a ring of yellow carotenoid around the entrance to the corolla throat is lost leaving the area white. *Lantana montevidense* appears to combine the two with the disappearance of yellow carotenoid pigment from the throat area being accompanied by production of purple anthocyanin. Experimental studies on color change in *L. camara* (Mohan Ram and Mathur, 1984) indicate, as we have just seen, that newly opened flowers are yellow but shade to orange and then to scarlet after pollination. Shriveled flowers are magenta having, presumably, reached maximum anthocyanin concentration. Thrips are attracted only by yellow flowers with the formation of anthocyanin beginning immediately following pollination. These workers demonstrated that an extract of pollen placed on receptive stigmas was capable of stimulating anthocyanin synthesis, but the nature of the stimulatory chemical(s) was not reported. Many species of lupine have yellow–white spots on their banner petals. Working with *Lupinus arizonicus* and *L. sparsiflorus*, Wainwright (1978) showed that the spots change from yellow–white to a deep purple–red within four hours following pollination, but that neither fertilization nor pollen tube growth was necessary; simple manipulation of the flower was sufficient to initiate color change. Studies of *L. texensis*, by Schaal and Leverich (1980), showed that flower color change begins on the fifth day after opening and by the sixth day the formerly pale spots have acquired the same color as the rest of the petal. The pigments produced are the same as those that provide the background color. Additional experiments showed that seed set was markedly reduced if six-day flowers were pollinated as compared to yield from flowers pollinated on the fifth day. Flowers on plants maintained in exclosures exhibited floral color change in the absence of visitors (insects or scientists). Several questions arise when one considers possible mechanisms that might be responsible for filling in the spots with anthocyanin. Is anthocyanin transported into the area from reserves in adjacent cells, or is pigment synthesized *de novo* in the spot cells? If it is the latter, then what factor or factors are involved in suppressing anthocyanin formation during the first five days? Or is it the entire flavonoid pathway that has been temporarily blocked? That other types of compounds are involved in flower color change, either directly or indirectly, can be seen in the changes that occur in aging *Fuchsia hybrida* flowers. Yazaki (1976) showed that the change from blue–violet in young flowers to purple–red in older ones was brought about by co-pigmentation of malvin with other flavonoids along with a change in vacuolar pH from 4.8 to 4.2 brought about by an increase in the concentration of organic acids.

The reader interested in delving into studies of floral color change in more detail should consult the work of Casper and La Pine (1984) on *Cryptantha humilis* (Boraginaceae), Gori (1989) on *Lupinus argenteus*, and Delph and Lively (1989) on *Fuchsia excorticata*. The *Fuchsia* study is particularly interesting in that it explains why pollinated flowers must be "protected" from further attention of the pollinators (bell birds). If the floral assembly is disturbed during the time it takes for pollen tube growth and fertilization to occur there is a risk that the newly formed embryo may be lost. Changing to a non-attractive color prevents further bird visits.

Flavonols, Pollen Germination, and Pollen Tube Growth

Some 15 years ago, Coe *et al.* (1981) described a maize (*Zea mays*) mutant that produced nonfunctional, white pollen rather than the yellow-colored pollen normally seen. Although there was some pollen germination and limited pollen tube growth, no seed was set. It was established that the condition is brought about by the existence of two stable recessive mutations in maize at loci that control chalcone synthase (CHS), which of course controls the first committed step in flavonoid biosynthesis. Its absence means that no carbon is entering the pathway. It was also noted in that paper that the mutant pollen lacked flavonoids, whereas normal maize pollen is known to accumulate them (Ceska and Styles, 1984). The mutant pollen-flavonoid connection was firmly established by L. P. Taylor and her associates at Washington State University in a study of chalcone synthase-deficient petunia (Taylor and Jorgensen, 1992). The absence of flavonoids in pollen of CHS-deficient petunia could be partially overcome by interaction with wild-type stigmas. Taylor and Jorgensen (1992) suggested that normal development of the male gametophyte requires flavonoid synthesis.

The observation that mutant pollen, ineffective in self pollination, is partially functional in contact with wild-type stigmas suggested that some pollen germination inducer was present in the stigmas. Extracts from wild-type stigmas or wild-type pollen when added to flavonoid-deficient pollen effectively restored pollen tube growth in culture conditions and full seed set *in vivo*. The active principal was isolated and determined to be the flavonol kaempferol (Mo *et al.*, 1992). Moreover, these workers determined that the amount of kaempferol per wild-type stigma was approximately 60 nanograms and with an estimated volume of 34 μL per stigma contained flavonoid at a concentration approximating 6 μM, an amount capable of eliciting a strong germination response. The control experiment with either mutant stigmas or anthers yielded no detectable flavonoid peak by HPLC analysis.

Subsequent studies, however, suggest that the aglycone was produced through the action of some glycosidase present in the stigmas (Pollak *et al.*, 1993). In that paper, it was shown that there was excellent concordance between CHS activity and the presence of flavonol glycosides in the stigmas. This hypothesis was based upon several lines of evidence, among which was the finding that no flavonol aglycones were detectable, while a substantial pool of glycosides was present. Furthermore, it was shown by HPLC analysis that glycosides are converted to aglycones in incubated stigma extracts. The presence of general glycosidase activity in the tissue was also noted. Since it is generally known that glycosylation is a terminal step in flavonoid biosynthesis in most systems, it was logical to ask whether mutant pollen might possess the capacity to glycosylate flavonoids. (The lack of CHS should not alter the presence of other flavonoid biosynthetic enzymes.) Two glycosyltransferase activities were detected in the pollen (Vogt and Taylor, 1995). The compounds obtained by incubating kaempferol or quercetin with enzyme preparations were identified as the 3-*O*-(2″-*O*-β-D-glucopyranosyl)-β-D-galactopyranosides), which are identical with the diglycosides known from wild-type pollen (Strack *et al.*, 1984; Zerback *et al.*, 1989). These diglycosides appear to be unique components of pollen since examination of roots, leaves, and corollas showed no trace of them, whereas monoglucosides and rutinosides are known.

Further observations on the nature of the glycosylation reactions in petunia pollen were also reported by Vogt and Taylor (1995). The absence of flavonol 3-*O*-galactosides and the difficulties encountered in solubilizing the enzymes suggested that the enzymes are closely associated with one another, possibly membrane-bound, and that the product of the first reaction, the 3-*O*-galactoside, is immediately converted to the diglycoside. This sort of sequential channeling is known (see review by Hrazdina and Jensen, 1992). This situation for glycosyltransferases is highly unusual; all other known glycosyltransferases are soluble enzymes. A recent paper by Taylor and Hepler (1997) reviews current knowledge on the subject.

The discovery that a widespread flavonol was intimately involved in pollen germination and pollen tube growth prompted the Washington State University group to investigate other flavonoids as potential moderators of these critical processes. The first series of experiments (Mo *et al.*, 1992) established that 3,5,7-trihydroxyflavone (galangin) and quercetin 3′-methyl ether (isorhamnetin) had activity equal to that of kaempferol (concentration for response of 1 μM). Quercetin and morin (3,5,7,2′,4′-pentahydroxyflavone) were next most active (at 10 μM level), with myricetin and fisetin (5-deoxyquercetin) active at 10 times that concentration. 3-Hydroxyflavone and dihydroquercetin (taxifolin) were only slightly active. Four flavones, three flavanones, the chalcone corresponding to naringenin, the isoflavone genistein, two anthocyanidins, and catechol showed no activity at all in the assay. A more detailed study of flavonols (Vogt *et al.*, 1995) showed kaempferol, kaempferol 4′-methyl ether, and isorhamnetin with equal activity (measured as 100% pollen germination), while quercetin, myricetin-3,5′-dimethyl ether, and myricetin-3,4′-dimethyl ether each brought about approximately 25% germination. It is noteworthy that methylation of either kaempferol or quercetin at the C-3 or C-7 hydroxyl groups effectively rendered the compounds inactive. Glycosylation had a similar effect; quercetin 3-*O*-rutinoside, kaempferol 7-*O*-neohesperidoside, and kaempferol 3-*O*-robinoside-7-*O*-rhamnoside were inactive. These latter workers also looked at the effect of modifying the flavonol A-ring. Again using kaempferol as standard, they found that 6-hydroxykaempferol had about 80% activity while 6-methoxykaempferol matched kaempferol in activity. By contrast, quercetin, its 6-hydroxy derivative (quercetagetin), and its 6-methoxy derivative (patuletin) had essentially equal activities. Galangin, the B-ring deoxyflavonol, and its 6-methoxy derivative were equal to kaempferol in activity while 6-hydroxygalangin was essentially inactive in the germination test. Flavonols with either 8-hydroxy or 8-methoxy substitution showed activity in the 15–20% germination range. Conclusive statements from structure-activity studies are always risky, but it seems clear that for maximal activity a flavonol must have hydroxyl groups at C-3, C-5, and C-7. Substitution by methyl or by a sugar moiety destroys the biological activity of the molecule, at least within the time it normally takes pollen to germinate (<1 hour). If kaempferol 7-methyl ether is allowed to remain with pollen for prolonged periods of time (>8 hours) some germination occurs. The authors (Vogt *et al.*, 1995) pointed out that this has no apparent physiological significance but may indicate that different levels of signal recognition exist. The C-4 carbonyl is necessary as indicated by the inactivity of catechol (3,5,7,3′,4′-pentahydroxyflavan) which lacks that functional group. It is tempting to suggest that the requirement for the carbonyl group and hydroxyl groups at C-3 and C-5 may have something to do with the potent hydrogen bonding possibilities associated with those functions. The delayed activity of kaempferol 7-methyl ether could be taken to indicate that, while it fits the surface of the receptor, it does not do so ideally. The nature of the B-ring substitution appears to be much less critical except that the system appears to be sensitive to the presence of more than one methoxy group. The fact that the unsubstituted flavonol (galangin) was highly active suggests a non-polar interaction between the B-ring and a nonpolar region on the receptor. Compounds substituted at C-6 give such variable results that any suggestions would be highly speculative. Vogt and coworkers (1995a) suggest that this part of the molecule may be involved somehow in precise structural recognition of the receptor surface.

The studies of flavonoid involvement in germination of maize and petunia suggest that the effect may be widespread in the plant kingdom. A further indication that this might be the case comes from the work of Ylstra and coworkers (1992) who examined development, germination, and tube growth in tobacco pollen. Anther-derived substances strongly stimulated tobacco pollen germination and seed set. Chromatographic examination of diffusates from pollen suggested that flavonoids are involved. When kaempferol, quercetin, or myricetin was added to the germination medium, pollen tube growth was strongly stimulated. Other tested flavonoids (2′,4′,6′,4-tetrahydroxychalcone, naringenin, and naringin) had no effect. The concentrations at which the flavonols were active (0.15–1.5 μM) suggest a signal function. Two recent reports suggest that the phenomenon is not universal, however (Ylstra *et al.*, 1996; Burbulis *et al.*, 1996). Both involved *Arabidopsis thaliana* lines that have mutations in the gene coding for chalcone synthase. What other secondary compounds might be involved in pollen germination is under active investigation by these groups.

Flavonoids as Messengers

One of the most important and best known biological associations involves the interaction of roots of certain flowering plants and nitrogen fixing bacteria. This is most commonly seen in interactions between legume species and members of the microbial genera *Azorhizobium*, *Bradyrhizobium*, and *Rhizobium* (often referred to collectively as Rhizobia) (Pueppke, 1996), but other flowering plant species and other nitrogen-fixers are known. A recent review by Pueppke (1996) discusses the genetic and biochemical basis of nodulation in detail. Other reviews of note are those of Rao (1990), which treats the broad subject of root flavonoids and their various functions, including nitrogen fixation, Phillips and Tsai (1992), which includes a general overview of the development of the field, and Heidstra and Bisseling (1996) in which *nod*-factor-induced host responses, along with mechanisms of *nod*-factor perception are discussed.

The first report of flavonoids as soil constituents appears to be that of Lundegardh and Stenlid (1944), but it was much more recently that the nature of these components was established. Studies of root exudates of soybean (*Glycine max*) and lentil (*Lens esculenta*) demonstrated the presence of polyphenolic compounds, some of which were identified as isoflavones (D'Arcy-Lameta, 1986; D'Arcy-Lameta and Jay, 1987). The second of these papers discussed the possibility that the isoflavonoids exerted some influence on the growth of microorganisms in the vicinity of roots. An observation of major significance, which appeared at about the same time, was that of Peters and coworkers (1986) who showed that the flavone luteolin could induce the expression of nodulation genes in *Rhizobium meliloti*. There soon followed reports of other plants and other flavonoids involved in the establishment of these symbioses. Compounds identified from various legume species include 4,4′-dihydroxy-2′-methoxychalcone, 7,4′-dihydroxyflavone, 7,4′-dihydroxyflavanone, apigenin 7-*O*-glucoside, luteolin, luteolin 3′-methyl ether (chrysoeriol), 7,4′-dihydroxyisoflavone (daidzein), 5,7,4′-trihydroxyisoflavone (genistein), anthocyanidin 3-*O*-glycosides, flavonol 3-*O*-glycosides, naringenin, and eriodictyol (see Phillips and Tsai, 1992, for further details).

The interactions between plant and bacterium can be quite complex as witnessed by the fact that certain flavonoids may attenuate the action of others. For example, there is evidence to suggest that the weak *nod*-gene inducer 7,4′-dihdyroxyflavone may compete with the stronger inducer luteolin for binding sites on the *nodD*-gene product in alfalfa plants (Hartwig *et al*., 1989). Germinating seeds of alfalfa release glucosidases that hydrolyze the sugar linkage in the inactive luteolin 7-*O*-glucoside to produce the active aglycone (Hartwig and Phillips, 1991). In addition to their gene inducing action, flavonoids can also act as chemotactic agents at concentrations in the nano- and micromolar ranges (Aguilar *et al*., 1988). Caetano-Anolles and coworkers (1988) have shown that mutation at the *nodD* locus eliminates the chemotactic response of *Rhizobium meliloti* to luteolin. Other activities of flavonoids in the rhizosphere include stimulation of bacterial growth, spore germination, hyphal growth, and hyphal branching.

Communication between plant root (or germinating seed) and microorganisms in the soil is not a one-way street. Once the microorganism has been "contacted" by the inducing flavonoid(s) other processes are set in motion. The nodulation genes direct the synthesis of Nod factors. These factors are lipooligosaccharides based upon a chain of β-1,4-linked *N*-acetyl-D-glucosamine units consisting of three to five units. The non-reducing end of the chain is linked to a fatty acid group, which varies with the species of *Rhizobium*. Other units may be attached as in the case of *R. leguminosarum* where an *O*-acetyl group is attached at the C-6 position on the non-reducing terminal sugar unit. In this *Rhizobium* species the fatty acid consists of 18 carbons with either one or four double bonds (Dénarié and Cullimore, 1993; Spaink *et al*., 1991; Spaink, 1992). Within one hour of contact (contact is five to 10 minutes) with Nod factor, root hairs begin to undergo deformation and by three hours about 80% of them have become deformed (Heidstra *et al*., 1994). During contact with root hairs the glucosamine oligomers are shortened. In addition to root hair deformation, *nod* factors cause depolarization of root hair membrane potential, induce nodulin gene expression, and initiate nodule primordia. A further appreciation of the complexity of this system can be gained by reading the work of Scheres and coworkers (1990) on the sequential induction of nodulin gene expression in developing pea nodules. The elucidation of the nodulation process represents one of the most interesting cases of biochemical detective work to grace the plant biochemical/physiological literature in recent years.

Flavonoids as Oviposition Stimulants

One of the most complex aspects of plant–insect interactions involves the oviposition response of insects to plant chemicals. Harborne (1988, p. 179) lists examples involving simple terpenes (e.g., methyl eugenol), coumarins (e.g., bergapten), sterols, glucosinolates (e.g., sinigrin), and several flavonoids. A closer look at flavonoids as oviposition stimulators shows that several classes have been implicated and that they do not necessarily function alone. The intricacy of these systems can be appreciated by looking at the work of Feeny and colleagues (1988) on oviposition of the black swallowtail butterfly (*Papilio polyxenes*) on carrot (*Daucus carota*). Two compounds were isolated and identified as *trans*-chlorogenic acid and luteolin 7-*O*-(6″-*O*-malonyl)-β-D-glucopyranoside. Neither compound was active alone but together they accounted for 70% of the activity of the original plant extract. On standing, extracts of carrot lost their stimulatory activity, which was found to result from loss of the malonic acid function from the flavonoid. Other studies have shown that the *C*-glycosylflavone vicenin-2, rutin, narirutin (naringenin 7-*O*-rutinoside), and hesperidin are stimulants for *P. xuthus* (Oshugi *et al*., 1985; Nishida *et al*., 1987) while naringin and hesperidin are stimulants for *P. protenor* (Honda, 1986, 1990).

The involvement of flavonoids take on a cryptic nature in some, if not most, situations. For example, extracts of *Heterotropa aspera* (Aristolochiaceae) placed on filter papers strongly stimulated oviposition behavior of *Luehdordia japonica*, a swallowtail specifically associated with this plant species. Nishida (1994) reported identification of isorhamnetin 3-*O*-glucosyl-(1 → 6)-galactoside 7-*O*-glucoside from extracts of the plant. He reported that the flavonoid itself did not have any stimulatory activity by itself but when it was mixed with other unidentified components of the plant ovipositioning behavior was again observed. A convincing account of the complexity of the phenomenon comes from the work of Honda (1990) on the chemistry of *Citrus* plants that host *Papilio protenor*. Ovipositional behavior was elicited by the synergistic action of (−)-stachydrine (*N*,*N*-dimethylproline), D-(−)-quinic acid, (−)-synephrine (a phenylethylamine derivative), and L-(−)-proline. The effectiveness of this mixture was increased by the addition of the flavanone glycosides naringin and hesperidin, both of which are known *Citrus* constituents. Sucrose, glucose, and inositols, which also occur in *Citrus* species, did not have any effect upon the butterflies.

The combination effect was observed again in the case of oviposition stimulators from the azuki bean, *Vigna angularis* (Fabaceae). This crop plant is seriously challenged by the azuki bean weevil, *Callosobruchus chinensis*, which lays its eggs on young and immature pods. Several studies (Tebayashi *et al.*, 1995 and work cited therein) have shown the existence of three compounds responsible for initiating ovipositing behavior, D-catechin, taxifolin, and quercetin 7-*O*-β-D-glucoside. In a comparison of effectiveness, it was shown that these compounds were active in the descending order: D-catechin > quercetin 7-*O*-glucoside > taxifolin. The mixture of the three compounds in their natural proportions (63 : 19 : 71 μg per g bean, respectively) was the most effective.

A recent study of oviposition behavior of the monarch butterfly, *Danaus plexippus*, visiting the milkweed, *Asclepias curassovica* (Asclepiadaceae), was described by Haribal and Renwick (1996). These workers isolated six quercetin glycosides the following of which were found to be effective oviposition stimulants: quercetin 3-*O*-(2″,6″-α-L-dirhamnosyl)-β-D-galactopyranoside, quercetin 3-*O*-(2″,6″-α-L-dirhamnosyl)-β-D-glucopyranoside, quercetin 3-*O*-β-D-glucopyranosyl-(1 → 6)-β-D-galactopyranoside, and quercetin 3-*O*-rutinoside. In most, if not all, of the cases discussed above, the stimulatory flavonoids were active only in the presence of some other compound or compounds. The *Danaus*/*Asclepias* interaction is a case where flavonoids are active in the absence of other plant constituents.

Flavonoids, Butterflies and Moths

The presence of flavonoids, and other plant chemicals, in butterflies has been recognized for many years (Harborne, 1967, pp. 123–125). Apparently, the first definitive work on these pigments in butterflies was that of Morris and Thomson (1963a,b, 1964); earlier reports had been by and large based upon color tests and not upon rigorous chemical analysis. Morris and Thomson's work involved the Marble White butterfly, *Melanargea galathea*, from which they obtained tricin (5,7,4′-trihydroxy-3′,5′-dimethoxyflavone), free and as a glycoside, and a glycoside of orientin. It is interesting to learn that E. C. Bate-Smith had tentatively identified tricin in this animal on the basis of chromatographic comparisons (footnote in Morris and Thomson, 1963a).

The lycaenid butterfly *Polyommatus bellargus* has the capacity to sequester derivatives of the *C*-glucosylflavone isovitexin selectively from leaves of *Coronilla varia* (Fabaceae), which contains both isovitexin and isoörientin derivatives (Geuder *et al.*, 1977). It was suggested by those workers that the sequestered flavonoids are stored in the animal's wings and serve as visual mate attractants. The concentration of flavonoids is much higher in the wings of females. It was also noted that animals raised in the laboratory and those collected in the wild have similar wing flavonoid chemistry. Small amounts of kaempferol and quercetin *O*-glycosides were also detected in the animals' wings. It was suggested that these compounds must be of animal or microorganismal origin since *Coronilla* apparently does not accumulate them.

In a detailed study of several swallowtail butterflies, Wilson (1986) observed that the zebra swallowtail (*Eurytides marcellus* Cr.) selectively sequesters quercetin 3-*O*-glucoside from a mixture of quercetin 3-*O*-glucoside, quercetin 3-*O*-rutinoside, and quercetin 3-*O*-rutinoside-7-*O*-glucoside present in its larval food plant *Asimina triloba* (Annonaceae). In all, 10 of 27 other papilionid species also have the capacity to sequester flavonoids from their respective food plants. In all of these there was a clear choice for kaempferol and quercetin 3-*O*-glycosides and 7-*O*-glycosides. In a detailed study of flavonoids in the chalkhill blue butterfly (*Lysandra coridon*), Wilson (1987) discussed such factors as larval food source, selective metabolism and sequestering of flavonoids, and how the compounds might be involved in complex interactions between the butterflies and other organisms. The interested reader might wish to consult a recent paper by Trigo and coworkers (1996) that discusses patterns of pyrrolizidine alkaloids in butterflies belonging to Ithomiinae (Lepidoptera, Nymphalidae).

The larvae of the small ermine moth *Yponomeuta malinellus* feed on species of apple (*Malus* spp.) as a result of which they encounter considerable amounts of the dihydrochalcone phloridzin (2′,4′,6′,4-tetrahydroxydihydrochalcone 2′-*O*-glucoside) long known to be a constituent of the root bark of the plant. Concentrations can reach 12% in dry root bark (Williams, 1966). Two specimens of *Y. malinellus* larvae were shown by Fung (1988) to accumulate 0.03 and 0.26% dry weight of the glucoside and 0.05% of the aglycone. It appears that all of the sequestered flavonoid is eliminated (metabolized?) before pupation since neither pupae nor imagines possessed detectable amounts. Fung speculated that the compound(s) serve as chemical defensive agents in the larval stage of the insect. This argument is supported by findings of Schafer and coworkers (1983) that phloridzin is both repellent and toxic to red-winged blackbirds.

Flavonoids as Defensive Agents – Combined Insect and Fungal Attack

Combined attacks by two or more parasitic organisms on certain plant species are a common phenomenon that can have disastrous consequences. Physical injury by an attacking insect can provide a means of access for opportunistic pathogenic microorganisms. A case in point involves beech bark disease which affects *Fagus sylvatica* in Europe and *F. grandifolia* in North America. This affliction arises from the combined attack of scale-forming *Nectria coccinea*, which penetrates the bark, and the offending fungus *Cryptococcus fagisuga*, which uses the wound as access to the bark parenchyma. (Note: the Latin names were reversed in the original paper.) Neither organism is lethal alone. Dübeler and coworkers (1997) have shown that infection of *F. sylvatica* results in major changes in certain phenolic components of the host. The concentrations of catechin, *cis*-isoconiferin, and *cis*-syringin did not change in response to infection. The concentration of *cis*-coniferin dropped, while there were significant increases in the concentrations of (2*R*,3*R*)-(+)-dihydroquercetin 3-*O*-glucoside, (2*R*,3*R*)-(+)-dihydroquercetin 3-*O*-xyloside, and (2*S*,3*S*)-(−)-dihydroquercetin

3-*O*-glucoside. Coniferin is 3-methoxy-4-hydroxycinnamyl alcohol 4-*O*-glucoside; syringin is 3,5-dimethoxy-4-hydroxycinnamyl alcohol 4-*O*-glucoside. In iso-coniferin the glucose is attached at the aliphatic hydroxyl group instead of at the phenolic group.

Flavonoids as Defensive Agents – Allelopathy and Phytoalexins

A wide variety of secondary plant products, including flavonoids, have been identified as defensive agents that protect plants against a number of other, "attacking," organisms in Nature. The attack may involve microorganisms, including viruses, bacteria, or pathogenic fungi, invertebrate or vertebrate herbivores, or the encroachment of another plant (of the same or other species). There are two general types of response to threatening situations. One involves allelopathy, wherein some compound or compounds serve to discourage (or at least reduce the impact of) an attacking or otherwise invasive organism. The second involves the formation of some compound, a phytoalexin, in response to infection by a pathogenic microorganism.

The term allelopathy was coined by Molisch (1937) to identify biochemical interactions between plants of all sorts, including microorganisms. Both stimulatory and inhibitory interactions were included in his definition. There was a tendency, for a time, to include only negative effects in defining allelopathy, but the current definition includes both negative and positive effects. Rice (1984) noted that the distinction is an artificial one since many organic compounds can be either stimulatory or inhibitory depending upon their concentrations. The field has been reviewed three times in great detail by Rice (1974, 1984, 1995) and has been discussed in greater or lesser details by many other authors, e.g., Harborne (1988), Rao (1990), Inderjit (1996). Several examples will demonstrate the widespread occurrence of this phenomenon in the plant kingdom as well as the variety of flavonoids that are involved.

Pluchea lanceolata (Asteraceae) is an aggressive weedy species widespread in semiarid India (Inderjit and Dakshini, 1994). Vegetative reproduction via stolons is rapid and fragments remain viable long after a field has been cleared. Contributing to the success of this plant are several phenolic compounds that have been shown to interfere with germination of other species. Hesperidin (5,7,3′-trihydroxy-4′-methoxyflavanone 7-*O*-rutinoside) [7-7], dihydrokaempferol 3-*O*-arabinoside [7-8], and formononetin 7-*O*-glucoside [7-9] have been identified as components of the soil surrounding roots of the weed (Inderjit and Dakshini, 1991, 1992). The 1994 paper by these workers described the quantitative determination of chlorogenic acid, formononetin 7-*O*-glucoside, methylated coumarins, phloroglucinol, and phenol from aqueous leachate of the roots of *P. lanceolata*. The leach solutions from roots were, for the most part, active in repressing root and shoot growth of seedlings of radish, mustard, wheat, and pea. Although we do not know from these studies what effect the *Pluchea* compounds would have on the germination and growth of natural species in the region, it is abundantly clear that potential crop plants, as represented by the test species, are sensitive to the weed's phenolic compounds.

One of the reasons put forward to explain the long history and survival success of ferns involves their capacity to produce secondary compounds with allelopathic

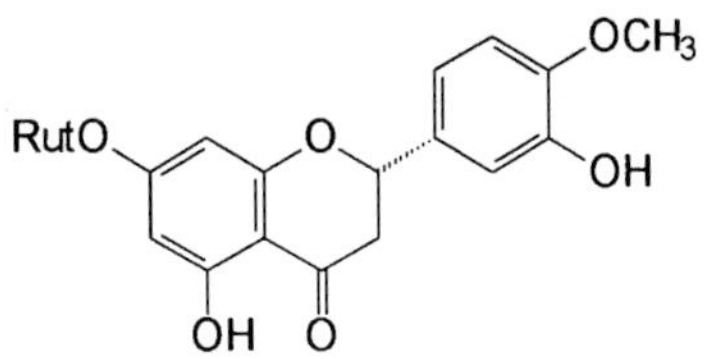

potential. A test of this hypothesis was described by Star (1980) using *Pityrogramma calomelanos* on the Hawaiian Islands. Three of the major components of the frond exudate of this species were identified as 2′,6′-dihydroxy-4′-methoxychalcone [7-10], the corresponding dihydrochalcone, and 3,5-dihydroxy-7-methoxyflavone (izalpin) [7-11]. All three compounds inhibited spore germination and gametophyte growth at most concentrations with the dihydrochalcone being the most active. At low concentrations, 5×10^{-7} M for the chalcone and 5×10^{-4} M for izalpin, spore germination was enhanced. Star discussed the inhibitory nature of the exudate compounds in terms of the influence they might have on plant growth in the field. *Pityrogramma*

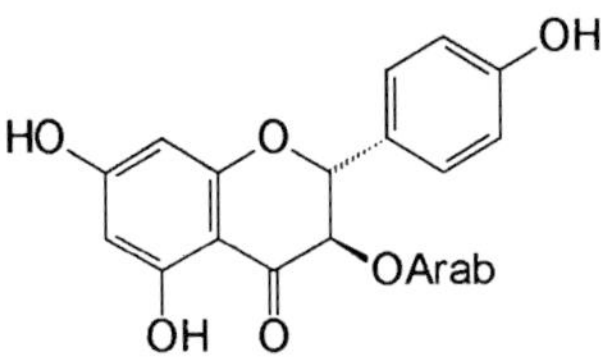

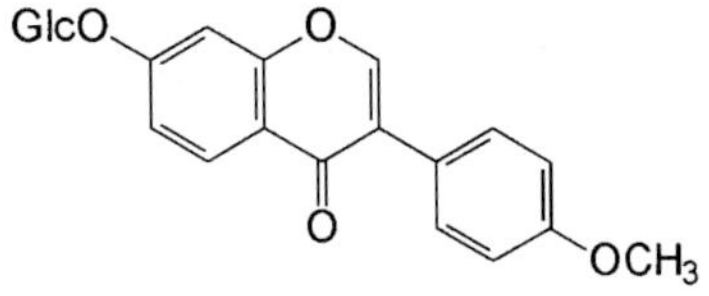

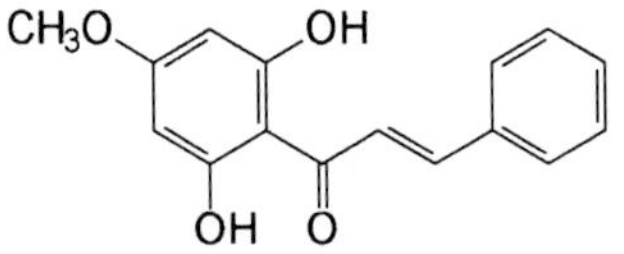

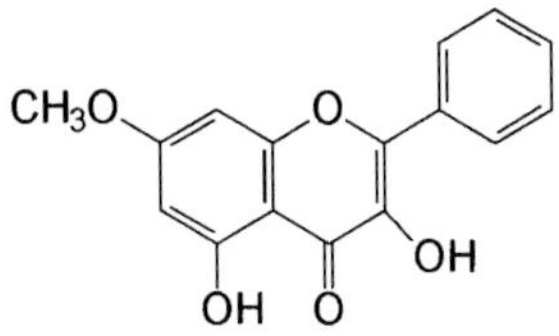

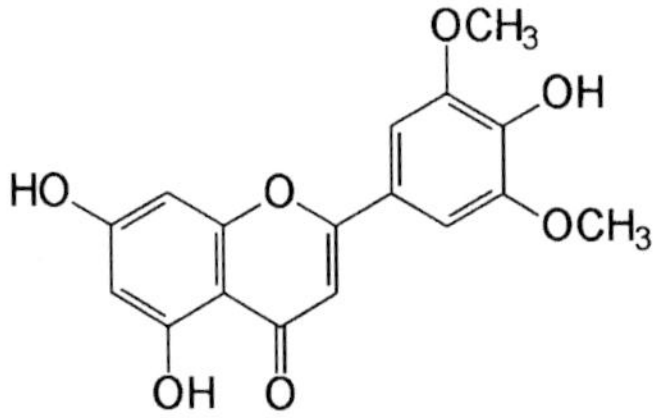

calomelanos grows in open arrays with mature plants occurring at intervals of 2–3 meters. There is little in the way of accompanying vegetation. Star suggested that the distance between individual mature plants reflects the existence of zones of inhibition owing to the presence of inhibitory flavonoids in the soil.

An often cited example of apparent allelopathy involves the heathland of northwestern Spain, an area that had been cleared of forests in preparation for use in agriculture. The area now supports extensive growths of ericaceous shrubs. The result of this change in vegetation is that native grasses and other herbaceous species are at a serious disadvantage. That species of *Erica* are involved in this phenomenon gains considerable weight from several experimental studies. Ballister and coworkers (1979) demonstrated that aqueous extracts of the aerial parts of *E. australis* and *E. arborea* inhibited root and hypocotyl growth of red clover. Further, a layer of leaves or flowers of either of these species spread on soil in a greenhouse significantly inhibited the growth of red clover. No efforts to examine the chemical components of these plants was made as part of those experiments. Such work was done, however, by Carballeira (1980) who identified 15 phenolic compounds from extracts of various parts of *E. australis* plants as well as in the associated soil. Nine of the compounds were hydroxybenzoic or hydroxycinnamic acids, two were identified as coumarins, and one as the simple phenol orcinol, 1,3-dihydroxybenzene. The other three were shown to be kaempferol, quercetin, and myricetin. A problem exists in situations of this sort with regard to the origin of the phenolic acids and simple phenols. Are these compounds derived from the phenolic fraction of the plant or are they derived from more complex molecules, the flavonoids as a case in point, though the action of soil microorganisms? Degradation of the dihydrochalcone phloridzin into simpler phenolic compounds, such as phloroglucinol and cinnamic acid derivatives, is known to occur in the case of the soil sickness associated with old apple orchards (Börner, 1959).

Rice (1984) devoted a chapter to the question of secondary compounds of seeds and their possible role in preventing decay of the seed before germination. We were reminded in the introduction to that chapter that some seeds must have exceptional abilities to cope with potential infecting organisms since they may lie dormant for many years before germination. Many of the types of compound seen in other allelopathic systems have been identified as seed components. The examples listed in Rice's chapter included tricin (5,7,4′-trihydroxy-3′,5′-dimethoxyflavone) [7-12] in seeds of *Orobanche* (Orobanchaceae), quercetin and myricetin in seeds of *Trifolium*, rutin in seeds of *Brassica*, and apigenin glycosides in celery seeds. Many other phenolic compounds, including coumarins, simple phenols, and a variety of phenolic acids are also known components of seeds all or any of which may serve as antibiotics.

Flavonoids also play a significant part in defending plants against attack by pathogenic organisms. The idea that plants have the capacity to produce antibiotic substances was first suggested by Müller and Börger (1940) who were studying the infection of potatoes with virulent and avirulent strains of *Phytophthora*. The term "phytoalexin" was coined for these defensive compounds. The initial definition of a phytoalexin required that it be produced *de novo* in the plant, that is, it could not exist preformed in the plant nor could it be produced from some other compound already present in the plant. Among other things, the antibiotic must be a discrete chemical substance, the reaction could only involve living cells, and the reaction must be non-specific in its action toward a given pathogen, although different pathogens could be more or less sensitive to the antibiotic substance produced (see Harborne, 1988, pp. 315–316 for further details). The classical proof of the phytoalexin theory came from the work of Cruickshank and Perrin (1960, 1964) who isolated and identified the first such agent from pea pods (*Pisum sativum*) that had been inoculated with *Sclerotina fructicola*. The compound identified, called "pisatin" by those workers, is a pterocarpan having structure [7-13]. Pisatin could also be detected in peas that had been infected with *Asclochyta pisi* under field conditions. It has been shown subsequently that a number of other legume species produce similar compounds, a case in point being "phaseolin" [7-14] from *Phaseolus vulgaris*. The occurrence of phytoalexins in legumes in general has been discussed in some detail by Ingham (1982). Another useful review of elicitors and defense gene activation can be found in a work by Templeton and Lamb (1988).

Production of flavonoid phytoalexins is not limited to legumes as we can appreciate by looking at the structures of several flavan derivatives produced by members of other plant families. *Narcissus pseudonarcissus* (Liliaceae, daffodil, was

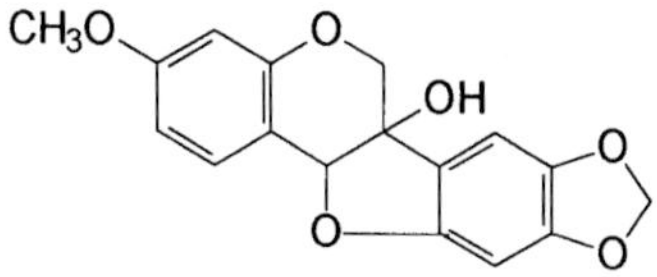

7-13

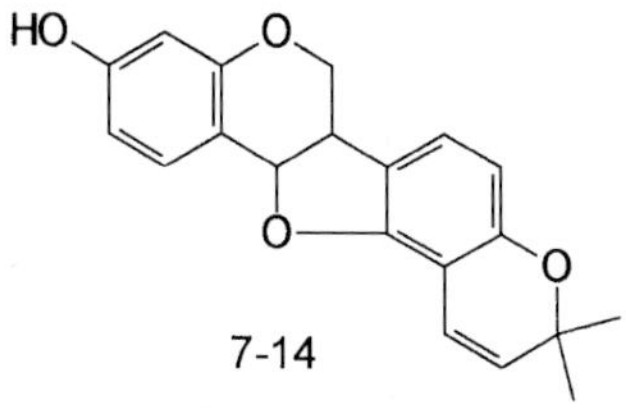

7-14

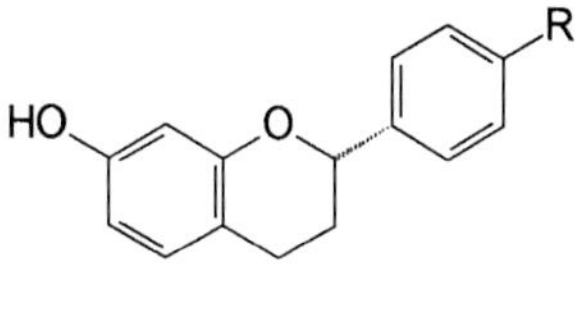

7-15 R = H

7-16 R = OH

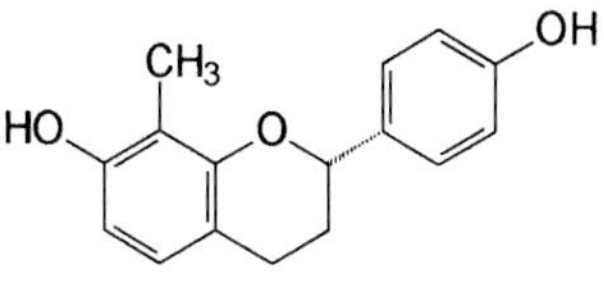

7-17

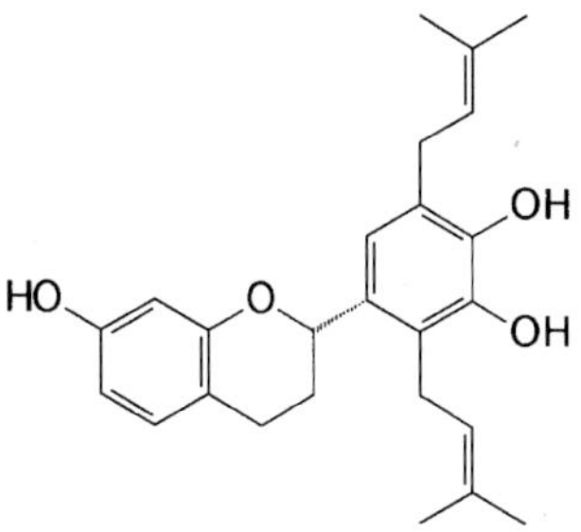

7-18

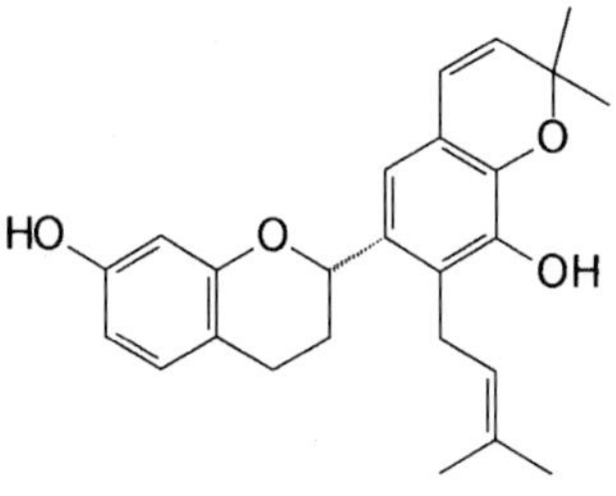

7-19

shown by Coxon and coworkers (1980) to produce 7-hydroxyflavan [7-15], 7,4′-dihydroxyflavan [7-16], and 7,4′-dihydroxy-8-*C*-methylflavan [7-17] in response to infection by *Botrytis cinerea.* Compound [7-16] along with its 4′-methyl ether were identified in shoots of *Broussonetia papyrifera* (Moraceae) that had been physically wounded by (Takasugi *et al*., 1980). Later work on the cortex of *B. papyrifera* (Ikuta *et al*., 1985) added 7,3′,4′-trihydroxy-2′,5′-di-*C*-prenylflavan [7-18] and the cyclized derivative (kazinol B) [7-19] to the list of flavan phytoalexins. Based upon positive antimicrobial activity of *Psiadia trinervia* (Asteraceae) extracts against *Cladosporium cucumerinum* and *Bacillus cereus*, Wang and coworkers (1989) isolated and identified 13 flavonol 3-methyl ethers from the plant. In addition, these workers prepared 29 other flavonoid derivatives by permethylation, selective demethylation, and acetylation. Testing of all 42 compounds against the two microbial species revealed that the antifungal activity of the original extracts could be ascribed to 5,3′-dihydroxy-3,7,4′-trimethoxyflavone (ayanin) [7-20], 5,3′-dihydroxy-3,6,7,4′-tetramethoxyflavone (casticin) [7-21], 5,3′,4′-trihydroxy-3,6,7-trimethoxyflavone (chrysosplenol D) [7-22], and 5,7,4′-trihydroxy-3,8-dimethoxyflavone [7-23]. Four compounds showed antibacterial activity: chrysosplenol D, kaempferol 3-methyl ether (isokaempferide) [7-24], 5,7,4′-trihydroxy-3,3′-dimethoxyflavone (quercetin 3,3′-dimethyl ether) [7-25], and 5,7,4′-trihydroxy-3,8-dimethoxyflavone [7-23]. The possibility that these compounds were active because of their lipid solubility, and hence, ease of passing through membranes, was discounted by the observation that the lipid soluble permethylated compounds lacked significant biological activity. The presence of phenolic hydroxyl groups appear to be required for activity. The most active compounds, however, were 4′-hydroxy-3,5,7-trimethoxyflavone acetate [7-26], 3′,4′-dihydroxy-3,5,7-trimethoxyflavone diacetate, and 7,4′-dihydroxy-3,5,8,3′-tetramethoxyflavone diacetate. Deacetylation after entry into the organism would release the active phenols.

A recent structure-activity study using mycelial growth of *Verticillium albo-atrum* as an indicator of biological activity was described by Picman and her associates (1995). The most active compounds (complete inhibition of growth) were the totally unsubstituted flavonoids flavanone and flavone. Substitution does not seem to be the major factor in this interaction since naringenin and hesperetin, with three and four substitutions, respectively, and morin with five, are also active. Glycosylated flavonoids were not active, which is in general agreement with the idea that lipophilic molecules are more effective than those with more polar substituents (O'Neill and Mansfield, 1982; Weidenborner *et al*., 1990).

Peach, plum, and sweet cherry trees are prone to growth of perennial bark cankers caused by *Cytospora persoonii* (imperfect stage of *Leucostoma persoonii*). The sour cherry, *Prunus cerasus*, however, is more resistant to fungal attack. In order to see if some constituents of sour cherry bark accounted for this resistance Geibel (1995) undertook a study of the fungistatic properties of bark flavonoids. He tested flavanones, flavones, and isoflavones with various substitution patterns, a dozen compounds in all. Naringenin showed complete fungistatic effects at a concentration of 2.5 mM, 5,7-dihydroxyflavone (chrysin) was active at 1.0 and 2.5 mM, naringenin 7-methyl ether (sakuranetin) was not active at 0.1 mM but was active at the higher concentrations. The polar nature of compounds appears to have an effect as judged by the observation that both 5,4′-dihydroxy-7-methoxyisoflavone (prunetin) [7-27] and 5-hydroxy-7-methoxyflavone (tectochrysin) [7-28] were more active than their 5-*O*-glucosides.

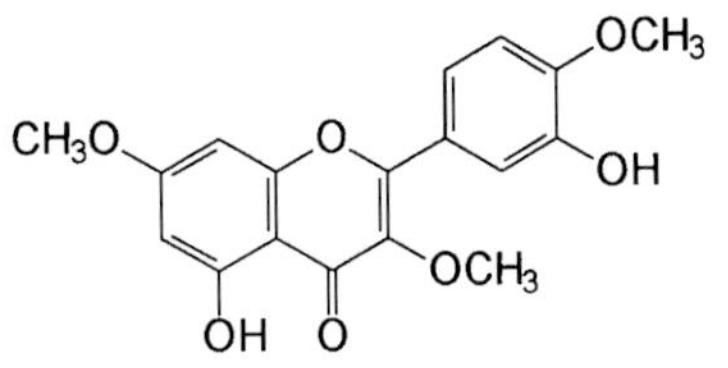

7-20

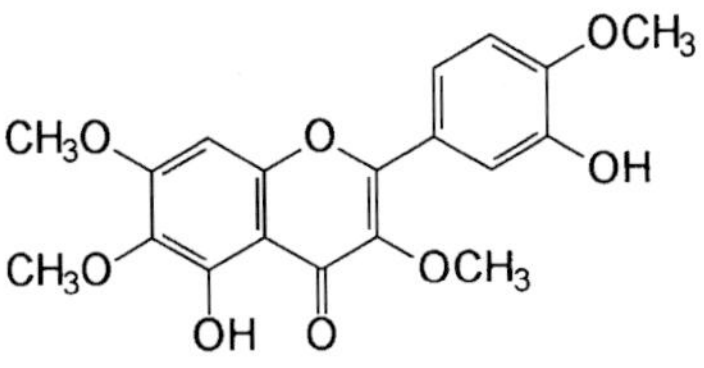

7-21

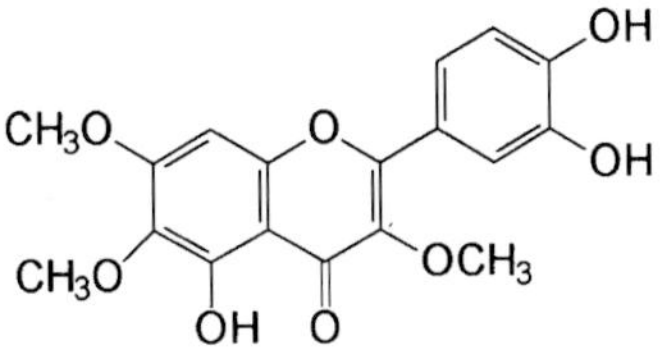

7-22

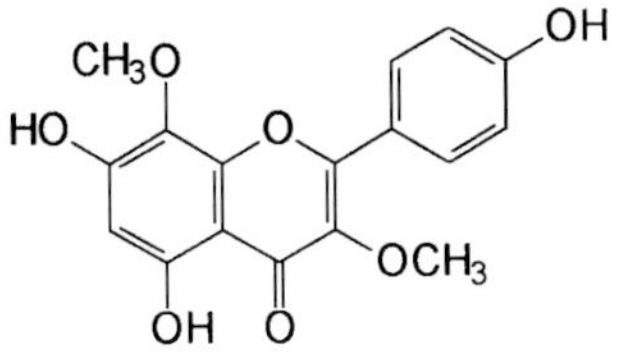

7-23

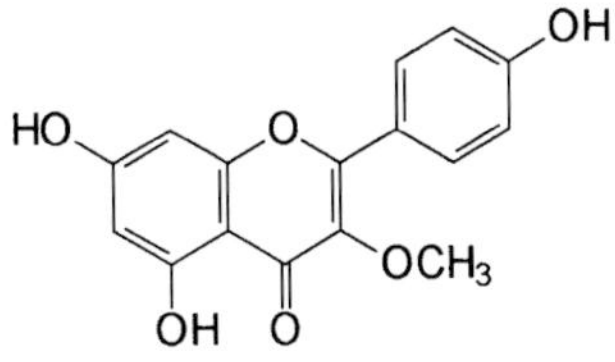

7-24

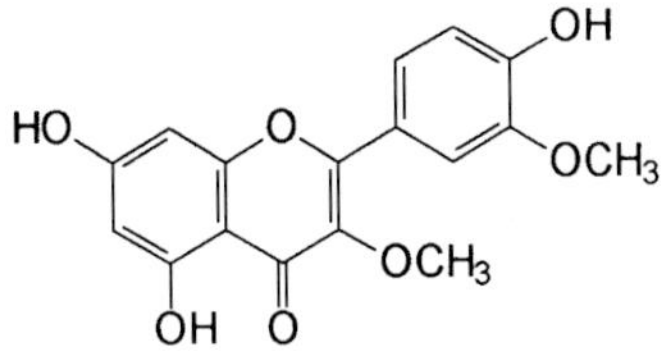

7-25

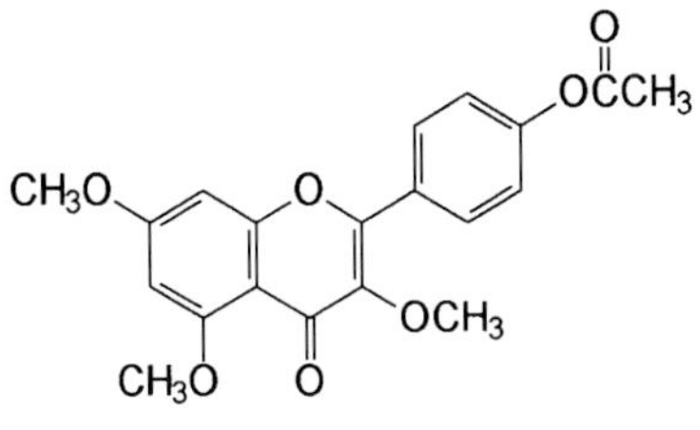

7-26

It should be pointed out that phytoalexin production does not necessarily depend upon infection with a pathogenic fungus. Other factors can also bring about the response although the magnitude of response may not be as large as when microorganisms are involved. Phytoalexins may be produced in response to temperature shock, UV radiation, wounding, or treatment of the plant with inorganic compounds such as mercuric chloride or cupric chloride. A recent article by Hanawa and associates (1991) demonstrated the capacity of *Iris pseudoacorus* to produce stress metabolites in response to cupric chloride treatment. In a large scale experiment these workers soaked 125 g of perforated leaves of the iris in a liter of 3 mM cupric

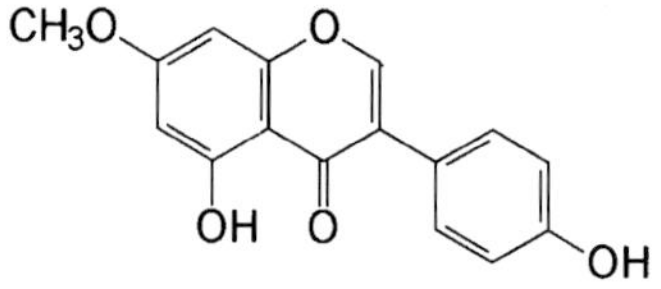

7-27

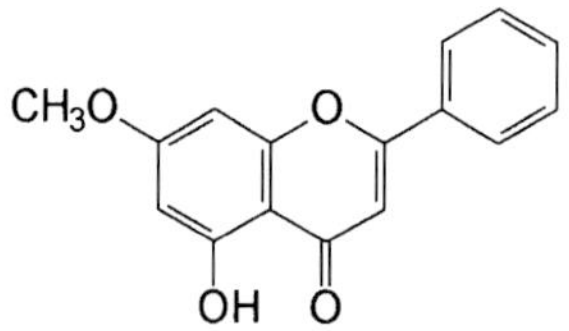

7-28

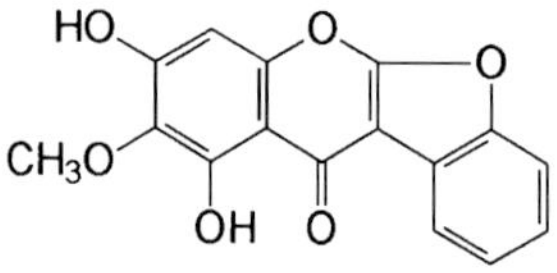

7-29

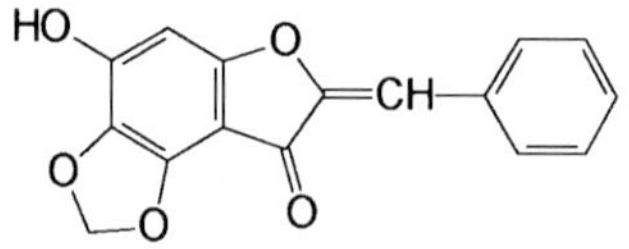

7-30

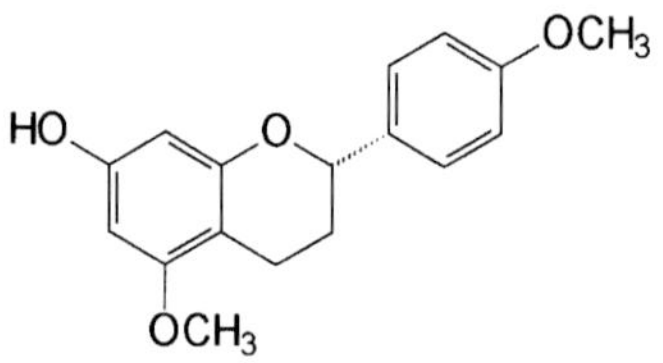

7-31

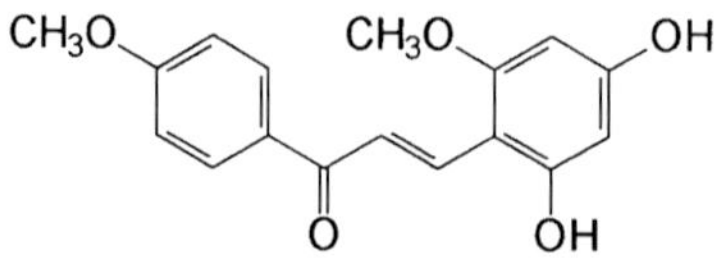

7-32

chloride solution for 24 hours. Sixteen phenolic compounds were isolated and identified 10 of which were simple isoflavones. The other six were shown to be coumaronochromones represented by [7-29], which is known as ayamenin A. Some of the identified compounds exhibited fungitoxic activity indicating that the plant's response is a general one and not linked directly with attack by a pathogenic fungus.

The work just described on iris (Hanawa *et al*., 1991) was the first report of stress metabolites from a member of Iridaceae. That the capacity to produce phytoalexins is widespread in the plant kingdom is seen in another recent study, this time involving a member of Cactaceae. *Cephalocereus senilis*, the old-man-cactus, responded to challenge by chitin, either in whole plants or in liquid suspension culture, by producing a light sensitive pigment shown to be an aurone. Pare and coworkers (1991) identified 4,5-methylenedioxy-6-hydroxyaurone [7-30] as the active compound. The compound, named cephalocerone by those workers, was shown to be active against bacterial cultures including a species of *Erwinia* known to produce cactus rot in injured plants.

Most of the known instances of phytoalexin production involve an attack of a microorganism on a higher plant. This is not always the case, however, as has been demonstrated by Ghosal and coworkers (1986) in a study of parasitism of the grass *Imperata cylindrica* on *Pancratium biflorum* (Amaryllidaceae). The main roots of the grass were found to penetrate the fleshy leaves of the bulbs of *P. biflorum*. Examination of infected bulbs showed a necrotic zone surrounding the invading grass. Immediately adjacent to the necrotic zone were cells filled with a red substance which yielded a number of compounds including several flavonoid derivatives. Two of these compounds were identified as flavans, the known 7-hydroxy-4′-methoxyflavan and the newly reported 7-hydroxy-5,4′-dimethoxyflavan [7-31]. The retrochalcone 2,4-dihydroxy-6,4′-dimethoxychalcone [7-32], was also newly reported. Five alkaloids, two of them new, were also identified. It is useful to remind the reader that flavans have been implicated as phytoalexins in other plants, e.g. *Narcissus* (see above).

Insofar as some plants have the capacity to make antibiotic substances without having been infected with a pathogen some term other than phytoalexin would seem to be in order. VanEtten and coworkers (1994) addressed this issue in a letter to the editor of *The Plant Cell*. The term "phytoanticipin" was suggested (by J. W. Mansfield) to accommodate compounds that possess antibiotic properties and are constituitively present in the plants in question. The definition suggested in that letter was: "Phytoanticipins are low molecular weight, antimicrobial compounds that are present in plants before challenge by microorganisms or are produced after infection solely from preexisting constituents." It was pointed out by those authors that the distinction between a phytoanticipin and a phytoalexin rests with its production and not with its chemical structure and that it is possible for the same compound to be both a phytoanticipin and a phytoalexin. The example cited involves the isoflavonoid derivative maackiain, which is released from a preexisting glucoside through the action of a glucosidase released from injured plant cells. In this case maackiain would be classified as a phytoanticipin. If the formation of maackiain were brought about by infection by microorganisms it would then be classified as a phytoalexin.

Considering the complexity of some of the systems developed in plants to protect themselves from attacks by microorganisms, it should not come as any surprise that some microorganisms fight back. The capacity of fungi to metabolize phenolic compounds has been appreciated for many years (Towers, 1964). A more recent treatment (Harborne, 1988, pp. 324–327) lists several situations where pathogenic fungi convert plant defensive compounds to less or non-toxic products. Recently described examples of this phenomenon can be found in works by Tahara and colleagues (1997) and by Soby and coworkers (1997). In the former, 6-prenylnaringenin is converted to a variety of oxygenated products by cultures of *Aspergillus flavus*, while in the latter work the phytoalexin maackiain [7-33] is oxidized in stages to 6,6a-dihydroxymaackiain [7-34] by *Colletrotrichum gloeosporioides*. Another approach to combating toxic compounds would involve interference with an entire pathway by some means. This sort of interaction is modeled in the response of cultured alfalfa cells to addition of both fungal elicitor and the methylation inhibitor tubericidin. In this system, the cells accumulate flavones rather than isoflavones (Daniell *et al*., 1997). Obviously, if the plant's antifungal agent were an isoflavonoid, the capacity to inhibit the isoflavonoid pathway would confer a significant adaptive advantage to the micro-organism.

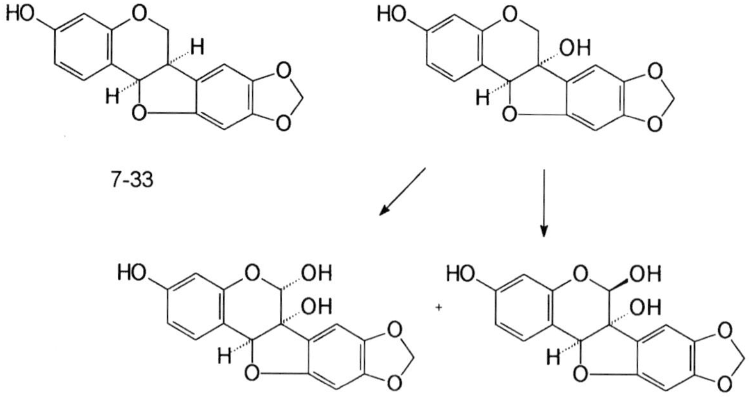

7-33

7-34

Miscellaneous Flavonoid Involvements

A description of the effect of polyphenolic compounds on the control of nitrogen release from pine litter suggests that these substances play an important role in recycling valuable nutrients (Northup *et al*., 1995). Although the polyphenolic compounds were not identified, it seems reasonable to suspect that proanthocyanidins and possibly dihydroflavonols, both of which are well represented in species of *Pinus*, could be involved.

Fleck and Woolfenden (1997) asked an interesting question in their title: "Can acorn tannin predict scrub-jay caching behavior?" Using four oak species, these workers found that cached acorns exhibited higher protein precipitating ability (ppa) than non-cached acorns. Protein precipitating ability is generally taken as an indication of higher tannin content. The higher ppa values correlate well with lower insect damage leading to the suggestion that the scrub jays (*Aphelocoma coerulescens*) select for better quality acorns, that is, acorns with higher tannin content.

CHAPTER EIGHT

HUMAN USES OF FLAVONOIDS

INTRODUCTION

In this chapter we will examine some of the uses to which flavonoids have been put by humankind. Flavonoids have applications in the food industry where some compounds have been found useful as sweetening agents, others as food colors (Marakis, 1982), while many others figure significantly as flower pigments where they contribute to the garden and cut flower industry. The involvement of flavonoids in agricultural situations, although often useful, is not always so as evidenced by the detrimental estrogenic effect of certain isoflavonoids on stock. Many flavonoids have been found to have antiviral, antibacterial, or antifungal properties. Many flavonoids, both natural and synthetic, have been tested as potential medicinal agents against human diseases including malaria and HIV. Owing to the size of the literature dealing with the testing of flavonoids for potential medicinal use, we only have space for an overview of the subject. Sufficient references to the literature have been provided, however, to provide the interested reader access to the broader picture.

Medicinal Uses of Flavonoids

The use of plants, in one form or another, for medicinal purposes long predates the recorded history of humankind. Knowledge of the usefulness of certain plants, perhaps because they alleviated toothache or simply because they tasted good, and unhealthy properties of others, was passed from generation to generation by word of mouth. By the time of the earliest herbals (Arber, 1938) a very considerable medicinal folklore had accumulated, differing, naturally, as one journeyed from one part of the world to another. There was much in the way of superstition, to be certain, with a measure of power resting in the hands of those who knew the properties, both healthful and not, of the local flora. Despite the superstitions, there was a significant amount of pharmacological truth to the use of many species. Examples of this are well documented. The alleviation of toothache by chewing on a willow (*Salix*) twig is based on the presence of salicylic acid derivatives, which has given rise to present day use of acetyl salicylic acid (ASA or aspirin) [8-1] and its many synthetic variants to alleviate minor pain. Natural medicine in India included preparations of the roots of *Rauwolfia serpentina* (Apocynaceae) that were given to ease those individuals possessed of madness. Studies eventually led to the discovery of the tryptophan-based alkaloids responsible for the tranquilizing properties of the plant and in time to the development of clinically important tranquilizer drugs, e.g. reserpine [8-2]. The Efik people of West Africa, who live near the Calabar Coast

(Calabar is in southern Nigeria near the border with Cameroon), use the seeds of *Physostigma venenosa* (Fabaceae), the Calabar bean, in ordeals to determine guilt or innocence. It was a simple matter; the guilty died, the innocent survived. Since the alkaloids present in Calabar beans are known to be toxic, a guilty verdict was almost guaranteed! Innocence was more a matter of being fortunate enough to vomit (preparations are very bitter) before the toxic chemicals did their work. It was ultimately determined that the active principle in the beans is the alkaloid physostigmine [8-3]. An interesting account of the discovery of the active compound and its development into a clinically useful ophthalmic drug can be found in John Mann's (1992) engaging little book entitled *Murder, Magic and Medicine.*

Much effort is being expended at the present time to assess the reliability and pharmacological basis of aboriginal uses of plants (Taniguchi *et al.*, 1978; Gottlieb, 1982; Taniguchi and Kubo, 1993). University programs in ethnobotany and ethnopharmacology are becoming more common while long-standing research programs in pharmacognosy in schools of pharmaceutical sciences continue. Most of the disciplines have specialist journals that treat the chemistry as well as the more "ethno" aspects of the subject. These studies have given rise more recently to matters of intellectual property rights of the peoples who have used plants for medicinal purposes for generations (perhaps for millennia). This subject has been examined with sensitivity in a recent review entitled "Changing strategies in natural products chemistry" (Cordell, 1995). This review also covers more technical aspects of natural product chemistry, such as new analytical methodology, and is a valuable guide to the recent literature (468 references).

Flavonoids have figured prominently in the search for physiologically active natural products. Examples of the attention given to flavonoids in general include a review by Havsteen (1983) on the pharmacological potential of flavonoids, two collections of essays on a broad array of medicinal uses of flavonoids by Cody *et al.* (1986, 1988), a book by Das (1989) entitled *Flavonoids in Biology and Medicine*, and a thorough treatment of the impact of flavonoids on mammalian biology by Middleton and Kandaswami (1994) in the most recent volume of *The Flavonoids – Advances in Research* series. Other recent reviews include those of Rao (1990), who lists members of 50 medicinally useful genera whose roots have yielded flavonoids, a timely addition to the *Advances in Phytochemistry* series entitled "Phytochemistry of Medicinal Plants" (Arnason *et al.*, 1995), and Dakora (1995) who lists many genera from the African flora that have been used for a variety of medicinal purposes by native peoples. Dakora concluded (p. 9) by saying, "...it is most likely that the efficacy of these indigenous medicinal legumes and non-legumes against physiological disorders is due to active flavonoid ingredients." Although this enthusiasm is to be lauded, the reader should be cautioned that there is actually very little support for generalizations of this sort. It is entirely likely that a significant fraction of the species listed could be active because of alkaloids or, for that matter, other types of natural products. Nonetheless, flavonoids have been shown to possess remarkable physiological activities in mammalian systems. In most cases, the mechanism of action is not known with any degree of certainty, but the growing sophistication of analytical tools, notably X-ray analysis of flavonoid-enzyme interactions, continues to reveal ever increasing details. The following examples, which include naturally

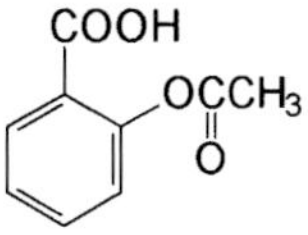

8-1

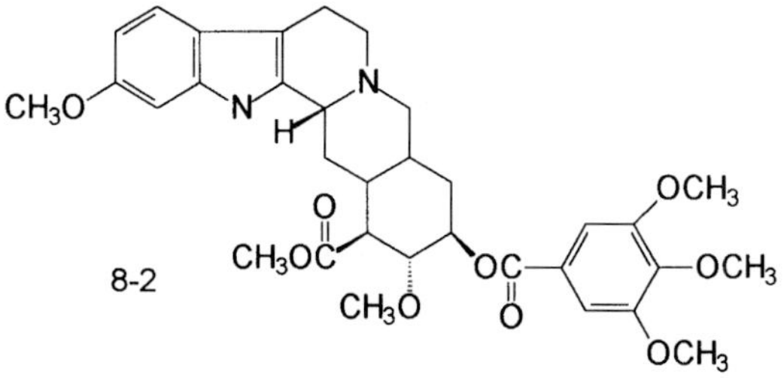

8-2

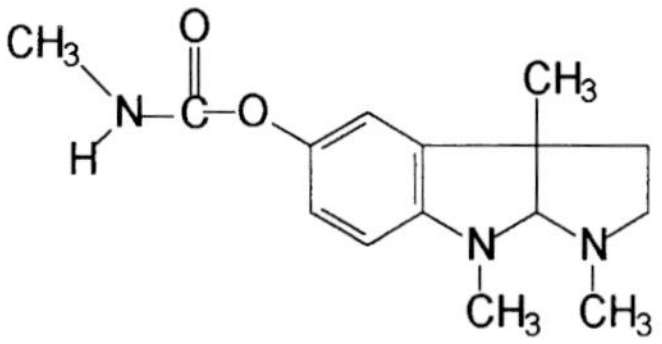

8-3

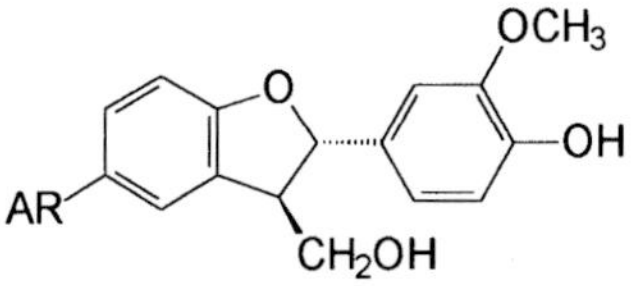

8-5

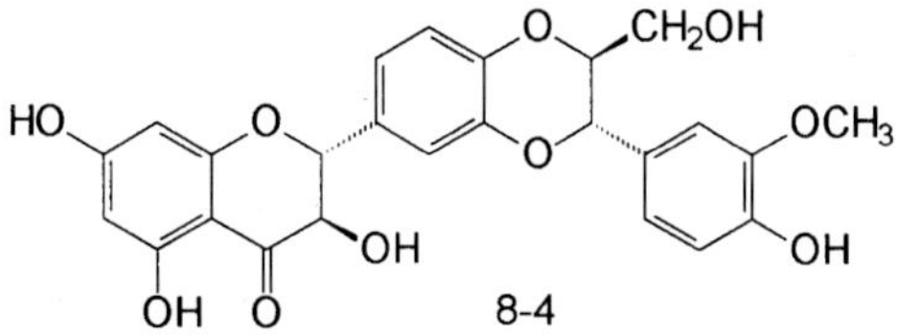

8-4

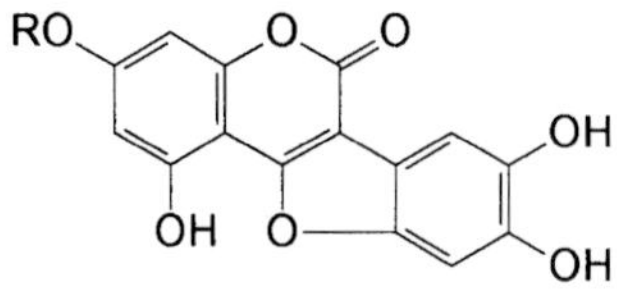

8-6 R = CH_3

8-7 R = H

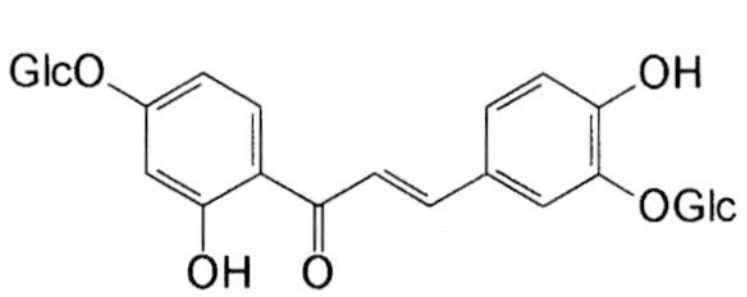

8-8

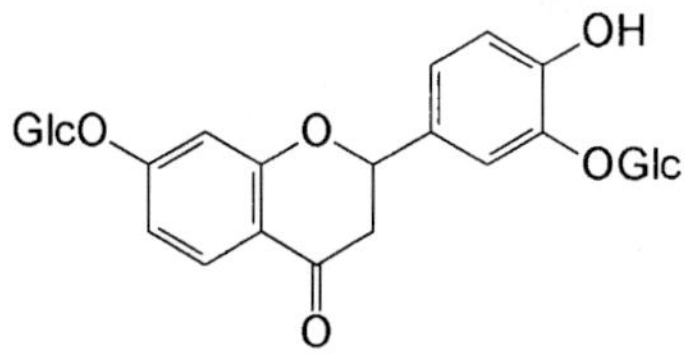

8-9

occurring as well as synthetic flavonoid derivatives, are just a sampling of the information available in the literature. We start with an excellent example of the discovery of a useful drug obtained from a plant long used in European folk medicine.

Seeds of the milk thistle, *Silybum marianum*, have long been used as a remedy for liver disease (Morazzoni *et al*., 1995). As Wagner (1986) pointed out, the discovery of the active flavanolignans of *S. marianum* came about, not as a result of systematic pharmacological screening, but from investigation of the use of this plant in folk medicine. Several active compounds have been identified from the plant representative of which are silybin [8-4] and silychristin [8-5]. The potent effectiveness of these compounds in preventing liver damage has been dramatically demonstrated by the work of Tuchweber and colleagues (1979) who showed that a single dose of silybin given to Swiss mice abolished structural changes and reduced leakage of liver enzymes into the bloodstream normally caused by the highly toxic compound phalloidin from *Amanita phalloides*. Other experimental studies of antihepatotoxic flavonoids are reviewed by Middleton and Kandaswami (1994). Wagner (1986) described studies of other plants as sources of liver protecting drugs taken from Indian folk medicine, viz. *Butea monosperma* (Fabaceae), *Eclipta alba*, and *Wedelia calendulaea* (both Asteraceae). *Eclipta alba* preparations are considered the best medication against cirrhosis of the liver in the Indian Ayurvedic system of medicine (Barua *et al.*, 1970). Studies of extracts of *Eclipta* showed that the active ingredients are wedelolactone [8-6] and demethylwedelolactone [8-7], both of which are also known constituents of *Wedelia* (from which they obviously derive their name). The compounds responsible for the antihepatotoxic activity of *Butea* extracts were identified as the well known chalcone diglucoside [8-8] and the isomeric flavanone diglucoside [8-9].

Before we leave the subject of flavanolignans, it is interesting to note that compounds closely related to the *Silybum* compounds have been identified from other species and shown to possess significant biological activity. Afifi and colleagues (1993) reported the new flavonolignan [8-10], which they named "sinaiticin," the structurally related hydnocarpin [8-11], and the two common flavones luteolin and chrysoeriol from *Verbascum sinaiticum*. All of the compounds exhibited dose-dependent cytotoxicity when tested against the murine lymphocytic leukemia P-388 cell line. Hydnocarpin has also been shown to be active against several human tumor lines and Ehrlich ascites tumors in mice as well as to exhibit anti-inflammatory and hypolipidemic activities (cited by Afifi *et al.*, 1993).

In a biological activity-directed study of *Melicope triphylla* (Rutaceae), using the P-388 cell line, Hou and coworkers (1994) obtained three flavonoids, the known 4′-hydroxy-3,5,7,3′-tetramethoxyflavone [8-12] and 3,5,8,3′,4′-pentamethoxy-6,7-methylenedioxyflavone [8-13], both of which were active, and the new 3,5-dihydroxy-6,7,8-trimethoxy-3′,4′-methylenedioxyflavone [8-14], which was inactive. The simple 5,7-dihydroxyflavanone (pinocembrin) 7-*O*-rhamnosylglucoside [8-15], obtained from the fern *Onychium japonicum*, has also been shown to be active against the P-388 cell line (Xu *et al.*, 1993). That both compounds [8-15] and the *O*-methylated flavones are active against the same cell line represents the sort of challenge that studies of structure-activity relationships regularly face. In a situation such as this, where one active compound is non-polar and the other is highly polar, one would suspect different modes of action.

Several workers have shown that certain flavonoids are active against HIV at one stage or another of infection. For example, extracts of *Scutellaria baicalensis* var. *georgi* showed activity for which 5,6,7-trihydroxyflavone 7-*O*-glucoside (baicalin) [8-16] was responsible (Li *et al.*, 1993). A year earlier, Baylor and colleagues (1992) had shown baicalin to inhibit human T-cell leukemia virus type 1 (HTLV-1). In an examination of the efficacy of flavones as inhibitors of HIV-1 proteinase, it was

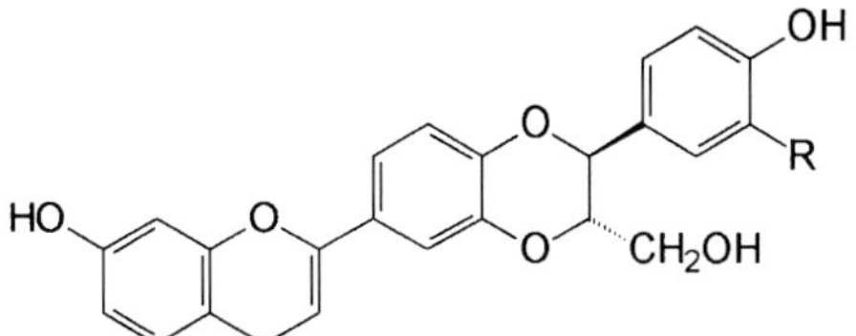

8-10 R = H
8-11 R = OCH_3

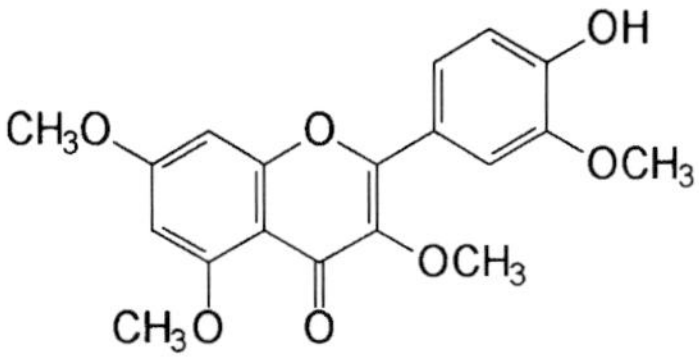

8-12

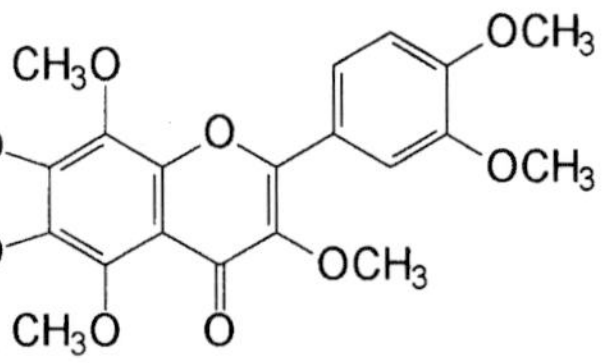

8-13

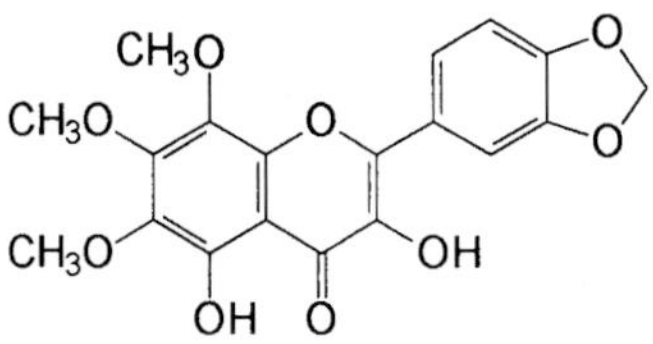

8-14

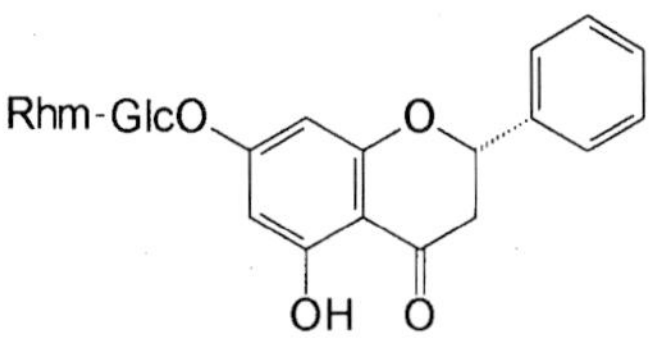

8-15

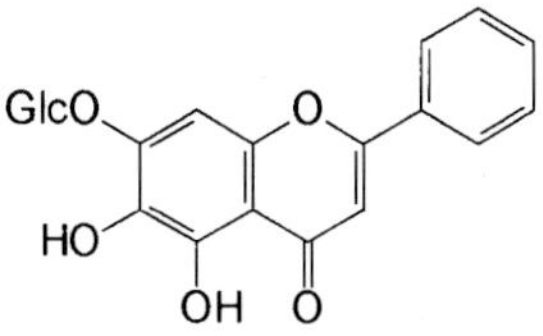

8-16

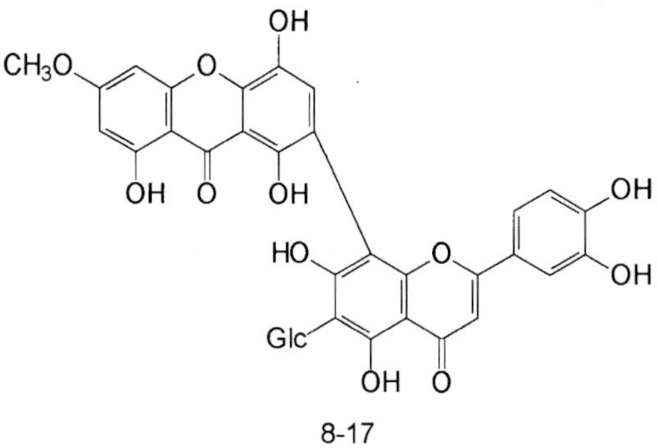

8-17

concluded that they have only modest capacity to bind to the active site (Brinkworth *et al.*, 1992). Wang and coworkers (1994), however, showed that the novel flavone-xanthone dimer [8-17] from *Swertia franchetiana* inhibited HIV-reverse transcriptase. The isoflavones genistein [8-18] and orobol [8-19] have been shown to induce mammalian topoisomerase II dependent DNA cleavage with activities comparable to those of known antitumor agents. Quercetin and fisetin (5-deoxyquercetin) had activities similar to that of the drug Adriamycin. Flavonoids lacking action included naringenin, dihydroquercetin, dihydromyricetin, and the isoflavone daidzein (5-deoxygenistein) (Yamashita *et al.*, 1990).

Thyroxine (3,5,3′,5′-tetraiodothyronine, T_4) [8-20] and 3,5,3′-triiodothyronine (T_3) [8-21] are important in mammalian systems owing to their effects upon other hormone systems, upon rates of metabolism, and upon growth. They are biosynthesized by successive iodination of tyrosine residues in thyroglobin followed by structural rearrangements to yield the diphenyl ether system. The extrathyroidal control of these compounds is controlled by removal of iodine catalyzed by iodothyronine monoiododeiodinase (ITHD) in the liver. Certain plants used in traditional medicine are known to have antihormonal activity. That flavonoids may be the active principles in these preparations received support from studies by Auf'mkolk and coworkers (1986) that showed 4,6,4′-trihydroxy- [8-22] and 4,6,3′,4′-tetrahydroxyaurones [8-23] were potent inhibitors of rat liver microsomal type I iodothyronine deiodinase ($IC_{50} = 0.5\,\mu M$). Computer studies showed that the conformation of the aurone molecule closely approximated that of the thyroid hormones. This observation suggests the possible use of aurones as experimental probes for the study of the deiodinating enzyme(s) as well as, and, more significantly, for the production of medicinally useful agents.

A more recent study showed that a synthetic flavonoid EMD-21388, which is 6,4′-dihydroxy-3-methyl-3′,5′-dibromoflavone [8-24], was the most potent inhibitor of thyroid hormone 5′-deiodination tested (Köhrle *et al.*, 1989). The crystal and molecular structure of EMD-21388 have been determined by X-ray analysis (Cody *et al.*, 1990). Note the similarity of the hormone and the synthetic flavonoid.

Several additional examples from the recent literature show the diversity of effects that have been demonstrated for flavonoids. For example, morin [8-25] and fisetin [8-26] were shown to inhibit the oxidative modification of low density lipoproteins (LDL) by macrophages, with $IC_{50} = 1\,\mu M$ (de Whalley *et al.*, 1990). Quercetin and gossypetin were effective at concentrations of $2\,\mu M$. Those authors concluded by saying that flavonoids "…may therefore be natural anti-atherosclerotic components of the diet, although this will depend to a large extent on their pharmacokinetics." This is a useful warning to bear in mind especially in view of the widespread belief that some plant preparations have almost magical properties. It is almost certain that many do contain potentially useful compounds, but the issues of effective dose, duration of effect, transport to site of action, life-time of the effective agent, and other factors, must be considered before the claims can be said to have any pharmacological validity.

Flavonoids have been intensively studied for their potential as anti-inflammatory agents. One of the more interesting recent findings in studies of human immune response and inflammatory reactions involved identifying nitric oxide (NO) as a major participant. It has now been demonstrated that certain flavonoids, some natural, some synthetic, have a significant inhibitory effect on the production of nitric oxide (Krol *et al.*, 1995). Flavone *per se* was an effective inhibitor and the isoflavone genistein (5,7,4′-trihydroxyisoflavone) [8-18] was also active, but the most potent inhibitors were amino substituted flavones. The most active within that group was 3′-amino-4′-hydroxyflavone [8-27]. Those workers suggest that flavonoids might find a role in helping to moderate the immune reaction through their capacity to inhibit nitric oxide production.

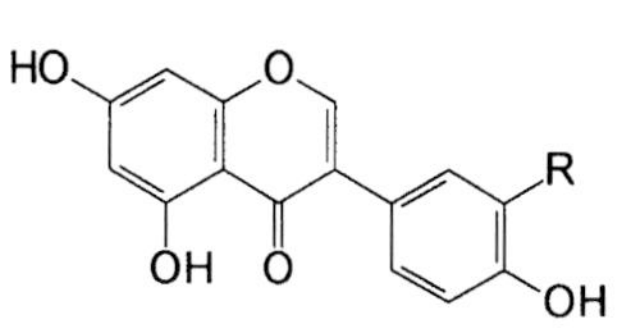

8-18 R = H
8-19 R = OH

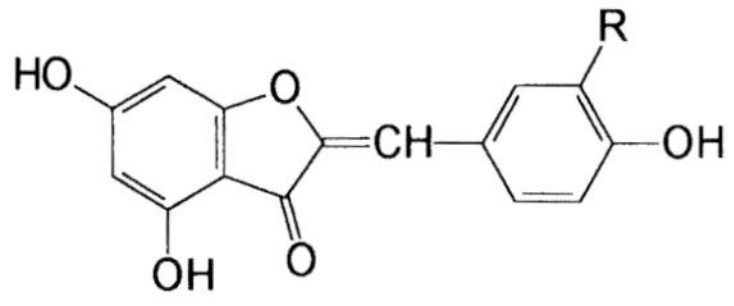

8-22 R = H
8-23 R = OH

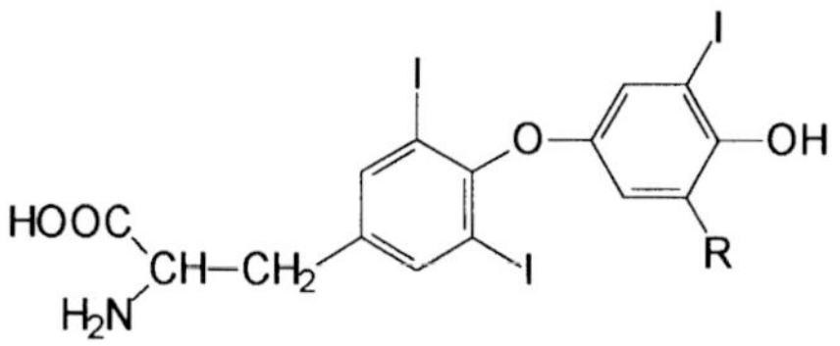

8-20 R = I
8-21 R = H

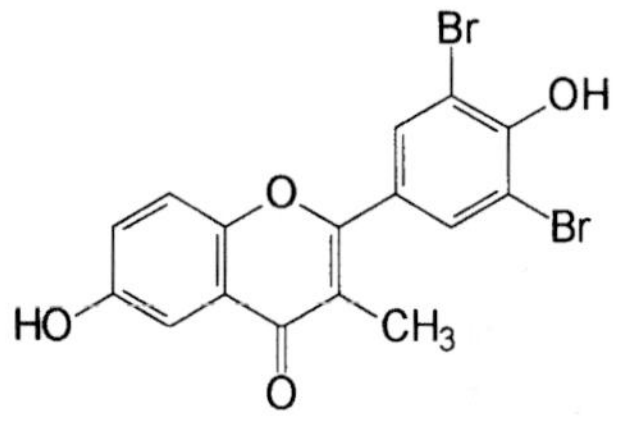

8-24

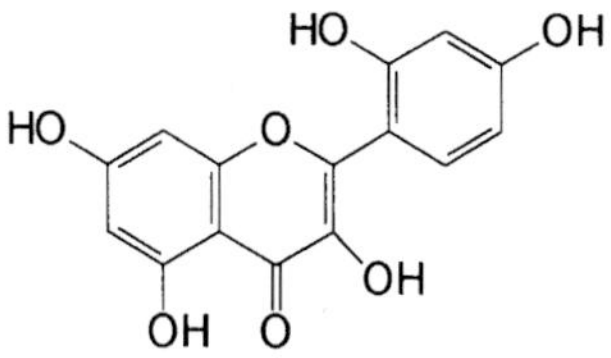

8-25

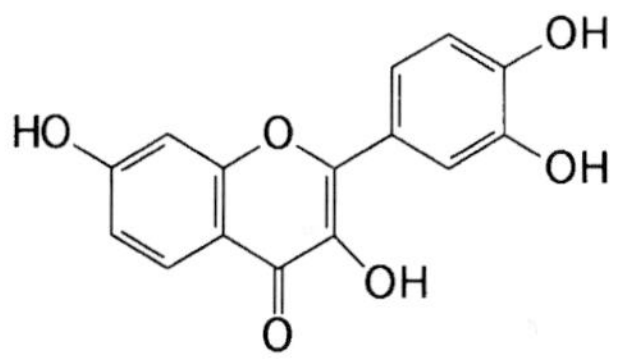

8-26

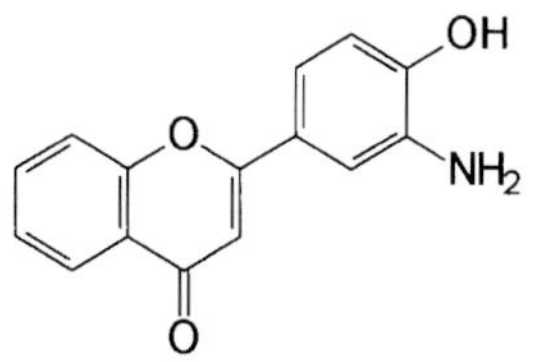

8-27

Among other interesting observations are the report that kaempferol and quercetin efficiently suppress oxygen-induced cytotoxicity at concentrations of 5 μM (Nakayama *et al.*, 1993). Several other flavonoids were also active but only at much higher concentrations. Quercetin has also been shown to be a potent and specific inhibitor of the P-form of phenolsulfotransferase in humans (Walle *et al.*, 1995). This enzyme is active in modifying phenolic compounds preparatory to elimination;

quercetin would have the effect of slowing down the process and prolonging the residence time in the body of desirable drugs. The most widespread quercetin derivative, rutin, has been found effective in reducing toxic effects of iron-overload in experimental animals (Afanas'ev *et al*., 1995). The flavonoid is thought to complex iron atoms thus preventing them from catalyzing the conversion of superoxide ions to harmful hydroxyl radicals. An interesting observation, also involving the maintenance of effective levels of drugs within the body, came from the work of Bailey and collaborators (1994) who speculated that flavonoids were responsible for the inhibition by grapefruit juice of the cytochrome P450-catalyzed oxidation of drugs. Two recent articles that address the issue of dietary antioxidants are those of Gordon (1996) and Rice–Evans and colleagues (1997), which covers recent work in the field and provides comparisons of the effectiveness of various flavonoids as they occur in tea, wine, and other products. An interesting example of the chemical modification of flavonoids in the search for antiradical and antioxidant agents came from the recent studies of Metodiewa and coworkers (1997) who compared *N, N*-diethylaminoethyl ethers of flavanone oximes with rutin. The compounds, which lacked A-ring substitution but had different substitutions on the B-ring, were sufficiently more active than rutin to suggest that compounds of this sort could be developed into useful antioxidants and radioprotective agents.

Many additional examples of flavonoid interactions with mammalian systems can be found in the review by Middleton and Kandaswami (1994). In addition to flavonoid involvement in the immune response (e.g., Middleton and Drzewiecki, 1982; Middleton *et al*., 1981), their antihepatotoxic effects, and their capacity to interact with animal hormone systems, the topics covered include effects of flavonoids on smooth muscle, their anti-oxidative properties, and cancer-related subjects. Many of these physiological activities are likely caused by flavonoids binding to enzymes in one way or another. Middleton and Kandaswami (1994) devote a large portion of their review to descriptions of flavonoid interactions with mammalian enzyme systems. In order to demonstrate the exceptional versatility of flavonoids in this regard, several examples have been brought together in Table 8.1. Also worth mentioning specifically is a study of the biological activity of C-prenylflavonoids obtained from species of mulberry (Lin and Shieh, 1992). Lin and coworkers (1993) demonstrated that several of these compounds showed strong inhibition of arachidonic acid- and collagen-induced platelet aggregation with IC_{50} values of 12.5–14.4 μM, respectively. The most active compound was shown to be "cyclocommunin" which is based on 5,7,2′,4′-tetrahydroxyflavone with prenyl groups attached at C-6 and C-3. The C-3 group is further cyclized to the 2′-oxygen.

Several flavonoids and flavonoid derivatives have been found to possess anti-mitotic behavior which suggests possible use as anticancer drugs. A recent example of such work comes from a study of *Fissistigma lanuginosum* described by Alias and coworkers (1995). Extracts of this annonaceous species exhibited inhibition of the conversion of tubulin into microtubules as well as cytotoxic behavior toward (KB) human nasopharyngeal carcinoma cells. Biological activity-directed fractionation led to the isolation of the known 2′,5′-dihydroxy-3′,4′,6′-trimethoxychalcone "pedicin," as well as two compounds derived by condensation of pedicin with a ten-carbon isoprenoid unit.

Table 8.1 A Sampling of flavonoid–enzyme interactions

Enzyme	Comment	Reference
Adenosine receptors	Many flavonoids	Ji *et al*., 1995
Aldose reductase	*C*-Methylflavanone glycoside	Kadota *et al*., 1994
AMP-dependent protein kinase	Prenylated isoflavone	Wang *et al*., 1997
Ca^{++}-transporting ATPase	Structure-activity study	Bennett *et al*., 1981
DNA topoisomerase	Structure-activity study	Constantinou *et al*., 1995
β-Glucuronidase	8-Hydroxytricetin 7-*O*-glucuronide	Kawasaki *et al*., 1988
Glutathione reductase	Structure-activity study	Elliott *et al*., 1992
Histidine decarboxylase	3-Isopropoxy-4′-methoxyflavone-6-carboxylic acid	Pfister *et al*., 1980
	5,7,3′,4′-Tetrahydroxy-3-methoxyflavan	Parmer *et al*., 1984
	Isoflavones	Umezawa *et al*., 1975
Hyaluronidase	Tannins, flavones, kaempferol, silybin	Kuppusamy *et al*., 1990
5-Lipoxygenase	Many flavonoids	Laughton *et al*., 1991
Phospholipase A_2	Several flavonoids	Fawzy *et al*., 1988
Protein kinase C	Fisetin, luteolin, quercertin active	Ferriola *et al*., 1989
Tyrosinase	Kaempferol and isorhamnetin glycosides	Kubo and Yokokawa, 1992

In a recent paper, Zhu and coworkers (1997) address the question, "Are plant polyphenols biologically active compounds or just non-selective binders of protein?" Twenty phenolic compounds including proanthocyanidins and gallic acid/ellagic acid derivatives were assessed for their capacity to inhibit specific ligands at 16 receptor sites. Although the polyphenolic compounds exhibited a wide range of activities, some of them showed specific activities at the receptor level that cannot be explained solely on the basis of polyphenol-protein binding. They concluded that routine removal of tannins from plant extracts prior to testing for biological activity may result in missing compounds of potential drug value.

Flavonoids as Antiviral Agents

Flavonoids have been found to be active against a variety of animal and plant viruses. Among the animal viruses are poliovirus (Castillo *et al*., 1986), herpes simplex (Kaul *et al*., 1985), and pseudorabies virus (Mucsi and Pragai, 1985). Additional examples can be found in articles by Selway (1986) and Vlietinck *et al*. (1986), which appear in a collection of reports entitled *Plant Flavonoids in Biology and Medicine* edited by V. Cody *et al*. (1986). These workers discuss the activity of synthetic as well as naturally occurring flavonoid derivatives.

The first report of flavonoids as agents against plant viruses was apparently that of Verma (1973) who showed that they were active against potato virus X (PVX). The effects of a range of flavonoids on tobacco mosaic virus (TMV) and potato virus X were carried out by French and coworkers (1991) and French and Towers (1992). It was thought that flavonoids reduced TMV infectivity by weakening interactions between virus coat proteins leading to increased susceptibility to host RNAases (French *et al*., 1991).

A recent study of the action of flavonoids on tomato ringspot nepovirus (ToRSV), using *Chenopodium quinoa* as host plant, addressed the question of structure-activity relationships of the active compounds (Malhotra *et al*., 1996). These workers tested 37 compounds at concentrations of 10 and 50 $\mu g\,ml^{-1}$. Compounds found to be ineffective in inhibiting germination of spores of ToRSV included kaempferol, quercetagetin, 5-hydroxyflavone (primuletin), apigenin, apigenin 4′-methyl ether (acacetin), catechin, pelargonidin chloride, apigenin 7,4′-disulfate, kaempferol 7,4′-disulfate, 5,7,4′-trihydroxyisoflavone (genistein), and 2′-hydroxygenistein. Twenty-two compounds showed activities ranging from 3-hydroxyflavone ($16 \pm 2.3\%$ reduction in spore germination) to the highest, quercetin 7,4′-dimethyl ether ($76 \pm 6.2\%$). The activities for quercetin ($73 \pm 2.2\%$) and for dihydroquercetin ($23 \pm 2.2\%$) show the effect of saturation of the C2-C3 double bond which substantially alters the conformation of the heterocyclic ring and, likely, the compound's capacity to bind to some protein or membrane surface feature. Addition of an hydroxyl group to C-6, forming quercetagetin eliminates activity, while moving the hydroxyl group, relative to quercetin, from C-3′ to C-2′, to form morin, reduces the activity to about 28%. Straightforward correlations with structure are clearly difficult when one sees that kaempferol has no appreciable activity and yet differs from quercetin only by the absence of an hydroxyl group in the B-ring. Unfortunately, 3,5,7,3′-tetrahydroxyflavone ("isokaempferol") was not available for testing. However, 3,7,3′-trihydroxy-4′-methoxyflavone, which does have the 3′-hydroxyl function, is one of the more active compounds (ca. 68%). We have already noted that the most active compound tested was quercetin 7,4′-dimethyl ether. Adding to the difficulty in arriving at a clear-cut relationship between flavonoid ring structure and substitution pattern was that group's finding that aurones, which have five-membered heterocyclic rings, and chalcones, with no heterocyclic ring, were also active. In an effort to sort out some of the cryptic structure-activity relationships observed in that study, Onyilagha and coworkers (1997) studied the activity of 21 chalcones against ToRSV. Although several compounds exhibited activity against the virus (at 5 $\mu g\,ml^{-1}$) it was not possible to determine the specific structural features responsible. These studies exemplify the problems of structure-activity research; there is rarely a shortage of seemingly contradictory observations.

Flavonoids as Antibacterial and Antifungal Agents

One of the standard approaches for determining potential medicinal use of flavonoids is to screen them for activity against a range of viruses, bacteria, and pathogenic fungi. We saw the results of some viral studies above. Many flavonoids have been found to have antibacterial properties. To most people in the screening-for-medicinal-properties business detecting antibacterial activity and identifying the compound or compounds responsible is the motivating force. There is little or, very often, no interest in determining whether the observed activity is of any significance to the plant. We examined some cases in Chapter Seven where such activity was part of a plant's defense against disease. In this section, however, we will look only at cases where potentially useful antibacterial agents were the target. The range of organisms against which certain flavonoids are active is quite impressive and in many cases, not trivial. *Staphylococcus aureus*, *E. coli*, and *Bacillus* species are frequently used as screening organisms. The critical importance of these searches for antibacterial agents can be appreciated by the finding of *C*-alkylflavanones from *Sophora exigua* that were active against methicillin-resistant *Staphylococcus aureus* (Iinuma *et al.*, 1994b). The importance of such finds cannot be over emphasized in view of the increasing resistance of many strains of pathogenic bacteria to routinely used antibiotics. Table 8.2 summarizes a number of recent searches for antibacterial flavonoids. In addition to the more common pathogenic bacteria, note that several studies have used cariogenic bacteria as screening agents.

There are many reports in the literature documenting antifungal activity of flavonoids. A sampling of organisms and compounds that have been shown to be active is presented in Table 8.3.

As part of a study of phytoalexin analogues and their potential use as biocides, Laks (1987) synthesized a series of long-chain alkyl derivatives of flavonoids and tested them against a variety of wood-destroying fungi and both Gram-positive and Gram-negative bacteria. The compounds were synthesized by a variation of the thiolytic cleavage reaction that has been used for analysis of proanthocyanidin oligomers (condensed tannins). Treatment of the oligomers [8-28], obtained from pine and eastern hemlock bark, with alkyl sulfides yielded epicatechin 4-alkylsulfides of the general type shown as [8-29], where $n=5$, 7, 9, 11, and 15. The most active of these was the decane ($n=9$) derivative when tested against a variety of fungi and both Gram-positive and Gram-negative bacteria. Activity ranged from 10 ppm against some fungi (*Scytalidium lignicola*) and Gram-positive bacteria (e.g., *Bacillus cereus*,

Table 8.2 Flavonoids as antibacterial agents

Microorganism	Compound	Plant source	Reference
Cariogenic *Streptococcus*	Isoflavanones	*Swartzia polyphylla*	Osawa *et al.*, 1992
Oral bacteria	Isoflavanones	*Erythrina* × *bidwilli*	Iinuma *et al.*, 1992
Oral bacteria	Isoflavones	*Ormosia monosperma*	Iinuma *et al.*, 1994a
Staphylococcus aureus, *Streptococcus faecalis*, *Bacillus subtilis*, *Mycobacterium smegmatis*	Isoflavanones	*Desmodium canum*	Delle Monache *et al.*, 1996
Staph. aureus, *B. subtilis*, *Trichophyton mentagrophytes*, *Mycobacterium intracellulare*, *Cryptococcus neoformans*	Dihydroflavonol	*Petalostemum purpureum*	Hufford *et al.*, 1993
B. subtilis, *E. coli*	Chalcones, Dihydrochalcones	*Myrica serrata*	Gafner *et al.*, 1996
Staph. aureus	*C*-Alkylflavanones	*Sophora exigua*	Iinuma *et al.*, 1994b
Staph. aureus, *E. coli*	Isoflavone galactoside	*Cnestis ferruginea*	Parvez and Rahman, 1992
Staph. aureus	Isoflavanone	*Erythrina eriotricha*	Nkengfack *et al.*, 1995
Staph. aureus (MRSA)	Kaempferol diacyl rhamnoside	*Pentachondra pumila*	Bloor, 1995
Bacillus cereus	3-Methyl flavonols	*Psiadia trinervia*	Wang *et al.*, 1989

Table 8.3 Flavonoids as antifungal agents

Microorganism	Flavonoid	Plant source	Reference
Cladosporium cucumerinum	Chalcones, Dihydrochalcones	*Myrica serrata*	Gafner *et al.*, 1996
Un-named in reference	Flavans, flavone	*Dracaena cochinchinensis*	Wang *et al.*, 1995
Candida albicans	Isoflavone galactoside	*Cnestis ferruginea*	Parvez and Rahman, 1992
Alternaria tenuissima	Chalcones	*Echinops echinatus*	Singh and Pandey, 1993
Aspergillus fumigatus	Isoflavone	*Erythrina sigmoidea*	Nkengfack *et al.*, 1994
Candida albicans, *Clad. cucumerinum*	Flavans	*Mariscus psilostachys*	Garo *et al.*, 1996
Prophyromonas gingivalis, *Actinomyces*	Isoflavones	*Ormosia monosperma*	Iinuma *et al.*, 1994

Micrococcus luteus) to values over 500 ppm for Gram-negative bacteria. It was suggested in that paper that the activity of these compounds may arise from their structural similarity to compounds such as the known phytoalexin kievitone [8-30] here presented to emphasize the spatial position of the *C*-alkyl function.

Flavonoids have been implicated as the compounds responsible for the toxic action of several plant species against a variety of pests other than the microorganisms discussed and listed above. For example, *Parkia clappertoniana*, the West African locust bean, possesses antimolluscicidal properties against *Biomphalaria glabrata* the snail responsible for transmission of schistosomiasis. Two flavones were isolated, identified, and tested for toxicity against the snail (Lemmich *et al.*, 1996). 7-hydroxy-3,8,4′-trimethoxyflavone [8-31] proved to be inactive, but 2′-hydroxy-3,7,8,4′,5′-pentamethoxyflavone [8-32] was effective (80% kill) at a concentration of 25 ppm. Other toxic flavonoids include a quercetin 3-*O*-triglycoside from *Engelhardtia colebrookiana* that was shown by Uniyal and coworkers (1992) to be poisonous to fish, and a series of compounds with greater or lesser activity against the European corn-borer *Ostrinia nubialis* (Abou-Zaid *et al.*, 1993). These later workers showed that quercetin, quercetin 3-*O*-glycosides, and rhamnetin were most effective, catechin somewhat less so, and pinocembrin least active of the compounds tested.

Flavonoids Acting as Animal Hormones

Several isoflavonoid derivatives produced in high yield by field legumes may have serious effects upon grazing livestock owing to their capacity simulate the action of animal hormones. Shutt (1976) estimated that over a million ewes per year fail to lamb because of isoflavonoid-rich pasture legumes. Compare the structures of formononetin [8-33] and coumestrol [8-34], both widely occurring and highly estrogenic isoflavonoids, with the known animal hormone oestron [8-35] and the synthetic molecule diethylstilboestrol [8-36]. A recent paper by Ruh and coworkers (1995) showed that naringenin was only weakly estrogenic in tests in the female rat and using MCF-7 human breast cancer cells.

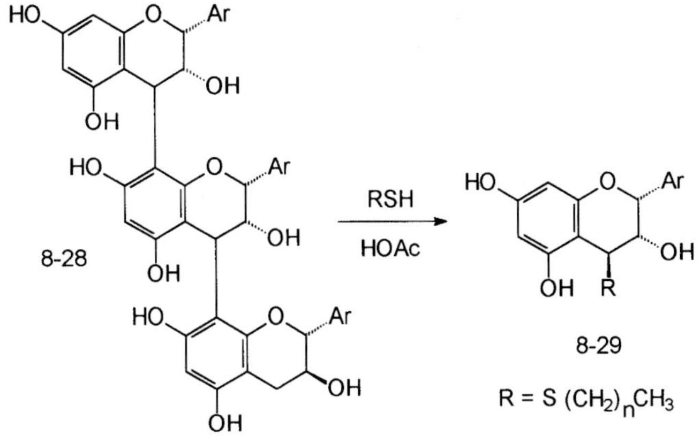

8-28 —RSH / HOAc→ 8-29

$R = S\,(CH_2)_nCH_3$

$n = 5, 7, 9, 11, 15$

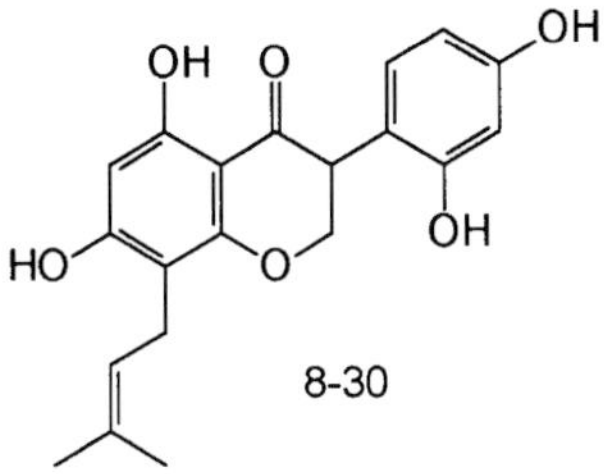

8-30

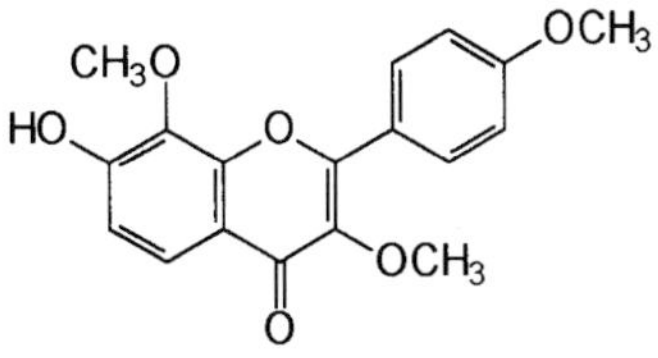

8-31

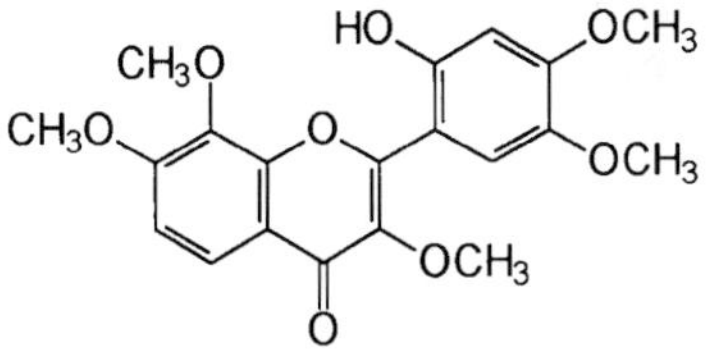

8-32

An added problem in assessing the toxicity of isoflavonoids (this applies to other compounds as well) is the confounding feature that some can be metabolized within the animal to a more active form. This is seen in the conversion of formononetin to the isoflavan equol [8-37] by demethylation, reduction of the C2-C3 double bond, and elimination of the carbonyl group. In instances where subsequent metabolism converts a dietary compound to an active or more active compound, the starting material, formononetin in this case, is referred to as a proestrogen.

Dakora (1995), in his timely review of useful exploitation of natural flavonoids, points out an interesting distinction that seems to exist among naturally occurring isoflavonoids. He notes that the isoflavonoids involved in *nod*-gene induction in

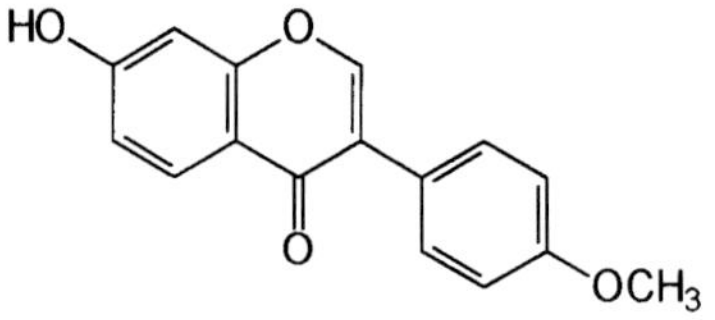

8-33

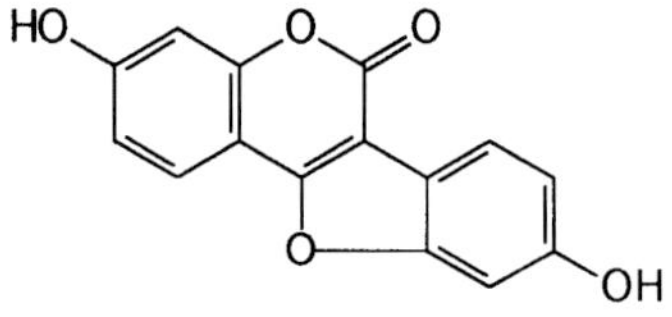

8-34

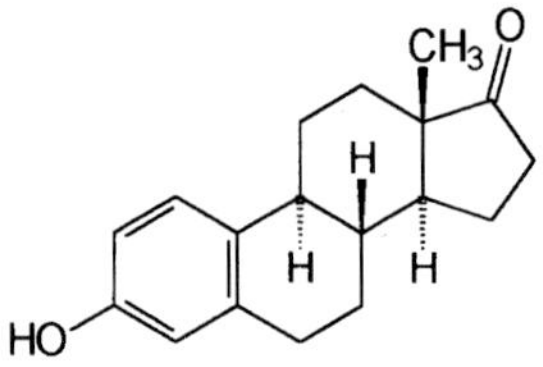

8-35

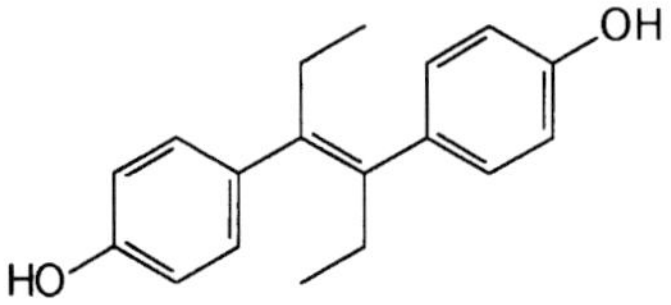

8-36

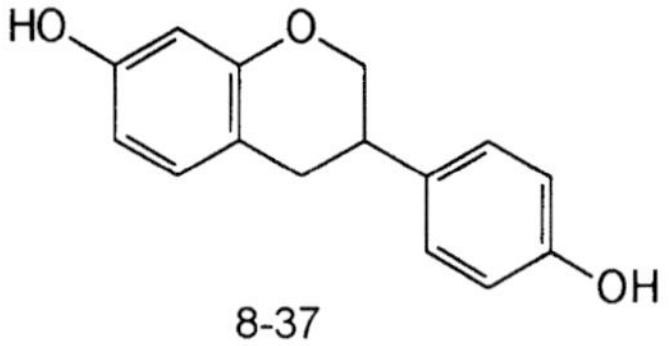

8-37

Rhizobium and related microorganisms do not appear to possess the oestrogen-like activity of formononetin and its relatives. He goes on to suggest that selection for lines of field legumes with reduced capacity to make the harmful isoflavonoids without interfering with biosynthesis of the *nod*-gene inducing compounds could result in increased production of both wool and animal protein.

Flavonoids and Foods

Flavonoids are a normal and regular part of the food we eat. Are they beneficial, harmless, or do they have undesirable effects of which most people are unaware? The answers to all of these questions is almost certainly yes. The benefits may only be the pleasure of a particular flavor, and the undesirable effects may be minuscule with normal levels of consumption, but the effects are certainly real. The suggestion that some dietary flavonoids have neither positive nor negative effects is probably open to debate, but not here. A general overview of flavonoids in foodstuffs written by Tony Swain (1962) is worth reading in order to gain an overall appreciation of the impact these compounds make in food science. A more recent article by Butler (1989) discusses the effects of condensed tannins on animal nutrition. The nutritional value of flavonoids has been reviewed in some depth by Roger (1988) in an article that addresses the importance of flavonoids as antiscorbutic (anti-scurvy) elements of foods. The conclusion that flavan-3-ols are of particular significance is important in view of the long standing debate on the efficacy of "vitamin-P" and "bioflavonoids" as essential dietary elements.

We will start with an example wherein flavonoids make a major contribution to flavor, in particular, the gentle bitterness of the grapefruit. An examination of the structural features responsible for this characteristic of grapefruit and fruits of related *Citrus* species will take us, quite unexpectedly, into the realm of flavonoids as sweetening agents. We will also look at the contribution that flavonoids make to other common beverages, wine, beer, and tea.

In the first case studies in Chapter Four, brief mention was made of the difference in taste between different glycosidic derivatives of the *Citrus* flavanones. The 7-*O*-neohesperidosides (2-α-L-rhamnosylglucosides) of naringenin, isosakuranetin, eriodictyol, and hesperetin are all very bitter in contrast to the corresponding 7-*O*-rutinosides (6-α-L-rhamnosylglucosides) which have no taste. Structures, common names, and a measure of bitterness using quinine as a standard, are presented in Table 8.4 (adapted from Horowitz and Gentili, 1969).

The finding that all of the *Citrus* neohesperidosides were bitter while all of the rutinosides were tasteless, gave a clear message that the structure of the diglycoside plays a key role in determining taste characteristics. Modifications of the B-ring plays only a minor role in determining intensity of bitterness. Structure-activity studies done by Horowitz and Gentili (1969) established to a significant degree what features were required for bitterness, and their work led to a most unexpected result. Structural modifications that have been investigated include loss or modification of large pieces of the parent compound and conversions to other flavonoids. The first, and obvious, large piece to be examined involved the nature of the saccharide portion of the molecule. Since movement of the rhamnose substituent from position-2 on the glucose to position-6 results in loss of bitterness there must be some feature of the 2-linkage that is important. Konishi and coworkers (1983) have shown that movement of the rhamnose to position-3 or 4 of the glucose gives compounds that are only slightly bitter. Loss of rhamnose from either naringin or neohesperidin, leaving just the 7-*O*-glucosides, did not result in loss of bitterness, so the neohesperidose entity *per se* is not required. Removal of both sugars resulted in complete loss of taste, so a sugar at position-7 is clearly a structural requirement.

The next large structural alteration involved removal of the B-ring. (Recall that B-ring substitution has comparatively little effect on taste.) The bitter flavanones

Table 8.4 Bitterness of *Citrus* flavanones (from Horowitz and Gentili, 1969)

Common Names			
Glycoside	Aglycone	Aglycone structure	Relative bitterness
Naringin	Naringenin	5,7,4′-Trihydroxyflavanone	20
Poncirin	Isosakuranetin	5,7-Dihydroxy-4′-methoxyflavanon	20
Neoeriocitrin	Eriodictyol	5,7,3′,4′-Tetrahydroxyflavanone	<2
Neohesperidin	Hesperetin	5,7,3′-Trihydroxy-4′-methoxyflavanone	2
Quinine HCl			100

were degraded to phloroacetophenone 4′-*O*-neohesperidoside [8-38], which is intensely bitter. Degradation all the way to the phloroglucinol level [8-39] removes any trace of bitterness. This suggests that the carbonyl function is required for binding to the tasting site. The 4-carbonyl-5-hydroxyl combination in flavonoids and the *o*-hydroxyl-carbonyl combination in acetophenones are known to complex metal ions which suggests at least one of the possible interactions of these compounds with the bitter taste site.

The second category of structural modifications involved conversion of the bitter flavanone glycosides into other types of flavonoids. Whereas the flavone diglycosides

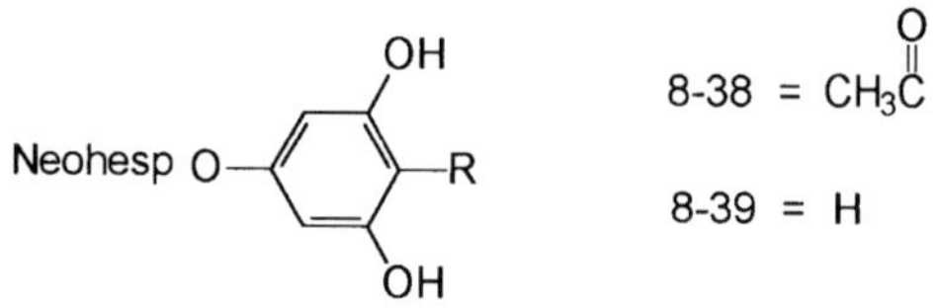

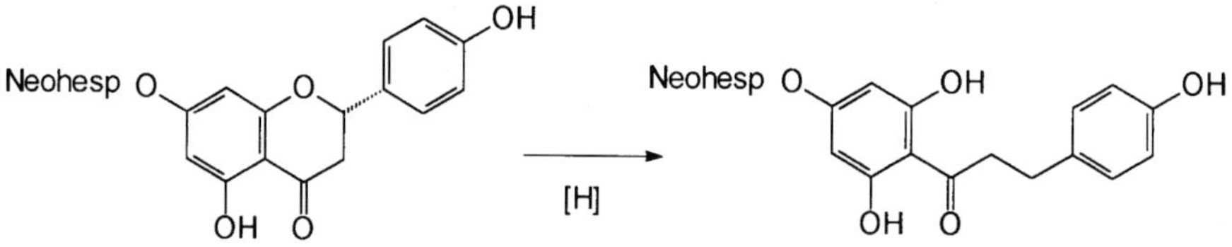

Neohesp O OH R OH

8-38 = $CH_3C(=O)$

8-39 = H

Neohesp O O OH OH O [H] Neohesp O OH OH OH O

8-40

8-41

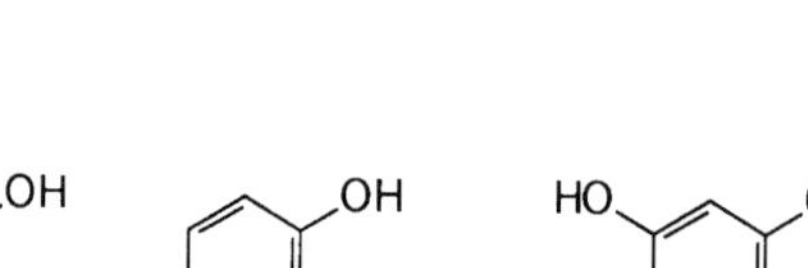

HO OH OCH_3 HOOC OH OH O

8-42

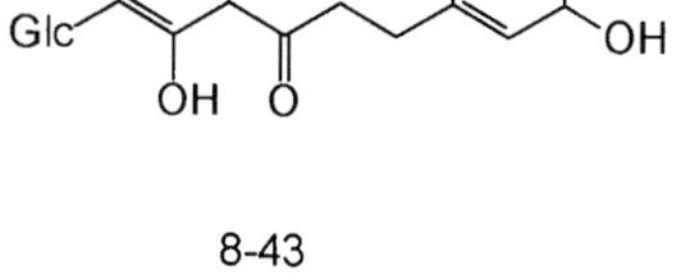

HO OH OH Glc OH OH O

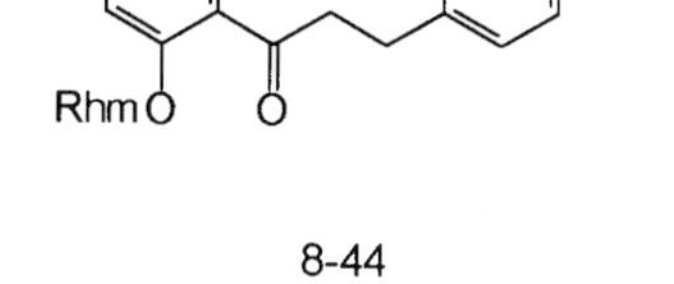

8-43

HO OH OH RhmO O

8-44

corresponding to the bitter flavanone diglycosides are not themselves bitter, they have the capacity to reduce the bitterness of the latter compounds. One of the ways in which this might be accomplished would be by competitive binding of the flavone derivative at the tasting site. Flavones had the same effect when tested using bitter compounds other than the flavanones (Horowitz, 1986). Probably the most striking effect observed in the transformation experiments came from the reduction of the bitter naringin to the corresponding dihydrochalcone according to the sequence of reactions [8-40] to [8-41]. The product, 2′,4′,6′,4-tetrahydroxydihydrochalcone 4′-*O*-neohesperidoside, or naringin dihydrochalcone, was intensely sweet! Interestingly, from a structure-activity point of view, only naringin and neohesperidin yielded intensely sweet dihydrochalcones; the others were either only slightly sweet or still retained a bitter taste.

Manipulation of B-ring substituents afforded some additional interesting observations. Whereas hesperidin dihydrochalcone, with the 3-hydroxy-4-methoxy arrangement, was intensely sweet, the isomeric compound with the 3-methoxy-4-hydroxy arrangement had no taste at all. Adding another hydroxyl group to the B-ring to give the 3,5-dihydroxy-4-methoxy arrangement abolished taste despite the presence of the active 3-hydroxy-4-methoxy grouping. Replacing the 4-*O*-methoxy group by *C*-methyl resulted in loss of taste, while replacing the 4-*O*-methyl group by 4-*O*-*n*-propyl produced a compound that was among the sweetest tested. Whitelaw and Daniel (1991) synthesized a number of dihydrochalcones with a variety of substituents, both electron donating and electron withdrawing, and subjected them to sensory evaluation. The most powerful sweetener was found to be 3′-carboxyhesperetin dihydrochalcone [8-42], which was shown to be about 3400 times sweeter than a 6% aqueous solution of sucrose. The lingering aftertaste characteristic of dihydrochalcone sweeteners remains a problem, however. Additional studies of the effects of modifying the sugar portion of the dihydrochalcones have been done, again with quite variable results. The interested reader should consult the article by Horowitz (1986) for details.

The leaves and stems of the South African shrubby legume *Aspalathus linearis* are used to prepare a beverage known as "rooibos tea." The tea is considered choice by many owing to its flavor, its mineral and ascorbic acid content, its lack of caffeine, its low tannin content, and its naturally sweet taste. Rooibos tea is currently exported to a number of countries. In view of the known phenolic chemistry of the plant itself, Rabe and coworkers (1994) undertook a study of the phenolic constituents of the beverage. Compounds identified included hydroxybenzoic and cinnamic acids, the flavones luteolin and chrysoeriol, quercetin derivatives, several *C*-glycosylflavones, and the *C*-glucosyldihydrochalcone aspalathin [8-43]. Since aspalathin is known to possess a sweet taste, it is probably the compound responsible for the sweetness in the tea. The sweet taste of dihydrochalcones has been known since the late 19th century when Rennie (1886) determined that glycyphyllin, the bitter-sweet principle of *Smilax glycyphylla*, (Smilacaceae) was 2′,4′,6′,4-tetrahydroxydihydrochalcone 2′-*O*-α-L-rhamnoside [8-44]. It is interesting to note, again on the subject of structure-activity relationships, that replacement of the rhamnose with glucose to give the 2′-*O*-β-D-glucoside eliminates any element of sweetness from the compound. This glucoside, called phloridzin, is quite bitter to taste.

Rhizomes of the fern *Selliguea feei* (syn. *Polypodium feei*) is used to make a tea in western Java that is used as a treatment for rheumatism, and as a male tonic. Baek and coworkers (1993) undertook a study of the rhizomes in order to determine the chemical nature of their sweet-bitter components. A new proanthocyanidin, named "selliguean A," was isolated and shown to be very sweet by a panel of tasters, who also reported that a 0.5% solution was pleasant tasting and not astringent (many proanthocyanidins are intensely astringent). In semi-quantitative terms, selliguean A was judged to be about 35 times sweeter than a 2% aqueous solution of sucrose. The structure of selliguean A was determined to be a trimer based on (+)-afzelechin] [8-45] and (+)-epiafzelechin, which are the (2*R*:3*S*)- and (2*S*:3*S*) isomers of 3,5,7,4′-tetrahydroxyflavan. (+)-Epiafzelechin is epimeric to (+)-afzelechin at C-3. The structure of the trimer was established as epiafzelechin-($4\beta \rightarrow 8$, $2\beta \rightarrow O \rightarrow 7$)-epiafzelechin-($4\beta \rightarrow 8$)-afzelechin and is shown as [8-46]. Other reportedly sweet proanthocyanidins have been obtained from the leaves of the fern *Arachnioides sporadosora* (Tanaka *et al*., 1991), and from root bark of *Cinnamomum sieboldii* (Morimoto *et al*., 1985). The sweet compounds from *Cinnamomum* bark were trimers, whereas the two tetramers and a pentamer that also were present in the plant were not sweet but did have the astringent properties normally associated with condensed tannins. The structure of one of the trimers is shown in full as [8-47]. The other compound differs only in the stereochemistry of the terminal unit, the partial structure of which is shown in [8-48]. The sweet compound from *Arachnioides* has the same upper and middle units as does the *Cinnamomum* compound, but is distinguished by the existence of a rare lactone structure that is part of the terminal unit. The partial structure of this compound, showing the terminal unit, appears as [8-49].

Red Wine

Flavonoids play an extremely important role in the production and enjoyment of several widely used beverages, in particular, tea, beer ands its relatives, and wine. One of the most intensely studied subjects, and the one we'll look at first, involves the flavonoids of grapes and how they influence the ultimate quality of wine. Although representatives of many different classes of flavonoids are present in grapes, two stand out above all others in the influence they have on the process of wine making and the ultimate quality of the product, anthocyanins and proanthocyanidins (tannins in most general literature). Anthocyanins, of course, are responsible for the color of grapes, while the proanthocyanidins are responsible for the astringency. The astringency is masked more or less completely by the high concentration of sugar in ripe grapes, but with the disappearance of the sugars through fermentation, it becomes significant. We will concern ourselves exclusively with red grapes in this discussion; vinification, flavor, and storage problems associated with wine made from white grapes involve a different set of problems. (In fact, the aim is to remove as much polyphenolic matter as possible from white wines in order to prevent discoloration through oxidation.) For balanced treatments of all aspects of grape and wine chemistry, the interested reader should consult works by Webb (1974) and Jackson (1994), both of which have extensive bibliographies.

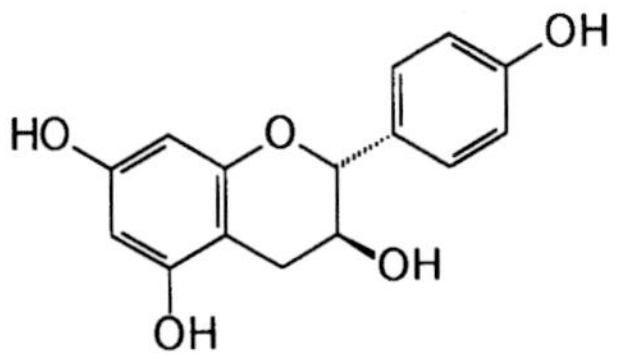

8-45

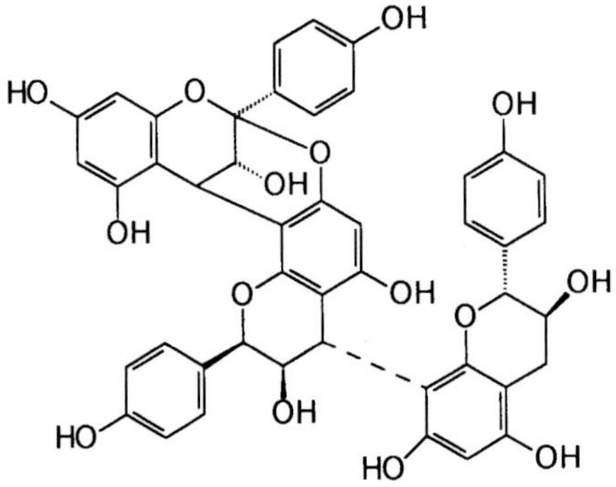

8-46

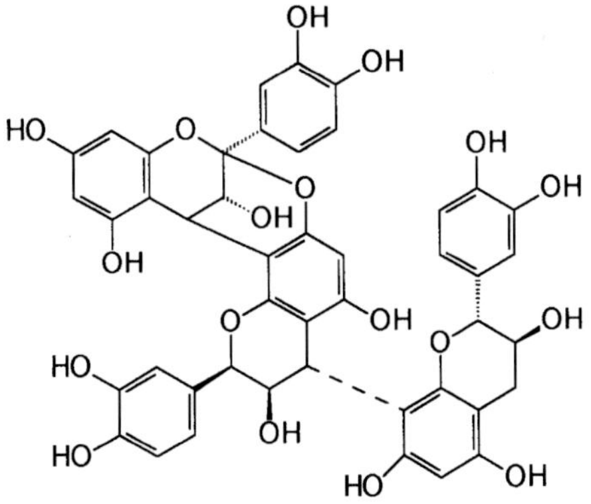

8-47

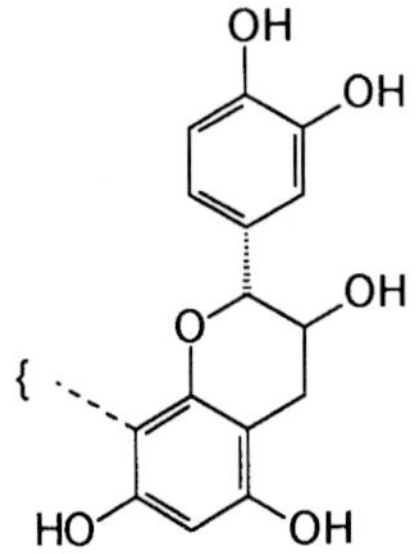

8-48

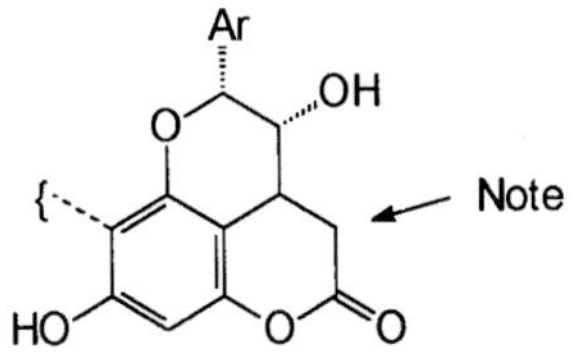

8-49

The anthocyanin chemistry of grapes is a complex one with the five aglycones petunidin, cyanidin, peonidin, delphinidin, and malvidin present in various mixtures of 3-*O*-mono- and 3,5-di-*O*-monoglucosides some of which are acylated, the most common acid being *p*-coumaric (Ribéreau-Gayon, 1974). Different species of *Vitis* accumulate different arrays of anthocyanins such that the pigment profiles of these taxa can be used to identify them. This is important for two reasons: (1) to identify juice from non vinifera grapes; and (2) to identify contaminating species in grape hybrids. Both of these contaminations lead to lower quality products and are strictly controlled. Other monomeric flavonoids present in grapes are glycosides of the common flavonols kaempferol, quercetin, and myricetin, and the common flavan-3-ols catechin, epicatechin, and gallocatechin (Jackson, 1994). Proanthocyanidins consisting of from two to five subunits occur in reasonably high concentration in the skins and seeds (pips) of red wine grapes. The major components of this fraction consists of procyanidin and prodelphinidin roughly in the ratio of 4 : 1 (Haslam, 1989, p. 197). The chemical modification of these oligomeric proanthocyanidins (e.g., oxidation) and the interaction between them and the anthocyanidins are extremely important processes that define the nature of finished product. For early studies of proanthocyanidin chemistry in aging wine the interested reader should consult the work of Somers (1971).

Two easily detectable changes occur in red wine as it undergoes the aging process: (1) there is a change in its color; and (2) there is a dramatic lessening of astringency. As we shall see, these changes do not necessarily occur independently of one another. Without going into mechanistic details, we can say in general that addition of monomeric proanthocyanidin units to oligomeric proanthocyanidins leads to increased molecular weight eventually reaching a point where precipitation of the polymer begins to occur (Haslam, 1980, 1989). The precipitation of polymeric material draws the equilibrium away from soluble proanthocyanidins, with concomitant decrease in astringency. In a set of experiments to test the fate of proanthocyanidins in wine, Haslam (1980) added measured quantities of oligomeric procyanidins to various wines and observed the formation of precipitates. When white wine was used, turbidity became evident after 15 days and precipitate began to form. After 48 days, solids representing 50–70% of the added polymer had accumulated. Quantitative analysis of the precipitate indicated that it accounted for 75–80% of the theoretical amount of cyanidin available in the original procyanidin polymer. A parallel experiment demonstrated that a solution of procyanidin polymer and either (+)-catechin or (−)-epicatechin remained clear for 48 days. The slow polymerization occurring in the procyanidin-augmented wine appears to come about by the acid-catalyzed bond-breaking and bond-making reactions characteristic of proanthocyanidins. Similar results were observed when red wine was treated with polymeric procyanidins. Changes in anthocyanidin content of the augmented wine and in the red wine itself were evident over time. In unadulterated red wine there was a reduction of pigment of approximately 20% over 120 days, while the reduction in the experimental solutions was in the range 10–15%. The change of color of red wines toward the tawny that occurs over time, seems to be due, to some extent at least, to the involvement of anthocyanins in the polymerization reaction. In time, these polymers will precipitate resulting in a reduced concentration of both anthocyanins and soluble proanthocyanidins. In this way, the wine undergoes a reduction in astringency and color change synchronously. Kinetic studies of some of these process have been described (Beart *et al.*, 1985).

Haslam (1989) also discusses another mechanism whereby polymerization of proanthocyanidins may be facilitated. The agent in this process is acetaldehyde

(CH_3CHO) whose formation in the system by partial oxidation of ethanol by available oxygen is easily rationalized (Ribéreau-Gayon, 1974). In this process, the carbonyl carbon would serve as a target for nucleophilic attack by a phenolic ring carbon followed by reaction with a phenolic ring carbon of another molecule (the reaction resembles the process by which phenol-formaldehyde resins are produced). In the aging of wine, then, the acetaldehyde molecule serves admirably well as a cross-linking agent. The result of these processes, other than the obvious precipitation of high molecular weight polymers, can be seen in the average size of polymers which changes from 500–700 for young wines to 2000–3000 for old wines (Ribéreau-Gayon, 1974, p. 65).

Recent experimental studies by Escribano-Bailón *et al.* (1996) have addressed the issue of reactions between proanthocyanidins and anthocyanins, as well as the involvement of acetaldehyde in flavonoid oligomerization in the aging of red wines. These workers used (+)-catechin [8-50] as a tannin (proanthocyanidin) model and 3,4′-dimethoxy-7-hydroxyflavylium chloride [8-51] and 5,7-dihydroxy-3,4′-dimethoxyflavylium chloride [8-52] as model anthocyanidins. The reaction between catechin and two equivalents of [8-51/52] to form the dimer [8-53] was extremely slow, four months being required to yield enough product to allow structural studies. By contrast, catechin plus flavylium [8-52/51] in the presence of acetaldehyde yielded easily isolable amounts of dimer [8-54] in only a week. Evidence was also presented to suggest that further oligomerization may occur to form a product that consists of two catechin and two flavylium units. These tetramers might represent products of early stages of the polymerization and precipitation that occur in the aging of red wine.

The importance of the polymerization process, and thus the status of the "tannins" in an aging wine, can hardly be overestimated. This can be appreciated by the regular references to tannins in the rather poetic language used in descriptions of wines in trade literature. This is clearly seen in a selection of quotations from the January 31, 1997 issue of *Wine Spectator*, in which recent vintages of red wine are assessed by experts as to their "drinkability." The first two come from J. Laube's comments on the California Zinfandel vintage of 1994; the rest from an article by H. Steiman on 1994 Pinot Noirs from Oregon and Washington: "Smooth in texture to the finish, where the tannins weigh in." "Opens to reveal more depth and richness before the tannins kick in." "... superbly balanced, it has the polish to be drinkable now, and the fine-grained tannins to hold well past 2000." "Tannins are smooth and polished." "Firm tannins need some time; try in 2001." "Fine tannins are present but not intrusive." "... delicious and packed with flavors just barely held back by firm tannins. Promises to be a winner." It is clear from most of these descriptions that tannins, which are variously described as "firm," "fine grained," or "smooth and polished," have some negative features as well, especially if they become "intrusive." This may sound like so much advertising verbiage to many, but there is a very serious message underlying the whole exercise. The status of these polymers plays a major role in determining when a wine is ready to be drunk with maximum pleasure (absence of undesirable taste sensations), when it is likely to be at its best and, in many instances, whether it will continue to improve with age. Every wine has its time; sometimes we just have to wait!

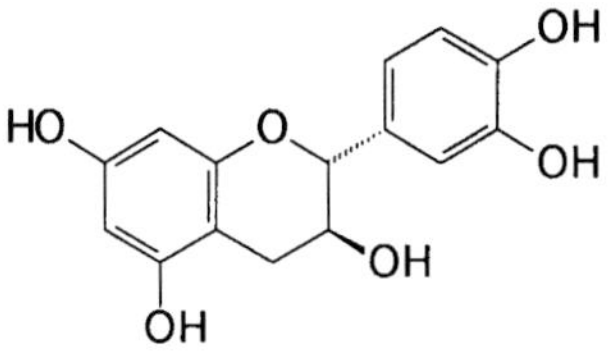

8-50

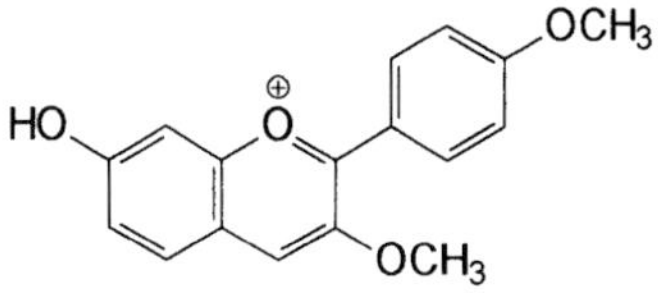

8-51

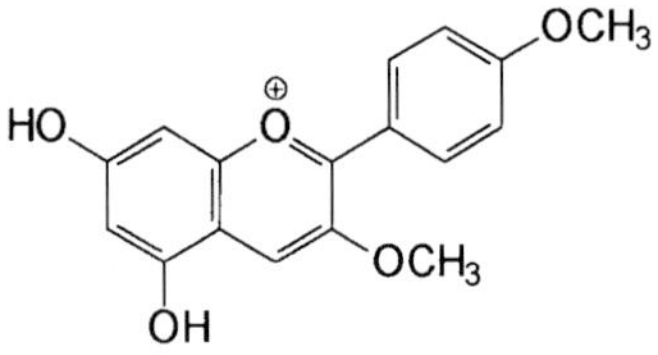

8-52

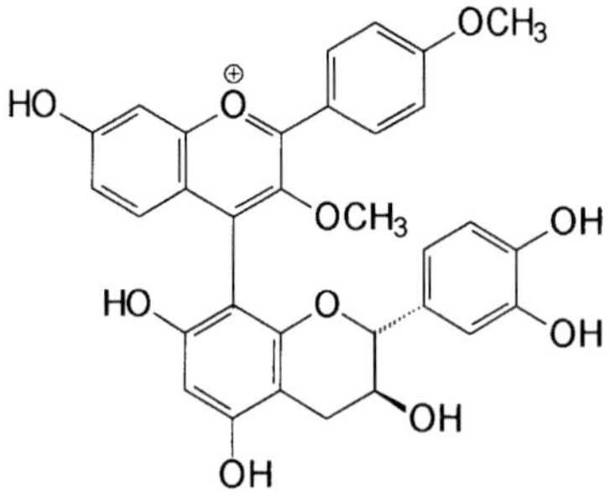

8-53

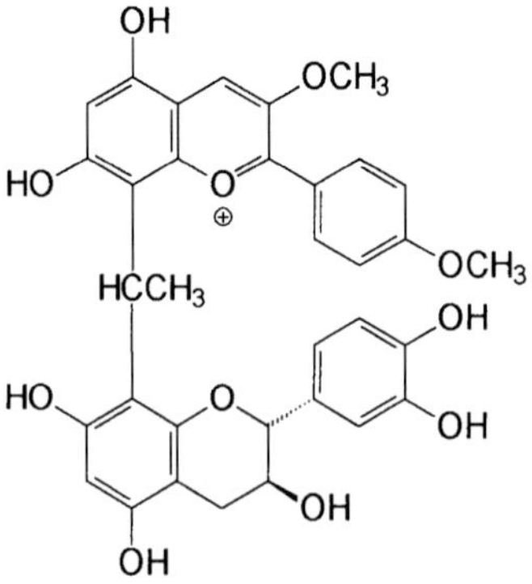

8-54

Beers, Lagers, and Ales

One of the problems that plagues brewers is the formation of hazes in their products. Hazes may be permanent or they may appear when the products are chilled. The mixture of material precipitated can be very complex, and can include proteins, carbohydrates, polyphenolic materials and even inorganic substances. Whatever the composition of a haze may be, it obviously reduces the value of the product and must be removed. One of the major contributors to the haze problem is the presence of the polyphenolic compounds the majority of which are proanthocyanidins (Haslam, 1989). An ideal solution to the problem would be to remove these components at source rather than having to resort to some physical means of separation (e.g., filtration) at a later stage of manufacture. This approach has been realized through the use of barley mutants that do not have the capacity to biosynthesize proanthocyanidins. Papers by Van Wettstein and colleagues (1977) and Erdal (1986) address the issue of proanthocyanidin-free barley.

Tea

One of the most widely consumed beverages in the world derives from the leaves of *Camellia sinensis* (originally described by Linnaeus as *Thea sinensis*). Many species belong to the genus, including the familiar flowering camellia (*C. japonica*), but it is *C. sinensis* that ranks as the world's most important caffeine beverage (Mabberley, 1987). Green tea consists of leaves that have been steamed to destroy enzymes and then dried; it is not a separate species as once thought. The black tea of commerce is produced through the action of oxidative enzymes on the substantial amount of polyphenolic material in the leaves. This process, commonly referred to as a "fermentation," although it does not involve any microbial activity, results in the formation of two types of products, the theaflavins and the thearubigens. In order to understand where these products come from it is necessary to comment briefly on the main polyphenolic compounds present in tea leaves prior to any processing. Haslam (1989) informs us that the "tea flush" consists of 3–4% caffeine and up to 30% of polyphenolic compounds the principle components of which are the flavan-3-ols (−)-epicatechin gallate [8-55], (−)-epigallocatechin [8-56], and (−)-epigallocatechin gallate [8-57].

During the drying stage of tea production cellular integrity of the leaves is disrupted bringing the flavan-3-ol derivatives, which had been stored in the vacuoles, into contact with oxidases released from the chloroplasts. Roberts (1962) indicates in his description of the process, that the oxidase has rather specific affinities. It will oxidize the simple phenols catechol (1,2-dihydroxybenzene) and pyrogallol (1,2,3-trihydroxybenzene), but it acts much more quickly on flavanols and proanthocyanidins. Oxygen is required and, as expected for an oxidase whose substrates all contain vicinal hydroxyl groups, *ortho*-quinones are involved. In the presence of ascorbic acid oxygen absorption ceases and no colored products are formed. Addition of ascorbic acid at the end of the "fermentation" process has no effect on the color or taste of the products. These observations are characteristic of processes involving *ortho*-quinone intermediates. Carbon dioxide is also produced suggesting that significant carbon–carbon breakage occurs. Theaflavins appear to be the major product of the preliminary oxidation reactions. A typical member of this family of dimers appears as [8-58]. Gallic acid conjugates of these compounds are also known.

It is important to point out that not all "tea" preparations, including those made from *Camellia sinensis*, can be consumed without some risk. Morton (1989) has had a continuing interest in the high incidence of esophageal cancer associated with continued use of high tannin food stuffs, teas, and masticatories. Occasional use is not a problem, but prolonged and repeated use of such products as strong black tea, the stimulant drink maté, made from dried leaves of *Ilex paraguariensis*, the masticatore khat (*Catha edulis*, Celastraceae), and betel (*Areca catechu*, Arecaceae, syn. Palmae) can result in increased risk of mouth and throat cancers. Adding of milk to common tea acts to neutralize the undesirable effects of tannins through complexation

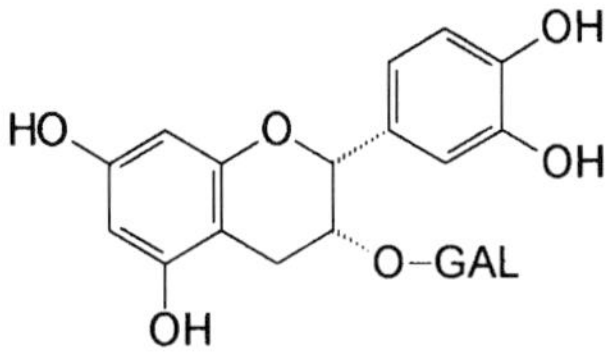

8-55

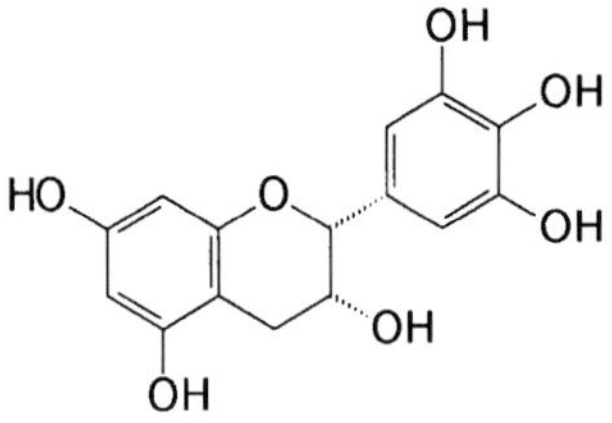

8-56

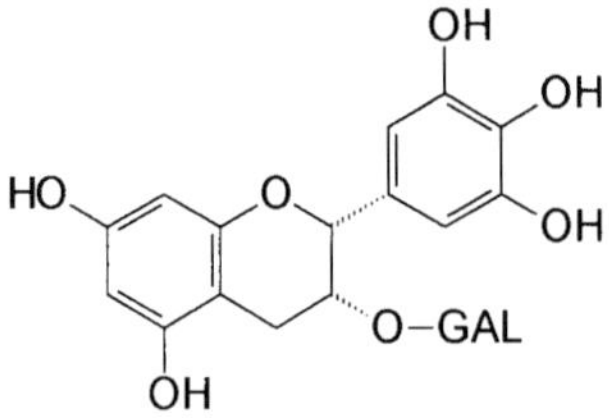

8-57

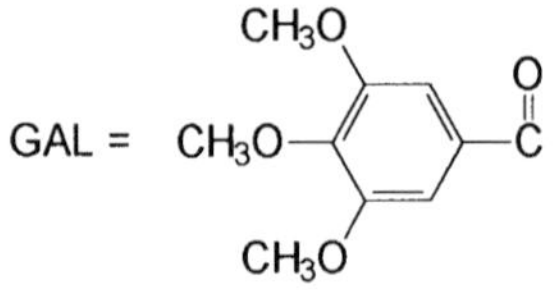

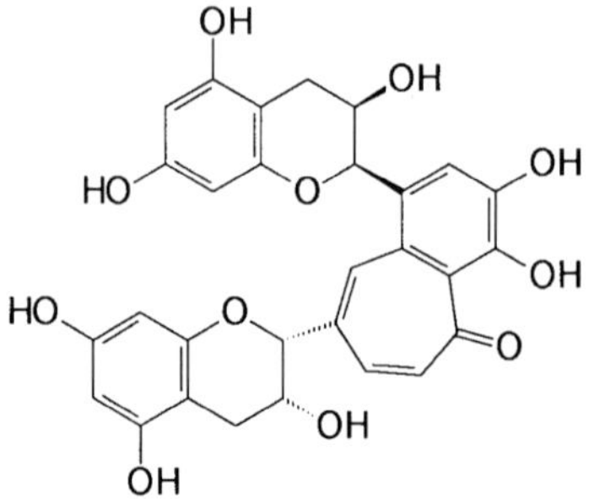

GAL =

8-58

with the milk proteins. It is interesting to note that although betel nut users have elevated risk of mouth and throat cancers, the incidence of stomach cancers in these people is much reduced, presumably as a result of the tannins' capacity to react with nitrosamines.

Flavonoids in Honey and Bee Pollen

Tomás-Barberán and colleagues (1993a) have developed a high performance liquid chromatography (HPLC) protocol for the study of honey and related materials.

Since the quality of a honey depends to a large extent upon its floral origin, it would be desirable to have an analytical system capable of providing both qualitative and quantitative information. The procedure is based on the analysis of flavonoids of these materials and comparison of the profiles with those of plants known or suspected of being involved. This method has the potential of helping to establish the geographical and botanical origins of the species whose nectar contributes to the product. There is a significant commercial preference for monofloral honeys, such as those derived from citrus, rosemary, *Calluna* (heather), or white clover, as opposed to multifloral honeys. Theoretically, the presence of adulterants could also be documented by an analytical system of the sort under investigation. The cited study described attempts to maximize systems that could be used for analysis of citrus and rosemary honey. One of the systems developed worked well for the separation of hesperetin, the marker for citrus honey, while another system worked best for the detection of apigenin, which, within this narrow sampling, was the characteristic flavonoid of rosemary honey. A companion study (Tomás-Barberán *et al.*, 1993b) demonstrated that HPLC methods could be equally useful for analysis of components of beeswax. The method is capable of resolving quite complex mixtures and clearly showed differences among the flavonoid profiles of *Apis mellifera* beeswax collected in "La Alcarria" region of Spain, wax scales (as they are secreted by the bees), and honey from the same region of Spain. The concentrations (in $\mu g\,g^{-1}$ material) of a sampling of the flavonoids in these sources as reported are: for 5,7-dihydroxyflavanone (pinocembrin), 2.3, 14.7, and 2.9; for 5,7-dihydroxyflavone (chrysin), 10.0, 88.7, and 2.3; and for 5-hydroxy-7-methoxyflavone (tectochrysin), 19.6, 22.2, and 0.6, respectively for beeswax, wax scales, and honey.

A recent note (Campos *et al.*, 1996) reported that the flavonoid/phenolic profiles of individual bee pollens appear to be species specific. The major plant species contributing to a given pollen can be determined by their phenolic profiles even when they cannot be ascertained by microscopic examination of the pollen itself. The test organisms examined in this study were four species of *Salix* (willow) that were easily distinguished by the HPLC system. It was also determined that the flavonoid pollen profiles remain unchanged for at least a century!

Markham and Campos (1996) have recently described otherwise rare flavonol glycosides as components of bee pollen from *Ranunculus sardous* and *R. raphanistrum* and from *Ulex europaeus*. All were derivatives of herbacetin (8-hydroxykaempferol). The compound obtained from *R. sardous* was determined to be 7-*O*-methylherbacetin 3-*O*-sophoroside, which is a new compound, that from *R. raphanistrum*, 8-*O*-methylherbacetin 3-*O*-sophoroside, and that from *Ulex*, 8-*O*-methylherbacetin 3-*O*-glucoside. While these workers were not necessarily concerned with using these compounds as markers for nectar sources, their unusual nature suggests that they might well be useful for that purpose. It is interesting to note the occurrence of herbacetin derivatives in such unrelated plant groups as *Ranunculus* (Ranunculaceae) and *Ulex* (Fabaceae). It is based on more than curious interest, however, to echo the observation of Markham and Campos that two of the flavonols exist as 2-*O*-glucosylglucosides (sophorosides) and that flavonol sophorosides may play a significant role in pollen tube growth (see Chapter Seven).

Leather Tanning

One of the oldest activities involving polyphenolic compounds is the conversion of animal hides into leather. In his chapter on tanning, Bliss (1989) pointed out that there are written records of tanning that go back over 10,000 years. The process of tanning involves treating hides with substances that protect the molecular form of the collagen fibers of which the hides are composed (Haslam, 1989). The steps in accomplishing this, again as described by Haslam (1989), involve first the binding of polyphenolic molecules with the amide linkages and polar amino acid residues of the collagen, through hydrogen bonding, followed by packing additional polyphenolic molecules in the remaining spaces. Inorganic substances, such as chromium or aluminum salts, are also used to tan hides and the use of tannic substances and aluminum in combination is also employed. The reader will have noticed that the generic term "polyphenolic" has been used so far. We have not specified whether we are dealing with condensed tannins, in other words, proanthocyanidins, or with hydrolyzable tannins which are often very complex carbohydrate esters of gallic acid. Both types have the capacity to act as tanning agents, although the exact nature of the product will be different under different conditions. Bliss (1989, Table 1) lists several common sources of vegetable tannins, without specifying the chemical type: species of *Rhus* (sumac) and *Schinopsis* (quebracho) both Anacardiaceae, *Acacia* (wattle), *Castanea* (chestnut) and *Quercus* (oak) (Fagaceae), *Eucalyptus* (Myrtaceae), *Terminalia* (the myrobalans) (Combretaceae), and *Caesalpinia* (Fabaceae). The major polyphenolic compounds presumed to be responsible for the tanning capacity of some of these species are hydrolyzable tannins rather than proanthocyanidins. Reference to Haslam's (1989) Table 3.1 shows species of *Rhus*, *Caesalpinia* (providing tanning substances called algarobilla, divi-divi, and tara), *Castanea*, *Quercus*, and *Terminalia* (myrobalans) as sources of complex gallic acid derivatives. On the other hand, quebracho and quebracho-like tannins, from species of *Rhus* and *Schinopsis*, are proanthocyanidins (Haslam, 1989, p. 79). This apparent nomenclatural problem is of historical origin, the term "tannin" having come into use to identify any substance that was capable of converting animal hides to leather. The term was well and truly entrenched in the language before chemists established that two quite different classes of compounds were involved.

Flavonoids and Flower Color

Flower colors have played a central role in the study of flavonoids from the early recognition that anthocyanins were responsible for the various hues, through their use as visible markers for plant genetic studies, to the present day where their manipulation, either by plant breeders or molecular geneticists, has had a significant impact on the horticultural community. In this section we will look at a few situations where some aspect of flavonoid chemistry or biochemistry is under investigation from a very specific commercial point of view. Some are purely analytical; others are manipulative in the extreme.

One of the most popular of garden plants is the rose. The intense interest in developing new color forms has led to the production of an almost unmanageable array of varieties (variety is used here in the general rather than the taxonomic sense). As Olivier and colleagues (1995) point out in a brief historical review of the problem, several methods have been used to study the diversity within *Rosa* including the classical comparative morphological approach (Roberts, 1977), the use of meiotic figures (Lata, 1981, 1982a,b), isozyme analysis (Kuhns and Fretz, 1978a,b), and, more recently, the use of DNA restriction fragment analysis (Hubbard *et al*., 1992). Attempts to use flavonoids as indicators of identity of and relationships among rose varieties was undertaken by Asen (1982) who used HPLC methods for quantitative identification of rose cultivars. This work was followed by studies by the group at Lyon on a number of *Rosa* hybrids (Biolley *et al*., 1992, 1993). A more recent study (Raymond *et al*., 1995) has taken the study a step further. HPLC methods were used to resolve and identify 17 rose petal flavonoids followed by multivariate statistical analysis and preparation of a graphical two-dimensional array (F1 × F2 factorial correspondence analysis). The result was a reasonably clear-cut grouping of rose varieties around the founder species of modern roses: *R. chinensis, R. gallica*, and *R. moschata*. It was thus possible to group modern varieties into closely related families and, moreover, to provide a means of estimating likely ancestors of varieties whose background is questionable or unknown. An example will serve to illustrate the power of their methods. *Rosa x alba* is considered by some authorities to have been the result of hybridization between *R. canina* and *R. gallica* both of which represent presumed ancestral gene pools. The flavonoid profiles of these two species show distinct differences with kaempferol and quercetin 3-*O*-sophorosides and quercetin 3-*O*-rhamnoside present as major components of *R. canina* while the former two compounds appear as only trace constituents at most and the latter compound as only a minor constituent of *R. gallica*. *Rosa gallica*, on the other hand, is characterized by the presence of kaempferol 3-*O*-glucoside and 3-*O*-glucuronide and quercetin 3-*O*-arabinoside as major components. The combination of all of these flavonol glycosides in *R. x alba* is taken as evidence that it combines the flavonoid genetic capabilities of both species. There are sufficient reports in the flavonoid literature suggesting that flavonoid biosynthetic capabilities are inherited in an additive fashion to consider the present hypothesis a creditable one. Other examples of the use of flavonoids as diagnostic indicators of hybridization were discussed in Chapter Three.

In addition to the practical advantage of developing a method that allows identification of the likely ancestors of modern rose varieties, this study also points out the critical importance of quantitative flavonoid information in discerning relationships among very closely related organisms, hybrids and cultivars in this particular situation.

Other genera of horticultural interest that have been the subject of flavonoid analysis include the Transvaal daisy (*Gerbera jamesonii*, Asteraceae). Asen (1984) studied 18 varieties in detail showing that pelargonidin and cyanidin 3-*O*-malonylglucosides were the major pigments that contributed to flower color. The unacylated anthocyanins were present as minor components. Accompanying the anthocyanins were apigenin and luteolin 7-*O*- and 4′-*O*-glucosides, kaempferol and quercetin

3-*O*- and 4′-*O*-glucosides, and their respective malonyl derivatives. In view of subsequent studies by many other workers (see section on anthocyanins in Chapter Two) the flavones and flavonols in *Gerbera* are almost certainly involved in copigment formation and thus make very significant contributions to the flower color variation seen in this species.

The coupling of detailed knowledge of flavonoid genetics with sophisticated manipulations of the genetic material itself has made it possible to modify the flavonoid biosynthetic pathway in such a way that plants can be induced to produce novel compounds. Even more drastically, the production of certain compounds can be suppressed completely. Two advantages accrue to plant breeders who take advantage of these techniques. First, it is possible to place the normal range of colors of a plant in the best possible genetic background, which might involve such features as cold hardiness, stature, or disease resistance. And secondly, it becomes possible to engineer novel flower colors (Mol *et al.*, 1989a,b; Courtney-Gutterson, 1993). An example of this approach involves altering the flavonoid biosynthetic capacities of the common petunia. The dihydroflavonol reductase (DFR) of petunia has the peculiar property of not being able to accommodate dihydrokaempferol as substrate while being able to reduce both dihydroquercetin and dihydromyricetin. This means that natural, or wild-type, petunia can not make pelargonidin-based anthocyanins but can make cyanidin derivatives (from dihydroquercetin) and delphinidin derivatives (from dihydromyricetin). Petunia was transformed with the DNA coding for the dihydroflavonol reductase gene (*DFR*) from corn, which yields an enzyme that does accept dihydrokaempferol as substrate. The result was a line of transformed petunias that exhibited orange-red flowers unlike any known from non-transformed "normal" petunias (Forkmann, 1989, 1991; Linn *et al.*,1990; Meyer *et al.*, 1987, 1989, 1992).

Another example of engineered color comes from work of Markham (1996) on the genetic transformation of lisianthus (*Eustoma grandiflorum*, Gentianaceae). Normally, this plant produces anthocyanins with galactose at position-3 and glucose at position-5 yielding flowers that range from white to shades of cream, pink, mauve, and purple. Transformation with DNA for flavonoid-3-*O*-glucosyltransferase: UDP-glucose from the snapdragon (*Antirrhinum majus*) yielded a lisianthus in which about 33% of the total petal pigments were new compounds, all of which had glucose attached at position-3. The new compounds were identified as delphinidin 3-*O*-β-D-(6-*O*-α-L-rhamnosylglucoside)-5-*O*-β-D-[6-*E*(and *Z*)-*p*-coumaroyl]glucoside and the diglycoside lacking rhamnose. The transformed plants exhibited shades of color not known from the natural species.

It is also possible to "alter" flower color by eliminating it all together. This situation exists in Nature, of course, being caused by the absence of functional chalcone synthase (CHS), or some other enzyme on the pathway leading to anthocyanidin biosynthesis. The same effect can be achieved experimentally by transforming a plant in such a way that the CHS doesn't function normally, e.g., by transformation with sense or anti-sense CHS constructs (van der Krol *et al.*, 1988, 1990). In some instances white flowers result, while in others there is a low level of pigmentation owing to the natural existence of duplicate copies of the CHS gene in the plant. In the following example, yellow pollen was eliminated from a line of petunias by getting rid of the flavonoid that accumulated and was responsible for the color. The yellow-pollen line had recessive alleles for chalcone isomerase (CHI) which means, of course, that chalcone accumulates yielding the pigmented tissue. Transformation of the yellow-pollen line with DNA from a wild-type petunia resulted in synthesis of chalcone isomerase with consequent conversion of chalcone to flavanone, which is colorless (van Tunen *et al*., 1990).

In a number of plants, perhaps a majority, flower color requires the interaction of anthocyanins with flavone or flavonol glycoside co-pigments. Holton and coworkers (1993b) transformed petunia in such a way that the flavonol synthase gene no longer functioned normally. The result was a change in color of the flowers from red to magenta directly attributable to the absence of the co-pigment molecules. As we saw in Chapter Two, the structural forms in which an anthocyanin may exist can also be influenced by the pH of the vacuolar medium. Chuck and coworkers (1993) have shown that modification of flower color can be achieved by altering the genetic control of vacuolar pH.

REFERENCES

Abe, F., Iwase, Y., Yamauchi, T., Yahara, S., and Nohara, T. (1995) Flavonol sinapoyl glycosides from leaves of *Thevetia peruviana*. *Phytochemistry*, **40**, 557–581.

Abou-Zaid, M. M., Beninger, C. W., Arnason, J. T., and Nozzolillo, C. (1993) The effect of one flavone, two catechins and four flavonols on mortality and growth of the European corn borer. *Biochem. Syst. Ecol.*, **21**, 415–420.

Abou-Zaid, M., Dumas, M., Chauret, D., Watson, A., and Thompson, D. (1997) *C*-Methyl flavonols from the fungus *Colletotrichum dematium* f. sp. *epilobii*. *Phytochemistry*, **45**, 957–961.

Abraham, R. J., Takagi, S., Rosenstein, R. D., Shiono, R., Wagner, H., Hörhammer, L., Seligmann, O., and Farnsworth, N. R. (1970) The structure of silydianin, an isomer of silymarin (silybin), by X-ray analysis. *Tet. Lett.*, 2675–2678.

Adams, R. P. (1972) Numerical analyses of some common errors in chemosystematics. *Brittonia*, **24**, 9–21.

Adams, R. P. (1974) On "numerical chemotaxonomy" revisited. *Taxon*, **23**, 336–338.

Adekenov, S. M., Aituganov, K. A., and Golovtsov, N. I. (1987) Artausin, a new sesquiterpene lactone from *Artemisia austriaca*. *Khim. Prir. Soedin.*, 148–149.

Afanas'ev, I. B., Ostrachovitch, E. A., Abramova, N. E., and Korkina, L. G. (1995) Different antioxidant activities of bioflavonoid rutin in normal and iron-overloading rats. *Biochem. Pharmacol.*, **50**, 627–635.

Afifi, M. S. A., Ahmed, M. M., Pezzuto, J. M., and Kinghorn, A. D. (1993) Cytotoxic flavonolignans and flavones from *Verbascum sinaiticum* leaves. *Phytochemistry*, **34**, 839–841.

Aft, H. (1965) Chemistry of dihydroquercetin. II. Reaction of partially acetylated polyhydroxyflavanones with N-halosuccinimides. *J. Org. Chem.*, **30**, 897–901.

Aguilera, A., Alcantara, A., Marinas, J. M., and Siniestra, J. V. (1987) Barium hydroxide as the catalyst in organic reactions. Part XIV. Mechanism of the Claisen-Schmidt condensation. *Can. J. Chem.*, **65**, 1165–1171.

Ahmad, M. (1986) Naturally occurring acetoside from *Buddleja davidii*. *J. Pharm. Univ. Karachi*, **4**, 65–68.

Ahmad, S. and Razaq, S. (1971) A new approach to the synthesis of symmetrical biflavones. *Tet. Lett.*, **48**, 4633–4636.

Ahmad, S. and Razaq, S. (1976) New synthesis of biflavones of cupressuflavone series. *Tetrahedron*, **32**, 503–506.

Ahmad, V. U. and Ismail, N. (1991) 5-Hydroxy-3,6,7,2′,5′-pentamethoxyflavone from *Inula grantioides*. *Phytochemistry*, **30**, 1040–1041.

Ahmad, V. U., Shah, M. G., Mohammad, F. V., Ismail, N., and Noorwala, M. (1991) Macrophylloside, a flavone glucoside from *Primula macrophylla*. *Phytochemistry*, **30**, 4206–4208.

Ahmad, V. U., Shah, M. G., Noorwala, M., and Mohammad, F. V. (1992) Isolation of 3,3′-dihydroxychalcone from *Primula macrophylla. J. Nat. Prod.*, **55**, 956–958.

Ahmed, M., Datta, B. K., and Rouf, A. S. S. (1990) Rotenoids from *Boerhaavia repens. Phytochemistry*, **29**, 1709–1710.

Ahmed, M., Datta, B. K., Rouf, A. S. S., and Jakupovic, J. (1991) Flavone and α-santalene derivatives from *Polygonum flaccidum. Phytochemistry*, **30**, 3155–3156.

Airy-Shaw, H. K. (1964) Diagnoses of new families, new names, etc. for the seventh edition of Willis's "Dictionary". *Kew Bulletin*, **18**, 249–273.

Ali, M. A. and Kagan, J. (1974) The biosynthesis of flavonoid pigments: on the incorporation of phloroglucinol and phloroglucinyl cinnamate into rutin in *Fagopyrum esculetum. Phytochemistry*, **13**, 1479–1482.

Ali, S. M. and Ilyas, M. (1986) Biomimetic approach to biflavonoids: Oxidative coupling of 2′-hydroxychalcones with iodine in alkaline methanol. *J. Org. Chem.*, **51**, 5415–5417.

Alias, Y., Awang, K., Hadi, A. H. A., Thoison, O., Sévenet, T., and Pais, M. (1995) An antimitotic and cytotoxic chalcone from *Fissistigma lanuginosum. J. Nat. Prod.*, **58**, 1160–1166.

Allais, D. P., Chulia, A. J., Kaouadji, M., Simon, A., and Delage, C. (1995) 3-Desoxycallunin and 2″-acetylcallunin, two minor 2,3-dihydroflavonoid glucosides from *Calluna vulgaris. Phytochemistry*, **39**, 427–430.

Allan, H. H. (1961) *Flora of New Zealand*, Government Printer, Wellington.

Almtorp, G. T., Bachmann, T. L., and Torssell, K. B. G. (1991) Syntheses of flavonodis via the isoxazoline route. *Acta Chem. Scand.*, **45**, 212–215.

Alston, R. E. (1967) Biochemical systematics. In *Evolutionary Biology*, T. Dobzhansky, M. K. Hecht, and W. C. Steere, eds., Appleton-Century-Crofts, New York, 197–305.

Alston, R. E. and Turner, B. L. (1959) Applications of paper chromatography to systematics: Recombination of parental biochemical components in a *Baptisia* hybrid population. *Nature* (*London*), **184**, 285–286.

Alston, R. E. and Turner, B. L. (1962) New techniques in analysis of complex natural hybridization. *Proc. Nat. Acad. Sci., U.S.A.*, **48**, 130–137.

Alston, R. E. and Turner, B. L. (1963a) *Biochemical Systematics*, Prentice-Hall, Englewood Cliffs, New Jersey.

Alston, R. E. and Turner, B. L. (1963b) Natural hybridization among four species of *Baptisia* (Leguminosae). *Amer. J. Bot.*, **50**, 159–173.

Amrhein, N. (1979) Biosynthesis of cyanidin in buckwheat hypocotyls. *Phytochemistry*, **18**, 585–589.

Andersen, W. K., Omar, A. A., and Cristensen, S. B. (1987) Isorhamnetin 3-(2,6-dirhramnosylgalactoside)-7-rhamnoside and 3-(6-rhamnosylgalactoside)-7-rhamnoside from *Rhazya stricta. Phytochemistry*, **26**, 291–294.

Andersen, O. M. and Fossen, T. (1995) Anthocyanins with an unusual acylation pattern from stem of *Allium victorialis. Phytochemistry*, **40**, 1809–1812.

Anthonsen, T., Falkenberg, I., Laake, M., Midelfart, A., and Mortensen, T. (1971) Some unusual flavonoids from *Myrica gale L. Acta Chem. Scand.*, **25**, 1929–1930.

Araki, Y. and Cooper-Driver, G. (1993) Changes during gametophyte development in *Matteuccia struthiopteris* from the cinnamte to the chalcone synthase pathway. *Phytochemistry*, **34**, 83–84.

Arber, A. (1938) *Herbals*, Cambridge University Press, Cambridge.

Arisawa, M., Shiojima, M., Bai, H., Hayashi, T., Tezuko, Y., Taga, T., Miwa, Y., Kikuchi, T., and Morita, N. (1992) Chrysograyanone, a novel chromone derivative from *Chrysosplenium grayanum* Maxim. *Tet. Lett.*, **33**, 5977–5980.

Arnason, J. T., Mata, R., and Romeo, J. T. (1995) Phytochemistry of Medicinal Plants. Recent Advances in Phytochemistry, Plenum Press, New York, 364.

Arnone, A., Camarda, L., Merlini, L., Nasini, G., and Taylor, D. A. H. (1981) Isoflavonoid constitutents of the West African red wood *Baphia nitida*. *Phytochemistry*, **20**, 799–801.

Arriaga-Giner, F. J., Wollenweber, E., and Hradetzky, D. (1986) New flavonoids from the exudate of *Baccharis bigelovii* (Asteraceae). *Z. Naturforsch.*, **41c**, 946–948.

Arya, R., Babu, V., Ilyas, M., and Nasim, K. T. (1992) Myricetin 3′-rhamnoside-3-galactoside from *Buchanania lanzan* (Anacardiaceae). *Phytochemistry*, **31**, 2569–2570.

Asakawa, Y., Toyota, M., and Takemoto, T. (1978) Seven new bibenzyls and a dihydrochalcone from *Radula variabilis*. *Phytochemistry*, **17**, 2005–2010.

Asen, S. (1982) Identification of flavonoid chemical markers in roses and their High Pressure Liquid Chromatographic resolution and quantitation for cultivar identification. *J. Amer. Soc. Hort. Sci.*, **107**, 744–750.

Asen, S. (1984) HPLC analysis of flavonoid chemical markers in petals from Gerbera flowers as an adjunct for cultivar and germ plasm identification. *Phytochemistry*, **23**, 2523–2526.

Asen, S., Norris, K. H., and Stewart, R. N. (1971) Effect of pH and concentration of the anthocyanin-flavonol copigment complex on the color of 'Better times' roses. *J. Amer. Soc. Hort. Sci.*, **96**, 770–773.

Asen, S., Stewart, R. N., and Norris, K. H. (1977) Anthocyanin and pH involved in the color of 'Heavenly Blue' morning glory. *Phytochemistry*, **16**, 1118–1119.

Atsatt, P. R. (1988) Are vascular plants 'inside-out' lichens? *Ecology*, **69**, 17–23.

Atsatt, P. R. (1991) Fungi and the origin of land plants. In *Symbiosis as a Source of Evolutionary Innovation*, L. Margulis and R. Fester (eds.), MIT press, Cambridge, MA, 301–315.

Auf'mkolk, M., Koehrle, J., Hesch, R.-D., and Cody, V. (1986) Inhibition of rat liver iodothyronine deiodinase. Interaction of aurones with the iodothyronin ligand-binding site. *J. Biol. Chem.*, **261**, 11623–11630.

Augustyn, J. A. N., Bezuidenhoudt, B. C. B., Swanepoel, A., and Ferreira, D. (1990) Enantioselective synthesis of flavonoids. Part 2. Poly-oxygenated α-hydroxy-dihydrochalcones and circular dichroic assessment of their absolute configuration. *Tetrahedron*, **46**, 4429–4442.

Averett, J. E., and Bohm, B. A. (1986) Eight flavonol glycosdies in *Pyrola* (Ericaceae). *Phytochemistry*, **25**, 1995–1996.

Ayabe, S., Udagawa, A., and Furuya, T. (1988a) Stimulation of chalcone synthase activity by yeast extract in cultured *Glycyrrhiza echinata* cells and 5-deoxyflavanone formation by isolated protoplasts. *Plant Cell. Rep.*, **7**, 35–38.

Ayabe, S., Udagawa, A., and Furuya, T. (1988b) NAD(P)H-dependent 6′-deoxychalcone synthase activity in *Glycyrrhiza echinata* cells induced by yeast extract. *Arch. Biochem. Biophys.*, **261**, 458–462.

Baek, N.-I., Chung, M.-S., Shamon, L., Kardono, B. S., Tsauri, S., Padmawinata, K., Pezzuto, J. M., Soejarto, D. D., and Kinghorn, A. D. (1993) Selligueain A, a novel highly sweet proanthocyanidin from the rhizomes of *Selliguea feei*. *J. Nat. Prod.*, **56**, 1532–1538.

Baetcke, K. P. and Alston, R. E. (1968) The composition of a hybridizing population of *Baptisia sphaerocarpa* and *Baptisia leucophaea*. *Evolution*, **22**, 157–165.

Bailey, D. G., Arnold, J. M. O., and Spence, J. D. (1994) Grapefruit juice and drugs: How significant is the interaction? *Clinical Pharmacokinetics*, **26**, 91–98.

Bajaj, K. L., De Luca, V., Khouri, H., and Ibrahim, R. K. (1983) Purification and properties of flavonol ring-B glucosyltransferase from *Chrysosplenium americanum*. *Plant Physiol.*, **72**, 891–896.

Baker, W. and Ollis, W. D. (1956) Developments in the synthesis of isoflavones. *Sci. Proc. Royal Dub. Soc.*, **27**, 119–128.

Baker, W., Chaddenson, J., Harborne, J. B., and Ollis, W. D. (1953a) A new synthesis of isoflavones. Part I. *J. Chem. Soc.*, 1852–1860.

Baker, W., Harborne, J. B., and Ollis, W. D. (1953b) A new synthesis of isoflavones. Part II. 5 : 7 : 2′-trihydroxyisoflavone. *J. Chem. Soc.*, 1860–1864.

Baker, W., Finch, A. C. M., Ollis, W. D., and Robinson, K. W. (1959) Biflavonyls, a new class of natural product. The structure of ginkgetin, isoginkgetin and sciadopitysin. *Proc. Chem. Soc.*, 91–92.

Bakhtiar, A., Gleye, J., Moulis, C., and Fouraste, I. (1994) *O*-Glycosyl-*C*-glycosylflavones from *Galipea trifoliata*. *Phytochemistry*, **35**, 1593–1594.

Bakker, J., Bridle, P., Honda, T., Kuwano, H., Saito, N., Terahara, N., and Timberlake, C. F. (1997) Identification of an anthocyanin occurring in some red wines. *Phytochemistry*, **44**, 1375–1382.

Balas, L. and Vercauteren, J. (1994) Extensive high-resolution reverse 2D NMR analysis for the structural elucidation of procyanidin oligomers. *Mag. Res. Chem.*, **32**, 386–393.

Balas, L., Vercauteren, J., and Laguerre, M. (1995) 2D NMR Structure elucidation of proanthocyanidins: the special case of the catechin-(4a-8)-catechin-(4a-8)-catechin trimer. *Mag. Res. Chem.*, **33**, 85–94.

Ballester, A., Vieitez, A. M., and Vieitez, E. (1979) The allelopathic potential of *Erica australis* L. and *E. arborea* L. *Bot. Gazz.*, **140**, 433–436.

Ballio, A., Dittrich, S., and Marini-Betolo, G. B. (1953) The constitution of the aglycone of cernoside, the yellow pigment of *Oxalis cernua*. *Gazz. Chim. Ital.*, **83**, 224–230.

Balza, F., Jamieson, L., and Towers, G. H. N. (1985) Chemical constituents of the aerial parts of *Artemisia dracunculus*. *J. Nat. Prod.*, **48**, 339–340.

Balza, F., Muir, A. D., and Towers, G. H. N. (1985) 3′-Hydroxypsilotin, a minor phenolic glycoside from *Psilotum nudum*. *Phytochemistry*, **24**, 529–531.

Barber, G. A. (1962) Enzymatic glycosylation of quercetin to rutin. *Biochemistry (A.C.S.)*, **1**, 463–468.

Barrett, M. L., Scutt, A. M., and Evans, F. J. (1986) Cannflavin A and B, prenylated flavones from *Cannabis sativa*. *Experientia*, **42**, 452–453.

Barrett, S. C. H. (1987) Mimicry in Plants. *Scientific American*, 76–83.

Barron, D. and Ibrahim, R. K. (1988) Hydrochloric acid and aryl-sulphatase as reagents for UV-spectral detection of 3- and 4′-sulphated flavonoids. *Phytochemistry*, **27**, 2335–2338.

Barron, D., Varin, L., Ibrahim, R. K., Harborne, J. B., and Williams, C. A. (1988) Sulphated flavonoids – an update. *Phytochemistry*, **27**, 2375–2395.

Bar-Yosef, O., and Kislev, M. (1989) Early farming communities in the Jordan Valley. In *The Evolution of Plant Exploitation*, D. Harris and G. C. Hillman (eds.), Unwin Hyman, London, 621–631.

Barz, W. and Grisebach, H. (1966) Significance of 3,4′,5,7-tetrahydroxyflavone (dihydrokaempferol) for the biosynthesis of isoflavones. *Z. Naturforsch.*, **21b**, 47–52.

Bass, R. J. (1976) Synthesis of chromones by cyclization of 2-hydroxyphenyl ketones with boron trifluoride-diethyl ether and methanesulfonyl chloride. *J. Chem. Soc., Chem. Commun.*, 78–79.

Bate-Smith, E. C. (1954) Leuco-anthocyanins. 1. Detection and identification of anthocyanins formed from leuco-anthocyanins in plant tissues. *Biochem. J.*, **58**, 122–125.

Bate-Smith, E. C. (1962) The phenolic constituents of plants and their taxonomic significance. I. Dicotyledons. *Bot. J. Linnean Soc.*, **58**, 95–173.

Bate-Smith, E. C. (1964) Paper chromatography of phenolics. In *Methods in Polyphenol Chemistry*, J. B. Pridham (ed.), The Macmillan Company, New York, 73–79.

Bate-Smith, E. C. and Geissman, T. A. (1951) Benzalcoumaranones. *Nature* (*London*), **167**, 688.

Bate-Smith, E. C. and Richens, R. H. (1973) Flavonoid chemistry and taxonomy of *Ulmus*. *Biochem. Syst.*, **1**, 141–146.

Batschauer, A., Rocholl, M., Kaiser, T., Nagatani, A., Furuya, M., and Schäfer, E. (1996) Blue and UV-A light-regulated *CHS* expression in *Arabidopsis* independent of phytochrome A and phytochrome B. *The Plant Journal*, **9**, 63–69.

Baudouin, G., Tillequin, F., Koch, M., Vuilhorgne, M., Lallemand, J.-Y., and Jacquemin, H. (1983) Isolement, structure et synthèse de la vochysine, pyrrolidinoflavanne de *Vochysia guaianensis*. *J. Nat. Prod.*, **46**, 681–687.

Baylor, N. W., Fu, T., Yan, Y. D., and Ruscetti, F. W. (1992) Inhibition of human T-cell leukemia virus by the plant flavonoid baicalin (7-glucuronic acid, 5,6-dihydroxyflavone). *J. Infectious Diseases*, **165**, 433–437.

Beale, G. H. (1940) The genetics of *Verbena*. I. *J. Genet.*, **40**, 337–358.

Beale, G. H., Price, J. R., and Scott-Moncrieff, R. (1940) The genetics of *Verbena*. II. Chemistry of the flower color variations. *J. Genet.*, **41**, 65–74.

Beart, J. E., Lilley, T. H., and Haslam, E. (1985) Polyphenol interactions. Part 2. Covalent binding of procyanidins to protein during acid-catalysed decomposition; observations on some polymeric proanthocyanidins. *J. Chem. Soc., Perkin Trans. 2*, 1439–1443.

Beerhues, L. and Wiermann, R. (1988) Chalcone synthases from spinach (*Spinacia oleracea* L.). I. Purification, peptide patterns, and immunological properties of different forms. *Planta*, **173**, 532–543.

Beerhues, L., Robenek, H., and Wiermann, R. (1988) Chalcone synthases from spinach (*Spinacia* oleracea L.). II. Immunofluorescence and *immunogold* localization. *Planta*, **173**, 544–553.

Beerhues, L., Forkmann, G., Schöpker, H., Stotz, G., and Wiermann, R. (1989) Flavanone 3-hydroxylase and dihydroflavonol oxygenase activities in anthers of *Tulipa*. The significance of the tapetum fraction in flavonoid metabolism. *J. Plant Physiol.*, **133**, 743–746.

Begley, M. J., Crombie, L., Rossiter, J., Sanders, M., and Whiting, D. A. (1986) Prochirality transfer in the enzymic conversion of (*E*)-[4′-^{13}C]-(6a*S*,12a*S*)-rot-2′-enoic acid into (6a*S*,12a*S*)-deguelin: restricted stereoselectivity in an electrocyclisation. *J. Chem. Soc., Chem. Commun.*, 353–356.

Behnke, G.-D. (1976) Ultrastructure of sieve-element plastids in Caryophyllales (Centrospermae), evidence for the delimitation and classification of the order. *Plant Syst. Evol.*, **126**, 31–54.

Beld, M. G. H. M., Martin, C., Huits, H., Stuitje, A. R., and Gerats, A. G. M. (1989) Flavonoid synthesis in *Petunia*: partial characterisation of dihydroflavonol 4-reductase genes. *Plant Mol. Biol.*, **13**, 491–502.

Beltrami, E., de Bernardi, M., Fronza, G., Mellerio, G. Vidari, F., and Vita-Finzi, P. (1982) Coatline A and B, two *C*-glucosyl-α-hydroxydihydrochalcones from *Eysenhardtia polystachya*. *Phytochemistry*, **21**, 2931–2933.

Bendz, G., Martensson, O., and Terenius, L. (1962) Moss pigments. I. The anthocyanins of *Bryum cryophilum* O. Mart. *Acta Chem. Scand.*, **16**, 1183–1190.

Bendz, G. and Santesson, J. (1973) *Chemistry in Botanical Classification*, Academic Press, New York.

Bennett, J. P., Gomperts, B. D., and Wollenweber, E. (1981) Inhibitory effects of natural flavonoids on secretion from mast cells and neutrophils. *Arzneimittel-Forsch.*, **31**, 433–437.

Bennett, M., Burke, A. J., and O'Sullivan, W. I. (1996) Aspects of the Algar-Flynn-Oyamada (AFO) reaction. *Tetrahedron*, **52**, 7163–7178.

Bentham, G. (1846) *Prodromus*, **10**, 432.

Benveniste, I., Gabriac, B., and Durst, F. (1986) Purification and characterization of the NADPH-cytochrome P-450 (cytochrome c) reductase from higher-plant microsomal fraction. *Biochem. J.*, **235**, 365–373.

Beyerinck, M. W. (1885) Die Galle von *Cecidomyia poae* an *Poa nemoralis*. Entstehung normaler Wurzeln in Folge der Wirkung einer Gallenthieres. *Bot. Ztg., Leipzig*, **43**, 305–315, 321–331. (Cited by Onslow, 1925.)

Bezuidenhout, S. C., Bezuidenhoudt, B. C. B., and Ferreira, D. (1988) α-Hydroxydihydrochalcones and related 1,3-diarylpropan-2-ones from *Xanthocercis zambesica*. *Phytochemistry*, **27**, 2329–2334.

Bhandari, P., Van Bruggen, N., Crombie, L., and Whiting, D. A. (1989) Mechanism and stereochemistry of the enzymic conversion of prenyl to chromen structures, as effected by deguelin cyclase. *J. Chem. Soc., Chem. Commun.*, 982–984.

Bhandari, P., Crombie, L., Daniels, P., Holden, I., Van Bruggen, N., and Whiting, D. A. (1992) Biosynthesis of the A/B/C/D-ring system of the rotenoid amorphigenin by *Amorpha fruticosa* seedlings. *J. Chem. Soc., Perkin Trans. 1*, 839–849.

Bidhendi, G. N. and Bannerjee, N. R. (1989) Syntheses of 5,6,7,4′-tetramethoxyflavanone and 5,6,7-trimethoxy-3′,4′-methylenedioxyflavone. *Ind. J. Chem.*, **28B**, 352–353.

Bilia, A. R., Catalano, S., De Simone, F., Morelli, I., and Pizza, C. (1991) An acetylated flavanone glucoside from leaves of *Pyracantha coccinea*. *Phytochemistry*, **30**, 3830–3831.

Biolley, J. P., Jay, M., and Barbe, J. P. (1992) Chemometric approach (flavonoids) in an automatic recognition of modern rose cultivars. *Biochem. Syst. Ecol.*, **20**, 697–705.

Biolley, J. P., Jay, M., and Viricel, M. R. (1993) Flavonoid diversity and metabolism in 100 *Rosa* x *hybrida* cultivars. *Phytochemistry*, **35**, 413–419.

Birch, A. J. (1957) Biosynthetic relations of some natural phenolic and enolic compounds. *Advances in the Chemistry of Organic Natural Products*, **14**, 186–216.

Birch, A. J. (1962) Biosynthesis of flavonoids and anthocyanins. In *The Chemistry of Flavonoid Compounds*, T. A. Geissman (ed.), The Macmillan Company, New York, 618–625.

Birch, A. J. and Donovan, F. W. (1953) Studies in relation to biosynthesis. I. Some possible routes to derivatives of orcinol and phloroglucinol. *Australian J. Chem.*, **6**, 360–368.

Birch, A. J., Clark-Lewis, J. W., and Robertson, A. V. (1957) Relative and absolute stereochemistry of catechins and epicatechins. *J. Chem. Soc.*, 3586–3594.

Birch, A. J., Donovan, F. W., and Moewus, F. (1953) Biogenesis of flavonoids in *Chlamydomonas eugametuos*. *Nature* (*London*), **172**, 902.

Bird, A. E. and Marshall, A. C. (1969) Structure of chlorflavonin. *J. Chem. Soc., C*, 2418–2420.

Blank, F. (1947) The anthocyanin pigments of plants. *The Botanical Reviews*, **13**, 241–317.

Blasdale, W. C. (1945) Composition of the solid secretion produced by *Primula denticulata*. *J. Amer. Chem. Soc.*, **67**, 491–493.

Blaskó, G., Xun, L., and Cordell, G. A. (1988) Studies in the Thymelaeaceae. V. 2′-Hydroxyflavone from *Daphnopsis sellowiana*: Isolation and synthesis. *J. Nat. Prod.*, **51**, 60–65.

Blechert, S., Brodshelm, W., Hölder, S., Kammerer, L., Kutchan, T. M., Mueller, M. J., Xia, Z.-Q., and Zenk, M. H. (1995) The octadecanoid pathway: signal molecules for the regulation of secondary pathways. *Proc. Natl. Acad. Sci. USA*, **92**, 4099–4105.

Bless, W. and Barz, W. (1988) Isolation of pterocarpan synthase, the terminal enzyme of pterocarpan phytoalexin biosynthesis in cell suspension cultures of *Cicer arietinum*. *FEBS Lett.*, **235**, 47–50.

Bliss, E. D. (1989) Using tannins to produce leather. In *Chemistry and Significance of Condensed Tannins*, R. W. Hemingway and J. J. Karchesy (eds.), Plenum Press, New York, 493–502.

Bloom, M. and Vickerey, R. K. (1973) Pattern partitioning of floral flavonoids in the *Mimulus luteus* complex. *Phytochemistry*, **12**, 165–167.

Bloor, S. J. (1995) An antimicrobial kaempferol-diacyl-rhamnoside from *Pentachondra pumila*. *Phytochemistry*, **38**, 1033–1035.

Blume, D. E., Jaworski, J. G., and McClure, J. W. (1979) Uridinediphosphaeglucose: isovitexin 7-*O*-glucosyltransferase from barley protoplasts: subcellular localization. *Planta*, **146**, 199–202.

Boeswinkel, F. D. (1984) Ovule and seed structure in Datiscaceae. *Acta Bot. Neerlandica*, **33**, 419–429.

Bohlmann, F. and Ates (Gören), N. (1984) Three prenylated flavonoids *from Helichrysum athrixiifolium. Phytochemistry*, **23**, 1338–1341.

Bohlmann, F. and Misra, L. N. (1984) Weitere Phloroglucin- und α-Pyrone-Derivate aus *Helichrysum* Arten. *Planta Medica*, **50**, 271.

Bohlmann, F. and Suwita, A. (1979) Weitere Phloroglucin-Derivate aus *Helichrysum*-Arten. *Phytochemistry*, **18**, 2046–2049.

Bohlmann, F., Mahanta, P. K., and Zdero, C. (1978) Neue Chalkon-Derivate aus Sudafrikanischen *Helichrysum*-Arten. *Phytochemistry*, **17**, 1935–1937.

Bohlmann, F., Zdero, C., Abraham, W.-R., Suwita, A., and Grenz, M. (1980) Neue Diterpene und neue Dihydrochalkon-Derivate sowie weitere Inhaltsstoffe aus *Helichrysum*-Arten. *Phytochemistry*, **19**, 873–879.

Bohm, B. A. (1975a) Chalcones, aurones and dihydrochalcones. In *The Flavonoids*, J. B. Harborne, T. J. Mabry, T. J., and H. Mabry (eds.), Chapman and Hall, London, 442–504.

Bohm, B. A. (1975b) Flavanones and dihydroflavonols. In *The Flavonoids*. J. B. Harborne, T. J. Mabry and H. Mabry (eds.), Chapman and Hall, London, 560–631.

Bohm, B. A. (1979) Flavonoids of *Tolmiea menziesii*. *Phytochemistry*, **18**, 1079–1080.

Bohm, B. A. (1982) The minor flavonoids. In *The Flavonoids: Advances in Research*, J. B. Harborne and T. J. Mabry (eds.), Chapman and Hall, London, 313–416.

Bohm, B. A. (1987) Intraspecific Flavonoid Variation. *The Botanical Reviews*, **53**, 197–279.

Bohm, B. A. (1988a) Flavonoid systematics of the Datiscaceae. *Biochem. Syst. Ecol.*, **16**, 151–155.

Bohm, B. A. (1988b) The minor flavonoids. In *The Flavonoids: Advances in Research since 1980*, J. B. Harborne and T. J. Mabry (eds.), Chapman and Hall, London, 329–388.

Bohm, B. A. (1989) Chalcones and Aurones. In *Methods in Plant Biochemistry. Vol. 1. Plant Phenolics*, J. B. Harborne (ed.), Academic Press, New York, 237–274.

Bohm, B. A. (1994) The minor flavonoids. In *The Flavonoids: Advances in Research since 1986*, J. B. Harborne (ed.), Chapman and Hall, London, 387–440.

Bohm, B. A. and Chan, J. (1992) Flavonoids and affinities of Greyiaceae with a discussion of the occurrence of B-ring deoxyflavonoids in dicotyledonous families. *Syst. Bot.*, **17**, 272–281.

Bohm, B. A. and Collins, F. W. (1979) Flavonoids of some species of *Chrysosplenium. Biochem. Syst. Ecol.*, **7**, 195–201.

Bohm, B. A. and Glennie, C. W. (1969) The isolation of 2′,4, 4′-trihydroxydihydrochalcone from extracts of *Viburnum davidii. Phytochemistry*, **8**, 905.

Bohm, B. A. and Stuessy, T. E. (1995) Flavonoid chemistry of Barnadesioideae (Asteraceae). *Syst. Bot.*, **20**, 22–27.

Bohm, B. A., Chalmers, G., and Bhat, U. G. (1988) Flavonoids and the relationship of *Itea* to the Saxifragaceae. *Phytochemistry*, **27**, 2651–2653.

Bohm, B. A., Donovan, L. S., and Bhat, U. G. (1985a) Flavonoids of some spcies of *Bergenia, Francoa, Parnassia* and *Lepuropetalon. Biochem. Syst. Ecol.*, **13**, 75–77.

Bohm, B. A., Nicholls, K. W., and Bhat, U. G. (1985b) Flavonoids of the Hydrangeaceae Dumortier. *Biochem. Syst. Ecol.*, **13**, 441–445.

Bohm, B. A., Herring, A., Nicholls, K. W., Bohm, L. R., and Ornduff, R. (1989) A six-year study of flavonoid distribution in a population of *Lasthenia californica* (Asteraceae). *Amer. J. Bot.*, **76**, 157–163.

Bohm, B. A., Reid, A., Devore, M., and Stuessy, T. F. (1995) Flavonoid chemistry of Calyceraceae. *Can. J. Bot.*, **73**, 1962–1965.

Bohm, B. A., Saleh, N. A. M., and Ornduff, R. (1974) The flavonoids of *Lasthenia* (Compositae). *Amer. J. Bot.*, **61**, 551–561.

Bold, H. C. (1973) *Morphology of Plants*. Harper and Rowe, New York.

Bolwell, G. P., Bell, J. N., Cramer, C. L., Schuch, W., and Dixon, R. A. (1985) L-Phenylalanine ammonia-lyase from *Phaseolus vulgaris*: Characterization and differential induction of multiple forms from elicitor-treated cell suspension cultures. *Eur. J. Biochem.*, **149**, 411–419.

Bolwell, G. P., Bozak, K., and Zimmerlin, A. (1994) Plant cytochrome P450. *Phytochemistry*, **37**, 1491–1506.

Börner, H. (1959) The apple replant problem. I. The excretion of phlorizin from apple root residues. *Contrib. Boyce Thompson Inst.*, **20**, 39–56.

Bose, P. K., Chakrabarti, P., and Sanyal, A. K. (1971) A direct conversion of chalcones to flavones. *J. Indian Chem. Soc.*, **48**, 1163–1164.

Boss, P. K., Davies, C., and Robinson, S. P. (1996) Analysis of the expression of anthocyanin pathway genes in developing *Vitis vinifera* L. cv Shiraz grape berries and the implications for pathway regulation. *Plant Physiol.*, **111**, 1059–1066.

Boyle, T. H. and Stimart, D. P. (1989) Anatomical and biochemical factors determining ray floret color of *Zinnia angustifolia*, *Z. elegans*, and their interspecific hybrids. *J. Amer. Soc. Hort. Sci.*, **114**, 499–505.

Braconnot, H. (1807) Observations sur le *phytolacca*, vulgaire raisin d'amerique. *Annales de Chimie, Paris*, **62**, 71–90. (Cited by Onslow, 1925.)

Bradley, W. and Robinson, R. (1928) A synthesis of pyrylium salts of anthocyanidin type. XVIII. The synthesis of malvidin chloride by means of *O*-benzoylphloroglucinaldehyde. *J. Chem. Soc.*, 1541–1569.

Brehm, B. G. and Krell, D. (1975) Flavonoid localization in epidermal papillae of flower petals: a specialized adaptation for ultraviolet absorption. *Science*, **190**, 1221–1223.

Bremer, K. and Jansen, R. K. (1992) A new subfamily of the Asteraceae. *Ann. Missouri Bot. Gard.*, **79**, 414–416.

Bremer, K. (1994) *Asteraceae. Cladistics & Classification*, Timber Press, Portland, Oregon.

Breslow, D. S. and Hauser, C. R. (1940) Condensations. XI. Condensations of certain active hydrogen compounds effected by boron trifluoride and aluminum chloride. *J. Am. Chem. Soc.*, **62**, 2385–2388.

Bridle, P., Loeffles, R. S. T., Timberlake, C. F., and Self, R. (1984) Cyanidin 3-malonylglucoside in *Cichorium intybus*. *Phytochemistry*, **23**, 2968–2969.

Brinkworth, R. I., Stoermer, M. J., and Fairlie, D. P. (1992) Flavones are inhibitors of HIV-1 proteinase. *Biochem. Biophys. Res. Commun.*, **188**, 631–637.

Britsch, L. (1990) Purification and characterization of flavone synthase I, a 2-oxoglutarate-dependent desaturase. *Arch. Biochem. Biophys.*, **282**, 152–160.

Britsch, L. and Grisebach, H. (1986) Purification and characterization of (2*S*)-flavanone 3-hydroxylase from *Petunia hybrida*. *Eur. J. Biochem.*, **156**, 569–577.

Britsch, L., Heller, W., and Grisebach, H. (1981) Conversion of flavanone to flavone, dihydroflavonol and flavonol with an enzyme system from cell cultures of parsley. *Z. Naturforsch.*, **36c**, 742–750.

Brown, W. H. (1938) The bearing of nectaries on the phylogeny of flowering plants. *Proc. Amer. Philosoph. Soc.*, **79**, 549–595.

Brouillard, R. and Dangles, O. (1994) Flavonoids and flower colour. In *The Flavonoids: Advances in Research since 1986*, J. B. Harborne (ed.), Chapman and Hall, London, 565–588.

Bruno, M., Savona, G., Lamartina, L., and Lentini, F. (1985) New flavonoids from *Bonannia graeca* (L.) Halscsy. *Heterocycles*, **23**, 1147–1153.

Bu, X. and Li, Y. (1996) Synthesis of exiguaflavanone K and (±) leachianone G. *J. Nat. Prod.*, **59**, 968–969.

Bucar, F., Kartnig, T., Paschek, G., Winter, E., and Schubert-Zsilavecz, M. (1995) Flavonoid glycosides from *Lycopus europaeus*. *Planta Medica*, **61**, 489.

Buchholz, J. T. and Gray, N. E. (1948) A taxonomic revision of the genus *Podocarpus*. I. The sections of the genus and their subdivisions with special reference to leaf anatom. *J. Arnold Arboretum*, **29**, 49–63.

Bunyapraphatsara, N., Blasko, G., and Cordell, G. A. (1989) Hortensin, an unusual flavone from *Millingtonia hortensis*. *Phytochemistry*, **28**, 1555–1556.

Burbulis, I. E., Iacobucci, M., and Shirley, B. W. (1996) A null mutation in the first enzyme of flavonoid biosynthesis does not affect male sterility in *Arabidopsis*. *Plant Cell*, **8**, 1013–1025.

Butler, L. G. (1989) Effects of condensed tannin on animal nutrition. In *Chemistry and Significance of Condensed Tannins*, R. W. Hemingway and J. J. Karchesy (eds.), Plenum Press, New York, 391–402.

Byrne, L. T., Cannon, J. R., Gawad, D. H., Joshi, B. S., Skelton, B. W., Toia, R. F., and White, A. H. (1982) The crystal structure of (*S*)-(-)-6-bromo-5, 7-dihydroxy-8-methyl-2-phenyl-2,3-dihydro-4*H*-1-benzopyran-4-one [(-)-6-bromo-cryptostrobin] and a C-13 N.M.R. study of cryptostrobin and related substances. Revision of the structures of the natural products lawinal, unonal, 7-*O*-methylunonal and isounonal. *Australian J. Chem.*, **35**, 1851–1859.

Caetano-Anolles, G., Crist-Estes, D. K., and Bauer, W. D. (1988) Chemotaxis of *Rhizobium meliloti* to the plant flavone luteolin requires functional nodulation genes. *J. Bacteriol.*, **170**, 3164–3169.

Caldwell, M. M., Robberecht, R., and Flint, S. D. (1983) Internal filters: Prospects for UV-acclimation in higher plants. *Physiol. Plantarum*, **58**, 445–450.

Calloway, N. O. and Green, L. D. (1937) Reactions in the presence of metallic hydrides. I. β-Unsaturated ketone formation as a side reaction in Friedel-Crafts acylations. *J. Am. Chem. Soc.*, **59**, 809–811.

Calzada, F., Mata, R., Bye, R., and Linares, E. (1990) A retrochalcone from *Anredera scandens*. *Phytochemistry*, **29**, 2737–2738.

Camarda, L., Merlini, L., and Nasini, G. (1982) Synthesis of 2′,4′-dimethoxy-3′,6,7-trihydroxy-3-isoflavene, a constituent of *Baphia nitida*. *Gazz. Chim. Ital.*, **112**, 289–291.

Camm, E. L. and Towers, G. H. N. (1973) Phenylalanine ammonia lyase. *Phytochemistry*, **12**, 961–973.

Campbell, E. O., Markham, K. R., Moore, N. A., Porter, L. J., and Wallace, J. W. (1979) Taxonomic and phylogenetic implications of comparative flavonoid chemistry of species in the family Marchantiaceae. *J. Hattori Bot. Lab.*, No. 45, 185–199.

Campos, M., Markham, K. R., and Da Cunha A. P. (1996) Quality assessment of bee-pollens using flavonoid/phenolic profiles. *Polyphenols Commun.*, **96**, 53–54.

Cannon, J. R. and Martin, P. F. (1977) The flavanones of *Agonis spathulalta* (Myrtaceae). *Australian J. Chem.*, **30**, 2099–2101.

Carballeira, A. (1980) Phenolic inhibitors in *Erica australis* L. and the associated soil. *J. Chem. Ecol.*, **6**, 593–596.

Cardona, M. L., Garcia, B., Pedro, J. R., and Sinisterra, J. F. (1990) Flavonoids, flavonolignans and a phenylpropanoid from *Onopordum corymbosum*. *Phytochemistry*, **29**, 629–631.

Carew, D. P. and Krueger, R. J. (1976) Anthocyanidins of *Catharanthus roseus* callus cultures. *Phytochemistry*, **15**, 442.

Carle, R., Dölle, B., Müller, W., and Baumeister, U. (1993) Thermospray liquid chromatography/mass spectrometry (TSP LC/MS): analysis of acetylated apigenin 7-glucosides from *Chamomilla recutita*. *Die Pharmazie*, **48**, 304–306.

Casper, B. B. and La Pine, T. R. (1984) Changes in corolla color and other floral characteristics in *Cryptantha humilis* (Boraginaceae): Cues to discourage pollinators? *Evolution*, **38**, 128–141.

Ceska, O. and Styles, E. D. (1984) Flavonoids from *Zea mays* pollen. *Phytochemistry*, **23**, 1822–1823.

Chadenson, M., Hauteville, M., and Chopin, J. (1972) Synthesis of 2,5-dihydroxy-7-methoxyflavanone, cyclic structure of the benzoyl(2,6-dihydroxy-4-methoxybenzoyl)methane from *Populus nigra* buds. *J. Chem. Soc., Chem. Commun.*, 107–108.

Chandra, S. and Babber, S. (1987) Synthesis of 5,4′-dihydroxy-7,8,3′,5′-tetramethoxyflavone and two new isomeric pentaoxygenated flavanones isolated from *Lepidium sativum & Vitex negundo*. *Ind. J. Chem.*, **26B**, 82–84.

Chantrapromma, K., Pakawatchai, C., Skelton, B. W., White, A. H., and Worapatamasri, S. (1989) 5-Hydroxy-7-methoxy-2-phenyl-4H-1-benzopyran-4-one (tectochrysin) and 2,5-dihydroxy-7-methoxy-2-phenyl-2,3-dihydro-4H-1-benzopyran-4-one. Isolation from *Uvaria rufus* and X-ray structures. *Australian J. Chem.*, **42**, 2289–2293.

Charrier, B., Leroux, C., Kondorosi, A., and Ratet, P. (1996) The expresion of alfalfa *flavanone 3-hydroxylase* promoter-*gus* fusion in *Nicotiana benthamiana* correlates with the presence of flavonoids detected *in situ*. *Plant Mol. Biol.*, **30**, 1153–1168.

Chaudhuri, S. K., Fullas, F., Wani, M. C., Wall, M. E., Tucker, J. C., Beecher, C. W. W., and Kinghorn, A. D. (1996) Two isoflavones from the bark of *Petalostemon purpurescens*. *Phytochemistry*, **41**, 945–946.

Chibber, S. S. and Sharma, R. P. (1979) 5-Hydroxy-7-methoxyisoflavone from seeds of *Derris robusta*. *Planta Medica*, **36**, 379–380.

Chmiel, E., Sütfeld, R., and Wiermann, R. (1983) Conversion of phloroglucinol-type chalcones by purified chalcone isomerase from tulip anthers and from *Cosmos* petals. *Biochem. Physiol. Pflanz.*, **178**, 139–146.

Chopin, J., Bouillant, M. L., Wagner, H., and Galle, K. (1974) Endgültige Struktur von Schaftosid aus *Silene schafta*. *Phytochemistry*, **13**, 2583–2586.

Chopin, J. and Bouillant, M. L. (1975) *C*-Glycosylflavonoids. In *The Flavonoids*, J. B. Harborne, T. J. Mabry, and H. Mabry (eds.), Chapman and Hall, London, 632–691.

Chopin, J., Hautelville, M., Joshi, B. S., and Gawad, D. H. (1978) A novel example of a natural 2,5-dihydroxyflavanone from *Unona lawii*. *Phytochemistry*, **17**, 332–334.

Chuck, G., Robbins, T., Nijjar, C., Ralston, E., Courtney-Gutterson, N., and Dooner, H. K. (1993) Tagging and cloning of a petunia flower color gene with the maize transposable element *Activator*. *Plant Cell*, **5**, 371–378.

Chung, M.-I., Lai, M.-H., Yen, L.-H., Wu, R.-R., and Lin, C.-N. (1997) Phenolics from *Hypericum geminiflorum*. *Phytochemistry*, **44**, 943–947.

Clark, R. (1984) *J.B.S. The Life and Work of J. B. S. Haldane*, Oxford University Press, Oxford.

Clark-Lewis, J. W. (1968) Flavan derivatives. XXV. Mass spectra of 3-hydroxy-flavanones, flavan-3-ols and flavan-3,4-diols. *Australian J. Chem.*, **21**, 3025–3054.

Clevenger, S. (1964) A new anthocyanidin in *Impatiens*. *Can. J. Biochem.*, **42**, 154–155.

Clifford, H. T. and Stephenson, W. (1975) *An Introduction to Numerical Classification*, Academic Press, New York.

Climent, M. J., Garcia, H., Iborra, S., Miranda, M. A., and Primo, J. (1989) Photosensitized dehydrogenation of flavanones to flavones using 2,4,6-triphenyl-pyrylium tetrafluoroborate (TPT). *Heterocycles*, **29**, 115–121.

Climent, M. J., Corma, A., Iborra, S., and Primo, J. (1995) Base catalysis for fine chemicals production: Claisen-Schmidt condensation on zeolites and hydrotal-cites for the production of chalcones and flavanones of pharmaceutical interest. *J. Catal.*, **151**, 60–66.

Cody, V., Middleton Jr., E., and Harborne, J. B. (1986) Plant Flavonoids in Biology and Medicine. In *Progress in Clinical and Biological Research*, N. Back, G. J. Brewer, V. P. Eijsvoogel, R. Grover, K. Hirschorn, S. S. Kety, S. Udenfriend, and J. W. Uhr (eds.), Alan R. Liss, Inc., New York, 592.

Cody, V., Middleton Jr., E., Harborne, J. B., and Beretz, A. (1988) Plant Flavonoids in Biology and Medicine. In *Progress in Clinical and Biological Research*, N. Back, G. J. Brewer, V. P. Eijsvoogel, R. Grover, K. Hirschorn, S. S. Kety, S. Udenfriend, and J. W. Uhr (eds.), Alan R. Liss Inc., New York.

Cody, V., Luft, J., and McCourt, M. (1990) Conformational analysis of flavonoids: crystal and molecular structure of 3′,5′-dibromo-3-methyl-6,4′-dihydroxyflavone (1 : 2) triphenylphosphine complex. *Struct. Chem.*, **2**, 601.

Cody, V. and Luft, J. R. (1994) Conformational analysis of flavonoids: crystal and molecular structure of morin hydrate and myricetin (1:2) triphenylphosphine oxide complex. *J. Mol. Struct.*, **317**, 89–97.

Coe, E. H., McCormick, S. M., and Modena, S. A. (1981) White pollen in maize. *J. of Heredity*, **72**, 318–320.

Coen, E. S., Carpenter, R., and Martin, C. (1986) Transposable elements generate novel patterns of gene expression in *Antirrhinum majus*. *Cell*, **47**, 285–296.

Cole, J. R., Torrance, S. J., Wiedhopf, R. M., Arora, S. K., and Bates, R. B. (1976) Uvaretin, a new antitumor agent from *Uvaria acuminata* (Annonaceae). *J. Org. Chem.*, **41**, 1852–1855.

Collie, J. N. and Myers, W. S. (1893) The formation of orcinol and other condensation products from dehydracetic acid. *J. Chem. Soc.*, 122–128.

Collie, J. N. (1893) The production of naphthalene derivatives from dehydracetic acid. *J. Chem. Soc.*, 329–337.

Collie, J. N. and Hilditch, T. P. (1907) An isomeric change of dehydracetic acid. *J. Chem. Soc.*, 787–789.

Collie, J. N. and Chrystall, E. R. (1907) The production of orcinol derivatives from the sodium salt of ethyl acetoacetate by the action of heat. *J. Chem. Soc.*, 1802–1806.

Collie, J. N. (1907) Derivatives of multiple keten group. *J. Chem. Soc.*, 1806–1813.

Collins, F. W., Bohm, B. A., and Wilkins, C. K. (1975) Flavonol glycoside gallates from *Tellima grandiflora*. *Phytochemistry*, **14**, 1099–1102.

Collins, F. W., De Luca, V., Ibrahim, R. K., Voirin, B., and Jay, M. (1981) Polymethylated flavonols of *Chrysosplenium americanum*. I. Identification and enzymatic synthesis. *Z. Naturforsch.*, **36c**, 730–736.

Combes, G., Vassort, P., and Winternitz, F. (1970) Structure de la rubranine, chalcone isolee de l'*Aniba rosaeodora* Ducke. *Tetrahedron*, **26**, 5981–5992.

Constantinou, A., Mehta, R., Runyan, C., Rao, K., Vaughn, A., and Moon, R. (1995) Flavonoids as DNA topoisomerase antagonists and poisons: structure-activity relationships. *J. Nat. Prod.*, **58**, 217–225.

Cordell, G. A. (1995) Changing strategies in natural product chemistry. *Phytochemistry*, **40**, 1585–1612.

Corma, A., Climent, M. J., Garcia, H., and Primo, J. (1989) Design of synthetic zeolites as catalysts in organic reactions: acylation of anisole by acid chlorides or carboxylic acids over acid zeolites. *Appl. Catal.*, **49**, 109–123

Cornuz, G., Wyler, H., and Lauterwein, J. (1981) Pelargonidin 3-malonylsophoroside from the red Iceland poppy, *Papaver nudicaule*. *Phytochemistry*, **20**, 1461–1462.

Correa, J., Cervera, M. L., and Maniero, R. M. (1971) Tithonine synthesis. *Rev. Lationoamer. Quim.*, **2**, 67–68.

Courtney-Gutterson, N. (1993) Molecular breeding for color, flavor and fragrance. *Scient. Hort.*, **55**, 141–160.

Coxon, F. T., O'Neill, T. M., Mansfield, J. W., and Porter, A. E. A. (1980) Identification of three hydroxyflavan phytoalexins from daffodil bulbs. *Phytochemistry*, **19**, 889–891.

Cramer, C. L., Edwards, K., Dron, N., Liang, X., Dildine, S. L., Bolwell, G. P., Dixon, R. A., Lamb, C. J., and Schuch, W. (1989) Phenylalanine ammonia-lyase gene organization and structure. *Plant Mol. Biol.*, **12**, 367–383.

Crawford, D. J. (1975) Systematic relationships in the narrow-leaved species of *Chenopodium* of the western United States. *Brittonia*, **27**, 279–288.

Crawford, D. J. (1976) Variation in the seed protein profiles of *Chenopodium fremontii*. *Biochem. Syst. Ecol.*, **4**, 169–172.

Crawford, D. J. (1979) Flavonoid chemistry and angiosperm evolution. *The Botanical Reviews*, **44**, 431–456.

Crawford, D. J. (1990) *Plant Molecular Systematics. Macromolecular Approaches*, Wiley Interscience, New York.

Crawford, D. J. and Levy, M. (1978) Flavonoid profile affinities and genetic similarity. *Syst. Bot.*, **3**, 369–373.

Crawford, D. J. and Mabry, T. J. (1978) Flavonoid chemistry of *Chenopodium fremontii*. Infraspecific variation and systematic implications at the interspecific level. *Biochem. Syst. Ecol.*, **6**, 189–192.

Crawford, D. J. and Stuessy, T. F. (1981) The taxonomic significance of anthochlors in the subtribe Coreopsidinae (Compositae, Heliantheae). *Amer. J. Bot.*, **68**, 107–117.

Creaser, C. S., Koupai-Abyazani, M. R., and Stephanson, G. R. (1989) Capillary column gas chromatography of methyl and trimethylsilyl derivatives of some naturally occurring flavonoid aglycones and other phenolics. *J. Chromatog.*, **478**, 415–421.

Creaser, C. S., Koupai-Abyazani, M. R., and Stephenson, G. R. (1991a) Origin and control of multi-peak formation in the analysis of trimethylsilyl derivatives of flavanone aglycones by capillary column gas chromatography. *J. Chromatog.*, **586**, 323–328.

Creaser, C. S., Koupai-Abyazani, M. R., and Stephanson, G. R. (1991b) Mass spectra of trimethylsilyl derivatives of naturally occurring flavonoid aglycones and chalcones. *Org. Mass Spectrom.*, **26**, 157–160.

Creaser, C. S., Koupai-Abyazani, M. R., and Stephanson, G. R. (1992) Gas-chromatographic-mass spectrometric characterization of flavanones in citrus and grape juices. *Analyst*, **117**, 1105–1109.

Creelman, R. A. and Mullet, J. E. (1997) Biosynthesis and action of jasmonates in plants. *Annu. Rev. Plant Physiol. Mol. Biol.*, **48**, 355–381.

Crespo-Irizar, A., Fernandez, M., Perales, A., Gonzalez, A. G., and Gutierrez Ravelo, A. (1992) Crystal structure of a proanthocyanidin derivative and conformation by molecular mechanics. *Spectroscopy* (*Amsterdam*), **10**, 17–24.

Crombie, L., Green, C. L., and Whiting, D. A. (1968) Biosynthesis of rotenoids: the origin of C-6 and C-6a. *J. Chem. Soc.* (*C*), 3029–3032.

Crombie, L., Dewick, P. M., and Whiting, D. A. (1973) Biosynthesis of rotenoids. Chalcone, isoflavone, and rotenoid steps in the formation of amorphigenin by *Amorpha fruticosa* seedlings. *J. Chem. Soc., Perkin Trans. 1*, 1285–1294.

Crombie, L., Holden, I., Kilbee, G. W., and Whiting, D. A. (1982) Biosynthesis of rotenoids by *Amorpha fruticosa*: Sequence, specificity, and stereochemistry in the development of the hemiterpenoid segment. *J. Chem. Soc., Perkin Trans. 1*, 789–797.

Crombie, L. (1984) Rotenoids and their biosynthesis. *Nat. Prod. Reports*, **1**, 3–19.

Crombie, L., Rossiter, J., and Whiting, D. A. (1986) Biosynthetic origin of the 2,2-dimethylchroman ring: Formation of deguelin by a cyclase enzyme from *Tephrosia vogellii*. *J. Chem. Soc., Chem. Commun.*, 352–353.

Cronquist, A. (1981) *An Integrated System of Classification of Flowering Plants*, Columbia University Press, New York.

Crowden, R. K. (1982) Pseudobase of malvidin 3-rhamnoside-5-glucoside in *am* mutants of *Pisum sativum*. *Phytochemistry*, **21**, 2989–2990.

Cruickshank, I. A. M. and Perrin, D. R. (1960) Isolation of a phytoalexin from *Pisum sativum* L. *Nature* (*London*), **187**, 799–800.

Cruickshank, I. A. M. and Perrin, D. R. (1964) Pathological function of phenolic compounds in plants. In *Biochemistry of Phenolic Compounds*, J. B. Harborne (ed.), Academic Press, London, 511–544.

Da-Cunha, E. V. L., Dias, C., Barbarosa-Filho, J. M., and Gray, A. I. (1996) Eryvellutinone, an isoflavanone from the stem bark of *Erythrina vellutina*. *Phytochemistry*, **43**, 1371–1373.

Dagne, E. and Bekele, A. (1990) *C*-Prenylated isoflavones from *Millettia ferruginea*. *Phytochemistry*, **29**, 2679–2682.

Dahlgren, R. M. T. and Clifford, H. T. (1982) *The Monocotyledons, A Comparative Study*, Academic Press, London.

Dahlgren, R. M. T. and Rasmussen, F. N. (1983) Monocotylendon evolution: characters and phylogenetic estimation. *Evol. Biol.*, **16**, 255–395.

Dahlgren, R. M. T., Clifford, H. T., and Yeo, P. F. (1985) *The Families of the Monocotyledons*, Springer, Berlin.

Dakora, F. D. (1995) Plant flavonoids: Biological molecules for useful exploitation. *Australian J. Plant Physiol.*, **22**, 87–99.

Dale, J. A. and Mosher, H. S. (1973) Nuclear magnetic resonance enantiomer reagents. Configurational correlations via nuclear magnetic resonance shifts of diastereomeric mandelate, *O*-methylmandelate, and α-methoxy-α-trifluromethylphenylacetate (MTPAS) esters. *J. Amer. Chem. Soc.*, **95**, 512–519.

Dangl, J. L. (1992) Chalcone synthase. In *Genes Involved in Plant Defense*, T. Boller and F. Meins, Jr. (eds.), Springer, Vienna, 303–326.

Dangles, O., Saito, N., and Brouillard, R. (1993) Anthocyanin intramolecular copigment effect. *Phytochemistry*, **34**, 119–124.

Danieli, B., De Bellis, P., Carrea, G., and Riva, S. (1989) Enzyme-mediated acylation of flavonol monoglycosides. *Heterocycles*, **29**, 2061–2064.

Danieli, B., De Bellis, P., Carrea, G., and Riva, S. (1990) Enzyme-mediated regioselective acylations of flavonoid disaccharide monoglycosides. *Helv. Chim. Acta*, **73**, 1837–1845.

Danieli, B., Bertario, A., Carrea, G., Redigolo, B., Secundo, F., and Riva, S. (1993) Chemo-enzymatic synthesis of 6″-*O*-(3-arylprop-2-enol) derivatives of the flavonol glucoside isoquercitrin. *Helv. Chim. Acta*, **76**, 2981–2991.

Danieli, B. and Riva, S. (1994) Enzyme-mediated regioselective acylation of polyhydroxylated natural products. *Pure & Appl. Chem.*, **66**, 2215–2218.

Danieli, B. and Riva, S. (1996) Enzyme-mediated regioselective modification of natural glycosides. *Korean J. Med. Chem.*, **6**, 303–308.

Danieli, B., Luisetti, M., Sampognaro, G., Carrea, G., and Riva, S. (1997) Regioselective acylation of polyhydroxylated natural compounds catalyzed by *Candida antarctica* lipase B (Novozym 435) in organic solvents. *J. Mol. Catal. B: Enzymatic*, 78, **3**, 193–201.

Daniell, T., O'Hagan, D., and Edwards, R. (1997) Alfalfa cell cultures treated with a fungal elicitor accumulate flavone metabolites rather than isoflavones in the presence of the methylation inhibitor tubericidin. *Phytochemistry*, **44**, 285–291.

Darmat, A., Carnat, A.-P., Chavignon, O., Heitz, A., Wylde, R., and Lamaison, J.-L. (1995) Luteolin 7-diglucuronide, the major flavonoid compound from *Aloysia triphylla* and *Verbena officinalis*. *Planta Medica*, **61**, 490.

Das, B. and Chakravarty, A. K. (1993) Three flavone glycosides from *Gelonium multiflorum*. *Phytochemistry*, **33**, 493–496.

Das, N. P. (1989) *Flavonoids in Biology and Medicine III*, National University of Singapore, Singapore.

Daumer, K. (1958) Blumenfarben, wie sie die Bienen sehen. *Z. Vergl. Physiol.*, **41**, 49–110.

Davidson, C. (1973) An anatomical and morphological study of Datiscaceae. *Aliso*, **8**, 49–110.

Davidson, C. (1976) Anatomy of xylem and phloem of the Datiscaceae. *Los Angeles County Museum Contrib. Sci.*, **280**, 1–28.

Dawidar, A. M., Metwally, M. A., Abou-Elgahab, M., and Abdel-Mogib, M. (1989) Chemical constituents of two *Centaurea* species. *Die Pharmazie*, **44**, 735–736.

de Alleluia, I. B., Braz Filho, R., Gottlieb, O. R., Magalhaes, E. G., and Marques, R. (1978) (−)-Rubranine from *Aniba rosaeodora*. *Phytochemistry*, **17**, 517–521.

de Candolle, A. (1830) *Prodromus*, **4**, 434.

de Candolle, A. (1845) *Prodromus*, **9**, 12.

De Koninck, L. (1835) *Ann. Chem. Pharm.* **15,** 75, 258. (Cited by Shimokoriyama, 1962.)

de Laire, G. and Tiemann, F. (1893) *Chem. Ber.*, **26**, 2010. (Cited by Onslow, 1925.)

de Laubenfels, D. J. (1969) A revision of the Malesian and Pacific rainforest conifers. I. Podocarpaceae, in part. *J. Arnold Arboretum*, **50**, 274–369.

de Laubenfels, D. J. (1978) The genus *Prumnopitys* (Podocarpaceae) in Malesia. *Blumea*, **24**, 189–190.

De Luca, V. and Ibrahim, R. K. (1982) Characterization of three distinct flavonol *O*-methyltransferases from *Chrysosplenium americanum*. *Phytochemistry*, **21**, 1537–1540.

De Luca, V., Brunet, G., Kouri, H., Ibrahim, R. K., and Hrazdina, G. (1982) Flavonol 3-*O*-methyltransferase in plant tissues. *Z. Naturforsch.*, **37c**, 134–135.

De Luca, V. and Ibrahim, R. K. (1985a) Enzymatic synthesis of polymethylated flavonols in *Chrysosplenium americanum*. I. Partial purification and some properties of *S*-adenosyl-L-methionine: flavonol 3-, 6-, 7-, and 4′-*O*-methyltransferases. *Arch. Biochem. Biophys.*, **238**, 596–605.

De Luca, V. and Ibrahim, R. K. (1985b) Enzymatic synthesis of polymethylated flavonols in *Chrysosplenium americanum*. II. Substrate interaction and product inhibition studies of flavonol 3-, 6-, and 4′-O-methyltransferases. *Arch. Biochem. Biophys.*, **238**, 606–618.

de Vlaming, P. and Kho, K. F. F. (1976) 4,2′,4′,6′-Tetrahydroxychalcone in pollen of *Petunia hybrida. Phytochemistry*, **15**, 348–349.

de Whalley, C. V., Rankin, S. M., Hoult, J. R. S., Jessup, W., and Leake, D. S. (1990) Flavonoids inhibit the oxidative modification of low density lipoproteins by macrophages. *Biochem. Pharmacol.*, **39**, 1743–1750.

Dean, H. F. and Nierenstein, M. (1925) Attempts to synthesize myricetin. *J. Amer. Chem. Soc.*, **47**, 1676–1684.

Dean, F. M. (1963) *Naturally Occurring Oxygen Ring Compounds*, Butterworths, London.

Dean, F. M., Monkhe, T. V., Mulholland, D. A., and Taylor, D. A. H. (1993) An isoflavanoid from *Aglaia ferruginaea,* an Australian member of the Meliaceae. *Phytochemistry*, **34**, 1537–1539.

Dedaldechamp, F., Uhel, C., and Macheix, J.-J. (1995) Enhancement of anthocyanin synthesis and dihydroflavonol reductase (DFR) activity in response to phosphate deprivation in grape cell suspensions. *Phytochemistry*, **40**, 1357–1360.

Delle-Monache, F., Labbiento, L., Marta, M., and Lwande, W. (1986) 4β-Substituted flavans from *Tephrosia hildebrandtii. Phytochemistry*, **25**, 1711–1713.

Delle Monache, G., Botta, B., Vinciguerra, V., de Mello, J. F., and Chiappeta, A. d. A. (1996) Antimicrobial isoflavanones from *Desmodium canum. Phytochemistry*, **41**, 537–544.

Delph, L. F. and Lively, C. M. (1989) The evolution of floral color change: Pollinator attraction versus physiological constraints in *Fuchsia excorticata. Evolution*, **43**, 1252–1262.

Dement, W. A. and Raven, P. H. (1974) Pigments responsible for ultraviolet patterns in flowers of *Oenothera* (Onagraceae). *Nature* (*London*), **252**, 705–706.

Denford, K. E. (1981) Chemical subdivisions within the genus *Arctostaphylos* based on flavonoid profiles. *Experientia*, **37**, 1287–1288.

Dénarié, J. and Cullimore, J. (1993) Lipo-oligosaccharide nodualtion factors: a new class of signalling molecules mediating recognition and morphogenesis. *Cell*, **74**, 951–954.

Desrochers, A. M. and Bohm, B. A. (1995) Biosystematic study of *Lasthenia californica* (Asteraceae). *Syst. Bot.*, **20**, 65–84.

DeVore, M. L. (1994) Systematic studies of Calyceraceae, Ph.D. Dissertation, Ohio State University, Columbus, Ohio.

DeVore, M. L. and Stuessy, T. F. (1995) The time and the place of origin of the Compositae, with comments on the Calyceraceae and the Goodeniaceae. In *Advances in Compositae Systematics*, D. J. N. Hind, G. V. Pope, and C. Jeffrey (eds.), Royal Botanic Gardens, Kew, England, 23–40.

Dewick, P. M. (1982) Isoflavonoids. In *The Flavonoids: Advances in Research*, J. B. Harborne and T. J. Mabry (eds.), Chapman & Hall, London, 535–640.

Dewick, P. M. and Steele, M. J. (1982) Biosynthesis of the phytoalexin phaseolin in *Phaseolus vulgaris. Phytochemistry*, **21**, 1599–1603.

Dewick, P. M. (1988) Isoflavonoids. In *The Flavonoids: Advances in Research since 1980*, J. B. Harborne (ed.), Chapman & Hall, London, 125–209.

Dewick, P. M. (1994) Isoflavonoids. In *The Flavonoids: Advances in Research since 1986*, J. B. Harborne (ed.), Chapman & Hall, London, 117–238.

Dillon, M. O., Mabry, T. J., Besson, E., Bouillant, M. L., and Chopin, J. (1976) New flavonoids from *Flourensia cernua. Phytochemistry*, 15, 1085–1086.

Dittrich, H., Kutchan, T. M., and Zenk, M. H. (1992) The jasmonate precursor, 12-oxo-phytodienoic acid, induces phytoalexin synthesis in *Petroselinum crispum* cell cultures. *FEBS Letters*, **309**, 33–36.

Dominguez, X. A. and Torre, B. (1974) Two pentamethoxylated flavonoids from *Gymnosperma glutinosum. Phytochemistry*, **13**, 1624–1625.

Dominguez, X. A., Gutierrez, M., and Aragon, R. (1976) Isolation of baileyolin, a tumor inhibitory and antibiotic sesquiterpene lactone from *Baileya multiradiata. Planta Medica*, **30**, 356–359.

Dominguez, E., Lete, E., Villa, M.-J., Igartua, A., Sotomayer, N., Arietta, J. M., Berisa, A., Labeaga, L., Orjales, A., Germain, G., and Nastopoulos, V. (1991) Synthesis, crystal structure determination, and pharmacological activity of 7,8,3′,4′-tetramethoxyisoflavone. *J. Heterocycl. Chem.*, **28**, 1885–1889.

Donnelly, D. M. X. (1975) Neoflavonoids. In *The Flavonoids*, J. B. Harborne, T. J. Mabry, and H. Mabry (eds.), Chapman & Hall Ltd., London, 801–865.

Donnelly, D. M. X. and Sheridan, M. H. (1988) Neoflavonoids. In *The Flavonoids: Advances in Research since 1980*, J. B. Harborne (ed.), Chapman & Hall, London, 212–232.

Donnelly, D. M. X., Finet, J.-P., and Rattigan, B. A. (1993a) Organolead-mediated arylation of allyl β-ketoesters: A selective synthesis of isoflavanones and isoflavones. *J. Chem. Soc., Perkin Trans. 1*, 1729–1735.

Donnelly, D. M. X., Fitzpatrick, B. M., O'Reilly, B. A., and Finet, J.-P. (1993b) Aryllead mediated synthesis of isoflavanone and isoflavone derivatives. *Tetrahedron*, **49**, 7967–7976.

Donnelly, D. M. X. and Boland, G. (1994) Neoflavonoids. In *The Flavonoids: Advances in Research since 1986*, J. B. Harborne (ed.), Chapman & Hall, London, 239–258.

Donnelly, D. M. X. and Boland, G. M. (1995) Isoflavonoids and neoflavonoids: Naturally occurring *O*-heterocycles. *Nat. Prod. Reports*, **11**, 321–338.

Dooner, H. K. (1983) Coordinate genetic regulation of flavonoid biosynthetic enzymes in maize. *Mol. Gen. Genet.*, **189**, 136–141.

Dorn, P. S. and Bloom, W. L. (1984) Anthocyanin variation in an introgressive complex in *Clarkia. Biochem. Syst. Ecol.*, **12**, 311–314.

Doshi, A. G., Soni, P. A., and Ghiya, B. J. (1986) Oxidation of 2′-hydroxychalcones. *Ind. J. Chem.*, **25B**, 759.

Douglas, C., Hoffmann, H., Schulz, W., and Hahlbrock, K. (1987) Structure and elicitor- or u.v.-light-stimulated expression of two 4-coumarate:CoA ligase genes in parsley. *The EMBO J.*, **6**, 1189–1195.

Douglas, C. J. (1996) Phenylpropanoid metabolism and lignin biosynthesis: from weeds to trees. *Trends in Plant Science*, **1**, 171–178.

Downie, S. R. and Palmer, J. D. (1994) A chloroplast FDNA phlylogeny of the Caryophyllales based on structural and inverted repeat restriction site variation. *Syst. Bot.*, **19**, 236–252.

Drewes, S. E. (1968) Mass spectra of derivatives of some flavan-3,4-diols. *J. Chem. Soc., C.*, 1140–1148.

Drewes, S. E. and Roux, D. G. (1965) Condensed tannins. Optically active diastereoisomers of (+)-mollisacacedin by epimerization. *Biochem. J.*, **94**, 482–487.

Drewes, S. E., Taylor, C. W., and Cunningham, A. B. (1992) (+)-Afzelechin 3-rhamnoside from *Cassipourea gerrardii*. *Phytochemistry*, **31**, 1073–1075.

Dreyer, D. L. and Jones, K. C. (1981) Feeding deterency of flavonoids and related phenolics towards *Schizaphis graminum* and *Myzus persicae*: Aphid feeding deterrents in wheat. *Phytochemistry*, **20**, 2489–2493.

Dübeler, A., Voltmer, G., Gora, V., Lunderstädt, J., and Zeeck, A. (1997) Phenols from *Fagus sylvatica* and their role in defence against *Cryptococcus fagisuga*. *Phytochemistry*, **45**, 51–57.

Ducrey, B., Wolfender, J. L., Marston, A., and Hostettmann, K. (1995) Analysis of flavonol glycosides of thirteen *Epilobium* species (Onagraceae) by LC-UV and thermospray LC-MS. *Phytochemistry*, **38**, 129–137.

Duncan, T. and Stuessy, T. F. (1984) *Cladistics: Perspectives on the Reconstruction of Evolutionary History*. Columbia University Press, New York, 312.

Dübeler, A., Voltmer, G., Gora, V., Lunderstädt, J., and Zeeck, A. (1997) Phenols from *Fagus sylvatica* and their role in defence against *Cryptococcus fagisuga*. *Phytochemistry*, **45**, 51–57.

du Preez, I. C. and Roux, D. G. (1970) Novel flavan-3,4-diols from *Acacia cultriformis*. *J. Chem. Soc., C.*, 1800–1804.

Eagle, A. (1982) *Eagle's trees and shrubs of New Zealand*, Collins, Auckland.

Ebel, J. and Hahlbrock, K. (1982) Biosynthesis. In *The Flavonoids: Advances in Research*, J. B. Harborne, T. J. Mabry, and H. Mabry (eds.), Chapman & Hall, London, 641–679.

Ebel, J., Barz, W., and Grisebach, H. (1970) Biosynthesis of acacetin in *Robinia pseudoacacia*: incorporation of multiple labelled *p*-methoxycinnamic acid. *Phytochemistry*, **9**, 1529–1534.

Eisner, T., Silberglied, R. E., Aneshansley, D., Carrel, J. E., and Howland, H. C. (1969) Ultaviolet vide-viewing: the television camera as an insect eye. *Science*, **166**, 1172–1174.

Eisner, T., Eisner, M., Hyypio, P. A., Aneshanslely, D., and Silberglied, R. F. (1973) Plant taxonomy: ultraviolet patterns of flowers visible as fluorescent patterns in pressed herbarium specimens. *Science*, **179**, 486–487.

El Tohamy, S. F., Azzam, S. M., El Gohary, H. M., and Fathy, M. M. (1993) Phytochemical study of *Lonchocarpus speciosus* Bolus grown in Egypt. *Egyptian J. of Pharm. Sci.* **34**, 309–318.

El-Ansari, M. A., Barron, D., Abdalla, M. F., Saleh, N. A. M., and LeQuéré, J. L. (1991) Flavonoid constituents of *Stachys aegyptica*. *Phytochemistry*, **30**, 1169–1173.

El-Ansari, M. A., Nawwar, M. A., and Saleh, N. A. M. (1995) Stachysetin, a diapigenin-7-glucoside-*p*,*p*′-dihydroxytruxinate from *Stachys aegyptiaca*. *Phytochemistry*, **40**, 1543–1548.

El-Sayed, N. H., Abu Dooh, A. M., El-Khrisy, E. A. M., and Mabry, T. J. (1992) Flavonoids of *Cassia italica*. *Phytochemistry*, **31**, 2187.

El-Sayed, N. H., El-Khrisy, E. A. M., Khadiga, M. A., and Mabry, T. J. (1991) Flavonoids of *Conyza linifolia*. *Rev. Latinoamericana Quim.*, **22**, 89–90.

Elliott, A. J., Schreiber, S. A., Thomas, C., and Pardini, R. S. (1992) Inhibition of glutathione reductase by flavonoids. A structure-activity study. *Biochem. Pharmacol.*, **44**, 1603–1608.

Endress, R. (1972) Versuche zum Vorkommen von Chalkonen als natürliche Zwischenstufen im Biosyntheseweg der Flavonoide. *Z. Pflanzenphysiol.*, **67**, 188–194.

Engler, A. (1891) Saxifragaceae. In *Die natürlichen Pflanzenfamilien*, A. Engler and K. Prantl (eds.), Engelmann, Leipzig, 41.

Engler, A. (1928) Saxifragaceae. In *Die natürlichen Pflanzenfamilien*, A. Engler and K. Prantl (eds.), Engelmann, Leipzig, 74.

Erdal, K. (1986) Proanthocyanin-free barley-malting and brewing. *J. Inst. Brewing*, **92**, 220–224.

Etter, M. C., Urbanczuk, Z., Baer, S., and Barbara, P. F. (1986) The crystal structurers and H-bond properties of three 3-hydroxyflavone derivatives. *J. Mol. Struct.*, **144**, 155–167.

Everest, A. E. (1915) Recent chemical investigations of the anthocyanin pigments and their bearing upon the production of these pigments in plants. *J. Genetics*, **4**, 361–367.

Faegri, L. and van der Pijl, L. (1971) *The Principles of Pollination Ecology*, Pergamon Press, Oxford.

Fang, N., Leidig, M., and Mabry, T. M. (1985a) Highly oxygenated flavonoid aglycones from *Gutierrezia grandis*. *Phytochemistry*, **24**, 2693–2698.

Fang, N., Leidig, M., and Mabry, T. M. (1985b) Six 2′-hydroxyflavonols *from Gutierrrezia microcephala*. *Phytochemistry* **24**, 3029–3034.

Farkas, L. and Pallos, L. (1959) Final proof of structure and synthesis of coreopsin. *Chem. Ber.*, **92**, 1263–1270.

Farkas, L. and Pallos, L. (1967) Natürlich vorkommende Auronglykoside. *Progress in the Chemistry of Organic Natural Products*, **25**, 150–174.

Farkas, L., Major, A., Pallos, L., and Várady, J. (1958) Acylation of active methylene groups. II. Synthesis of pseudobaptigenin and formononetin and some related isoflavones. *Chem. Ber.*, **91**, 2858–2861.

Farkas, L., Strelisky, J., and Vermes, B. (1969) Ring isomerization of flavones. III. Synthesis of eupatorin, a flavone from *Eupatorium semiserratum*. *Chem. Ber.*, **102**, 112–117.

Farkas, L., Gottseguent, A., and Nógrádi, M. (1974) Synthesis of sophorol, violanone, lonchocarpan, claussiquinone, philenopteron, leiocalycin and some other natural isoflavonoids by the oxidative rearrangement of chalcones with thallium (III) nitrate. *J. Chem. Soc., Perkin Trans. 1*, 305–312.

Fawzy, A. A., Vishwanath, B. S., and Franson, R. C. (1988) Inhibition of human non-pancreatic phosphokinases A2 by retinoids and flavonoids. Mechanism of action. *Agents and Actions*, **25**, 394–400.

Feeny, P., Sachdev, K., Rosenberry, L., and Carter, M. (1988) Luteolin 7-*O*-(6″-*O*-malonyl)-β-D-glucoside and *trans*-chlorogenic acid: Oviposition stimulants for the black swallowtail butterfly. *Phytochemistry*, **27**, 3439–3448.

Ferrari, F., Messama, I., and Sant'Ana, A. E. G. (1991) Two new isoflavonoids from *Boerhaavia coccinea*. *J. Nat. Prod.*, **54**, 597–598.

Ferreira, D. and Bekker, R. (1996) Oligomeric proanthocyanidins: Naturally occurring *O*-heterocycles. *Nat. Prod. Reports*, **13**, 411–433.

Ferreres, F., Tomas-Lorente, F., Tomas-Barberan, F. A., Rivera, D., and Obon, C. (1989) Biochemical identification of *Sideritis serrata* X *S. bourgaeana* hybrids by HPLC analyses of flavonoids. *Z. Naturforsch.*, **44c**, 568–572.

Ferriola, P. C., Cody, V., and Middleton Jr., E. (1989) Protein kinase C inhibition by plant flavonoids. Kinetic mechanisms and structure-activity relationships. *Biochem. Pharmacol.*, **38**, 1617–1624.

Fiebig, M. and Wagner, H. (1984) Neue antihepatotoxisch wirksame Flavonolignane aus einer weissblühenden *Silybum*-Varietät. *Planta Medica*, **50**, 310–313.

Fieser, L. F. and Fieser, M. (1961) *Advanced Organic Chemistry*. Reinhold Publishing Corp., New York.

Figueiredo, P., Ehhabiri, M., Toki, K., Saito, N., Dangles, O., and Brouillard, R. (1996) New aspects of anthocyanin complexation. Intramolecular copigmentation as a means for colour loss? *Phytochemistry*, **41**, 301–308.

Filhol, E. (1860) Nouvelles recherches sur les matieres colorantes végétales. *C. R. Acad. sci., Paris*, **50**, 1182–1185. (Cited by Onslow, 1925.)

Finnemore, H. (1910) Examination of a species of *Prunus*. *Pharm. J.*, **85**, 604. (Cited by Onslow, 1925.)

Fischer, E. and Nouri, O. (1917) Synthesis of phloretin and preparation of the nitriles of phenylcarboxylic acids. *Chem. Ber.* **50**, 611–623.

Fischer, D., Ebenau-Jehle, C., and Grisebach, H. (1990) Purification and characterization of pterocarpan synthase from elicitor-challenged soybean cell cultures. *Phytochemistry*, **29**, 2879–2882.

Fischer, S., Böttcher, U., Reuber, S., Anhalt, S., and Weissenböck, G. (1995) Chalcone synthase in the liverwort *Marchantia polymorpha*. *Phytochemistry*, **39**, 1007–1012.

Fisher, E. and Nouri, O. (1917) Synthesis of phloretin. *Sitzb. kgl. preuss. Akad.*, 982–989.

Fleak, L. S. (1971) Biosystematics of the *Lupinus sericeus* complex, Ph.D. Dissertation, University of Missouri, Columbia.

Fleak, L. S. and Dunn, D. B. (1971) Nomenclature of the *Lupinus sericeus* L. complex (Papilionaceae). *Trans. Missouri Acad. Sci.* **5**, 85–88.

Fleck, D. C. and Woolfenden, G. E. (1997) Can acorn tannin predict scrub-jay caching behavior? *J. Chem. Ecol.*, **23**, 793–806.

Fleischer, T. C., Waigh, R. D., and Waterman, P. G. (1997) Bisabolene sesquiterpenes and flavonoids from *Friesodielsia enghiana*. *Phytochemistry*, **44**, 315–318.

Fliegmann, J., Schröder, G., Schanz, S., Britsch, L., and Schröder, J. (1992) Molecular analysis of chalcone and dihydropinosylvin synthase from Scots pine (*Pinus sylvestris*), and differential regulation of these and related enzyme activities in stressed plants. *Plant Mol. Biol.*, **18**, 489–503.

Forkmann, G. (1977) Anthocyanin pigments in *Callistephus chinensis*. *Phytochemistry*, **16**, 299–301.

Forkmann, G. (1980) The B-ring hydroxylation pattern of intermediates of anthocyanin synthesis in pelargonidin- and cyanidin-producing lines of *Matthiola incana*. *Planta*, **148**, 157–161

Forkmann, G., Heller, W., and Grisebach, H. (1980) Anthocyanin biosynthesis in flowers of *Matthiola incana*: flavanone 3-hydroxylase and flavonoid 3′-hydroxylase. *Z. Naturforsch.*, **35c**, 691–695.

Forkmann, G. and Dangelmayr, B. (1980) Genetic control of chalcone isomerase activity in flowers of *Dianthus caryophyllus*. *Biochem. Genet.*, **18**, 519–527.

Forkmann, G. (1989) Gene-enzyme relations and genetic manipulation of anthocyanin biosynthesis in flowering plants. In *The Genetics of Flavonoids*, D. E. Styles, G. A. Gavazzi, and M. L. Racchi (eds.), Edizioni Unicopli, Milano, 49–60.

Forkmann, G. (1991) Flavonoids as flower pigments: The formation of the natural spectrum and its extension by genetic engineering. *Plant Breeding*, **106**, 1–26.

Forkmann, G. (1994) Genetics of flavonoids. In *The Flavonoids: Advances in Research since 1990*. J. B. Harborne (ed.), Chapman and Hall, London, 536–564.

Forkmann, G. and Stotz, G. (1984) Selection and characterisation of flavanone 3-hydroxylase mutants of *Dahlia, Streptocarpus, Verbena* and *Zinnia*. *Planta*, **161**, 261–265.

Forkmann, G., de Vlaming, P., Spribille, R., Wiering, H., and Schram, A. W. (1986) Genetic and biochemical studies on the conversion of dihydroflavonols to flavonols and flavones of *Petunia hybrida*. *Z. Naturforsch.*, **41c**, 179–186.

Franca, N. C., Diaz Diaz, P. P., Gottlieb, O. R., and de Paula Rosa, B. (1974) Flavans from *Iryanthera* species. *Phytochemistry*, **13**, 1631–1632.

Franke, A. and Markham, K. R. (1989) Quercetin-3-*O*-α-[2-*O*-*p*-hydroxybenzoyl-4-*O*-*p*-coumaroylrhamnopyranoside], an aglycone-like flavonol glycoside from *Libocedrus bidwillii*. *Phytochemistry*, **28**, 3566–3568.

Franz, G. and Grün, M. (1983) Chemistry, occurrence and biosynthesis of *C*-glycosyl compounds in plants. *Planta Medica*, **47**, 131–140.

French, C. J. and Towers, G. H. N. (1992) Inhibition of infectivity of potato virus X by flavonoids. *Phytochemistry*, **31**, 3017–3020.

French, C. J., Elder, M., Leggett, F., Ibrahim, R. K., and Towers, G. H. N. (1991) Flavonoids inhibit infectivity of tobacco mosaic virus. *Can. J. Plant Pathol.*, **13**, 1–6.

Frey-Wyssling, A. (1938) The origin of secondary plant products (trans.). *Naturwissen.*, **26**, 624–628.

Friebolin, H. (1993) *Basic One- and Two-Dimensional NMR Spectroscopy*, Becconsall, J. K., translator, VCH Publishers, New York.

Fritsch, H. and Grisebach, H. (1975) Biosynthesis of cyanidin in cell cultures of *Haplopappus gracilis*. *Phytochemistry*, **14**, 2437–2442.

Fritze, K., Staiger, D., Czaja, I., Walden, R., Schell, J., and Wing, D. (1991) Developmental and UV light regulation of the snapdragon chalcone synthase promoter. *Plant Cell*, **3**, 893-905.

Fritzmeier, K.-H., Cretin, C., Kombrink, E., Rohwer, F., Taylor, J., Scheel, D., and Hahlbrock, K. (1987) Transient induction of phenylalanine ammonia-lyase and 4-coumarate:CoA ligase mRNAs in potato leaves infected with virulent or avirulent races of *Phytophthora infestans*. *Plant Physiol.*, **85**, 34–41.

Froemel, S., de Vlaming, P., Stotz, G., Wiering, H., Forkmann, G., and Schram, A. W. (1985) Genetic and biochemical studies on the conversion of flavanones to dihydroflavonols in flowers of *Petunia hybrida. Theor. Appl. Genet.*, **70**, 561–568.

Fröst, S., Holm, G., and Asker, S. (1975) Flavonoid patterns and the phylogeny of barley. *Hereditas*, **79**, 133–142.

Fröst, S. and Holm, G. (1975) Variation of flavonoid patterns in *Hordeum spontaneum* and *H. agriocrithon. Hereditas*, **80**, 167–172.

Fröst, S., Harborne, J. B., and King, L. (1977) Identification of the flavonoids in five chemical races of cultivated barley. *Hereditas*, **85**, 163–168.

Fuentes, A., Marinas, J. M., and Siniestra, J. V. (1987) Catalytic synthesis of chalcones under interfacial solid–liquid conditions with ultrasound. *Tet. Lett.*, **28**, 4541–4544.

Fujii, S., Yamagata, Y., Jin, G. Z., and Tomita, K. (1994) Novel molecular conformation of (R,S)-hesperetin in anhydrous crystal. *Chem. Pharm. Bull.*, **42**, 1143–1145.

Fukai, T., Hano, Y., Hirakura, K., Nomura, T., and Uzawa, J. (1985) Structure of a novel 2-arylbenzofuran derivative and two flavone derivatives from the cultivated mulberry tree (*Morus lhou* Koidz.). *Chem. Pharm. Bull.*, **33**, 4288–4295.

Fukasawa-Akada, T., Kung, S.-d., and Watson, J. C. (1996) Phenylalanine ammonia-lyase gene structure, expression, and evolution in *Nicotiana. Plant Mol. Biol.*, **30**, 711–722.

Fukui, H., Matsumoto, T., Nakamura, S., Nakayama, M., and Horie, T. (1968) The synthesis of jaceidin. *Experientia*, **24**, 108–109.

Fung, S. Y. (1988) Food plant derived dihydrochalcones in *Yponomeuta malinellus. Proc. Koninklijke Nederlandse Akad. Wetenschappen*, **91**, 217–221.

Furuya, M. A., Galston, A. W., and Stowe, B. B. (1962) Isolation from peas of co-factor and inhibitors of indole-3-acetic acid oxidase. *Nature* (*London*), **193**, 456–457.

Gaffield, W. (1996) Absolute configuration of the *cis*-dihydroquercetin-3-*O*-β-L-rhamnosides isoastilbin and neoastilbin. *Chem. Pharm. Bull.*, **44**, 1102–1103.

Gafner, S., Wolfender, J.-L., Mavi, S., and Hostettmann, K. (1996) Antifungal and antibacterial chalcones from *Myrica serrata. Planta Medica*, **62**, 67–69.

Galston, A. W. (1969) Flavonoids and photomorphogenesis in peas. In *Perspectives in Phytochemistry*, J. B. Harborne and T. Swain (eds.), Academic Press, New York, 193–204.

Gandhidasan, R., Neelakantan, S., Raman, P. V., and Devaraj, S. (1987) Components of the galls on the leaves of *Pongamia glabra*: Structures of pongagallone-A and pongagallone-B. *Phytochemistry*, **26**, 281–283.

Garo, E., Maillard, M., Antus, S., Mavi, S., and Hostettmann, K. (1996) Five flavans from *Mariscus psilostachys. Phytochemistry*, **43**, 1265–1269.

Gauthier, A., Gulick, P. J., and Ibrahim, R. K. (1996) cDNA cloning and characterization of a 3′/5′-*O*-methyltransferase for partially methylated flavonols from *Chrysosplenium americanum. Plant Mol. Biol.*, **32**, 1163–1169.

Gautier, J., Cave, A., Kunesch, G., and Polonsky, J. (1972) On the biosynthesis of neoflavonoids. *Experientia*, **28**, 759–761.

Geibel, M. (1995) Sensitivity of the fungus *Cytospora persoonii* to the flavonoids of *Prunus cerasus. Phytochemistry*, **38**, 599–601.

Geiger, H. (1994) Biflavonoids and triflavonoids. In *The Flavonoids: Advances in Research since 1986*, J. B. Harborne (ed.), Chapman and Hall, London, 95–115.

Geiger, H. and Markham, K. R. (1992) Campylopusaurone, an auronoflavanone biflavonoid from the mosses *Campylopus clavatus* and *Campylopus holomitrium*. *Phytochemistry*, **31**, 4325–4328.

Geiger, H. and Quinn, C. (1975) Biflavonoids. In *The Flavonoids*, J. B. Harborne, T. J. Mabry, and H. Mabry (eds.), Chapman and Hall, London, 692–742.

Geiger, H. and Quinn, C. (1982) Biflavonoids. In *The Flavonoids: Advances in Research*, J. B. Harborne and T. J. Mabry (eds.), Chapman and Hall, London, 505–534.

Geiger, H. and Quinn, C. (1988) Biflavonoids. In *The Flavonoids: Advances in Research since 1980*, J. B. Harborne (ed.), Chapman and Hall, London, 99–124.

Geiger, H., Seeger, T., Hahn, H., Zinsmeister, D., Markham, K. R., and Wong, H. (1993) ^{1}H NMR assignments in biflavonoid spectra by proton-detected C-H correlation. *Z. Naturforsch.*, **48c**, 821–826.

Geiger, H., Voigt, A., Seeger, T., Zinsmeister, H.-D., López-Sáez, J.-A., Pérez-Alonso, M.-J., and Velasco-Negeruela, A. (1995) Cyclobartramiatriluteolin, a unique triflavonoid from *Bartramia stricta*. *Phytochemistry*, **39**, 465–467.

Geilbel, M. and Feucht, W. (1991) Flavonoid 5-glucosides from *Prunus cerasus* bark and their characteristic weak glycosidic bonding. *Phytochemistry*, **30**, 1519–1521.

Geissman, T. A. and Heaton, C. D. (1943) Anthochlor pigments. IV. The pigments of *Coreopsis grandiflora* Nutt. I. *J. Amer. Chem. Soc.*, **65**, 677–683.

Geissman, T. A. and Heaton, C. D. (1944) Anthochlor pigments. V. The pigments of *Coreopsis grandiflora* Nutt. II. *J. Amer. Chem. Soc.*, **66**, 486–487.

Geissman, T. A. and Hinreiner, E. (1952a,b) Theories of the biogenesis of flavonoid compounds (Parts I & II). *The Botanical Reviews*, **18**, 77–164; 165–244.

Geissman, T. A. and Moje, W. (1951) Anthochlor pigments. VIII. The pigments of *Coreopsis grandiflora*. *J. Amer. Chem. Soc.*, **73**, 5765–5768.

Geissman, T. A. and Swain, T. (1957) Biosynthesis of flavonoid compounds in higher plants. *Chem. & Ind.* (*London*), 984.

Gerats, A. G. M., Farcy, E., Wallroth, M., Groot, S. P. C., and Schram, A. (1984) Control of anthocyanin synthesis in *Petunia hybrida* by multiple allelic series of the genes *An1* and *An2*. *Genetics*, **106**, 501–508.

Gerats, A. G. M., Vrijlandt, E., Wallroth, M., and Schram, A. (1985) The influence of the genes *An1*, *An2* and *An4* on the activity of the enzyme UDP glucose : flavonoid 3-*O*-glucosyltransferase in flowers of *Petunia hybrida*. *Biochem. Genetics*, **23**, 591–598.

Ghosal, S., Kumar, Y., Chakrabarti, D. K., Lal, J., and Singh, S. K. (1986) Parasitism of *Imperata cylindrica* on *Pancriatium biflorum* and the concomitant chemical changes in the host species. *Phytochemistry*, **25**, 1097–1102.

Giannasi, D. E. (1988) Flavonoids and evolution in the dicotyledons. In *The Flavonoids: Advances in Research since 1980*, J. B. Harborne (ed.), Chapman and Hall, London, 479–504.

Giannasi, D. E. (1979) Systematic aspects of flavonoid biosynthesis and evolution. *The Botanical Reviews*, **44**, 399–429.

Giannasi, D. E. and Niklas, K. J. (1977) Flavonoid and other chemical constituents of fossil Miocene *Celtis* and *Ulmus* (Succor Creek Flora). *Science*, **197**, 765–767.

Giannasi, D. E. and Niklas, K. J. (1981) Comparative paleobiochemistry of some fossil and extant Fagaceae. *Amer. J. Bot.*, **68**, 762–770.

Giannasi, D. E. and Crawford, D. J. (1986) Biochemical Systematics. II. A Reprise. In *Evolutionary Biology*, M. K. Hecht, B. Wallace, and G. T. Prance (eds.), Plenum Publishing Co., 25–248.

Gibbs, R. D. (1974) *Chemotaxonomy of Flowering Plants*, McGill-Queens University Press, Montreal.

Glässgen, W. E., Wray, V., Strack, D., Metzger, J. W., and Seitz, H. U. (1992) Anthocyanins from cell suspension cultures of *Daucus carota*. *Phytochemistry*, **31**, 1593–1601.

Glennie, C. W. and Harborne, J. B. (1971) Flavone and flavonol 5-glucosides. *Phytochemistry*, **10**, 1325–1329.

Gluchoff-Fiasson, F., Favre-Bonvin, L., and Fiasson, J. L. (1991) Glycosides and acylated glycosides of isoëtin from European species of *Hypochoeris*. *Phytochemistry*, **30**, 1673–1675.

Goda, Y., Shimizu, T., Kato, Y., Nakamura, M., Maitani, T., Yamada, T., Terahara, N., and Yamaguchi, M. (1997) Two acylated anthocyanins from purple sweet potato. *Phytochemistry*, **44**, 183–186.

Goldsmith, T. H. and Bernard, G. D. (1974) The visual system of insects. In *The Physiology of Insecta*, M. Rockstein (ed.), Academic Press, New York, 165–272.

Goldsmith, T. H. (1980) Hummingbirds see near ultraviolet light. *Science*, **207**, 786–788.

Gomez, F., Quijano, L., Calderon, J. S., and Rodriguez, C. (1985) Prenylflavones from *Tephrosia watsoniana*. *Phytochemistry*, **24**, 1057–1059.

Gomez-Garibay, F., Chilpa, R. R., Quiano, L., Calderon Pardo, J. S., and Rios Castillo, T. (1990) Methoxy furan auranols with fungistatic activity from *Lonchocarpus castilloi*. *Phytochemistry*, **29**, 459–463.

Gonzalez, A. G., Bermejo, J., Estevez, F., and Velazquez, R. (1983) Phenolic derivatives from *Artemisia glutinosa*. *Phytochemistry*, **22**, 1515–1516.

Goodwin, R. S., Rosler, K.-H., A., Mabry, T. J., and Varma, S. (1984) Flavonoids from *Brickellia glutinosa*. *J. Nat. Prod.*, **47**, 711–714.

Goppelsroeder, F. (1901) *Capillaranalyse beruhend auf Capillaritäts- und Adsorptionserscheinungen mit dem Schlusskapitel: Das Emporsteigen der Farbstoffe in Pflanzen*, Basel. (Cited by Onslow, 1925.)

Gordon, M. H. (1996) Dietary antioxidants in disease prevention. *Nat. Prod. Reports*, **12**, 269–273.

Gorham, J. (1990) Phenolic compounds, other than flavonoids, from bryophytes. *Proc. Phytochem. Soc.*, **29**, 143–159.

Gori, D. F. (1989) Floral color change in *Lupinus argenteus* (Fabaceae): Why should plants advertise the location of unrewarding flowers to pollinators. *Evolution*, **43**, 870–881.

Goris, A. and Canal, H. (1935a) 2′,6′-Dihydroxy-4′-methoxy-β-phenylpropiophenone, extracted from the oil of *Populus balsamifera*. *Compt. rend. Acad. Sci.*, **201**, 1435–1437.

Goris, A. and Canal, H. (1935b) Synthesis of 2′,6′-dihydroxy-4′-methoxy-β-phenylpropiophenone obtained from the oil of *Populus balsamifera. Compt. rend. Acad. Sci.*, **201**, 1520–1521.

Gornall, R. J. and Bohm, B. A. (1978) Angiosperm flavonoid evolution: A reappraisal. *Syst. Bot.*, **3**, 353–368.

Gornall, R. J., Bohm, B. A., and Dahlgren, R. (1979) The distribution of flavonoids in the angiosperms. *Bot. Notiser*, **132**, 1–30.

Goto, T. (1987) Structure, stability and color variation of natural anthocyanins. *Progress in the Chemistry of Organic Natural Products*, **5**, 113–158.

Goto, T. and Kondo, T. (1991) Structure and molecular association of anthocyanins. Variation of flower colors. *Angew. Chem., Int. Edn. Engl.* **30**, 17–33.

Gottlieb, O. R. (1982a) Ethnopharmacology *versus* chemosystematics in the search for biologically active principles in plants. *J. Ethnopharmacol.*, **6**, 227–238.

Gottlieb, O. R. (1982b) *Micromolecular Evolution, Systematics and Ecology*. Springer-Verlag, Berlin.

Graham, A. (1963) Outline of the origin and historical affinities between Asia and eastern North America. In *Floristics and Paleofloristics of Asia and Eastern North America*, A. Graham (ed.), Elsevier Publ. Co., Amsterdam, 1–16.

Grande, M., Piera, F., Cuenca, A., Torres, P., and Bellido, I. S. (1985) Flavonoids from *Inula viscosa. Planta Medica*, **51**, 414–419.

Grayer, R. J., Bryan, S. E., Veitch, N. C., Goldstone, F. J., Paton, A., and Wollenweber, E. (1996) External flavones in sweet basil, *Ocimum basilicum*, and related taxa. *Phytochemistry*, **43**, 1041–1047.

Greenaway, W., Scaysbrook, T., and Whatley, F. R. (1988) Phenolic analysis of bud exudate of *Populus lasiocarpus* by GC/MS. *Phytochemistry*, **27**, 3513–3515.

Greenaway, W., English, S., and Whatley, F. R. (1990a) Phenolic composition of bud exudates of *Populus deltoides. Z. Naturforsch.*, **45c**, 587–593.

Greenaway, W., Davidson, C. G., Scaysbrook, T., May, J., and Whatley, F. R. (1990b) Hybrid origin of *Populus* x *jackii* confirmed by gas chromatography-mass spectrometry analysis of its bud exudate. *Z. Naturforsch.*, **45c**, 594–598.

Greenaway, W., May, J., Scaysbrook, T., and Whatley, F. R. (1992a) Analysis of phenolics of bud exudate of *Populus violascens* by GC-MS. *Z. Naturforsch.* **47c**, 773–775.

Greenaway, W., May, J., and Whatley, F. R. (1992b) Analysis of phenolics of bud exudate of *Populus laurifolia* by GC-MS. *Z. Naturforsch.* **47c**, 776–778.

Greenham, J., Williams, C., and Harborne, J. B. (1995) Identification of lipophilic flavonols by a combination of chromatographic and spectral techniques. *Phytochem. Anal.* **6**, 211–217.

Grew, N. (1682) The anatomy of plants, with an idea of a philosophical history of plants, and several other lectures, read before the Royal Society, London. (Cited by Onslow, 1925.)

Griffiths, A. J. F. and Ganders, F. R. (1983) *Wildflower Genetics-A Field Guide for British Columbia and the Pacific Northwest*. Flight Press, Vancouver.

Gripenberg, J. (1962) Flavones. In *The Chemistry of Flavonoid Compounds*, T. A. Geissman (ed.), Macmillan, New York, 406–440.

Grisebach, H. (1957) The biogenesis of cyanidins. I. Experiments with acetate -1-^{14}C and acetate-2-^{14}C. *Z. Naturforsch.*, **12b**, 227–231.

Grisebach, H. and Brandner, G. (1962) Einbau des 2′,4,4′,6′-tetrahydroxychalkon-2′-glucosid-[β-C-14] in Isoflavone. *Experientia*, **18**, 400–401.

Grisebach, H. and Doerr, N. (1960) Biogenesis of isoflavones. II. Mechanism of rearrangement. *Z. Naturforsch.*, **15b**, 284–286.

Grisebach, H. and Grambow, H. J. (1968) Biosynthesis of flavonoids – XV. Occurrence and biosynthesis of flavonoids in *Datisca cannabina*. *Phytochemistry*, **7**, 51–56.

Griesbach, R. J. and Kamo, K. K. (1996) The effect of induced polyploidy on the flavonols of *Petunia* 'Mitchell'. *Phytochemistry*, **42**, 361–363.

Grün, M. and Franz, G. (1981) *In vitro* biosynthesis of the *C*-glycosidic bond in aloin. *Planta*, **152**, 562–564.

Guinaudeau, H., Seligmann, O., Wagner, H., and Neszmely, A. (1981) Faralatroside and faratroside, two flavonol triglycosides from *Colubrina faralaotra*. *Phytochemistry*, **20**, 1113–1116.

Gupta, B. K., Gupta, G. K., Dhar, K. L., and Atal, C. K. (1980) A *C*-formylated chalcone from *Psoralea corylifolia*. *Phytochemistry*, **19**, 2034–2035.

Gupta, D. and Singh, J. (1991) Flavonoid glycosides from *Cassia alata*. *Phytochemistry*, **30**, 2761–2763.

Gupta, S. R., Seshadri, T. R., and Good, G. R. (1975) A new formylated chalkone, 5′-formyl-2′,4-dihydroxy-4′-methoxychalkone, from the seeds of *Psoralea corylifolia*. *Ind. J. Chem.*, **13B**, 632.

Gupta, S. R., Seshadri, T. R., and Good, G. R. (1977) The structure and synthesis of neobavachalcone, a new component of *Psoralea corylifolia*. *Phytochemistry*, **16**, 1995–1997.

Gustafsson, M. H. G., and Bremer, K. (1995) Morphology and phylogenetic interrelationships of the Asteraceae, Calyceraceae, Campanulaceae, Goodeniaceae, and related families (Asterales). *Amer. J. Bot.*, **82**, 250–265.

Guthrie, J. L. and Rabjohn, N. (1957) Some reactions effected by means of bromomagnesium t-alkoxides. *J. Org. Chem.*, **22**, 176–179.

Günther, H. (1995) *NMR Spectroscopy*, Günther, H., translator, John Wiley & Sons, New York.

Hadj-Mahammed, M. and Meklati, B. Y. (1987) Qualitative determination of polymethoxylated flavones in Valencia orange peel oil and juice by LC-UV/MS and LC-MS techniques. *Lebensm. Wiss. Technol.*, **20**, 111–114.

Hagmann, M., Heller, W., and Grisebach, H. (1983) Induction and characterization of a microsomal flavonoid-3′-hydroxylase from parsley cell cultures. *Eur. J. Biochem.*, **134**, 547–549.

Hagmann, M. L. and Grisebach, H. (1984) Enzymatic rearrangement of flavanone to isoflavone. *FEBS Lett.*, **175**, 199–202.

Hahlbrock, K. (1976) Regulation of the enzymes of phenylpropanoid metabolism in relation to specific growth stages of plant cell suspension cultures. *Physiol. Veg.*, **14**, 207–213.

Hahlbrock, H. and Grisebach, H. (1975) Biosynthesis of flavonoids. In *The Flavonoids*, J. B. Harborne, T. J. Mabry, and H. Mabry (eds.), Chapman and Hall, London, 866–915.

Hahlbrock, K. and Grisebach, H. (1979) Enzymic controls in the biosynthesis of lignin and flavonoids. *Annu. Rev. Plant Physiol.*, **30**, 105–130.

Hahlbrock, K. and Ragg, H. (1975) Light-induced changes of enzyme activities in parsley cell suspension cultures. Effects of inhibitors of RNA and protein synthesis. *Arch. Biochem. Biophys.*, **166**, 41–46.

Hahlbrock, K. and Scheel, D. (1989) Physiology and molecular biology of phenylpropanoid metabolism. *Annu. Rev. Plant Physiol. Plant Mol. Biol.*, **40**, 347–369.

Hahlbrock, K., Ebel, J., Ortmann, R., Sutter, A., Wellmann, E., and Grisebach, J. (1971) Regulation of enzyme activities related to the biosynthesis of flavonoid glycosides in cell suspension cultures of parsley (*Petroselinum hortense*). *Biochim. Biophys. Acta*, **244**, 7–15.

Hahn, H., Seeger, T., Geiger, H., Zinsmeister, H. D., Markham, K. R., and Wong, H. (1995) The first biaurone, a triflavone and biflavonoids from two *Aulacomnium* species. *Phytochemistry*, **40**, 573–576.

Hakamatsuka, T., Ebizuka, Y., and Sankawa, U. (1991) Induced isoflavonoids from copper chloride-treated stems of *Pueraria lobata. Phytochemistry*, **30**, 1481–1482.

Hakamatsuka, T., Noguchi, H., Ebizuka, Y., and Sankawa, U. (1988) Deoxychalcone synthase from cell suspension cultures of *Pueraria lobata. Chem. Pharm. Bull.*, **36**, 4225–4228.

Hakamatsuka, T., Mori, K., Ishida, S., Choi, K. A., Hashim, M. F., Noguchi, H., Ebizuka, Y., and Sankawa, U. (1994) P-450-Dependent oxidative aryl migration in isoflavonoid biosynthesis. *Tennen Yuki Kagobutsu Toronkai Koen Yoshishu*, 36th, 196–203.

Hanawa, F., Tahara, S., and Mizutani, J. (1991) Isoflavonoids produced by *Iris pseudacorus* leaves treated with *Cupric chloride. Phytochemistry*, **30**, 157–163.

Hano, Y., Mitsui, P., and Nomura, T. (1990) Constituents of the Moraceae plants. VI. Two new prenylaurones, antiarones A and B, from the root bark of *Antiaris toxicaria* Lesch. *Heterocycles*, **30**, 1023–1030.

Hano, Y., Mitsui, P., Nomura, T., Kawai, T., and Yoshida, Y. (1991) Two new dihydrochalcone derivatives, antiarones J and K, from the root bark of *Antiaris toxicaria. J. Nat. Prod.*, **54**, 1049–1055.

Hans, N. and Grover, S. K. (1993) An efficient conversion of 2′-hydroxychalcones to flavones. *Synth. Commun.*, **23**, 1021–1023.

Hansen, H. V. (1992) Studies in the Calyceraceae with a discussion of its relationship to Compositae. *Nordic J. Bot.*, **12**, 63–75.

Hanson, K. R. and Havir, E. A. (1972a) Phenylalanine Ammonia-lyase. In *The Enzymes, 3rd Ed.*, P. D. Boyer (ed.), Academic Press, New York, 75–166.

Hanson, K. R. and Havir, E. A. (1972b) Phenylalanine Ammonia-lyase. In *Recent Advances in Phytochemistry, Vol. 4*, V. C. Runeckles and J. E. Watkin (eds.), Appleton-Century-Crofts, New York, 45–85.

Hanson, K. R. and Havir, E. A. (1981) Phenylalanine Ammonia-lyase. In *Secondary Plant Products*, E. E. Conn (ed.), Academic Press, New York, 577–625.

Hao, H., Handong, S., and Shouxun, Z. (1996) Flavonoids from *Isodon oresbius. Phytochemistry*, **42**, 1247–1248.

Harada, N., Ono, H., Uda, H., Parveen, M., Din, K. N., Achari, B., and Dutta, P. K. (1992) Atropisomerism in natural products. Absolute stereochemistry of biflavone, (-)-4′,4‴,7,7″-tetra-*O*-methylcupressuflavone, as determined by the theoretical calculation of CD spectra. *J. Amer. Chem. Soc.*, **114**, 7687–7692.

Harborne, J. B. (1962) Plant polyphenols. 5. Occurrence of azalein and related pigments in flowers of *Plumbago* and *Rhododendron* species. *Arch. Biochem. Biophys.*, **96**, 171–178.

Harborne, J. B. (1964) Ultraviolet spectroscopy of polyphenols. In *Methods in Polyphenol Chemistry*, J. B. Pridham (ed.), The Macmillan Company, New York, 13–36.

Harborne, J. B. (1965) Plant polyphenols-XIV. Characterization of flavonoid glycosides by acidic and enzymic hydrolyses. *Phytochemistry*, **4**, 107–120.

Harborne, J. B. (1966) Comparative biochemistry of flavonoids. II. 3-Desoxyanthocyanins and their systematic distribution in ferns and gesnerads. *Phytochemistry*, **5**, 589–600.

Harborne, J. B. (1967a) *Comparative Biochemistry of the Flavonoids*. Academic Press, New York.

Harborne, J. B. (1967b) Comparative biochemistry of the flavonoids. IV. Correlations between chemistry, pollen morphology and systematics in the family Plumbaginaceae. *Phytochemistry*, **6**, 1415–1428.

Harborne, J. B. (1967c) Comparative biochemistry of the flavonoids. VI. Flavonoid patterns in the Bignoniaceae and the Gesneriaceae. *Phytochemistry*, **6**, 1643–1651.

Harborne, J. B. (1970) *Phytochemical Phylogeny*, Academic Press, New York.

Harborne, J. B. (1977) Flavonoids and the evolution of the angiosperms. *Biochem. Syst. Ecol.*, **5**, 7–22.

Harborne, J. B. (1988) *Introduction to Ecological Biochemistry*, Academic Press, New York.

Harborne, J. B. (1989) General procedures and measurement of total phenolics. In *Plant Phenolics*, Harborne, J. B. (ed.), Academic Press, New York, 1–28.

Harborne, J. B. (1991) Revised structures for three isoëtin glycosides, yellow flower pigments in *Heywoodiella*. *Phytochemistry*, **30**, 1677–1678.

Harborne, J. B. and Boardley, M. (1985) The widespread occurrence in Nature of anthocyanins as zwitterions. *Z. Naturforsch.*, **40c**, 305–308.

Harborne, J. B. and Geissman, T. A. (1956) Anthochlor pigments. XII. Maritimein and marein. *J. Amer. Chem. Soc.*, **78**, 829–832.

Harborne, J. B. and Smith, D. M. (1978) Anthochlors and other flavonoids as honey guides in the Compositae. *Biochem. Syst. Ecol.*, **6**, 287–291.

Harborne, J. B. and Turner, B. L. (1984) *Plant Chemosystematics*, Academic Press, New York.

Harborne, J. B. and Williams, C. A. (1973) A chemotaxonomic survey of flavonoids and simple phenols in leaves of the Ericaceae. *Bot. J. Linnean Soc.*, **66**, 37–54.

Harborne, J. B. and Williams, C. A. (1975) Flavone and flavonol glycosides. In *The Flavonoids*, J. B. Harborne, T. J. Mabry, and H. Mabry (eds.), Chapman and Hall, London, 376–441.

Harborne, J. B. and Williams, C. A. (1982) Flavone and flavonol glycosides. In *The Flavonoids: Advances in Research*, J. B. Harborne and T. J. Mabry (eds.), Chapman and Hall, New York, 261–301.

Harborne, J. B. and Williams, C. A. (1988) Flavone and flavonol glycosides. In *The Flavonoids: Advances in Research since 1986*, J. B. Harborne (ed.), Chapman and Hall, New York, 303–328.

Harborne, J. B., Boulter, D., and Turner, B. L. (1971) *Chemotaxonomy of the Leguminosae*, Academic Press, New York.

Harborne, J. B., Williams, C. A., and Wilson, K. L. (1985) Flavonoids in leaves and inflorescences of Australian Cyperaceae. *Phytochemistry*, **24**, 751–766.

Harborne, J. B., Williams, C. A., Greenham, J., and Eagles, J. (1994) Variations in the lipophilic and vacuolar flavonoids of the genus *Vellozia. Phytochemistry*, **35**, 1475–1480.

Haribal, M. and Renwick, J. A. A. (1996) Oviposition stimulants for the monarch butterfly: Flavonol glycosides from *Asclepias curassavica. Phytochemistry*, **41**, 139–144.

Harker, C. L., Ellis, T. H. N., and Coen, E. S. (1990) Identification and genetic regulation of the chalcone synthase multigene family in pea. *Plant Cell*, **2**, 185–194.

Harlan, J. R. (1976) *Hordeum vulgare*. In *Evolution of Crop Plants*, N. W. Simmonds (ed.), Longman, London, 93–98.

Harley, R. M. and Reynolds, T. (1992) *Advances in Labiate Science*, Royal Botanic Gardens, Kew, UK, 568.

Harrison, B. J. and Stickland, R. G. (1974) Precursors and genetic control of pigmentation. 2. Genotype analysis of pigment controlling genes in acyanic phenotypes in *Antirrhinum majus. Heredity*, **33**, 112–115.

Hartwig, U. A., Maxwell, C. A., Joseph, C. M., and Phillips, D. A. (1989) Interactions among flavonoid *nod* gene inducers released from alfalfa seeds and roots. *Plant Physiol.*, **91**, 1138–1142.

Hashidoko, Y., Tahara, S., and Mizutani, J. (1986) Isoflavonoids of yellow lupine. Part I. New complex isoflavones in the root of yellow lupine (*Lupinus luteus* L. cv. Barpine). *Agricul. Biol. Chem.*, **50**, 1797–1800.

Hashim, M. F., Hakamatsuka, T., Ebizuka, Y., and Sankawa, U. (1990) Reaction mechanism of oxidative rearrangement of flavanone in isoflavone biosynthesis. *FEBS Lett.*, **271**, 219–222.

Haslam, E. (1974) *The Shikimate Pathway*. John Wiley & Sons, New York.

Haslam, E. (1980) *In vino veritas*: Oligomeric procyanidins and the ageing of red wine. *Phytochemistry*, **19**, 2577–2582.

Haslam, E. (1982) Proanthocyanidins. In *The Flavonoids: Advances in Research*, J. B. Harborne and T. J. Mabry (eds.), Chapman and Hall, London, 417–447.

Haslam, E. (1989) *Plant polyphenols. Vegetable tannins revisited*. Cambridge University Press, Cambridge.

Hatayama, K. and Komatsu, M. (1971) Studies on the constituents of *Sophora* species. V. Constituents of the root of *Sophora angustifolia* Sieb. & Zucc. (2). *Chem. Pharm. Bull.*, **19**, 2126–2131.

Havir, E. A., Reid, P. D., and Marsh, J. H.V. (1971) L-Phenylalanine ammonia-lyase (maize). *Plant Physiol.*, **48**, 130–136.

Havsteen, B. (1983) Flavonoids, a class of natural products of high pharmacological potency. *Biochemical Pharmacology*, **32**, 1141–1148.

Hawkes, J. G. (1968) (ed.) *Chemotaxonomy and Serotaxonomy*, Academic Press, New York.

Hawkes, J. G., Lester, R. N., and Skelding, A. D. (1979) *The Biology and Taxonomy of the Solanaceae*, Academic Press, New York, 738.

Hayashi, K. (1962) The anthocyanins. In *The Chemistry of Flavonoids Compounds*, T. A. Geissman (ed.), The Macmillan Company, New York, 248–285.

He, K., Timmermann, B. N., Aladesanmi, A. J., and Zeng, L. (1996) A biflavonoid from *Dysoxylum lenticellulare* Gillespie. *Phytochemistry*, **42**, 1199–1201.

Hedin, P. A. and Phillips, V. A. (1991) Chemical ionization (methane) mass spectrometry of sugars and their derivatives. *J. Agric. Food Chem.*, **39**, 1106–1109.

Hedin, P. A. and Phillips, V. A. (1992) Electron impact mass spectral analysis of flavonoids. *J. Agric. Food Chem.*, **40**, 607–611.

Hegnauer, R. (1962, 1963, 1964, 1966, 1969, 1973, 1986, 1989, 1990, 1992, 1994) *Chemotaxonomie der Pflanzen*, Vols. I-XIa, Birkhäuser Verlag, Basal.

Hegnauer, R. (1986) Phytochemistry and plant taxonomy – an essay on the chemotaxonomy of higher plants. *Phytochemistry*, **25**, 1519–1535.

Heidstra, R., Geurts, R., Franssen, H., Spaink, H. P., van Kammen, A., and Bisseling, T. (1994) Root hair deformation activity of nodulation factors and their fate on *Vicia sativa*. *Plant Physiol.*, **105**, 787–797.

Heidstra, R. and Bisseling, T. (1996) Nod-factor-induced host responses and mechanisms of Nod factor perception. *New Phytologist*, **133**, 25–43.

Heimler, D., Pieroni, A., Mittempergher, L., and Buzzini, P. (1993) The use of flavonoid glycosides for the identification of elm hybrids. *Can. J. Forest Res.*, **23**, 611–616.

Heller, W. and Forkmann, G. (1988) Biosynthesis. In *The Flavonoids: Advances in Research since 1980*, J. B. Harborne (ed.), Chapman and Hall, London, 399–425.

Heller, W. and Forkmann, G. (1994) Biosynthesis. In *The Flavonoids: Advances in Research since 1986*, J. B. Harborne (ed.), Chapman and Hall, London, 499–535.

Heller, W. and Hahlbrock, K. (1980) Highly purified 'flavanone synthase' from parsley catalyzes the formation of naringenin chalcone. *Arch. Biochem. Biophys.*, **200**, 617–619.

Heller, W., Britsch, L., Forkmann, G., and Grisebach, H. (1985) Leucoanthocyanidins as intermediates in anthocyanidin biosynthesis in flowers of *Matthiola incana* R. Br., *Planta*, **163**, 191–196.

Hemmingway, R. W., Foo, L. Y., and Porter, L. J. (1982) Linkage isomerism in trimeric and polymeric 2,3-*cis*-procyanidins. *J. Chem. Soc.*, *Perkin 1*, 1209–1216.

Hennion, F., Fiasson, J. L., and Gluchoff-Fiasson, K. (1994) Morphological and phytochemical relationships between *Ranunculus* species from Iles Kerguelen. *Biochem. Syst. Ecol.*, **22**, 533–542.

Hess, D. (1964) Der Einbau Methylgruppen-markierter Ferulasäure und Sinapinsäure in die Anthocyane von *Petunia hybrida*. *Planta*, **60**, 568–581.

Heywood, V. H. (1971) *The Biology and Chemistry of the Umbelliferae*, Academic Press, New York, 438.

Heywood, V. H. (1973) The role of chemistry in plant systematics. *Pure and Applied Chem.*, **34**, 355–375.

Heywood, V. H., Harborne, J. B., and Turner, B. L. (1977) *The Biology and Chemistry of the Compositae*, Academic Press, New York, 1189.

Hilger, A. (1879) Über den Farbstuff der Familie der Caryophyllinen. *Landw. Versuchstat., Berlin*, **23**, 456–461. (Cited by Onslow, 1925.)

Hitchcock, C. L., Cronquist, A., Ownbey, M., and Thompson, J. W. (1961) *Vascular Plants of the Pacific Northwest*, University of Washington Press, Seattle.

Hoffmann, B. and Hölzl, J. (1988) Weiter acylierte Chalkone aus *Bidens pilosa. Planta Medica*, **54**, 450–451.

Hoffmann, B. and Hölzl, J. (1989) Acylated compounds from *Bidens pilosa. Planta Medica*, **55**, 108–110.

Holowczak, J., Kuc, J., and Williams, E. G. (1960) Metabolism *in vitro* of phloridzin and other host compounds by *Venturia inaequalis. Phytopathology*, **50**, 640.

Holton, T. A. and Cornish, E. C. (1995) Genetics and biochemistry of anthocyanin biosynthesis. *Plant Cell*, **7**, 1071–1083.

Holton, T. A., Brugliera, F., Lester, D. R., Tanaka, Y., Hyland, C. D., Menting, J. G. T., Lu, C.-Y., Farcy, E., Stevenson, T. W., and Cornish, E. C. (1993a) Cloning and expression of cytochrome P450 genes controlling flower colour. *Nature (London)*, **366**, 276–279.

Holton, T. A., Brugliera, F., and Tanaka, Y. (1993b) Cloning an expression of flavonol synthase from *Petunia hybrida. Plant J.*, **4**, 1003–1010.

Honda, K. (1986) Flavanone glycosides as oviposition stimulants in a papilionid butterfly, *Papilio protenor. J. Chem. Ecol.*, **12**, 1999–2010.

Honda, K. (1990) Identification of host-plat chemicals stimulating oviposition by swallowtail butterfly, *Papilio protenor. J. Chem. Ecol.*, **16**, 325–337.

Hörhammer, L., Wagner, H., and Gloggengiesser, F. (1958) A new type of flavone glycoside. I. Isolation of a luteolin and apigenin glycoside from *Polygonum orientale. Arch. Pharm.*, **291**, 126–137.

Horie, T., Kourai, H., Osaka, H., and Nakayama, M. (1982) Studies of the selective *O*-alkylation and dealkylation of flavonoids. V. The synthesis of 5,6-dihydroxy-3,4′,8-trimethoxyflavone and a revised structure for the flavone from *Conyza stricta. Bull. Chem. Soc. Japan*, **55**, 2933–2940.

Horie, T., Kouri, H., Tsukayama, M., Masumura, M., and Nakayama, M. (1985) Studies of the selective *O*-alkylation and dealkylation of flavonoids. VII. Partial dealkylation of 5,6,7-trioxygenated flavones and synthesis of pectolinarigenin and its analogues. *J. Pharm. Soc. Japan*, **105**, 232–240.

Horie, T., Tsukayama, M., Yamada, T., Miura, I., and Nakayama, M. (1986) Three flavone glycosides from *Citrus sudachi. Phytochemistry*, **25**, 2621–2624.

Horie, T., Tsukayama, M., Kawamura, Y., and Seno, M. (1987) Studies of the selective *O*-alkylation and dealkylation of flavonoids. X. Selective demethylation of 7-hydroxy-3,5,8-trimethoxyflavones with aluminum halide in acetonitrile or ether. *J. Org. Chem.*, **52**, 4702–4709.

Horie, T., Kawamura, Y., Yamamoto, H., Kitou, T., and Yamashita, K. (1995) Studies of the selective *O*-alkylation and dealkylation of flavonoids. XVII.

Synthesis of 5,8-dihydroxy-6,7-dimethoxyflavones and revised structures for some natural flavones. *Phytochemistry*, **39**, 1201–1210.

Horie, T., Sasagawa, M., Torii, F., Kawamura, Y., and Yamashita, K. (1996a) Studies of the selective *O*-alkylation and dealkylaltion of flavonoids. XX. A convenient method for synthesizing 5,6,7-trihydroxyisoflavones and 5,6-dihydroxy-7-methoxyisoflavones. *Chem. Pharm. Bull.*, **44**, 486–491.

Horie, T., Kitou, T., Kawamura, Y., and Yamashita, K. (1996b) Studies of the selective *O*-alkylation and dealkylation of flavonoids. XXI. A convenient method for synthesizing 3,5,7-trihydroxy-6,8-dimethoxy- and 5,7-dihydroxy-3,6,8-trimethoxyflavones. *Bull. Chem. Soc. Jpn.*, **69**, 1033–1041.

Horovitz, A. (1976) Edaphic factors and flower colour distribution in the Anemoneae (Ranunculaceae). *Plant Syst. Evol.*, **126**, 239–242.

Horovitz, A. and Cohen, Y. (1972) Ultaviolet reflectance characteristics in flowers of crucifers. *Amer. J. Bot.*, **59**, 706–713.

Horowitz, R. M. (1964) Relations between the taste and structure of some phenolic glycosides. In *Biochemistry of Phenolic Compounds*, J. B. Harborne (ed.), Academic Press, New York, 545–571.

Horowitz, R. M. (1986) Taste effects of flavonoids. In *Plant Flavonids in Biology and Medicine*, V. Cody, E. Middleton Jr., and J. B. Harborne (eds.), Alan R. Liss, Inc., New York, 163–175.

Horowitz, R. M. and Gentilli, B. (1964) Structure of vitexin and isovitexin. *Chem. Ind.* (*London*), 498–499.

Horowitz, R. M. and Gentilli, B. (1969) Taste and structure in phenolic glycosides. *J. Agric. and Food Chem.*, **17**, 696–700.

Hoshino, T. (1991) An approximate estimate of self-association constants and the self-stacking conformation of malvin quinonoidal bases studied by proton NMR. *Phytochemistry*, **30**, 2049–2055.

Hoshino, T., Matsumoto, U., and Goto, T. (1981) Self-association of some anthocyanins in neutral aqueous solution. *Phytochemistry*, **20**, 1971–1976.

Hosokawa, K., Fukushi, E., Kawabata, J., Fujii, C., Ito, T., and Yamamura, S. (1995) Three acylalted cyanidin glucosides in pink flowers of *Gentiana*. *Phytochemistry*, **40**, 941–944.

Hossain, M. A. and Islam, A. (1993) Synthesis of some derivatives of ovalichalcone and ovalichalcone-A. *J. Bangladesh Chem. Soc.*, **6**, 199–207.

Hostettmann, K. and Hostettmann, M. (1982) Isolation techniques for flavonoids. In *The Flavonoids: Advances in Research*, J. B. Harborne (ed.), Chapman & Hall, London, 1–18.

Hou, R.-S., Duh, C.-Y., Wang, S.-K., and Chang, T.-T. (1994) Cytotoxic flavonoids from leaves of *Melicope triphylla*. *Phytochemistry*, **35**, 271–272.

Howles, P. A., Sewalt, V. J. H., Paiva, N. L., Elkind, Y., Bate, N. J., Lamb, C., and Dixon, R. A. (1996) Overexpression of L-phenylalanine ammonia-lyase in transgenic tobacco plants reveals control points for flux into phenylpropanoid biosynthesis. *Plant Physiol.*, **112**, 1617–1624.

Hrazdina, G. and Jensen, R. A. (1992) Spatial organization of enzymes in plant metabolic pathways. *Annu. Rev. Plant Physiol. Plant Mol. Biol.*, **43**, 241–267.

Hubbard, M., Kelly, J., Rajapakse, S., Abbot, A., and Ballard, R. (1992) Restriction fragment length polymorphism in rose and their use for cultivar identification. *HortScience*, **27**, 172–173.

Hufford, C. D. and Lasswell Jr., W. L. (1976) Uvaretin and isouvaretin, two novel cytotoxic *C*-benzylflavanones from *Uvaria chamae*. *J. Org. Chem.*, **41**, 1297–1300.

Hufford, C. D. and Lasswell Jr., W. L. (1978a) C-13 NMR studies of *C*-benzylated flavanones. *Lloydia*, **41**, 151–155.

Hufford, C. D. and Lasswell Jr., W. L. (1978b) Antimicrobial activities of constituents of *Uvaria chamae*. *Lloydia*, **41**, 156–160.

Hufford, C. D. and Oguntimein, B. O. (1980) Dihydrochalcones from *Uvaria angolensis*. *Phytochemistry*, **19**, 2036–2038.

Hufford, C. D., Jia, Y., Croom, J., E.M., Muhammed, I., Okunade, A. L., Clark, A. M., and Rogers, R. D. (1993) Antimicrobial compounds from *Petalostemon purpureum*. *J. Nat. Prod.*, **56**, 1878–1889.

Hufford, C. D., Oguntimein, B. O., and Baker, J. K. (1981) New flavonoid and coumarin derivatives of *Uvaria afzelii*. *J. Org. Chem.*, **46**, 3073–3078.

Huke, M. and Gorlitzer, K. (1969) N.M.R. spectra of 3-coumaranones, aurones and some sulphur analogs. *Arch. Pharm.*, **302**, 423–434.

Hutchinson, A., Taper, C. D., and Towers, G. H. N. (1959) Studies of phloridzin in *Malus*. *Can. J. Biochem. Physiol.*, **37**, 901–910.

Hutchinson, J. (1959) *The Families of Flowering Plants*, Oxford University Press, Oxford, England.

Hutchinson, J. (1967) *The Genera of Flowering Plants*, Oxford University Press, Oxford, England.

Ibrahim, R. K. (1992) Immunolocalization of flavonoid conjugates and their enzymes. *Rec. Advan. Phytochem.*, **26**, 25–61.

Ibrahim, R. K., De Luca, V., Khouri, H., Latchinian, L., Brisson, L., and Charest, P. M. (1987) Enzymology and compartmentation of polymethylated flavonol glucosides in *Chrysosplenium americanum*. *Phytochemistry*, **26**, 1237–1245.

Ibrahim, R. K., Latchinian, L., and Brisson, L. (1989) Biogenesis and localization of polymethylaed flavonoids in cell walls of *Chrysosplenium americanum*. In *Plant Cell Wall Polymers, Biogenesis and Biodegradation*, N. G. Lewis and M. G. Paice (eds.), American Chemical Society, Washington, DC, 122–136.

Ichino, K. (1989) Two flavonoids from two *Lindera umbellata* varieties. *Phytochemistry*, **28**, 955–956.

Ichino, K., Tanaka, H., and Ito, K. (1988a) Two novel flavonoids from the leaves of *Lindera umbellata* var. *lancea* and *L. umbellata*. *Tetrahedron*, **44**, 3251–3260.

Ichino, K., Tanaka, H., and Ito, K. (1989) Isolation and structures of two new flavonods form *Lindera umbellata.*, *Chem. Lett.* 363–366.

Ichino, K., Tanaka, H., Ito, K., Tanaka, T., and Mizuno, M. (1988b) Synthesis of helilandin B, pashanone, and their isomers. *J. Nat. Prod.*, **51**, 906–914.

Iinuma, M. and Mizuno, M. (1989) Natural occurrence and synthesis of 2′-oxygenated flavones, flavonols, flavanones and chalcones. *Phyrochemistry*, **28**, 681–694.

Iinuma, M., Tanaka, T., and Matsuura, S. (1984a) Synthetic studies on the flavone derivatives. XIII. Synthesis of flavones with tetramethoxyl groups in ring B. *Chem. Pharm. Bull.*, **32**, 3354–3360.

Iinuma, M., Iwashima, K., Tanaka, T., and Matsuura, S. (1984b) Synthesis of 5,6′-dihydroxy-2′,3′,4′,6,7-pentamethoxyflavone. *Chem. Pharm. Bull.*, **32**, 4217–4219.

Iinuma, M., Iwashima, K., and Matsuura, S. (1984c) Synthetic studies on flavone derivatives. XIV. Synthesis of 2′,4′,5′-trioxygenated flavones. *Chem. Pharm. Bull.*, **32**, 4935–4941.

Iinuma, M., Roberts, M. F., Matlin, S. A., Stacey, V. E., Timmermann, B. N., Mabry, T. J., and Brown, R. (1985) Synthesis and revised structure of the flavone brickellin. *Phytochemistry*, **24**, 1367–1368.

Iinuma, M., Matoba, Y., Tanaka, T., and Mizuno, M. (1986a) Flavonoid syntheses. I. Synthesis and spectroscopic properties of flavones with two hydroxy and five methoxy groups at C-2′,3′,4′,5,6,6′,7 and C-2′,3,4′,5,5′,6,7. *Chem. Pharm. Bull.*, **34**, 1656–1662.

Iinuma, M., Tanaka, T., Mizuno, M., and Mabry, T. J. (1986b) Flavonoid syntheses. IV. Syntheses of 2′,3,4′,5,5′,6,7,8-octaoxygenated flavones. *Chem. Pharm. Bull.*, **34**, 2228–2230.

Iinuma, M., Tanaka, T., Ito, K., and Mizuno, M. (1987) Flavonoid syntheses. V. Synthesis of flavonoids with three hydroxy and four methoxy groups and their spectral properties. *Chem. Pharm. Bull.*, **35**, 660–667.

Iinuma, M., Ohyama, M., Tanaka, T., Mizuno, M., and Hong, S.-K. (1992) Three 2′,4′,6′-trioxygenated flavanones in roots of *Echinosophora koreensis*. *Phytochemistry*, **31**, 2855–2858.

Iinuma, M., Tanaka, T., Mizuno, M., Yamamoto, H., Kobayashi, Y., and Yonemori, S. (1992) Phenolic constituents in *Erythrina* X *bidwilli* and their activity against oral microbial organisms. *Chem. Pharm. Bull.*, **40**, 2749–2752.

Iinuma, M., Okawa, Y., Tanaka, T., Ho, F.-C., Kobayashi, Y., and Miyauchi, K.-I. (1994a) Anti-oral microbial activity of isoflavonoids in root bark of *Ormosa monosperma*. *Phytochemistry*, **37**, 889–891.

Iinuma, M., Tsuchiya, H., Sato, M., Yokohama, J., Ohyama, M., Masayoshi, O., Ohkawa, Y., Tanaka, T., Fujiewara, S., and Fujii, T. (1994b) Flavanones with potent antibacterial activity against methicillin-resistant *Staphylococcus aureus*. *J. Pharm. Pharmacol.*, **46**, 892–895.

Ikeshiro, Y. and Konoshima, M. (1972) Synthesis of (+/−)-hepta-*O*-methylfukugetin and hepta-*O*-methylsaharanflavone. *Tet. lett.*, **43**, 4383–4386.

Ikuta, J., Hano, Y., and Nomura, T. (1985) Constituents of the cultivated mulberry tree. XXXI. Components of *Broussonetia papyrifera* (L.) Vent. 2. Structures of two new isoprenylated flavans, kazinols A and B. *Heterocycles*, **23**, 2835–2842.

Ikuta, J., Hano, Y., Nomura, T., Kawakami, Y., and Sato, T. (1986) Components of *Broussonetia kazinoki* Sieb. I. Structure of two new isoprenylated flavans and five new isoprenylated 1,3-diphenylpropane derivatives. *Chem. Pharm. Bull.*, **34**, 1968–1979.

Imafuku, K., Honda, M., and McOmie, J. F. W. (1987) Cyclodehydration of 2′-hydroxychalcones with DDQ: A simple route for flavones and aurones. *Synth. Commun.* (2), 199–201.

Imperato, F. (1989) Bracteatin from the fern *Asplenium kaulfussii. Chim. Ind.* (*Milan*), **71**, 86.

Imperato, F. (1992) Kaempferol 3-(2″-*p*-coumaroylrhamnoside)-7-rhamnoside from *Cheilanthes fragrans. Phytochemistry*, **31**, 3291–3292.

Imperato, F. (1993) 3,6,8-Tri-*C*-xylosylapigenin from *Asplenium viviparum. Phytochemistry*, **33**, 729–730.

Imperato, F. (1995) Flavonol glycosides from *Pteridium aquilinum. Phytochemistry*, **40**, 1801–1802.

Imperato, P. (1996a) A polyglycine chain attached to a flavonol glycoside from *Pteridium aquilinum. Amer. Fern J.*, **86**, 127–128.

Imperato, F. (1996b) Kaempferol 3-*O*-(5″-feruloylapioside) from *Pteridium aquilinum. Phytochemistry*, **43**, 1421–1423.

Inderjit and Dakshini, K. M. M. (1991) Hesperetin 7-rutinoside (hesperidin) and taxifolin 3-arabinoside as germination an growth inhibitors in soils associated with the weed *Plucheae lanceolata* (DC.) C. B. Clarke (Asteraceae). *J. Chem. Ecol.*, **17**, 1585–1591.

Inderjit and Lakshini, K. M. M. (1992) Formononetin 7-*O*-glucoside (ononin), an additional growth inhibitor in soils associated with the weed *Pluchea lanceolata* (DC.) C. B. Clarke (Asteraceae). *J. Chem. Ecol.*, **18**, 713–718.

Inderjit and Dakshini, K. M. M. (1994) Allelopathic potential of the phenolics from the roots of *Pluchea lanceolata. Physiol. Plantarum*, **92**, 571–576.

Inderjit. (1996) Allelopathy. *The Botanical Reviews*, **62**, 186–202.

Ingham, J. L. (1982) Phytoalexins from the Leguminosae. In *Phytoalexins*, J. A. Bailey and J. W. Mansfield (eds.), Blackie, Glasgow, 21–80.

Ishikura, N. and Yamamoto, E. (1980) Some factors involved in the blue color of 'heavenly blue' morning glory flowers. Studies on the flower color of morning glory. Part II. *Nippon Nogei Kagaku Kaishi*, **54**, 637–643.

Ishikura, N. and Yang, Z.-q. (1994) Multiple forms of flavonol *O*-glucosyltransferase in young leaves of *Euonymus alatus* forma *ciliato-dentatus. Phytochemistry*, **36**, 1139–1145.

Islam, A. and Hossain, M. A. (1993) Synthesis of 3-*O*-methyl-6-*C*-methyl myricetin. *J. Bangladesh Acad. Sci.*, **17**, 173–178.

Itokawa, H., Sato, K., and Takeya, K. (1981) Structure of isoagastachoside and agastachin, new glucosylflavones isolated from *Agastache rugosa. Chem. Pharm. Bull.*, **29**, 1777–1779.

Iwagawa, T., Kawasaki, J.-I., Hase, T., Sako, S., Okubo, T., Ishida, M., and Kim, M. (1990) An acylated flavonol glycoside from *Lasiobema japonica. Phytochemistry*, **29**, 1013–1014.

Iwashina, T., Matsumoto, S., and Yoshida, Y. (1993) Apigenin 7-rhamnoside-4′-glucosylrhamnoside from *Asplenium normale. Phytochemistry*, **32**, 1629–1630.

Iyer, P. R. and Iyer, C. S. R. (1989) Synthesis of robustone, isorobustone and 4′-*O*-methyl alpinum isoflavone: decarboxylative rearrangement of angular isoflavone carboxylic acids. *J. Nat. Prod.*, **52**, 711–715.

Jackson, R. S. (1994) *Wine Science. Principles and Applications*, Academic Press, San Diego.

Jadhav, G. V. and Kulkarni, U. G. (1951) Borax as a new condensing agent for the preparation of chalcones. *Curr. Sci.*, **20**, 42–43.

Jain, A., Bhartiya, H. P., and Vishwakarma, A. N. (1982) A chalcone glycoside from the heartwood of *Shorera robusta. Phytochemistry*, **21**, 957.

Jain, A. C., Nayyar, N. K., and Arya, P. (1986) Aromatic benzylation. Part V. Synthesis of nuclear benzylated chalcones and dihydrochalcones from gallacetophenone and β-resacetophenone as analogues of uvaretin and isourvaretin. *Ind. J. Chem.*, **25B**, 259–263.

Jain, A. C. and Nayyar, N. K. (1987) Synthesis of (+/−)-sativone and (+/−)-dihydrodaidzein. *Ind. J. Chem.*, **26B**, 136–139.

Jain, A. C. and Prasad, A. K. (1989) Synthesis of naturally occurring pongagallone-A, pongagallone-B and their analogues. *Ind. J. Chem.*, **28B**, 193–194.

Jain, A. C., Kumar, A., and Sharma, N. K. (1991) Synthesis of isoflavanones from isoflavones by reduction with sodium hydrogen telluride. *Ind. J. Chem.*, **30B**, 290–291.

Jansen, R. K. and Palmer, J. D. (1987) A chloroplast DNA inversion marks an ancient evolutionary split in the sunflower family (Asteraceae). *Proc. Natl. Acad. Sci. U.S.A.*, **84**, 5818–5822.

Jaques, U., Kessmann, H., and Barz, W. (1987) Accumulation of phenolic compounds and phytoalexins in sliced and elicitor-treated cotyledons of *Cicer arietinum* L. *Z. Naturforsch.*, **42c**, 1171–1178.

Jay, M. (1994) *C*-Glycosylflavonoids. In *The Flavonoids: Advances in Research since 1986*, J. B. Harborne (ed.), Chapman and Hall, London, 57–93.

Jay, M., De Luca, V., and Ibrahim, R. K. (1983) Meta-Methylation of flavonol rings A (8-) and B (3′-) is catalysed by two distinct *O*-methyltransferases in *Lotus corniculatus. Z. Naturforsch.*, **38c**, 413–417.

Jeandet, P., Sbaghi, M., Bessis, R., and Meunier, P. (1995) The potential relationship of stilbene (resveratrol) synthesis to anthocyanin content in grape berry skins. *Vitis*, **34**, 91–94.

Jha, H. C., Zilliken, F., and Breitmaier, E. (1981) Isoflavone synthesis with 1,3,5-triazine. *Angew. Chem.*, **93**, 129–130.

Ji, X.-d., Melman, N., and Jacobson, K. A. (1995) Interactions of flavonoids and other phytochemicals with adenosine receptors. *J. Med. Chem.*, **39**, 781–788.

Jiang, J., Zhou, R., Meng, Z., and Li, N. (1994a) A new flavanone from *Matteuccia orientalis. Zhongguo Yaoke Daxue Xuebao*, **25**, 199–201.

Jiang, J., Zhou, R., Wang, L., and Li, N. (1994b) Chemical constituents of *Matteuccia intermedia* C. Chr. *Zhongguo Yaoke Daxue Xuebao*, **25**, 265–266.

Johow, F. (1884) Über die Beziehungen einiger Eigenschaften der Laubblätter zu den Standortsverhältnissen. *Bot. Ztg., Leipzig*, **15**, 282–310. (Cited by Onslow, 1925.)

Johns, S. R., Russel, J. H., and Hefferman, M. L. (1965) Ficine, a novel flavonoidal alkaloid. *Tet. Lett.* (24), 1987–1991.

Jones, D. H. (1984) Phenylalanine ammonia-lyase: Regulation of its induction, and its role in plant development. *Phytochemistry*, **23**, 1349–1359.

Jonsson, L. M. V., Aarsman, M. E. G., Schram, A. W., and Bennink, G. J. (1982) Methylation of anthocyanins by cell-free extracts of flower buds of *Petunia hybrida*. *Phytochemistry*, **21**, 2457–2459.

Jonsson, L. M. V., de Vlaming, P., Wiering, H., Aarsman, M. E. G., and Schram, A. W. (1983a) Genetic control of anthocyanin *O*-methyltransferase activity in flowers of *Petunia hybrida*. *Theor. Appl. Genet.*, **66**, 349–355.

Jonsson, L. M. V., Aarsman, M. E. G., Poulton, J. E., and Schram, A. W. (1983b) Properties and genetic control of four methyltransferases involved in methylation of anthocyanins in flowers of *Petunia hybrida*. *Planta*, **160**, 174–179.

Jorgensen, R. (1993) The origin of land plants: a union of alga and fungus advanced by flavonoids? *BioSystems*, **31**, 193–207.

Jungblut, T. P., Schnitzler, J.-P., Heller, W., Hertkorn, N., Metzger, J. W., and Szymczak, J. W. S. (1995a) Structures of UV-B induced sunscreen pigments of the Scots pine (*Pinus sylvestris* L.). *Angew. Chem.*, *International Ed. English*, **34**, 312–314.

Jungblut, T. P., Schnitzler, J.-P., Hertkorn, N., Metzger, J. W., Heller, E., and Sandermann, J. H. (1995b) Acylated flavonoids as plant defense compounds against environmentally relevant UV-B radiation in Scots pine seedlings. *Curr. Topics Plant Physiol.*, 266–267.

Jurd, L. (1962) Spectral properties of flavonoid compounds. In *The Chemistry of Flavonoid Compounds*, T. A. Geissman (ed.), The Macmillan Company, New York, 107–155.

Jurd, L. (1972) Anthocyanidins. In *Recent Progress in the Chemistry of Flavylium Salts in Structural and Functional Aspects of Phytochemistry*, V. R. Runeckles (ed.), Academic Press, New York, 135–145.

Jury, S. L., Reynolds, T., Cutler, D. F., and Evans, F. J. (1987) *The Euphorbiales*, Academic Press, New York, 326.

Kadota, S., Lami, N., Tezuka, Y., and Kikuchi, T. (1988) Boeravinone A and B, new rotenoid analogues from *Boerhaavia diffusa* Linn. *Chem. Pharm. Bull.*, **36**, 834–836.

Kadota, S., Lami, N., Tezuka, Y., and Kikuchi, T. (1989) Constituents of the roots of *Boerhaavia diffusa* L. Examination of sterols and structures of new rotenoids, boeravinones A and B. *Chem. Pharm. Bull.*, **37**, 3214–3220.

Kadota, S., Basnet, P., Hase, K., and Namba, T. (1994) Matteuoriente A and B, two new and potent aldose reductase inhibitors from *Matteuccia orientalis* (Hook.) Trev. *Chem. Pharm. Bull.*, **42**, 1712–1714.

Kagal, S. A., Karmarkar, S. S., and Venkataraman (1956) Synthetical experiments in the chromone group. XXXII. A synthesis of tectorigenin. *Proc. Indian Acad. Sci.*, **44A**, 36–41.

Kajiyama, K., Demizu, S., Hirage, Y., Kinoshita, K., Koyama, K., Takahashi, K., Tamura, Y., Okada, K., and Kinoshita, T. (1992) New prenylflavones and dibenzoylmethane from *Glycyrrhiza inflata*. *J. Nat. Prod.*, **55**, 1197–1203.

Kametaka, T. and Perkin, A. G. (1910) Carthamine. I. *J. Chem. Soc.*, **97**, 1415–1427.

Kamsteeg, J., van Brederode, J., and van Nigtevecht, G. (1978) Identification, properties and genetic control of UDP-glucose:cyanidin-3-rhamnoside-(1-6)-glucoside-5-*O*-glucosyl transferase isolated from petals of red campion (*Silene dioica*). *Biochem. Genetics*, **16**, 1059–1071.

Kamsteeg, J., van Brederode, J., and van Nigtevecht, G. (1979) Properties and genetic control of UDP-L-rhamnose:anthocyanin 3-*O*-glucoside, 6-*O*-rhamnosyl transferase from petals of red campion, *Silene dioica*. *Phytochemistry*, **18**, 659–660.

Kamsteeg, J., van Brederode, J., and van Nigtevecht, G. (1980) Genetical and biochemical evidence that the hydroxylation pattern of the anthocyanin B-ring in *Silene dioica* is determined at the *p*-coumaroyl-coenzyme A stage. *Phytochemistry*, **19**, 1459–1462.

Kamsteeg, J., van Brederode, J., Verschuren, P. M., and van Nigtevecht, G. (1981) Identification, properties and genetic control of *p*-coumaroyl coenzyme A 3-hydroxylase isolated from petals of *Silene dioica*. *Zeit. Pflanzenphysiol.*, **102**, 435–442.

Kaouadji, M. (1986) Grenoblone, nouvelle oxodihydrochalcone des bourgeons de *Platanus acerifolia*. *Journal of Natural Products*, **49**, 500–503.

Kaouadji, M. (1990) Polyphenols from *Platanus acerifolia*. Part 7. Further nonpolar flavonoids from *Platanus acerifolia* buds. *Phytochemistry*, **29**, 1348–1350.

Kaouadji, M., Morand, J.-M., and Gilly, C. (1986) 4-Hydroxygrenoblone, another uncommon *C*-prenylated flavonoid from *Platanus acerifolia* leaves. *J. Nat. Prod.*, **49**, 508–510.

Karabourniotis, G., Kyparissis, A., and Manetas, Y. (1993) Leaf hairs of *Olea europaea* protect underlying tissues against ultraviolet-B radiation damage. *Environ. Exp. Bot.*, **33**, 341–345.

Karabourniotis, G., Papadopoulos, K., Papmarkou, M., and Manetas, Y. (1992) Ultraviolet-B radiation absorbing capacity of leaf hairs. *Physiol. Plant.*, **86**, 414–418.

Karikome, H., Ogawa, K., and Sashida, Y. (1992) New acylated glucosides of chalcones from the leaves of *Bidens frondosa*. *Chem. Pharm. Bull.*, **40**, 689–691.

Kasai, R., Hirono, S., Chou, W.-H., Tanaka, O., and Chen, F.-H. (1988) Sweet dihydroflavonol rhamnoside from leaves of *Engelhardtia chrysolepis*, a Chinese folk medicine, Hung-qi. *Chem. Pharm. Bull.*, **36**, 4167–4170.

Kashiwada, Y., Morita, M., Nonaka, G.-i., and Nishioka, I. (1990) Tannins and related compounds. XCI. Isolation and characterization of proanthocyanidins with an intermolecularly double-linked unit from the fern *Dicranopteris pedata* Houss. *Chem. Pharm. Bull.*, **38**, 856–860.

Kaul, T. N., Middleton Jr., E., and Ogra, P. L. (1985) Antiviral effect of flavonoids on human viruses. *J. Med. Virology*, **15**, 71–79.

Kaval, A. A. and Shah, N. M. (1962) *J. Sci. Ind. Res., Sec. B*, **21**, 234.

Kawano, N. (1962) Recent advances in the chemistry of biflavones. In *Chemistry of Natural and Synthetic Colouring Matters*, T. S. Gore, B. S. Joshi, S. V. Sunthankar, and B. D. Tilak (eds.), Academic Press, New York, 177–185.

Kawasaki, M., Hayashi, T., Arisawa, M., Morita, N., and Berganza, L. H. (1988) 8-Hydroxytricetin 7-glucuronide, a β-glucuronidase inhibitor from *Scoparia dulcis*. *Phytochemistry*, **27**, 3709–3711.

Kay, Q. O. N. and Daoud, H. S. (1981) Pigment distribution, light reflection and cell structure in petals. *Bot. J. Linnean Soc.*, **83**, 57–84.

Kellam, S. J., Mitchell, K. A., Blunt, J. W., Munro, M. H. G., and Walker, J. R. L. (1993) Luteolin and 6-hydroxyluteolin glycosides from *Hebe stricta*. *Phytochemistry*, **33**, 867–869.

Kerscher, F. and Franz, G. (1987) Biosynthesis of vitexin and isovitexin: enzymatic synthesis of the *C*-glucosylflavones vitexin and isovitexin with an enzyme preparation from *Fagopyrum esculentum* M. seedlings. *Z. Naturforsch.*, **42c**, 519–524.

Kerscher, F. and Franz, G. (1988) Isolation and some properties of an UDP-glucose: 2-hydroxyflavanone-6(or 8)-*C*-glucosyltransferase from *Fagopyrum esculentum* M. cotylendons. *J. Plant Physiol.*, **132**, 110–115.

Kevan, P. G. (1972) Floral colors in the high arctic with reference to insect-flower relations and pollination. *Can. J. Bot.*, **50**, 2289–2316.

Kevan, P. G. (1978) Floral coloration, its colorimetric analysis and significance in anthecology. In *The Pollination of Flowers by Insects*, A. J. Richards (ed.), Academic Press, New York, 51–78.

Khalilullah, M., Sharma, V. M., and Rao, P. S. (1993) Ramosismin, a new prenylated chalcone from *Crotalaria ramosissima. Fitoterapia*, **64**, 232–234.

Khan, M. S. Y., Khan, S. U., Khan, C. I. Z., and Parthasarathy, M. R. (1985a) Synthesis of biflavonoids. Part I. *J. Indian Chem. Soc.*, **62**, 310–312.

Khan, M. S. Y., Khan, C. I. Z., and Khan, S. U. (1985b) Synthesis of biflavonoids. Part II. Oxidative coupling of phenolic ketones with ferric chloride-dimethylformamide: proof for inter-flavonyl linkage. *J. Indian Chem. Soc.*, **62**, 335–337.

Khan, M. S. Y., Khan, M. H., and Javed, K. (1990) Attempted synthesis of 3,3′-linked biflavonoids. *Ind. J. Chem.*, **29B**, 1101–1106.

Khan, I. Z., Aqil, M., Ogarawu, V. C., and Rahman, F. A. (1994) Selective demethylation of the 5-position in flavonoids. *Chem. Environ. Res.*, **3**, 125–127.

Khan, I. Z. and Aqil, M. (1995) Synthesis of biflavonoids. Part III. *Ultra Sci. Phys. Sci.*, **7**, 143–146.

Khanna, M. S., Singh, O. V., Garg, C. P., and Kapoor, R. P. (1992) Oxidation of flavanones using thallium (III) salts: A new route for the synthesis of flavones and isoflavones. *J. Chem. Soc., Perkin Trans. 1*, 2565–2568.

Kho, K. F. F., Bennink, G. J. H., and Wiering, H. (1975) Anthocyanin synthesis in a white flowering mutant of *Petunia hybrida* by a complementation technique. *Planta*, **127**, 271–279.

Kho, K. F. F., Bolsman-Louwen, A. C., Vuik, J. C., and Bennink, G. J. H. (1977) Anthocyanin synthesis in a white flowering mutant of *Petunia hybrida*. II. Accumulation of dihydroflavanol intermediates in white flowering mutants; uptake of intermediates in isolated corollas and conversion into anthocyanins. *Planta*, **135**, 109–118.

Khouri, H. E. and Ibrahim, R. K. (1987) Resolution of five position-specific flavonoid *O*-methyltransferases by fast protein liquid chromatofocusing. *J. Chromatog.*, **407**, 291–297.

Khouri, H. E., Tahara, S., and Ibrahim, R. K. (1988a) Partial purification, characterization, and kinetic analysis of isoflavone 5-*O*-methyltransferase from yellow lupin roots. *Arch. Biochem. Biophys.*, **262**, 592–598.

Khouri, H. E., De Luca, V., and Ibrahim, R. K. (1988b) Enzymatic synthesis of polymethylated flavonols in *Chrysosplenium americanum*. III. Purification and kinetic analysis of *S*-adenosyl-L-methionine: 3-methylquercetin 7-*O*-methyltransferase. *Arch. Biochem. Biophys.*, **265**, 1–7.

Kijima, H., Ide, T., Otsuka, H., and Takeda, Y. (1995) Alangiflavoside, a new flavonol glycoside from the leaves of *Alangium premnifolium*. *J. Nat. Prod.*, **58**, 1753–1755.

Kikuchi, Y., Miyaichi, Y., Yamaguchi, Y., Kizu, H., Tomimori, T., and Vetschera, K. (1991) Studies on the constituents of *Scutellaria* species. XIV. On the constituents of the roots and of the leaves of *S. alpina* L. *Chem. Pharm. Bull.*, **39**, 199–201.

King, R. M. and Kranz, V. E. (1975) Ultraviolet reflectance patterns in the Asteraceae. I. Local and cultivated species. *Phytologia*, **31**, 66–85.

Kinghorn, A. D. (1987) Biologically active compounds from plants with reputed medicinal and sweetening properties. *J. Nat. Prod.*, **50**, 1009–1024

Kleinehollenhorst, G., Behrens, H., Pegels, G., Srunk, N., and Wiermann, R. (1982) Formation of flavonol 3-*O*-diglycosides and flavonol 3-*O*-triglycosdies by enzyme extracts from anthers of *Tulipa* cv. Apeldoorn. *Z. Naturforsch.*, **37c**, 587–599.

Knobloch, K.-H. and Hahlbrock, K. (1975) Isoenzymes of *p*-coumarate: CoA ligase from cell suspension cultures of *Glycine max*. *Eur. J. Biochem.*, **52**, 311–320.

Knobloch, K.-H. and Hahlbrock, K. (1977) 4-Coumarate:CoA ligase from cell suspension cultures of *Petroselinum hortense* Hoffm. Partial purification, substrate specificity, and further properties. *Arch. Biochem. Biophys.*, **184**, 237–248.

Knogge, W. and Weisenböck, G. (1984) Purification, characterization, and kinetic mechanism of *S*-adenosyl-L-methionine: vitexin 2″-*O*-rhamnoside 7-*O*-methyltransferase of *Avena sativa* L. *Eur. J. Biochem.*, **140**, 113–118.

Kny, L. (1892) Zur physiologischen Bedeutung des Anthocyans. *Congresso Botanico Internazionale*. (Cited by Onslow, 1925.)

Kochs, G., and Grisebach, H. (1986) Enzymic synthesis of isoflavones. *Eur. J. Biochem.*, **155**, 311–318.

Kochs, G. and Grisebach, H. (1987) Induction and characterization of a NADPH-dependent flavone synthase from cell cultures of soybean. *Z. Naturforsch.*, **42c**, 343–348.

Koeppen, B. H. (1962) Flavone *C*-glycosides. The interrelation of orientin and homoorientin. *S. African J. Lab. Clin. Med.*, **8**, 125–126.

Koeppen, B. H. (1964) *Z. Naturforsch.*, **19b**, 173.

Koeppen, B. H. and Roux, D. G. (1965) Aspalatin, a novel *C*-glycosylflavonoid from *Aspalathus linearis*. *Tet. Lett.*, 3497–3503.

Koes, R. E., Spelt, C. E., Mol, J. N. M., and Gerats, A. G. M. (1987) The chalcone synthase multigene family of *Petunia hybrida* (V30): sequence homology, chromosomal localization and evolutionary aspects. *Plant Mol. Biol.*, **10**, 375–385.

Koes, R. E., Spelt, C. E., and Mol, J. N. M. (1989) The chalcone synthase multigene family of *Petunia hybrida* (V30): differential, light-regulated expression during flower development and UV light induction. *Plant Mol. Biol.*, **12**, 213–225.

Komissarenko, N. F., Chernobai, V. T., and Derkach, A. I. (1988) Flavonoids of inflorescences of *Calendula officinalis*. *Khim. Prir. Soedn.*, 795–801.

Kondo, T., Yoshida, K., Nakagawa, A., Kawai, T., Tamura, H., and Goto, T. (1992) Structural basis of blue-colour development in flower petals from *Commelina communis*. *Nature* (*London*), **358**, 515–518.

Kondo, T., Yamashiki, J., Kawahori, K., and Goto, T. (1989a) Structure of lobelinin A and B, novel anthocyanins acylated with three and four different organic acids, respectively. *Tet. Lett.*, **30**, 6055–6058.

Kondo, T., Yoshikane, M., Yoshida, K., and Goto, T. (1989b) Structure of anthocyanins in scarlet, purple and blueb flowers of *Salvia. Tet. Lett.*, **30**, 6729–6732.

Konishi, F., Esaki, S., and Kamiya, S. (1983) Synthesis and taste of flavanone and dihydrochalcone glycosides containing 3-*O*-α-L-rhamnopyranosyl-D-glucopyranose or 4-*O*-α-L-rhamnopyranosyl-D-glucopyranose in the sugar moiety. *Agric. Biol. Chem.*, **47**, 265–275.

Koukol, J. and Conn, E. E. (1961) The metabolism of aromatic compounds in higher plants. IV. Purification and properties of the phenylalanine deaminase of *Hordeum vulgare*. *J. Biol. Chem.*, **236**, 2692–2698.

Koupai-Abyazani, M. R., Creaser, C. S., and Stephanson, G. R. (1992) Separation and identification of flavone, flavonol, isoflavone and flavanone aglycones by capillary column gas chromatography. *Phytochem. Anal.*, **3**, 80–84.

Koupai-Abyazani, M. R., McCallum, J., Muir, A. D., Bohm, B. A., Towers, G. H. N., and Gruber, M. Y. (1993a) Developmental changes in the composition of proanthocyanidins from leaves of sainfoin (*Onobrychis viciifolia* Scop.) as determined by HPLC analysis. *J. Agric. Food Chem.*, **41**, 1066–1070.

Koupai-Abyazani, M. R., McCallum, J., Muir, A. D., Lees, G. L., Bohm, B. A., Towers, G. H. N., and Gruber, M. Y. (1993b) Purification and characterization of a proanthocyanidin polymer from seed of alfalfa (*Medicago sativa* cv. Beaver). *J. Agric. Food. Chem.*, **41**, 565–569.

Köhrle, J., Fang, S. L., Yang, Y., Irmscher, K., Hesch, R. D., Pino, S., Alex, S., and Braverman, L. E. (1989) Rapid effects of the flavonoid EMD-21388 on serum thyroid hormone binding and thyrotropin regulation in the rat. *Endocrinology*, **125**, 532–537.

König, W. A., Krauss, C., and Zähner, H. (1977) Stoffwechselprodukte von Mikroorganism. 6-Chlorgenistein und 6,3′-Dichlorgenistein. *Helv. Chim. Acta*, **60**, 2071–2078.

Köster, J. and Barz, W. (1981) UDP-Glucose: isoflavone 7-*O*-glucosyltransferase from roots of chick pea (*Cicer arietinum* L.). *Arch. Biochem. Biophys.*, **212**, 98–104.

Kreuzaler, F. and Hahlbrock, K. (1972) Enzymatic synthesis of aromatic compounds in higher plants: Formation of naringenin (5,7,4′-trihydroxyflavanone) from *p*-coumaroyl coenzyme A and malonyl coenzyme A. *FEBS Lett.*, **28**, 69–72.

Kreuzaler, F. and Hahlbrock, K. (1973) Flavonol glycosides from illuminated cell suspension cultures of *Petroselinum hortense*. *Phytochemistry*, **12**, 1149–1152.

Krishna Prasad, A. V., Kapil, R. S., and Popli, S. P. (1986) Synthesis of (±)-isomedicarpin, (±)-homopterocarpin and tuberostan: A novel entry of 'hydrogenative cyclisation' into pterocarpans. *J. Chem. Soc., Perkin Trans. 1*, 1561–1563.

Krishnan, S. K., Murti, V. V. S., and Seshadri, T. R. (1966) Synthesis of hinokiflavone pentamethyl ether. *Curr. Sci.*, **35**, 64–65.

Krishnamurty, H. G. and Prasad, J. S. (1977) A new synthesis of isoflavones using 'active formate'. *Tet. Lett.*, 3071–3072.

Krishnamurty, H. G. and Sathyanarayana, S. (1986) Catalytic transfer hydrogenation, a chemo-selective reduction of isoflavones to isoflavanones. *Synth. Commun.*, **16**, 1657–1663.

Krishnamurty, H. G. and Sathyanarayana, S. (1989) Catalytic transfer hydrogenation, a facile conversion of hydroxyflavanones into hydroxydihydrochalcones. *Synth. Commun.*, **19**, 119–123.

Krol, W., Czuba, Z. P., Threadgill, M. D., Cunningham, B. D. M., and Pietsz, G. (1995) Inhibition of nitric oxide (NO) production in murine macrophages by flavones. *Biochem. Pharmacol.*, **50**, 1031–1035.

Kubitzki, K. and Gottlieb, O. R. (1984) Phytochemical aspects of angiosperm origin and evolution. *Acta Bot. Neerland.*, **33**, 457–468.

Kubo, I. and Yokokawa, Y. (1992) Two tyrosinase inhibiting flavonol glycosides from *Buddleia coriacea. Phytochemistry*, **31**, 1075–1077.

Kubo, I., Kim, M., and Naoki, H. (1987) New insect growth inhibitory flavan glycosides from *Viscum tuberculatum. Tet. Lett.*, **28**, 921–924.

Kugler, H. (1943) Hummeln als Blütenbesucher. *Ergeb. Biol.*, **19**, 143–323. (Cited by Onslow, 1925.)

Kuhn, B., Forkmann, G., and Seyffert, W. (1978) Genetic control of chalcone-flavanone isomerase activity in *Callistephus chinensis* (L.) Nees. *Planta*, 138, 199–203.

Kuhns, L. J. and Fretz, T. A. (1978a) Distinguishing rose cultivars by polyacrylamide gel electrophoresis. I. Extraction and storage of protein and active enzyme from rose leaves. *J. Am. Soc. Hort. Sci.*, **103**(4), 503–508.

Kuhns, L. J. and Fretz, T. A. (1978b) Distinguishing rose cultivars by polacrylamide gel electrophoresis. II. Isozyme variation among cultivars. *J. Am. Soc. Hort. Sci.*, **103**, 509–516.

Kulanthaivel, P. and Benn, M. H. (1986) A new truxillate and some flavonoid esters from the leaf gum of *Traversia baccharoides* Hook. f. *Can. J. Chem.*, **64**, 514–519.

Kulkarni, M. M., Rojatkar, S. R., and Nagasampagi, B. A. (1987) Four 6-hydroxyflavonols from *Blumea malcomii. Phytochemistry*, **26**, 2079–2083.

Kunesch, G. and Polonsky, J. (1967) Biosynthesis of neoflavonoids: callophyllolide (4-phenylcoumarin). *Chem. Commun.*, 317–318.

Kuntze, O. (1877) Die Schutzmittel der Pflanzen gegen Thiere und Wetterungunst. *Bot., Ztg., Leipzig*, Supplementheft. (Cited by Onslow, 1925.)

Kunz, S., Burkhardt, G., and Becker, H. (1994) Riccionidins A and B, anthocyanidins from the cell walls of the liverwort *Ricciocarpos natans. Phytochemistry*, **35**, 233–235.

Kuppusamy, U. R., Khoo, H. E., and Das, N. P. (1990) Structure-activity studies of flavonoids as inhibitors of hyaluronidase. *Biochem. Pharm.*, **40**, 397–401.

Kuroki, G. and Poulton, J. E. (1981) The *para-O*-methylation of apigenin to acacetin by cell-free extracts of *Robinia pseudoacacia* L. *Z. Naturforsch.*, **36c**, 916–920.

Kurosawa, K., Ollis, W. D., Redman, B. T., Sutherland, I. O., Oliveira, A. B., Gottlieb, O. R., and Alves, H. M. (1968) Natural occurrence of isoflavans and isoflavanquinones. *Chem. Commun.* (20), 1263–1264.

Kuroyanagi, M., Sato, M., Veno, A., and Nishi, K. (1987) Flavonoids of *Andrographis paniculata. Chem. Pharm. Bull.*, **35**, 4429–4435.

Kvist, L. P. and Pedersen, J. A. (1986) Distribution and taxonomic implications of some phenolics in the family Gesneriaceae determined by EPR spectroscopy. *Biochem. Syst. Ecol.*, **14**, 385–405.

La Duke, J. and Crawford, D. J. (1979) Character compatibility and phyletic relationship in several closely related species of *Chenopodium* of the western United States. *Taxon*, **28**, 307–314.

Laflamme, P., Khouri, H., Gulick, P., and Ibrahim, R. K. (1993) Enzymatic prenylation of isoflavones in white lupin. *Phytochemistry*, **34**, 147–151.

Lai, S. M. F., Orchison, J. J. A., and Whiting, D. A. (1989) A new synthetic approach to the rotenoid ring system. *Tetrahedron*, **45**, 5895–5906.

Laks, P. E. (1987) Flavonoid biocides: Phytoalexin analogues from condensed tannins. *Phytochemistry*, **26**, 1617–1621.

Lal, J. B. and Dutt, S. (1935) Chemical examination of *Butea frondosa* flowers. Isolation of a crystalline glucoside of butin. *J. Indian Chem. Soc.*, **12**, 262–267.

Lam, J. and Wrang, P. (1975) Flavonoids and polyacetylenes in *Dahlia tenuicaulis*. *Phytochemistry*, **14**, 1621–1623.

Lammers, T. G. (1992) Circumscription and phylogeny of the Campanulales. *Ann. Missouri. Bot. Gard.*, **79**, 388–413.

Lang, W., and Potrykus, I. (1971) Lichtmikroskopische und pigmentanalytische Untersuchungen zum Ablauf der Petalenentwicklung von *Torenia baillonii*. *Z. Pflanzenphysiol.*, **65**, 1–12.

Larson, R. L. and Coe, E. H. (1977) Gene-dependent flavonoid glucosyltransferase in maize. *Biochem. Genetics*, **15**, 153–156.

Larson, R. L. (1989) Flavonoid 3′-*O*-methylation by a *Zea mays* L. preparation. *Biochem. Physiol. Pflanzen*, **184**, 453–460.

Lata, P. (1981) Cytological studies in the genus *Rosa*. I. Mitotic analysis. *Caryologia*, **34**, 409–417.

Lata, P. (1982a) Cytological studies in the genus *Rosa*. II. Meiotic analysis of ten species. *Caryologia*, **47**, 631–637.

Lata, P. (1982b) Cytological studies in the genus Rosa. III. Meiotic analysis of ten cultivars. *Cytologia*, **47**, 639–647.

Latchinian, L. and Ibrahim, R. K. (1989) Characterization of a monoclonal antibody specific to a flavonol 2′-*O*-glucosyltransferase. *Biochem. Cell Biol.*, **67**, 210–213.

Laughton, M. J., Evans, P. J., Moroney, M. A., Hoult, J. R. S., and Halliwell, B. (1991) Inhibition of mammalian 5-lipoxygenase and cyclo-oxygenase by flavonids and phenolic dietary additives. Relationship to antioxidant activity and to iron ion reducing ability. *Biochem. Pharmacol.*, **42**, 1673–1681.

Laux, D. O., Stefani, G. M., and Gottlieb, O. R. (1985) The chemistry of Brazilian Leguminosae. Part 62. Bausplendin, a dimethylenedioxyflavone from *Bauhinia splendens*. *Phytochemistry*, **24**, 1085–1088.

Lawrence, W. J. C. and Price, J. R. (1940) The genetics and chemistry of flower colour variation. *Biol. Reviews*, **15**, 35–58.

Lawrence, W. J. C. and Scott-Moncrieff, R. (1935) The genetics and chemistry of flower colour in *Dahlia*: a new theory of specific pigmentation. *J. Genetics*, **37**, 299–315.

Lee, K.-H., Anuforo, D. C., Huang, E.-S., and Piantadosi, C. (1972) Antitumor agents. I. Angustibalin, a new cytotoxic sesquiterpene lactone from *Balduina angustifolia* (Pursh.) Robins. *J. Pharm. Sci.*, **61**, 626–628.

Lee, S.-S., Tsai, F.-Y., and Chen, I.-S. (1995) Chemical constituents from *Berchemia formosana*. *J. Chin. Chem. Soc.*, **42**, 101–105.

Lee, D., Ellard, M., Wanner, L. A., Davis, K. R., and Douglas, C. J. (1995) The *Arabidopsis thaliana* 4-coumarate: CoA ligase (*4CL*) gene: stress and developmentally regulated expression and nucleotide sequence of its cDNA. *Plant Mol. Biol.*, **28**, 871–884.

Lee, D. and Douglas, C. J. (1996) Two divergent members of a tobacco 4-coumarate: coenzyme A ligase (4CL) gene family. *Plant Physiol.*, **112**, 193–205.

Leeuwenberg, A. J. M. (1980) *Polypremum. Die Natürlich. Pflanzenfam.* 28b.

Lemmich, E., Adewunmi, C. O., Furu, P., Kristensen, A., Larsen, L., and Olsen, C. E. (1996) 5-Deoxyflavones from *Parkia clappertoniana*. *Phytochemistry*, **42**, 1011–1013.

Levai, A. and Sebok, P. (1992) New procedures for the preparation of isoflavones with unsubstituted Ring A. *Synth. Commun.*, **22**, 1735–1750.

Levy, M. and Levin, D. A. (1971) The origin of novel flavonoids in *Phlox* allotetraploids. *Proc. Natl. Acad. Sci., U.S.A.*, **68**, 1627–1630.

Levy, M. and Levin, D. A. (1974) Novel flavonoids and reticulate evolution in the *Phlox pilosa-P. drummondii* complex. *Amer. J. Bot.*, **61**, 156–167.

Levy, M. and Levin, D. A. (1975) The novel flavonoid chemistry and phylogenetic origin of *Phlox floridana*. *Evolution*, **29**, 487–499.

Levy, M. (1976) Altered glycoflavone expression in induced autotetraploids of *Phlox drummondii*. *Biochem. Syst. Ecol.*, **4**, 249–254.

Levy, M. (1977) Minimum biosynthetic-step indices as measures of comparative flavonoid affinity. *Syst. Bot.*, **2**, 89–98.

Li, B.-Q., Fu, T., Yan, Y.-D., Baylor, N. W., Ruscetti, R. W., and Kung, H.-F. (1993) Inhibition of HIV infection by baicalin, a flavonoid compound purified from Chinese herbal medicine. *Cell. Mol. Biol. Res.*, **39**, 119–124.

Li, J., Ou-Lee, T.-M., Raba, R., Amundson, R. G., and Last, R. L. (1993) *Arabidopsis* flavonoid mutants are hypersensitive to UV-B radiation. *The Plant Cell*, **5**, 171–179.

Li, R., Fang, N., and Mabry, T. J. (1987) Flavonoids from *Gutierrezia texana* var. *texana*. *Phytochemistry*, **26**, 2831–2833.

Li, W.-K., Pan, J.-Q., Lü, M.-J., Xiao, P.-G., and Zhang, R.-Y. (1996) Anhydroicaritin 3-*O*-rhamnosyl(1–>2)rhamnoside from *Epimedium koreanum* and a reappraisal of other rhamnosyl(1–>2, 1–>3 and 1–>4)rhamnoside structures. *Phytochemistry*, **42**, 213–216.

Li, X.-C., Cai, L., and Wu, C. D. (1997) Antimicrobial compounds from *Ceanothus americanus* against oral pathogens. *Phytochemistry*, **46**, 97–102.

Liang, X., Dron, M., Schmid, J., Dixon, R. A., and Lamb, C. J. (1989) Developmental and environmental regulation of a phenylalanine ammonia-lyase-β-glucuronidase gene fusion in transgenic tobacco. *Proc. Natl. Acad. Sci. USA*, **86**, 9284–9288.

Light, R. J. and Hahlbrock, K. (1980) Randomization of the flavonoid A ring during biosynthesis of kaempferol from [1,2-^{13}C]-acetate in cell suspension cultures of parsley. *Z. Naturforsch.*, **35c**, 717–721.

Lin, Y. L., Ou, J. C., Chen, C. F., and Kuo, Y. H. (1991) Flavonoids from the roots of *Scutellaria luzonica* Rolfe. *J. Chinese Chem. Soc.*, **38**, 619–623.

Lin, C.-N. and Shieh, W.-L. (1992) Pyranoflavonoids from *Artocarpus communis*. *Phytochemistry*, **31**, 2922–2924.

Lin, C.-N., Shieh, W.-L., Ko, F.-N., and Teng, C.M. (1993) Antiplatelet activity of some prenylflavonoids. *Biochem. Pharmacol.*, **45**, 509–512.

Linn, F., Heidmann, I., Saedler, H., and Meyer, P. (1990) Epigenetic changes in the expression of the maize *A1* gene in *Petunia hybrida*: Role of numbers of integrated gene copies and state of methylation. *Mol. Gen. Genet.*, **222**, 329–336.

List, P. H. and Freund, B. (1968) Geruchsstoffe der Stinkmorchel, *Phallus impudicus* L. 18 Mitt. über Pilzinhaltsstoffe. *Plant Medica* (*Suppl.*), 123.

Litkei, G., Gulácsi, K., Antus, S., and Blaskó, G. (1995) Cyclodehydration of 2′-hydroxychalcones with hypervalent iodine reagent: a new synthesis of flavones. *Liebig's Ann.*, 1711–1715.

Little, H. N. and Bloch, K. (1950) The utilization of acetic acid from the biological synthesis of cholesterol. *J. Biol. Chem.*, **183**, 33–46.

Liu, K. C.-S., Yang, S. L., Roberts, M. F., and Phillipson, J. D. (1989) Flavonol glycosides with acetyl substitution from *Kalanchoe gracilis*. *Phytochemistry*, **28**(10), 2813–2818.

Liu, Q., Liu, M., Mabry, T. J., and Dixon, R. A. (1994) Flavonol glycosides from *Cephalocereus senilis*. *Phytochemistry*, **36**, 229–231.

Locksley, H. D. (1973) The chemistry of biflavanoid compounds. *Progress in the Chemistry of Organic Natural Products*, **30**, 207–312.

Lois, R. and Hahlbrock, K. (1992) Differential wound activation of members of the phenylalanine ammonia-lyase and 4-coumarate:CoA ligase gene families in various organs of parsley plants. *Z. Naturforsch.*, **47c**, 90–94.

Lois, R. (1994) Accumulation of UV-absorbing flavonoids induced by UV-B radiation in *Arabidopsis thalinana* L. I. Mechanisms of UV-resistance in *Arabidopsis*. *Planta*, **194**, 498–503.

Lois, R. and Buchanan, B. B. (1994) Severe sensitivity to ultraviolet radiation in an *Arabidopsis* mutant deficient in flavonoid accumulation. II. Mechanisms of UV-resistance in *Arabidopsis*. *Planta*, **194**, 504–509.

Lowry, B., Lee, D., and Hebant, C. (1980) The origin of land plants: a new look at an old problem. *Taxon*, **29**, 183–197.

Lowry, J. B. (1972) Anthocyanins of some Malaysian members of the Gesneriaceae. *Phytochemistry*, **11**, 3267–3269.

Lowry, J. B. (1973) *Rhabdothamnous solandri*. Phytochemical results. *N. Z. J. Bot.*, **11**, 555–560.

Lu, T. S., Saito, N., Yokoi, M., Shigihara, A., and Honda, T. (1991) An acylated peonidin glycoside in the violet-blue flowers of *Pharbitis nil*. *Phytochemistry*, **30**, 2387–2390.

Lu, T. S., Saito, N., Yokoi, M., Shigihara, A., and Hondo, T. (1992a) Acylated pelargonidin glycosides in the red-purple flowers of *Pharbitis nil*. *Phytochemistry*, **31**, 289.

Lu, T. S., Saito, N., Yokoi, M., Shigihara, A., and Hondo, T. (1992b) Acylated peonidin glycosides in the violet-blue cultivars of *Pharbitis nil*. *Phytochemistry*, **31**, 659–663.

Lu, C.-M. and Lin, C.-N. (1993) Two 2′,4′,6′-Trioxygenated flavanones form *Artocarpus heterophyllus*. *Phytochemistry*, **33**, 909–911.

Ludwig, F. (1889) Extranuptiale Saftmale bei Ameisenpflanzen. *Humboldt*, **8**, 294–297. (Cited by Onslow, 1925.)

Ludwig, F. (1891) Die Beziehungen zwischen Pflanzen an Schnecken. *Beihefte z. Bot. Centralbl., Cassel*, **1**, 35–39. (Cited by Onslow, 1925.)

Lundegardh, H. and Stenlid, G. (1944) On the exudation of nucleotides and flavanones from living roots. *Ark. Bot.*, **31A**, 1–27.

Mabberley, D. J. (1987) *The plant-book*, Cambridge University Press, Cambridge, England.

Mabry, T. J. (1973) Is the order Centrospermae monophyletic? In *Chemistry in Botanical Classification*, G. Bendz and J. Santesson (eds.), Academic Press, New York, 275–285.

Mabry, T. J. (1977) The order Centrospermae. *Ann. Missouri Bot. Gard.*, **64**, 210–220.

Mabry, T. J. (1980) Betalains. In *Secondary Plant Products*, E. A. Bell and B. V. Charlwood (eds.), Springer-Verlag, New York, 513–533.

Mabry, T. J., Markham, K. R., and Thomas, M. B. (1970) *The systematic identification of flavonoids*, Springer-Verlag, New York.

Macaire-Princep. (1828) Mémoire sur la coloration automnale des feuilles. *Mém. Soc. Phys., Genève*, **4**, 43–53. (Cited by Onslow, 1925.)

Macchiati, L. (1899) Osservazioni sui nettarii estranuziali del *Prunus laurocerasus*. *Boll. Soc. bot. ital., Firenze*, 144–147. (Cited by Onslow, 1925.)

Madhavi, F. L., Juthangkoon, S., Lewen, K., Berber-Jimenez, M. D., and Smith, M. A. L. (1996) Characterization of anthocyanins from *Ajuga pyramidalis* cell cultures. *J. Agric. Food. Chem.*, **44**, 1170–1176.

Mahal, H. S., Rai, H. S., and Venkataraman, K. (1934) Synthetical experiments in the chromone grouo. Part XII. Synthesis of 7-hydroxyisoflavone and of α- and β-naphthaisoflavone. *J. Chem. Soc.*, 1120–1122.

Mahesh, V. B. and Seshadri, T. R. (1955) Iodine oxidation (dehydrogenation) of hydroxyflavanones. *J. Sci. Indust. Res. India*, **14B**, 608–609.

Mahmoud, E. N. and Waterman, P. G. (1985) Flavonoids from the stem bark of *Millettia hemsleyana*. *Phytochemistry*, **24**, 369–371.

Major, A., Nógradi, M., Vermes, B., and Kajtar-Peredy, M. (1988) Synthesis of the natural isoflav-3-enes haginin A, B and D. *Liebig's Annalen*(6), 555–558.

Makrandi, J. K. and Seema (1989) An efficient procedure for cyclization of 2′-hydroxychalcones into flavones. *Chem. & Ind.* (*London*), 607.

Malan, E. (1993) A flavonol with a tetrasubstituted B-ring from *Distemonanthus benthamianus*. *Phytochemistry*, **32**, 1631–1632.

Malan, E. and Swinny, E. (1990) Flavonoids and isoflavonoids from the heartwood of *Virgilia oroboides*. *Phytochemistry*, **29**, 3307–3309.

Malan, E., Sireeparsad, A., Swinny, E., and Ferreira, D. (1997) The structure and synthesis of a 7,8,4′-trihydroxyflavan-epioritin dimer from *Acacia caffra*. *Phytochemistry*, **44**, 529–531.

Malhotra, B., Onyilagha, J. C., Bohm, B. A., Towers, G. H. N., James, D., Harborne, J. B., and French, C. J. (1996) Inhibition of tomato ringspot virus by flavonoids. *Phytochemistry*, **43**, 1271–1276.

Mallik, U. K., Saha, M. M., and Mallick, A. K. (1989) Cyclodehydrogenation of 2′-hydroxychalcones and dehydrogenation of flavanones using nickel peroxide. *Ind. J. Chem.*, **28B**, 970–972.

Malterud, K. E., Bremnes, T. E., Faegri, A., Moe, T., and Dugstad, E. K. S. (1985) Flavonoids from the wood of *Salix capra* as inhibitors of wood-destroying fungi. *J. Nat. Prod.*, **48**, 559–563.

Manickam, M., Ramanathan, M., Jahromi, M. A. F., Chansouria, J. P. N., and Ray, A. B. (1997) Antihyperglycemic activity of phenolics from *Pterocarpus marsupium*. *J. Nat. Prod.*, **60**, 609–610.

Mann, J. (1992) *Murder, Magic and Medicine*, Oxford University Press, Oxford.

Manners, G. D. and Jurd, L. (1979) Additional flavanoids in *Gliricidia sepium*. *Phytochemistry*, **18**, 1037–1042.

Marakis, P. (1982) Anthocyanins as Food Colors. In *Food Science and Technology*, G. F. Stewart, B. S. Schweigert, and J. Hawthorn (eds.), Academic Press, New York.

Marchelli, R. and Vining, L. C. (1973) The biosynthetic origin of chlorflavonin, a flavonoid antibiotic from *Aspergillus candidus*. *Can. J. Biochem.*, **51**, 1624–1629.

Marco, J. A., Adell, J., Barbera, O., Strack, D., and Wray, V. (1989) Two isorhamnetin triglycosdies from *Anthyllis sericea*. *Phytochemistry*, **28**, 1513–1516.

Marco, J. A., Barbera J. O., Rodriguez, S., Domingo, C., and Adell, J. (1988) Flavonoids and other phenolics from *Artemisia hispanica*. *Phytochemistry*, **27**, 3155–3159.

Markham, K. R. (1982) *Techniques of Flavonoid Identification*, Academic Press, New York.

Markham, K. R. (1984) The structures of amentoflavone glycosides isolated from *Psilotum nudum*. *Phytochemistry*, **23**, 2053–2056.

Markham, K. R. (1975) Isolation techniques for flavonoids. In *The Flavonoids*, J. B. Harborne, T. J. Mabry, and H. Mabry (eds.), Chapman & Hall, London, 1–44.

Markham, K. R. (1988) Distribution of flavonoids in the lower plants and its evolutionary significance. In *The Flavonoids: Advances in Research since 1980*, J. B. Harborne (ed.), Chapman and Hall, London, 427–468.

Markham, K. R. (1989a) A reassessment of the data supporting the structures of *Blumea malcolmii* flavonols. *Phytochemistry*, **28**, 243–244.

Markham, K. R. (1989b) Flavones, flavonols and their glycosides. In *Methods in Plant Biochemistry. Vol. 1, Plant Phenolics*, J. B. Harborne (ed.), Academic Press, New York, 197–235.

Markham, K. R. (1996) Novel anthocyanins produced in petals of genetically transformed lisianthus. *Phytochemistry*, **42**, 1035–1038.

Markham, K. R. and Campos, M. (1996) 7- And 8-*O*-methylherbacetin-3-*O*-sophorosides from bee pollens and some structure/activity observations. *Phytochemistry*, **43**, 763–767.

Markham, K. R. and Geiger, H. (1994) H-1 nuclear magnetic resonance spectroscopy of flavonoids and their glycosides in hexadeuterodimethylsulfoxide.

In *The Flavonoids: Advances in Research Since 1986*, J. B. Harborne (ed.), Chapman & Hall, London, 441–497.

Markham, K. R. and Mabry, T. J. (1975) Ultraviolet-visible and proton magnetic resonance spectroscopy of flavonoids. In *The Flavonoids*, J. B. Harborne, T. J. Mabry, and H. Mabry (eds.), Chapman and Hall, London, 45–77.

Markham, K. R. and Porter, L. J. (1978) Production of an aurone by bryophytes in the reproductive phase. *Phytochemistry*, **17**, 159–160.

Markham, K. R., Geiger, J., and Jaggy, H. (1992) Kaempferol-3-*O*-glucosyl(1-2) rhamnoside from *Ginkgo biloba* and a reappraisal of other gluco(1-2, 1-3 and 1-4)rhamnoside structures. *Phytochemistry*, **31**, 1009–1011.

Markham, K. R., Hammett, K. R. W., and Ofman, D. J. (1992) Floral pigmentation in two yellow-flowered *Lathyrus* species and their hybrid. *Phytochemistry*, **31**, 549–554.

Markham, K. R., Mitchell, K. A., and Boase, M. R. (1997) Malvidin-3-*O*-glucoside-5-*O*-(6-acetylglucoside) and its colour manifestation in 'Johnson's Blue' and other 'blue' geraniums. *Phytochemistry*, **45**, 417–423.

Markham, K. R., Webby, R. F., Molloy, B. P. J., and Vilain, C. (1989) Support from flavonoid glycoside distribution for the division of *Dacrydium sensu lato*. *N. Z. J. Bot.*, **27**, 1–11.

Markham, K. R., Webby, R. F., Whitehouse, L. A., Molloy, B. P. J., Vilain, C., and Mues, R. (1985a) Support from flavonoid glycoside distribution for the division of *Podocarpus* in New Zealand. *N. Z. J. Bot.*, **23**, 1–13.

Marquart, L. C. (1835) *Die Farben der Blüthen. Eine chemisch-physiol. Abhandlung*, Bonn. (Cited by Onslow, 1925.)

Marston, A., Slacanin, I., and Hostettmann, K. (1990) Centrifugal partition chromatography in the separation of natural products. *Phytochem. Anal.*, **1**, 3–17.

Martin, M. and Dewick, P. M. (1980) Biosynthesis of pterocarpan, isoflavan and coumestan metabolites of *Medicago sativa*: the role of an isoflav-3-ene. *Phytochemistry*, **19**, 2341–2346.

Martin, C. and Gerats, T. (1993) The control of flower coloration. In *The Molecular Biology of Flowering*, B. R. Jordan (ed.), CAB International, Wallingford, 219–255.

Martinez, M. and Swain, T. (1977) Variation in flavonoid patterns in relation to chromosome changes in *Gibasis schiedeana*. *Biochem. Syst. Ecol.*, **5**, 37–43.

Mastenbroek, O., Maas, J. W., van Brederode, J., and van Nigtevecht, G. (1982) The geographic distribution of flavone glycosylation genes in *Silene pratensis* (Rafn.) Godron & Gren. (Caryophyllaceae). *Genetica*, **59**, 139–144.

Mastenbroek, O., Hogeweg, P., van Brederode, J., and van Nigtevecht, G. (1983a) A pattern analysis of the geographical distribution of flavone-glycosylating genes in *Silene pratensis*. *Biochem. Syst. Ecol.*, **11**, 91–96.

Mastenbroek, O., Prentice, H. C., Kamps-Heinsbroek, R., van Brederode, J., Niemann, G. J., and van Nigtevecht, G. (1983b) Geographic trends in flavone-glycosylation genes and seed morphology in European *Silene pratensis* (Caryophyllaceae). *Plant Syst. Evol.*, **141**, 257–271.

Masterova, I., Uhrin, D., and Tomko, J. (1987) Lilaline-A flavonoid alkaloid from *Lilium candidum*. *Phytochemistry*, **26**, 1844–1845.

Matern, U., Potts, J. R. M., and Hahlbrock, K. (1981) Two flavonoid-specific malonyltransferases from cell suspension cultures of *Petroselinum hortense*: partial purification and some properties of malonyl-coenzyme A: flavone/flavonol 7-*O*-glycoside malonlyltransferase and malonyl-coenzyme A: flavone 3-*O*-glucoside malonyltransferase. *Arch. Biochem. Biophys.*, **208**, 233–241.

Matern, U., Feser, C., and Hammer, D. (1983a) Further characterization and regulation of malonyl-coenzyme A: flavonoid glucoside malonyltransferases from parsley cell suspension cultures. *Arch. Biochem. Biophys.*, **226**, 206–217.

Matern, U., Heller, W., and Himmelspach, K. (1983b) Conformational changes in apigenin 7-*O*-(6-*O*-malonylglucoside), a vacuolar pigment from parsley, with solvent composition and protein concentration. *Eur. J. Biochem.*, **133**, 439–448.

Matern, U., Reichenbach, C., and Heller, W. (1986) Efficient uptake of flavonoids into parsley (*Petroselinum hortense*) vacuoles rerquires acylated glycosides. *Planta*, **167**, 183–189.

Matlawska, I., Sikorska, M., and Kowalewski, Z. (1991) Flavonoid compounds in the herb *Arabis caucasica*. II. Heteroside derivatives of kaempferol. *Acta Polonica Pharm.*, **48**, 31–34.

Matsui, S., Suzuki, T., and Nakamura, M. (1984) Distribution of flower pigments in perianth of *Vandeae* orchids. *Research Bulletin of the Faculty of Agriculture, Gifu University*, (49), 361–369.

Matsui, S. and Nakamura, M. (1988) Distribution of flower pigments in perianth of *Cattleya* and allied genera. I. Species. *J. Japanese Soc. Hort. Sci.*, **57**, 222–232.

Matsui, S. (1988) Distribution of flower pigments in perianth of *Cattleya* and allied genera. II. Hybrids. *Research Bulletin of the Faculty of Agriculture, Gifu University*, (53), 393–402.

Matsuura, S., Kunii, T., and Matsuura, A. (1973) Synthetic studies of the flavone derivatives. I. Syntheses of cirsiliol and cirsileneol. *Chem. Pharm. Bull.*, **21**, 2757–2759.

Maule, A. J. and Ride, J. P. (1983) Cinnamate 4-hydroxylase and hydroxycinnamate : CoA ligase in wheat leaves infected with *Botrytis cinerea*. *Phytochemistry*, **22**, 1113–1116.

Maumené, E. J. (1882) Sur l'oenocyanine. *C. R. Acad. Sci., Paris*, **95**, 924. (Cited by Onslow, 1925.)

Maxwell, C. A., Harrison, M. J., and Dixon, R. A. (1993) Molecular characterization and expression of alfalfa isoliquiritigenin 2′-*O*-methyltransferase, an enzyme specifically involved in the biosynthesis of an inducer of *Rhizobium meliloti* modulation genes. *The Plant Journal*, **4**, 971–981.

Mazza, L. and Guarna, A. (1980) An improved synthesis of 1,3-diphenyl-2-buten-1-ones (β-methylchalcones). *Synthesis* (1), 41–44.

Mc Cormick, S. (1978) Pigment synthesis in maize aleurone from precursors fed toanthocyanin mutants. *Biochem. Genet.*, **16**, 777–785.

McCormick, S., Robson, K., and Bohm, B. A. (1986) Flavonoids of *Wyethia angustifolia* and *W. helenioides*. *Phytochemistry*, **25**, 1723–1726.

McCrea, K. D. and Levy, M. (1983) Photographic visualization of floral colors as perceived by honeybee pollinators. *Amer. J. Bot.*, **70**, 369–375.

McDougal, K. M. and Parks, C. R. (1986) Environmental and genetic components of flavonoid variation in red oak, *Quercus rubra*. *Biochem. Syst. Ecol.*, **14**, 291–298.

McGhie, T. K. (1993) Analysis of sugarcane flavonoids by capillary zone electrophoresis. *J. Chromatog.*, **634**, 107–112.

McGhie, T. K. and Markham, K. R. (1994) Separation of flavonols by capillary electrophoresis: the effect of structure on electrophoretic mobility. *Phytochem. Anal.*, **5**, 121–126.

McInnes, A. G., Yoshida, S., and Towers, G. H. N. (1965) A phenolic glycoside from *Psilotum nudum*. *Tetrahedron*, **21**, 2939–2946.

McPhee, J. (1991) *Oranges*, Macfarlane Walter & Ross, Toronto.

Mears, J. A. (1980a) Flavonoid diversity and geographic endemism in *Parthenium*. *Biochem. Syst. Ecol.*, **8**, 361–370.

Mears, J. A. (1980b) The flavonoids of *Parthenium* L. *Journal of Natural Products*, **43**, 708–715.

Mellon, F. A., Chapman, J. R., and Pratt, J. A. E. (1987) Thermospray liquid chromatography-mass spectrometry in food and agricultural research. *J. Chromatogr.*, **394**, 209–222.

Menadue, Y. and Crowden, R. K. (1983) Morphological and chemical variation in populations of *Richea scoparia* and *R. angustifolia* (Epacridaceae). *Australian J. Bot.*, **31**, 73–84.

Merlin, J.-C., Statoua, A., and Brouilard, R. (1985) Investigation of the *in vivo* organization of anthocyanins using resonance Raman microspectrometry. *Phytochemistry*, **24**, 1575–1581.

Merlin, J.-C., Statoua, A., Cornard, J.-P., Saidi-Idrissi, M., and Brouillard, R. (1994) Resonance Raman spectroscopic studies of anthocyanins and anthocyanidins in aqueous solutions. *Phytochemistry*, **35**, 227–232.

Meselhy, M. R., Kadota, S., Momose, Y., Hattori, M., and Namba, T. (1992) Tinctoramine, a novel calcium antagonist N-containing quinochalcone *C*-glucoside from *Carthamus tinctorius* L. *Chem. Pharm. Bull.*, **40**, 3355–3357.

Messana, I., Ferrari, F., and Goulart Sant'ana, A. E. (1986) Two 12a-hydroxyrotenoids from *Boerhaavia coccinea*. *Phytochemistry*, **25**, 2688–2689.

Messanga, B., Tih, R. G., Sondengam, B.-L., Martin, M.-T., and Bodo, B. (1994) Biflavonoids from *Ochna calodendron*. *Phytochemistry*, **35**, 791–794.

Metodiewa, D., Kochman, A., and Karolczak, S. (1997) Evidence for antiradical and antioxidant properties of four biologically active *N,N*-diethylaminoethyl ethers of flavanone oximes: A comparison with natural polyphenolic flavonoid (rutin)n action. *Biochem. Mol. Biol. Internat.*, **41**, 1067–1075.

Meyer, P., Heidmann, I., Forkmann, G., and Saedler, H. (1987) A new *Petunia* flower colour generated by transformation of a mutant with a maize gene. *Nature* (*London*), **330**, 677–678.

Meyer, P., Linn, F., Heidmann, I., and Saedler, H. (1989) Engineering a new flower colour variety of *Petunia*. In *Plant Gene Transfer*, C. J. Lamb and R. N. Beachy (eds.), Alan R. Liss Inc., New York, 319–326.

Meyer, P., Heidmann, I., Meyer, Z. A., Neidenhof, I., and Saedler, H. (1992) Endogenous and environmental factors influence 35S promoter methylation of a maize *A1* gene construct in transgenic petunia and its colour phenotype. *Mol. Gen. Genet.*, **231**, 345–352.

Middleton Jr., E. and Drzewiecki, G. (1982) Effects of flavonoids and transitional metal cations on antigen-induced histamine release from human basophils. *Biochem. Pharmacol.*, **31**, 1449–1453.

Middleton Jr., E. and Kandaswami, C. (1994) The impact of plant flavonoids on mammalian biology: implications for immunity, inflammation and cancer. In *The Flavonoids: Advances in Research since 1986*, J. B. Harborne (ed.), Chapman and Hall, London, 619–652.

Middleton Jr., E., Drzewiecki, G., and KrishnaRao, D. G. (1981) Quercetin: an inhibitor of antigen-induced human basophil histamine release. *J. Immunol.*, **127**, 546–550.

Miles, D. H., de Medeiros, J. M. R., Chittawong, V., Hedin, P. A., Swithenbank, C., and Lidert, Z. (1991) 3′-Formyl-2′,4′,6′-trihydroxydihydrochalcone from *Psidium acutangulum*. *Phytochemistry*, **30**, 1131–1132.

Minami, H., Okubo, A., Kodama, M., and Fukuyama, Y. (1996) Highly oxygenated isoflavones from *Iris japonica*. *Phytochemistry*, **41**, 1219–1221.

Mishima, H., Kurabayashi, M., and Hirai, K. (1971) The total synthesis of silymarin (silybin). *Ann. Sankyo Res. Lab.*, **23**, 70–88.

Misirlioglu, H., Stevens, R., and Meikle, T. (1978) Synthesis of dihydrochalcones of *Myrica gale*. *Phytochemistry*, **17**, 2015–2019.

Miyaichi, Y., Kizu, H., Tomimori, T., and Lin, C.-C. (1989) Studies on the constituents of *Scutellaria* species. XI. On the flavonoid constituents of the aerial parts of *Scutellaria indica* L. *Chem. Pharm. Bull.*, **37**, 794–797.

Miyano, M. and Matsui, M. (1958) Partialsynthese des Rotenons und Dihydrorotenons. *Chem. Ber.*, **91**, 2044–2049.

Mizobuchi, S. and Sato, Y. (1985) A new flavanone with antifungal activity isolated from hops. *Rep. Rers. Lab. Kirin Brew. Co.*, **28**, 33–38.

Mizuno, M., Kanie, Y., Iinuma, M., Tanaka, T., and Lang, F. A. (1991a) Two flavonol glycosides, hexandrasides C and D, from the underground parts of *Vancouveria hexandra*. *Phytochemistry*, **30**, 2765–2768.

Mizuno, M., Kato, M., Iinuma, M., Tanaka, T., Kimura, A., Ohashi, H., and Sakai, H. (1987) Acylated luteolin glucoside from *Salix gilgiana*. *Phytochemistry*, **26**, 2418–2420.

Mizuno, M., Kyotani, Y., Iinuma, M., Tanaka, T., Kojima, H., and Iwatsuki, K. (1991b) Mearnsetin 3,7-dirhamnoside from *Asplenium antiquum*. *Phytochemistry*, **30**, 2817–2818.

Mizuno, M., Matsuura, N., Tanaka, T., Iinuma, M., and Ho, F.-C. (1991c) Four flavonoids in the roots of *Euchresta formosana*. *Phytochemistry*, **30**, 3095–3097.

Mo, Y., Nagel, C., and Taylor, L. P. (1992) Biochemical complementation of chalcone synthase mutants defines a role for flavonols in functional pollen. *Proc. Natl. Acad. Sci. U.S.A.*, **89**, 7213–7217.

Moewus, F. (1954a) On inherited and adapted rutin-resistance in *Chlamydomonas*. *Biol. Bull.*, **107**, 293.

Moewus, F. (1954b) "Hormone control in the life cycle of the green alga *Chlamydomonas engametuos*." *International Botanical Congress*, 46–47.

Mohan, P. and Joshi, T. (1989) Two anthochlor pigments from heartwood of *Pterocarpus marsupium*. *Phytochemistry*, **28**, 2529–2530.

Mohan Ram, H. Y. and Mathur, G. (1984) Flower color changes in *Lantana camara*. *J. Expt. Bot.*, **35**, 1656–1662.

Mol, J. N. M., Schram, A. W., de Vlaming, P., Gerats, A. G. M., Kreuzaler, F., Hahlbrock, K., Reif, H. J., and Veltkamp, E. (1983) Regulation of flavonoid gene expression in *Petunia hybrida*: description and partial characterization of a conditional mutant in chalcone synthase gene expression. *Mol. Gen. Genet.*, **192**, 424–429.

Mol, J. N. M., Stuitje, A., Gerats, A., van der Krol, A., and Jorgnsen, R. (1989a) Saying it with genes: molecular flower breeding. *Trends in Biotechnology*, **7**, 148–153.

Mol, J. N. M., Stuitje, A. R., and van der Krol, A. (1989b) Genetic manipulation of floral pigmentation genes. *Plant Mol. Biol.*, **13**, 287–294.

Molisch, H. (1889) Über den Farbenwechsel anthokyanhältiger Blätter bei rasch eintretenden Tode. *Bot. Ztg., Leipzig*, **47**, 17–23. (Cited by Onslow, 1925.)

Molisch, H. (1937) *Der Einfluss einer Pflanze auf die andere-Allelopathie*. Fischer, Jena.

Money, T. (1970) Biogenetic-type synthesis of phenolic compounds. *Chem. Reviews*, **70**, 553–560.

Momose, T., Abe, K., and Yoshitama, K. (1977) 5-*O*-Methylcyanidin 3-glucoside from leaves of *Egeria densa*. *Phytochemistry*, **16**, 1321.

Montero, J. L. and Winternitz, F. (1973) Structure de la rubranine chalcone isolee du bois de rose *Aniba rosaeodora*-II. *Tetrahedron*, **29**, 1243–1252.

Morazzoni, P. and Bombardelli, E. (1995) *Silybum marianum* (*Carduus marianus*). *Fitoterapia*, **66**, 3–42.

Moreira, I. C., Sobrinho, D. C., De Carvalho, M. G., and Braz-Filho, R. (1994) Isoflavanone dimers hexaspermone A, B and C from *Ouratea hexasperma*. *Phytochemistry*, **35**, 1567–1572.

Mori, K. and Kisida, H. (1988) Synthesis of pterocarpans. I. Synthesis of both enantiomers of pterocarpin. *Liebigs Ann. Chem.*, 721–723.

Mori, K. and Kisida, H. (1989) Synthesis of pterocarpans. II. Synthesis of both the isomers of pisatin. *Liebigs Ann. Chem.*, 35–39.

Morimoto, S., Nonaka, G.-i., and Nishioka, I. (1985) Tannins and related compounds. XXXV. Proanthocyanidins with a doubly linked unit from the root bark of *Cinnamomum sieboldii* Meisner. *Chem. Pharm. Bull.*, **33**, 4338–4345.

Morin, P., Villard, F., Dreux, M., and André, P. (1993) Borate complexation of flavonoid *O*-glycosides in capillary electrophoresis. II. Separation of flavonoid-3-*O*-glycosides differing in their sugar moiety. *J. Chromatog.*, **628**, 161–169.

Moriyama, M., Tahara, S., Ingham, J. L., and Mizutani, J. (1993) Isoflavonoid alkaloids from *Piscidia erythrina*. *Phytochemistry*, **32**, 1317–1325.

Morris, S. J. and Thomson, R. H. (1963a) Flavonoid pigments of the marbled white butterfly *Melanargia galathea* Seltz, *Tet. Lett.*, **2**, 101–104.

Morris, S. J. and Thomson, R. H. (1963b) The flavonoid pigments of the marbled white butterfly *Melanargia galathea* Seltz. *J. Insect Physiol.*, **9**, 391–399.

Morris, S. J. and Thomson, R. H. (1964) The flavonoid pigments of the small heath butterfly, *Coenonympha pamphilus* L. *J. Insect Physiol.*, **10**, 377–383.

Morton, J. F. (1989) Tannin as a carcinogen in bush-tea: tea, maté, and khat. In Chemistry and Significance of Condensed Tannins, R. W. Hemingway and J. J. Karchesy (eds.), Plenum Press, New York, 403–416.

Moustafa, E. and Wong, E. (1967) Purification and properties of chalcone–flavanone isomerase from soya bean seed. *Phytochemistry*, **6**, 625–632.

Moyano, E., Martinez-Garcia, J. F., and Martin, C. (1996) Apparant redundancy in *myb* gene function provides gearing for the control of flavonoid biosynthesis in *Antirrhinum* flowers. *The Plant Cell*, **8**, 1519–1532.

Möhle, B., Heller, W., and Wellmann, E. (1985) UV-Induced biosynthesis of quercetin 3-*O*-beta-D-glucuronide in dell cell cultures. *Phytochemistry*, **24**, 465–467.

Mucsi, I. and Prágai, B. M. (1985) Inhibition of virus multiplication and alteration of cyclic-AMP level in cell cultures by flavonoids. *Experientia*, **41**, 930–931.

Mukherjee, R. K., Fujimoto, Y., and Kakinuma, K. (1994) 1-(ω-Hydroxyfatty-acyl)glycerols and two flavanols from *Cinnamomum camphora. Phytochemistry*, **37**, 1641–1643.

Muller, D. and Fleury, J.-P. (1991) A new strategy for the synthesis of biflavonoids via arylboronic acids. *Tet. Lett.*, **32**, 2229–2232.

Müller, K. O. and Börger, H. (1941) *Arb. Biol. Abt. (Ansl. Reichstanst.), Berlin*. **23**, 189–231. (Cited by Harborne, 1988.)

Murakami, S. and Robinson, R. (1928) A synthesis of pyrylium salts of anthocyanidin type. Part XVII. The synthesis of peonidin chloride by means of *O*-benzoylphloroglucinaldehyde. *J. Chem. Soc.*, 1537–1541.

Murray, B. G. and Williams, C. A. (1973) Polyploidy and flavonoid synthesis in *Briza media* L. *Nature* (*London*), **243**, 87–88.

Murray, B. G. and Williams, C. A. (1976) Chromosome number and flavonoid synthesis in *Briza* L. (Gramineae). *Biochem. Genet.*, **14**, 897–904.

Musgrave, A. and van den Ende, H. (1987) How *Chlamydomonas* court their partners. *TIBS*, **12**, 470–473.

Nagai, M., Dohi, J., Morihara, M., and Sakurai, N. (1995) Diarylheptanoids from *Myrica gale* var. *tomentosa* and revised structure of porson. *Chem. Pharm. Bull.*, **43**, 1674–1677.

Nair, P. M. and Vining, L. C. (1965) Cinnamic acid hydroxylase in spinach. *Phytochemistry*, **4**, 161–168.

Nakano, J., Uchida, K., and Fujimoto, Y. (1989) An efficient total synthesis of AC-5-1: novel 5-liopxygenase inhibitor isolated from *Artocarpus communis. Heterocycles*, **29**, 427–430.

Nakashima, M., Vichnewski, W., Diaz, J. G., and Herz, W. (1994) Two flavones from *Graziela mollissima. Phytochemistry*, **37**, 285–286.

Nakayama, T., Yamada, M., Osawa, T., and Kawakiski, S. (1993) Suppression of active oxygen-induced cytotoxicity by flavonoids. *Biochem. Pharmacol.*, **45**, 265–267.

Nakazawa, K. (1959) Synthesis of ginkgetin tetramethyl ether. *Chem. Pharm. Bull.*, **7**, 748–749.

Nakazawa, K. and Ito, M. (1963) Synthesis of ring-substituted flavonoids and allied compounds. X. Synthesis of ginkgetin. *Chem. Pharm. Bull.*, **11**, 283–288.

Narayana Rao, M., Krupadanam, G. L. D., and Srimannarayana, G. (1994) Four isoflavones and two 3-aryl coumarins from stems of *Derris scandens. Phytochemistry*, **37**, 267–269.

Narkhede, D. D., Iyer, P. R., and Iyer, C. S. R. (1989) Synthesis of (+/−)-neorautane. *J. Nat. Prod.*, **52**, 502–505.

Narkhede, D. D., Iyer, P. R., and Iyer, C. S. R. (1990) Total synthesis of (+/−)-leiocarpin and (+/−)-isohemileiocarpin. *Tetrahedron*, **46**, 2031–2034.

Nawwar, M. A. M., El-Mousallamy, A. M. D., Barakat, H. H., Buddrus, J., and Linsheid, M. (1989) Flavonoid lactates from leaves of *Marrubium vulgare. Phytochemistry*, **28**, 3201–3206.

Neish, A. C. (1960) Biosynthetic pathways of aromatic compounds. *Ann. Rev. Plant Physiol.*, **11**, 55–80.

Neu, R. (1957) Chelate von Diarylborsauren mit aliphatischen Oxyalkylaminen als Nachweisreagenz für den Nachweis von Oxyphenyl-benz-o-pyronen. *Naturwissenschaften*, **44**, 181.

Nicholls, K. W. and Bohm, B. A. (1982) Quantitative flavonoid variation in *Lupinus sericeus. Biochem. Syst. Ecol.*, **10**, 225–231.

Nicholls, K. W., Bohm, B. A., and Wells, E. F. (1986) Flavonoids of *Mitella, Bensoniella* and *Conimitella. Can. J. Bot.*, **64**, 525–530.

Nicholls, K. W. and Bohm, B. A. (1987) Flavonoids of *Lupinus* sects. *Platycarpos* and *Lupinellus* (Fabaceae). *Syst. Bot.*, **12**, 320–323.

Niemann, G. J. (1988) Distribution and evolution of the flavonoids in gymnosperms. In *The Flavonoids: Advances in Research since 1980*, J. B. Harborne (ed.), Chapman and Hall, London, 469–478.

Niesbach-Klösgen, U., Barzen, E., Bernhardt, J., Rohde, W., Schwarz-Sommer, Z., Reif, H. J., Wienand, U., and Saedler, H. (1987) Chalcone synthase genes in plants: A tool to study evolutionary relationships. *J. Mol. Evol.*, **26**, 213–225.

Niklas, K. J. and Giannasi, D. E. (1977a) Flavonoids and other chemical constituents of fossil Miocene *Zelkova* (Ulmaceae). *Science*, **196**, 877–878.

Niklas, K. J. and Giannasi, D. E. (1977b) Geochemistry and thermolysis of flavonoids. *Science*, **197**, 767–769.

Niklas, K. J. and Giannasi, D. E. (1978) Angiosperm paleobiochemistry of the Succor Creek flora (Miocene) Oregon, USA. *Amer. J. Bot.*, **65**, 943–952.

Nilsson, M. (1959) Structure of ceroptene. *Acta Chem. Scand.*, **13**, 750–757.

Nishida, R. (1994) Oviposition stimulant of a Zeryntiine swallowtail butterfly, *Luehdorfia japonica. Phytochemistry*, **36**, 873–877.

Nkengfack, A. E., Kouam, J., Vouffo, T. W., Meyer, M., Tempesta, M. S., and Fomum, Z. T. (1994a) An isoflavanone and a coumestan from *Erythrina sigmoidea. Phytochemistry*, **35**, 521–526.

Nkengfack, A. E., Vouffo, T. W., Vardamides, J. C., Fomum, Z. T., Bergendorff, J. C., and Sterner, O. (1994b) Sigmoidins J and K, two new prenylated isoflavonoids from *Erythrina sigmoidea. J. Nat. Prod.*, **57**(8), 1172–1177.

Nkengfack, A. E., Vouffo, T. W., Fomum, Z. T., Meyer, M., Bergendorff, O., and Sterner, O. (1994c) Prenylated isoflavanone from the roots of *Erythrina sigmoidea*. *Phytochemistry*, **36**, 1047–1051.

Nkengfack, A. E., Vardamides, J. C., Fomum, Z. T., and Meyer, M. (1995) Prenylated isoflavanone from *Erythrina eriotricha*. *Phytochemistry*, **40**, 1803–1808.

Nkunya, M. H. H., Waibel, R., and Achenback, H. (1993) Three flavonoids from the stem bark of the antimalarial *Uvaria dependens*. *Phytochemistry*, **34**, 853–856.

Nomura, T. (1988) Phenolic compounds of the mulberry tree and related plants. *Progress in the Chemistry of Organic Natural Products*, **53**, 87–201.

Nonaka, G.-i., Muta, M., and Nishioka, I. (1983) Myricatin, a galloyl flanonol sulfate and prodelphinidin gallates from *Myrica rubra*. *Phytochemistry*, **22**, 237–241.

Nondek, L. and Malek, J. (1980) Kinetics of condensation of benzaldehyde and its derivatives with acetone and methylethyl ketone catalyzed by aluminum oxide. *Collect. Czech. Chem. Commun.*, **45**, 1813–1819.

Norbaek, R., Christensen, L. P., Bojesen, G., and Brandt, K. (1996) Anthocyanins in Chilean species of *Alstroemeria*. *Phytochemistry*, **42**, 97–100.

Norbedo, C., Ferraro, G., and Coussio, J. D. (1984) Flavonoids from *Achyrocline flaccida*. *Phytochemistry*, **23**, 2698–2700.

Northup, R. R., Yu, Z., Dahlgren, R. A., and Vogt, K. A. (1995) Polyphenol control of nitrogen release from pine litter. *Nature* (*London*), **377**, 227–229.

O'Neill, T. M. and Mansfield, J. W. (1982) Antifungal activity of hydroxyflavan and other flavonoids. *Trans. British Mycological Soc.*, **79**, 229–237.

O'Reilly, C., Shepherd, N. S., Perreira, A., Schwarz-Sommer, Z., Bertram, I., Robertson, D. S., Peterson, P. A., and Saedler, H. (1985) Molecular cloning of the *a1* locus of *Zea mays* using the transposable elements En and Mu. *EMBO Journal*, **4**, 877–882.

Obara, H., Onodera, J., Yusa, K., Tsuchiya, M., and Matsuba, S. (1989) Synthesis of ceratiolin, a constituent of *Ceratiola ericoides*. *Bull. Chem. Soc. Japan*, **62**, 3371–3372.

Ogiso, A. and Kashida, I. (1972) Flavonoids of *Leucothoë keiskei*. *Phytochemistry*, **11**, 3545.

Ogundaini, A., Farah, M., Perera, P., Samuelsson, G., and Bohlin, L. (1996) Isolation of two new antiinflammatory biflavanoids from *Sarcophyte piriei*. *J. Nat. Prod.*, **59**, 587–590.

Okamura, H., Mimura, A., Niwano, M., Takahara, Y., Yasuda, H., and Yosida, H. (1993) Two acylated flavonol glycosides from *Eucalyptus rostrata*. *Phytochemistry*, **33**, 512–514.

Okorie, D. A. (1977) New benzyldihydrochalcone from *Uvaria chamae*. *Phytochemistry*, **16**, 1591–1594.

Ollis, W. D. (1962) The isoflavonoids. In *The Chemistry of Flavonoid Compounds*, T. A. Geissman (ed.), The Macmillan Company, New York, 353–405.

Onodera, J.-i., Saito, T., and Obara, H. (1979) Hydrolysis of carthamin. *Chem. Lett.*, 1327–1330.

Onodera, J.-i., Obara, H., Osone, M., Maruyama, Y., and Sato, S. (1981) The structure of safflomin-A, a component of safflower yellow. *Chem. Lett.*, 433–436.

Onslow, M. W. (1925) *The Anthocyanin Pigments of Plants*, Cambridge University Press, Cambridge.

Onslow, M. W. (1931) The chemical effect of a Mendelian factor for flower colour. *Nature (London)*, **128**, 373–374.

Onyilagha, J. C., Malhotra, B., Elder, M., French, C. J., and Towers, G. H. N. (1997) Comparative studies of inhibitory activities of chalcones on tomato ringspot virus (ToRSV). *Can. J. Plant Path.*, **19**, 133–137.

Ornduff, R. (1966) A biosystematic survey of the goldfield genus *Lasthenia* (Compositae: Helenieae). *University of California Publications in Botany*, **40**, 1–92.

Ortmann, R., Sandermann, J. H., and Grisebach, H. (1970) Transfer of apiose from UDP-apiose to 7-*O*-(β-D-glucosyl)-apigenin and 7-*O*-(-β-glucosyl)-chrysoeriol with an enzyme preparation from parsley. *FEBS Lett.*, **7**, 164–166.

Osawa, K., Yasuda, H., Maruyama, T., Morita, H., Takeya, K., and Itokawa, H. (1992) Isoflavanones from the heartwood of *Swarzia polyphylla* and their antibacterial activity against cariogenic bacteria. *Chem. Pharm. Bull.*, **40**, 2970–2974.

Österdahl, B. G. (1979) *Acta Chemica Scandinavica*, **33B**, 400. (Cited by Markham, 1988.)

Painuly, P. and Tandon, J. S. (1983) Two 3-*C*-methylflavone glycosides from *Eugenia kurzii*. *Phytochemistry*, **22**, 243–245.

Pale, E., Kouda-Bonafas, M., Nacro, M., Vanhaelen, M., Vanhaelen-Fastré, R., and Ottinger, R. (1997) 7-*O*-Methylapigeninidin, an anthocyanidin from *Sorghum caudatum*. *Phytochemistry*, **45**, 1091–1092.

Pallares, E. S. and Garza, H. M. (1949) *Arch. Biochem.*, **21**, 377. (Cited by Harborne, 1966.)

Pare, P. W., Dmitrieva, N., and Mabry, T. M. (1991) Phytoalexin aurone induced in *Cephalocereus senilis* liquid-suspension culture. *Phytochemistry*, **30**, 1133–1135.

Parker, W. H. (1976) Comparison of numerical taxonomic methods used to estimate flavonoid similarities in the Limnanthaceae. *Brittonia*, **28**, 390–399.

Parmar, N. S., Hennings, G., and Gulati, O. P. (1984) Histidine decarboxylase inhibition: a novel approach towards the development of an effective and safe gastric anti-ulcer drug. *Agents and Actions*, **15**, 494–499.

Parmar, V. S., Jain, R., and Singh, S. (1987) Synthesis of 2-(3,4-dihydroxyphenyl)-5-hydroxy-3,6,7,8-tetramethoxy-1-benzopyran-4-one, the cytotoxic principle of *Gutierrezia resinosa*. *J. Chem. Res. Synop.* (12), 404–405.

Parmar, V. S., Gupta, S., and Sharma, V. K. (1989) Synthesis of 2,3-dihydro-6,7-dimethoxy-2-(3,4-methylenedioxyphenyl)-[1]-benzopyran-4-one and its dehydro derivative. *Ind. J. Chem.*, **28B**, 268–269.

Parthasarathy, M. R., Ranganathan, K. R., and Sharma, D. K. (1977) A novel synthesis of hexa-*O*-methyl-6,6″-biapigenin. *Ind. J. Chem.*, **15B**, 942–944.

Parthasarathy, M. R. and Gupta, S. (1984) Oxidative coupling of phloroacetophenone dimethyl ether, resacetophenone and resacetophenone monomethyl ether using silica-bound ferric chloride. *Ind. J. Chem.*, **23B**, 227–230.

Partridge, S. M. (1950) Partition chromatography and its application to carbohydrate studies. In Biochemical Society Symposium No. 3 (London), 52–61.

Partridge, S. M. and Westfall, R. G. (1947) Filter paper partition chromatography of sugars. I. General description and application to the qualitative analysis of sugars in apple juice, egg white and fetal blood of sheep. *Biochem. J.*, **42**, 238–250.

Parvez, M. and Rahman, A. (1992) A novel antimicrobial isoflavone galactoside from *Cnestis ferruginia* (Connaraceae). *J. Chem. Soc. Pakistan*, **14**, 221–223.

Pelletier, M. K. and Shirley, B. W. (1996) Analysis of flavanone 3-hydroxylase in *Arabidopsis* seedlings. Coordinate regulation with chalcone synthase and chalcone isomerase. *Plant Physiol.*, **111**, 339–345.

Pelter, A. and Foot, S. (1976) A new convenient synthesis of isoflavones. *Synthesis*, **5**, 326.

Pelter, A. and Hänsel, R. (1968) The structure of silymarin (*Silybum* substance E6), the first flavonolignan. *Tetrahedron Letters*, 2911.

Pelter, A., Bradshaw, J., and Warren, R. F. (1971) Oxidation experiments with flavonoids. *Phytochemistry*, **10**, 835–850.

Pelter, A., Hänsel, R., and Kaloga, M. (1977) The structure of silychristin. *Tet. Lett.* (51), 4547–4548.

Pelter, A., Ward, R. S., and Ashdown, D. H. J. (1978) The synthesis of mono-, di-, and trihydroxyisoflavones. *Synthesis*, 843.

Perkin, A. G. and Everest, A. E. (1918) *The Natural Organic Colouring Matters*, Longmans, Green, London.

Peters, N. K., Frost, J. W., and Long, S. R. (1986) A plant flavone, luteolin, induces expression of *Rhizobium meliloti* nodulation genes. *Science*, **233**, 977–980.

Petersen, F. and Fairbrothers, D. E. (1979) *Amphipterygium* – an amentiferous member of the Anacardiaceae. *Phytochem. Bull.*, **12**, 28–29.

Pfister, J. R., Wymann, W. E., Schuler, M. E., and Roszkowski, A. P. (1980) Inhibition of histamine-induced gastric secretion by flavone 6-carboxylic acids. *J. Med. Chem.*, **23**, 335–338.

Phillips, D. A. and Tsai, S. M. (1992) Flavonoids as plant signals in rhizosphere microbes. *Mycorrhiza*, **1**, 55–58.

Piattelli, M. (1981) The Betalains: Structure, Biosynthesis, and Chemical Taxonomy. In *Secondary Plant Products*, E. E. Conn (ed.), Academic Press, New York, 557–575.

Picman, A. K., Schneider, E. F., and Picman, J. (1995) Effect of flavonoids on mycelial growth of *Verticillium albo-atrum*. *Biochem. Syst. Ecol.*, **23**, 683–693.

Platnick, N. I. and Funk, V. A. (1983) *Advances in Cladistics*, Columbia University Press, New York, 218.

Pollak, P. E., Vogt, T., Mo, Y., and Taylor, L. P. (1993) Chalcone synthase and flavonol accumulation in sigmas and anthers of *Petunia hybrida*. *Plant Physiol.*, **102**, 925–932.

Pollock, H. G., Vickery, Jr., R. K., and Wilson, K. G. (1967) Flavonoid pigments in *Mimulus cardinalis* and its related species. I. Anthocyanins. *Amer. J. Bot.*, **54**, 695–701.

Polonsky, J. (1955) Chemical constitution of calophyllolide. IV. Alkaline degradation of calophyllolide: isolation of 5-hydroxy-7-methoxy-4-phenylcoumarin. *Bull. Soc. Chim. France*, 541–549.

Porter, L. J. (1988) Flavans and proanthocyanidins. In *The Flavonoids: Advances in Research since 1980*, J. B. Harborne (ed.), Chapman and Hall, London, 21–62.

Potts, J. R. M., Wecklych, R., and Conn, E. E. (1974) The 4-hydroxylation of cinnamic acid by sorghum microsomes and the requirement for cytochrome P-450. *J. Biol. Chem.*, **249**, 5019–5026.

Poulton, J. E. and Kauer, M. (1977) Identification of an UDP-glucose: flavonol 3-*O*-glucosyl-transferase from cell suspension cultures of soybean (*Glycine max* L.). *Planta*, **136**, 53–59.

Poulton, J. E. (1981) Transmethylation and demethylation reactions in the metabolism of secondary plant products. In *The Biochemistry of Plants*, E. E. Conn (ed.), Academic Press, New York, 667–723.

Prakash Rao, C., Prashant, A., and Krupadanam, G. L. D. (1996) Two prenylated isoflavans from *Millettia racemosa*. *Phytochemistry*, **41**, 1223–1224.

Pratt, D. D. and Robinson, R. (1922) A synthesis of pyrylium salts of anthocyanidin type. *J. Chem. Soc.*, **121**, 1577–1585.

Pridham, J. B. (1964) Methods in Polyphenol Chemistry, The Macmillan Company, New York.

Proctor, M. and Yeo, P. (1972) *The Pollination of Flowers*, Taplinger, New York.

Pueppke, S. G. (1996) The genetic and biochemical basis for nodulation of legumes by Rhizobia. *Critical Reviews in Biotechnology*, **16**, 1–51.

Puri, B. and Seshadri, T. R. (1955) Survey of anthoxanthins. Part IX. Isolation and constitution of palasitrin. *J. Chem. Soc.*, 1589–1592.

Quijano, L., Calderon, J. S., Gomez Garbary, F., Escobar, E., and Rios, T. (1987) Further polysubstituted flavones from *Ageratum houstonianum*. *Phytochemistry*, **26**, 2075–2078.

Quinn, C. J. (1982) Taxonomy of *Dacrydium* Sol. ex Lamb. emend. de Laub. (Podocarpaceae). *Australian J. Bot.*, **30**, 311–320.

Rabe, C., Steenkamp, J. A., Joubert, E., Burger, J. F. W., and Ferreira, D. (1994) Phenolic metabolites from rooibos tea (*Aspalathus linearis*). *Phytochemistry*, **35**, 1559–1565.

Raguenet, H., Barron, D., and Mariotte, A.-M. (1996) Total synthesis of 8-(1,1-dimethylallyl)apigenin. *Heterocycles*, **43**, 277–285.

Raju, V. S., Subbaraju, G. V., Manchas, M. S., Kaluza, Z., and Bose, A. K. (1992) Synthesis of kukulkanins A and B-methoxychalcones from *Mimosa tenuifolial*. *Tetrahedron*, **48**, 8347–8352.

Rall, S. and Hemleben, V. (1984) Characterization and expression of chalcone synthase in different genotypes of *Matthiola incana* R. Br. during flower development. *Plant Mol. Biol.*, **3**, 137–145.

Ramesh, P. and Yuvarajan, C. R. (1995) Coromandelin, a new isoflavone apioglucoside from the leaves of *Dalbergia coromaneliana*. *J. Nat. Prod.*, **58**, 1240–1241.

Ranganathan, K. R. and Seshadri, T. R. (1973) New flavono-lignan from *Hydnocarpus wightiana*. *Tet. Lett.*, 3481–3482.

Ranjeva, R., Faggion, R., and Boudet, A. (1975a) Métabolisme des composés phénoliques des tissus de *Pétunia*. II. Étude des système enzymatique d'activation des acides cinnamiques. Mise en évidence de deux formes de la "cinnamoyl-coenzyme A ligase". *Physiol. Veg.*, **13**, 725–734.

Ranjeva, R., Boudet, A. M., Harada, H., and Marigo, G. (1975b) Phenolic metabolism in *Petunia* tissues: characteristic responses of enzymes involved in different steps of polyphenol synthesis to different hormonal influences. *Biochem. Biophys. Acta*, **399**, 23–30.

Ranjeva, R., Boudet, A. M., and Faggion, R. (1976) Phenolic metabolism in *Petunia* tissue. IV. Properties of *p*-coumarate : coenzyme A ligase isoenzymes. *Biochemie*, **58**, 1255–1262.

Rao, A. S. (1990) Root flavonoids. *The Botanical Reviews*, **56**, 1–84.

Rasoanaivo, P., Ratsimamanga-Urverg, S., Messana, I., De Vincente, Y., and Galeffi, C. (1990) Cassinopin, a kaempferol trirhamnoside from *Cassinopsis madagascariensis*. *Phytochemistry*, **29**, 2040–2043.

Rauter, A. P., Bronco, I., Tostao, Z., Pais, M. S., Gonzalez, A. G., and Bermejo, J. B. (1989) Flavonoids from *Artemisia campestris* subsp. *maritima*. *Phytochemistry*, **28**, 2173–2175.

Raymond, P., Biolley, J. P., and Jay, M. (1995) Fingerprinting the selection process of ancient roses by means of floral phenolic metabolism. *Biochem. Syst. Ecol.*, **23**, 555–565.

Real, L. (ed.) (1983) *Pollination Biology*. Academic Press, New York.

Reddy, G. M., Britsch, L., Salamini, F., Saedler, H., and Ronde, W. (1987) The *A1* (anthocyanin-1) locus in *Zea mays* encodes dihydroquercetin reductase. *Plant Science*, **52**, 7–13.

Reid, A. R. and Bohm, B. A. (1993) Vacuolar and exudate flavonoids of New Zealand *Cassinia*. *Biochem. Syst. Ecol.*, **22**, 501.

Reinold, S., Hauffe, K. D., and Douglas, C. J. (1993) Tobacco and parsley 4-coumarate : coenzyme A ligase genes are temporally and spatially regulated in a cell type-specific manner during tobacco flower development. *Plant Physiol.*, **101**, 373–383.

Rennie, E. H. (1886) Glycyphyllin, the sweet principle of *Smilax glycyphylla*. *J. Chem. Soc.*, **49**, 857.

Repcak, M. and Martonfi, P. (1995) The variability pattern of apigenin glucosides in *Chamomilla recutita* diploid and tetraploid cultivars. *Die Pharmazie*, **50**, 696–699.

Reynolds, T. M., Robinson, R., and Scott-Moncrieff, R. (1934) Experiments on the synthesis of anthocyanins. Part XXII. Isolation of an anthocyanin of *Salvia patens* termed delphin, and its synthesis. *J. Chem. Soc.*, 1235–1243.

Ribéreau-Gayon, P. (1982) The anthocyanins of grapes and wines. In *Anthocyanins as Food Colors*, P. Markakis (ed.), Academic Press, New York, 209–244.

Ribéreua-Gayon, P. (1974) The chemistry of red wine color. In *Chemistry of Winemaking*, A. D. Webb (ed.), American Chemical Society, Washington, D.C., 50–87.

Rice, E. L. (1974) *Allelopathy*, Academic Press, New York.

Rice, E. L. (1984) *Allelopathy*, Academic Press, New York.

Rice, E. L. (1995) *Biological Control of Weeds and Plant Diseases: Advances in Applied Allelopathy*, University of Oklahoma Press, Norman, OK.

Rice-Evans, C. A., Miller, N. J., and Paganga, G. (1997) Antioxidant properties of phenolic compounds. *Trends in Plant Sci.*, **2**, 152–159.

Richards, A. J. (ed.) (1978) *The Pollination of Flowers by Insects*, Academic Press, New York.

Richards, J. H. and Hendrickson, J. B. (1964) *The Biosynthesis of Steroids, Terpenes, and Acetogenins*. W. A. Benjamin, Inc., New York.

Richards, M., Bird, A. E., and Munden, J. E. (1969) Chlorformin, a new antifungal antibiotic. *J. Antibiotics*, **22**, 388–389.

Richardson, M. (1978) Flavonols and *C*-glycosylflavonoids of the Caryophyllales. *Biochem. Syst. Ecol.*, **6**, 283–286.

Richardson, P. M. and Young, D. A. (1982) The phylogenetic content of flavonoid point scores. *Biochem. Syst. Ecol.*, **10**, 251–256.

Rieseberg, L. H. and Schilling, E. E. (1985) Floral flavonoids and ultraviolet patterns in *Viguiera* (Compositae). *Amer. J. Bot.*, **72**, 999–1004.

Rieseberg, L. H. and Soltis, D. E. (1987) Flavonoids of Miocene *Platanus* and its extant relatives. *Biochem. Syst. Ecol.*, **15**, 109–112.

Riva, S., Chopinear, J., Kieboom, A. P. G., and Klibanov, A. M. (1988) Protease-catalyzed regioselective esterification of sugars and related compounds in anhydrous dimethylformamide. *J. Amer. Chem. Soc.*, **110**, 584–589.

Riva, S., Danieli, B., and Luisetti, M. (1996) A two-step efficient chemoenzymatic synthesis of flavonoid glycoside malonates. *J. Nat. Prod.*, **59**, 618–621.

Robberecht, R. (1989) Environmental photobiology. In *The Science of Photobiology*, K. C. Smith (ed.), Plenum Press, New York, 135–154.

Robberecht, R. and Caldwell, M. M. (1983) Protective mechanisms and acclimation to solar ultraviolet-B radiation in *Oenothera stricta*. *Plant, Cell and Environ.*, **6**, 477–485.

Roberts, A. V. (1977) Relationship between species in the genus *Rosa*, section *Pimpinellifoliae*. *Bot. J. Linnean Soc.*, **74**, 309–328.

Roberts, E. A. H. (1962) Economic importance of flavonoid substances: tea fermentation. In *The Chemistry of Flavonoid Compounds*, T. A. Geismann (ed.), The Macmillan Company, New York, 468–512.

Roberts, E. A. H. and Williams, D. M. (1958) Phenolic substances of manufactured tea. III. Ultraviolet and visible absorption spectra. *J. Sci. Food Agric.*, **9**, 217–223.

Roberts, M. F., Timmermann, B. N., Mabry, T. J., Brown, R., and Matlin, S. A. (1984) Brickellin, a novel flavone from *Brickellia veronicaefolia* and *B. chlorolepis*. *Phytochemistry*, **23**, 163–165.

Robertson, A. and Robinson, R. (1928) CC. A synthesis of pyrylium salts of anthocyanidin type. Part XV. The synthesis of cyanidin chloride by means of *O*-benzoylphloroglucinaldehyde. *J. Chem. Soc.*, 1526–1532.

Robertson, A., Robinson, R., and Sugiura, J. (1928) A synthesis of pyrylium salts of anthocyanidin type. Part XVI. The synthesis of pelargonidin chloride by means of *O*-benzoylphloroglucinaldehyde. *J. Chem. Soc.*, 1533–1537.

Robertson, A. (1933) Experiments on the synthesis of rotenone and its derivatives. Part III. The dehydrorotenone nucleus. *J. Chem. Soc.*, 489–493.

Robinson, G. M. and Robinson, R. (1933) *Biochem. J.*, **27**, 206–212.

Robinson, G. M., Robinson, R., and Todd, A. R. (1934) Experiments on the synthesis of anthocyanins. Part XIX. 5-Glucosidylapigeninidin, believed to be identical with gesnerin, an anthocyanin of *Gesnera fulgens. J. Chem. Soc.*, 809–813.

Robinson, R. (1936) The formation of anthocyanin in plants. *Nature* (*London*), **137**, 172.

Robinson, R. (1962) Synthesis in the brazilin group. In *Chemistry of Natural and Synthetic Colouring Matters*, T. S. Gore, B. S. Joshi, S. V. Sunthankar, and B. D. Tilak (eds.), Academic Press, New York, 1-11.

Roger, C. R. (1988) The nutritional incidence of flavonoids: some physiological and metabolic considerations. *Experientia*, **44**, 725–733.

Rohlf, F. J. (1992) NTSYS-pc. Numerical Taxonomy and Multivariate Analysis System, Exeter Software, Setauket, New York.

Rolfs, C.-H. and Kindl, H. (1984) Stilbene synthase and chalcone synthase. Two different constitutive enzymes in cultured cells of *Picea excelsa. Plant Physiol.*, **75**, 489–492.

Romeo, J. T., Saunders, J. A., and Barbosa, P. (eds.) (1996) *Phytochemical Diversity and Redundancy in Ecological Interactions*. Recent Advances in Phytochemistry, Plenum Press, New York, 319.

Rosemann, D., Heller, W., and Sandermann, J. H. (1991) Biochemical plant responses to ozone. II. Induction of stilbenme biosynthesis in Scots pine (*Pinus sylvestris* L.) seedlings. *Plant Physiol.*, **97**, 1280–1286.

Rosemann, D., Heller, W., and Sandermann, H. (1991) Biochemical plant responses to ozone. II. Induction of stilbene biosynthesis in Scots Pine (*Pinus sylvestris* L.) seedlings. *Plant Physiol.*, **97**, 1280–1286.

Rosenheim, O. (1920) Observations on anthocyanins. I. The anthocyanins of the young leaves of the grape vine. *Biochem. J.*, **14**, 178–188.

Rösler, J., Krekel, F., Amrhein, N., and Schmid, J. (1997) Maize phenylalanine ammonia-lyase has tyrosine ammonia-lyase activity. *Plant Physiol.*, **113**, 175–179.

Rossouw, W., Hundt, A. F., Steenkamp, J. A., and Ferreira, D. (1994) Oligomeric flavanoids. Part 17. Absolute configurations of flavan-3-ols and 4-aryl-3-ols via the Mosher method. *Tetrahedron*, **50**, 12477–12488.

Roussis, V., Ampofo, S. A., and Wiemer, D. F. (1987) Flavanones from *Lonchocarpus minimiflorus. Phytochemistry*, **26**, 2371–2375.

Rubin, B., Penner, D., and Saetler, A. W. (1983) Induction of isoflavonoid production in *Phaseolus vulgaris* L. leaves by ozone, sulfur dioxide and herbacide stress. *Environ. Toxicol. Chem.*, **2**, 295–306.

Rubery, P. H. and Jacobs, M. (1988) Auxin transport and its regulation by flavonoids. In *Plant Growth Substances* [Int. Conf. Plant Growth Subst.] 13th, 428–440.

Ruh, M. F., Zacharewski, T., Connor, K., Howell, J., Chen, I., and Safe, S. (1995) Naringenin: A weakly estrogenic bioflavonoid that exhibits antiestrogenic activity. *Biochem. Pharmacol.*, **50**, 1485–1493.

Runemark, H. (1968) Critical comments on the use of statistical methods in chemotaxonomy. *Bot. Notiser*, **121**, 29–43.

Russell, D. W. and Conn, E. E. (1967) The cinnamic acid 4-hydroxylase of pea seedlings. *Arch. Biochem. Biophys.*, **122**, 256–258.

Russell, D. W., Conn, E. E., Sutter, A., and Grisebach, H. (1968) Hydroxylation-induced migration and retention of tritium on conversion of 4-tritiocinnamic acid to 4-hydroxycinnamic acid by an enzyme from pea seedlings. *Biochem. Biophys. Acta*, **170**, 210–213.

Russell, D. W. (1971) The metabolism of aromatic compounds in higher plants. X. Properties of the cinnamic acid 4-hydroxylase of pea seedlings and some aspects of its metabolic and developmental control. *J. Biol. Chem.*, **246**, 3870–3878.

Ryan, F. J. (1955) Attempt to reproduce some of Moewus' experiments in *Chlamydomonas* and *Polytoma*. *Science*, **122**, 470.

Saini, T. R., Pathak, V. P., and Khanna, R. N. (1983) Glabrachromene II, a minor constituent of seeds of *Pongamia glabra*. *J. Nat. Prod.*, **46**, 936.

Saito, N. and Harborne, J. B. (1992) Correlations between anthocyanin type, pollinator and flower colour in the Labiatae. *Phytochemistry*, **31**, 3009–3015.

Saito, N., Tatsuzawa, F., Hongo, A., Win, K. W., Yokoi, M., Shigihara, A., and Honda, T. (1996a) Acylated pelargonidin 3-sambubioside-5-glucosides in *Matthiola incana*. *Phytochemistry*, **41**, 1613–1620.

Saito, N., Tatsuzawa, F., Kasahara, K., Yokoi, M., Iida, S., Shigihara, A., and Honda, T. (1996b) Acylated peonidin glycosides in the slate flowers of *Pharbitis nil*. *Phytochemistry*, **41**, 1607–1611.

Saito, N., Tatsuzawa, F., Yoda, K., Yokoi, M., Kasahara, K., Iida, S., Shigihara, A., and Honda, T. (1995) Acylated cyanidin glycosides in the violet-blue flowers of *Ipomoea purpurea*. *Phytochemistry*, **40**, 1283–1289.

Saito, N., Toki, K., Özden, S., and Honda, T. (1996c) Acylated delphinidin glycosides in the blue-violet flowers of *Consolida armeniaca*. *Phytochemistry*, **41**, 1599–1605.

Saito, N., Yokoi, M., Yamaji, M., and Hondo, T. (1985) Anthocyanidin glycosides from the flowers of *Alstroemeria*. *Phytochemistry*, **24**, 2125–2126.

Saito, T., Shibata, S., Sankawa, U., Furuya, T., and Ayabe, S. (1975) Biosynthesis of echinatin. New biosynthetical scheme of retrochalcone. *Tet. Lett.*, 4463–4466.

Saleh, N. A. M. (1976) Chromatographic differentiation of tamarixetin and isorhamnetin by spraying. *J. Chromatog.*, **124**, 174.

Santhosh, K. C. and Balasubramanian, K. K. (1992) Ligand coupling route to isoflavanones and isoflavones. *J. Chem. Soc., Chem. Commun.*, 224–225.

Santos, H., Turner, D. L., Lima, J. C., Figueiredo, P., Pina, F. S., and Maçanita, A. L. (1993) Elucidation of the multiple equilibria of malvin in aqueous solution by one- and two-dimensional NMR. *Phytochemistry*, **33**, 1227–1232.

Sathyanarayana, S. and Krishnamurty, A. G. (1988) Corroborative studies on the highly efficient preparation of 2′-hydroxychalcones using partially dehydrated barium hydroxide catalyst. *Curr. Sci.*, **57**, 1114–1116.

Sato, S., Obara, H., Takeuchi, H., Tawaraya, T., Endo, A., and Onodera, J.-I. (1995) Syntheses of 2,4,6-trihydroxy-, 2,4,6-trihydroxy-5-methyl- and 2,4,5,6-tetrahydroxy-substituted 3-(3-phenylpropionyl)benzaldehydes and their bacterial activity. *Phytochemistry*, **38**, 491–493.

Sauer, J. D. (1994) *Historical Geography of Crop Plants – A Select Roster*, CRC Press, Boca Raton.

Saxena, S., Makrandi, J. K., and Grover, S. K. (1986) A diagnostic cleavage of aurones and chalcones with alkaline hydrogen peroxide in the presence of triethylbenzylammonium chloride (TEBA). *Ind. J. Chem.*, **25B**, 473–477.

Saxena, V. K. and Bhadoria, B. K. (1990) 3′-Prenyl-4′-methoxyisoflavone-7-*O*-β-D-(2″-*O*-*p*-coumaroyl)glucopyranoside, a novel phytoestrogen from *Sopubia delphinifolia*. *J. Nat. Prod.*, **53**, 62–65.

Scalbert, A. (1993) *Polyphenolic Phenomena*. INRA Editions, Paris.

Schaal, B. A. and Leverich, W. J. (1980) Pollination and banner markings in *Lupinus texensis* (Leguminosae). *Southwestern Naturalist*, **25**, 280–282.

Schafer, Jr., E. W., Bowles, Jr., W. A., and Hurlbut, J. (1983) The acute oral toxicity, repellency, and hazard potential of 998 chemicals to one or more species of wild and domestic birds. *Arch. Environm. Contam. Toxicol.*, **12**, 355–382.

Scheres, B., van Engelen, F., van der Knaap, E., van de Wiel, C., van Kammen, A., and Bisseling, T. (1990) Sequential induction of nodulin gene expression in the developing pea nodule. *The Plant Cell*, **2**, 687–700.

Schilling, E. E. (1989) External flavonoid aglycones of *Viguiera* series *Viguiera* (Asteraceae: Heliantheae). *Biochem. Syst. Ecol.*, **17**, 535–538.

Schilling, E. E. and Panero, J. L. (1988) Flavonoids of *Viguiera* series *Brevifoliae*. *Biochem. Syst. Ecol.*, **16**, 417–418.

Schmid, J., Doerner, P. W., Clouse, S. D., Dixon, R. A., and Lamb, C. J. (1990) Developmental and environmental regulation of a bean chalcone synthase promoter in transgenic tobacco. *The Plant Cell*, **2**, 619–631.

Schrall, R. and Becker, H. (1977) Callus- und Suspensioskulturen von *Silybum marianum*. II. Umsetzung von Flavonoiden mit Coniferylalkohol zu Flavonolignanen. *Planta Medica*, **32**, 27–32.

Schramm, A. W., Timmermann, A. W., de Vlaming, P., Jonsson, L. M. V., and Bennink, G. J. H. (1981) Glucosylation of flavonoids in petals of *Petunia hybrida*. *Planta*, **153**, 459–461.

Schramm, A. W., Al, E. J. M., Douma, N., Jonsson, L. M. V., de Vlaming, P., Kool, A., and Bennink, G. J. H. (1982) Cell wall localization of dihydroflavonol-glucoside β-glucosidase in flowers of *Petunia hybrida*. *Planta*, **155**, 162–165.

Schuda, P. F. and Price, W. A. (1987) Total synthesis of isoflavones: jamaicin, calopogonium isoflavone B, pseudobaptigenin, and maxima substance-B. Friedel-Crafts acylation reactions with acid-sensitive substrates. *J. Org. Chem.*, **52**, 1972–1979.

Schulz, M. and Weissenböck, G. (1988) Three specific UDP-glucuronate: flavone-glucuronosyltransferases from primary leaves of *Secale cereale*. *Phytochemistry*, **27**, 1261–1267.

Scogin, R. (1992) The distribution of acetoside among Angiosperms. *Biochem. Syst. Ecol.*, **20**, 477–480.

Scogin, R. and Zakar, K. (1976) Anthochlor pigments and floral UV patterns in the genus *Bidens*. *Biochem. Syst. Ecol.*, **4**, 165–167.

Scogin, R., Young, D. A., and Jones, J., C.E. (1977) Anthochlor pigments and pollination biology. II. The ultraviolet floral pattern of *Coreopsis gigantea* (Asteraceae). *Bull. Torrey Bot. Club*, **104**, 155–159.

Scogin, R. (1978) Floral UV-absorption patterns and anthochlor pigments in the Asteraceae. *The Southwestern Naturalist*, **23**, 371–374.

Scogin, R. (1983) Natural variation and regulation of floral ultraviolet absorption in *Coreropsis* and *Bidens*. *Aliso*, **10**, 443–448.

Scogin, R. and Freeman, C. E. (1987) Floral anthocyanins of the genus *Penstemon*: Correlations with taxonomy and pollination. *Biochem. Syst. Ecol.*, **15**, 355–360.

Scogin, R. and Romo-Contreras, V. (1992) Familial assignment of *Polypremum*: Evidence from phenolic chemistry. *Biochem. Syst. Ecol.*, **20**, 787–788.

Scogin, R. (1992) The distribution of acetoside among angiosperms. *Biochem. Syst. Ecol.*, **20**, 477–480.

Scora, R. W. (1964) Dependency of pollination on patterns in *Monarda* (Labiatae). *Nature* (*London*), **204**, 1011–1012.

Scora, R. (1976) Floral UV patterns and anthochlor pigments in the genus *Coreropsis* (Asteraceae). *Aliso*, **8**, 429–431.

Scott-Moncrieff, R. (1936) A biochemical survey of some mendelian factors for flower colour. *J. Genetics*, **32**, 117. (Cited by Stafford, 1990.)

Seabra, R. M., Andrade, P. B., Ferreres, F., and Moreira, M. M. (1997) Methoxylated aurones from *Cyperus capitatus*. *Phytochemistry*, **45**, 839–840.

Seabra, R. M., Moreira, M. M., Costa, M. A. C., and Paul, M. I. (1995) 6,3′,4′-Trihydroxy-4-methoxy-5-methylaurone from *Cyperus capitatus*. *Phytochemistry*, **40**, 1579–1580.

Seeger, T., Geiger, H., and Zinsmeister, H. D. (1991) Bartramiaflavone, a macrocyclic biflavonoid from the moss *Bartramia pomiformis*. *Phytochemistry*, **30**, 1653–1656.

Seeger, T., Geiger, H., and Zinsmeister, H. D. (1992) Isolation and structure elucidation of bartramia-triluteolin, bartramic acid and biflavonoids from the moss *Bartramia pomiformis*. *Z. Naturforsch.*, **47c**, 525–530.

Seeger, T., Voigt, A., Geiger, H., Zinsmeister, H. D., Schilling, G., and López-Sáez, J.-A. (1995) Isomeric triluteolins from *Bartramia stricta* and *Bartramia pomiformis*. *Phytochemistry*, **40**, 1531–1536.

Seetharaman, J. and Rajan, S. (1992) Structure of 6-hydroxyflavone. *Acta Cryst.*, **C48**, 1714–1715.

Seetharaman, J. and Rajan, S. (1995) Structure of 2′-hydroxyflavone. *Z. Krystallogr.*, **210**, 104–106.

Seigler, D. S. and Wollenweber, E. (1983) Chemical variation in *Notholaena standleyi*. *Amer. J. Bot.*, **70**, 790–798.

Seikel, M. K. and Geissman, T. A. (1950) Anthochlor pigments. VII. The pigments of yellow *Antirrhinum majus*. *J. Amer. Chem. Soc.*, **72**, 5725–5730.

Seikel, M. K. and Mabry, T. J. (1965) New type of glycoflavone from *Vitex lucens*. *Tet. Lett.* (16), 1105–1109.

Seitz, U., Bonn, G., Öfner, P., and Popp, M. (1991) Isotachophoretic analysis of flavonoids and phenolcarboxylic acids of relevance to phytopharmaceutical industry. *J. Chromatog.*, **559**, 499–504.

Selway, J. W. T. (1986) Antiviral activity of flavones and flavans. In *Plant Flavonoids in Biology and Medicine*, V. Cody, J. Middleton, E., and J. B. Harborne (eds.), Alan R. Liss, Inc., New York, 521–536.

Seshadri, T. R. (1951) Biochemistry of natural pigments (exclusive of heme pigments and carotenoidsd). *Ann. Revs. Biochem.*, **20**, 487–512.

Seshadri, T. R. (1962a) Isolation of flavonoid compounds from plant materials. In *The Chemistry of Flavonoid Compounds*, T. A. Geissman (ed.), The Macmillan Company, New York, 6–33.

Seshadri, T. R. (1962b) Interconversions of flavonoid compounds. In *The Chemistry of Flavonoid Compounds*, T. A. Geissman (ed.), The Macmillan Company, New York, 156–196.

Setchell, K. D. R., Welsh, M. B., and Lim, C. K. (1987) High performance liquid chromatographic analysis of phytoestrogens in soy protein preparations with ultraviolet, electrochemical and thermospray mass spectrometric detection. *J. Chromatogr.*, **386**, 315–323.

Sharaf, M. (1996) Isolation of an acacetin tetraglycoside from *Peganum harmala. Fitoterapia*, **67**, 294–296.

Sharaf, M., El-Ansari, M. A., Matlin, S. A., and Saleh, N. A. M. (1997) Four flavonoid glycosides from *Peganum harmala. Phytochemistry*, **44**, 533–536.

Sharma, D. K., Ranganathan, K. R., Parthasarathy, M. R., Bhushan, B., and Seshadri, T. R. (1979) Flavonolignans from *Hydnocarpus wrightiana. Planta Medica*, **37**, 79–83.

Sharma, V. M. and Rao, P. S. (1992) A prenylated chalcone from the roots of *Tephrosia spinosa. Phytochemistry*, **31**, 2915–2916.

Shawl, A. S., Mengi, N., Misra, L. N., and Vishwapaul. (1988) Irispurinol, a 12a-hydroxyrotenoid from *Iris spuria. Phytochemistry*, **27**, 3331–3332.

Sheahan, J. J. (1996) Sinapate esters provide greater UV-B attenuation than flavonoids in *Arabidopsis thaliana* (Brassicaceae). *Amer. J. Bot.*, **83**, 679–686.

Shi, Y.-p., Li, Y., and Zhang, H.-c. (1992) Chemical constituents of *Artemisia subdigitata* Mattf. *Gaodeng Xuexiao Huaxue Xuebao*, **13**, 1258–1261.

Shibata, K., Shibata, Y., and Kasiwagi, I. (1919) Anthocyanins: Color variation in anthocyanins. *J. Amer. Chem. Soc.*, **41**, 208–209.

Shibata, S. and Yamazaki, M. (1958) The biogenesis of rutin. *Pharm. Bull. (Tokyo)*, **5**, 501–502.

Shimokoriyama, M. (1962) Flavanones, Chalcones and Aurones In *The Chemistry of Flavonoid Compounds*, T. A. Geissman (ed.), The Macmillan Co., New York.

Shimokoriyama, M. and Hattori, S. (1953) Anthochlor pigments of *Cosmos sulphureus, Coreopsis lanceolata* and *C. saxicola. J. Amer. Chem. Soc.*, **75**, 1900–1904.

Shin, W., Kim, S., and Chun, K. S. (1987) Crystal structure of hesperetin hydrate. *Acta Crystallogr., Sect. C.*, **C43**, 1946.

Shirataki, Y., Endo, M., Yokoe, I., and Komatsu, M. (1983) Studies on the constituents of *Sophora* species. XVIII. Constituents of the root of *Sophora tomentosa* L. (3). *Chem. Pharm. Bull.*, **31**, 2859–2863.

Shirataki, Y., Komatsu, M., Yokoe, I., and Manaka, A. (1981) Studies on the constituents of *Sophora* species. XVI. Constituents of the root of *Euchresta japonica* Hook. f. ex Regel. *Chem. Pharm. Bull.*, **29**, 3033–3036.

Shirley, B. W., Kubasek, W. L., Storz, G., Bruggemann, E., Koornneef, M., Ausubel, F. M., and Goodman, H. M. (1995) Analysis of *Arabidopsis* mutants deficient in flavonoid biosynthesis. *The Plant Journal*, **8**, 659–671.

Shirley, B. W. (1996) Flavonoid biosynthesis: 'new' functions for an 'old' pathway. *Trends in Plant Science*, **1**, 377–382.

Shoja, M. (1990) 5-Hydroxyflavone. *Acta Cryst.*, **C46**, 517–519.

Shoukry, M. M., Darwish, N. A., and Morsi, M. A. (1982) Synthesis of 2′,3-dimethoxy-3′,6,7-trihydroxy-3-isoflavene, a constituent of *Baphia nitida. Gazz. Chim. Ital.*, **112**, 289–291.

Shriner, R. L., Fuson, R. C., and Curtin, D. Y. (1956) *The Systematic Identification of Organic Compounds*, John Wiley & Sons, New York.

Shute, J. L., Jourdan, P. S., and Mansell, R. L. (1979) UDP-Glucose: glucosyltransferase activity involved in the biosynthesis of flavonol triglucosides in *Pisum sativum* L. seedlings. *Z. Naturforsch.*, **34c**, 738–741.

Shutt, D. A. (1976) The effects of plant oestrogens on animal reproduction. *Endeavour*, **35**, 110–113.

Siaens, E., De Keukeleire, D., and Verzele, M. (1977) The metal ion catalysed oxidation of hexahydrolupulone. *Tetrahedron*, **33**, 423–426.

Siddiqui, S. A. and Sen, A. B. (1971) Hypolaetin 7-glucoside from *Juniperus macropoda. Phytochemistry*, **10**, 434–435.

Silverstein, R. M., Bassler, G. C., and Morrill, T. C. (1981) *Spectrometric Identification of Organic Compounds*, John Wiley & Sons, New York.

Silverstein, R. M., Bassler, G. C., and Morrill, T. C. (1991) *Spectrometric Identification of Organic Compounds*, John Wiley & Sons, New York.

Singh, K. N. and Pandey, V. B. (1993) Antifungal flavonoids of *Echinops echinatus. Oriental Journal of Chemistry*, **9**, 149–151.

Singh, O. V. and Kapil, R. S. (1993) A general method for the synthesis of isoflavones by oxidative rearrangement of flavanones using thallium (III) perchlorate. *Ind. J. Chem.*, **32B**, 911–915.

Singh, S., McCallum, J., Gruber, M. Y., Towers, G. H. N., Muir, A. D., Bohm, B. A., Koupai-Abyazani, M. R., and Glass, A. D. M. (1997) Biosynthesis of flavan-3-ols by leaf extracts of *Onobrychis viciifolia. Phytochemistry*, **44**, 425–432.

Sink, K. C. (ed.) (1984) *Petunia*. Springer Verlag, New York.

Skaltsa, H. and Shammas, G. (1988) Flavonoids from Lippia citriodora. *Planta Medica*, **54**, 465.

Skaltsa, H., Verykokidou, E., Harvala, C., Karabouriotis, G., and Manetas, Y. (1994) UV-B Protective potential and flavonoid content of leaf hairs of *Quercus ilex. Phytochemistry*, **37**, 987–990.

Slimestad, R. and Hostettmann, K. (1996) Characterization of phenolic constituents from juvenile and mature needles of Norway spruce by means of HPLC-MS. *Phytochemical Anal.*, **7**, 42–48.

Slimestad, R., Nerdal, W., Francis, G. W., and Andersen, O. M. (1993) Myricetin 3,4′-diglucoside and kaempferol derivatives from needles of Norway spruce, *Picea abies. Phytochemistry*, **32**, 179–181.

Smiley, C. J. and Huggins, L. M. (1981) *Pseudofagus idahoensis*, N. Gen. et Sp. (Fagaceae) from the Miocene Clarkia Flora of Idaho. *Amer. J. Bot.*, **68**, 741–761.

Smith, H. G. and Read, J. (1923) Glucoside occurring in the timber of the "red ash," *Alphitonia excelsa* Reiss. *J. Proc. Roy. Soc. N. S. Wales*, **56**, 253–259.

Smith, D. M. (1980) Flavonoid analysis of the *Pityrogramma triangularis* complex. *Bull. Torrey Bot. Club*, **107**, 134–145.

Smith, D. M., Craig, S. P., and Santarosa, J. (1971) Cytological and chemical variation in *Pityrogramma triangularis*. *Amer. J. Bot.*, **58**, 292–299.

Smith, L. C. (1944) *Lupinus*, Stanford University Press, Stanford.

Smith, P. M. (1976) *The Chemotaxonomy of Plants*, Edward Arnold, London.

Soby, S., Bates, R., and Van Etten, H. (1997) Oxidation of the phytoalexin maackiain to 6,6a-dihydroxymaackiain by *Colletotrichum gloeosporioides*. *Phytochemistry*, **45**, 925–929.

Sokol, R. R. and Sneath, P. H. A. (1963) *Principles of Numerical Taxonomy*, Freeman, San Francisco.

Soltis, D. E. (1980a) Flavonoids of *Sullivantia*: Taxonomic implications at the generic level within the Saxifraginae. *Biochem. Syst. Ecol.*, **8**, 149–151.

Soltis, D. E. (1980b) Karyotypic relationships among species of *Boykinia, Heuchera, Mitella, Sullivantia, Tiarella*, and *Tolmiea* (Saxifragaceae). *Syst. Bot.*, **5**, 17–29.

Soltis, D. E. (1984) Autopolyploidy in *Tolmiea menziesii* (Saxifragaceae). *Amer. J. Bot.*, **71**, 1171–1174.

Soltis, D. E. and Bohm, B. A. (1986) Flavonoid chemistry of diploid and tetraploid cytotypes of *Tolmiea menziesii* (Saxifragaceae). *Syst. Bot.*, **11**, 20–25.

Soltis, P. S. and Soltis, D. E. (1986) Anthocyanin content in diploid and tetraploid cytotypes of *Tolmiea menziesii* (Saxifragaceae). *Syst. Bot.*, **11**, 32–34.

Soltis, D. E., Bohm, B. A., and Nesom, G. L. (1983) Flavonoid chemistry of cytotypes in *Galax* (Diapensiaceae). *Systematic Botany*, **8**, 15–23.

Soltis, P. E., Soltis, D. E., and Doyle, J. J. R. (1992) *Molecular Systematics of Plants*. Chapman & Hall, London.

Somers, T. C. (1971) The polymeric nature of wine pigments. *Phytochemistry*, **10**, 2175–2186.

Sorauer, P. (1886) *Handbuch der Pflanzenkrankheiten*, Berlin. Vol. 1, 324. (Cited by Onslow, 1925.)

Spaink, H. P. (1992) Rhizobial lipo-oligosaccharides: answers and questions. *Plant Mol. Biol.*, **20**, 977–986.

Spaink, H. P., Sheely, D. M., Van Brussel, A. A. N., Glushka, J., York, W. S., Tak, T., Geiger, O., Kennedy, E. P., Reinhold, V. N., and Lughtgenberg, B. J. J. (1991) A novel highly unsaturated fatty acid moiety of lipo-oligosaccharide signals determines host specificity of *Rhizobium*. *Nature*, **354**, 125–130.

Sparvoli, F., Martin, C., Scienza, A., Gavazzi, G., and Tonelli, C. (1994) Cloning and molecular analysis of structural genes involved in flavonoid and stilbene biosynthesis in grape (*Vitis vinifera* L.). *Plant Mol. Biol.*, **24**, 743–755.

Spongberg, S. A. (1972) The genera of Saxifragaceae in the southeastern United States. *J. Arnold Arboretum*, **53**, 409–498.

Sprengel, C. K. (1793) *Das endeckte Geheimniss der Natur im Bau und in der Befruchtung der Blumen*, Vieweg, Berlin.

Spribille, R. and Forkmann, G. (1981) Genetic control of chalcone synthase activity in flowers of *Matthiola incana* R. Br. *Z. Naturforsch.*, **36c**, 619–624.

Spribille, R. and Forkmann, G. (1982) Genetic control of chalcone synthase activity in flowers of *Antirrhinum majus*. *Phytochemistry*, **21**, 2231–2234.

Spribille, R. and Forkmann, G. (1984) Conversion of dihydroflavonols to flavonols with enzyme extracts from flower buds of *Matthiola incana* R. Br. *Z. Naturforsch.*, **39c**, 714–719.

Srivastava, S. K. and Gupta, H. O. (1983) A new flavanone from *Adina cordifolia*. *Planta Medica*, **48**, 58–59.

Stafford, H. A. (1965) Flavonoids and related phenolic compounds produced in the first internode of *Sorgham vulgare* in darkness and in light. *Plant Physiol.*, **40**, 130–138.

Stafford, H. A. (1990) *Flavonoid Metabolism*. CRC Press, Boca Raton, Florida.

Stafford, H. A. (1991) Flavonoid evolution: An enzymatic approach. *Plant Physiol.*, **96**, 680–685.

Stafford, H. A. (1994) Anthocyanins and betalains: evolution of the mutually exclusive pathways. *Plant Science*, **101**, 91–98.

Stafford, H. A. and Lester, H. H. (1982) Enzymic and nonenzymic reduction of (+)-dihydroquercetin to its 3,4-diol. *Plant Physiol.*, **70**, 695–698.

Stapleton, A. E. and Walbot, V. (1994) Flavonoids can protect maize DNA from the induction of ultraviolet radiation damage. *Plant Physiol.*, **105**, 881–889.

Star, A. E. (1980) Frond exudate flavonoids as allelopathic agents in *Pityrogramma*. *Bull. Torrey Bot. Club*, **107**, 146–153.

Star, A. E. and Mabry, T. J. (1971) Flavonoid frond exudates from two Jamaican ferns, *Pityrogramma tartarea* and *P. calomelanos*. *Phytochemistry*, **10**, 1217–1218.

Stern, W. L. (1952) The comparative anatomy of the xylem and the phylogeny of the Julianaceae. *Amer. J. Bot.*, **39**, 220–229.

Steyns, J. M. and van Brederode, J. (1986) Variation in the substrate specificity of allozymes catalyzing flavone *O*-glucoside biosynthesis in *Silene* plants. *Biochemical Genetics*, **24**, 349–360.

Stich, K. and Forkmann, G. (1988a) Biosynthesis of 3-deoxyanthocyanins with flower extracts from *Sinningia cardinalis*. *Phytochemistry*, **27**, 785–789.

Stich, K. and Forkmann, G. (1988b) Studies on columnidin biosynthesis with flower extracts from *Columnea hybrida*. *Z. Naturforsch.*, **43c**, 311–314.

Stich, K., Ebermann, R., and Forkmann, G. (1988) Effect of cytochrome P-450 specific inhibitors on the activity of flavonoid 3′-hydroxylase and flavone synthase II in certain plants. *Phyton* (*Austria*), **28**, 237–247.

Stich, K., Eidenberger, T., Wurst, F., and Forkmann, G. (1992) Flavonol synthase activity and the regulation of flavonol and anthocyanin biosynthesis during flower development in *Dianthus caryophyllus* L. (carnation). *Z. Naturforsh.*, **47c**, 553–560.

Stickland, R. G. and Harrison, B. J. (1974a) Pathways to colour. *J. Royal Hort. Soc.*, 526–528.

Stickland, R. G. and Harrison, B. J. (1974b) Precursors and genetic control of pigmentation. 1. Induced biosynthesis of pelarganodin, cyanidin and delphinidin in *Antirrhinum majus*. *Heredity*, **33**, 108–112.

Stothers, J. B. (1972) Carbon-13 NMR spectroscopy. In *Organic Chemistry: A Series of Monographs*, *Vol. 24*. Academic Press, New York.

Stotz, G. and Forkmann, G. (1981) Oxidation of flavanones to flavones with flower extracts of *Antirrhinum majus* (snapdragon). *Z. Naturforsch.*, **36c**, 737–741.

Stotz, G. and Forkmann, G. (1982) Hydroxylation of the B-ring of flavonoids in the 3′-and 5′-position with enzyme extracts from flowers of *Verbena hybrida*. *Z. Naturforsch.*, **37c**, 19–23.

Stotz, G., de Vlaming, P., Wiering, H., Schram, A. W., and Forkmann, G. (1985) Genetic and biochemical studies on flavonoid 3′-hydroxylation in flowers of *Petunia hybrida*. *Theor. Appl. Genet.*, **70**, 300–305.

Stöckigt, J. and Zenk, M. H. (1975) Chemical syntheses and properties of hydroxycinnamoyl-coenzyme A derivatives. *Z. Naturforsch.*, **30c**, 352–358.

Strack, D. and Wray, V. (1994) The anthocyanins. In *The Flavonoids: Advances in Research since 1986*, J. B. Harborne (ed.), Chapman and Hall, London, 1–22.

Strack, D., Busch, E., and Klein, E. (1989) Anthocyanin patterns in European orchids and their taxonomic and phylogenetic relevance. *Phytochemistry*, **28**, 2127–2139.

Strack, D., Meurer, B., Wray, V., Grotjahn, L., Austenfeld, F. A., and Wiermann, R. (1984) Quercetin 3-glucosylgalactoside from pollen of *Corylus avellana*. *Phytochemistry*, **23**, 2970–2971.

Stuessy, T. F. (1990) *Plant Taxonomy*, Columbia University Press, New York.

Stuessy, T. F. and Crawford, D. J. (1983) Flavonoids and phylogenetic reconstruction. *Plant Syst. Evol.*, **143**, 83–107.

Stuessy, T. F., Foland, K. A., Sutter, J. F., and Silva, O. M. (1984) Botanical and geological significance of potassium-argon dates from the Juan Fernandez Islands. *Science*, **225**, 49–51.

Stuppner, H. and Müller, E. P. (1994) Rare flavonoid aglycones from *Flourensia retinophylla*. *Phytochemistry*, **37**, 1185–1187.

Sturgeon, K. B. and Mitten, J. B. (1980) Cone color polymorphism associated with elevation iln white fir, *Abies concolor*, in southern Colorado. *Amer. J. Bot.*, **67**, 1040–1045.

Styles, D. E., Gavazzi, G. A., and Racchi, M. L. (eds.) (1989) *The Genetics of Flavonoids*. Edizioni Unicopli, Milan.

Sullivan, G. R., Dale, J. A., and Mosher, H. S. (1973) Correlation of configuration and F-19 chemical shifts of α-methoxy-α-trifluoromethylphenylacetate derivatives. *J. Org. Chem.* **38**, 2143–2147.

Sultana, S. and Ilyas, M. (1987) Chemical investigation of *Macaranga indica* Wight. *Indian J. Chem.*, **26B**, 801–802.

Suresh, R. V., Iyer, C. S. R., and Iyer, P. R. (1985) Synthesis of erythrinin-A, a naturally occurring pyranoisoflavone. *Tetrahedron*, **41**, 2479–2482.

Sutter, A. and Grisebach, G. (1973) UDP-Glucose flavonol 3-*O*-glucosyltransferase from cell suspension cultures of parsley. *Biochem. Biophys. Acta*, **309**, 289–295.

Sutter, A., Ortmann, R., and Grisebach, H. (1972) Purification and properties of an enzyme from cell suspension cultures of parsley catalyzing the transfer of D-glucose from UDP-D-glucose to flavonoids. *Biochem. Biophys. Acta*, **258**, 71–87.

Sutter, A., Poulton, J., and Grisebach, H. (1975) Oxidation of flavanone to flavone with cell-free extracts from parsley leaves. *Arch. Biochem. Biophys.*, **170**, 547–556.

Sütfeld, R. and Wiermann, R. (1980) Chalcone synthesis with enzyme extracts from tulip anther tapetum using a biphasic enzyme assay. *Arch. Biochem. Biophys.*, **201**, 64–72.

Sütfeld, R. and Wiermann, R. (1981) Purification of chalcone synthase from tulip anthers and comparison with the synthase from *Cosmos* petals. *Z. Naturforsch.*, **36c**, 30–34.

Swain, T. (1962) Economic importance of flavonoid compounds: foodstuffs. In *The Chemistry of Flavonoid Compounds*, T. A. Geissman (ed.), The Macmillan Company, New-York, 513–552.

Swain, T. (1973) *Chemistry in Evolution and Systematics*, Butterworth & Co., London.

Swain, T. (1977) Secondary compounds as protective agents. *Annu. Rev. Plant Physiol.*, **28**, 479–501.

Sweigard, J. A., Matthews, D. E., and VanEtten, H. D. (1986) Synthesis of the phytoalexin pisatin by a methyltransferase from pea. *Plant Physiol.*, **80**, 277–279.

Swensen, S. M., Mullin, B. C., and Chase, M. W. (1994) Phylogenetic affinities of Datiscaceae based on an analysis of nucleotide sequences from the plastid *rbc*L gene. *Syst. Bot.*, **19**, 157–168.

Tabak, A. J. H., Schram, A. W., and Bennink, G. J. H. (1981) Modification of the B-ring during flavonoid synthesis in *Petunia hybrida*: Effect of hydroxylation gene *Hf1* on dihydroflavonol intermediates. *Planta*, **153**, 462–465.

Tahara, S., Moriyama, M., Ingham, J. L., and Mizutani, J. (1993) Isoflavone atropisomers from *Piscidia erythrina*. *Phytochemistry*, **34**, 545–552.

Tahara, S., Tanaka, M., and Barz, W. (1997) Fungal metabolism of prenylated flavonoids. *Phytochemistry*, **44**, 1031–1036.

Tak, H., Fronczek, F. R., and Fischer, N. H. (1993) Ceratiolin from *Ceratiola ericoides*. *Acta Crystallographica*, **C49**, 1990–1992.

Takagi, M., Funahashi, S., Ohta, K., and Nakabayashi, T. (1980) Phyllospadine, a new flavonoidal alkaloid from the sea grass Phyllospadix iwatensis. *Agric. Biol. Chem.*, **44**, 3019–3020.

Takahashi, Y., Saito, K., Yanagiya, M., Ikura, M., Hikichi, K., Matsumoto, T., and Wada, M. (1984) Chemical constituents of safflower yellow-B, a quinochalcone *C*-glycoside from the flower petals of *Carthamus tinctorius* L. *Tet. Lett.*, **25**, 2471–2474.

Takahashi, H., Sasaki, T., and Ito, M. (1987) New flavonoids isolated from infected sugarbeet root. *Bull. Chem. Soc. Japan*, **60**, 2261–2262.

Takasugi, M., Kumagai, Y., Nagao, S., Masamune, T., Shirata, A., and Takahashi, K. (1980) Studies of phytoalexins in Moraceae. 6. The co-occurrence of flavans and 1,3-diphenylpropane derivatives in wounded paper mulberry. *Chem. Lett.*, 1459–1460.

Takeda, K., Kariuda, M., and Itoi, H. (1985a) Blueing of sepal colour of *Hydrangea macrophylla*. *Phytochemistry*, **24**, 2251–2254.

Takeda, K., Kubota, R., and Yagioka, C. (1985b) Copigments in the blueing of sepal colour of *Hydrangea macrophylla*. *Phytochemistry*, **24**, 1207–1209.

Takeda, K., Sato, S., Kobayashi, H., Kanaitsuka, Y., Ueno, M., Kinoshita, T., Tazaki, H., and Fujimori, T. (1994) The anthocyanin responsible for purplish blue flower colour of *Aconitum chinense*. *Phytochemistry*, **36**, 613–616.

Takeda, K., Yamaguchi, S., Iwata, K., Tsujino, Y., Fujimori, T., and Husain, S. Z. (1996) A malonylated anthocyanin and flavonols in the blue flowers of *Meconopsis*. *Phytochemistry*, **42**, 863–865.

Takeda, K., Yamashita, T., Takahashi, A., and Timberlake, C. F. (1990) Stable blue complexes of anthocyanin-aluminium 3-*p*-coumaroyl- or 3-caffeoyl-quinic acid involved in the blueing of *Hydrangea* flower. *Phytochemistry*, **29**, 1089–1091.

Takhtajan, A. (1986) *Floristic Regions of the World*, University of California Press, Berkeley.

Talapatra, B., Deb, T., and Talapatra, S. (1986) Condensation of phenols and cinnamic acids in presence of polyphosphoric acid: a novel biogenetic-type oxidative self-cyclization of *p*-methoxycinnamic acid to 7-methoxycoumarin. *Ind. J. Chem.*, **25B**, 1122–1125.

Tamara, H., Hayashi, Y., Sugisawa, H., and Kondo, T. (1994) Structure determination of acylated anthocyanins in Muscat Bailey A grapes by homonuclear Hartmann-Hahn (HOHAHA) spectroscopy and liquid chromatography-mass spectrometry. *Phytochem. Anal.*, **5**, 190–196.

Tamura, H., Kondo, T., and Goto, T. (1986) The composition of commelinin, a highly associated metalloanthocyanin present in the blue flower petals of *Commelina communis*. *Tet. Lett.*, **27**, 1801–1804.

Tanaguchi, M. and Kubo, I. (1993) Ethnobotanical drug discovery based on medicine men's trials in the African savanna: Screening of East African plants for antimicrobial activity II. *J. Nat. Prod.*, **56**, 1539–1546.

Tanaguchi, M., Chapya, A., Kubo, I., and Nakanishi, K. (1978) Screening of East African plants for antimicrobial activity. I. *Chem. Pharm. Bull.*, **26**, 2910–2913.

Tanaka, H., Hiroo, M., Ichino, K., and Ito, K. (1989) Total synthesis of silychristin, an antihepatotoxic flavonolignan. *Chem. Pharm. Bull.*, **37**, 1441–1445.

Tanaka, T., Iinuma, M., and Mizuno, M. (1986) Spectral properties of 2′-oxygenated flavones. *Chem. Pharm. Bull.*, **34**, 1667–1671.

Tanaka, T., Kawamura, K., Hohda, H., Yamasaki, K., and Tanaka, O. (1982) Glycosides of the leaves of *Symplocos* spp. (Symplocaceae). *Chem. Pharm. Bull.*, **30**, 2421–2423.

Tanaka, N., Orii, R., Ogasa, K., Wada, H., Murakami, T., Saiki, Y., and Chen, C.-M. (1991) Chemical and chemotaxonomical studies of ferns. LXXX. Proanthocyanidins of *Arachnioides sporadosora* Nakaike and *A. exilis* Ching. *Chem. Pharm. Bull.*, **39**, 55–59.

Tanrisever, N., Franczek, F. R., Fischer, N. H., and Williamson, G. B. (1987) Ceratiolin and other flavonoids from *Ceratiola ericoides*. *Phytochemistry*, **26**, 175–179.

Tatsuzawa, F., Saito, N., Yokoi, M., Shigihara, A., and Honda, T. (1994) An acylated cyanidin glycoside in the red-purple flowers of x *Laeliocattleya* cv. mini purple. *Phytochemistry*, **37**, 1179–1183.

Taylor, E. C., Conley, R. A., Johnson, D. K., McKillop, A., and Ford, M. E. (1980) Thallium inorganic synthesis. 57. Reaction of chalcones and chalcone ketals with thallium (III) trinitrate. *J. Org. Chem.*, **45**, 3433–3436.

Taylor, L. P. and Hepler, P. K. (1997) Pollen germination and tube growth. *Annu. Rev. Plant Physiol. Mol. Biol.*, **48**, 461–491.

Taylor, L. P. and Jorgensen, R. (1992) Conditional male fertility in chalcone synthase-deficient petunia. *J. Heredity*, **83**, 11–17.

Tebayashi, S.-i., Matsuyama, S., Suzuki, T., Kuwahara, Y., Nemoto, T., and Fujii, K. (1995) Quercimeritrin: The third oviposition stimulant of the azuki bean weevil from the host azuki bean. *J. Pesticide Sci.*, **20**, 299–305.

Templeton, M. D. and Lamb, C. J. (1988) Elicitors and defence gene activation. *Plant, Cell and Environment*, **11**, 395–401.

Terahara, N., Oda, M., Matsui, T., Osajima, Y., Saito, N., Toki, N., and Honda, T. (1996) Five new anthocyanins, ternatins A3, B4, B3, B2, and D2 from *Clitoria ternatea* flowers. *J. Nat. Prod.*, **59**, 139–144.

Terahara, N., Saito, N., Honda, T., Toki, K., and Osajima, Y. (1990a) Acylated anthocyanins of *Clitoria ternatea* flowers and their acyl moieties. *Phytochemistry*, **29**, 949–953.

Terahara, N., Saito, N., Honda, T., Toki, K., and Osajima, Y. (1990b) Further structural elucidation of the anthocyanin deacylternatin from *Clitoria ternatea. Phytochemistry*, **29**, 3686–3687.

Terashima, K., Aqil, M., and Niwa, M. (1995) Garcinianin, a novel biflavonoid from the roots of *Garcinia kola. Heterocycles*, **41**, 2245–2250.

Tétényi, P. (1970) *Infraspecific Chemical Taxa of Medicinal Plants.* Akadémiai Kiadó, Budapest.

Teusch, M. (1986) UDP-Xylose:anthocyanidin 3-O-glucosexylosyltransferase from petals of *Matthiola incana.* R. Br. *Planta*, **169**, 559–563.

Teusch, M. and Forkmann, G. (1987) Malonyl-coenzyme A: anthocyanidin 3-glucoside malonyltransferase from flowers of *Callisptephus chinensis. Phytochemistry*, **26**, 2181–2183.

Teusch, M., Forkmann, G., and Seyffert, W. (1986a) Genetic control of UDP-glucose: anthocyanin 5-*O*-glucosyltransferase from flowers of *Matthiola incana* R. Br. *Planta*, **168**, 586–591.

Teusch, M., Forkmann, G., and Seyffert, W. (1986b) UDP-Glucose: anthocyanidin/flavonol 3-*O*-glucosyltransferase in enzyme preparations from flower extracts of genetically defined lines of *Matthiola incana* R. Br. *Z. Naturforsch.*, **41c**, 699–706.

Thompson, J. F., Honda, S. I., Hunt, G. E., Krupka, R. M., Morris, C. J., Powell, J. L. E., Silberstein, O. O., Towers, G. H. N., and Zacharias, R. M. (1959) Partition chromatography and its use in the plant sciences. *The Botanical Reviews*, **25**, 1–263.

Thompson, W. R., Meinwald, J., Aneshansley, D., and Eisner, T. (1972) Flavonols: Pigments responsible for ultraviolet absorption on nectar guide. *Science*, **177**, 528–530.

Thomsen, I. and Torssell, K. B. G. (1988) Use of nitrile oxides in synthesis. Novel synthesis of chalcones, flavanones, flavones and isoflavones. *Acta Chem. Scand.*, **B42**, 303–308.

Toki, K., Saito, N., Kawano, K., Lu, T. S., Shigihara, A., and Honda, T. (1994a) An acylated delphinidin glycoside in the blue flowers of *Evolvulus pilosus. Phytochemistry*, **36**, 609–612.

Toki, K., Saito, N., Iimura, K., Suzuki, T., and Honda, T. (1994b) (Delphinidin 3-gentiobiosyl)(apigenin 7-glucosyl) malonate from the flowers of *Eichhornia crassipes*. *Phytochemistry*, **36**, 1181–1183.

Toki, K., Saito, N., Kuwano, H., Shigihara, A., and Honda, T. (1995a) Acylated pelargonidin 3,7-diglycosides from pink flowers of *Senecio cruentus*. *Phytochemistry*, **38**, 1509–1512.

Toki, K., Saito, N., Terahara, N., and Honda, T. (1995b) Pelargonidin 3-glucoside-5-acetylglucoside in *Verbena* flowers. *Phytochemistry*, **40**, 939–940.

Toki, K., Takeuchi, M., Saito, N., and Honda, T. (1996) Two malonylated anthocyanidin glycosides in *Ranunculus asiaticus*. *Phytochemistry*, **42**, 1055–1057.

Tomás-Barberán, F. A., Ferreres, F., Blázquez, M. A., Garcia-Viguera, C., and Tomás-Lorente, F. (1993a) High-performance liquid chromatography of honey flavonoids. *J. Chromatog.*, **634**, 41–46.

Tomás-Barberán, F. A., Ferreres, F., and Tomás-Lorente, F. (1993b) Flavonoids from *Apis mellifera* beeswax. *Z. Naturforsch.*, **48c**, 68–72.

Tomas-Barberan, F. A., Garcia-Viguera, C., and Bridle, P. (1996c) Is capillary electrophoresis the HPLC of the 90s in polyphenol analysis? *Polyphenols Actualites*, No. 15(July), 15–17.

Tomas-Barberan, F. A., Grayer-Barkmeijer, R. J., Gil, M. I., and Harborne, J. B. (1988) Distribution of 6-hydroxy-, 6-methoxy- and 8-hydroxyflavone glycosides in the Labiatae,the Scrophulariaceae and related families. *Phytochemistry*, **27**, 2631–2645.

Tomás-Barberán, F. A., Iniesta-Sanmartin, E., Ferreres, F., and Tomás-Lorente, F. (1990) High performance liquid chromatography, thin layer chromatography and ultraviolet behaviour of flavone aglycones with unsubstituted B rings. *Phytochem. Anal.*, **1**, 44–47.

Tomlin, D. W. and Cantrell, J. S. (1990) Structure of chloroflavanone. *Acta Cryst.*, **C46**, 519–521.

Towers, G. H. N. (1964) Fungal metabolism of phenolic compounds. In *Biochemistry of Phenolic Compounds*, Harborne, J. B. (ed.), Academic Press, London, 249–294.

Trigo, J. R., Brown, Jr., K. S., Henriques, S. A., and Barata, L. E. S. (1996) Qualitative patterns of pyrrolizidine alkaloids in Ithomiinae butterflies. *Biochem. Syst. Ecol.*, **24**, 181–188.

Tryon, R. (1962) Taxonomic fern notes. II. *Pityrogramma* (including *Trismeria*) and *Anogramma*. *Contrib. Gray Herb.*, **189**, 52–76.

Tryon, R. M. (1969) Taxonomic problems in the geography of North American ferns. *BioScience*, **19**, 790–795.

Tryon, R. M. (1972) Endemic areas and geographic speciation in tropical American ferns. *Biotropica*, **4**, 121–131.

Tryon, R. M. and Tryon, A. F. (1973) Geography, spores and evolutionary relationships in the cheilanthoid ferns. *Bot. J. Linnean Soc., Supplement*, **67**, 145–153.

Tryon, R. M. and Tryon, A. F. (1982) *Ferns and Allied Plants, with Special Reference to Tropical America*, Springer-Verlag, New York.

Tse, A. and Towers, G. H. N. (1967) The occurrence of psilotin in *Tmesipteris*. *Phytochemistry*, **6**, 149.

Tsukayama, M., Kawamura, Y., Tamaki, H., and Horie, T. (1991) Synthesis of parvisoflavones A and B. *Chem. Pharm. Bull.*, **39**, 1704–1706.

Tuchweber, B., Sieck, R., and Trost, W. (1979) Prevention by silybin of phalloidin-induced acute hepatotoxicity. *Toxicol. Appl. Pharmacol.*, **51**, 265–275.

Tuntiwachwuttikul, P., Pakawatchai, C., Patrick, V. A., Reutrakul, V., Skelton, B. W., and White, A. H. (1988) Flavonoid derivatives of *Boesenbergia*. *J. Sci. Soc. Thailand*, **14**, 301–307.

Turner, B. L. and Alston, R. E. (1959) Segregation and recombination of chemical constituents is a hybrid swarm of *Baptisia laevicaulis* X *B. viridis*. *Amer. J. Bot.*, **46**, 678–686.

Turner, R. B., Lindsey, D. L., Davis, D. D., and Bishop, R. D. (1975) Isolation and identification of 5,7-dimethoxyisoflavone, an inhibitor of *Aspergillus flavus* from peanuts. *Mycopathologia*, **57**, 39–40.

Ueda, S., Nomura, T., Fukai, T., and Matsumoto, J. (1982) Kuwanon J, a new Diels-Alder adduct, and chalcomoracin from callus culture of *Morus alba* L. *Chem. Pharm. Bull.*, **30**, 3042–3045.

Uhlmann, A. and Ebel, J. (1993) Molecular cloning and expression of 4-coumarate: coenzyme A ligase, an enzyme involved in the resistance response of soybean (*Glycine max* L.) against pathogen attack. *Plant Physiol.*, **102**, 1147–1156.

Umezawa, H., Tobe, H., Shibamoto, N., Nakamura, F., Nakamura, K., Matsuzaki, M., and Takeuchi, T. (1975) Isolation of isoflavones inhibiting DOPA decarboxylase from fungi and streptomyces. *Journal of Antibiotics*, **28**, 947–952.

Underhill, E. W., Watkin, J. E., and Neish, A. C. (1957) Biosynthesis of quercetin in buckwheat. Part I. *Can. J. Biochem. Physiol.*, **35**, 219–228.

Uniyal, S. K., Badoni, V., and Sati, O. P. (1992) A new piscicidal flavonoid glycoside from *Engelhardtia colebrookiana*. *International J. Pharmacog.*, **30**, 209–212.

Urzua, A. and Cuadra, P. (1989) Flavonoids from the resinous exudate of *Gnaphalium robustum*. *Bol. Soc. Chil. Quim.*, **34**, 247–251.

Uyar, T., Malterud, K. E., and Anthonsen, T. (1978) Two new dihydrochalcones from *Myrica gale*. *Phytochemistry*, **17**, 2011–2013.

Valant-Vetschera, K. M. (1985) *C*-Glycosylflavones as an accumulation Tendency: A critical review. *The Botanical Reviews*, **51**, 1–51.

Valdebenito, H. A., Stuessy, T. F., and Crawford, D. J. (1990a) Synonymy in *Peperomia berteroana* (Piperaceae) results in biological disjunction between Pacific and Atlantic Oceans. *Brittonia*, **42**, 121–124.

Valdebenito, H. A., Stuessy, T. F., and Crawford, D. J. (1990b) A new biogeographic connection between islands in the Atlantic and Pacific Oceans. *Nature* (*London*), **347**, 549–550.

Valdebenito, H. A., Stuessy, T. F., Crawford, D. J., and Silva, O. M. (1992) Evolution of *Peperomia* (Piperaceae) in the Juan Fernandez Islands. *Plant Syst. Evol.*, **182**, 107–119.

van Brederode, J. and van Nigtevecht, G. (1975) Dominance relationships between allelic glycosyltransferase genes in *Melandrium*: An enzyme-kinetic approach. *Theor. Appl. Genet.*, **46**, 353–358.

van Brederode, J., Kamps-Heinsbroek, R., and Steyns, J. (1987) Serological and genetical studies on the evolution of substrate specificity of flavone glycosyltransferase genes in *Silene*. *Experientia*, **43**, 202–205.

van der Krol, A. R., Lenting, P. E., Veenstra, J., van der Meer, I. M., Koes, R. E., Gerats, A. G. M., Mol, J. N. M., and Stuitje, A. R. (1988) An anti-sense chalcone synthase gene in transgenic plants inhibits flower pigmentation. *Nature* (*London*), **333**, 866–869.

van der Krol, A. R., Mur, L. A., Beld, M., Mol, J. N. M., and Stuitje, A. R. (1990) Flavonoid genes in petunia: Addition of a limited number of gene copies may lead to a suppression of gene expression. *Plant Cell*, **2**, 291–299.

van der Meer, I. M., Stuitje, A. R., and Mol, J. N. M. (1992) Factors controlling expression of phenylalanine ammonia lyase. In *Control of Plant Gene Expression*, D. P. S. Verma (ed.), CRC Press, Boca Raton, FL, 125–155.

van Eldik, G. J., Reijnen, W. H., Ruiter, R. K., van Herpen, M. M. A., Schrauwen, J. A. M., and Wullems, G. J. (1997) Regulation of flavonol biosynthesis during anther and pistil development, and during pollen tube growth in *Solanum tuberosum*. *The Plant Journal*, **11**, 105–113.

Van Sumere, C., Fache, P., Vande Casteele, K., De Cooman, L., Everaert, E., De Loose, R., and Hutsebaut, W. (1993) Improved extraction and reversed phase – high performance liquid chromatographic separation of flavonoids and the identification of *Rosa* cultivars. *Phytochem. Anal.*, **4**, 279–292.

van Tunen, A. J., Koes, R. E., Spelt, C. E., van der Krol, A. R., Stuitje, A. R., and Mol, J. N. M. (1988) Cloning of two chalcone flavanone isomerase genes from *Petunia hybrida*: Coordinate light regulated and differential expression of flavonoid genes. *EMBO Journal*, **7**, 1257–1263.

van Tunen, A. J., Hartman, S. A., Mur, L. A., and Mol, J. N. M. (1989) Regulation of chalcone flavanone isomerase (CHI) gene expression in *Petunia hybrida*: the use of alternative promoters in corolla, anthers and pollen. *Plant Mol. Biol.*, **12**, 539–551.

van Tunen, A. J., Mur, L. A., Brouns, G. S., Rienstra, J.-D., Koes, R. E., and Mol, J. N. M. (1990) Pollen- and anther-specific *chi* promoters from petunia: tandem promoter regulation of the *chiA* gene. *Plant Cell*, **2**, 393–401.

Vande Casteele, K., De Pooter, H., and Van Sumere, C. F. (1976) Gas chromatographic separation and analysis of trimethylsilyl derivatives of some naturally occurring non-volatile phenolic compounds and related substances. *J. Chromatog.*, **121**, 49–63.

Vande Casteele, K., Geiger, H., and Van Sumere, C. F. (1982) Separation of flavonoids by reversed-phase high performance liquid chromatography. *J. Chromatog.*, **240**, 81–94.

VanEtten, H. D., Mansfield, J. W., Bailey, J. A., and Farmer, E. E. (1994) Two classes of plant antibiotics: Phytoalexins versus "Phytoanticipins". *Plant Cell*, **6**, 1191–1192.

Vwit, M., Geiger, H., Czygan, F.-C., and Markham, K. R. (1990) Malonylated flavone 5-O-glucosides in the barren sprouts of *Equisetum arvense*. *Phytochemistry*, **29**, 2555–2560.

Veitch, N. C. and Stevenson, P. C. (1997) 2-Methoxyjudaicin, an isoflavene from the roots of *Cicer bijugum. Phytochemistry*, **44**, 1587–1589.

Venkata Rao, E., Rajendra Prasad, Y., and Ganapaty, S. (1992) Three prenylated isoflavones from *Milletia auriculalta. Phytochemistry*, **31**, 1015–1017.

Venkataraman, K. (1959) Flavones and Isoflavones. *Advances Chem. Org. Nat. Prod.*, **17**, 1–69.

Venkataraman, K. (1962) Methods for determining the structures of flavonoid compounds. In *The Chemistry of Flavonoid Compounds*, T. A. Geissman (ed.), The Macmillan Company, New York, 70–106.

Verma, V. S. (1973) Study on the effect of flavonoids on the infection of potato virus X. *Zbl. Bakt. Abt. II*, **128S**, 467–472.

Vlietinck, A. J., Vanden Berge, D. A., Van Hoof, L. M., Vrijsen, R., and Boeyé, A. (1986) Antiviral activity of 3-methoxyflavones. In *Plant Flavonoids in Biology and Medicine*, V. Cody, E. Middleton Jr., and J. B. Harborne (eds.), Alan R. Liss, Inc., New York, 537–540.

Vogt, T. and Taylor, L. P. (1995b) Flavonol 3-*O*-glycosyltransferases associated with petunia pollen produce gametophyte-specific flavonol diglycosides. *Plant Physiol.*, **108**, 903–911.

Vogt, T., Wollenweber, E., and Taylor, L. P. (1995a) The structural requirements of flavonols that induce pollen germination of conditionally male fertile *Petunia. Phytochemistry*, **38**, 589–592.

Voirin, B. and Lebreton, P. (1967) Chemotaxonomic investigation of vascular plants. Presence of 6-methylchrysin in the fern *Lonchitis tisserantii. Bull. Soc. Chim. Biol.*, **49**, 1402–1405.

Voirin, B., Jay, M., and Hauteville, M. (1975) Isoetine, nouvelle flavone isolée de *Isoetes delilei* et de *Isoetes durieui. Phytochemistry*, **14**, 257–259.

Voirin, B., Rasamoelisendra, R., Favre-Bonvin, J., Adriantsiferana, M., and Rabesa, Z. (1986) 6,8-Di-C-Methyldihydromyricetin from *Alluaudia humbertii. Phytochemistry*, **25**, 560–556.

von Kostanecki, S. and Tambor, J. (1895) *Ber.*, **28**, 2302. (Cited by Onslow, 1925.)

von Mohl, H. (1838) Recherces sur la coloration hibernale der feuilles. *Bul. Acad. roy., Bruxelles*, **20**, 197–235. (Cited by Onslow, 1925.)

von Wettstein, D., Jende-Strid, B., Ahrenst-Larsen, B., and Sorensen, J. A. (1977) Biochemical mutant in barley renders chemical stabilization of beer superfluous. *Carlsberg Res. Commun.*, **42**, 341–351.

Vowinkel, E. (1975) Die Struktur des Sphagnorubins. *Chem. Ber.*, **108**, 1166–1181.

Wagner, H. (1964) Infrared spectroscopy of flavonoids. In *Methods in Polyphenol Chemistry*, Pridham, J. B. (ed.), The Macmillan Company, New York, 37–48.

Wagner, H. (1986) Antihepatotoxic flavonoids. In *Plant Flavonoids in Biology and Medicine*, V. Cody, E. Middleton Jr., and J. B. Harborne (eds.), Alan R. Liss, Inc., New York, 545–558.

Wagner, H. and Farkas, L. (1975) Synthesis of Flavonoids. In *The Flavonoids*, J. B. Harborne, T. J. Mabry, and H. Mabry (eds.), Chapman and Hall, London, 127–213.

Wagner, H., Aurnhammer, G., Hörhammer, L., Farkas, L., and Nógrádi, M. (1968) Endgültige Konstitionsaufklärung und Synthese von Narirutin, Didymin, Rhoifolin, Poncirin und Fortunellin. *Tet. Lett.*, 1635–1639.

Wagner, H., Chari, V. M., Seitz, M., and Riess-Maurer, I. (1978) The structure of silychristin-a C-13 NMR study. *Tet. Lett.* (4), 381–384.

Wagner, H., Höer, R., Murakami, T., and Farkas, L. (1973) Isolierung, Strukturaufklärung und Synthese von 4′,5,7-Trihydroxy-3′,6-dimethoxyflavon-7-mono-β-D-glucopyranosid (Jaceosid), einem neuen Flavonglycosid aus den Wurzeln von *Centaurea jacea* L. *Chem. Ber.*, **106**, 20–27.

Wainwright, C. M. (1978) The floral biology and pollination ecology of two desert lupines. *Bull. Torrey Bot. Club*, **105**, 24–38.

Walker, J. C. (1923) Disease resistance to onion smudge. *J. Agric. Res.*, **24**, 1019–1039.

Wallace, J. W. and Grisebach, H. (1973) The *in vivo* incorporation of a flavanone into *C*-glycosylflavones. *Biochem. Biophys. Acta*, **304**, 837–841.

Wallace, J. W. and Markham, K. R. (1978) Apigenin and amentoglavone glycosides in the Psilotaceae and their phylogenetic significance. *Phytochemistry*, **17**, 1313–1317.

Wallace, J. W., Mabry, T. J., and Alston, R. E. (1969) On the biosynthesis of flavone *O*-glycosides and *C*-glycosides in the Lemnaceae. *Phytochemistry*, **8**, 93–99.

Walle, T., Eaton, E. A., and Walle, U. K. (1995) Quercetin, a potent and specific inhibitor of the human P-form phenolsulfotransferase. *Biochem. Pharmacol.*, **50**, 731–734.

Wallet, J.-C., Gaydou, E. M., and Baldy, A. (1989) Structure of 2-(2,6-dimethoxyphenyl)-4*H*-1-benzopyran-4-one (2′,6′-dimethoxyflavone). *Acta Cryst.*, **C45**, 512–515.

Wallet, J.-C., Gaydou, E. M., Jaud, J., and Baldy, A. (1990) Structures of 2-(2-methoxyphenyl)-4*H*-1-benzopyran-4-one (1) and 5,7-dimethoxy-2-(2,4-dimethoxyphenyl)-4*H*-1-benzopyran-4-one (2) (2′-methoxyflavone and 2′,4′,5,7-tetramethoxyflavone). *Acta Cryst.*, **C46**, 1536–1540.

Wallet, J. C., Cody, V., Wojtczak, A., and Blessing, R. H. (1993) Structural and conformational studies on bio-active flavonoids. Crystal and molecular structure of a complex formed betwen 2′,6′-dimethoxyflavone and *ortho*-phosphoric acid: a model for flavone-nucleotide interactions. *Anti-Cancer Drug Res.*, **8**, 325–332.

Walton, E. and Butt, V. S. (1970) The activation of cinnamate by an enzyme from leaves of spinach beet (*Beta vulgaris* L. ssp. *vulgaris*). *J. Expt. Bot.*, **21**, 887–891.

Walton, E. and Butt, V. S. (1971) The demonstration of cinnamoyl-CoA synthetase activity in leaf extracts. *Phytochemistry*, **10**, 295–304.

Wandji, J., Fomum, Z. T., Tillequin, F., Baudouin, G., and Koch, M. (1994) Epoxyisoflavones from *Erythrina senegalensis*. *Phytochemistry*, **35**, 1573–1577.

Wang, B. H., Ternai, B., and Polya, G. (1997) Specific inhibition of cyclic AMP-dependent protein kinase by warangalone and robustic acid. *Phytochemistry*, **44**, 787–796.

Wang, J., Li, X., Jiang, D., Ma, P., and Yang, C. (1995) Chemical constituents of dragon's blood rersin from *Dracaena cochinchinensis* in Yunnan and their antifungal activity. *Yunnan Zhiwu Yanjiu*, **17**, 336–340.

Wang, J.-N., Hou, C.-Y., Liu, Y.-L., Lin, L. Z., Gil, R. R., and Cordell, G. A. (1994) Swertifrancheside, an HIV-reverse transcriptase inhibitor and the first flavone-xanthone dimer, from *Swertia franchetiana. J. Nat. Prod.*, **57**, 211–217.

Wang, M., Li, J., and Liu, W. (1987) Two flavanones from the root bark of *Lespedeza davidii. Phytochemistry*, **26**, 1218–1219.

Wang, Y., Hamburger, M., Gueho, J., and Hostettmann, K. (1989) Antimicrobial flavonoids from *Psiadia trinervia* and their methylated and acetylated derivatives. *Phytochemistry*, **28**, 2323–2327.

Warburton, W. K. (1954) Isoflavones. *Quart. Revs.* (*London*), **8**, 67–87.

Waser, N. M. (1983) The adaptive nature of floral traits: Ideas and evidence. In *Pollination Biology*, L. Real (ed.), Academic Press, New York, 241–285.

Waterman, P. G. and Grundon, M. F. (1983) *Chemistry and Chemical Taxonomy of the Rutales*, Academic Press, New York, 464.

Waterman, P. G. and Mole, S. (1994) *Analysis of Phenolic Plant Metabolites*, Blackwell Scientific Publications, Oxford.

Watkin, J. E., Underhill, E. W., and Neish, A. C. (1957) Biosynthesis of quercetin in buckwheat. Part II. *Can. J. Biochem. Physiol.*, **35**, 229–237.

Watson, J. T. (1985) *Introduction to Mass Spectrometry*, Raven Press, New York.

Watson, S. (1873) Revision of the extra-tropical North American species of the genera *Lupinus, Potentilla* and *Oenothera. Proc. Amer. Acad. of Arts*, **8**, 517–618.

Wähälä, K. and Hase, T. A. (1989) Hydrogen transfer reduction of isoflavones. *Heterocycles*, **28**, 183–186.

Weatherby, C. A. (1920) Varieties of *Pityrogramma triangularis. Rhodora*, **22**, 113–120.

Webb, A. D. (1974) Chemistry of Winemaking. In *Advances in Chemistry Series* (ACS), R. F. Gould (ed.), American Chemical Society, Washington, D.C.

Webby, R. F. and Markham, K. R. (1990) Flavonol 3-*O*-triglycosides from *Actinidia* species. *Phytochemistry*, **29**, 289–292.

Weidenborner, M., Hindorf, H., Jha, H. C., and Tsotsonos, P. (1990) Antifungal activity of flavonoids against storage fungi of the genus *Aspergillus. Phytochemistry*, **29**, 1103–1105.

Weimarck, G. (1972) On "numerical chemotaxonomy". *Taxon*, **21**, 615–619.

Weiring, H. and de Vlaming, P. (1984) Genetics of flower and pollen colours. *Monographs on Theoretical and Applied Genetics*, K. C. Sink (ed.), Springer Verlag, Berlin, 49–75.

Weiss, M. R. (1995) Floral color change: A wide spread functional convergence. *Amer. J. Bot.*, **82**, 167–185.

Weitz, S. and Ikan, R. (1977) Bracteatin from the moss *Funaria hygrometrica. Phytochemistry*, **16**, 1108–1109.

Welle, R. and Grisebach, H. (1988) Isolation of a novel NADPH-dependent reductase which coacts with chalcone synthase in the biosynthesis of 6′-deoxychalcone. *FEBS Letters*, **236**, 221–225.

Welle, R., Schröder, G., Schiltz, E., Grisebach, H., and Schröder, J. (1991) Induced plant responses to pathogen attack. *Eur. J. Biochem.*, **196**, 423–430.

Wengenmayer, H., Ebel, J., and Grisebach, H. (1974) Purification and properties of a *S*-adenosyl-L-methionine: isoflavone 4′-*O*-methyltransferase from cell suspension cultures of *Cicer arietinum. Eur. J. Biochem.*, **50**, 135–143.

Wheldale, M. (1909a) On the nature of anthocyanin. *Proc. Phil. Soc., Cambridge*, **15**, 137–161.

Wheldale, M. (1909b) The colours and pigments of flowers with special reference to genetics. *Proc. Royal Soc., London*, **81**, 44–60. (Cited by Onslow, 1925.)

Wheldale, M. (1914) Our present knowledge of the chemistry of the Mendelian factors for flower colour. *J. Genetics*, **4**, 109–120.

Wheldale, M. (1915) Our present knowledge of the chemistry of the Mendelian factors for flower-colour. Part 2. *J. Genetics*, **4**, 369–376.

Whitelaw, M. L. and Daniel, J. R. (1991) Synthesis and sensory evaluation of ring-substituted dihydrochalcone sweeteners. *J. Agric. Food Chem.*, **39**, 44–51.

Wiering, H. and de Vlaming, P. (1984) Inheritance and biochemistry of pigments. In *Petunia*, K. C. Sink (ed.), Springer-Verlag, Berlin, 49–76.

Wiermann, R. (1972) Aktivität der Chalkon-Flavanone Isomerase und Akkumulation von phenylpropanoiden Verbinungen in Antheren. *Planta*, **102**, 55–60.

Wigand, A. (1862) Einige Sätze über die physiologische Bedeutung des Gerbstoffes und der Pflanzenfarbe. *Bot. Ztg., Leipzeig*, **20**, 121–125. (Cited by Onslow, 1925.)

Wightman, F., Schneider, E. A., and Thimann, K. (1980) Hormonal factors controlling the initiation and development of lateral roots. *Physiol. Plant*, **49**, 304–314.

Wilbert, S. M., Schenske, D. W., and Bradshaw, J. H. D. (1997) Floral anthocyanins from two monkeyflower species with different pollinators. *Biochem. Syst. Ecol.*, **25**, 437–443.

Williams, A. H. (1966) Dihydrochalcones. In *Comparative Phytochemistry*, T. Swain (ed.), Academic Press, New York, 297–307.

Williams, A. H. (1967) A dibenzoylmethane glucoside from *Malus* leaf. *Chem. & Ind.*, 526–527.

Williams, C. A. (1979) The leaf flavonoids of the Orchidaceae. *Phytochemistry*, **18**, 803–813.

Williams, C. A. and Harborne, J. B. (1988) Distribution and evolution of flavonoids in the monocotyledons. In *The Flavonoids: Advances in Research since 1980*, J. B. Harborne (ed.), Chapman and Hall, London, 505–524.

Williams, C. A. and Harborne, J. B. (1989) Isoflavonoids. In *Methods in Plant Biochemistry. Vol. 1. Plant Phenolics*, J. B. Harborne (ed.), Academic Press, New York, 421–449.

Williams, C. A. and Harborne, J. B. (1994) Flavone and flavonol glycosides. In *The Flavonoids: Advances in Research since 1986*, J. B. Harborne (ed.), Chapman and Hall, London, 337–385.

Williams, C. A. and Murray, B. G. (1972) Flavonoid variation in the genus *Briza*. *Phytochemistry*, **11**, 2507–2512.

Williams, C. A., Harborne, J. B., and Goldblatt, P. (1986) Correlations between phenolic patterns and tribal classification in the family Iridaceae. *Phytochemistry*, **25**, 2135–2154.

Williams, C. A., Harborne, J. B., and Mathew, B. (1988) A chemical appraisal via leaf flavonoids of Dahlgren's Liliiflorae. *Phytochemistry*, **27**, 2609–2629.

Willstätter, R. and Everest, A. E. (1913) Anthocyan I. The dye of the corn flower. *Liebig's Ann.*, **401**, 189–232. (Cited by Onslow, 1925.)

Willstätter, R. and Zechmeister, L. (1914) Synthesis of pelargonidin (transl.). *Sitzber. preuss. Akad. Wiss., Physik.-math. Kl.*, **34**, 886–893. (Cited by Onslow, 1925.)

Wilson, A. (1986) Flavonoid pigments in swallowtail butterflies. *Phytochemistry*, **25**, 1309–1313.

Wilson, A. (1987) Flavonoid pigments in chalkhill blue (*Lysandra coridon* Poda) and other lycaenid butterflies. *J. Chem. Ecol.*, **13**, 473–493.

Wilson, R. D. (1979) Chemotaxonomic studies in the Rubiacceae: I. Methods for the identification of hybridization in the genus *Coprosma* J.R. et G. Forst. using flavonoids. *N. Z. J. Bot.*, **22**, 195–200.

Woldu, Y. and Abegaz, B. (1990) Isoflavonoids from *Salsola somalensis. Phytochemistry*, **29**, 2013–2015.

Wolf, S. J. and Denford, K. E. (1984) *Arnica gracilis* (Compositae), a natural hybrid between *A. latifolia* and *A. cordifolia. Syst. Bot.*, **9**, 12–16.

Wolf, S. K. and Whitkus, R. (1987) A numerical analysis of flavonoid variation in *Arnica* subgen. *Austromontana* (Asteraceae). *Amer. J. Bot.*, **74**, 1577–1584.

Wolfender, J.-L., Maillard, M., and Hostettmann, K. (1994) Thermospray liquid chromatography-mass spectrometry in phytochemical analysis. *Phytochem. Anal.*, **5**, 153–182.

Wolfender, J.-L. and Hostettmann, K. (1995) Application of liquid chromatography-mass spectrometry to the investigation of medicinal plants. In *Phytochemistry of Medicinal Plants*, J. T. Arnason, R. Mata, and J. T. Romeo (eds.), Plenum Press, New York, 189–215.

Wollenweber, E. (1982a) Flavonoid aglycones as constituents of epicuticular layers in ferns. In *The Plant Cuticle*, D. F. Cutler, K. L. Alvin, and C. E. Price (eds.), Academic Press, London, 215–224.

Wollenweber, E. (1982b) Flavones and flavonols. In *The Flavonoids: Advances in Research*, J. B. Harborne and T. J. Mabry (eds.), Chapman and Hall, London, 189–260.

Wollenweber, E. (1984) The systematic implication of flavonoids secreted by plants. In *Biology and Chemistry of Plant Trichomes*, E. Rodriguez, P. L. Healey, and I. Mehtal (eds.), Plenum Press, New York, 53–69.

Wollenweber, E. (1986) Flavonoid aglycones as leaf exudate constituents in higher plants. In *Flavonoids and Bioflavonoids*, L. Farkas, M. Gabor, and F. Kallay (eds.), Elsevier Sci. Publ., Amsterdam, 155–169.

Wollenweber, E. (1994) Flavones and Flavonols. In *The Flavonoids-Advances in Research since 1986*, J. B. Harborne (ed.), Chapman & Hall, New York, 259–335.

Wollenweber, E. and Dietz, V. H. (1979) A table of mass spectral parent ions as an aid in flavonoid analysis. *Phytochem. Bull.*, **12**, 48–52.

Wollenweber, E. and Dietz, V. H. (1980) Flavonoid patterns in the farina of goldenback and silverback ferns. *Biochem. Syst. Ecol.*, **8**, 21–33.

Wollenweber, E. and Dietz, V. H. (1981) Occurrence and distribution of free flavonoid aglycones in plants. *Phytochemistry*, **20**, 869–932.

Wollenweber, E. and Jay, M. (1988) Flavones and flavonols. In *The Flavonoids: Advances in Research since 1980*, J. B. Harborne (ed.), Chapman and Hall, 233–302.

Wollenweber, E. and Kohorst, G. (1981) Epicuticular leaf flavonoids from *Eucalyptus* species and from *Kalmia latifolia*. *Z. Naturforsch*, **36c**, 913–915.

Wollenweber, E. and Mann, K. (1986) Neue Flavonoide aus Primelmehl. *Biochem. Physiol. Pflanzen*, **181**, 665–669.

Wollenweber, E. and Seigler, D. S. (1982) Flavonoids from the exudate of *Acacia neovernicosa*. *Phytochemistry*, **21**, 1063–1066.

Wollenweber, E. and Wiermann, R. (1979) On the pigmentation of the pollen of *Nothofagus antartica*. *Z. Naturforsch.*, **34c**, 1289–1291.

Wollenweber, E., Mann, K., Iinuma, M., Tanaka, T., and Mizuno, M. (1988a) 8,2′-Dihydroxyflavone from *Primula pulverulenta*. *Phytochemistry*, **27**, 1483–1486.

Wollenweber, E., Mann, K., Iinuma, M., Tanaka, T., and Mizuno, M. (1988b) 2′,5′-Dihydroxyflavone and its 5′-acetate – novel compounds from the farinose exudate of *Primula*. *Z. Naturforsch.*, **43c**, 305–307.

Wollenweber, E., Marx, D., Favre-Bonvin, J., Voirin, B., and Kaouadji, M. (1988c) 3-Methoxy flavones with uncommon B-ring substitution from two species of *Notholaena*. *Phytochemistry*, **27**, 2673–2676.

Wollenweber, E., Arriaga-Giner, F. J., Rumbero, A., and Greenaway, W. (1989) New phenolics from *Baccharis* leaf exudate. *Z. Naturforsch.*, **44c**, 727.

Wollenweber, E., Armbruster, S., and Roitman, J. N. (1994) A herbacetin methyl ether from the farinose exudate of a *Pentagramma triangularis* hybrid. *Phytochemistry*, **37**, 455–456.

Wong, E. (1965) Flavonoid biosynthesis in *Cicer arietinum*. *Biochim. Biophys. Acta*, **111**, 358–363.

Wong, E. (1966) Occurrence and biosynthesis of 4′,6-dihydroxyaurone in soybean. *Phytochemistry*, **5**, 463–467.

Wong, E. (1968) The role of chalcones and flavanones in flavonoid biosynthesis. *Phytochemistry*, **7**, 1751–1758.

Wong, E. (1975) The isoflavonoids. In *The Flavonoids*, J. B. Harborne, T. J. Mabry, and H. Mabry (eds.), Chapman & Hall, London, 743–800.

Wong, E. and Grisebach, H. (1969) Further studies on the role of chalcone and flavanone in biosynthesis of flavonoids. *Phytochemistry*, **8**, 1419–1426.

Wong, E. and Wilson, J. M. (1972a) the oxidation of chalcone catalyzed by peroxidase. *Phytochemistry*, **11**, 875.

Wong, S.-M., Konno, C., Oshina, Y., Pezzuto, J. M., Fong, H. H. S., and Farnsworth, N. R. (1987) Irisones A and B: Two new isoflavones from *Iris missouriensis*. *J. Nat. Prod.*, **50**, 178–180.

Wu, J. H., Liao, S. X., Liang, H. Q., and Mao, S. L. (1994) Isolation and identification of flavones from *Desmos cochinchinensis* Lour. *Yaoxue Xuebao*, **29**, 621–623.

Wu, L. J., Miyase, T., Veno, A., Kuroyanagi, M., Noro, T., and Fukushima, S. (1985) Studies on the constituents of *Sophora flavescens* Aiton. II. *Chem. Pharm. Bull.* **33**, 3231–3236.

Xie, J., Wang, L., Liu, C., and Ge, D. (1986) Synthesis of expectorant principle of natural flavanone, "matteucinol". *Zhongguo Yixue Kexueyuan Xuebao*, **8**, 84–87.

Xu, Y., Kubo, I., and Ma, Y. (1993) A cytotoxic flavanone glycoside from *Onychium japonicum*: structure of onychin. *Phytochemistry*, **33**, 510–511.

Yamaguchi, M.-A., Maki, T., Ohishi, T., and Ino, I. (1995) Succinyl-coenzyme A: anthocyanidin 3-glucoside succinyltransferase in flowers of *Centaurea cyanus*. *Phytochemistry*, **39**, 311–313.

Yamaguchi, M.-A., Kawanobu, S., Maki, T., and Ino, I. (1996) Cyanidin 3-malonylglucoside and malonyl-coenzyme A: anthocyanidin malonyltransferase in *Lactuca sativa* leaves. *Phytochemistry*, **42**, 661–663.

Yamashita, Y., Kawada, S.-z., and Nakano, H. (1990) Induction of mammalian topoisomerase II dependent DNA cleavage by non-intercalative flavonoids, genistein and orobol. *Biochem. Pharmacol.*, **39**, 737–744.

Yatskievych, G., Windham, M. D., and Wollenweber, E. (1990) A reconsideration of the genus *Pityrogramma* (Adiantaceae) in western North America. *American Fern Journal*, **80**, 9–17.

Yazaki, Y. (1976) Co-pigmentation and color change with age in petals of *Fuchsia hybrida*. *Botanical Magazine (Tokyo)*, **89**, 45–57.

Ylstra, B., Touraev, A., Moreno, R. M. B., Stöger, E., Tunen, v. A. J., Vicente, O., Mol, J. N. M., and Heberle-Bors, E. (1992) Flavonols stimulate development, germination, and tube growth of tobacco pollen. *Plant Physiol.*, **100**, 902–907.

Ylstra, B., Muskens, M., and Van Tunen, A. J. (1996) Flavonols are *not* essential for fertilization in *Arabidopsis thaliana*. *Plant Mol. Biol.*, **32**, 1155–1158.

Yoshida, K., Kameda, K., and Kondo, T. (1993) Diglucuronoflavones from purple leaves of *Perilla ocimoides*. *Phytochemistry*, **33**, 917–919.

Young, D. A. (1976) Flavonoid chemistry and phylogenetic relationships of the Julianaceae. *Syst. Bot.*, **1**, 149–162.

Young, D. A. (1979) Heartwood flavonoids and the infrageneric relationships of *Rhus* (Anacardiaceae). *Amer. J. Bot.*, **66**, 502–510.

Young, D. A. (1981) The usefulness of flavonoids in angiosperm phylogeny: Some selected examples. In *Phytochemistry and Angiosperm Phylogeny*, D. A. Young and D. S. Seigler (eds.), Praeger Publishers, New York, 205–232.

Young, D. A. and Seigler, D. S. (1981) *Phytochemistry and Angiosperm Phylogeny*, Praeger Publishers, New York, 295.

Youngs, R. L. (1955) The xylem anatomy of *Orthopterygium* (Julianaceae). *Tropical Woods*, **101**, 29–43.

Yu, S., Fang, N., and Mabry, T. M. (1988) Flavonoids from *Gymnosperma glutinosum*. *Phytochemistry*, **27**, 171–177.

Yuan, H. Q. and Zuo, C. X. (1992) Chemical constituents of *Cynanchum thesioides*. *Yaoxue Xuebao*, **27**, 589–594.

Zähringer, U., Schaller, E., and Grisebach, H. (1981) Induction of phytoalexin synthesis in soybean. Structure and reactions of naturally occurring and enzymatically prepared prenylated pterocarpans from elicitor-treated cotyledons and cell cultures of soybean. *Z. Naturforsch.*, **36c**, 234–241.

Zeng, L., Fukai, T., Nomura, T., Zhang, R.-Y., and Lou, Z.-C. (1992) Phenolic constituents of *Glycyrrhiza* species. 9. Five new isoprenoid-substituted flavonoids, glyasperins F, G, H, I and J from the roots of *Glycyrrhiza aspera*. *Heterocycles*, **34**, 1813–1828.

Zeng, J.-F., Li, G.-L., Xu, X., and Zhu, D.-Y. (1996) Two isoprenoid-substituted isoflavans from roots of *Maackia tenuifolia. Phytochemistry*, **43**, 893–896.

Zerback, R., Bokel, M., Geiger, H., and Hess, D. (1989) A kaempferol 3-glucosylgalactoside and further flavonoids from pollen of *Petunia hybrida. Phytochemistry*, **28**, 897–899.

Zhang, Y. Y., Li, X., Gou, Y. Z., Harigaya, Y., Onda, M., Hashimoto, K., Ikeya, Y., Okada, M., and Maruno, M. (1994) Studies on the constituents of the roots of *Scutellaria planipes. Chinese Chem. Lett.*, **5**, 851–854.

Zhang, F.-J., Lin, G.-Q., and Huang, Q.-C. (1995) Synthesis, resolution, and absolute configuration of optically pure 5,5″-dihydroxy-4′,4‴,7,7″-tetramethoxy-8,8″-biflavone and its derivatives. *J. Org. Chem.*, **60**, 6427–6430.

Zhang, X.-H. and Chiang, V. L. (1997) Molecular cloning of 4-coumarate : coenzyme A ligase in loblolly pine and the roles of this enzyme in the biosynthesis of lignin in compression wood. *Plant Physiol.*, **113**, 65–74.

Zheng, S., Sun, L. S. X., and Yi, W. (1996) Flavonoid constituents from *Mosla chinensis* Maxim. *Indian J. Chem.*, **35B**, 392–394.

Zhu, M., Phillipson, J. D., Greengrass, P. M., Bowery, N. E., and Cai, Y. (1997) Plant polyphenols: Biologically active compounds or non-selective binders to protein. *Phytochemistry*, **44**, 441–447.

Zinsmeister, H. D. and Mues, R. (1990) *Bryophytes. Their Chemistry and Chemical Taxonomy*, Clarendon Press, Oxford, 470.

Zohary, D. and Hopf, M. (1988) *Domestication of Plants in the Old World*, Clarendon Press, Oxford.

INDEX

Index to general topics (t = data in table; s = data in scheme; syn = synthesis)

B

D

I

Index of families and higher taxa (t = data in table)

A

B

C

D

E

Index of plant genera (t = table)

B

C

Index of animal genera

Index of insect genera

Index of microorganism genera